Negative Strand Viruses and the Host Cell

Negative Strand Viruses and the Host Cell

Edited by

B. W. J. MAHY

and

R. D. BARRY

Division of Virology
Department of Pathology
University of Cambridge
Cambridge, England.

1978

ACADEMIC PRESS

London New York San Francisco

A Subsidiary of Harcourt Brace Jovanovich, Publishers

ACADEMIC PRESS INC. (LONDON) LTD.
24/28 Oval Road,
London NW1

United States Edition published by
ACADEMIC PRESS INC.
111 Fifth Avenue
New York, New York 10003

Library of Congress Catalog Card Number: 77-93203
ISBN: 0-12-465350-2

Printed in Great Britain
by Galliard (Printers) Limited, Great Yarmouth

CONTRIBUTORS

ABRAHAM, G. *Division of Virology, National Institute for Medical Research, Mill Hill, London N.W.7.*

ALBRECHT, P. *Department of Health, Education and Welfare, Public Health Service, Food and Drug Administration, Bureau of Biologics, Division of Virology, Bethesda, Maryland, 20014, U.S.A.*

ALLEN, R. *Department of Microbiology, The University of Texas Health Science Center, Southwestern Medical School, Dallas, Texas, U.S.A.*

ALMOND, J.W. *Division of Virology, Department of Pathology, University of Cambridge, Laboratories Block, Addenbrooke's Hospital, Hills Road, Cambridge, England.*

APOSTOLOV, K. *Department of Virology, Royal Postgraduate Medical School, London, England.*

ARMSTRONG, M.A. *Department of Microbiology and Immunobiology, The Queen's University of Belfast, Belfast, N. Ireland.*

ATANASIU, P. *Rage Recherche Rhabdovirus, Institut Pasteur, Paris, France.*

AULAKH, G.S. *Department of Health, Education and Welfare, Public Health Service, Food and Drug Administration, Bureau of Biologics, Division of Virology, Bethesda, Maryland, 20014, U.S.A.*

BACHMEYER, H. *Sandoz Forschungsinstitut, A-1235, Vienna, Austria.*

BALL, L.A. *Microbiology Section, Biological Sciences Group, University of Connecticut, Storrs, Connecticut 06268, U.S.A.*

BANERJEE, A.K. *Roche Institute of Molecular Biology, Nutley, New Jersey, 07110, U.S.A.*

BARRETT, T. *Division of Virology, Department of Pathology, University of Cambridge, Laboratories Block, Addenrbooke's Hospital, Hills Road, Cambridge, England.*

BARRY, R.D. *Division of Virology, Department of Pathology, University of Cambridge, Laboratories Block, Addenbrooke's Hospital, Hills Road, Cambridge, England.*

BEAN, W.J., Jr. *Division of Virology, St. Jude Children's Research Hospital, 332 N. Lauderdale, P.O. Box 318, Memphis, Tennessee 38101, U.S.A.*

BEARE, A.S. *M.R.C. Common Cold Research Unit, Harvard Hospital, Salisbury, Wiltshire, England.*

BENEKE, J. *Department of Bacteriology, Iowa State University, Ames, Iowa 50011, U.S.A.*

BENNETT, J.C. *Department of Microbiology and Division of Clinical Immunology and Rheumatology, University of Alabama Medical Center, University of Alabama in Birmingham,*

Birmingham, Alabama, U.S.A.

BERK, W. *Institut für Virologie der Justus-Liebig-Universitat Giessen, Frankfurter Str. 107, 63 Giessen/Lahn, Germany.*

BHOWN, A.S. *Department of Microbiology and Division of Clinical Immunology and Rheumatology, University of Alabama Medical Center, University of Alabama in Birmingham, Birmingham, Alabama, U.S.A.*

BISHOP, D.H.L. *Department of Microbiology, The Medical Center, University of Alabama in Birmingham, Birmingham, Alabama, 35294, U.S.A.*

BOSCH, F.X. *Institut für Virologie an der Veterinarmedizinischen Fakultat der Justus Liebig-Universitat Giessen, 63 Giessen, Frankfurter Strasse 87, Germany.*

BOULOY, M. *Viral Ecology Unit, Institut Pasteur, Paris, France.*

BRAMBILLA, R. *Research Department, Pharmaceuticals Division, CIBA-GEIGY Limited, Basle, Switzerland.*

BRATT, M.A. *Department of Microbiology, University of Massachusetts Medical School, Worcester, Mass. 01605, U.S.A.*

BRONI, B. *The Institute for Cancer Research, Fox Chase Cancer Center, Philadelphia, Pa 19111, U.S.A.*

BROWN, F. *Animal Virus Research Institute, Pirbright, Woking, Surrey, England.*

BROWNSON, J.M. *Division of Virology, Department of Pathology, University of Cambridge, Laboratories Block, Addenbrooke's Hospital, Hills Road, Cambridge, England.*

BUCHMEIER, M.J. *Department of Immunopathology, Scripps Clinic and Research Foundation, La Jolla, California 92037, U.S.A.*

BURDICK, J. *Department of Bacteriology, Iowa State University, Ames, Iowa 50011, U.S.A.*

BUSSEREAU, F. *Bat. 400, Universite Paris-Sud, 91405 Orsay-Cedex, France.*

CALIGUIRI, L.A. *Department of Microbiology and Immunology, Albany Medical College of Union University, Albany, New York, U.S.A.*

CAREY, N.H. *G.D. Searle & Company Limited, Research Division, P.O. Box 53, High Wycombe, Buckinghamshire, HP12 4HL, England.*

CASH, P. *M.R.C. Virology Unit, Institute of Virology, Church Street, Glasgow, G11 5JR, Scotland.*

CHOPPIN, P.W. *The Rockefeller University, New York, N.Y.10021, U.S.A.*

CLARK, H.F. *The Wistar Institute of Anatomy and Biology, 36th Street at Spruce, Philadelphia, Pennsylvania 19104, U.S.A.*

CLEWLEY, J.P. *Department of Microbiology, The Medical Center, University of Alabama in Birmingham, Birmingham, Alabama, 35294, U.S.A.*

CONTRIBUTORS

CLINKSCALES, C.W. *Department of Microbiology and Immunology, Oral Roberts University School of Medicine & Dentistry, Tulsa, Oklahoma 74171, U.S.A.*

CLUXTON, D.H. *Russell Sage College, Troy, New York, U.S.A.*

COLLINS, P.L. *Microbiology Section, Biological Sciences Group, University of Connecticut, Storrs, Connecticut 06268, U.S.A.*

COLONNO, R.J. *Roche Institute of Molecular Biology, Nutley, New Jersey, 07110, U.S.A.*

COMPANS, R.W. *Department of Microbiology, The Medical Center, University of Alabama in Birmingham, Birmingham, Alabama, 35294, U.S.A.*

CONTENT, J. *Department of Virology, Institut Pasteur du Brabant, Brussels 1040, Belgium.*

CONTI, G. *Instituto di Microbiologia, Scuola di Medicina, Universita degli Studi di Parma, Parma 43100, Italy.*

COX, N.J. *World Health Organization Influenza Center, Center for Disease Control, Atlanta, Georgia 30333, U.S.A.*

CRICK, J. *Animal Virus Research Institute, Pirbright, Woking, Surrey, GU24 ONF, England.*

DERMOTT, E. *Department of Microbiology and Immunobiology, The Queen's University of Belfast, Belfast, N. Ireland.*

DIERKS, P.M. *Department of Microbiology, University of Virginia School of Medicine, Charlottesville, Virginia 22901, U.S.A.*

DOEL, T.R. *Animal Virus Research Institute, Pirbright, Woking, Surrey, GU24 ONF, England.*

DUBOVI, E.J. *Department of Microbiology, University of Virginia, Charlottesville, Virginia 22901, U.S.A.*

DUNLAP, R.C. *Department of Health, Education and Welfare, Public Health Service, Food and Drug Administration, Bureau of Biologics, Division of Virology, Bethesda, Maryland, 20014, U.S.A.*

DURAND, D.P. *Department of Bacteriology, Iowa State University, Ames, Iowa 50011, U.S.A.*

EMERSON, S.U. *Department of Microbiology, University of Virvinia School of Medicine, Charlottesville, Virginia 22901, U.S.A.*

ERMINE, A. *Rage Recherche Rhabdovirus, Institut Pasteur, Paris, France.*

ERON, L. *Department of Health, Education and Welfare, Public Health Service, Food and Drug Administration, Bureau of Biologics, Division of Virology, Bethesda, Maryland, 20014, U.S.A.*

FELLNER, P. *G.D. Searle & Company Limited, Research Division, P.O. Box 53, High Wycombe, Buckinghamshire, HP12 4HL, England.*

FIELDS, B.N. *Department of Microbiology and Molecular Genetics, Harvard Medical School, Boston, Massachusetts 02115, U.S.A.*

CONTRIBUTORS

FISHER, D. *Department of Biochemistry and Chemistry, Royal Free Hospital School of Medicine, University of London, 8 Hunter Street, London WC1N 1BP, England.*

FLAMAND, A. *Bât. 400, Université Paris-Sud, 91405 Orsay-Cedex, France.*

FRANZE-FERNANDEZ, M.T. *Departamento de Quimica Biologica, Facultad de Farmacia y Bioquimica, Universidad de Buenos Aires, Argentina.*

FRASER, K.B. *Department of Microbiology and Immunobiology, The Queen's University of Belfast, Belfast, N. Ireland.*

FREEMAN, G.J. *Department of Microbiology and Molecular Genetics, Harvard Medical School, Boston, Massachusetts 02115, U.S.A.*

FUKAI, K. *Research Institute for Microbial Diseases, Osaka University, Suita City, Osaka, Japan.*

GARD, G.P. *Veterinary Research Station, Glenfield, N.S.W., Australia.*

GENTSCH, J. *Department of Microbiology, The Medical Center, University of Alabama in Birmingham, Birmingham, Alabama, 35294, U.S.A.*

GENTY, N. *Universite Paris XI, Institut de Microbiologie, Centre d'Orsay, 91405 Orsay-Cedex, France.*

GERSTEIN, H. *Department of Microbiology and Immunology, Albany Medical College of Union University, Albany, New York, U.S.A.*

GHARPURE, M. *Department of Microbiology and Immunobiology, The Queen's University of Belfast, Belfast, N. Ireland.*

GIMENEZ, H.B. *M.R.C. Virology Unit, Institute of Virology, Church Street, Glasgow, G11 5JR, Scotland.*

GLIMP, T. *Department of Microbiology, The University of Texas Health Science Center, Southwestern Medical School, Dallas, Texas, U.S.A.*

GOTTLIEB, C. *Departments of Microbiology and Immunology, Medicine and Biochemistry, Washington University School of Medicine, St. Louis, Mo. 63110, U.S.A.*

GOULD, E.A. *Departments of Biochemistry and Microbiology, The Queen's University of Belfast, Belfast, N. Ireland.*

GRAVES, M.C. *The Rockefeller University, New York, N.Y. 10021, U.S.A.*

GREGORIADES, A. *Public Health Research Institute of The City of New York, Inc., New York, N.Y. 10016, U.S.A.*

HALL, W.W. *Institute for Virology and Immunobiology, University of Würzburg, Versbacher Landstrasse 7, 8700 Würzburg, W. Germany.*

HALLINAN, T. *Department of Biochemistry and Chemistry, Royal Free Hospital School of Medicine, University of London, 8 Hunter Street, London WC1N 1BP, England.*

CONTRIBUTORS

HALLUM, J.V. *Department of Microbiology and Immunology, University of Oregon Health Sciences Center, Portland, Oregon 97201, U.S.A.*

HARMS, E. *Institut für Virologie der Justus-Liebig-Universitat Giessen Frankfurter Str. 107, 63 Giessen/Lahn, Germany.*

HART, C.A. *Department of Medical Microbiology, The New Medical School, Liverpool University, Pembroke Place, P.O. Box 147, Liverpool L69 3BX, England.*

HAY, A.J. *Division of Virology, National Institute for Medical Research, Mill Hill, London N.W.7., England.*

HERRING, L. *The Institute for Cancer Research, Fox Chase Cancer Center, Philadelphia, Pa. 19111, U.S.A.*

HERRLER, G. *Institut für Medizinische Mikrobiologie der Techn., Universitat München, West Germany.*

HICKS, J.T. *Department of Health, Education and Welfare, Public Health Service, Food and Drug Administration, Bureau of Biologics, Division of Virology, Bethesda, Maryland, 20014, U.S.A.*

HIGHTOWER, L.E. *Microbiology Section, Biological Sciences Group, University of Connecticut, Storrs, Connecticut 06268, U.S.A.*

HOLLAND, J.J. *Department of Biology, University of California, San Diego, La Jolla, California 92093, U.S.A.*

HORISBERGER, M.A. *Research Department, Pharmaceuticals Division, CIBA-GEIGY Limited, Basle, Switzerland.*

HOSAKA, Y. *Research Institute for Microbial Diseases, Osaka University, Suita City, Osaka, Japan.*

HUANG, A.S. *Department of Microbiology and Molecular Genetics, Harvard Medical School, Boston, Massachusetts 02115, U.S.A.*

INGLIS, S.C. *Division of Virology, Department of Pathology, University of Cambridge, Laboratories Block, Addenbrooke's Hospital, Hills Road, Cambridge, England.*

JOHNSON, L.D. *Lahoratory of Molecular Biology, Molecular Virology Section, NINCDS, National Institutes of Health, Bethesda, Maryland 20014, U.S.A.*

KANG, C.Y. *Department of Microbiology, The University of Texas Health Science Center, Southwestern Medical School, Dallas, Texas, U.S.A.*

KAWAI, A. *Institute for Virus Research, Kyoto University, Kyoto 606, Japan.*

KELLER, P. *Department of Physiology, CMDNJ-Rutgers Medical School, Piscataway, New Jersey, U.S.A.*

KENDAL, A.P. *World Health Organization Influenza Center, Center for Disease Control, Atlanta, Georgia 30333, U.S.A.*

KIESSLING, W.R. *Institute for Virology and Immunobiology, University of Würzburg, Versbacher Landstrasse 7, 8700 Würzburg, West Germany.*

CONTRIBUTORS

KINGSBURY, D.W. *Division of Virology, St. Jude Children's Research Hospital, P.O. Box 318, Memphis, Tennessee 38101, U.S.A.*

KLENK, H.-D. *Institut für Virologie der Justus-Liebig-Universitat Giessen, Frankfurter Str. 107, 63 Giessen/Lahn, Germany.*

KO, K.K. *Research Institute for Microbial Diseases, Osaka University, Suita City, Osaka, Japan.*

KOLAKOFSKY, D. *Department of Microbiology, University of Utah Medical Center, Salt Lake City, Utah 84132, U.S.A.*

KORNFELD, S. *Departments of Microbiology and Immunology, Medicine and Biochemistry, Washington University School of Medicine, St. Louis, Mo. 63110, U.S.A.*

KORT, L. *Department of Microbiology, University of Utah Medical Center, Salt Lake City, Utah 84132, U.S.A.*

KRUG, R.M. *Memorial Sloan-Kettering Cancer Center, New York, U.S.A.*

LAM, T. *The Ontario Cancer Institute and Department of Medical Biophysics, University of Toronto, Toronto, Canada.*

LAMB, R.A. *The Rockefeller University, New York, N.Y.10021, U.S.A.*

LANDSBERGER, F.W. *The Rockefeller University, New York, N.Y. 10021, U.S.A.*

LAZZARINI, R.A. *Laboratory of Molecular Biology, Molecular Virology Section, NINCDS, National Institutes of Health, Bethesda, Maryland 20014, U.S.A.*

LEAVITT, R. *Departments of Microbiology and Immunology, Medicine and Biochemistry, Washington University School of Medicine, St. Louis, Mo. 63110, U.S.A.*

LENARD, J. *Department of Physiology, CMDNJ-Rutgers Medical School, Piscataway, New Jersey, U.S.A.*

LEPPERT, M. *Department of Microbiology, University of Utah Medical Center, Salt Lake City, Utah 84132, U.S.A.*

LEUNG, W-C. *Department of Pathology, McMaster University, Hamilton, Ontario, Canada.*

LUCY, J.A. *Department of Biochemistry and Chemistry, Royal Free Hospital School of Medicine, University of London, 8 Hunter Street, London WC1N 1BP, England.*

LYLES, D.S. *The Rockefeller University, New York, N.Y. 10021, U.S.A.*

MAASSAB, H.F. *The University of Michigan, School of Public Health, Ann Arbor, Michigan 48104, U.S.A.*

MADANSKY, C.H. *Department of Microbiology and Molecular Genetics, Harvard Medical School, Boston, Massachusetts 02115, U.S.A.*

MAHY, B.W.J. *Division of Virology, Department of Pathology, University of Cambridge, Laboratories Block, Addenbrooke's Hospital, Hills Road, Cambridge, England.*

CONTRIBUTORS

MARK, G.E. *The Institute for Cancer Research, Fox Chase Cancer Center, Philadelphia, Pa. 19111, U.S.A.*

MARTIN, S.J. *Departments of Biochemistry and Microbiology, The Queen's University of Belfast, Belfast, N. Ireland.*

MATSUMOTO, S. *Institute for Virus Research, Kyoto University, Kyoto 606, Japan.*

McGEOCH, D.J. *Division of Virology, Department of Pathology, University of Cambridge, Laboratories Block, Addenbrooke's Hospital, Hills Road, Cambridge, England.*

MEIER-EWERT, H. *Institut für Medizinische Mikrobiologie der Techn. Universitat München, West Germany.*

MELLON, M.G. *Department of Microbiology, University of Virginia School of Medicine, Charlottesville, Virginia 22901, U.S.A.*

METZEL, P.S. *Department of Microbiology, University of Illinois, Urbana, Illinois 61801, U.S.A.*

MEULEN V. ter *Institute for Virology and Immunobiology, University of Würzburg, Versbacher Landstrasse 7, 8700 Würzburg, W. Germany.*

MONTO, A.S. *University of Michigan, School of Public Health, Ann Arbor, Michigan 48104, U.S.A.*

MOORE, A. *Department of Microbiology and Immunobiology, The Queen's University of Belfast, Belfast, N. Ireland.*

MOORE, N.F. *Unit of Invertebrate Virology, South Parks Road, Oxford OX1 3RX, England.*

NAGAI, Y. *Department of Virology, Cancer Research Institute, Nagoya University School of Medicine, Nagoya, Japan.*

NAGASHIMA, K. *Institute of Virology and Immunobiology, University of Würzburg, Versbacher Landstrasse 7, 8700 Würzburg, W. Germany.*

NAKAMURA, K. *Department of Microbiology, University of Alabama Medical Center, Birmingham, Alabama 35294, U.S.A.*

NORRBY, E. *Department of Virology, Karolinska Institutet School of Medicine, S.B.L., Stockholm 1, Sweden.*

OLDSTONE, M.B.A. *Department of Immunopathology, Scripps Clinic and Research Foundation, La Jolla, California 92037, U.S.A.*

ORLICH, M. *Institut für Virologie, Justus-Liebig-Universitat, Giessen, Germany.*

OXFORD, J.S. *Division of Virology, National Institute for Biological Standards and Control, Holly Hill, London, NW3 6RB, England.*

PALESE, P. *Mount Sinai School of Medicine of CUNY, New York, N.Y. 10029, U.S.A.*

PARSONS, J.T. *Department of Microbiology, University of Virginia School of Medicine, Charlottesville, Virginia 22901, U.S.A.*

CONTRIBUTORS

PATZER, E.J. *Department of Microbiology, University of Virginia School of Medicine, Charlottesville, Virginia 22901, U.S.A.*

PELUSO, R.W. *The Rockefeller University, New York, N.Y. 10021, U.S.A.*

PENNINGTON, T.H. *Department of Virology and MRC Virology Unit, Institute of Virology, University of Glasgow, Glasgow, Scotland.*

PERRAULT, J. *Department of Microbiology and Immunology, Washington University School of Medicine, St. Louis, Missouri 63110, U.S.A.*

PETTERSSON, R.F. *Department of Virology, University of Helsinki, Haartmaninkatu 3, SF-00290, Helsinki 29, Finland.*

PLOTCH, S.J. *Memorial Sloan-Kettering Cancer Center, New York, U.S.A.*

POSTE, G. *Department of Experimental Pathology, Roswell Park Memorial Institute, Buffalo, New York 14263, U.S.A.*

PRINGLE, C.R. *M.R.C. Virology Unit, Institute of Virology, University of Glasgow, Glasgow G11 5JR, Scotland.*

PRITCHARD, D.G. *Department of Microbiology, University of Alabama Medical Center, Birmingham, Alabama 35294, U.S.A.*

RACANIELLO, V.R. *Mount Sinai School of Medicine of CUNY, New York, U.S.A.*

RANKI, M. *Department of Virology, University of Helsinki, Haartmaninkatu 3, SF-00290, Helsinki 29, Finland.*

RAO, D.D. *Department of Microbiology and Molecular Genetics, Harvard Medical School, Boston, Massachusetts 02115, U.S.A.*

REEVE, P. *Sandoz Forschungsinstitut, A-1235, Vienna, Austria.*

REICHMANN, M.E. *Department of Microbiology, University of Illinois, Urbana, Illinois 61801, U.S.A.*

RIMA, B.K. *Departments of Biochemistry and Microbiology, The Queen's University of Belfast, Belfast, N. Ireland.*

ROBERTSON, B.H. *Department of Microbiology and Division of Clinical Immunology and Rheumatology, University of Alabama Medical Center, University of Alabama in Birmingham, Birmingham, Alabama, U.S.A.*

ROHDE, W. *Institut für Virologie der Justus-Liebig-Universitat Giessen, Frankfurter Str. 107, 63 Giessen/Lahn, Germany.*

ROSE, J.K. *Department of Biology and Center for Cancer Research, Massachusetts Institute of Technology, Cambridge, Massachusetts 02139, U.S.A.*

ROTT, R. *Institut für Virologie der Justus-Liebig-Universitat Giessen, Frankfurter Str. 107, 63 Giessen/Lahn, Germany.*

ROY, P. *Department of Microbiology, The Medical Center, University of Alabama in Birmingham, Birmingham, Alabama 35294, U.S.A.*

CONTRIBUTORS

SAKURAI, T. *Research Institute for Microbial Diseases, Osaka University, Suita City, Osaka, Japan.*

SANBORN, M. *Department of Bacteriology, Iowa State University, Ames, Iowa 50011, U.S.A.*

SAWA, M.I. *Department of Virology, Royal Postgraduate Medical School, London, England.*

SCHEID, A. *The Rockefeller University, New York, N.Y. 10021, U.S.A.*

SCHILD, G.C. *Division of Virology, National Institute for Biological Standards and Control, Holly Hill, London NW3 6RB, England.*

SCHITO, G.C. *Instituto di Microbiologia, Facolta' di Medicina E Chirurgia, Universita' Degli Studi di Parma, Italy.*

SCHLESINGER, S. *Departments of Microbiology and Immunology, Medicine and Biochemistry, Washington University School of Medicine, St. Louis, Mo. 63110, U.S.A.*

SCHNITZLEIN, W.M. *Department of Microbiology, University of Illinois, Urbana, Illinois 61801, U.S.A.*

SCHOLTISSEK, C. *Institut für Virologie der Justus-Liebig-Universitat, Giessen, Frankfurter Str. 107, 63 Giessen/Lahn, Germany.*

SCHULMAN, J.L. *Mount Sinai School of Medicine of the City University of New York, 100th Street and Fifth Avenue, New York City, New York, U.S.A.*

SEMLER, B.L. *Department of Biology, University of California, San Diego, La Jolla, California 92093, U.S.A.*

SHAW, M.W. *Department of Microbiology, University of Alàbama Medical Center, Birmingham, Alabama 35294, U.S.A.*

SHIRODARIA, P.V. *Department of Microbiology, The Queen's University of Belfast, Belfast, N. Ireland.*

SILVER, S.M. *The Rockefeller University, New York, N.Y. 10021, U.S.A.*

SKEHEL, J.J. *Division of Virology, National Institute for Medical Research, Mill Hill, London NW7, England.*

SMITH, J.C. *G.D. Searle & Company Limited, Research Division, P.O. Box 53, High Wycombe, Buckinghamshire HP12 4HL, England.*

SMITH, M.D. *Microbiology Section, Biological Sciences Group, University of Connecticut, Storrs, Connecticut 06286, U.S.A.*

SPRAGUE, J.A. *Department of Health, Education and Welfare, Public Health Service, Food and Drug Administration, Bureau of Biologics, Division of Virology, Bethesda, Maryland 20014, U.S.A.*

SPRING, S.B. *Laboratory of Infectious Diseases, National Institutes of Health, Bethesda, Maryland 20014, U.S.A.*

STALLCUP, K.C. *Department of Microbiology and Molecular Genetics,*

CONTRIBUTORS

Harvard Medical School, Boston, Massachusetts 02115, U.S.A.
STANNERS, C.P. *The Ontario Cancer Institute and Department of Medical Biophysics, University of Toronto, Toronto, Canada.*
SZILAGYI, J.F. *M.R.C. Virology Unit, Institute of Virology, University of Glasgow, Glasgow G11 5JR, Scotland.*
TAYLOR, J.M. *The Institute for Cancer Research, Fox Chase Cancer Center, Philadelphia, Pa. 19111, U.S.A.*
TESTA, D. *Roche Institute of Molecular Biology, Nutley, New Jersey, 07110, U.S.A.*
THOMAS, G.P. *Division of Virology, Department of Pathology, University of Cambridge, Laboratories Block, Addenbrooke's Hospital, Hills Road, Cambridge, England.*
TOMASZ, J. *The Institute of Biophysics, Szeged, Hungary.*
TRUANT, A.L. *Department of Microbiology and Immunology, University of Oregon Health Sciences Center, Portland, Oregon 97201, U.S.A.*
TYRRELL, D.L.J. *Division of Infectious Diseases, Department of Medicine, University of Alberta, Edmonton, Alberta, Canada.*
UZGIRIS, E.E. *General Electric Research and Development Center, Schenectady, New York, U.S.A.*
VALCAVI, P. *Instituto di Microbiologia, Facolta' di Medicina E Chirurgia, Universita' Degli Studi di Parma, Italy.*
VEZZA, A.C. *Department of Microbiology, The Medical Center, University of Alabama in Birmingham, Birmingham, Alabama 35294, U.S.A.*
WAGNER, R.R. *Department of Microbiology, University of Virginia, Charlottesville, Virginia 22901, U.S.A.*
WEBSTER, R.G. *Division of Virology, St. Jude Children's Research Hospital, 332 N. Lauderdale, P.O. Box 318, Memphis, Tennessee 38101, U.S.A.*
WECHSLER, S.L. *Department of Microbiology and Molecular Genetics, Harvard Medical School,Boston, Massachusetts 02115, U.S.A.*
WIT, L. de *Department of Virology, Institut Pasteur du Brabant, Brussels 1040, Belgium.*
WOLSTENHOLME, A.J. *Division of Virology, Department of Pathology, University of Cambridge, Laboratories Block, Addenbrooke's Hospital, Hills Road, Cambridge, England.*
WUNNER, W.H. *The Wistar Institute of Anatomy and Biology, 36th Street at Spruce, Philadelphia, Pennsylvania 19104, U.S.A.*
WYNNE, L.R. *Department of Microbiology, The Medical Center, University of Alabama in Birmingham, Birmingham, Alabama 35294, U.S.A.*

PREFACE

Having convened two successful symposia, on large RNA-containing viruses in 1969, and on negative strand viruses in 1973, it was perhaps inevitable that after a further four year interval we should organize a third. We were hardly prepared, however, for the overwhelming response which resulted in this third Cambridge virus symposium, held in King's College from August 1st to 5th, 1977, becoming by far the most comprehensive of the three meetings. Since 1973, our understanding of the molecular biology of the group of viruses which we then termed 'Negative Strand Viruses' (the influenza, parainfluenza, and rhabdoviruses) has expanded enormously. In addition, we now recognize two further groups, the arenaviruses (which include viruses such as lymphocytic choriomeningitis and Lassa fever) and the bunyaviruses (a large group of insect-borne viruses) as having a negative-stranded genome.

The title of the symposium emphasises the fact that an intimate association with normal cell processes is, for many of these viruses, an essential feature of the replication cycle. It is also becoming apparent that the cell itself may regulate those properties of the virus most closely associated with pathogenesis, such as the ability to persist without killing the host cell, and the production of defective interfering forms.

The volume contains eighty-five papers, mostly based on talks given at the symposium, which offer an up-to-date and full account of current knowledge of these viruses. The papers are grouped into nine sections which at the meeting were chaired respectively by D.W. Kingsbury (Nature of the Genome), F. Brown and C. Scholtissek (Structure and synthesis of virus-specified products), D.H.L. Bishop, B.W.J. Mahy and D. Kolakofsky (Genome transcription and translation), R.D. Barry (Aspects of virus replication), P.W. Choppin (Host cell modification), R.R. Wagner (Defective virus formation), R. Rott (Aspects of virulence), M.A. Bratt (Genetics), and R.W. Compans (Membranes and assembly).

Our thanks are due to all those who attended and contributed to the success of the meeting and to all the authors for providing us with such an excellent series of manuscripts from which to compile this book. We are particularly indebted to our secretaries, Brenda Wood who helped with the meeting and then cheerfully embarked on the unenviable task of typing this book, and Mary Wright who was able to lift some of the work from all our shoulders. We would like to thank Pat Davies, who single-handedly produced the index, Jim Robertson who helped with the art-work, and Tom Barrett, Jenny Brownson, Steve Inglis, Stuart Nichol, Paul Thomas, and Adrian Wolstenholme, who helped with the proof-reading. The unfailing guidance of Mrs. Judy Sibley, of Academic Press, is also gratefully acknowledged.

Finally, we would like to acknowledge the financial support, so necessary to the success of this meeting, which was provided by Beecham Pharmaceuticals Research Division; The Boots Company Ltd.; Imperial Chemical Industries Ltd., Pharmaceuticals Division; Geigy Pharmaceuticals; Philips-Duphar B.V.; Pye-Unicam Ltd.; Recherche et Industrie Therapeutiques; Sandoz Forschungsinstitut; and G.D. Searle & Co. Ltd.

B.W.J. Mahy
R.D. Barry

Division of Virology,
Department of Pathology,
University of Cambridge,
Laboratories Block,
Addenbrooke's Hospital,
Hills Road
Cambridge, CB2 2QQ,
England.

CONTENTS

STRUCTURE AND FUNCTION OF THE INFLUENZA VIRUS GENOME

R.D. BARRY, J.W. ALMOND, D.J. McGEOCH
S.C. INGLIS and B.W.J. MAHY

Division of Virology, Department of Pathology, University of Cambridge, Laboratories Block, Addenbrookes Hospital, Hills Road, Cambridge, England.

Early attempts to define the size of isolated influenza virus RNA indicated that it was smaller than anticipated, and probably consisted of a number of components (9, 10). These data supported the notion that "recombination" is the result of reassortment of independent genetic elements (2, 5, 15). Subsequent analysis of influenza virus RNA by acrylamide gel electrophoresis partially resolved at least six components (4, 8, 31). Even so, for more than a decade there remained the doubt that the apparent segmentation of the genome might be artifactual. However, the almost simultaneous application by a number of laboratories of a greatly improved fractionation procedure for influenza virus RNA, consisting of acrylamide gel electrophoresis in the presence of high levels of urea (13), rapidly established that the genome consists of a definite number of individual RNA segments. It was found that RNA from avian influenza A (20, 29) and various strains of human influenza A (3, 22, 25, 27) could be fractionated into eight species by electrophoresis. The separate identity of each species was established by two-dimensional oligonucleotide fingerprinting; this demonstrated that each species has a distinct nucleotide sequence (20). Furthermore the molecular weights and molar representation of the individual RNA segments suggest that individual virus particles contain on average 8 distinct genes (20).

Until recently there have also been doubts and uncertainties concerning the number and nature of the gene products of the influenza virus genome. For instance, it was reported that the high molecular weight, structural component associated with polymerase activity (P protein) consisted of one (6, 19, 30) or two polypeptides (33). Using high resolution polyacrylamide gel electrophoresis, we detected eight distinct gene products in chick embryo cells infected with fowl plague (influenza A) virus (16). These comprise three polymerase-associated polypeptides (P_1,P_2,P_3), haemagglutinin (HA), nucleoprotein (NP), neuraminidase (NA), membrane polypeptide (M), and a non-structural polypeptide (NS). A comparison of the electrophoretic separation of the virus-induced polypeptides of two strains of avian influenza A, Rostock and Dobson, is shown in Fig.1. The three P polypeptides were detected both in virions and infected cells, and their separate identity was established by two-dimensional tryptic peptide mapping. One continuing uncertainty concerns the smaller, non-structural protein - NS_2 (32), for which we do not find an RNA of corresponding molecular weight. In our experience, NS_2 is not found consistently in productive infections of chick cells by FPV, although it is regularly found in FPV-infected BHK cells; it is still uncertain whether

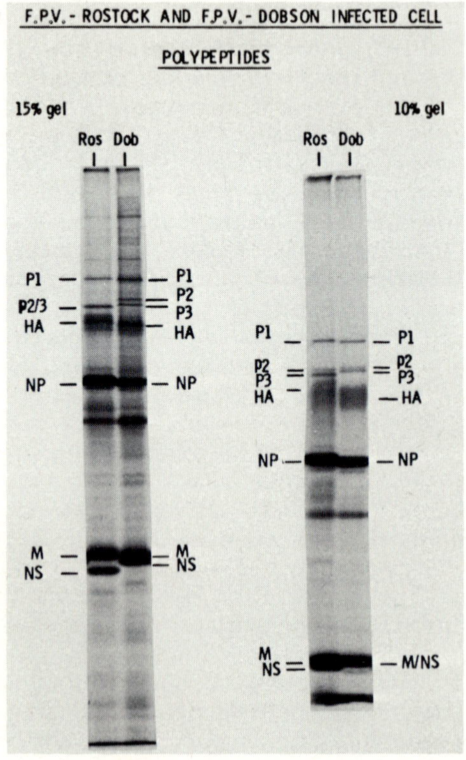

Fig.1.

NS_2 is a primary gene product.

The maximum polypeptide-coding capacity of each of the 8 virion RNA segments has been estimated, and they agree closely in decreasing order of size with the virion proteins as follows, P_1, P_2, P_3, HA, NP, NA, M and NS (20). It has been emphasized more recently however (1, 21) that the apparent size of RNA molecules does not always correlate with the size of the proteins for which they code. It is essential to establish a direct relation between gene and product before definite assignment of coding functions is possible. In this paper we consider two procedures that have enabled us to determine the polypeptide coding potential of each of the 8 RNA segments of FPV.

IN VITRO TRANSLATION

Influenza virus is a negative strand virus, that is, the virion RNA is complementary to the RNA with messenger activity (11, 14). In cell-free systems, such as wheat-germ, RNA extracted from virions does not induce synthesis of recognizable virus proteins (7, 12, 26, 34). However, RNA extracted from infected cells has been reported to direct the synthesis of up to 5 virus-specific polypeptides (7, 12, 26). We have found that cytoplasmic RNA extracts from chick cells infected with FPV can be translated in the wheat germ cell-free protein-synthesising system, to produce at least 7 polypeptides; 6 had electrophoretic mobilities corresponding to P_1, P_2, P_3, NP, M and NS, while the tryptic peptide map of the seventh polypeptide indicated that it was the non-glycosylated (HA_0) form of the HA glycoprotein (17).

We have used this system to identify the protein coding functions of the genome segments, using a technique involving hybridization of individual segments to mixtures of viral mRNAs. After hybridization, the species of mRNA complementary in sequence to the genome RNA segment is converted into a double-stranded form and so can no longer be translated in the cell-free system. Synthesis of the protein coded by that segment is therefore specifically reduced.

A preparation of purified FPV RNA, containing some ^{32}P-labelled RNA, was electrophoresed on a 2% acrylamide, 0.6%

Fig. 1. *Electrophoresis of FPV Dobson and FPV Rostock virus-infected cell polypeptides on 15% and 10% polyacrylamide slab gels. CEF monolayers were pulse-labelled for 15 min with ^{35}S methionine at 4 h p.i., and whole cell lysates were subjected to electrophoresis. Migrational differences were observed in all virus-specific polypeptides except the Matrix (M) on at least one gel.*

agarose slab gel for 16 hr at 180V, and the individual segments were detected by autoradiography. Under these circumstances, all segments were clearly separated with the exception of segments 1, 2, 3 (which were subsequently treated as a single band) and were extracted from the gel (17). Each segment was then annealed separately, and in excess to a preparation of total cytoplasmic RNA from FPV infected cells. Hybridization was performed in dimethyl sulphoxide, using a method based on that described by Ito and Joklik (18). The RNAs were precipitated, dissolved in water, and added directly to the wheat germ cell-free system. Comparison of the cell-free products of translation of infected cell RNA which had been annealed either in the presence or absence of each genome segment indicated which segment carried the genetic information for each virus polypeptide. These data have been presented fully elsewhere (17), and the results are summarised in Table I, where the relative proportions of virus polypeptides synthesized *in vitro* are shown in the presence or absence of annealing.

TABLE I

Viral polypeptides synthesised *in vitro* in response to infected cell RNA pre-annealed with individual genome segments

Treatment of infected cell RNA	Relative amounts of radioactivity in viral polypeptides[a]			
	HA_o	NP	M	NS
Unannealed	0.14	0.52	1.0	0.52
Self-annealed	0.15	0.40	1.0	0.45
Annealed with				
Segments 1-3	0.14	0.38	1.0	0.44
Segment 4	0.01	0.33	1.0	0.43
Segment 5	0.08	0.06	1.0	0.46
Segment 6	0.16	0.28	1.0	0.45
Segment 7[b]	0.12	0.35	0.38	0.49
Segment 8	0.15	0.32	1.0	0.05

[a]Results were obtained by microdensitometer scanning of autoradiogram tracks. Peaks corresponding to viral polypeptides were cut out and weighed. For each track the sizes of the viral polypeptide peaks were normalized to a value of 1.0 for the M protein, except for track 7[b].

[b]Viral polypeptide peak sizes in this track were normalized to an average value of 0.35 for NP.

The most abundant virus polypeptide synthesised *in vitro* was M. Annealing of cell RNA with genome segment 7 resulted in a large reduction in the synthesis of the M polypeptide relative to control, while synthesis of the other virus specific polypeptides was unaffected. Similarly, annealing of infected cell RNA with genome segments 4, 5 and 8 specifically reduced the synthesis of polypeptides HA_0, NP, and NS respectively. Annealing with segment 5 appeared also to reduce synthesis of HA_0, although to a much smaller degree; this is probably not significant since it has not been consistently observed in other experiments. The results obtained for annealing the pooled segments 1-3 are not shown because the P polypeptides are not synthesised in large quantities. However, when the experiment was repeated on a larger scale (17) it was apparent that segments 1-3 suppressed synthesis of all 3 P polypeptides. We have been unable to detect the synthesis of NA polypeptide in this cell-free system but since all 7 other virus specific polypeptides have been ascribed to genome segments, we conclude that segment 6 codes for the NA polypeptide.

ANALYSIS OF RECOMBINANTS

An unexpected bonus provided by the successful electrophoretic separation of influenza RNA genome segments is that, under standard conditions, various strains have different gel migration patterns for some or all of their RNA segments (21). A detailed analysis of the RNA's of the PR8 and HK human influenza A strains, for instance, indicated that each gene had a characteristic migration rate, and that the parental derivation of all 8 genes of PR8/HK recombinants could thus be determined readily (22, 23, 28, 29).

We have developed a method for assigning ts mutants of FPV to individual RNA segments, also based on the analysis of recombinants. We have isolated 29 ts mutants of the Rostock strain of FPV and have assigned them to 6 recombination groups by analysis of recombination frequencies obtained from pairwise crosses. For each recombination group, a series of recombinants was produced using the FPV strain Dobson as the other parent. Recombinants containing predominantly Rostock genes can be obtained by pre-irradiation with U.V. light of the Dobson parent (1). The Rostock gene bearing the ts lesion was detected in each case by comparing the electrophoretic migration of the RNA's and proteins of the recombinants contain the Dobson gene equivalent to that carrying the ts mutation, and its product can usually be detected amongst the virus-induced proteins found in infected cells.

An example of the method, described in detail elsewhere (1)

is presented briefly here. The group IV mutant, ts 47, possessed the unusual feature that its NS protein migrated significantly faster than the corresponding wild-type virus. A series of recombinants (labelled Di 47c or Di 47) were obtained, and the gel migration of their RNA segments, and induced proteins was determined. The findings are summarized in Table II, when the parental origin of each component is indicated. It can be seen that with the identifiable polypeptides, only NS is consistently of Dobson origin. Thus Rostock ts 47, and other members of recombination group IV, has a ts non-structural polypeptide which is encoded by RNA segment 8.

When other recombination groups were analysed in this way, it was possible to determine for each the RNA segment that carried the ts mutation, and the ts protein for which it coded. These results are summarized in Table III. Only in the case of the NA and M polypeptides have we been unable, by this method, to assign a corresponding gene.

Despite the generally close agreement in size and coding capacity between the RNA segments and their products, we have encountered several anomalies.

1. There is not an absolute correspondence between size of the RNA segment and the size of the protein for which it codes. For example, Rostock RNA segment 1 codes for P_2 (Table III).

2. The RNA segments do not migrate in the same order in different strains of virus. We find that Dobson segment 5 is equivalent to Rostock segment 6.

3. The P proteins do not migrate in the same order in the two strains examined. This is illustrated in Fig.2, where the migration patterns of the P proteins of Rostock and Dobson are diagramatically shown. Recombinants involving the P_2 and P_3 proteins were found to contain either the P_2 from both parents and no P_3 or the P_3 from both parents and no P_2. This indicates that the Rostock P_2 polypeptide is functionally equivalent to Dobson P_3, and Rostock P_3 is equivalent to Dobson P_2.

CONCLUSIONS

By means of the two types of experimental approach outlined above we have been able to assign gene functions to all 8 RNA segments of the Rostock strain of FPV. This is illustrated in Fig.3. With the exception of genes for the P polypeptides, there is a direct correlation in size between RNA segments and their products. However, segment 1 codes for P_2, segment 2 for P_1 and segment 3 for P_3. With the Dobson strain, our assignment of coding functions to the three largest genes was less definite because of the difficulty in separating them reproducably. However we tentatively concluded that genome segment 1 codes for P_3, and genes 2 and 3 together code for P_1 and P_2. The difference

TABLE II

Rostock *ts* 47 - U.V. Dobson *ts* i Recombinants

Protein	Proteins								Segment	RNA							
	Di 47c				Di 47					Di 47c				Di 47			
	1	3	4	6	2	6	9	12		1	3	4	6	2	6	9	12
P_1	R	R	R	R	R	R	R	R	1	R	R	R	R	R	R	R	R
P_2	R	R	R	R	R	R	R	R	2	?	?	?	?	?	?	?	?
P_3	R	R	R	R	R	R	R	R	3	R	R	R	R	R	R	R	R
HA	R	R	R	R	R	R	R	R	4	R	R	R	R	R	R	R	R
NP	R	R	R	R	R	R	D	R	5	R	R	R	R	R	R	D	R
NA	-	-	-	-	-	-	-	-	6	R	R	R	R	R	R	R	R
M	?	?	?	?	?	?	?	?	7	R	R	R	R	R	R	R	R
NS	D	D	D	D	D	D	D	D	8	D	D	D	D	D	D	D	D

TABLE III

Characterization of temperature sensitive mutants of influenza A/FPV Rostock virus mutants

Groups	RNA segment	Defective protein
I	5	NP
II	2	P_1
III	1	P_2
IV	8	NS
V	3	P_3
VI	4	HA

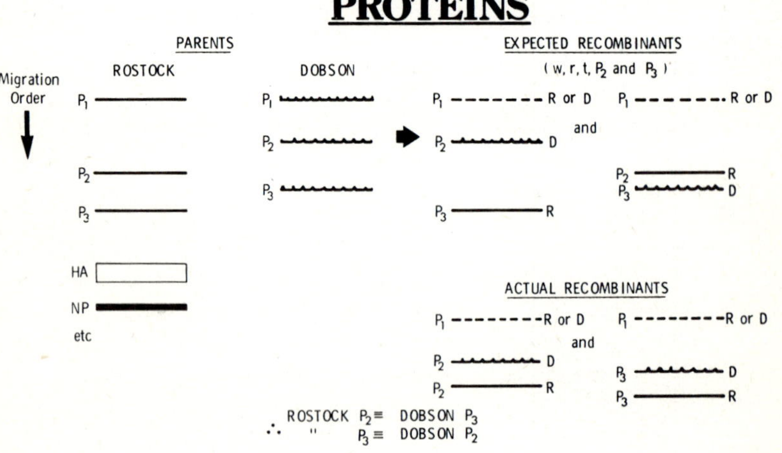

Fig. 2. *Diagrammatical representation of the gel migration of the P proteins of the Rostock and Dobson strains of FPV, and the expected and actual migration pattern obtained in the analysis of recombinants of FPV Rostock recombination group III and Dobson.*

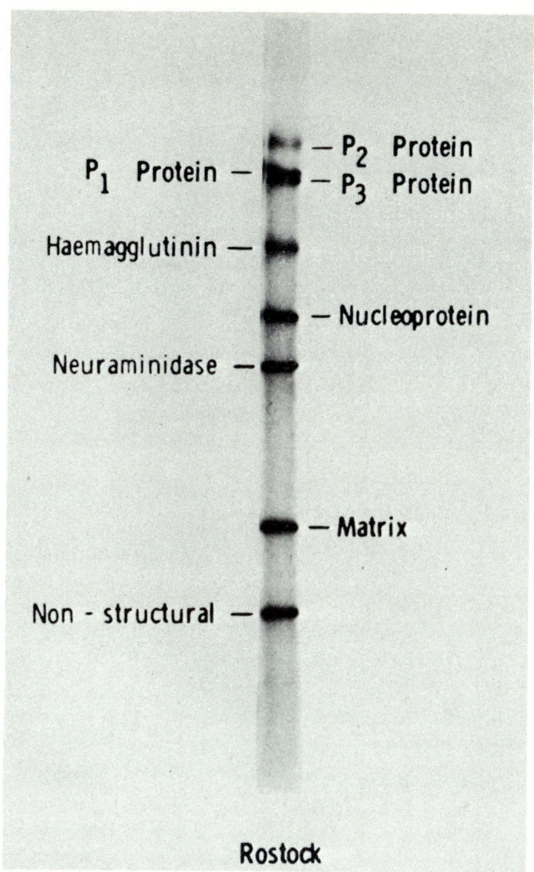

Fig. 3. *Assignment of coding activities to the separated RNA segments of influenza A/FPV Rostock*

---

in migration order of the P proteins and some of the RNA segments poses difficulties for the comparison of the ts phenotypes and of RNA segments between different strains of influenza virus A change in the nomenclature of the 3 P proteins is an obvious necessity.

REFERENCES

1. Almond, J.W., McGeoch, D. and Barry, R.D. (1977). *Virology* 81, 62.
2. Barry, R.D. (1961). *Virology* 14, 398.
3. Bean, W.J.Jr. and Simpson, R.W. (1976). *J. Virol.* 18, 365.
4. Bishop, D.H.L., Obijeski, J.F. and Simpson, R.W. (1971). *J. Virol.* 8, 74.

5. Burnet, F.M. (1956). *Science* 123, 1101.
6. Compans, R.W., Klenk, H.-D., Caliguiri, L.A. and Choppin, P.W. (1970). *Virology* 42, 880.
7. Content, J. (1976). *J. Virol.* 18, 604.
8. Content, J. and Duesberg, P.H. (1971). *J. Mol. Biol.* 62, 273.
9. Davies, P. and Barry, R.D. (1966). *Nature, Lond* .211, 384.
10. Duesberg, P.H. and Robinson, W.S. (1967). *J. Mol. Biol.* 25, 383.
11. Etkind, P.R. and Krug, R.M. (1974). *Virology* 62, 38.
12. Etkind, P.R. and Krug, R.M. (1975). *J. Virol.* 16, 1464.
13. Floyd, R.W., Stone, M.P. and Joklik, W.K. (1974). *Anal. Biochem.* 59, 599.
14. Glass, S.E., McGeoch, D. and Barry, R.D. (1975). *J. Virol.* 16, 1435.
15. Hirst, G.K. (1962). *Cold Spring Harbor Symp. Quant. Biol.* 27, 303.
16. Inglis, S.C., Carroll, A.R., Lamb, R.A. and Mahy, B.W.J. (1976). *Virology* 74, 489.
17. Inglis, S.C., McGeoch, D.J. and Mahy, B.W.J. (1977). *Virology* 78, 522.
18. Ito, Y. and Joklik, W.K. (1972). *Virology* 50, 189.
19. Klenk, H.-D., Scholtissek, C. and Rott, R. (1972). *Virology* 49, 723.
20. McGeoch, D., Fellner, P. and Newton, C. (1976). *Proc. Nat. Acad. Sci USA* 73, 3045.
21. Palese, P. (1977). *Cell* 10, 1.
22. Palese, P. and Schulman, J.L. (1976). *J. Virol.* 17, 876.
23. Palese, P. and Schulman, J.L. (1976). *Proc. Nat. Acad. Sci. USA* 73, 2142.
24. Palese, P., Ritchey, M.B. and Schulman, J.L. (1977). *Virology* 76, 114.
25. Pons, M.W. (1976). *Virology* 69, 789.
26. Ritchey, M.B. and Palese, P. (1976). *Virology* 72, 410.
27. Ritchey, M.B., Palese, P. and Kilbourne, E.D. (1976). *J. Virol.* 18, 738.
28. Ritchey, M.B., Palese, P. and Schulman, J.L. (1976). *J. Virol.* 20, 307.
29. Scholtissek, C., Harms, E., Rohde, W., Orlich, M. and Rott, R. (1976). *Virology* 74, 332.
30. Schulze, I. (1970). *Virology* 42, 890.
31. Skehel, J.J. (1971). *J. gen. Virol.* 11, 103.
32. Skehel, J.J. (1972). *Virology* 49, 23.
33. Skehel, J.J. and Schild, G.C. (1971). *Virology* 44, 396.
34. Tekamp, P.A. and Penhoet, E.E. (1976). *J. Virol.* 18, 812.

STUDIES ON THE GENOME STRUCTURE OF INFLUENZA A VIRUSES

W. Rohde, E. Harms, and C. Scholtissek

Institut für Virologie der Justus-Liebig-Universität Giessen
Frankfurter Str. 107, 63 Giessen/Lahn
Federal Republic Germany.

INTRODUCTION

The influenza virus genome consists of single-stranded RNA which can be resolved into eight distinct segments (genes) by polyacrylamide gel electrophoresis (PAGE) in the presence of urea (3). Using ^{32}P-labeled individual genes of various prototype strains and the complementary RNA (cRNA) of specific recombinants, we have been able to assign gene functions to six of the eight genes of fowl plague virus (FPV) (6). Comparative analysis of RNA patterns and protein fingerprints of recombinants with known gene constellation has completed the correlation between gene and gene functions for FPV and virus N. Part of these data will be presented in this publication. Furthermore, while the recombinant strains used in these studies have emerged through reassortment of total genes, we have obtained evidence that intragenic recombination may also take place. Recombinant strains derived from the "partial heterozygote" B 19/N 4 carry a nucleocapsid protein (NP) gene that contains both sequences of FPV and virus N. These recombinants show an abnormal replication pattern in that chick embryo fibroblasts (CEF) infected under standard conditions produce fully infectious virus as well as noninfectious particles. Some of the biochemical properties of these particles will be described.

THE GENOME MAP OF FPV AND VIRUS N

The recombinant strains 8/19/10 and 19/Eq 2 used in these studies have previously been shown to contain all genes from FPV with the exception of segments 5 and 8 which are derived from virus N and the equine 2 virus, respectively (6). As a consequence, tryptic peptide maps of the corresponding gene products should indicate which proteins are coded for by segments 5 and 8. Fig.1 summarises such comparative analysis.

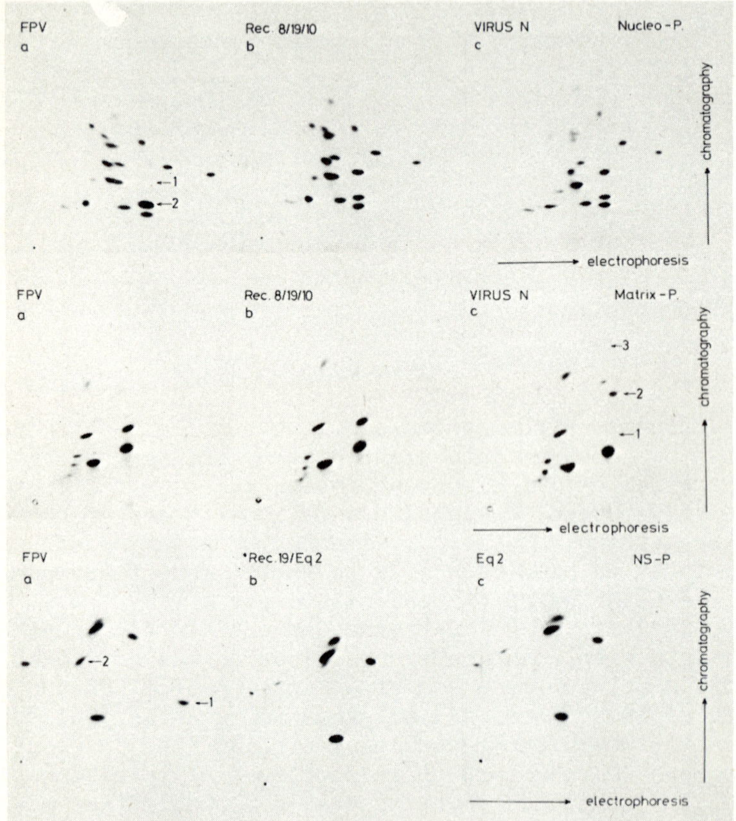

Fig.1. *Comparative tryptic peptide maps of the NP, M, and NS (NS 1) proteins from the wild-type strains FPV (Rostock), virus N, and equine 2 as well as the recombinant strains 8/19/10 and 19/Eq 2. Details on the preparation and analysis of tryptic digests will be presented elsewhere. Upper panel: Analysis of the NP protein; middle panel: analysis of the M protein; lower panel: analysis of the NS (NS 1) protein. Arrows indicate major map differences.*

While the matrix protein M is FPV-specific (middle panel) and thus represents the gene product of segment 7 (4, 6), tryptic fingerprints of the NP and NS proteins are identical with the ones obtained for the polypeptides from virus N and the equine

2 virus (upper and lower panel). Segment 8 is much too small to code for the NP protein; therefore, segment 5 codes for NP, while segment 8 codes for the NS protein. These results complete the map of the FPV genome; virus N has an identical gene order, while in equine 2 virus the gene order of the neuraminidase (segment 5) and NP (segment 6) genes is reversed.

ANALYSIS OF THE NP GENE OF THE RECOMBINANT 413 1, 1

During our studies on gene constellation and gene expression, recombinant strains were isolated which give rise to peculiar virus-host cell interaction (see below) and which are characterized by a particular NP gene. The recombinant strain 413 1, 1 used in the following studies as a representative virus strain, was obtained by sequential passage of the "partial heterozygote" B 19/N 4 isolated after double infection of CEF at 40° with virus N and FPV *ts* 19, a mutant temperature-sensitive in the NP protein (5). The gene constellation of 413 1, 1 was determined by molecular hybridization to comprise the FPV-specific genes 1, 2, 4, 6, 7, 8; gene 3 is derived from virus N. Gene 5 seems to contain sequences of both virus N and FPV. When the ^{32}P-labeled segment 5 of 413 1, 1 is hybridized with the cRNA's of FPV, virus N, and 413 1, 1, hybridization with the cRNA's of both wild-type strains is significantly below 100% (72% for FPV; 84% for virus N) as compared to the RNase resistance obtained with a mixture of both cRNA's or with the cRNA of 413 1, 1. From the previously determined genetic relatedness between the NP genes of FPV and virus N (57% in the presence of formaldehyde; ref.6) 35% of the sequences in the 413 1, 1 NP are calculated to be FPV-specific, while 65% should be derived from virus N. Furthermore, thermal transition of the double-stranded RNA's formed with ^{32}P-labeled segment 5 from FPV and the cRNA's of FPV, virus N, and 413 1, 1 differs significantly; the T_m of the hybrid 413 1, 1/FPV is between the T_m of the heterologous hybrid virus N/FPV and homologous double strand FPV/FPV. However, no difference could be picked up in the T_1 fingerprints of the ^{32}P-labeled NP genes from virus N and 413 1, 1 (data not shown). The close relatedness of the virus N and 413 1, 1 genes is evident in the tryptic peptide map of the corresponding gene products. While the NP proteins from FPV and virus N differ substantially as indicated by arrows 1 and 2 (Fig.2, upper panel), a major difference for virus N and 413 1, 1 is only observed for spot 3. When this ^{35}S-methionine-labeled peptide is eluted and cochromatographed on chromobead columns in a pH gradient with tryptic digests of ^{3}H-methionine-labeled NP from virus N and FPV, none of the virus N-specific peptides elutes at the exact position of spot 3 (Fig.2, middle and lower panel). These data favour the idea

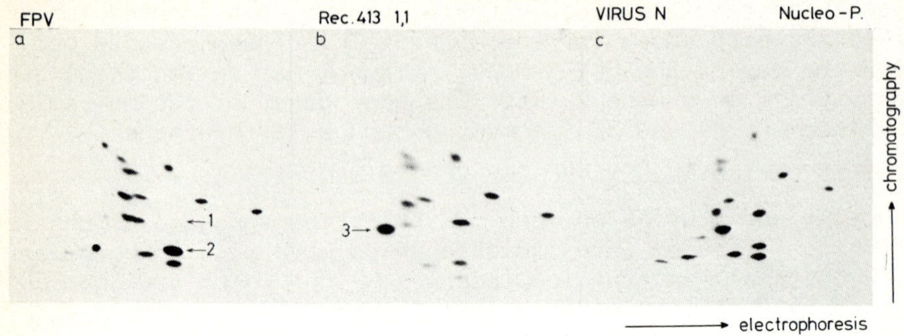

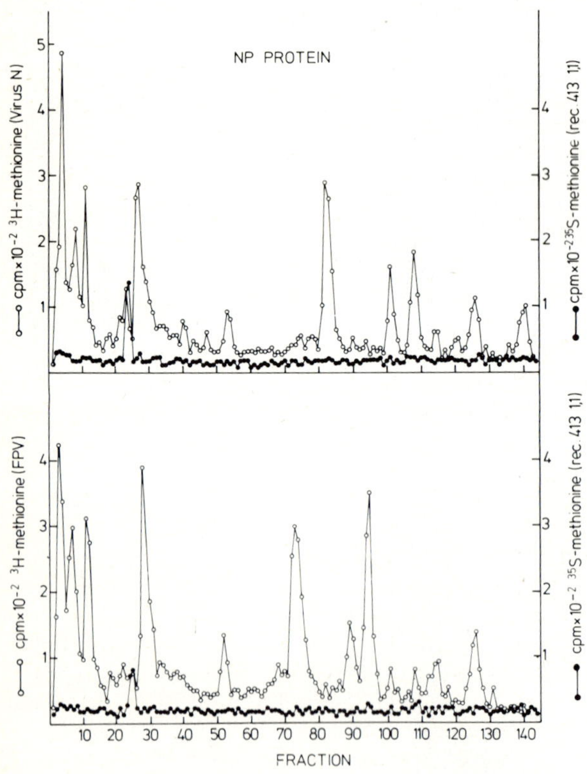

Fig.2. *Characterization of the NP proteins from FPV, virus N, and 413 1, 1. Upper panel: Two-dimensional peptide map. Middle and lower panel: Cochromatography of spot 3 with ^3H-methionine-labeled tryptic digests of the NP proteins from virus N and FPV.*

that the NP gene of 413 1, 1 is a result of intragenic recombination between the corresponding genes of FPV *ts* 19 and virus N.

CHARACTERIZATION OF INFECTIOUS AND NONINFECTIOUS PARTICLES

The recombinant strain 413 1, 1 and all virus strains carrying the NP gene as characterized above exhibit an unusual replication pattern in CEF. Fig.3a shows an equilibrium density gradient of particles produced in CEF by 413 1, 1 under standard conditions of replication. Two populations (H- and L-particles) separate due to differing buoyant densities. Using the glucosamine-label as an indicator for the presence of glycoproteins, almost all neuraminidase (NA) and hemagglutinin (HA) activity as well as infectivity resides within the H-population. Residual infectivity in the L-particle fraction can be abolished by adsorption to glutaraldehyde-fixed erythrocytes. In addition to the absence of virus-specific glycoproteins, SDS-PAGE analysis reveals that also M protein is absent in L-particles (Fig.3b). Autoradiograms of ^{35}S-methionine labeled proteins from L-particles separated on slab gels indicate that L-particles contain the proteins of the viral RNA polymerase. As with infectious H-particles, L-particles contain all eight RNA segments with a reduced amount of genes 1, 2, and 3 coding for the polymerase complex; furthermore, they do not differ qualitatively with regard to phospholipid composition (data not shown). Although they appear to be very fragile, spikeless particles similar in size and shape to H-particles can be visualized by electronmicroscopy. The data obtained with L-particles suggest that they consist of RNP's surrounded by a lipid bilayer.

SUMMARY AND DISCUSSION

Chick embryo cells infected with the recombinant strain 413 1, 1 produce two populations of virus and virus-like particles under standard conditions of infection. Biochemical analysis of the light fraction (L-population) suggests that these particles represent enveloped RNP's containing all eight RNA segments in a complex with the NP protein and viral RNA polymerase. The occurrence of an L-population has also been reported recently during replication of wild-type influenza strains at suboptimal temperatures (2). These particles are characterized by a reduced M protein content, while all other virus-specific proteins are present in seemingly normal amounts, and thus bear no resemblance to the ones described by us. However, CEF infected by FPV which has undergone serial undiluted passage through the embryonated egg (von Magnus virus) yields two populations of particles of high and low buoyant densities. By analogy with the

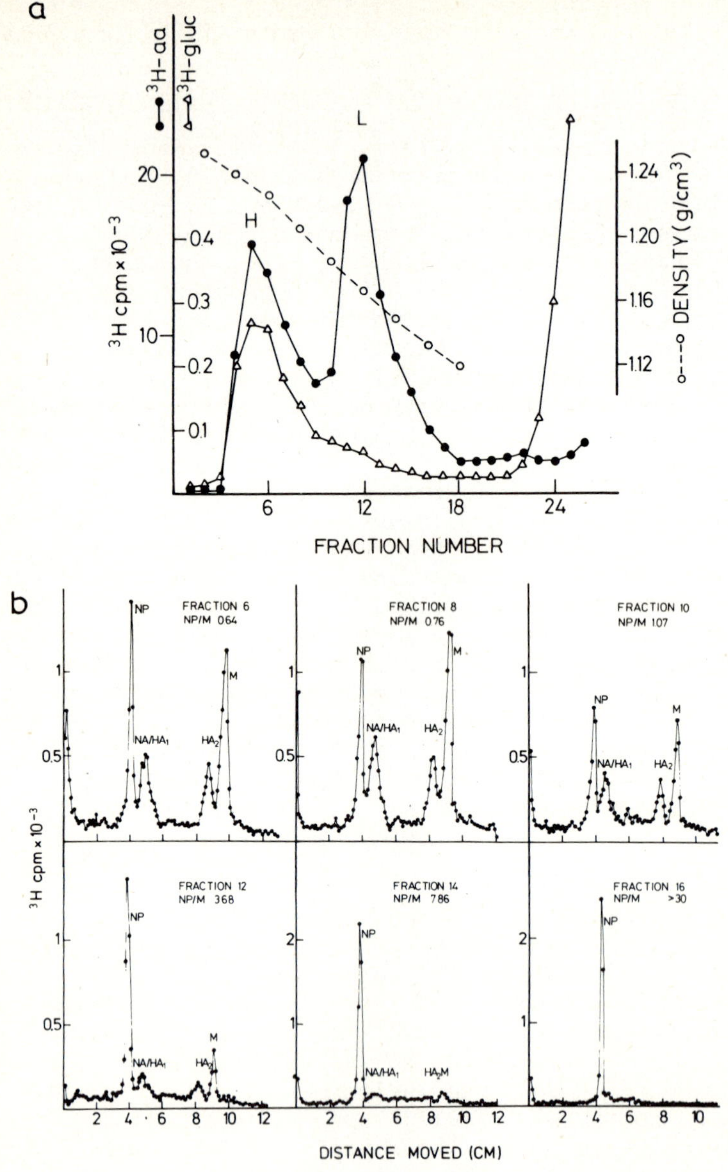

Fig.3. *Analysis of particles produced by CEF infected with the recombinant strain 413 1, 1. Conditions for infection, labeling and purification of virus have been described in detail (4). (a) Equilibrium density centrifugation. Virus and virus-like particles were banded in linear 20-70% (w/v) sucrose gradients*

L-fraction of 413 1, 1, von Magnus L-particles contain all eight RNA genes (segments 1, 2, and 3 in reduced amounts) and lack HA and NA as well as M, (Rohde, unpublished results).

The occurrence of L-particles in association with influenza strains carrying the NP gene as represented in the recombinant strain 413 1, 1 warrants further investigation. The biochemical data obtained for this gene favor the idea that it has resulted by recombination between the NP genes of FPV (*ts* 19) and virus N. We suggest that as a consequence of the production of an altered NP protein, the interaction of the NP protein with the RNA genome might not be optimal and that the formation of altered RNP's gives rise to the phenomena observed.

Genetic recombination in RNA viruses has been studied extensively by marker rescue in the case of poliovirus (1). The influenza virus system with the existence of single genes represents an opportunity for the direct demonstration of genetic recombination by biochemical techniques.

REFERENCES

1. Cooper, P.D. (1968). *Virology* 35, 584.
2. Kendal, A.P., Galphin, J.C., and Palmer, E.L. (1977). *Virology* 76, 186.
3. McGeoch, D., Fellner, P., and Newton, C. (1976). *Proc. Nat. Acad. Sci. U.S.A.* 73, 3045.
4. Rohde, W., Harms, E., and Scholtissek, C. (1977). *Virology* 79, 393.
5. Scholtissek, C., and Bowles, A. (1975). *Virology* 67, 576.
6. Scholtissek, C., Harms, E., Rohde, W., Orlich, M., and Rott, R. (1976). *Virology* 74, 332.

---

in TE (10 mM TRIS HCl pH 7.4; 1 mM EDTA), and aliquots withdrawn from each fraction for the determination of radioactivity. Populations labeled with ^{3}H-amino acids or ^{3}H-glucosamine were run in parallel gradients. (b) Polyacrylamide gel electrophoresis of ^{3}H-amino acid labeled viral proteins associated with different fractions of an equilibrium density gradient. Virus or virus-like particles were pelleted from the indicated fractions and subjected to SDS-PAGE in 15% polyacrylamide gels as described (4). The ratio NP/M was calculated by integration of radioactivity present in the position of the NP and M protein.

ASSIGNMENT OF GENE FUNCTIONS TO RNA SEGMENTS OF INFLUENZA A2-SINGAPORE AND GENETIC RELATEDNESS TO OTHER INFLUENZA STRAINS

C. SCHOLTISSEK, W. ROHDE, E. HARMS, AND R. ROTT

Institut für Virologie der Justus-Liebig-Universität Giessen
Frankfurter Str. 107, 63 Giessen/Lahn
Federal Republic Germany.

INTRODUCTION

Influenza viruses have a segmented genome in which each RNA segment carries the genetic information for the synthesis of one individual protein. The eight RNA segments can be separated by polyacrylamide gel electrophoresis using the method of Floyd *et al.* (2) (1, 4, 5, 7, 8, 10, 12).

The assignment of viral functions and/or proteins to the various RNA segments has been carried out by three different methods: (1) by molecular hybridization with the complementary RNA (cRNA) of specific recombinants (10, 12, see also chapter 2). (2) by comparison of the electrophoretic migration rates of the RNA segments and proteins of recombinant viruses with those of their parent strains (6, 9) and (3) by *in vitro* translation of cRNA into virus-specific proteins (3). During these studies it was found that the order of the migration rates of the segments 5 and 6 coding for the nucleocapsid protein (NP) and neuraminidase (NA), respectively, are reversed when certain influenza strains are compared (6, 10).

In this communication two different topics will be treated: (1) it will be shown that the gene order (migration rates) of segments 1 and 2 also are reversed relative to each other, when certain strains are compared; (2) by comparison of the base sequence homologies of the RNA segments of the various prototype influenza strains, it can be deduced that certain strains differ from each other by only a number of point mutations, while others differ as a result of reassortment (recombination).

ASSIGNMENT OF GENE FUNCTIONS TO RNA SEGMENTS OF THE INFLUENZA A2-SINGAPORE STRAIN

The separation of the RNA segments of ^{32}P-labelled vRNA of the A2-Singapore strain is shown in Fig.1. While segments 1, 2, 3, 4, 7, and 8 are well separated, segments 5 and 6 do not resolve under these conditions, and therefore were isolated as a mixture. The assignment of the various gene functions to the RNA segments of the A2-Singapore strain was achieved by molecular hybridization of the individual ^{32}P-labelled segments of the A2-strain with cRNA's of specific recombinants with fowl plague virus (FPV). The production and gene composition of these recombinants have been described recently (12). As can be seen in Table 1, the same RNase-resistant radioactivity found after hybridization with an individual A2-segment in the homologous system has been observed from the corresponding segment when present in the recombinant. If this segment is not present in the recombinant, then the radioactivity is the same as in the heterologous hybridization with FPV. Thus, the recombinant between A2 and the *ts* 90 mutant of FPV has segment 1 (A2-system) derived from the Singapore strain. Since *ts* 90 has a defect in the gene responsible for the transport of the polymerase complex from the nucleus to the cytoplasm (11, 12) segment 1 of A2 represents the transport gene. Investigating its recombinant with the mutant *ts* 3, it is shown that segment 2 represents the polymerase 1 gene. Thus the order of the migration rate of segments 1 and 2 of A2 is the reverse to the migration order of the corresponding segments of FPV (10). With PR8, the order of migration rates of segments 1 and 2 is the same as that found in A2, and different from that of FPV, whereas virus N resembles FPV in this respect (unpublished observations). This indicates that it is not possible to deduce the gene functions of these segments just by comparing the migration rates of the RNA segments of two individual strains. Except for segments 5 and 6 of A2, which have not yet been clearly separated and were therefore not used individually for molecular hybridization, the order of the other segments of A2 is the same as found for FPV. (Note that segments 2 and 3 are cross-contaminated by about 20%). The corresponding assignments are summarized in Fig.1.

BASE SEQUENCE HOMOLOGIES BETWEEN THE VARIOUS RNA SEGMENTS OF DIFFERENT INFLUENZA A PROTOTYPE STRAINS

Next, we determined and compared the base sequence homologies of the various RNA segments of individual influenza A prototype strains in order to explore whether it is possible to find a genetic basis for the taxonomy, taking into account how far one strain might differ from another one either by mutation or recombination (reassortment). We have shown recently (13) that the base sequence homologies between all individual RNA segments of the human A1 strain FM1, the human A0 strain PR8, and the swine influenza strain is extremely high (≥90%), while there is a dramatic break between the FM1 and A2 Singapore strain. This indicates that the FM1, PR8, and swine influenza strains are derived from each other only by a number of point mutations, while between FM1 and the human A2 strain, some kind of a recombination event must have occurred.

We therefore determined the base sequence homologies of the various RNA segments of A2 to the other influenza A prototype strains. It can be seen in Table 2 that concerning the PR8 strain and the FM1 strain segments 1, 7, and 8 of A2 are highly related, while segments 2, 3, and 4 are less related. Since the segment coding for the nucleocapsid protein (NP) is highly conserved among the influenza A strains (12) the difference in base sequence homology of the mixture of segments 5/6 is almost exclusively due to the gene coding for the neuraminidase (NA).

The difference in base sequence homology can be increased by heating the hybrids close (75°) to the melting point of the homologous hybrid in the presence of 1% formaldehyde. In this way mismatched regions of the heterologous hybrid molecules can be detected (12) (Table 2). From these data it can be concluded that at least 3 segments of FM1 and A2 are almost identical for their genetic information. Further, from other evidence (13) it is clear that the segments coding for NP are highly related. The other 4 segments are significantly less related, especially those coding for the haemagglutinin (HA, segment 4) and NA.

When the A2-strain is compared with the A3-Hong Kong strain, 7 out of 8 segments are genetically highly related if not identical. Only the segment coding for the HA exhibits a remarkably low base sequence homology (Table 2). On the other hand, it has been shown recently that there exists 80% base sequence homology between the equine 2- and Hong Kong strains in segment 4 (13) while there is little relatedness for the other segments of these latter strains (Table 2). Recently we have also determined the base sequence homologies between the labelled RNA segments of influenza virus strains duck Ukraine (Hav7Neq2) and Hong Kong. With segment 4 (HA) the homology was 90% while the other segments except for segments 3 and 6 (NA) showed homology comparable to the equine 2 strain (Scholtissek and von Hoyningen, unpublished observations). Other influenza A strains have been included in Table 2 for purpose of comparison.

CONCLUSION

In summary, we have assigned the various RNA segments of the

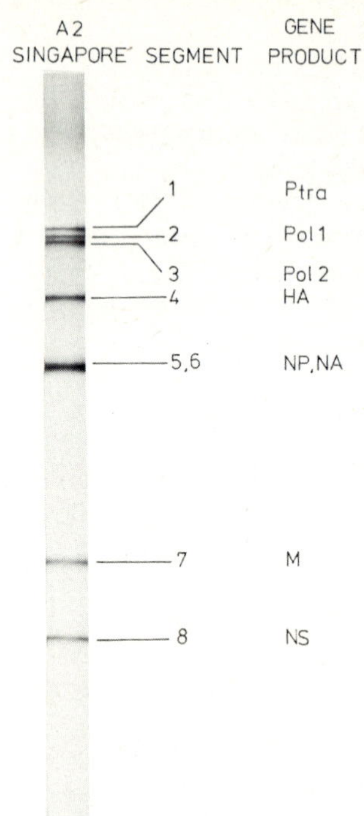

Fig.1. *Separation of the ^{32}P-labelled RNA segments of the A2-Singapore strain by polyacrylamide gel electrophoresis in 6 M urea according to Floyd et al. (2).*

Ptra = transport gene; Pol 1 = polymerase 1 gene; Pol 2 = polymerase 2 gene; HA = haemagglutinin gene; NP = nucleocapsid protein gene; NA = neuraminidase gene; M = matrix protein gene; NS = non-structural protein gene.

A2-Singapore strain to specific viral functions and/or proteins. The order of electrophoretic migration rates for segments 1 and 2 is the reverse of that seen with fowl plague virus.

From the data presented here and published recently (13) it can be concluded that the PR8-strain is derived from the swine influenza strain, and the FM1-strain from the PR8-strain by a number of point mutations, and not by recombination. On the other hand, a strong argument is presented here favouring the idea that A2 is derived from FM1 by recombination (4 segments exhibit almost 100% base sequence homology) and A3 from A2 also by recombination (7 segments show almost 100% base sequence homology). Segment 4 of the A3-Hong Kong-strain seems to be derived from the duck Ukraine strain, or these two strains might have a common ancestor for segment 4.

TABLE 1

Assignment of vRNA segments of A2-Singapore to gene functions using the technique of molecular hybridization with cRNA of specific recombinants of fowl plague virus.[a]

cRNA of recombinants with	Counts/5 min after hybridization and treatment with RNase ^{32}P-labelled segments of A2-Singapore						
	1	2	3	4	5/6	7	8
A2-Singapore (wild)	3600	4350	2880	6050	7850	6620	4950
FPV (wild)	600	1880	1100	1480	3900	4500	3000
*ts*3, Pol 1 (1)	580	<u>3800</u>	1400	1300	3680	4480	2900
*ts*90, Tra (2)	<u>3480</u>	2000	1230	1380	3670	4420	2850
*ts*263, Pol 2 (3)	700	2010	<u>2560</u>	1380	4000	4520	3060
HA (4)	600	1900	1120	<u>6000</u>	3900	4510	3080
*ts*113, NA (1+6)	570	<u>3780</u>	1110	1300	<u>7080</u>	4540	2880
NS (8)	570	1800	1040	1380	3740	4440	<u>4800</u>

[a]Hybrids with segments 1, 2, 3, 7, and 8 were heated to 75° in 1% formaldehyde prior to RNase digestion. The temperature-sensitive (*ts*) mutants have their defects in the genes listed. Pol 1 and Pol 2 are the two polymerase genes. Tra is the transport gene; NS is the gene of the non-structural protein. The figures in brackets represent the number of the segments (FPV-system, Scholtissek *et al* (1976)) of the recombinants derived from A2-Singapore. These segments are underlined in the A2-system.

TABLE 2

Percent base sequence homology between ^{32}P-labelled vRNA segments of A2-Singapore and cRNA of other prototype influenza A strains[a]

cRNA of	^{32}P-labelled segments of A2-Singapore													
	1		2		3		4		5/6		7		8	
	0°	75°	0°	75°	0°	75°	0°	75°	0°	75°	0°	75°	0°	75°
PR8 (A0)	96	-	72	-	75	-	24	-	56	-	94	-	95	-
FM1 (A1)	98	96	70	38	76	62	24	-	56	-	97	98	98	98
Singapore (A2)	100	100	100	100	100	100	100	-	100	100	100	100	100	100
Hong Kong (A3)	98	98	96	91	97	96	24	-	100	90	98	100	98	100
Swine	91	48	73	27	76	48	30	-	56	-	97	82	97	70
Equine 2	63	-	60	-	55	-	20	-	50	-	94	-	89	-
FPV	67	17	82	42	72	36	28	-	50	-	88	65	88	59
Virus N	73	20	86	63	74	43	22	-	50	-	88	66	40	-

[a]Hybrids were either heated to 75° in the presence of 1% formaldehyde (75°) or were not treated with formaldehyde (0°) prior to digestion with RNase.

It might be appropriate now to consider a new taxonomy of the influenza A strains on the basis of the genetic relatedness of their RNA segments. Such data would seem to be more meaningful and probably also more reliable to use in placing different strains into common subgroups.

REFERENCES

1. Bean, W.J. and Simpson, R.W. (1976). *J. Virol.* 18, 365.
2. Floyd, R.W., Stone, M.P. and Joklik, W.E. (1974). *Anal. Biochem.* 59, 599.
3. Inglis, S.C., McGeoch, D.J. and Mahy, B.W.J. (1977). *Virology* 78, 522.
4. McGeoch, D., Fellner, P. and Newton, C. (1976). *Proc. Nat. Acad. Sci. USA* 73, 3045.
5. Palese, P. and Schulman, J.L. (1976a). *J. Virol.* 17, 876.
6. Palese, P. and Schulman, J.L. (1976b). *Proc. Nat. Acad. Sci. USA* 73, 2142.
7. Pons, M.W. (1976). *Virology* 69, 789.
8. Ritchey, M.B., Palese, P. and Kilbourne, E.D. (1976a). *J. Virol.* 18, 738.
9. Ritchey, M.B., Palese, P. and Schulman, J.L. (1976b). *J. Virol.* 20, 307.
10. Rohde, W., Harms, E. and Scholtissek, C. (1977). *Virology* 79, 393.
11. Scholtissek, C. and Bowles, A.L. (1975). *Virology* 67, 576.
12. Scholtissek, C., Harms, E., Rohde, W., Orlich, M. and Rott, R. (1976). *Virology* 74, 332.
13. Scholtissek, C., Rohde, W., Harms, E. and Rott, R. (1977). *Virology* 79, 330.

THE GENES OF INFLUENZA VIRUS: ANALYSIS OF INFLUENZA B VIRUS STRAINS

VINCENT R. RACANIELLO AND PETER PALESE

Mt. Sinai School of Medicine of CUNY, New York, N.Y. 10029, USA.

A vast amount of biochemical information on influenza A viruses has accumulated since their discovery in 1931. One recent result of these efforts is the establishment of a genetic map for influenza A viruses which permits the correlation of each of the 8 RNA segments found in the virion with one virus specific polypeptide (4, 8, 17, 19, 23; for review see 16). The protein and RNA composition of another epidemiologically important myxovirus group, the influenza B viruses, has been studied less intensely. So far only four to six influenza B virus polypeptides have been described (3, 5, 9, 12, 15, 26), and very little is known about the RNAs of influenza B viruses (20).

This report describes experiments which lead to a characterization of the genome of influenza B viruses. We first confirm and extend the observation of Ritchey *et al.* (20) that the RNA of influenza B viruses consists of eight segments. Using established techniques, (20) the RNAs of several influenza B viruses isolated from 1940 to 1966 are shown to separate into eight bands in distinguishable patterns. Next, we present evidence which suggests that RNA 5 of influenza B viruses codes for the haemagglutinin (HA) polypeptide. (In contrast, RNA 4 of influenza A viruses so far studied carries the information for HA). Finally, studies on the protein composition of influenza B viruses are described. Gel electrophoresis of labelled viral proteins demonstrates the presence of two P proteins, HA, neuraminidase (NA), nucleoprotein (NP) and membrane polypeptide (M) in influenza B viruses. We can also identify, in influenza B virus infected cells, the presence of a previously unrecognized viral nonstructural (NS) protein. Search is currently in progress to locate an eighth virus specific protein, prsumably a third P protein.

INFLUENZA B VIRUSES CONTAIN EIGHT RNA SEGMENTS

Fig.1 is an autoradiogram of a polyacrylamide gel of the ^{32}P-labelled RNAs of four different influenza B viruses isolated from 1940 to 1966. The RNA of each isolate separates into eight bands. Counting the slowest moving RNA as #1, then bands 1, 2, and 3 of influenza B viruses cluster in a pattern similar to that of influenza A viruses. Bands 4 and 5 of B/Mass/66 virus do not separate well under the conditions used in this gel, while band 7 of B/GL/54 virus has labelled poorly in this preparation and is barely visible. It is interesting to note that the migration patterns of the RNAs of the various isolates are unique. For example, comparing the patterns of B/Lee and B/GL viruses, it can be seen (Fig.1) that RNAs 1, 3, 4, 5, 6, 7, and 8 migrate to different positions. (The relationship of this observation to possible genetic drift of influenza B viruses is not clear at this time.) In summary, like influenza A viruses, influenza B viruses possess a segmented RNA genome consisting of eight pieces. Interestingly, although influenza B viruses also undergo reassortment (1, 25) and possess a genome structure similar to that of influenza A viruses, a recombinant virus deriving genes from both an influenza A and an influenza B virus could not be isolated (Ritchey and Palese, unpublished experiments).

RNA 5 OF INFLUENZA B VIRUSES CODES FOR THE HEMAGGLUTININ POLYPEPTIDE

Fig.2a is an autoradiogram of the RNA patterns of B/Lee/40 and B/Maryland/59 viruses. Eight RNA segments are clearly separated for each virus in a distinguishable pattern. Fig. 2b shows the RNA pattern of a recombinant virus derived from these two parents selected in the presence of anti-B/Lee antiserum. Examination of RNA patterns indicates that this recombinant derives RNAs 1, 2, 3, 4, 6, 7, and 8 from B/Lee virus while only one RNA, RNA 5, is inherited from B/Maryland virus.

The result of hemagglutination inhibition tests for these viruses is shown in Table 1. The HA's of the 2 parent viruses cross-react to some extent, but the recombinant clearly resembles B/Maryland virus in its reactivity of the HA. This serologic information, considered with the RNA composition of the recombinant reported above, suggests that RNA 5 of influenza B viruses codes for the HA. In contrast, studies with different influenza A viruses (17) showed that the HA is coded for by RNA 4 in all strains examined.

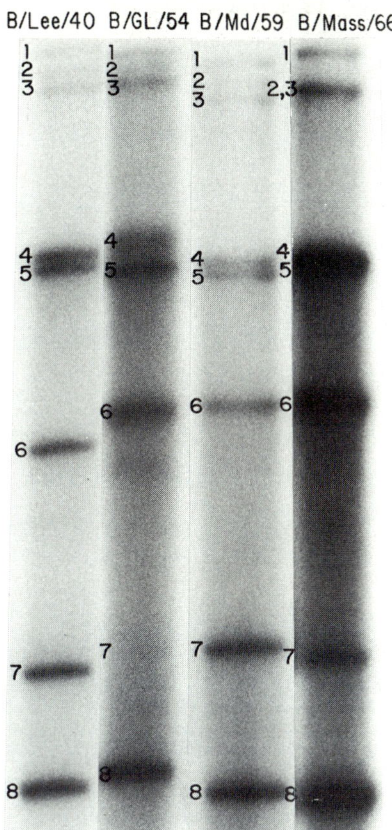

Fig.1. *RNA patterns of influenza B virus strains. RNA was labelled with ^{32}P in MDCK cells, extracted from purified virus and electrophoresed on 2.6% polyacrylamide - 6M urea gels as described previously (20). After drying, gels were exposed to Kodak XR-5 film. Electrophoresis at 120V for 14.5 hr, 26°C: Lane 1: RNA of B/Lee/40 virus. Lane 2: RNA of B/GL/54 virus. Lane 3: RNA of B/Maryland/59 virus. Lane 4: RNA of B/Massachusetts/66 virus.*

INFLUENZA B VIRUSES CODE FOR EIGHT VIRUS-SPECIFIC POLYPEPTIDES

In a previous report Choppin and collaborators (3) were able to detect HA,NP,NA, and M polypeptides in B/GL/1760/54 virus. The same report also demonstrated cleavage of HA into HA1 and HA2 *in vitro*. In addition, Oxford (15) found evidence for two P proteins in B/Lee/40 virus. We have also analyzed the proteins of influenza B viruses, and using B/Lee/40 virus, grown in MDCK cells in the absence of serum, we detect on 13%

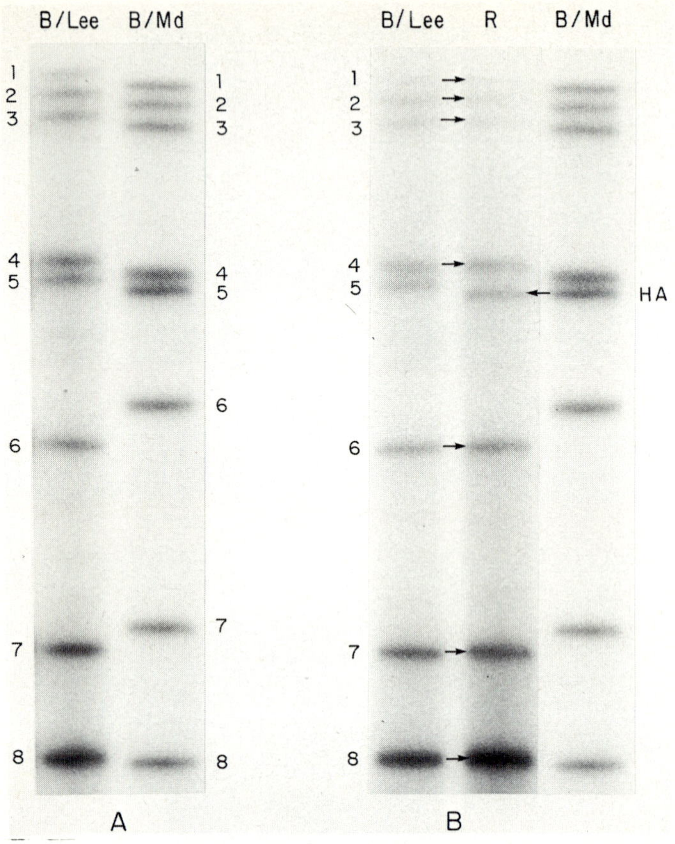

Fig.2. *RNA patterns of influenza B/Lee/40, B/Maryland/59 and recombinant virus R. A: Lane 1: RNA of B/Lee/40 virus. Lane 2: RNA of B/Maryland/59 virus. B: Lane 1: RNA of B/Lee/40 virus. Lane 2: RNA of recombinant R. Lane 3: RNA of B/Maryland/59 virus. Conditions are as described in Fig.1. Recombinant R was isolated as follows: an MDCK cell monolayer was mixedly infected with 1 PFU/cell each of B/Lee virus and B/Md viruses using described techniques (24). Cell supernatants were then plaqued in MDCK cells under agar containing hyperimmune anti-B/Lee antiserum diluted to 1/3600. Individual plaques were picked, grown up in MDCK cells, and their RNAs examined as described in Fig.1. Note that although antiserum to whole B/Lee virus was used, the Lee neuraminidase is present in the recombinant. Since influenza B virus neuraminidases are all antigenically similar (2), there is no selective advantage for recombinant viruses which contain the B/Md neuraminidase.*

TABLE 1

Haemagglutination - Inhibition Titres

Virus	Antiserum		
	B/Lee/40	B/Md/59	NRS***
B/Lee/40	1,280*	160	80
B/Md/59	320	1,280	80
R**	320	1,280	80

*reciprocal of highest serum dilution which causes inhibition of haemagglutination.

**recombinant virus described in Fig.2.

***normal rabbit serum.

polyacrylamide gels six protein peaks, four of which are glycosylated (Fig.3a). The identity of the peaks labelled as HA, NA, NP, HA1, HA2, and M is confirmed in part b of Fig.3. In this electropherogram the doubly labelled B/Lee virions were treated with 1 µg/ml of TPCK-trypsin for 20 min at 37°C. After treatment the first large glycosylated peak (HA) has nearly disappeared while the third glycosylated peak (HA1) increases in size. The HA2 peak does not change upon trypsin treatment; we do not know the reason for this unexpected finding but it is possible that HA2 is further degraded into smaller products. The presence of HA1 and HA2 in untreated virus suggests that B/Lee HA is partially cleaved in MDCK cells in the absence of serum. Under our conditions the second glycoprotein, most likely NA, migrates slightly slower than NP, in contrast to previous results in another system (3).

Analysis of the ^{35}S-methionine labelled proteins in purified viruses (B/Lee/40 and B/Md/59 viruses, lanes 1 and 2, Fig.4) on gradient polyacrylamide gels revealed, in addition to the HA, NP, and M proteins, the presence of two P proteins. The slower moving proteins P1 and P2 are barely visible in this exposure but can be readily detected in B/Lee virus infected cells (lane 8).

NS protein can be detected in virus-infected cells which migrates slower than the M protein (lanes 1 and 8). (In contrast, the M protein of influenza A viruses migrates slower than the NS protein under the same conditions - lane 7). It

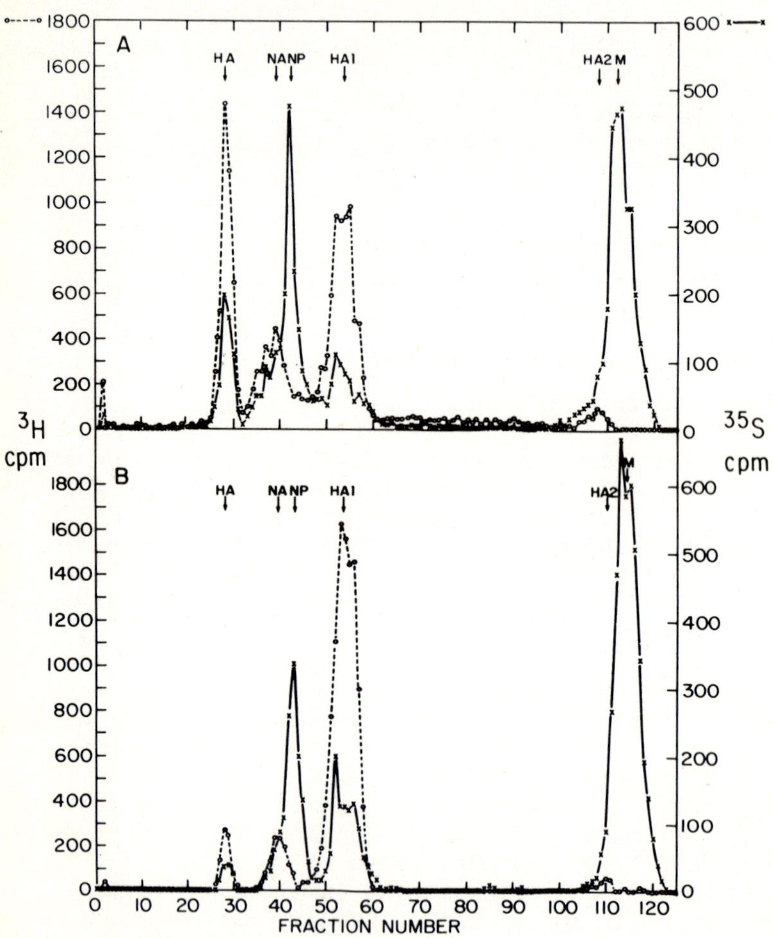

Fig.3. *Electrophoresis of purified influenza B/Lee/40 virus. Virus was doubly labelled with 10 μCi/ml ^{35}S-methionine and 90 μCi/ml ^{3}H-glucosamine in MDCK cells in the absence of serum, purified from supernatant fluids and processed for electrophoresis as described (22). Samples were electrophoresed on a 13% polyacrylamide gel containing 0.1% SDS (13) for 22 hours at 100V. After drying the gel was sliced and processed for scintillation counting as described (22). a: untreated virus. b:*

virus treated with TPCK-trypsin, 1 μg/ml, for 20 minutes at 37°C. Equal amounts of protein were added to each slot.

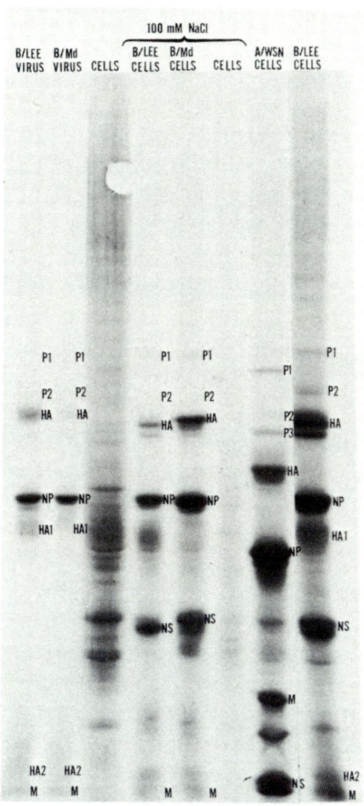

Fig.4. *Electrophoresis of influenza virus-specific proteins on a 5-13% gradient polyacrylamide gel. Purified influenza B/Lee/40 and B/Md/59 viruses were continuously labelled in MDCK cells with 10 μCi/ml ^{35}S-methionine. Pulse labelling of infected cell proteins for 15 minutes with ^{35}S-methionine, processing of samples and gel conditions have all been described (22). When labelling infected cell proteins in hypertonic medium, infected cells were washed six hours p.i. with warm Earle's salts containing 2% glucose (22) and 100 mM NaCl. After incubation for 15 minutes under 2.0 ml of the same medium, 40 μCi/ml of ^{35}S-methionine was then added for an additional 15 minutes. Infection of MDCK cells was done at 1-10 plaque forming units per cell. Lane 1: purified B/Lee/40 virus. Lane 2: purified B/Md/59 virus. Lane 3: uninfected cell proteins. Lane 4: cells infected with B/Lee virus in the presence of 100 mM NaCl. Lane 5: cells infected with B/Md virus in the presence of 100 mM NaCl. Lane 6: uninfected cells in the presence of 100 mM NaCl. Lane 7: cells infected with A/WSN virus. Lane 8: cells infected with*

B/Lee virus. Equal amounts of protein were added to each slot. Proteins are identified to the right of each slot.

should also be mentioned that the NS proteins of B/Lee virus and B/Maryland virus migrate differently (lanes 4 and 5).

In an attempt to reduce the host background in B/Lee virus-infected cells (lane 8), we made use of an observation of Nuss *et al.* (14) who reported that host protein synthesis is selectively depressed in virus-infected cells labelled in hypertonic medium. In lanes 3 and 6 of Fig.4 it can be seen that labelling the proteins of uninfected cells in the presence of 100 mM NaCl drastically reduces protein synthesis. Using hypertonic medium while labelling B/Lee and B/Maryland virus-infected cells it is possible to detect all virus-coded proteins while the host background is reduced (lanes 4 and 5). P2 protein cannot be seen well in this exposure. P1 and P2 proteins of both viruses migrate slower than the corresponding HA. Preliminary evidence based on peptide mapping indicates that the bands in B/Maryland and B/Lee virus infected cells which migrate slightly faster than the HA represent partially or unglycosylated HA.

In comparing the protein pattern of infected cells with that of purified virus for B/Lee and B/Maryland viruses (lanes 1, 2 and 4, 5), it was found that the HA of purified virus moves slower than the HA obtained from pulse-labelled infected cells. Presumably the slower moving band represents fully glycosylated HA of continuously labelled purified virus. In addition, migrational differences can be detected between the HAs of B/Lee and B/Maryland viruses (lanes 4 and 5).

It is of interest to note that influenza A (7, 11, 22) and also influenza B viruses code for 7 - 8 virus-specific proteins which differ in size. Comparing lanes 7 and 8 in Fig.4 it can be seen that the influenza B virus P1, P2, HA, NA, NP and NS proteins are all *larger* than their corresponding A virus proteins, while the M protein of influenza B viruses is *smaller* than the influenza A/WSN virus counterpart.

In summary, using established techniques of RNA electrophoresis on urea-polyacrylamide gels, we have examined various influenza B virus isolates. Our conclusion is that the segmented influenza B virus genome consists of eight pieces. By taking advantage of the fact that different influenza B virus isolates have different RNA patterns, we have identified a recombinant of influenza B/Lee/40 and B/Maryland/59 viruses which contains the HA from B/Md virus and which derives only one RNA, RNA 5, from B/Md virus. This result suggests that in influenza B viruses RNA 5 codes for the hemagglutinin. Finally, we have firmly identified seven virus-specific polypeptides in influenza B virus-infected cells. These polypeptides include P1, P2, HA, NA, NP, M, and a previously unrecognized NS protein. We also show that purified virus contains all virus specific polypeptides except for the NS protein. Our finding of eight RNA segments and seven virus-specific polypeptides has initiated work toward a complete map of the influenza B virus genome.

ACKNOWLEDGEMENT

For expert technical assistance we thank Ms. Marlene Lin. This work was supported by grants from the NIH (AI-11823), the NSF (PCM-76-11066) and the ACS (VC-234). V.R.R. was supported by the NIH training grant #5T01A100439-04.

REFERENCES

1. Beare, A.S. Sherwood, J.E., Callow, K.A., and Craig, J.W. (1977). *Infection and Immunity* 15, 347.
2. Chakraverty, P. (1972). *Bull. WHO* 46, 473.
3. Choppin, P.W., Lazarowitz, S.G., and Goldberg, A.R. (1975). *In* "Negative Strand Viruses" (R.D. Barry and B.W.J. Mahy, eds.) Academic Press, New York, pp.105-119.
4. Etkind, P.R., Buchhagen, D.L., Herz, C., Broni, B.A., and Krug, R.M. (1977). *J. Virol.* 22, 346.
5. Haslam, E.A., Hampson, A.W., Egan, J.A., and White, D.O. (1970a). *Virology* 42, 555.
6. Haslam, E.A., Hampson, A.W., Radiskevics, I., and White, D.O. (1970b). *Virology* 42, 566.
7. Inglis, S.C., Carroll, A.R., Lamb, R.A., and Mahy, B.W.J. (1976). *Virology* 74, 489.
8. Inglis, S.C., McGeoch, D.J., and Mahy, B.W.J. (1977). *Virology* 78, 522.
9. Kendal, A.P. (1975). *Virology* 65, 87.
10. Kilbourne, E.D. (1975). *In* "The Influenza Viruses and Influenza" (E.D. Kilbourne, ed.) Academic Press, New York, pp.1-13.
11. Lamb, R.A., and Choppin, P.W. (1976). *Virology* 74, 504.
12. Lazdins, I., Haslam, E.A., and White, D.O. (1972). *Virology* 49, 758.
13. Maizel, J.V. (1971). *In* "Methods in Virology" (K. Maramorosch and H. Koprowski, eds.)Academic Press, New York, pp.180-244.
14. Nuss, D.L., Oppermann, H., and Koch, G. (1975). *Proc. Nat. Acad. Sci. USA* 72, 1258.
15. Oxford, J.S. (1973). *J. Virol.* 12, 827.
16. Palese, P. (1977). *Cell* 10, 1.
17. Palese, P., and Schulman, J.L. (1976a). *Proc. Nat. Acad. Sci. USA* 73, 2142.

18. Palese, P., and Schulman, J.L. (1976b). *J. Virol.* 17, 876.
19. Palese, P., Ritchey, M.B., and Schulman, J.L. (1977). *Virology* 76, 114.
20. Ritchey, M.B., Palese, P., and Kilbourne, E.D. (1976a). *J. Virol.* 18, 738.
21. Ritchey, M.B., Palese, P., and Schulman, J.L. (1976b). *J. Virol.* 20, 307.
22. Ritchey, M.B., Palese, P., and Schulman, J.L. (1977). *Virology* 76, 122.
23. Scholtissek, C., Harms, E., Rohde, W., Orlich, M. and Rott, R. (1976). *Virology* 74, 332.
24. Schulman, J.L., and Palese, P. (1976). *J. Virol.* 20, 248.
25. Tobita, K., and Kilbourne, E.D. (1974). *J. Virol.* 13, 347.
26. Tobita, K., and Kilbourne, E.D. (1975). *Arch. Vir.* 47, 367.

COMPARATIVE STUDIES OF NUCLEOTIDE SEQUENCES WITHIN TWO INFLUENZA VIRUS GENES

*J.C. SMITH, *N.H. CAREY, *P. FELLNER,
**D. McGEOCH and **R.D. BARRY

**G.D. Searle & Co. Ltd., Research Division, P.O. Box 53, High Wycombe, Bucks., HP12 4HL.*

***Division of Virology, Department of Pathology, University of Cambridge, Cambridge, England.*

The influenza A virus genome consists of eight single-stranded RNA segments (2, 12 -14, 16). Each of these RNAs has a unique nucleotide sequence (12), and corresponds to an individual gene coding for a specific virus protein (8, 13).

This segmentation of the genome into its component genes presents an unusual opportunity to examine the sequences which may participate in controlling the transcription and replication of influenza virus, since although the coding regions of the genes must differ, parts of the non-coding sequences may be conserved if they are specifically interacting with the virus transcriptase or replicase.

We are therefore engaged in comparing the nucleotide sequences of the two smallest genes of an avian influenza A virus (fowl plague virus, Rostock strain (FPV)). These are: (i) Gene 7, assigned as coding for the matrix protein (8), with an approximate molecular weight of 3.2×10^5 daltons (about 1,000 nucleotides in length) (12). (ii) Gene 8, assigned as coding for non-structural protein I (8), with a molecular weight of about 2.8×10^5 daltons (850-900 nucleotides in length) (12).

The most satisfactory approach to sequence analysis of these genes would appear to be afforded by reverse transcription of them into DNA copies, which could then be analysed by recent rapid DNA sequencing methods (11). We are currently exploring this approach, but in the interim we have carried out some RNA sequencing studies directly on the genes, using classical methods (1, 18). However, we have experienced difficulty in obtaining the RNA species sufficiently labelled with ^{32}P, and this has limited the extent of our studies employing these methods. In this paper we report analyses on most of the unique oliognucleotides arising upon complete T_1 ribonuclease digestion of each gene, and the grouping of some of these products into larger blocks of sequence, usually partially resolved. We anticipate that this work will eventually permit us to map the relative positions of these unique oligonucleotides in each gene with sufficient precision to enable us to determine whether sequence homologies do exist close to the extremities of genes 7 and 8. This work should also guide us in synthesising complementary defined-sequence primers, for reverse transcription purposes.

PREPARATION OF ^{32}P-LABELLED VIRUS RNAs:

FPV was cultivated in primary chicken embryo fibroblasts, in low-phosphate medium 199 containing carrier-free ^{32}P-orthophosphate to a concentration of 2-3 mCi/ml, as described elsewhere (7). Usually 200-400 mCi of ^{32}P-orthophosphate were used, in all (the virus was grown at the Radiochemical Centre, Amersham). The labelled virus was purified according to Glass *et al.*, (7). The virus RNA was extracted and fractionated by gel electrophoresis through 2.2% acrylamide-0.6% agarose gels (5, 12). The RNA species were detected by autoradiography, and recovered from the gel by homogenisation and phenol extraction (12).

ANALYSIS OF COMPLETE T_1 RIBONUCLEASE DIGESTION PRODUCTS

Standard enzymatic digestion conditions and analysis procedures were employed, and will be summarised here.

The virus RNAs were subjected to complete digestion with T_1 ribonuclease (Sankyo), and the digests were fractionated either by bidimensional gel electrophoresis, largely according to De Wachter and Fiers (4) as previously described (6, 12) or by bidimensional electrophoresis on cellulose acetate and DEAE paper (18).

The T_1 ribonuclease digestion products were subjected to further complete hydrolysis with RNase U_2 (Sankyo), RNase A (Worthington), and CM-RNase A (pancreatic RNase carboxymethylated at position 41 lysine). Some oligonucleotides were partially digested with RNase A, to obtain further overlaps. The products arising from all of these enzymatic digests were fractionated by electrophoresis on DEAE-paper at pH 3.5 (0.5% pyridine, 5% acetic acid), and were further characterised either by alkaline hydrolysis, or digestion with RNase A to completion.

Partial RNase T_1 digestion, to obtain larger fragments of the RNAs, was carried out as follows: The RNA was dissolved in 0.2 ml of TMK buffer (0.025M Tris HCl, pH 7.5; 0.01M $MgCl_2$; 0.017M KCl), to a final concentration of 0.6 mg/ml (including carrier tRNA). The RNA solution was incubated at 37° for 20 min, and then chilled in ice for 10 min. RNase T_1 was then added to give an enzyme:substrate ratio of 1:12000 (mild digest) or 1:100 (strong digest), and hydrolysis was continued for 90 min at 0°. The digestion products were extracted with phenol, precipitated, and fractionated by gel electrophoresis. The large fragments (40-400 nucleotides) arising from mild digestion conditions were fractionated by electrophoresis through a 10% polyacrylamide slab (15). Products arising from strong partial digestion conditions were fractionated by bidimensional gel electrophoresis. The partial digestion products were then analysed by further complete digestion with RNases T_1 and A respectively. The RNase T_1 products were identified with previously characterised oligonucleotides by treatment with RNase A, and, conversely, the RNase A products were analysed by further digestion with RNase T_1.

RESULTS AND DISCUSSION

The RNA bands representing genes 7 and 8 were subjected to complete T_1 ribonuclease digestion and fractionated as shown in Fig.1.

We have analysed fully or partially 70 unique oligonucleotides from gene 7, out of an estimated total of roughly 85 which arise upon complete T_1 ribonuclease hydrolysis. We have also studied 57 products from a total of approximately 75 released from gene 8. The results of these analyses are set out in tables 1 and 2.

The analyses on these products taken together with the frequencies of occurrence of the small non-unique products indicate chain lengths of about 950 and 850 nucleotides for genes 7 and 8, respectively. However, this is likely to be a minimum value in each case, since our estimates of the frequencies of some unresolved isomeric oligonucleotides may be too low.

The numbers of unique T_1 ribonuclease products are higher than might be expected from RNAs of these lengths, presumably because of their low G contents (21.2% in gene 7 and 19.7% in gene 8).

The oligonucleotide sequences in so far as they have been determined do not reveal striking homologies between the genes. Certain homologies (up to 8 nucleotides) can be discerned, but their incidence is probably not statistically significant, in the absence of further information concerning their relative locations within the genes. However, this finding does not preclude the existence of appreciable sequence homologies

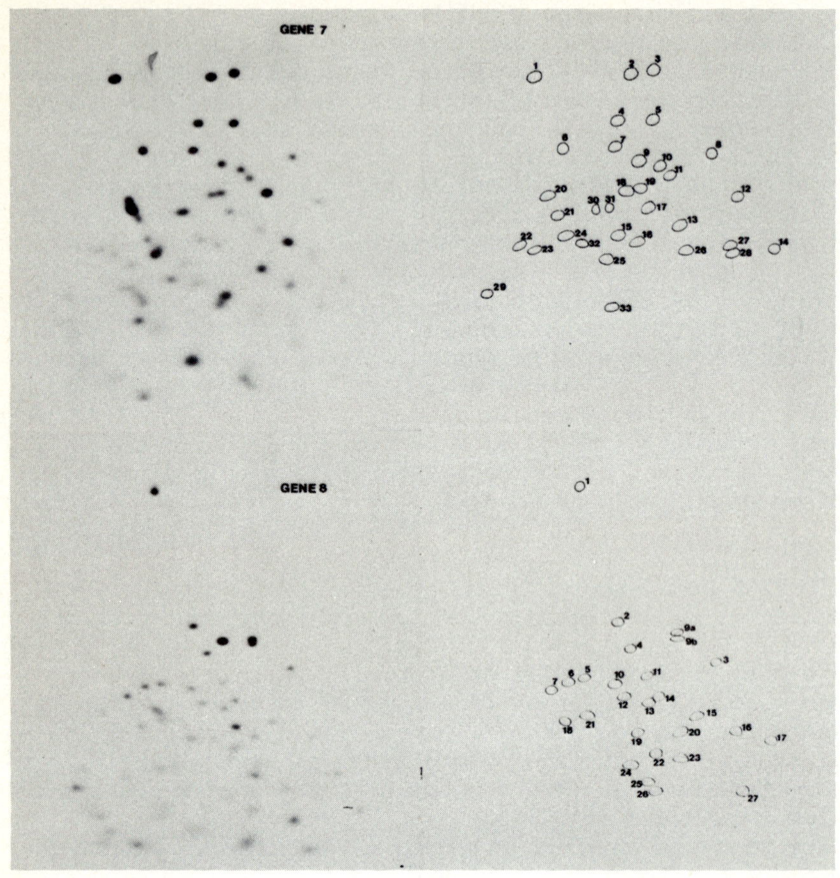

Fig.1. *Oligonucleotide fingerprints of influenza virus genes 7 and 8. The genes were digested with T_1 ribonuclease and fractionated as described in the text.*

between the two genes which would not be detected without more extensive analysis.

The oligonucleotide pppApGp arises from each gene in molar amounts, presumably from the 5'-terminus, and is also released upon T_1 ribonuclease digestion of the remaining six genes (our unpublished results), in agreement with previous reports concerning the 5'-terminus of influenza WSN RNAs (9, 19).

The predominant nucleoside at the 3'-terminus of influenza virus RNA has previously been reported to be uridine (10). Since we did not encounter any oligonucleotide amongst those fractionated by two-dimensional electrophoresis which lacked a 3'-terminal G residue, we subsequently attempted to isolate

TABLE 1

Unique T_1 oligomers from gene 7

No.	Oligomer
U_0G graticule	
7	ACCG
8	CACG
10a	ACAG
10b	AACG
11	ACCAG
12	CAAAG
13	ACCCAG
18	AAAAG
19	ACAAAG
g29	AAACAAG
	pppAG
U_1G graticule	
20a	(U,C)AG
20b	(U,AC)G
23a	CACUG
23b	(U,2C)AG
24	(U,3C)G
25a	AAUG
25b	AUAG
26	AAUCG
27a	AC(UC)AG
27b	(AC,U,C)AG
28	(U,3C)AG
29	AAUCAG
30	ACCAUCG
31	CCAAAUG
32	(AC,2C,U)AAG
g23	AAC(C,U)CAAG
g22	CC(AC,ACU)AG
g34	ACAAAAUG
U_2G graticule	
37	UUAG
39	AUUAG
40	UAUAG
41	UAACUG
42b	(AC,2U,C)G
43b	(2U,~3C)G
44	AUUAAG
45	CAAU(UC)G
g33	(2C,U)ACUG
g25	CAAUAAUG
46b	(AC,2U,3-4C)G
g32	AUCCCAAUG
g24	CCACCAU(U,C)G
g21	[AAC,C,(3C,U)]AUG
g20	UCA CAUCC ACAG A
U_3G graticule	
50	UUUG
51a	(AU,AU,U)G
51b	(3U,C)G
53	(3U,2C)G
g26	AUACUUG
g16	(U,3C)AAUUG
g15	CCAU(U,2C)AUG
g31	C[AC,AU(2U,2C)]AAG
g30	(AC,3U,~6C)G
g6	(AAUCC,AAU,AC)AUCAAG
g1	U[2-3U,AAAAC,2AAA C,AC, 4-5C]AG C(AU,2U,C)G
U_4G graticule	
g27	CAUUUUG
g28	(4U,C)AG
g13	(U,2C)(U,C)AUUG
g17	(C,UUU)AAC(2C,U)G
g18	CC(ACC,UUUUC)AG
g19	CAAAUUUUCAAG
U_5G graticule	
g14	UUAUUUG
g11	AU(C,AU,(3U,2C)AG
g10	C(AAAU,AU)ACAUUUG
g9	CCAUUACCAU(2U,C)G
g7	(4-5C,5U)AAACCG
U_6G graticule	
g12	(UUUUU,C)C(U,C)G
$U_{\geq 7}G$ graticule	
g8	UUUUUUAC(C,U)CAG
g4	(3C,2U)AU(4C,U)AUUUAG
g3	(AACAU,~6C,~9U)G
g2	[AAAU,AC(CU),AU(3U,C)AU (CU)]AC(CU)G

TABLE 2

Unique T_1 oligomers from gene 8

	U_0G graticule
76	CAAAG
8	AAACG
9	CCAAAG
10a	(A,C)CAAG
10b	ACACAG
	pppAG
	U_1G graticule
17b	(U,C)AG
18	UAAG
17a	(AU,C)G
19	(AC,U)G
20a	(AC,U,C)G
20b	(U,2C)AG
21	AAUG
22a	(U,C)AAG
22b	(U,AC)AG
23	(AAU,C)G
24	(AU,AC)G
25	AAUCCG
26	CAAAUCG
	U_2G graticule
28	UUAG
30	(2U,2C)G
31	(2U,3C)G
32	UAUAG
33	(U,C)AUG
g25	AU(U,C)CG
35	(AC,2C,2U)G
36	(AU,U,3C)G
g26	UUAAAG
g24	ACAAUUG
39	(AAAU,AU,C)G
g18	ACCCCAAUUG
g17	(AC,AAAAUC)ACUG
	U_3G graticule
41	CUUUG
42	(3U,C)AG
g23	ACUU(U,C)G
g22	(2U,C)AC(C,U)G
g19	ACUCAAUUG
g21	(2U,5C)AUCG
g6	[AU(3C,U),AUC,AC]AG
	U_4G graticule
48	(4U,C)G
g20	(2C,U)AU(2U,C)G
g13	(AAUCC,AUU)AUG
g12	(3C,U)ACUAUCUG
g10	ACAUC(3U,3C)AAG
g5	(4U,~8C)AG
	U_5G graticule
49	UUUUUG
g17	UUAUUUG
g16	CAUUUUUG
g15	CUAUUUUAAG
g14	[(4U,C),4C]AUG
g11	AUU(3U,5C)AG
	U_6G graticule
g4	C(A,CC)CAUCAUUU(2C,U,U)G
	$U_{\geq 7}G$ graticule
g3	(4U,C)AUCU(2U,C)G
g9a	UUUUUUU[AAAU,AUU,AUC,]AAG
g9b	UU[AAU(UC),AAU(2U,2C)]AUG
g2	(4C,2U)AU(2U,2C)AUUACUG
g1	1AAAU,AAU,AAC,2AU,AC 8C,~17U)G

TABLE 3

Gene 7 Partial T_1 digestion products

Section A

UAG[UUG,ACAAAAUG,UUUUUUAC(C,U)CAGC(AU,2U,C)G]

A-1

pppAG .... UAG[UUG,ACAAAAUG,UUUUUUAC(C,U)CAGC(AU,2U,C)G]

Section B

[AG,CACG,CG,UG,G]U(AAAAC,AAAC,AAAC,AC,2U,4-5C)AG

Section C

AAUCAGCAAU(U,C)GC(AC,U,C)AAG

Section D

(AC,C,U)AGGC(AC,5C,3U)G

Section E

ACCAUCGUCA↑CAUCC↑ACAG[(3U,~2C)G,(2C,U)ACUG,CUG,CCG]

A (bracketed under the two arrows)

Section F

AAGAC(C,U)[AAAU,AU(3U,C)]AU(C,U)AC(C,U)GCUUUUG

TABLE 4

Gene 8 Partial T_1 digestion products

Section A-1

GUCGGUCAAGGAAUGGGG[C(A,CC)CAUCAUUU(2C,2U)G,CAAAUCG,CU(U,C)G]

Section C

[CG, CCAAAG, UUUG, (AU,AC)G]

[CAG, AAAG, AUG, ACACAG]

[AAAG, 2UG,C,AG(2U,C)G,(AC,1-2C,3U)G]

UUG

(AAU,AU,1-2U,2C)G

[(4U,C)G,(AC,2U,~3C)G]

UUUUUG

the 3'-terminal product by means of the "diagonal" method of Dahlberg (3). In a single experiment we detected 1.0 moles of CpU arising upon RNase T_1 digestion of gene 7, and 1.1 moles of CpU from gene 8 (our unpublished results). This suggests that the 3'-terminal sequence of each gene is probably GCU, but this result must be further confirmed.

In tables 3 and 4 we report analyses of certain fragments of genes 7 and 8 arising upon partial T_1 ribonuclease hydrolysis. We have sequence data on six sections of gene 7 and two sections of gene 8, in the size range 20-50 nucleotides, derived from studies of relatively small products arising under strong conditions of partial digestion.

In addition, some analysis has been undertaken of a section of 80 nucleotides of gene 8, arising from mild conditions of partial hydrolysis in relatively good yield.

One of the sequences from gene 7 (section A) was detected within a somewhat larger fragment (approximately 50 nucleotides in length) which released an equimolar amount of pppAGU upon RNase A hydrolysis. This section therefore appears to be situated close to the 5'-terminus of the RNA molecule. It is noteworthy that the sequence UUUUUUA is present, and probably located 5-20 nucleotides from the 5'-terminus. We have also compiled oligonucleotide maps of six large sections of gene 8, in the size range 100-500 nucleotides (our unpublished results), and an oligonucleotide containing the sequence UUUUUUA is also present in a 100 nucleotide fragment which contains pppApGp, suggesting that this sequence is probably close to the 5'-terminus of both genes.

In transcription of procaryotic RNAs this sequence has been associated with termination of transcription (17) albeit in the opposite sense i.e. at the 3'-terminus of the transcribed RNA rather than in the template. It might nevertheless be in some way implicated in terminating transcription of the complementary mRNAs, and this notion is supported by the findings reported in chapter 31 which suggests that this sequence in genes 7 and 8 is not transcribed into mRNA, unlike any of the other unique oligonucleotides.

ACKNOWLEDGEMENT

J.C. Smith, N.H. Carey, P. Fellner wish to thank Dr. A.J. Hale for the provision of research facilities.

REFERENCES

1. Barrell, B.G. (1971). *In* "Procedures in Nucleic Acid Research, Volume 2" (Cantoni, G.L., and Davies, D.R., eds). Harper & Row, New York, p.751.

2. Bean, W.J., and Simpson, R.W. (1976). *J. Virol.* 18, 365.
3. Dahlberg, J.E. (1968). *Nature* 220, 548.
4. De Wachter, R., and Fiers, W. (1972). *Anal. Biochem.* 49, 184.
5. Floyd, R.W., Stone, M.P., and Joklik, W.K. (1974). *Anal. Biochem.* 59, 599.
6. Frisby, D.P., Newton, C., Carey, N.H., Fellner, P., Newman, J.F.E., Harris, T.J.R., and Brown, F. (1976). *Virology* 71, 379.
7. Glass, S.E., McGeoch, D., and Barry, R.D. (1975). *J. Virol.* 16, 1435.
8. Inglis, S.C., McGeoch, D.J., and Mahy, B.W.J. (1977). *Virology* 78, 522.
9. Krug, R.M., Morgan, M.A. and Shatkin, A.J. (1976). *J. Virol.* 20, 45.
10. Lewandowski, L.J., Content, J., and Leppla, S.H. (1971). *J. Virol.* 8, 701.
11. Maxam, A.M., and Gilbert, W. (1977). *Proc. Nat. Acad. Sci. U.S.A.* 74, 560.
12. McGeoch, D., Fellner, P., and Newton, C., (1976). *Proc. Nat. Acad. Sci. U.S.A.* 73, 3045.
13. Palese, P. (1977). *Cell* 10, 1.
14. Pons, M.W. (1976). *Virology* 69, 789.
15. Porter, A., Carey, N.H., and Fellner, P. (1972). *Nature* 248, 675.
16. Ritchey, M.B., Palese, P., and Kilbourne, E.D. (1976). *J. Virol.* 18, 738.
17. Rosenberg, M., De Crombrugghe, B., and Musso, R. (1976). *Proc. Nat. Acad. Sci. U.S.A.* 73, 717.
18. Sanger, F., Brownlee, G.G., and Barrell, B.G. (1965). *J. Mol. Biol.* 13, 373.
19. Young, R.J. and Content, J. (1971). *Nature New Biol.* 230, 140.

RIBOSOME RECOGNITION SITES IN VESICULAR STOMATITIS VIRUS MESSENGER RNA

JOHN K. ROSE

Department of Biology and Center for Cancer Research
Massachusetts Institute of Technology
Cambridge, Massachusetts 02139.

In contrast to prokaryotic mRNAs which commonly have multiple translation initiation sites, all eukaryotic mRNAs so far examined have only single active translation initiation sites, suggesting a fundamental difference between eukaryotic and prokaryotic mRNA translation (37). Also, prokaryotic mRNAs lack the capped and methylated 5'-termini which are found on most eukaryotic mRNAs (29). Nucleotide sequences from translation initiation sites in many mRNAs from prokaryotes have been determined in the past few years (see review in ref.34). Comparison of these sequences has suggested a model for initiation site selection involving base pairing between the 3'-end of the RNA from the small ribosome sub-unit with a sequence in the mRNA (30). By analogy with prokaryotic systems, structural studies on initiation sites are likely to prove valuable in understanding the factors which influence the efficiency of mRNA translation in eukaryotes.

Vesicular stomatitis virus (VSV) is an animal virus which produces five well characterized mRNAs which are complementary to the single-stranded RNA genome (5, 13, 14, 23). These mRNAs are known to be translated with equal efficiencies *in vivo* (36). I report here a procedure for obtaining translation initiation sites from the VSV mRNAs synthesized both in VSV infected cells and *in vitro* by the VSV virion-associated RNA polymerase (1, 25). The key feature of the method is the use of unfractionated mRNAs in large scale preparation of mRNA fragments which are protected from nuclease digestion in a ribosome-mRNA initiation complex. The protected sites are then completely separated from each other by exploiting differences in base composition and size on the two-dimensional gel electrophoretic system of De Wachter and Fiers (8). Using this procedure, which may be generally applicable to mixed mRNA populations, it was possible to obtain pure RNA fragments containing radioactivity sufficient for nucleotide sequence analysis.

SYNTHESIS AND RIBOSOME BINDING OF mRNA

Methods for synthesis of ^{32}P-labeled VSV mRNA *in vivo*, and *in vitro*, as well as β-elimination to remove 7-methylguanosine have been described previously (24, 25). Ribosome binding reactions (800 μl) contained 500 μl of rabbit reticulocyte extract, 1 mM ATP, 0.2 mM GTP, 5 mM DTT, 75 mM KCl, 1.25 mM $MgAc_2$ 20 mM methionine, 25 μg of hemin, 250 μg of anisomycin, and 25-50 μg (10^8-5 x 10^8) dpm of labeled mRNA. Incubation was for 2 min at 30°C followed by four-fold dilution into LSB (0.05 M NaCl, 0.02 M $MgCl_2$, 0.02 M Hepes, pH 7.5). Ribosome protected sites were obtained by addition of RNase T1 to a final concentration of 100 u/ml after the incubation. After 10 min at 30°C, RNase-treated reactions were diluted into 2 ml LSB (0°C) and layered on a 35 ml linear 15-50% (w/v) sucrose gradient. Centrifugation was for 10 hr in a Beckman SW27 rotor at 26,000 rpm (4°C). To obtain 40S complexes the same procedure was followed except that 0.4 mM β-γ-methylene GTP was substituted for GTP in the initial incubation.

FRACTIONATION AND SEQUENCE ANALYSIS OF PROTECTED SITES

Material recovered from 80S or 40S peaks was extracted with phenol, 1% SDS and 0.2 M sodium acetate, pH 5.2, followed by ethanol precipitation. The precipitated sample (usually 5 x 10^6 dpm and no more than 500 μg (RNA) was separated by two-dimensional gel electrophoresis as described previously (6, 8). Published techniques were applied to sequence analysis of the protected oligonucleotides eluted from the gels (2, 32).

ISOLATION OF RIBOSOME BINDING SITES

Previous work has shown that VSV mRNAs form an 80S initiation complex rapidly when added to a rabbit reticulocyte cell-free protein synthesizing system in the presence of anisomycin, an inhibitor of protein synthesis elongation (18, 24). Figs.1a, c show that mRNAs synthesized both *in vivo* and *in vitro* form this complex with high efficiency.

The method used for isolation of translation initiation sites from prokaryotic mRNAs has been formation of a ^{32}P-labeled mRNA-70S ribosome initiation complex, followed by digestion with ribonuclease under conditions where all sequences except those protected by the ribosomes are degraded (12, 33). The 70S com-

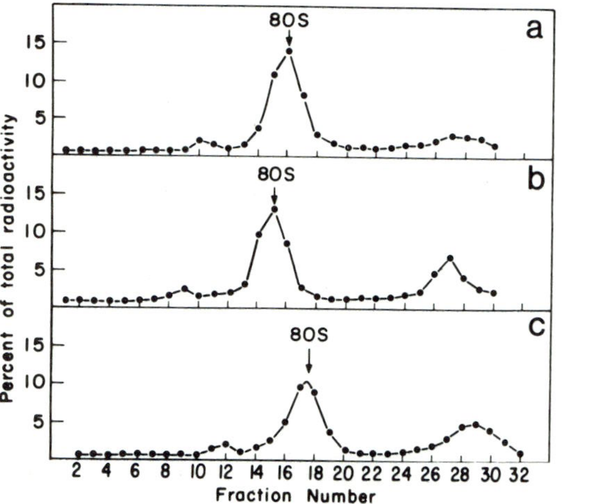

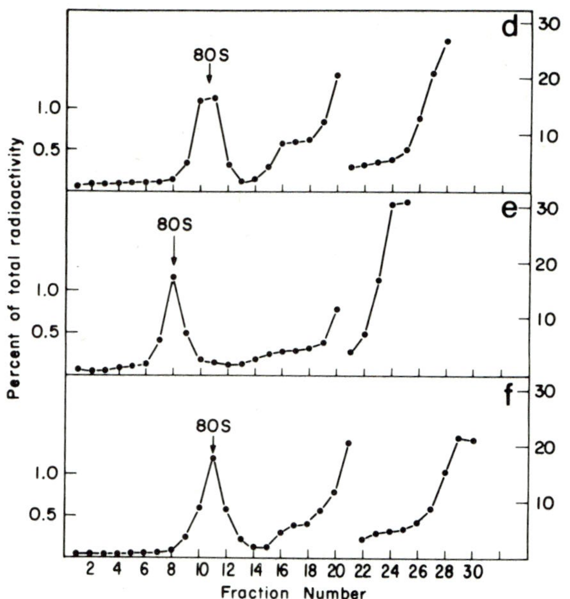

Fig.1. *Sucrose gradient sedimentation of ribosome-mRNA initiation complexes before and after RNase T1 digestion.*

Panels a, b and c show the ribosome binding of VSV mRNA synthesized in vivo, β-eliminated in vivo mRNA, and mRNA synthesized in vitro (α-^{32}P-GTP label) respectively. Panels d, e and f show the T1 RNase protected peak obtained from the respective mRNAs.

plex containing a specific sequence is then purified away from the degraded RNA by velocity gradient sedimentation. When this procedure was applied to initiation complexes formed with individual species of VSV mRNA, neither sufficient purity nor radioactivity could be obtained in the ribosome protected fraction to permit structural analysis. Therefore I developed a procedure for binding site isolation which eliminates mRNA purification steps and their inherent losses, and provides more highly purified sites suitable for sequence analysis.

Total VSV mRNA was labeled with $^{32}PO_4$ during an infection of BHK cells, or was labeled *in vitro* with α-^{32}P nucleoside triphosphates in reactions catalyzed by the virion-associated RNA polymerase. The RNA was then incubated in a reticulocyte cell-free protein synthesizing extract containing anisomycin. After incubation with RNase T1, the 80S peak of protected RNA was resolved by sedimentation (Fig.1d and f). The RNA obtained from the 80S peak was fractionated by two-dimensional gel electrophoresis, first at pH 3.5 in 6 M urea on an 8% polyacrylamide gel to separate the fragments by charge differences, and then in 22% acrylamide at pH 8.0 to separate the fragments by size. An autoradiogram of the region of the gel containing the complete set of ribosome protected fragments from mRNA synthesized *in vivo* is shown in Fig.2. RNA synthesized *in vitro* gave a similar pattern.

To determine the source of each spot, the experiment was repeated using individual VSV mRNAs purified by formamide-polyacrylamide gel electrophoresis (14, 23). Spots 1, 3 and 4 were obtained when N protein mRNA was used, whereas spot 6 was the only sequence derived from the G protein mRNA. Spots 2, 5, 8, 9 and 10 were all derived from the mixture of NS and M mRNAs which do not separate on formamide-polyacrylamide gels (14, 23). The minor sequence (spot 11) originates from the L mRNA. Because the VSV NS and M mRNAs have recently been separated and fingerprinted (10), it has been possible to assign spots 2 and 5 to the NS mRNA because they contain a large T1 oligonucleotide found in the NS mRNA fingerprint. The other major sequence (spots 9 and 10) derived from the mixture of NS and M mRNAs is presumably from the M mRNA because it is the other major sequence derived from this RNA mixture. The source of spot 7 was not identified. The next part of the analysis of the protected sites was elution of the radioactive material in each spot from the gel, followed by fingerprinting (as in Fig.3) of the products derived from each spot by complete digestion with RNase T1 or RNase A. Autoradiograms of these fingerprints as well as further sequence analysis showed that several spots fell into groups of related (overlapping sequences containing spots 1, 3 and 4 (N mRNA); 2 and 5 (NS mRNA);

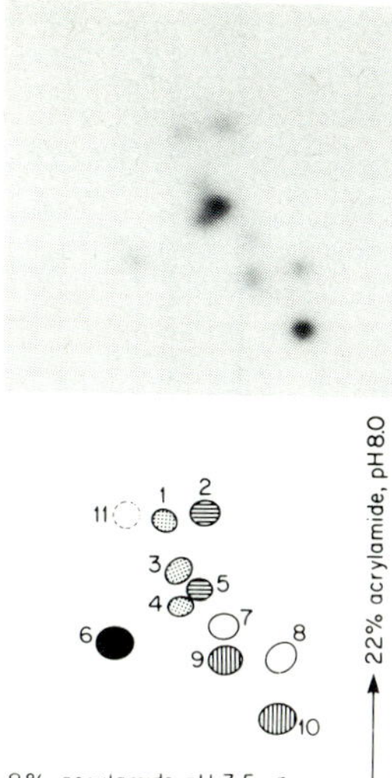

Fig.2. *Autoradiogram of a two-dimensional gel electrophoretic separation of the ribosome protected fragments from total VSV mRNA.*

RNA recovered from the ribosome protected peak (Fig.1d) was fractionated by two-dimensional gel electrophoresis. A five cm region of the first dimension extending from just slower than the xc dye to just slower than the bp dye (which was run 20 cm) contained all of the discrete bands of ribosome protected sites. The autoradiogram of the region of the second dimension (5 cm x 5 cm) containing the protected sites is shown (10 min exposure on Kodak XR-5 film). Spots 1, 3 and 4 originate from the N mRNA, 2 and 5 from the NS mRNA, 6 from the G mRNA, and 9 and 10 probably from the M mRNA, and spot 11 from the L mRNA. Spots 7 and 8 have not been assigned. The xc dye was coincident with spot 5.

and 9 and 10 (M mRNA). Spots 7 and 8 appeared to be unrelated to each other or to any of the other sequences and may be derived from internal protected sites in VSV mRNAs. Table 1 gives the quantitations, lengths, and mRNA assignments for the spots. The yields of the spots are approximately proportional to the molar amounts of the mRNAs (36).

SEQUENCE ANALYSIS OF THE RIBOSOME PROTECTED SITES

The fingerprints of T1 RNase digests of the shortest protected sites from the N and NS mRNAs (spots 1 and 2, Fig.2) and the only protected sites from the G and L mRNAs (spots 6 and 11, Fig.2) are shown in Fig.3a, b, c, and d. The sequences of oligonucleotides are indicated. The complete sequences of the

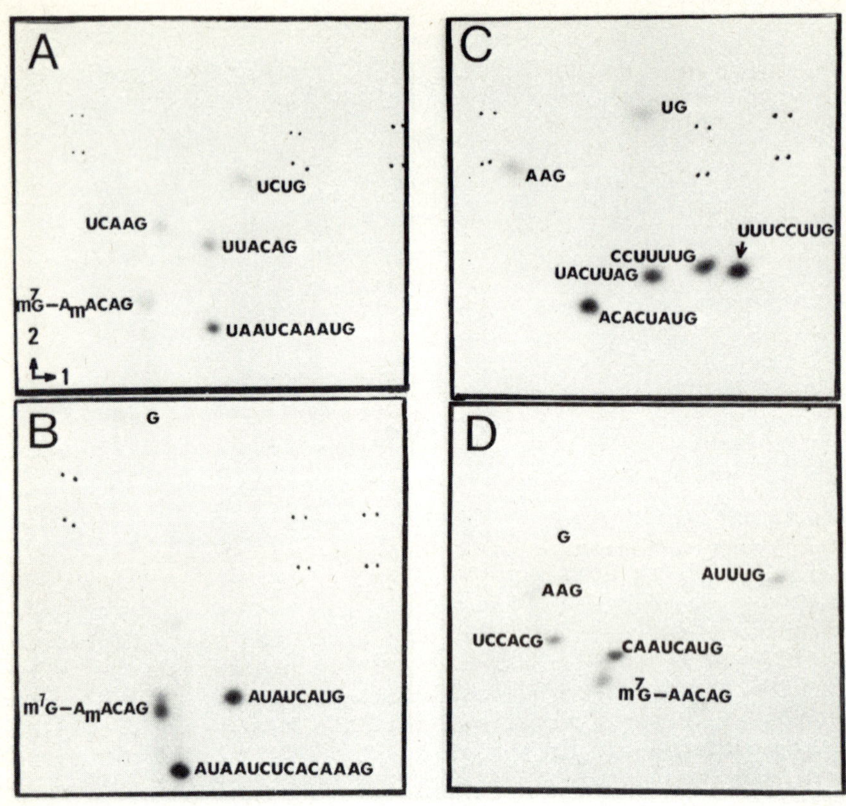

Fig.3. *Two-dimensional fingerprints of complete RNase T1 digests of ribosome binding sites.*

The first dimension was electrophoresis on cellulose acetate at pH 3.5 in 7 M urea followed by homochromatography on polyethyleneimine cellulose thin layer plates as described previously (2, 32). Panels A, B, C and D show the autoradiograms of fingerprints of spots 1, 2, 6 and 11 respectively.

binding sites are shown in Fig.4. These sequences were determined using mRNA synthesized both *in vivo* and *in vitro* using the standard procedures of further digestion with RNase A, RNase U_2, and RNase A after carbodiimide blocking of U residues (2). Further analyses of the products of these digestions by redigestion with RNase A or RNase T2 were used to establish the sequences. RNase C (28) was used in some cases to obtain partial products of oligopyrimidine tracts. Analyses of the RNase A products and T1 partial products of each site were used to order the T1 oligonucleotides. A more complete account of the sequencing of spots 1, 2 and 6 will be reported else-

TABLE 1

Quantitation of Ribosome Protected Sequences

mRNA	Spot no.	Counts per min.	Length
N	1	20,532	30*
	3	24,687	[34]*
	4	36,600	[36]
NS	2	27,002	28*
	5	46,231	[35]*
M	9	17,801	[35]
	10	36,624	[40]
G	6	25,510	35
L	11	4,722	28*
unassigned	7	14,878	[35]
unassigned	8	22,523	[37]

*Indicates sequence is capped; length is given excluding cap. Brackets indicate an estimated length based on gel mobility and fingerprint analysis.

where (22). Spot 11 (L mRNA) was obtained only in very small amounts from mRNA synthesized *in vivo* and *in vitro* consistent with the low level of L mRNA synthesis (23, 36). The complete sequence data for this site will be published elsewhere. The sequences of spots 9 and 10 (M mRNA) are not completed, although partial sequence data shows that they are uncapped, contain a single AUG codon, and are not closely related to the other sites.

When several related protected sites were obtained from a single mRNA (Fig.2 Table 1) the larger sites generally contained all of the sequences present in the smaller sites. The one exception observed was spot 4 from N mRNA which lacked only the 5'-terminal T1 oligonucleotide found in the smaller spots 1 and 3.

ANALYSIS OF 40S PROTECTED SITES.

When a 40S ribosome complex was formed with VSV mRNA and

the protected sites isolated by two-dimensional gel electrophoresis, as was done for the 80S sites, a pattern similar to that found for 80S sites was observed. Fingerprint analysis of the protected sites showed that the spots corresponding to 2 and 5 (NS mRNA), spot 10 (M mRNA) and spot 11 (L mRNA) were identical to those found in the 80S sites. In contrast, the predominant 40S site from the N mRNA is longer than any N site from the 80S complex (approximately 40 nucleotides) and contains the capped 5'-terminus. Also a longer site (approximately 50 nucleotides) from the G mRNA is protected in the 40S complex although it, like the M site, does not contain the capped 5'-terminus. Thus, in some mRNAs, 40S ribosomes and 80S ribosomes protect identical sequences, and in other cases a longer sequence is protected in the 40S site than in the 80S site. Also, not all mRNAs have the capped 5'-end in the site protected by 40S ribosomes.

DISCUSSION

Features of the sequences.

Legon (16) has recently reported the use of reticulocyte ribosomes to protect the translation initiation site in β-globin mRNA. The conditions used in the protection of β-globin mRNA are similar to those used here, and yielded a binding site encoding the NH_2-terminal sequence of the protein. The complete nucleotide sequence in this initiation site is now known from the complete sequence of the mRNA (9, 17) and does not show significant homology with all of the sites shown here, although the same sequence, AACAG, found in N, NS and L mRNAs also occurs in β-globin mRNA close to the initiation codon.

As was found for globin mRNA, each VSV site has a centrally located AUG initiation codon (Fig.4), and the predicted amino acids are shown starting from this position. The sequence predicted for the G protein is especially interesting because it contains a sequence of four hydrophobic amino acids leu-leu-tyr-leu which could form part of a labile "signal sequence" involved in insertion of the G protein into membranes (3).

This process might be involved in formation of the glycoprotein spikes which protrude from the viral membrane. The NH_2-terminus of the mature G protein isolated from virions has been reported to be alanine (27) suggesting that at least six amino acids are cleaved from the NH_2-terminus during maturation.

The nucleotide sequences of the N, NS, and L sites are identical to each other for 9 out of the first 11 positions extending from the capped 5'-end to the AUG codon (Fig.4). The high degree of sequence similarity in this region implies some important function(s), perhaps involving ribosome binding.

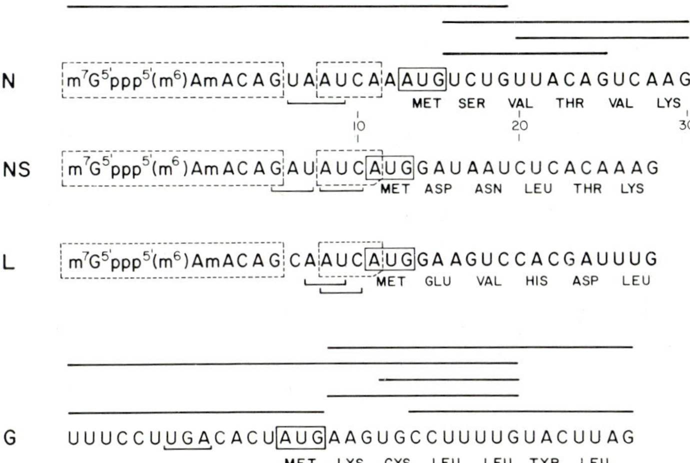

Fig.4. *Nucleotide sequences of ribosome binding sites from the VSV mRNAs specifying the N, NS, L and G proteins.*

The T1 partial products used to establish the sequences of spots 1 and 6 (N and G mRNA sites) are indicated by the solid lines above the sequences. The figure shows the regions of homology (dashed boxes) between N, NS and L and sequences complementary to the 3'-end of 18S rRNA (brackets under sequences).

Neither the capped 5'-end nor the conserved sequences are present in the 40S or 80S sites derived from G or M mRNAs although these mRNA share the 5'-sequence $m^7G^{5'}ppp^{5'}(m^6)AmAC$ in common with all VSV mRNAs (10). It is possible that the conserved sequence found in N, NS, and L mRNAs is also present in G and M mRNAs and determines the equal translation efficiencies of the mRNAs at a ribosome recognition stage prior to 40S complex formation. The observation that removal of the 5'-terminal 7-methylguanosine affects translation of all VSV mRNAs to about the same extent suggests an interaction of ribosomes with the 5' end (19). The failure of 40S or 80S ribosome complexes to protect the 5'-end of M and G mRNAs suggests that the interaction near the cap might occur at a stage prior to 40S complex formation. The protection of the capped 5'-end observed here for some VSV mRNAs and previously for reovirus mRNAs in the 40S complex (15) might be a consequence only of the proximity of the 5'-end to the translation initiation site.

Initiation site structure and the effect of hypertonic shock

Nuss and Koch (20) observed a differential reduction in translation of VSV mRNAs when infected cells were incubated in hypertonic medium. The structural studies reported here suggest an explanation for this effect. The three mRNAs for which translation was shown to be highly resistant to hypertonic initiation block (N, NS, and L) have the capped 5'-ends in the ribosome protected sites with initiation codons only 11 to 13 nucleotides from the capped 5'-ends. In contrast, the mRNAs for which translation was sensitive to hypertonic initiation block (G and M) do not have the capped end in the ribosome protected sites and have initiation codons at least 18 nucleotides from the 5'-end. Thus, the hypertonic block could affect a transition of the ribosome from the end of the mRNA to the initiation site. Alternatively, the resistance of N, NS, and L, could result from some feature of the conserved sequences in their ribosome recognition sites.

Differential effects of hypertonic shock on translation of cellular mRNA have also been observed (26) although most cellular mRNAs are sensitive to hypertonic initiation block. The differential effects on cellular mRNA translation might also be related to the proximity of the initiation codon to the 5'-end.

Lack of secondary structure and 18S rRNA complementarity

No extensive regions of secondary structure can be drawn for any of the protected sites although base pairing between positions 1-5 and 17-21 of the N protein mRNA could occur generating a loop containing the AUG. This structure would have minimal stability (35) but might be further stabilized by tertiary interactions or interaction with proteins.

The mechanism of initiation site selection in prokaryotes is thought to involve base pairing between the 3'-end of 16S rRNA and a sequence in the mRNA preceding the initiation codon (30, 34). The first complete initiation sequence established for a eukaryotic mRNA (the brome mosaic virus coat protein mRNA) has the sequence UAAU in the 9 nucleotides between the capped end and the initiation site (7). This sequence is also found in the VSV N protein mRNA and could pair with the 3'-end of eukaryotic rRNA, $3'_{HO}$AUUACUAG (31). The same sequence is not found in the NS and L mRNAs, however two triplets complementary to 18S rRNA are found in these sites. The site from the G mRNA contains only a single triplet complementary to the 3'-end of 18S rRNA. Because the minimal complementarities observed here would occur with 15-70% probability at random, these results do not support a model of eukaryotic initiation site interaction with the 3'-end of 18S rRNA, although additional sequences from the 3'-end of 18S rRNA might reveal more significant complementarities.

Initiation site selection in mRNA lacking 5'-terminal 7-methylguanosine

Considerable interest has recently been focused on the role of 5'-terminal 7-methylguanosine in translation of eukaryotic mRNAs which have this 5'-terminal modification (11, 29). The absence of this nucleotide at the 5'-end of VSV mRNA reduces translation 3-4 fold in the rabbit reticulocyte system and 10-20 fold in the wheat germ system (4, 19, 24). The defect in VSV mRNA lacking 7-methylguanosine appeared to be mainly in the rate of binding to reticulocyte ribosomes, rather than in the extent of binding (ref.24, Fig.1b of this paper).

Figs.1a, b, d and e show that VSV mRNA from which the 5'-terminal 7-methylguanosine was removed by periodate oxidation and β-elimination binds ribosomes to virtually the same extent as untreated mRNA and that both RNAs yield about two percent of the total radioactivity in the ribosome-protected 80S peak. It can be concluded that ribosomes found the proper initiation sites on both types of mRNA because two-dimensional fingerprints of RNase digests of the protected fraction showed the same major oligonucleotides from the N, NS, M and G mRNA binding sites including the normal and β-eliminated forms of the 5'-terminal T1 oligonucleotide. To obtain an estimate of the extent of proper recognition in the absence of 7-methylguanosine, the ratio of the radioactivity in the structures $m^7G^{5'}ppp^{5'}(m^6)Am$ and $ppp(m^6)$ to the other nucleotides was compared for the ribosome protected fractions and the total mRNA (Table 2). A 16-17 fold enrichment of the capped structure in the protected material was found for the control RNA and a smaller increase of 6-8 fold was seen for the triphosphate end from β-eliminated mRNA. Analysis of the fingerprints of the protected fractions suggests that this decrease in ribosome protection near the 5'-terminus results from increased protection of many minor non-initiation site sequences, but the cause of this decreased specificity is not yet clear. Although these experiments indicate that proper initiation site recognition can occur without 7-methylguanosine, they also suggest that results from ribosome binding and translation experiments performed *in vitro* should be interpreted with caution, because *in vitro* conditions may permit non-specific ribosome mRNA interactions which would not occur normally *in vivo*.

ACKNOWLEDGEMENTS

I thank Dr David Baltimore for providing encouragement,

TABLE 2

Quantitation of 5' Ends in Total and Ribosome Protected mRNA

		Fraction of total mRNA	Fraction of ribosome protected mRNA	Fold increase
Experiment I	$m^7G^{5'}ppp^{5'}(m^6)Am$ (control)	0.0020	0.0322	16.1
	$ppp^{5'}(m^6)Am$ (β-eliminated)	0.0019	0.0122	6.4
Experiment II	$m^7G^{5'}ppp^{5'}(m^6)Am$	0.0022	0.0378	17.2
	$ppp^{5'}(m^6)Am$	0.0018	0.0146	8.1

Total VSV mRNA (β-eliminated and control) synthesized *in vivo* and the ribosome protected fraction of each purified by sucrose gradient sedimentation (Fig.1), were digested with nuclease P1 and the 5'-terminal structures indicated were separated from the mononucleotides by ionophoresis at pH 3.5 as described previously (21). The positions of the nucleotides were identified by autoradiography, followed by excision from the paper and determination of radioactivity in a scintillation spectrometer. Approximately 10^6 total cpm were used for each determination, and the fraction of radioactivity in the 5'-ends relative to internal positions (mononucleotides) is given.

helpful suggestions, and the facilities in which this work was carried out. I also thank Dr Udy Olshevsky for gifts of reticulocyte extracts, Ralf Pettersson for careful instruction in the two-dimensional gel technique, Kahan Smith for assistance, and Marilyn Smith for preparing the manuscript. This work was supported by grant #AI-08388 to Dr David Baltimore from the U.S. National Institute of Allergy and Infectious Diseases and grant #CA-14051 from the U.S. National Cancer Institute.

REFERENCES

1. Baltimore, D., Huang, A.S. and Stampfer, M. (1970). *Proc. Nat. Acad. Sci. U.S.A.* 66, 572.
2. Barrell, B.G. (1971). *In* "Procedures in Nucleic Acid Research", vol.II, eds. Cantoni, G.L. and Davies, D.R. Harper and Row, New York, 751.
3. Blobel, G. and Dobberstein, B. (1975). *J. Cell Biol.* 67 835.
4. Both, G.W., Banerjee, A.K. and Shatkin, A.J. (1975). *Proc. Nat. Acad. Sci. U.S.A.* 72, 1189.
5. Both, G.W., Moyer, S.A. and Banerjee, A.K. (1975). *J. Virol.* 15, 1012.
6. Coffin, J.M. and Billeter, M.A. (1976). *J. Mol. Biol.* 100, 293.
7. Dasgupta, R., Shih, D.S., Saris, C. and Kaesberg, P. (1975). *Nature* 256, 624.
8. De Wachter, R. and Fiers, W. (1972). *Anal. Biochem.* 49, 184.
9. Efstratiadis, A., Kafatos, F.C. and Maniatis, T. (1977). *Cell* 10, 571.
10. Freeman, G.J., Rose, J.K., Clinton, G.M. and Huang, A.S. (1977). *J. Virol.* 21, 1094.
11. Griffin, B. (1976). *Nature* 263, 188.
12. Hindley, J. and Staples, D.H. (1969). *Nature* 224, 964.
13. Huang, A.S., Baltimore, D. and Stampfer, M. (1970). *Virology,* 42, 946.
14. Knipe, D., Rose, J.K. and Lodish, H.F. (1975). *J. Virol.* 15, 1004.
15. Kozak, M. and Shatkin, A.J. (1977). *J. Mol. Biol.* 112, 75.
16. Legon, S. (1976). *J. Mol. Biol.* 106, 37.
17. Lockard, R.E. and Rajbhandary, U.L. (1976). *Cell* 9, 747.
18. Lodish, H.F. (1971). *J. Biol. Chem.* 245, 7131.
19. Lodish, H.F. and Rose, J.K. (1977). *J. Biol. Chem.* 252, 1181.
20. Nuss, D.L. and Koch, G. (1976). *J. Virol.* 17, 283.
21. Rose, J.K. (1975). *J. Biol. Chem.* 250, 8098.
22. Rose, J.K. (1977). *Proc. Nat. Acad. Sci. U.S.A.*,
23. Rose, J.K. and Knipe, D. (1975). *J. Virol.* 15, 994.

24. Rose, J.K. and Lodish, H.F. (1976). *Nature* 262, 32.
25. Rose, J.K., Lodish, H.F. and Brock, M.L. (1977). *J. Virol.* 20, 324.
26. Saborio, J.L., Pong, S.-S. and Koch, G. (1974). *J. Mol. Biol.* 85, 195.
27. Schloemer, R.H. and Wagner, R.R. (1975). *J. Virol.* 16, 237.
28. Schmukler, M., Jewett, P.B. and Levy, C.C. (1975). *J. Biol. Chem.* 250, 2206.
29. Shatkin, A.J. (1976). *Cell* 9, 645.
30. Shine, J. and Dalgarno, L. (1974). *Proc. Nat. Acad. Sci. U.S.A.* 71, 1342.
31. Shine, J. and Dalgarno, L. (1974). *Biochem. J.* 141, 609.
32. Squires, C., Lee, F., Bertrand, K., Squires, C.L. Bronson, M.J. and Yanofsky, C. (1976). *J. Mol. Biol.* 103, 351.
33. Steitz, J.A. (1969). *Nature* 224, 957.
34. Steitz, J.A. and Jakes, K. (1975). *Proc. Nat. Acad. Sci. U.S.A.* 72, 4734.
35. Tinoco, I., Jr., Borer, P.N., Dengler, B., Levine, M.O. Uhlenbeck, P.C., Crothers, D.M. and Gralla, J. (1973). *Nature New Biol.* 246, 40.
36. Villarreal, L.P., Breindl, M. and Holland, J.J. (1976). *Biochemistry* 15, 1663.
37. Watson, J.D. (1976). *Molecular Biology of the Gene,* p.484, W.A. Benjamin Inc., Menlo Park, California.

7

GENETIC AND MOLECULAR STUDIES OF BUNYAVIRUSES

J. GENTSCH, J.P. CLEWLEY, L.R. WYNNE AND D.H.L. BISH

*Department of Microbiology, The Medical Center,
University of Alabama in Birmingham,
Birmingham, Alabama, 35294, USA.*

The Bunyaviridae are a family of RNA viruses which are unique by comparison to other enveloped, arthropod-borne, viruses because their virions contain a single-stranded, segmented RNA genome (12). Serological studies have indicated that some 90 bunyaviruses can be grouped into 11 serogroups with members of each group being serologically more closely related to other members of that group than to members of another group (1, 3, 7, 8, 12). However it has been shown that certain members of each of these groups are serologically related to one or more members of another group (1, 12, 13). The 11 groups of bunyaviruses constitute the Bunyamwera supergroup of viruses with Bunyamwera virus representing a prototype virus of the Bunyamwera group. Another group of viruses in the supergroup is the California encephalitis group whose members include California encephalitis, Inkoo, Jamestown Canyon, Jerry Slough, Keystone, La Crosse, Lumbo, Melao, San Angelo, snowshoe hare, Tahyna, and Trivittatus viruses (1). Of these 12 viruses Trivittatus, in addition to being serologically related to other members of the California encephalitis group, is related to Guaroa virus when hemagglutination inhibition (H.I.) and neutralization of infectivity tests are used (13). Guaroa virus is a member of the Bunyamwera group to which it has been shown to be related by complement fixation tests but not by H.I. or neutralization of infectivity tests (13).

Some fifty or more additional viruses are morphologically similar in many respects to Bunyamwera virus but serologically independent of members of the supergroup. Included in these other viruses is Uukuniemi virus. Although Uukuniemi virus is se and snowshoe hare viruses in that it has been ve a tripartite genome (4, 11), it differs in its nposition and surface structure (2, 9). However each hree viruses has a genome consisting of one large (L), m (M), and one small (S) RNA segment.

bility of temperature sensitive, conditional lethal of snowshoe hare bunyavirus to form recombinants at equency has been documented (5). In this paper we pro- vidence for the formation of recombinants *between* certain viruses. Evidence has also been obtained which indicates the S RNA alone codes for the viral nucleocapsid protein .

HE VIRION POLYPEPTIDES OF SNOWSHOE HARE AND LA CROSSE BUNYAVIRUSES.

The major virion polypeptides of snowshoe hare and La Crosse bunyaviruses include two external glycoproteins (G1, molecular weight = 110×10^3 and G2, molecular weight = 38×10^3) in equivalent molar proportions, as well as internal nucleocapsid protein (N, molecular weight = $24\text{-}21 \times 10^3$) which is associated with all three viral RNA segments (6, 9, 10). An additional minor polypeptide is also observed in virus preparations. This protein appears to have a large size (L, molecular weight = 180×10^3), although its function or virion location is not known (9). Whereas the sizes of the G1 (or G2) polypeptides of snowshoe hare and La Crosse viruses are similar to each other, the N protein of La Crosse virus appears to be slightly larger than that of snowshoe hare virus in that it has a slightly slower electrophoretic mobility (Fig.1, ref.6).

THE GENOME OF LA CROSSE AND SNOWSHOE HARE VIRUSES EACH CONSIST OF 3 UNIQUE, SINGLE-STRANDED, RNA SEGMENTS.

Ribonuclease T_1 digests of ^{32}P-labelled snowshoe hare and La Crosse L, M and S viral RNA species have provided oligonucleotide fingerprints which indicate that the three RNA species of each virus are unique (4). For example, for snowshoe hare virus, the S RNA contains oligonucleotides which are not found in the L or S RNA species and likewise the L RNA has oligonucleotides not found in the M or S RNA (Fig.2). It is of considerable interest that the L, M or S RNA species of snowshoe hare are, by fingerprinting, quite distinct from their La Crosse counterparts.

HIGH FREQUENCY RECOMBINATION WITH TEMPERATURE SENSITIVE, CONDITIONAL LETHAL, MUTANTS OF SNOWSHOE HARE VIRUS.

The initial studies which demonstrated the formation of wild type recombinants at high frequency when cells were coinfected

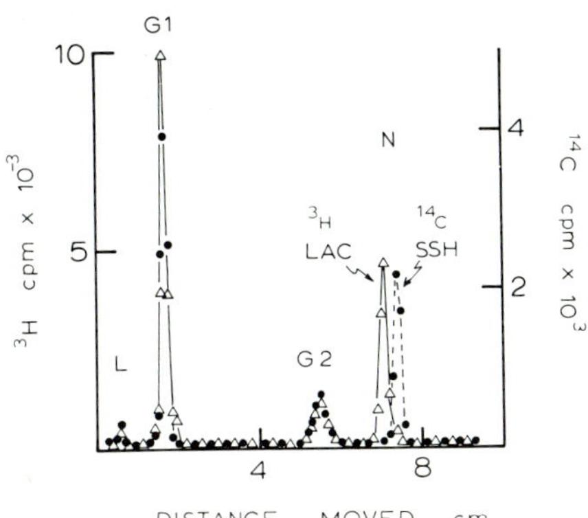

Fig.1. *Coelectrophoresis of the virus particle polypeptides of mixtures of ^{14}C-amino acid labelled snowshoe hare and ^{3}H-amino acid labelled La Crosse virus. Mixed samples of ^{14}C-amino acid labelled snowshoe hare and ^{3}H amino acid labelled La Crosse viruses were electrophoresed in 8% (w/v) polyacrylamide gels (6).*

with two temperature sensitive, conditional lethal, mutants of snowshoe hare virus (5) have been extended (Table 1). Most of the 47 *ts* mutants so far studied were isolated from mutageneses involving growth of infected cells in the presence of 50µg 5-fluorouracil per ml, although ten mutants (SSH ts #32, 33, 51, 52, 53, 54, 56, 59, 60, 61) were obtained as spontaneously occurring mutants (frequency 1.7%). Apart from two possible double mutants (SSH ts #14, 27), the snowshoe hare *ts* mutants can be categorized into two nonoverlapping recombination groups with Group I including 25 mutants (SSH ts #1, 3, 9, 12, 13, 15, 16, 20, 25, 28, 35, 36, 39, 41, 42, 47, 48, 49, 50, 52, 53, 54, 56, 58, 59 and 60), and Group II including 20 mutants (SSH ts #2, 7, 18, 21, 22, 23, 24, 26, 29, 30, 32, 33, 34, 37, 38, 40, 43, 51, 57 and 61). No recombination was observed between mutants within each group. Although three viral RNA segments have been identified for snowshoe hare virus only two recombination groups of mutants have been obtained. One possible reason for the lack of mutants representing a third recombination group might be the unequal target size of the three viral RNA species (L, molecular weight = 2.9×10^6, M, molecular weight = 1.6×10^6, and S, molecular weight = 0.41×10^6, ref.4). Another reason could be that mutations in the third RNA species may more frequently lead to lethal alterations in the gene product(s) than for the other two segments.

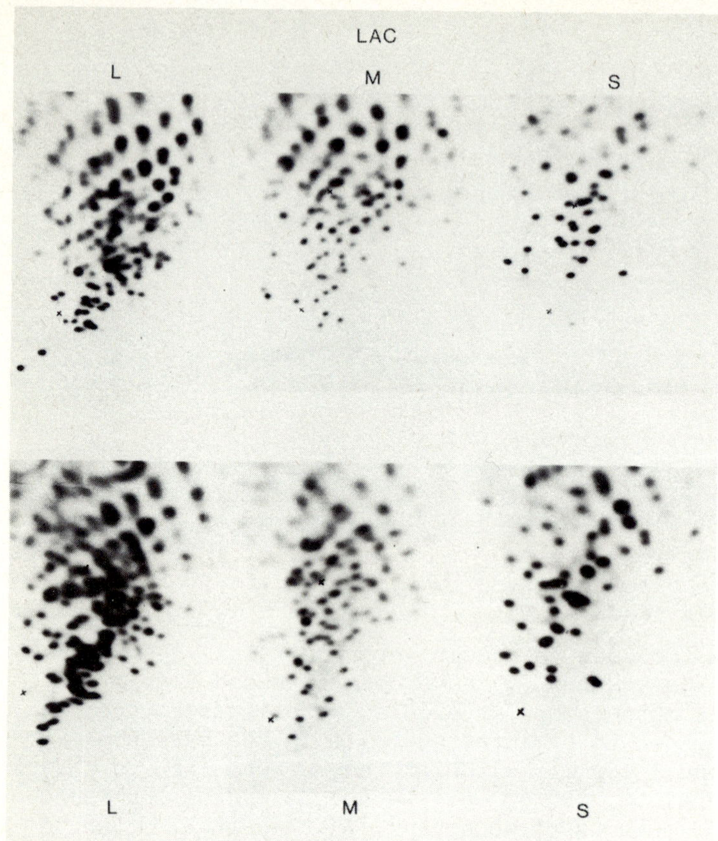

Fig.2. *Oligonucleotide fingerprints of the L, M, and S RNA species of La Crosse (LAC) and snowshoe hare (SSH) viruses. The oligonucleotide fingerprints of ribonuclease T_1 digests of the individual ^{32}P-labelled large (L), medium (M) and small (S) RNA species of La Crosse and snowshoe hare virus were obtained as described previously (4).*

A SEARCH FOR HETEROZYGOTES AMONG THE RECOMBINANT PROGENY OBTAINED FROM CROSSES INVOLVING SNOWSHOE HARE VIRUS *ts* MUTANTS.

The recombinant progeny, characterized as wild types from the point of view of their efficiency of plating at permissive (33°) and nonpermissive (39.8°) temperatures, could either represent wild type viruses with wild type genomes derived by parental segment reassortment, or heterozygotes containing in their virions duplicate RNA segments (*ts* and wild type) derived from both viruses, or aggregates of both parental viruses. Since for the latter two possibilities one might expect high

TABLE 1

Recombination analyses with snowshoe hare *ts* mutants.[a]

SSH ts mutant	% Recombinants with SSH ts mutants												
	1	2	3	7	9	12	13	14	15	16	18	20	21
1	.	4	0	7	0	0	0	0	0	0	14	0	21
2	.	.	17	0	18	4	1	0	16	9	0	7	0
3	.	.	.	24	0	0	0	0	0	0	14	0	2
7	.	.	.	.	5	4	3	0	5	5	18	4	0
9	.	.	.	.	.	0	0	0	0	0	5	0	7
12	.	.	.	.	.	.	0	0	0	0	25	0	4
13	.	.	.	.	.	.	.	0	0	0	9	0	6
14	.	.	.	.	.	.	.	.	0	0	0	0	0
15	.	.	.	.	.	.	.	.	.	0	13	0	6
16	.	.	.	.	.	.	.	.	.	.	6	0	2
18	.	.	.	.	.	.	.	.	.	.	.	2	0
20	.	.	.	.	.	.	.	.	.	.	.	.	2

Assignments: Group I: ts 1, ts 3, ts 9, ts 12, ts 13, ts 15, ts 16, ts 20. (8 mutants).
Group II: ts 2, ts 7, ts 18, ts 21. (4 mutants).
Unassigned: ts 14...Possible double mutant
[a]Recombination percentages are the average of 3 separate determinations and were calculated according to the formula:

$$\frac{\left[[AxB_{33}]_{39.8} - ([A_{33}]_{39.8} + [B_{33}]_{39.8})\right] \times 100 \times 2}{[AxB_{33}]_{33}}$$

frequency shedding of *ts* viruses on subsequent clonal analyses, this question was initially examined by determining the shedding rate of *ts* viruses from the 39.8° cloned wild type progeny of the SSH *ts* 2 x SSH *ts* 3 cross. The progeny from that cross were plated at 39.8° and 35 clones recovered. The efficiency of plating (E.O.P.) of virus in each clone was then determined by plating it out at 39.8° and 33°. All 35 clones had E.O.P. values of ~1 (PFU at 39.8° ÷ PFU at 33°). Six clones were subsequently replated at 33° and, per original clone, 50 plaques recovered and the E.O.P. of virus in these plaques determined. Of 300 clones analyzed, 298 had E.O.P. values of between 0.4 and 1.5 (i.e. comparable to that of wild type virus). The remaining two clones had E.O.P. values of less than 0.01. The plating efficiencies of these latter two viruses remained constant in subsequent titrations, suggesting that they were *de novo*, spontaneously produced, *ts* mutants. These results suggested that either the progeny virus from the original SSH *ts* 2 x SSH *ts* 3 cross are stable heterozygotes or that they represent wild type recombinants derived by reassortment of the competent wild type RNA species of the original parental mutants. Since we have observed that there is a spontaneous shedding of *ts* mutants from our wild type snowshoe hare stock virus at the rate of 1.7%, the two *ts* viruses probably reflect a spontaneous production of *ts* mutants from these clones.

Similar results were obtained with 13 putative recombinant clones which were recovered from a 33° plaque assay of the original SSH *ts* 2 x SSH *ts* 3 cross. Eleven of the 13 recombinants had plating efficiencies of between 0.5 and 1.0. When three of these clones were plated out at 33° and 50 plaques recovered for each, subsequent E.O.P. analyses of these 150 virus clones indicated that 145 had wild type E.O.P. values (~1), while 5 had low E.O.P. values (<0.01) and probably represented spontaneously produced *ts* mutants. Two of the 13 original clones from the SSH *ts* 2 x SSH *ts* 3 cross were found to have a low plating efficiency (~ 0.1). On subsequent clonal analyses these two clones were shown to contain predominantly temperature sensitive progeny (96 out of 100) with E.O.P. values of less than 0.01, suggesting that their original clones may have contained temperature sensitive viruses. The remaining 4 viruses had wild type E.O.P. values. Since the frequency of reversion to wild type of *ts* 2 and *ts*3 is less than 10^{-4} (5), the recovery of 4% wild type virus may indicate that the two original clones contained both *ts* 2 and *ts* 3 as well as recombinants. No further analyses were undertaken to investigate this possibility.

HIGH FREQUENCY RECOMBINATION WITH TEMPERATURE SENSITIVE, CONDITIONAL LETHAL, MUTANTS OF LA CROSSE VIRUS.

High frequency recombination has been obtained with some 20 *ts* mutants of La Crosse virus. The mutants have been categorized into two genetic recombination groups, Group I containing six mutants (LAC ts #9, 13, 16, 19, 20 and 23) and Group II containing 14 mutants (LAC ts #1, 2, 3, 4, 5, 10, 11, 12, 14, 15, 17, 18, 21 and 22). Although most of the mutants were isolated from stocks of virus grown in the presence of 5-fluorouracil, five were obtained as spontaneous mutants (LAC ts #19, 20, 21, 22 and 23).

RECOMBINATION BETWEEN TEMPERATURE SENSITIVE, CONDITIONAL LETHAL, MUTANTS OF SNOWSHOE HARE AND LA CROSSE BUNYAVIRUSES.

The results of recombination experiments between certain *ts* mutants of La Crosse and snowshoe hare viruses are shown in Table 2. All recombination experiments, their corresponding plaque assays, and other manipulations were performed in a Baker BioGard laminar flow hood in a P3 facility in the virology complex at the University of Alabama in Birmingham.

High frequency recombination was found for certain *ts* mutant crosses (e.g. SSH I-3 x LAC II-4). Clonal analyses of the plaques obtained from the 39.8° plaque assays gave efficiency of plating ratios of between 0.1 and 1.3 while the progenitor *ts* mutant efficiency of plating ratios were less than 0.0001.

TABLE 2

Recombination between snowshoe hare and La Crosse ts mutants

LAC ts mutants	% Recombinants with SSH ts mutants			
	SSH I-1	SSH I-3	SSH II-2	SSH II-18
LAC II-3	3.9	5.4	0	0
LAC II-4	6.5	6.2	0	0
LAC II-5	35.7	16.9	0	0

Recombination percentages were determined as described in Table 1.

OLIGONUCLEOTIDE FINGERPRINT ANALYSES OF THE L, M AND S RNA SPECIES OF THE LAC-SSH RECOMBINANTS

After pathogenicity tests in three model animals systems (3 week old mice, adult hamsters, adult guinea pigs) indicated that the LAC-SSH recombinants were no more pathogenic than the wild type snowshoe hare or La Crosse viruses, molecular analyses were undertaken to determine the genome composition of the recombinants.

The oligonucleotide fingerprints of the L, M and S RNA species of the recombinant obtained from the LAC II-4 x SSH I-1 cross are given in the upper part of Fig.3, while the fingerprints of the L, M and S RNA species of the recombinant obtained from the LAC II-5 x SSH I-3 cross are shown in the lower part of Fig.3. Comparison of the fingerprints of the two recombinants with those of the parent wild type viruses indicates that both recombinants have an SSH L RNA and a LAC M RNA. One recombinant has an SSH S RNA, while the other has a LAC S RNA.

No evidence was obtained for either recombinant having SSH M RNA or LAC L RNA species or both SSH and LAC S RNA species, indicating that neither recombinant is heterozygous with regard

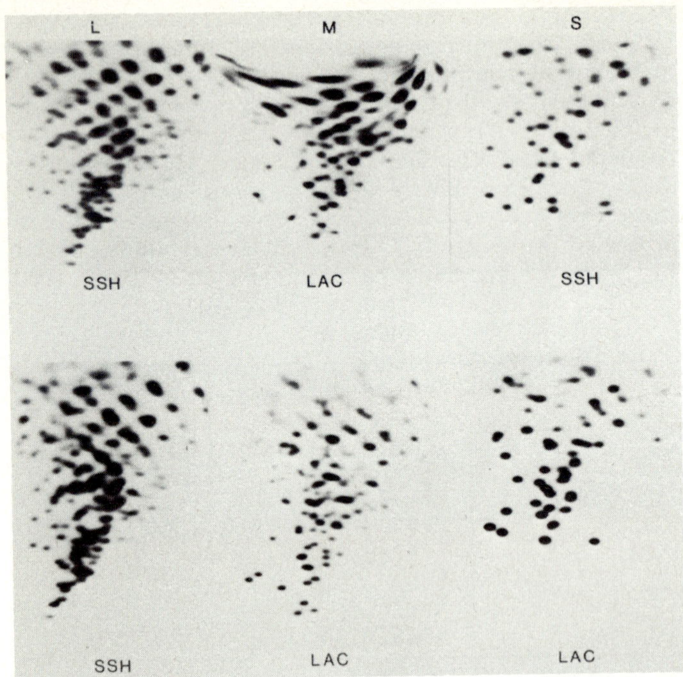

Fig.3. *Oligonucleotide fingerprint of the L, M, and S RNA species of the recombinants obtained from LAC II-4 x SSH I-1 and LAC II-5 x SSH I-3 crosses. The oligonucleotide fingerprints of the ribonuclease T_1 digests of the individual ^{32}P-labelled L, M and S RNA species of the recombinants derived from the (upper set) LAC II-4 x SSH I-1 cross, and (lower set) LAC II-5 x SSH I-3 cross were obtained as described previously (4). Comparison of the fingerprints with those of the progenitor wild type snowshoe hare and La Crosse viruses (Fig.2) indicate that the recombinant from the LAC II-4 x SSH I-1 cross has a SSH/LAC/SSH genotype and the recombinant from the LAC II-5 x SSH I-3 cross has a SSH/LAC/LAC genotype.*

to any single RNA size class. The results indicate that the recombinants were formed by segment reassortment rather than by intrasegment recombination since the fingerprints of all of the recombinant RNA species correspond totally to the RNA species of one or the other progenitor, and not partially to one progenitor and partially to the other. The recombinants can therefore be designated SSH/LAC/SSH (Fig.3 top), and SSH/LAC/LAC (Fig.3 bottom) to signify the origin of their respective L, M and S RNA species.

POLYPEPTIDE ANALYSES OF THE SSH/LAC/SSH AND SSH/LAC/LAC RECOMBINANTS INDICATE THAT THE S RNA CODES FOR N PROTEIN

The small sizes of the snowshoe hare or La Crosse S RNA species preclude their coding for the L or G1 polypeptides. Also it appears unlikely that the S RNA could code for both G2 and N.

Since the SSH/LAC/SSH and SSH/LAC/LAC recombinants differ only with respect to their S RNA species, the possibility that S RNA codes for N protein was examined by comparing the virion polypeptides of a ^{14}C-amino acid labelled SSH/LAC/SSH recombinant to those of a ^{3}H-leucine labelled SSH/LAC/LAC recombinant. The results, shown in Fig.4, indicate that SSH/LAC/LAC recombinant has an N protein with a slower electrophoretic mobility than that of the SSH/LAC/SSH recombinant. The G1 and G2 polypeptides of the two recombinants were indistinguishable. These results suggest therefore that the S RNA codes for N protein.

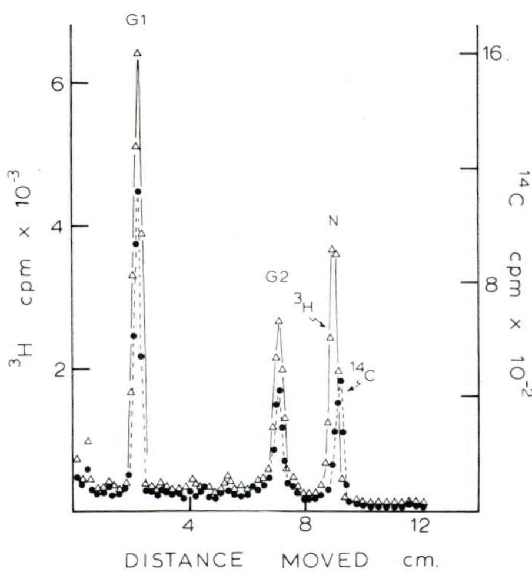

Fig.4. *Coelectrophoresis of the ^{14}C-amino acid labelled recombinant virus obtained from the LAC II-4 x SSH I-1 cross with the ^{3}H-leucine labelled recombinant virus obtained from the LAC II-5 x SSH I-3 cross. Virus particles of a ^{14}C-amino acid labelled preparation of the LAC II-4 x SSH I-1 recombinant and a ^{3}H-leucine labelled preparation of the LAC II-5 x SSH I-3 recombinant were mixed, dissociated by SDS and resolved by polyacrylamide gel electrophoresis (6, 9).*

DISCUSSION

The results presented in this paper indicate that the genomes of bunyaviruses consist of three unique RNA segments with the smallest RNA coding for the nucleocapsid protein. Analyses of recombinants formed from crosses involving Group II SSH mutants with Group I LAC mutants should indicate whether the L or M RNA codes for G1 and, or G2 proteins. Such recombinants have been obtained and will be analysed when their pathogenicity tests have been completed.

The number of new recombinant genotypes one could obtain from two bunyaviruses which can recombine with each other is 2^3-2 (e.g. LMS x lms giving, in addition to LMS and lms, the recombinants LMs, LmS, Lms, lmS, lMS and lMs, i.e. 6). So far we have obtained and characterized from La Crosse and snowshoe hare two recombinants, SSH/LAC/SSH and SSH/LAC/LAC. If all the bunyaviruses in the California group can recombine with each other then one might expect to generate 12^3-12 i.e. 1716 new virus genotypes. Whether recombination will occur between other bunyaviruses is not known. The potential for producing recombinants raises questions of whether recombinants can be detected in Nature, as well as their pathogenicity, host ranges and how they should be handled. We have elected to perform our recombination experiments in a P3 facility using the guidelines established in the U.S. for work on recombinant DNA. Pathogenicity tests, which will be reported elsewhere, have been undertaken before any large quantities of the recombinant viruses were produced for molecular analyses.

We conclude from the results recorded here that the snowshoe hare Group I mutants have a defect in their M RNA since only the M RNA of La Crosse virus was recovered in both recombinants. Likewise we conclude that the Group II mutants of La Crosse have a defect in their L RNA since only the L RNA of snowshoe hare virus was recovered in both recombinants. The fact that one recombinant had SSH S RNA and the other LAC S RNA suggests that the S RNA species were not defective for any of the progenitor mutants tested.

A nomenclature system (e.g. SSH/LAC/SSH) is proposed from the recombinant viruses which signifies the origin of the recombinant's respective L, M and S RNA segments. This system allows us to identify recombinants that may be produced from any three progenitor virus stocks.

ACKNOWLEDGEMENTS

This investigation was supported by Public Health service grant AI 13402 from the National Institute of Allergy and Infectious Diseases and by National Science Foundation grant

PCM 76-22218. We would like to express our appreciation to Gloria Robeson and Reginald Anderson for excellent technical assistance.

REFERENCES

1. Berge, T.O. (1975). *International Catalogue of arboviruses.* U.S. D.H.E.W. Publ. (CDC) 75-8301.
2. von Bonsdorff, C.-H. and Pettersson, R. (1975). *J. Virol.* 16, 1296.
3. Casals, J. (1971). *In* "Comparative Virology", (Maramorosch, K. and Kurstak, E., eds.) p.307. Academic Press: London and New York.
4. Clewley, J., Gentsch, J. and Bishop, D.H.L. (1977). *J. Virol.* 22, 459.
5. Gentsch, J. and Bishop, D.H.L. (1976). *J. Virol.* 20, 351.
6. Gentsch, J., Bishop, D.H.L. and Obijeski, J.F. (1977). *J. Gen. Virol.* 34, 257.
7. Murphy, F.A. (1975). *Proceedings of the U.S. Animal Health Association.* 425.
8. Murphy, F.A., Harrison, A.K. and Whitfield, S.G. (1973). *Intervirology* 1, 297.
9. Obijeski, J.F., Bishop, D.H.L., Murphy, F.A. and Palmer, E.L. (1976a). *J. Virol.* 19, 985.
10. Obijeski, J.F., Bishop, D.H.L., Palmer, E.L. and Murphy, F.A. (1976b). *J. Virol.* 20, 664.
11. Pettersson, R.F., Hewlett, M.J., Baltimore, D. and Coffin, J.M. (1977). *Cell* 11, 51.
12. Porterfield, J.S., Casals, J., Chumakov, M.P., Gaidamorich, S.Y., Hannoun, C., Holmes, I.H., Horzinek, M.C., Mussgay, M., Oker-Blom, N. and Russell, P.K. (1975/76). *Intervirology.* 6, 13.
13. Whitman, L. and Shope, R.E. (1962). *Amer. J. Trop. Med. Hyg.* 11, 691.

GENETIC AND MOLECULAR STUDIES OF ARENAVIRUSES

A.C. VEZZA, G.P. GARD*, R.W. COMPANS and D.H.L. BISHOP

Department of Microbiology, The Medical Center, University of Alabama in Birmingham, Birmingham, Alabama, 35294, U.S.A.

**On leave from the Veterinary Research Station, Glenfield, N.S.W., Australia.*

The arenaviruses are a family of enveloped, single-stranded RNA viruses whose members include the prototype lymphocytic choriomeningitis (LCM) virus, Lassa virus and the Tacaribe complex of viruses (Amapari, Junin, Latino, Machupo, Parana, Pichinde, Tacaribe and Tamiami), most of which have been isolated from South and Central America (16). Electron microscopic examinations of arenavirus infected cells, or negatively stained virus preparations, have revealed that arenavirus particles are spherical or pleomorphic (50 to 300 nm in diameter) with 10 nm club shaped surface projections (1, 9, 11). Characteristic electron dense granules have been observed in the interior of arenavirus particles. These have been shown by a variety of criteria to be host cell derived ribosomes (5, 6, 10, 14). What role if any the ribosomes have in developing an infection process is not known, nor is it known if the ribosomes are involved as obligate factors in the virus particle maturation. It is not certain that all arenavirus particles (infectious and, or noninfectious) possess ribosomes.

The genomes of Pichinde, LCM and Junin viruses are reported to consist of two or three RNA segments including a 31 *s*, 22 *s* and perhaps a 15 *s* RNA species (2, 5, 6, 10, 12, 13, 15). It has been shown that the 31 *s* and 22 *s* viral RNA species lack methylated nucleotides, they also lack a 3' polyadenosine sequence and are not infectious *per se* (7). This evidence, taken together with observations of a virion RNA-directed RNA polymerase capable of synthesizing viral complementary RNA, in the test tube has led Rawls and associates to postulate that arenaviruses are negative stranded viruses (7).

Three major structural polypeptides were identified in preparations of Pichinde virus by Ramos *et al*., (18). These included two glycoproteins (molecular weights = 72 x 10^3 and 34 x 10^3) and a nonglycosylated nucleocapsid protein (molecular weight = 72 x 10^3). A recent report indicates that Junin virus has two major viral polypeptides (a 64x10^3 dalton nucleoprotein and a 38x10^3 dalton glycoprotein), as well as several minor glycosylated and nonglycosylated polypeptides (19). For LCM, one large and five smaller polypeptides have been described (14).

In this paper we report structural analyses of Pichinde, Tacaribe and Tamiami viruses which indicate that Pichinde virus has two size classes of glycoproteins, while Tacaribe and Tamiami have only one. The glycoproteins for all three viruses are located on the external surface of the virus particles. Analyses of Pichinde viral RNA species indicate that although the virus has a segmented genome there is little, if any, 18*s* ribosomal RNA.

Preliminary genetic analyses suggest that high frequency wild type recombinants can be obtained from coinfections involving temperature sensitive, conditional lethal, Pichinde mutants. So far two nonoverlapping genetic recombination groups of mutants have been identified.

THE MAJOR STRUCTURAL POLYPEPTIDES OF PICHINDE VIRUS

When a preparation of ^{14}C-amino acid and ^{3}H-glucosamine labeled Pichinde virus was resolved on an 8% (w/v) polyacrylamide gel, two major size classes of viral polypeptides were identified. Both polypeptide species were labelled with the two isotopes (Fig.1a). Careful comparison of the distribution of the two labels indicated that not only was the ^{3}H-glucosamine label in the slower migrating band asymmetrical when compared to the ^{14}C label (unlike that of the faster band), but also the ^{14}C to ^{3}H label ratio was quite different for the two polypeptide bands. It will be shown later that the slower band consists of a nucleocapsid protein (N) and a large glycoprotein (G1), while the faster band contains a smaller glycoprotein (G2).

Coelectrophoresis of Pichinde viral polypeptides with the known molecular weight polypeptides of cytochrome C (molecular weight = 12x10^3), α-chymotrypsinogen A (molecular weight = 26x10^3), egg albumin (molecular weight = 43x10^3), or vesicular stomatitis L, G, N, and M proteins (respective molecular weights: 160 x 10^3, 65x10^3, 54x10^3, and 27x10^3), indicated that the N, G1 and G2 viral polypeptides of Pichinde have apparent molecular weights of 66x10^3, 64x10^3 and 38x10^3 respectively.

It has been shown recently that glycoproteins of various

enveloped viruses can be selectively labelled with ^{35}S-sulphate (8, 17). A preparation of ^{3}H-leucine and ^{35}S-sodium sulphate labelled Pichinde virus was grown in BHK-21 cells and the resulting purified virus preparation dissociated by SDS. When the polypeptides were resolved by polyacrylamide gel electrophoresis, no ^{35}S was observed in association with any of the major viral polypeptides. A similar result was obtained for Pichinde virus grown in MDBK cells. In experiments aimed at determining if any of the Pichinde viral polypeptides are phosphoproteins, a preparation of ^{32}P-phosphate labelled virus was dissociated by SDS and the viral polypeptides resolved on an 8% ($^{w}/v$) polyacrylamide gel. No ^{32}P label was associated with any of the major viral polypeptides.

Minor amounts of a 77×10^3 dalton and infrequently a 12×10^3 dalton nonglycosylated polypeptide have been observed in preparations of Pichinde virus. Their location or function have not been analysed further.

THE MAJOR STRUCTURAL POLYPEPTIDES OF TACARIBE AND TAMIAMI VIRUSES

Preparations of ^{3}H-glucosamine and ^{14}C-amino acid labelled Tamiami and Tacaribe viruses were dissociated by SDS and the polypeptides analysed by gel electrophoresis (Fig.2a, c). For both viruses a single major nonglycosylated polypeptide and a smaller glycosylated polypeptide (G) were obtained. As described below the nonglycosylated polypeptide is associated with the viral nucleocapsid(s) and is designated N. Coelectrophoresis of ^{14}C-amino acid labelled Pichinde viral polypeptides with ^{3}H -leucine labelled Tacaribe or Tamiami virus preparations, or ^{14}C-amino acid labelled Tamiami with ^{3}H-leucine labelled Tacaribe virus, indicated that the N polypeptides of Tacaribe and Tamiami viruses have apparent molecular weights of 68×10^3 and 66×10^3 respectively, while the G polypeptides have apparent molecular weights of 42×10^3 and 44×10^3 respectively. In Tacaribe virus preparations a slower moving, minor, 79×10^3 dalton polypeptide (P) was observed while Tamiami preparations contained a minor 77×10^3 dalton polypeptide (P). Neither minor polypeptide was glycosylated to any significant extent.

LOCATION OF THE VIRAL GLYCOPROTEINS

Glycoproteins of other enveloped viruses are located on the external surface of the virus particle and form the surface projections or spikes that are visible by electron microscopy. The characteristic surface spikes on Tacaribe virions are shown in Fig.3a. To establish the location of the glycoproteins of Pichinde, Tacaribe or Tamiami viruses, purified preparations of ^{14}C-amino acid and ^{3}H-glucosamine labelled viruses were treated

A. PICHINDE VIRUS
B. PICHINDE SPIKELESS PARTICLES
^{14}C cpm X 10^{-2}
^{3}H cpm X 10^{-2}
^{14}C
^{3}H
Slice No.

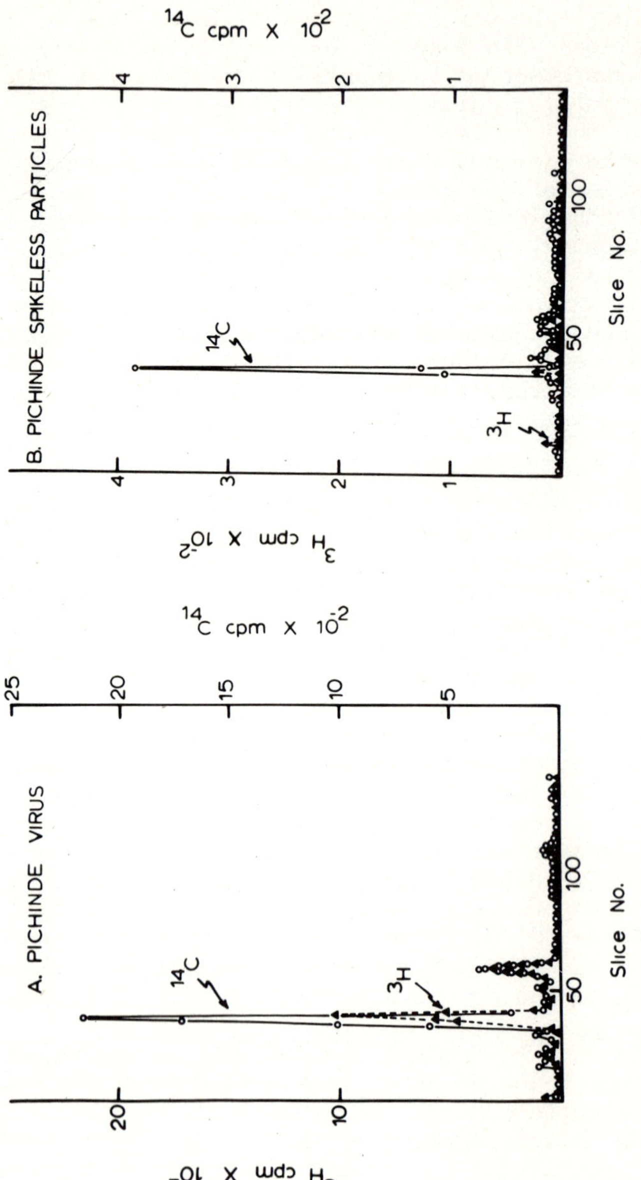

Fig.1

with proteases. The viruses were either incubated at 37° with α-chymotrypsin at a concentration of 1 mg/ml of 0.01 M phosphate buffer, pH7.0, for 2 hr (Tamiami), or 1 mg/ml for 2 hr followed by an additional 1 mg/ml for 1 hr (Tacaribe), or at 31° first for 1 hr with 10 mg pronase per ml and then with 10 mg bromelain and 0.02 ml of 0.1 M dithiothreitol per ml of the mixture for an additional 90 min (Pichinde). After electron microscopic identification to verify that the particles were essentially spikeless (for Pichinde see Fig.3b), the incubation mixtures were loaded over sucrose gradients and the particles repurified by centrifugation. Polyacrylamide gel electrophoresis of SDS dissociated spikeless particles (Fig.1b, 2b, d) revealed that for each virus the particles lacked most (Tacaribe, Tamiami), or all (Pichinde) of the viral glycoproteins. We conclude from these results that, like other enveloped viruses, the glycoproteins of arenaviruses are present on the external surface of the virus particles and are sensitive to protease removal.

IDENTIFICATION OF THE NUCLEOPROTEINS OF PICHINDE, TACARIBE AND TAMIAMI VIRUSES

A preparation of ^{32}P-labelled Pichinde was dissociated by 2% (wt/vol) Triton X100 in the presence of 1 M NaCl, loaded over a solution of 40% (wt/vol) CsCl, and centrifuged to equilibrium. In addition to acid soluble phospholipids recovered at the top of the gradient, a single peak of ^{32}P label was recovered at a density of 1.31 g/ml (Fig.4a). Eighty four percent of ^{32}P in that material was rendered acid soluble after incubation at 37° for 30 min with 5 μg pancreatic ribonuclease per ml. Similar results were obtained for ^{3}H-uridine labelled Tamiami virus. Electron microscopic examination of the material obtained from the peak fractions of the Tamiami ribonucleoprotein isolated by CsCl equilibrium centrifugation indicates that the structures consist of convoluted beaded strands 30-40 Å in diameter (Fig.3d), with globular subunits spaced at a

Fig.1. *Localization and characterization of Pichinde virus polypeptides. (A). A preparation of Pichinde virus labelled by ^{3}H-glucosamine and ^{14}C-amino acids, was dissociated by SDS and the distribution of radioactivity determined after resolution by polyacrylamide gel electrophoresis (4). A sample of the ^{3}H-glucosamine and ^{14}C-amino acid labelled Pichinde was digested with pronase and bromelain (see text) and repurified by centrifugation in a 20 to 70% sucrose gradient. After SDS lysis of the rebanded, protease treated, material followed by polyacrylamide gel electrophoresis, the distribution of radioactivity was determined (B).*

^{3}H dpm x 10^{-3}
^{14}C dpm x 10^{-3}
P N G
a
b
FRACTION NO.

^{3}H dpm x 10^{-3}
^{14}C dpm x 10^{-3}
N G
c.
d
FRACTION NO.

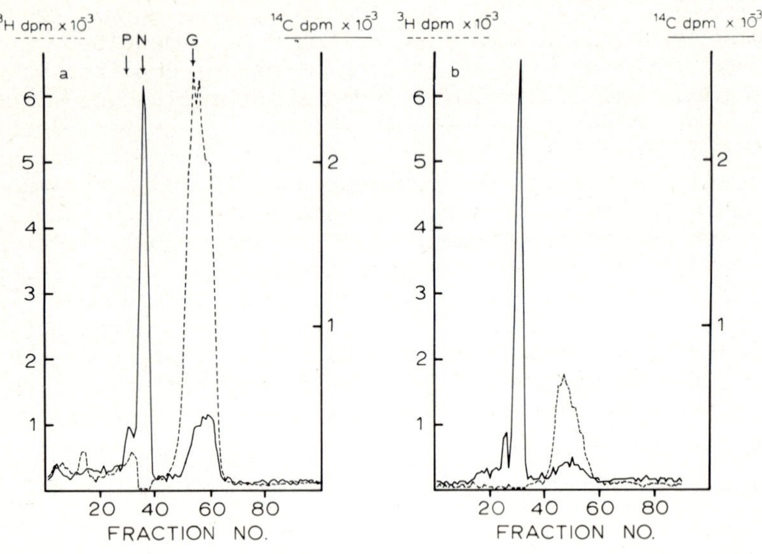

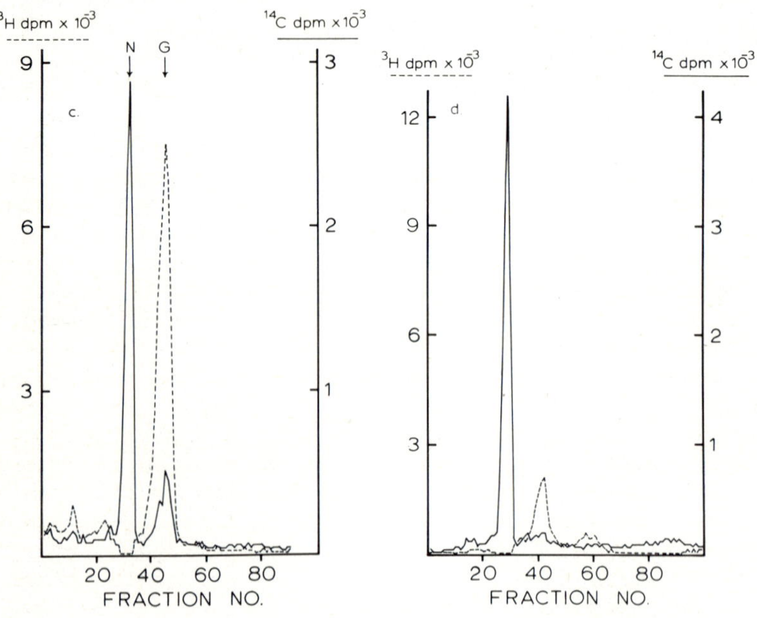

Fig. 2

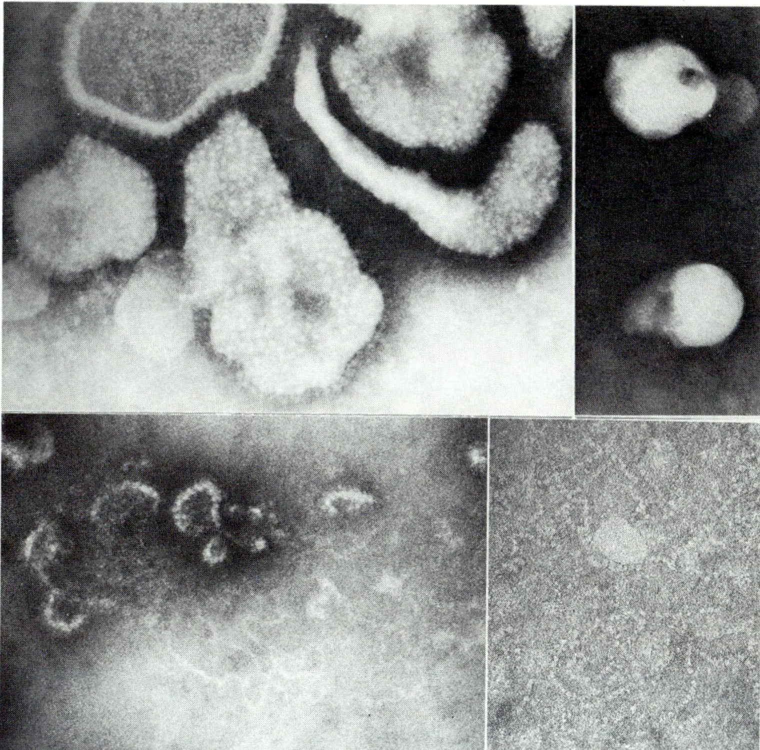

Fig.3. *Morphology of arenavirus particles and subviral components. A. Tacaribe virus particles negatively stained with sodium phosphotungstate. X240,000.B. Spikeless Pichinde virus particles obtained after protease treatment (see text). X180,000. C. Spontaneously disrupted Tacaribe virions revealing envelope fragments and strand-like components believed to be the internal ribonucleoproteins (arrows). X160,000. D. Beaded strand-like components seen in the peak fraction of Tamiami ribonucleoproteins after equilibrium centrifugation in CsCl, as described in Fig.4a. X240,000.*

Fig.2. *(a) Polyacrylamide gel electrophoresis of polypeptides of Tacaribe virions labelled with ^{3}H-glucosamine and ^{14}C-amino acids. (b) An aliquot of the Tacaribe virus preparation after incubation with α-chymotrypsin and rebanding in a sucrose gradient (see text). (c) Polyacrylamide gel electrophoresis of polypeptides of Tamiami virions labelled with ^{3}H-glucosamine and ^{14}C-amino acids. (d) An aliquot of the Tamiami virus preparation after incubation with α-chymotrypsin and rebanding in a sucrose gradient (see text).*

C. DEX.
N
^{14}C
^{3}H
D. PEG.
G1
G2
^{14}C cpm x 10^{-2}
^{3}H cpm x 10^{-2}
DISTANCE MOVED
(mm)

A. CsCl
^{32}P
FRACTION
n
B. Nucleocapsids
^{14}C
^{3}H
^{32}P cpm x 10^{-5}
^{3}H cpm x 10^{-3}
^{14}C cpm x 10^{-3}
DISTANCE MOVED
(mm)

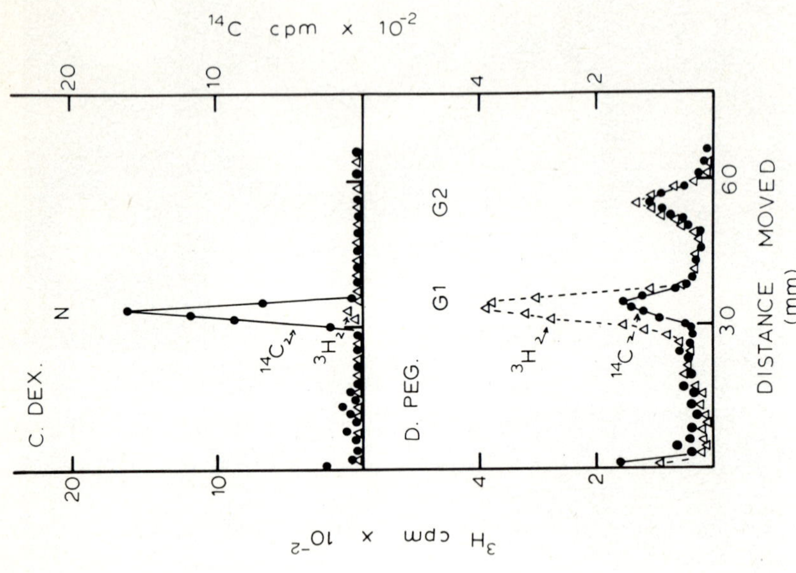

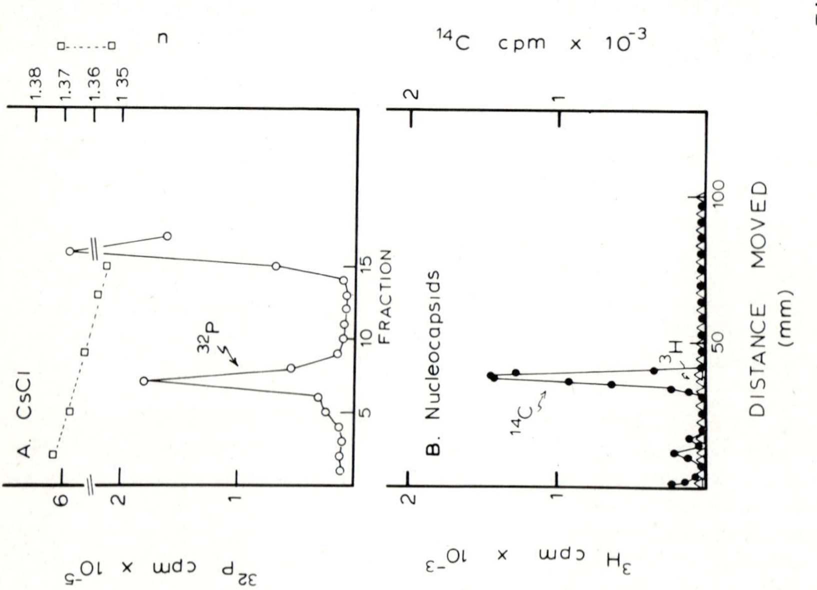

Fig.4

periodicity of ~ 45 Å. These strands are similar in appearance to those seen in preparations of virus containing spontaneously disrupted particles (Fig.3c).

When a preparation of ^{3}H-glucosamine and ^{14}C-amino acid labelled Pichinde virus was likewise treated with Triton X100 in the presence of 1 M NaCl and the products resolved by cesium chloride centrifugation, only ^{14}C label was present in the material recovered at a density of 1.31 g/ml. Both ^{14}C and ^{3}H radioactivity were recovered at the top of the gradient. The major polypeptide associated with the material recovered at a density of 1.31 g/ml was identified to be the 66×10^3 dalton viral species (Fig.4b). Similar results were obtained for Tamiami and Tacaribe viruses whereby their respective 66×10^3 and 68×10^3 dalton polypeptides were recovered in the nucleocapsid band, while for each virus only the glycoprotein was recovered at the top of the gradient. The minor polypeptides P of Tacaribe and Tamiami also appeared to be associated with the nucleocapsids of their respective viruses.

Dextran (DEX) - polyethylene glycol (PEG) phase separation of Triton dissociated virus preparations has been shown to be effective for separating viral nucleocapsids (DEX phase) from the solubilized viral components (PEG phase) for a variety of enveloped viruses (4). A preparation of ^{3}H-glucosamine and ^{14}C-amino acid labelled Pichinde virus was dissociated by Triton X100 in the presence of 1 M NaCl and resolved by PEG-DEX phase separation. The dextran phase was found to contain the ^{14}C-labelled 66×10^3 dalton polypeptide with no detectable ^{3}H label (Fig.4c), while the PEG phase contained the two glycosy-

---

Fig.4. *Identification of the major protein components of Pichinde nucleocapsids. A preparation of ^{32}P-labelled Pichinde virus, dissociated by Triton X100 in the presence of 1M NaCl, was resolved by CsCl gradient centrifugation (A), and the distribution of radioactivity and refractive indices of selected fractions determined. (B). Following a similar treatment of ^{3}H-glucosamine and ^{14}C-amino acid labelled Pichinde virus (see Fig.1A), the material recovered at a density of 1.31 g/ml by TCA precipitation, was dissociated by SDS and the polypeptides resolved by polyacrylamide gel electrophoresis. Another aliquot of the preparation of ^{3}H-glucosamine and ^{14}C-amino acid labelled Pichinde virus was dissociated by Triton X100 in the presence of 1 M NaCl and subjected to polyethylene glycol (PEG)-dextran (DEX) T 500 phase separation (4). The proteins recovered from the DEX phase (C) and PEG phase (D) were resolved by polyacrylamide gel electrophoresis. The polypeptide profile of untreated virus is given in Fig.1a.*

A
B
28 s
18 s
L
S
^{3}H
^{32}P
^{3}H cpm x 10^{-3}
^{3}H cpm x 10^{-5}
^{32}P cpm x 10^{-3}
DISTANCE MOVED cm.

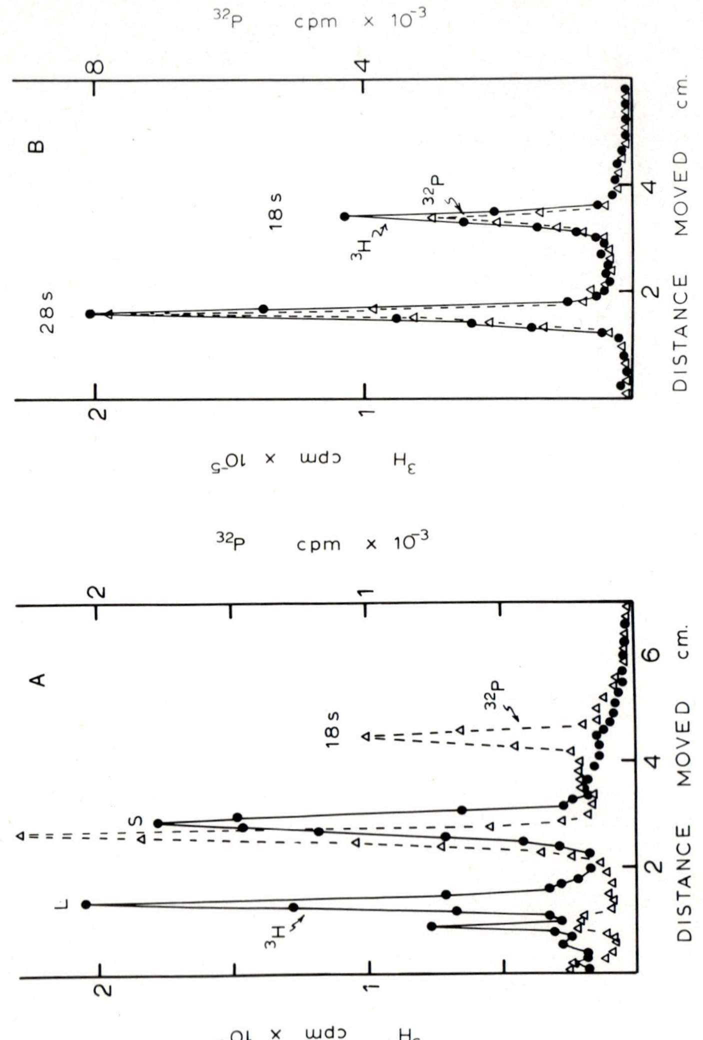

Fig.5

lated polypeptides (G1, molecular weight = 64×10^3; G2, molecular weight = 38×10^3) both labelled with ^{14}C and ^3H (Fig.4d). For both bands the ^{14}C and ^3H labels were symmetric with respect to each other although the G1 polypeptide had relatively more ^3H to ^{14}C label than the G2 polypeptide suggesting that, at least with respect to glucosamine, it is more heavily glycosylated.

We conclude from these results that the nucleocapsids of three arenaviruses, Pichinde, Tacaribe and Tamiami consist of RNA, a major nonglycosylated protein N ($66\text{-}68\times10^3$ daltons) with probably an associated minor polypeptide P ($77\text{-}79\times10^3$ daltons). Whereas Tacaribe, Tamiami and Pichinde all have a glycoprotein size class of approximately $38\text{-}44\times10^3$ daltons, Pichinde has an additional glycoprotein of about 64×10^3 daltons (Table 1).

THE RNA SPECIES ASSOCIATED WITH PICHINDE VIRUS PREPARATIONS

A freshly cloned stock of Pichinde virus was used to infect 80% confluent BHK-21 cell monolayers at an input multiplicity of infection of 5 PFU per cell. After addition of media containing ^3H-uridine, the cells were incubated at 35° for 3 days and the virus harvested from the supernatant fluids. The viral RNA was extracted, melted by heating at 100° for 30 sec, mixed with an aliquot of ^{32}P-labelled uninfected BHK ribosomal RNA, and resolved by polyacrylamide gel electrophoresis (3). The results obtained (Fig.5a) indicated that the Pichinde virus preparation contained two size classes of RNA, large RNA (L, molecular weight $\sim 3.2\times10^6$) and smaller RNA (S, molecular weight $\sim 1.6\times10^6$). Although the infected cells contained both ^3H labelled 28 *s* and 18 *s* ribosomal RNA, the viral RNA preparation essentially lacked ^3H labelled 18 *s* ribosomal RNA (Fig.5b). Whether any ^3H labelled 28 *s* ribosomal RNA was present could not be ascertained due to the similar (but not identical) electrophoretic mobility of the S RNA species to that of the 28 *s* marker ribosomal species. When cells were labelled one day prior to infection or one day after virus addition, similar results were obtained.

Fig.5. *A. The RNA species of Pichinde virus and infected cell preparations. A preparation of ^3H-uridine labelled Pichinde viral RNA was melted at 100° for 30 sec, then mixed with ^{32}P-labelled uninfected BHK 21 cell ribosomal RNA, and resolved by polyacrylamide gel electrophoresis as described in the text. B. The Pichinde infected cells obtained from the ^3H-uridine labelled infection which gave ^3H-labelled Pichinde virus, was extracted for RNA, mixed with ^{32}P-labelled uninfected cell ribosomal RNA, and resolved by polyacrylamide gel electrophoresis as described in the text.*

TABLE 1

Arenavirus structural polypeptides

PICHINDE

Protein species	% label recovered in viral polypeptides			Molecular weight	No. per average particle
	^{14}C-amino acids[a]	^{3}H-leucine[b]	^{3}H-glucosamine[a]		
P	ND	ND	0	77,000	<100
N	82±3	85±3	0	66,000	1530
G1			74±3	64,000	390
G2	11±2	10±3	19±2	38,000	440
other	7±2	5±2	7±2	-	-

G1 + G2 28% total viral protein mass (see text).

TACARIBE

Protein species	% label recovered in viral polypeptides		Molecular weight
	^{14}C-amino acids[a]	^{3}H-leucine[a]	
P	5±2	5±3	79,000
N	55±4	59±3	68,000
G	31±4	29±1	42,000
other	9±3	7±2	-

TAMIAMI

Protein species	% label recovered in viral polypeptides		Molecular weight
	^{14}C-amino acids[a]	^{3}H-leucine[b]	
P	3±1	4±1	77,000
N	69±2	68±2	66,000
G	21±2	23±1	44,000
other	7±2	5±2	-

[a] Average of 3 determinations, [b] Average of 2 determinations.

ESTIMATION OF THE NUMBER OF PICHINDE PROTEIN MOLECULES PER VIRION

In order to estimate the number of molecules of the major protein types per Pichinde virion, it is necessary to determine the virion RNA to protein ratio, the per cent of the virion protein in the N, G1 and G2 protein species, and the weight of viral RNA per virion.

A preparation of ^{3}H-uridine labelled virus (0.60 mg protein per ml, 1.67×10^{6} ^{3}H cpm per ml), was extracted for RNA and, from the optical density and radioactivity, the RNA specific activity was determined to be 8.69×10^{7} ^{3}H cpm per mg RNA. It was calculated that in the original virus suspension there was 0.019 mg RNA per 0.60 mg protein, i.e. an RNA to protein weight ratio of 1 to 31.6

The amount of virion proteins in N, G1 and G2 (Table 1), was calculated from (a) the proportions of ^{14}C amino acid label in the N plus G1 and G2 peaks (see Fig.1a), and (b) the ratio of ^{3}H-leucine label in the N plus G1 and G2 peaks for two preparations of ^{3}H-leucine labelled virus. The ^{14}C ratios of label in the polypeptides obtained from whole virus preparations were similar to the ^{3}H-leucine ratios (Table 1). The average of five different analyses gave results of 83 per cent of the virion polypeptides in N plus G1 and 11 per cent in G2 with 6 per cent in other polypeptides. The ratio of ^{14}C radioactivity in the G1 and G2 polypeptides recovered in the PEG phase of the PEG-dextran extraction was 60 to 40 respectively (see Fig.4d), while the ratio of ^{3}H-glucosamine label therein was 74 to 24. This corresponded well to the per cent of ^{3}H-glucosamine in the N plus G1, and G2 peaks of whole virions (Fig.1a). We cannot exclude the possibility that some N protein was recovered in the PEG phase, although using ^{3}H-uridine labelled virus, less than 5% of the ^{3}H label was recovered in the PEG phase. On the basis of these results we estimate that the per cent of virion protein in the N plus G1 peak that is G1 protein is 60/40 x 11 i.e. 16.5 per cent. The proportion of N protein in the N plus G1 peak would therefore correspond to 83-16.5 i.e. 66.5 per cent. The results obtained for the distribution of amino acid label between the N and G polypeptides of Tamiami and Tacaribe are also shown in Table 1 and correspond reasonably well to the Pichinde results.

On the basis that Pichinde virions contain only one unique L and one unique S RNA species, and no ribosomes, (sum molecular weight of viral RNA = 4.8×10^{6}), we calculate that one gram molecule of Pichinde virus will contain $4.8 \times 10^{6} \times 31.6$ g of protein (i.e. 1.52×10^{8} g). If there is 66.5% N protein (molecular weight = 66,000), 16.5% G1 protein (molecular weight = 64,000) and 11% G2 protein (molecular weight = 38,000), we calculate that there are of the order of 1530 molecules of N protein per virion (i.e. $1.52 \times 10^8 \times 0.665 \div 66,000$), 390 molecules of G1 protein and 440 molecules of G2 protein (Table 1). Since G1 contains more carbohydrate than G2, its molecular weight may be overestimated to a greater extent than that of G2 protein. This would mean that there would be proportionately more G1 to G2 protein molecules than calculated. Although we do not know to what extent the carbohydrate components of G1 and G2 influence their electrophoretic mobilities, these results indicate that the two polypeptides probably are present in similar molecular proportions.

HIGH FREQUENCY RECOMBINATION WITH TEMPERATURE SENSITIVE, CONDITIONAL LETHAL, MUTANTS OF PICHINDE VIRUS

High frequency recombination has been obtained with nine out of 12 initial temperature sensitive (ts), conditional lethal mutants of Pichinde virus (Table 2). The mutants have been categorized into two genetic recombination groups, Group I containing five mutants (ts #1, 10, 11, 12 and 13) and Group II containing four mutants (ts #2, 3, 5 and 9). All of the mutants were isolated from stocks of virus grown in the presence of 5-fluorouracil. Clonal analyses of the putative recombinants obtained from the ts 12 x 2, and ts 11 x 3 crosses indicate that, unlike the *ts* mutants which have efficiency of plating ratios of ~ 0.0001 (E.O.P. = PFU at 39.8° ÷ PFU at 35°), the recombinants have E.O.P. ratios of 0.2 - 1.0, i.e. similar to that of wild type virus.

These results agree with the postulate that Pichinde virus has a segmented genome with different gene products coded for by the different RNA species. Which RNA species corresponds to which group of mutants or viral polypeptide(s) remains to be determined.

DISCUSSION

We have determined the size and location of the major viral polypeptides of three arenaviruses, Pichinde, Tacaribe and Tamiami. All three viruses have a major nucleocapsid protein (N) of similar size ($66\text{-}68 \times 10^3$ daltons). All three viruses have a $38\text{-}44 \times 10^3$ dalton glycoprotein size class, with Pichinde having a second major glycoprotein of molecular weight 64×10^3. We do not know if the G1 and G2 glycoproteins of Pichinde are separate gene products or if G2 represents a cleavage derivative of G1. Nor do we know whether the $42\text{-}44 \times 10^3$ glycoproteins of Tacaribe or Tamiami virus represent a single glycoprotein species or multiple species of similar molecular weight. However growth of Tacaribe or Tamiami viruses in serum free medium

TABLE 2

Recombination analyses with Pichinde *ts* mutants[a]

Inoculum viruses	Incubation temp	48 hr yield assayed at: 35°	39.8°	% Recombinants
ts 1	35°	3.6×10^6	7.0×10^2	-
ts 2	35°	2.2×10^6	1.3×10^3	-
ts 5	35°	3.5×10^6	5.0×10^2	-
ts 1 x ts 2	35°	2.5×10^6	3.2×10^5	25
ts 1 x ts 5	35°	3.0×10^6	7.1×10^5	47
ts 2 x ts 5	35°	2.5×10^6	1.7×10^3	0

[a] % Recombinants = $$\frac{\left[(AB_{35})_{39.8} - [(A_{35})_{39.8} + (B_{35})_{39.8}]\right] \times 100 \times 2}{(AB_{35})_{35}}$$

% Recombinants with Pichinde ts mutants[b]

ts mutant	1	2	3	5	9	10	11	12	13
1	.	15	21	26	32	0	0	0	0
2	.	.	0	0	0	81	14	21	9
3	.	.	.	0	0	45	100	53	51
5	.	.	.	.	0	98	ND	ND	ND
9	.	.	.	.	.	54	ND	ND	ND
10	.	.	.	.	.	.	0	0	0
11	.	.	.	.	.	.	.	0	0
12	.	.	.	.	.	.	.	.	0

Assignments:

Group I ts 1, 10, 11, 12, 13 (5 mutants)

Group II ts 2, 3, 5, 9 (4 mutants)

[b] Average of 3 determinations. ND = Not done. Percentages greater than 50% may represent preferential multiplication of wild type recombinants during the 48 hr coinfection.

had no detectable effect on the viral polypeptide pattern nor did treatment of Pichinde virus preparations with a variety of proteolytic enzymes provide any evidence for G1 being cleaved to G2. Further experiments are required to determine whether the Tacaribe and Tamiami glycoproteins are unique and, or cleavage products of a larger polypeptide, and whether the two major glycoproteins of Pichinde have similar or unique amino acid sequences.

The results obtained for Junin virus by Segovia and De Mitri (19) correspond, at least for the major viral polypeptides, to the results we have obtained for Tacaribe and Tamiami viruses in that a single nucleocapsid protein and single major glycoprotein size class were observed. The major polypeptide observed for LCM may well correspond to the nucleocapsid protein found in the other arenavirus preparations although further experiments are required to determine the location and function of the other LCM viral polypeptides.

The function of the minor 77-79x10^3 dalton polypeptide species obtained in virus and nucleocapsid preparations of Tacaribe and Tamiami viruses, and also seen in preparations of Pichinde virus, is not known. Five minor polypeptides have been reported for LCM, two for Pichinde (18) and four for Junin virus (19); their location and function are unknown.

The ribonucleoproteins obtained from arenavirus particles appear to differ in morphology from the nucleocapsids of RNA viruses in other major groups. We have not been able to resolve helical symmetry in the arenavirus ribonucleoproteins.

The lack of labelled 18 *s* ribosomal RNA in the preparations of Pichinde virus that we have analysed raises the question of whether unlabelled ribosomes become preferentially incorporated into virus particles. Although prelabelling cells one day prior to infection has given us the same RNA profile as that shown in Fig.5a, we cannot totally exclude the possibility that unlabelled ribosomes are incorporated into Pichinde virus particles. Under conditions in which no 18 *s* ribosomal RNA was detected in Pichinde virus preparations both 28 *s* and 18 *s* ribosomal RNA species were found to be the predominant labelled RNA species in infected cells (Fig.5b). In a single experiment in which sufficient quantities of Pichinde virus were obtained for an optical determination of the content of 18 *s* RNA in the viral RNA preparation, scanning the viral RNA species which were resolved on a 2.4% gel at 280 nm did not detect 18 *s* ribosomal RNA when the two viral RNA peaks were plainly evident and 28 *s* and 18 *s* ribosomal RNA species were easily detected in a sample of cell RNA run on a parallel gel. We cannot exclude the possibility that 28 *s* ribosomal RNA is present in our Pichinde virus preparations. Oligonucleotide fingerprinting and

RNA complexity analyses will be needed in order to determine if labelled 28 *s* RNA is present and whether the L and S RNA species are unique or whether one represents a deletion derivative of the other. Using thin section electron microscopy we have had difficulty in obtaining infected BHK cells with detectable numbers of residual budding Pichinde virus particles, so that we have not been able to confirm by that technique whether our Pichinde virus preparations lack ribosomes.

The genetic evidence for the formation of wild type recombinants at high frequency from coinfections involving certain Pichinde *ts* mutants substantiates the theory that Pichinde has a segmented genome. Whether recombinants can be formed between different arenaviruses is not known.

ACKNOWLEDGEMENTS

This research was supported by a UAB Medical Center Faculty seed money Research grant to A.C.V. and D.H.L.B. and in part by Public Health Service Grants CA-18611, AI-13402 and AI-12680, and by Grants No. VC-149B from the American Cancer Society and PCM76-09711 from the National Science Foundation.

We thank Reginald Anderson and John Fitzpatrick for excellent technical assistance.

REFERENCES

1. Abelson, H.T., Smith, G.H., Hoffman, H.A. and Rowe, W.P. (1969). *J. Natl. Cancer Inst.* 42, 497.
2. Anon, M.C., Grau, O., Segovia, Z.M. and Franze-Fernandez, M.T. (1976). *J. Virol.* 18, 833.
3. Bishop, D.H.L. and Roy, P. (1971). *J. Mol. Biol.* 57, 513.
4. Bishop, D.H.L. and Roy, P. (1972). *J. Virol.* 10, 234.
5. Carter, M.F., Biswal, N. and Rawls, W.E. (1973a). *J. Virol.* 11, 61.
6. Carter, M.F., Biswal, N. and Rawls, W.E. (1973b). *J. Virol.* 12, 33.
7. Carter, M.F., Biswal, N. and Rawls, W.E. (1974). *J. Virol.* 13, 577.
8. Compans, R.W. and Pinter, A. (1975). *Virology* 66, 151.
9. Dalton, A.J., Rowe, W.P., Smith, G.H., Wilsnack, R.W. and Pugh, W.E. (1968). *J. Virol.* 2, 1465.
10. Farber, F.E. and Rawls, W.E. (1975). *J. Gen. Virol.* 26, 21.
11. Murphy, F.A., Webb, P.A. Johnson, K.M. Whitfield, S.G. Chappell, W.A. (1970). *J. Virol.* 6, 507.
12. Pedersen, I.R. (1970). *J. Virol.* 6, 414.
13. Pedersen, I.R. (1971). *Nature New Biol.* 234, 112.
14. Pedersen, I.R. (1973a). *In* "Lymphocytic choriomeningitis virus and other arenaviruses. (Lehmann-Grube, ed.),

p.13. Springer, Berlin.

15. Pedersen, I.R. (1973b). *J. Virol.* 11, 416.
16. Pfau, C.J., Bergold, G.H., Casals, J., Johnson, K.M., Murphy, F.A., Pedersen, I.R., Rawls, W.E., Rowe, W.P. Webb, P.A. and Weissenbacher, M.C. *Intervirology* 4, 207.
17. Pinter, A. and Compans, R.W. (1975). *J. Virol.* 16, 859.
18. Ramos, B.A., Courtney, R.J. and Rawls, W.E. (1972). *J. Virol.* 10, 661.
19. Segovia, Z.M. and De Mitri, M.I. (1977). *J. Virol.* 21, 579.

9

IDENTITY OF THE VIRAL PROTEIN RESPONSIBLE FOR SEROLOGIC CROSS REACTIVITY AMONG THE TACARIBE COMPLEX ARENAVIRUSES

M.J. BUCHMEIER and M.B.A. OLDSTONE

Department of Immunopathology,
Scripps Clinic and Research Foundation
La Jolla, California 92037.

INTRODUCTION

The Tacaribe complex of arenaviruses consists of 8 structurally similar enveloped RNA containing viruses, of which Pichinde virus is the best characterized (1, 4, 7). The Pichinde virion contains three major structural polypeptides, a 66,000 dalton polypeptide associated with the viral nucleocapsid and two surface glycopeptides with molecular weights of approximately 64,000 and 38,000 daltons (6, 9). Recent structural studies on two other Tacaribe complex viruses, Tacaribe and Tamiami, have revealed a similar nucleocapsid protein; however, these viruses apparently contain only a single major glycosylated protein (3).

Members of this virus group cross react extensively as judged both by complement fixation (CF) tests and by immunofluorescent staining of cytoplasmic antigen(s); however, they do not cross react in virus neutralization tests (8). This suggests that cells infected by the Tacaribe complex viruses possess group specific antigens internally but do not express these antigens on their external surfaces. Therefore, to define the biochemical basis for the serologic cross reactivity in terms of specific viral polypeptides, we used a sensitive radioimmune precipitation assay. The purpose was first to identify the CF antigen of Pichinde virus, and second to determine whether an antiserum known to react with heterologous Tacaribe complex viruses also reacts with the polypeptides of Pichinde virus.

MATERIALS AND METHODS

Antisera

Antiserum from guinea pigs immunized with Pichinde virus and monospecific antiserum against the purified CF antigen of Pichinde virus were prepared as described (1). Hyperimmune hamster antisera specific for the individual Tacaribe complex viruses produced at the Middle America Research Unit, in Panama, were obtained through the courtesy of Dr William E. Rawls, Hamilton, Ontario, Canada. Rabbit antiserum to either guinea pig or hamster IgG was prepared in this laboratory (5).

Virus

Pichinde virus strain AN 3739 was cultured, assayed and purified as reported (2). The virus was radiolabeled by adding 10 μCi/ml ^{35}S methionine or 5 μCi/ml ^{3}H glucosamine to infected cultures 28 hr after infection and harvesting the cultures after an additional 18-20 hr.

Radioimmune precipitation and gel analysis

The radiolabeled virus was disrupted and solubilized with 2% NP-40 and 20 μg/ml of ribonuclease-A (Sigma Chemical Co., St. Louis, Mo.) as follows. Approximately 45,000 cpm of ^{3}H glucosamine labeled virus and 30,000 cpm of ^{35}S methionine labeled virus in 100 μl volume was added to a tube containing 25 μl of 5% NP-40 and 25 μl of heat inactivated fetal calf serum. To this was added 0.4 ml of TNE buffer (0.01 M Tris HCl pH 7.4; 0.1 M NaCl, 0.001 M EDTA) and 10 μl of the test antiserum. The mixture was incubated first for 30 min at 37° C and then for 18 hr at 4°C. Antiserum directed against the appropriate species of IgG was then added in a quantity which precipitated greater than 95% of the IgG present, and the second incubation was continued for 30 min at 37° and 2 hrs at 4°. After centrifugation at 2500 x g for 20 min, the immune precipitates were collected, washed 3 times with TNE buffer, dried, and dissolved at 100° C for 2 min in 200 μl of a buffer consisting of 1% SDS, 1% 2-mercaptoethanol, and 0.5 M urea. Fifty μl of the dissolved precipitate was counted in a beta scintillation counter and 100 μl of the remaining portion was electrophoresed on a 15% acrylamide-0.1% SDS gel (1). Subsequently, the gels were cut into 1 mm sections, oxidized in 30% H_2O_2, and reduced with 15% ascorbic acid. The radioactivity in each piece was then determined by liquid scintillation counting. Corrections were made for ^{35}S crossover into the ^{3}H channel.

RESULTS

Identity of the CF antigen of Pichinde virus

To determine which of the viral polypeptides carried the CF antigenic determinant, we reacted Pichinde virus against immune serum directed against all the virion polypeptides, immune serum against the purified CF antigen,and non-immune serum, and by SDS polyacrylamide gels identified specific immune precipitated polypeptides. The results of this analysis are shown in Fig.1. Immune serum used in the top panel

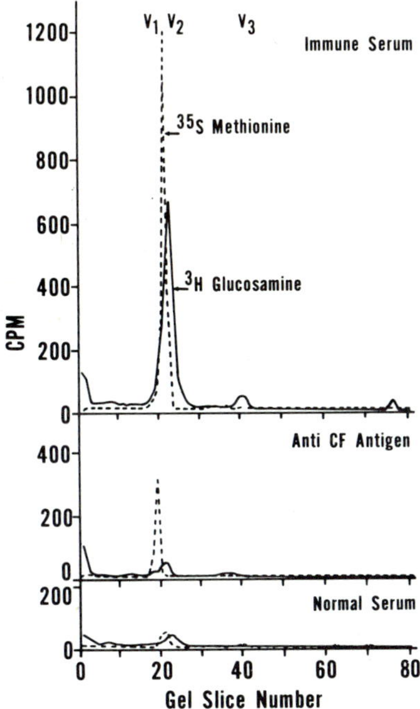

Fig.1. *Immune precipitation of Pichinde virus proteins by guinea pig serum containing antibodies against all the Pichinde virus polypeptides (top); monospecific antibody against purified soluble CF antigen (center); and nonimmune guinea pig serum control (bottom). Radiolabeled purified Pichinde virus was disrupted by incubation with 2% NP-40 and 20 μg/ml of RNase A and immune precipitated as described in Materials and Methods. After washing, the precipitates were electrophoresed on 15% acrylamide gels. The gels were sliced and counted; counts were corrected for crossover and plotted.*

recognized all 3 of the major Pichinde virus polypeptides (labeled V_1, V_2, V_3 by the convention of Ramos *et al.*, (6)). In contrast, monospecific antiserum to the purified CF antigen (middle panel) recognized only V_1, the nonglycosylated nucleocapsid protein. Both the anti CF and nonimmune sera (bottom panel) precipitated similar background levels of the V_2 glycoprotein. This demonstrated that the soluble CF antigen determinant is associated with the nucleocapsid protein and not the glycosylated proteins of Pichinde virus.

Identity of the cross reactive antigen among Tacaribe complex viruses

Hyperimmune hamster antisera against Pichinde, Amapari, Junin, and Tacaribe viruses demonstrating CF activity and a control nonimmune hamster serum were reacted by radioimmune precipitation with radiolabeled Pichinde virus proteins. Table 1 shows the amounts of glycoprotein and total protein precipitated. The homologous anti-Pichinde serum precipitated 56.7% of the ^3H glucosamine and 85.9% of the ^{35}S methionine label. Heterologous sera failed to precipitate significant amounts of glycoprotein label. In contrast, significant but varying amounts of the ^{35}S methionine labeled virus were immunoprecipitated by the different antisera (Table 1). Subsequently these immune precipitates were analyzed on gels and yielded the data shown in Fig.2. Antibodies to Pichinde virus clearly precipitated the three major viral proteins V_1, V_2 and V_3, but the various heterologous Tacaribe complex sera precipitated only the nonglycosylated nucleocapsid protein V_1. Study of Pichinde virus infected cells by immunofluorescence (data not shown) showed extensive cross reactivity between the various Tacaribe complex antisera and cytoplasmic antigens. In contrast, we were consistently unable to demonstrate any cross reactivity against antigens expressed on the surfaces of cells infected by Pichinde virus.

DISCUSSION

We have demonstrated that the antigen common to all the Tacaribe complex viruses is the viral nucleoprotein, V_1. Previous detection of cross reactivity among these viruses has been limited to antigens that are reactive in CF tests or cytoplasmic immunofluorescence assays (8). We used antisera specific for Pichinde, Amapari, Tacaribe and Junin viruses and all reacted with the V_1 nonglycosylated protein. These findings are consistent with the previous reports that neutralization tests (7), which probably involve recognition of surface glycoprotein(s), fail to show such cross reactivity and that the CF antigen purified from cells infected with Pichinde virus

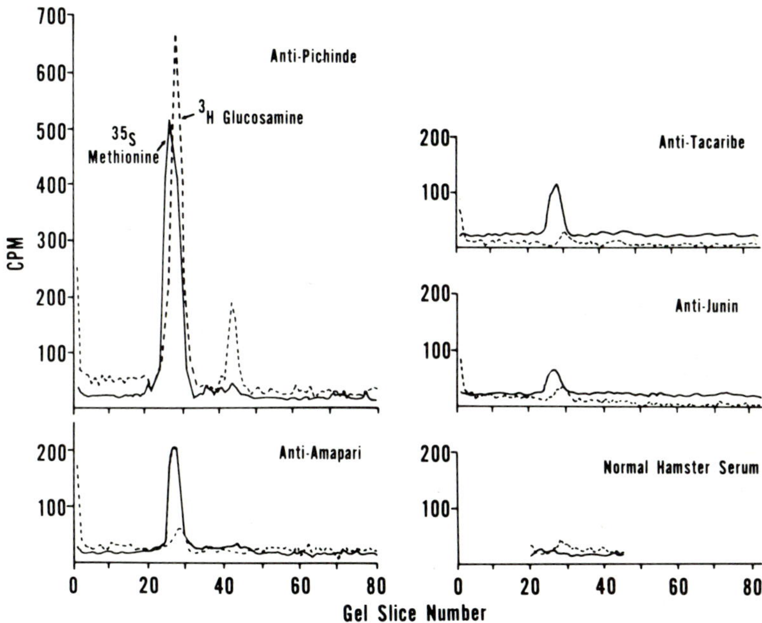

Fig.2. *Immune precipitation of Pichinde virus proteins by Tacaribe complex antisera. Pichinde virus was disrupted and reacted as described in the legend to Fig.1.*

is immunochemically identical to an internal component of the virion (1). Thus, it appears that the basis for serologic cross reactivity among the Tacaribe complex viruses lies in their similar nucleoproteins. Recent work by Gard *et al.*,(3) and by Vezza *et al.*,(9) has shown structural similarities among Pichinde, Tamiami and Tacaribe viruses. Undoubtedly, this nucleoprotein structure has been conserved during the evolution of the Tacaribe complex viruses. Using immunofluorescence assays on the surfaces of infected cells, we have been unable to demonstrate any significant cross reactivity among the Tacaribe complex of viruses or between Tacaribe viruses and lymphocytic choriomeningitis virus. Further, at the level of cytotoxic thymus (T) lymphocyte recognition, we have not found cross reactivity between lymphocytic choriomeningitis virus and Pichinde virus (unpublished data). Studies are currently under way to determine whether cross reactivity is detectable by cytotoxic T lymphocytes among Tacaribe complex viruses.

TABLE 1

Immune precipitation of radiolabeled Pichinde virus proteins by Tacaribe complex antisera

Antiserum	Percent of input cpm precipitated	
	^{3}H glucosamine	^{35}S methionine
Anti Pichinde	56.7[a]	85.9
Anti Amapari	3.9	30.0
Anti Junin	4.4	8.9
Anti Tacaribe	3.9	12.9
Nonimmune	2.6	2.8

[a]The percentages shown are the means of duplicate determinations.

ACKNOWLEDGEMENTS

This is manuscript number 1365 from the Department of Immunopathology, Scripps Clinic and Research Foundation, La Jolla, California 92037. This research was supported by U.S. Public Health Service grants AI 09484, AI 07007, NS 12428 and NIAID postdoctoral research fellowship award No.1 F32 AI05380 to M. Buchmeier.

The authors thank Jim Holmstoen for his technical assistance.

REFERENCES

1. Buchmeier, M.J., Gee, S.R. and Rawls, W.E. (1977). *J. Virol.* 22, 175.
2. Carter, M.F., Biswal, N. and Rawls, W.E. (1973). *J. Virol.* 11, 61.
3. Gard, G.P., Vezza, A.C., Bishop, D.H.L. and Compans, R.W. (1977). *Virology*,
4. Murphy, F.A., Webb, P.A., Johnson, K.M., Whitfield, S.G. and Chappell, W.A. (1970). *J. Virol.* 6, 507.
5. Oldstone, M.B.A. and Dixon, F.J. (1968). *Am. J. Path.* 52, 251.
6. Ramos, B.A., Courtney, R.J. and Rawls, W.E. (1972). *J. Virol.* 10, 661.
7. Rowe, W.P., Pugh, W.E., Webb, P.A. and Peters, C.J. (1970a). *J. Virol.* 5, 289.
8. Rowe, W.P., Murphy, F.A., Bergold, G.H., Casals, J., Hot-

chin, J., Johnson, K.M., Lehmann-Grube, F., Mims, C.A., Traub, E. and Webb, P.A. (1970b). *J. Virol.* 5, 651.
9. Vezza, A.C., Gard, G.P., Compans, R.W. and Bishop, D.H.L. (1977). *J. Virol.*, 23, 776.

SPATIAL RELATIONSHIPS OF THE PROTEINS OF VESICULAR STOMATITIS VIRUS

E.J. DUBOVI and R.R. WAGNER

Department of Microbiology,
University of Virginia,
Charlottesville, Va.

The mechanisms responsible for the maturation of enveloped viruses have yet to be clearly defined. Vesicular stomatitis (VS) virus, an enveloped rhabdovirus, contains three major proteins which may play key roles in the maturation process: glycoprotein (G protein), matrix protein (M protein), and the nucleocapsid protein (N protein). Of these three proteins, the M protein may be the critical element in the maturation process as it has been proposed that its synthesis is the rate-limiting step in virus assembly (3). If this hypothesis is correct, M protein must recognize and attach to regions of the cellular membrane which contain the G protein and must thereby provide a recognition site on the inner membrane surface for the attachment of the cytoplasmic nucleocapsid. The present studies were initiated in an attempt to define some of the spatial relationships of the viral proteins in the mature virion. These data could provide more insight into the maturation process by establishing which spatial relationships must be considered in any proposal for the dynamics of virus assembly.

CROSS-LINKING OF VS VIRAL PROTEINS

The approach used in these studies was to react the intact VS virion with bifunctional reagents which form covalently linked protein complexes. The complexes were separated by electrophoresis on cylindrical gels containing 5% acrylamide. Representative electropherograms of Coomassie blue stained gels are shown in Fig.1. Of the three reversible crosslinkers used, methyl-4-mercaptobutyrimidate (MMB) and dithiobis (succinimidyl propionate) (DTBSP) contain a disulphide bond which prohibits the use of reducing conditions prior to electrophoresis of the viral sample in the first dimension. Tartryl diazide (TDA) does not have this restriction but for purposes of comparison, electrophoresis for the gel scan shown in Fig. 1 was performed under non-reduced conditons.

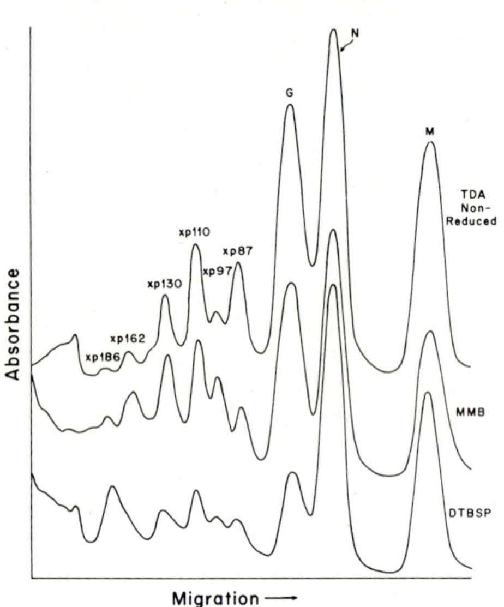

Fig.1. *Electrophoretic comparison of stained protein complexes (xp) produced by reaction of VS virions with three different cross-linking reagents. All cross-linking reactions were carried out with purified virus at a concentration of 2 mg of viral protein per ml of PBS. Virus was cross-linked with TDA (5 mM) in TEA buffer (pH 8.5) at 21°C for 15 min, MMB (2 mM) in TEA buffer (pH 8.0) at 21°C for 15 min and then* H_2O_2 *(2 mM) for 15 min, and DTBSP (1.25 x* 10^{-4}*M) in 5 mM phosphate buffer (pH 7.2) at 21°C for 15 min. All cross-linking reactions were terminated by the addition of methylamine (final concentration, 100 mM). Virus was dissociated in 2% SDS, electrophoresed at 2.5 mA/gel for 16 h and then at 6 mA/gel for 6 h, stained with Coomassie blue, and scanned at 620 nm. Cross-linked protein aggregates are designated by the prefix xp and ~ molecular weight (X*10^{-3}*); G, N, and M indicate positions of VS viral proteins. (Reprinted with permission from Dubovi and Wagner, 1977).*

Although these reagents are used under different conditions of pH and ionic strength, the patterns of cross-linking are very similar qualitatively. However, certain differences are noted in the amounts of some cross-linked complexes (xp). DTBSP, for example, generates larger amounts of xp186 than either TDA or MMB. Also, the amounts of xp162, xp110, and xp97 are reduced.

Cross-linking with MMB is achieved by oxidizing the free sulphydryl group on MMB, thus forming a disulphide bridge. This was routinely achieved by adding H_2O_2 to the virus sample following addition of MMB. In this case the reaction of MMB with the virus occurs under reducing conditions. In fact, without the addition of an oxidizing agent, the gel profile of the sample is indistinguishable from one reduced with β-mercaptoethanol. Cross-linking with MMB can also be achieved under oxidizing conditions, i.e. the disulphide bridge is formed by oxidation with H_2O_2 prior to addition of MMB to the virus sample. The gel profiles obtained under oxidizing or reducing conditions were indistinguishable. Oxidation of MMB using diamide, a thiol-oxidizing agent, also did not alter the pattern of cross-linked complexes.

EFFECT OF TEMPERATURE ON CROSS-LINKING

Standard conditions for cross-linking generally employed 21-22°C as the incubation temperature. Since the fluidity of lipid membranes is affected by temperature, it was necessary to determine the effect of temperature on the cross-linking patterns. Reaction mixtures were equilibrated at 0°, 21°, and 37°C followed by the addition of TDA at the appropriate temperature. No significant differences were noted in the gel profiles of virion proteins cross-linked at these three temperatures (Fig.2). DTBSP also was not influenced by the temperature of the reaction. These data suggest that the physical state and lateral mobility of the membrane lipids do not alter the cross-linking pattern.

EFFECT OF N-ETHYLMALEIMIDE (NEM) ON CROSS-LINKING

A potential problem inherent with cross-linking agents containing disulphide bonds is the possibility of the occurrence of mercaptan-disulphide interchange (6). This interchange can produce cross-linked complexes between proteins which do not lie adjacent to each other but undergo lateral diffusion. In order to alleviate this potential problem the virus was reacted with N-ethylmaleimide prior to cross-linking with DTBSP. This pretreatment should serve to block any free sulphydryl groups in the viral proteins. The pretreatment of VS virus with 10 mM NEM did not alter the cross-linking pattern produced by

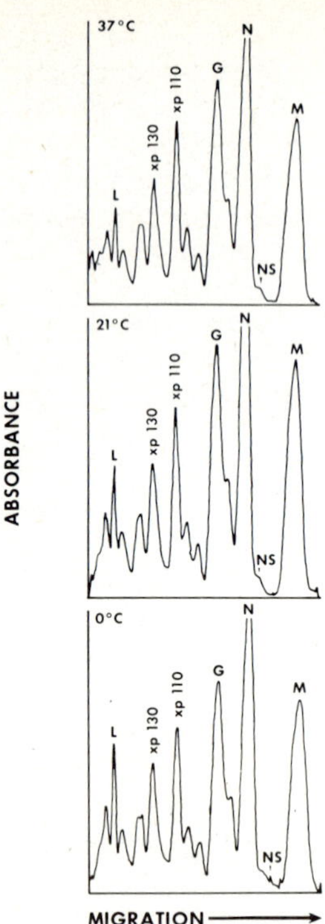

Fig.2. *Electrophoretic comparison of VS viral protein cross-linked at three different temperatures with TDA. Experimental conditions are as described in the legend to Fig.1 with the exception that the reactions were carried out at 0°, 21°, or 37°C.*

DTBSP (Fig.3). Therefore, mercaptan-disulphide interchange does not seem to be a major problem in cross-linking VS virus proteins.

CROSS-LINKING WITH PHENYLENEDIMALEIMIDE (PDM).

Previous work has demonstrated that the addition of oxidizing agents to intact virus results in the formation of a N-protein dimer (5). It was postulated that the N protein contains free sulphydryl groups in the completed virion. This possibility was examined using the sulphydryl cross-linking agent phenylenedimaleimide (Fig.4). Positive identification of the proteins in cross-linked complexes (designated xp) is much more difficult with PDM because the cross-linker can not be cleaved following electrophoretic separation of cross-linked complexes. However, some assignments can be made on the basis of glucosamine labelling and molecular weight estimates.

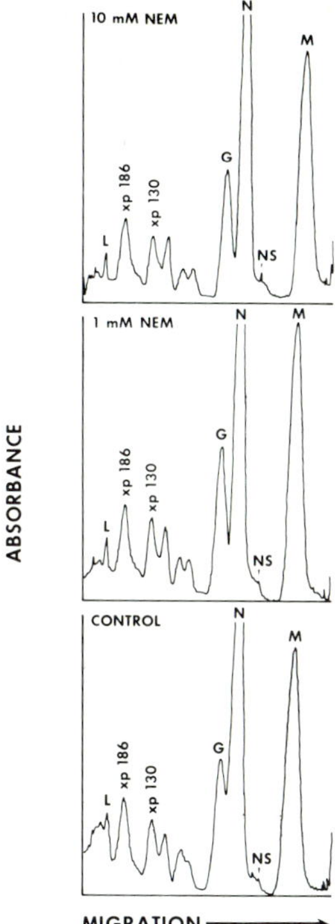

Fig.3. *Comparative analysis of the effect of NEM treatment on cross-linking of VS virus with DTBSP. Purified VS virus at 2 mg of viral protein per ml of PBS was treated with NEM at final concentrations of 10 mM and 1 mM for 1 hr at room temperature. The virus was pelleted at 165,000 x g for 30 min and the pellet resuspended at 2 mg viral protein per ml of 5 mM phosphate buffer pH 7.2. The virus was reacted with 0.5 mM DTBSP at 0°C as described in the legend to Fig.1.*

Analysis of ^{3}H-glucosamine-labelled virus cross-linked with 1.0 mM PDM showed that approximately 10% of the glucosamine label was associated with oligopeptide xp143 and 2-3% migrated between xp110 and xp96. It had been shown previously with other cross-linking agents that complexes in the xp130 to xp141 range, which are labeled with ^{3}H-glucosamine, consist of G-protein dimers (5). Confirmation of this identification is seen in Fig.4B. Triton-no salt treatment of VS virus removes only the G protein and cross-linking of the resulting viral particle does not produce xp143. The glucosamine label between xp110 and xp96 may represent a G-M heterodimer.

Absolute identification of M and N complexes is not possible using PDM as the cross-linker but certain complexes are con-

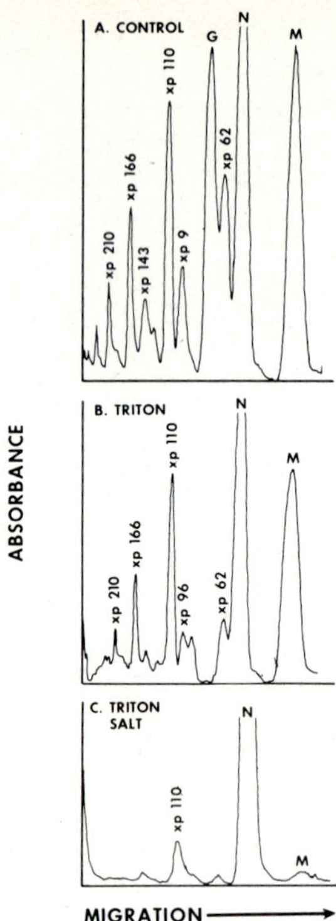

Fig.4. *Cross-linking of VS virus with phenylenedimaleimide. A. Purified virus (100 μg) was cross-linked with PDM (1 mM) in triethanolamine buffer (pH 8.0) at 37°C for 30 min. The reaction was terminated by the addition of β-mercaptoethanol and SDS (final conc 5% and 2%, respectively). B. Purified virus was treated with 2% Triton X-100 in 5 mM phosphate buffer for 1 hr at room temperature. The viral particles devoid of the glycoprotein were pelleted at 160,000 x g for 75 min. The pellets were resuspended in a volume of buffer equal to the volume of the original virus preparation. Cross-linking was as described for intact virus. C. Virus was treated with 2% Triton X-100 in NaCl (final conc 0.86 M) and the resulting nucleocapsids were crosslinked as described for intact virus.*

sistent with previous data. Complexes xp62 and xp96 appear to be M dimers and trimers, respectively. Removal of the G protein does not eliminate these complexes (Fig.4B), while a reduction of M protein results in a reduction of both complexes (Fig.4C).

The identification of N complexes follows a similar line of reasoning. Complexes xp110, xp166, and xp210 are consistent with the molecular weight estimates for the cross-linked N dimer, trimer, and tetramer, respectively. Removal of the G protein does not abolish these complexes, but removal of the M protein does result in a reduction in these complexes (Fig. 4B and 4C). If it is assumed that xp110, xp166 and xp210 are N protein complexes, then it must be concluded that the spatial arrangement of the N protein is altered in going from the coiled configuration of the nucleocapsid within the complete viral particle to the uncoiled structure of the free nucleocapsid (Fig.4A and 4C).

TWO-DIMENSIONAL ANALYSIS OF CROSS-LINKED VS VIRUS PROTEINS

More definitive identification of the VS virus proteins in cross-linked complexes can be ascertained by using two-dimensional electrophoretic analysis. This approach is feasible only when cleavable cross-linking agents are employed. MMB and DTBSP cross-linked complexes can be cleaved by reduction with β-mercaptoethanol while TDA cross-linked complexes can be cleaved by periodate oxidation. Fig.5 shows the results of the second-dimension analysis on slab gels of MMB, DTBSP, and TDA cross-linked VS virus proteins previously separated by first-dimension electrophoresis on cylindrical gels. Table 1 presents a summary of the data correlating the molecular weight estimates of the cross-linked and cleaved proteins separated by electrophoresis on first and second dimension gels.

The formation of homo-oligomers is readily seen with all three cross-linkers. The M protein does not appear to exist in a unique oligomeric complex since, in every situation examined, the amount of each cross-linked M-protein form progresses as follows: monomer>dimer>trimer>tetramer, etc. If a preferred M-protein oligomeric form existed, then one would expect a higher order oligomer to predominate when cross-linked as is seen with oligomeric enzymes (2, 4). The N protein is probably similar to M but structures greater than trimers have not been clearly identified. The results with the G protein are somewhat different. Both TDA and MMB produce G dimers and trimers but dimers predominate over trimers. However, DTBSP generates trimers in excess of dimers. This might suggest that the preferred structure of the G protein is a trimer.

The generation of hetero-oligomers is more difficult to ascertain. The existence of N-M heterodimers can be seen in TDA and MMB cross-linked samples (xp87). DTBSP cross-linked samples show a similar complex but generally in lower amounts, presumably due to a lesser degree of penetration by the hydrophobic DTBSP through the viral membrane. The existence of a G-M heterodimer is supported by the presence of G protein in the region of xp97 (Fig.5A and 5B). Definitive identification of this complex is hampered by the presence of the M trimer

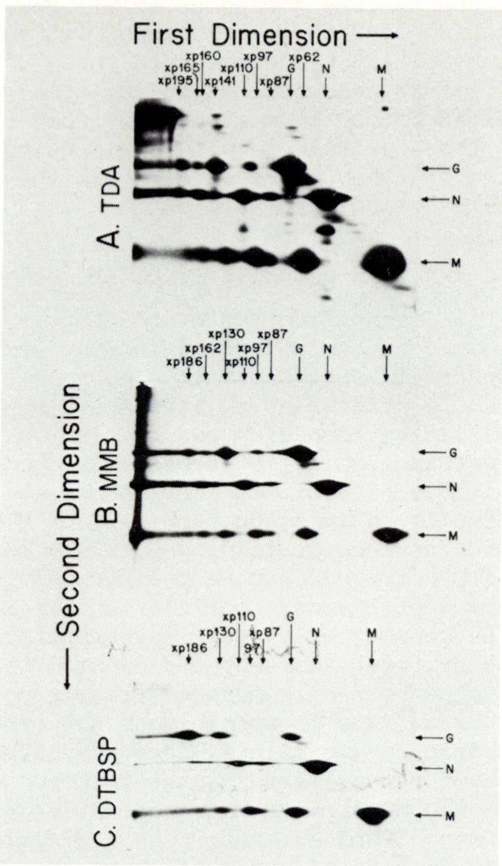

Fig.5. *Two-dimensional electrophoretic analysis of VS viral proteins cross-linked with (A) TDA, (B) MMB, and (C) DTBSP. Purified VS virions labeled with [^{35}S] methionine were treated with cross-linkers as described in the legend to Fig.1. Proteins extracted with SDS were then electrophoresed on cylindrical 5% polyacrylamide gels at 2.5 mA/gel for 16 hr and then at 6 mA/gel for 6 hr. First-dimension gels of TDA-cross-linked virus were soaked for 2.5 h in 15 mM sodium periodate in 20 mM TEA buffer (pH 7.6) and then for 2 hr in 0.125 M Tris (pH 6.8) containing 0.5% SDS and 3% glycerol. First-dimension gels of MMB- and DTBSP-cross-linked samples were soaked for 2 h in 1% 2-ME and 0.1% SDS. The second-dimension discontinuous slab gel consisted of a 10% resolving gel and a 5% stacking gel. The first-dimension gels were immobilized on top of each stacking gel with 1% agarose in 0.125 M Tris (pH 6.8) with 0.1% SDS and 1% 2-ME. Electrophoresis in the second dimension was at 75 to*

which migrates to the same region of the first dimension gel.

DISCUSSION AND SUMMARY

The spatial relationship of the major structural proteins of VS virus have been examined employing bifunctional reagents which can covalently link adjacent protein molecules. The intent of this study was to identify the unique spatial relationships of each protein and to apply these relationships to a scheme of virus maturation. Certain of these relationships have tentatively been identified and may shed some light on the structure of the VS virus particle.

The formation of N dimers by H_2O_2 and diamide treatment of intact virus as well as cross-linking with phenylenedimaleimide suggests that free sulphydryl groups exist on the N protein. The role of these free sulphydryl groups is unknown, but they do provide a means for monitoring the configurational changes of the nucleocapsid. In the coiled bullet-shaped form of the nucleocapsid, cross-linking of the N protein is readily achieved (Fig.4A) but in the uncoiled state, the N protein monomers are no longer in the same configuration since they cross-link very poorly (Fig.4C). Thus, it would appear that the N protein must undergo spatial rearrangements as the nucleocapsid is packaged during the maturation process.

The M protein does not appear to exist in a preferred oligomeric configuration unless all M monomers are aligned at equally spaced cross-linking distances. In the intact virus, it may be possible to serially cross-link M monomers so extensively as to produce a net-like structure of M proteins. In fact, virus highly cross-linked with PDM is relatively insensitive to the disruptive effects of Triton-high salt solubilizer (unpublished data). This observation is consistent with the report that formaldehyde fixed VS virus was resistant to disruption by detergent (1).

A key question in understanding the maturation of VS virus is whether there exists a direct interaction between the G and M proteins. This kind of interaction would provide the mechanism whereby the M protein recognizes areas of the cell membrane

---

100 V/gel and was stopped when the dye marker reached bottom. The slab gels were dried under vacuum, and autoradiography was performed with Kodak SB-54 X-ray film. Positions of first-dimension VS viral protein monomers and xp oligomers with estimated molecular weights are indicated by arrows at the top of each autoradiogram. Arrows at the side of the autoradiogram indicate the position after vertical migration of VS viral proteins G, N, and M on the second-eimension slab gel. (Reprinted with permission from Dubovi and Wagner, 1977).

TABLE I

Summary of VS viral protein components present in complexes formed by reacting intact VS virions with cross-linkers: TDA, MMB, DTBSP, and PDM

VS viral proteins in complex	Cross-linked protein (xp) complex (mol wt x 10^{-3})			
	TDA[a]	MMB[a]	DTBSP[a]	PDM[b]
M-M dimer	xp62	xp62	xp62	xp62
M-N dimer	xp87	xp87	xp87	ND[c]
M-M-M trimer	xp97	xp97	xp97	xp96
G-M dimer	xp97	xp97	ND	xp100
N-N dimer	xp110	xp110	xp110	xp110
G-G dimer	xp141	xp130	xp130	xp143
M-M-M-M tetramer	xp130	xp130	xp130	ND
N-N-N trimer	xp162	xp162	ND	xp166
G-G-G trimer	xp195	xp186	xp186	xp200

[a]Data obtained from first and second-dimension analysis of cross-linked VS viral protein complexes.

[b]Data from first-dimension analysis.

[c]ND, Not detected.

---

containing the G protein. The cross-linking data do suggest that such an interaction exists, but definitive proof is still lacking. All cross-linking agents employed to date have produced a glucosamine containing complex which migrates on polyacrylamide gels between 95,000 and 110,000 daltons, the region most likely to contain a G-M heterodimer. The hypothesis that the M protein may serve as a bridge between the G and N protein in the assembly of the VS virion is supported by the cross-linking data.

ACKNOWLEDGEMENTS

This study was supported by grant BMS72-02223 from the

National Science Foundation, grant VC-88 from the American Cancer Society, and Public Health Service grant AI-11112 from the National Institute of Allergy and Infectious Diseases. E.J. Dubovi is a postdoctoral fellow of the National Institute of Allergy and Infectious Diseases.

REFERENCES

1. Brown, F., Small, C.J. and Horzinek, M.C. (1974). *J. Gen. Virol.* 22, 455.
2. Carpenter, F.H. and Harrington, K.T. (1972). *J. Biol. Chem.* 247, 5580.
3. Cartwright, B. (1973). *J. Gen. Virol.* 21, 407.
4. Davies, G.E. and Stark, G.R. (1970). *Proc. Nat. Acad. Sci. USA* 66, 651.
5. Dubovi, E.J. and Wagner, R.R. (1977). *J. Virol.* 22, 500.
6. Lomant, A.J. and Fairbanks, G. (1976). *J. Mol. Biol.* 104, 243.

TRYPTIC PEPTIDE ANALYSIS OF THE STRUCTURAL PROTEINS OF VESICULAR STOMATITIS VIRUS

T.R. DOEL and F. BROWN

Animal Virus Research Institute,
Pirbright, Woking, Surrey.

The aim of the present work is to obtain an understanding of the variation within the VSV group of viruses at the molecular level and to relate such variations to the serological data (1). The glycoprotein carries the antigenic site responsible for the type specificity of the virus, whereas the group reactivity is associated with the ribonucleoprotein. The method we have used is that of tryptic peptide analysis of ^{125}I and ^{35}S- methionine containing peptides of the major viral proteins. The work is also intended to pave the way for limited sequencing studies if biologically interesting fragments of the major proteins of VSV can be readily obtained.

The viruses examined in the present study were New Jersey virus, Indiana virus and Brazil virus, which is regarded as a member of the Indiana serotype (2) although it can be readily distinguished from the classical strain by neutralization tests and by other properties. The major proteins of the viruses were either iodinated *in situ* and subsequently purified, or were purified from viruses grown in the presence of ^{35}S-methionine. The purified proteins were digested with trypsin and the peptides separated by electrophoresis and chromatography on thin layers of silica gel.

TRYPTIC PEPTIDE ANALYSIS OF G PROTEIN

Fig.1 shows the ^{125}-iodotyrosine peptides of the glycoproteins of the viruses. It can be seen that the maps differ considerably and that there are few, if any, peptides or groups of peptides which could reliably be considered common to the glycoproteins. Essentially the same results have been obtained

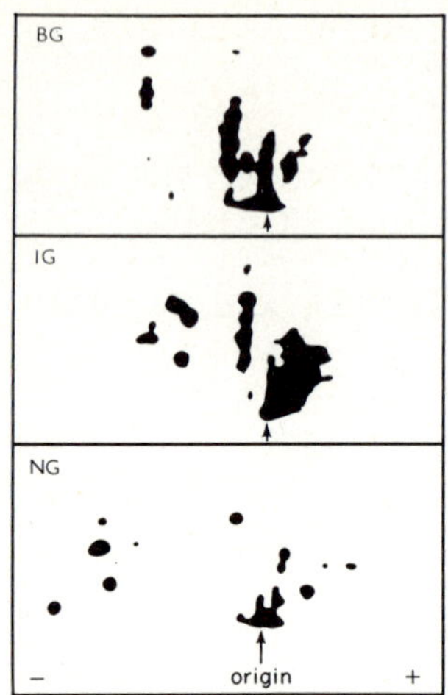

Fig.1. *Autoradiographs of* ^{125}I *tryptic peptide maps of the glycoproteins of Brazil virus (BG); Indiana virus (IG) and New Jersey virus (NG). Tryptic digests were spotted on to thin layers of silica gel and electrophoresed at 200 volts for 6 hours in pyridine: acetic acid: water (100:4:896), pH 6.5 (horizontal axis). For ascending chromatography (vertical axis) the sheets were placed in a tank containing* H_2O*: acetic acid: butan-1-ol (1:1:3) until sufficient resolution was achieved.*

with ^{35}S-methionine containing tryptic peptides. The observations with the peptide maps are entirely consistent with the serological experiments (1) which did not demonstrate significant levels of cross neutralization between the intact viruses and appropriate antisera. Although peptide differences were expected in view of the serological data, the results suggest very considerable variation among the primary structures of the glycoproteins. However, amino acid analyses (results not shown) do indicate that the glycoproteins are related and might be expected to contain some common sequences of amino acids. It is our intention to examine the glycoproteins in greater detail to see whether or not variability occurs randomly throughout the sequence.

TRYPTIC PEPTIDE ANALYSIS OF N PROTEIN

Fig.2 shows the ^{125}I peptide maps of the N proteins of the viruses. It is apparent that the N-proteins have similar although not identical peptide maps and share a significant

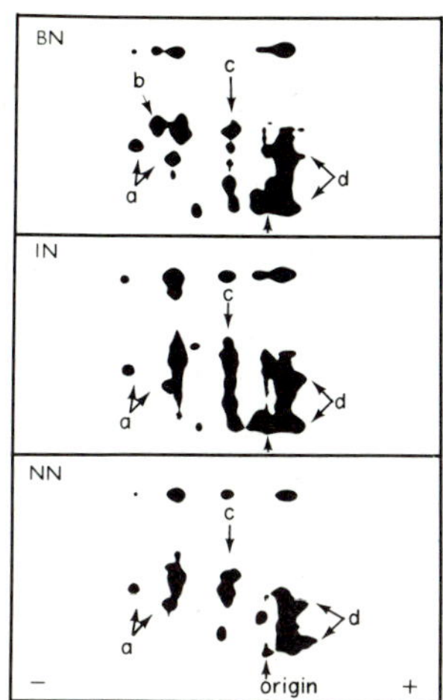

Fig.2. *Autoradiograph of the ^{125}I tryptic peptide maps of the N-proteins of Brazil virus (BN); Indiana virus (IN) and New Jersey virus (NN). Lettering and arrows are referred to in the text. The peptides were separated as in Fig.1.*

number of peptides. The horizontal line of peptides at the top of each autoradiograph run with the solvent front and may be artefacts. The group of peptides labelled 'a' can be recognised in each peptide map. Brazil N-protein is unique in possessing a particularly intense peptide 'b'. Both the vertical line of peptides 'c' and the group of peptides 'd' are also common to the three maps. New Jersey N-protein appears to contain fewer peptides in the group labelled 'c' and has a more obviously different pattern of peptides in group 'd'. Brazil and Indiana N-proteins are more closely related to each other than either are to New Jersey N-protein.

This observation was confirmed by the maps of ^{35}S-methionine containing tryptic peptides (Fig.3). The most striking differences between Indiana and Brazil N-proteins reside in peptides 'a', 'b' and 'c'. New Jersey N-protein is quite distinct from the other N proteins; particularly noticeable is its lack of peptides 'd', 'e' and 'f'.

Nevertheless, the N-proteins of Brazil, Indiana and New Jersey viruses are clearly related. This conclusion is in agreement with the work of Cartwright and Brown (1) who used detergent disrupted viruses and the appropriate antisera to show significant levels of cross reaction in neutralization tests. The present work indicates that Brazil and Indiana N-proteins, and therefore presumably the viruses, are more closely related to each other than either are to New Jersey N-protein.

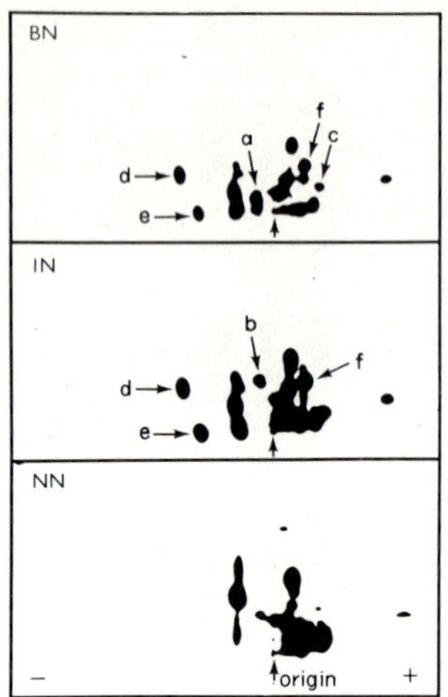

Fig.3. *Autoradiographs of the ^{35}S-methionine tryptic peptides of the N-proteins of Brazil virus (BN), Indiana virus (IN) and New Jersey virus (NN). Lettering and arrows are referred to in the text. The peptides were separated as in Fig.1.*

TRYPTIC PEPTIDE ANALYSIS OF M-PROTEIN

We have also examined the M-protein of the viruses by tryptic peptide analyses of ^{125}I and ^{35}S-methionine containing peptides (results not shown). Although the M-proteins are clearly related, there is little or no indication that the Brazil virus M-protein is more closely related to Indiana virus M-protein than it is to New Jersey virus M-protein.

DISCUSSION

Because of their ready differentiation by neutralization tests, a certain amount of variation among the glycoproteins was expected. However, the extent of structural variation indicated by the peptide maps suggests that very much more than the antigenic determinants vary among the glycoproteins. We interpret this to indicate that a large portion of the glycoprotein sequence has no strict function and that as a consequence, mutations which affect the region of the genome coding for the glycoprotein are less likely to be lethal than, for example, those which affect the regions coding for the N and M proteins.

The extent to which the sequence of the N-protein of the different viruses is conserved may be a reflection of its role. N-protein is intimately associated with the RNA so that the latter is protected from the action of ribonuclease. It is reasonable to assume that such an association makes strict demands on the sequence of the N-protein. The extent of sequence conservation within the M-proteins suggests that they too have an important role to play in the interaction of the nucleocapsid with the lipid membrane and it is interesting to note that Laver and Downie (3) also found very few differences between the peptide maps of the matrix proteins of type A influenza viruses from widely differing sources.

In view of the similarities observed in the tryptic peptide maps of the internal proteins of different strains of VSV, it would be interesting to examine the extent to which the internal proteins of serologically unrelated rhabdoviruses resemble those of VSV.

REFERENCES

1. Cartwright, B. and Brown, F. (1972). *J. Gen. Virol.* 16, 391.
2. Federer, K.E., Burrows, R. and Brooksby, J.B. (1967). *Research in Veterinary Science* 8, 103.
3. Laver, W.G. and Downie, J.C. (1976). *Virology* 70, 105.

PHOSPHOPROTEINS OF SPRING VIREMIA CARP VIRUS AND OTHER RHABDOVIRUSES

POLLY ROY and JON P. CLEWLEY

Department of Microbiology, The Medical Center,
University of Alabama in Birmingham,
Birmingham, Alabama 35294.

All animal rhabdoviruses so far studied have three or four major structural proteins including a single glycoprotein type, G, and single nucleocapsid protein, N and one or, for the rabies subgroup of viruses two, putative membrane proteins, M (2, 12). Although these protein species form the bulk of the virion proteins, minor amounts of other proteins have been observed, including a large protein, L, and phosphoproteins (e.g. the NS protein of vesicular stomatitis and related viruses). For VSV, the L and NS proteins are component parts of the virion transcriptase (3). All members of the VSV subgroup of rhabdoviruses appear to conform in protein composition to the prototype VSV Indiana (6). By contrast, rabies and the rabies subgroup viruses have a G, N and two M proteins, M1 and M2, as well as a large protein L (12). No separate NS phosphoprotein has been described for rabies virus although it has been reported that its nucleocapsid protein is phosphorylated (9). Other rabies subgroup rhabdoviruses are either similar in composition to rabies or in addition to a phosphorylated N protein have an M1 which is also phosphorylated (10). Whether their M1 functions as a transcriptase component like the NS protein of VSV, is not known. Spring viremia of carp virus, SVCV, which was isolated in 1974 from infected carp, has three major structural proteins G, N, and M, as well as the minor proteins, L and NS (10). Two phosphoproteins are present in SVCV virions, one is the NS protein and the other appears to be the virion nucleocapsid protein (10). In this paper we describe analyses of the structural proteins of SVCV and provide evidence that the two viral phosphoproteins are distingusihable from the bulk of the nucleocapsid protein. A similar result has been obtained for the phosphorylated protein of rabies virus.

ANALYSES OF THE PROTEINS OF SVCV

A preparation of purified Spring viremia of carp virus, grown in the presence of ^{3}H glucosamine, was dissociated with SDS and the proteins resolved by discontinuous polyacrylamide gel electrophoresis at pH 8.9 (Fig.1A). After staining with Coomassie brilliant blue, three major (G, N and M), and two minor (L and NS), viral polypeptides were identified. One of the major polypeptides was labelled by the ^{3}H-glucosamine and, according to the convention adopted for all rhabdoviruses, is designated G (12). When this virus preparation was treated with pronase and the residual spikeless particles repurified by sucrose gradient centrifugation, only the glycoprotein was absent (data not shown), indicating that, as for other rhabdoviruses, the glycoprotein constitutes the spikes seen on the external surface of the viral membrane. By comparison to known molecular weight protein standards, it was determined that the L, G, N, and M proteins of SVCV have respective molecular weights of 160,000, 85,000, 45,000, and 21,000. Since the NS protein is a phosphoprotein whose migration rate depends on the gel system used, its molecular weight was not estimated (see below).

In order to determine which of the virion proteins were phosphorylated, a purified preparation of ^{32}P labelled virus was extracted with phenol and the proteins separated from the labelled RNA by alcohol precipitation of the phenol phase. After a second alcohol precipitation to remove residual phospholipids, the proteins were resolved by electrophoresis on a continuous 8% polyacrylamide gel in 0.1 M phosphate buffer (pH 7.0). After staining with Coomassie brilliant blue, the three major and two minor virion polypeptides were identified (Fig.1B). It was noted that like certain other rhabdoviruses (6) the minor NS protein migrated faster than the nucleocapsid protein N. This is in contrast to its electrophoretic mobility in the discontinuous gel system shown in Fig.1A, where the NS protein migrated slower than the nucleocapsid protein N. When the distribution of labelled viral polypeptides was determined, it was found that two of the virion polypeptides were labelled with ^{32}P phosphate (Fig.1B). One of these polypeptides was the NS protein and the other appeared to be the nucleocapsid protein N. However visual comparison of the relative labelling of the two proteins indicated that, by comparison to the N protein, NS protein was obviously much more phosphorylated than N. Three possible explanations of these observations have been examined. One is that the NS protein has more phosphate

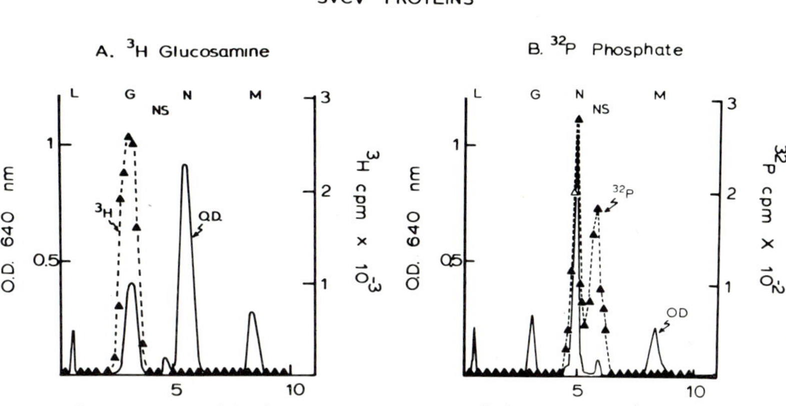

Fig.1. *The proteins of SVCV. Spring viremia of carp virus preparations labelled by (A) ^3H-glucosamine or (B) ^{32}P-phosphate were prepared for gel electrophoresis, (see text), and resolved on discontinuous gel electrophoresis at pH 8.9 (A), or by continuous gel electrophoresis at pH 7.0 (B) as described by Obijeski et al., (6). After staining with Coomassie brilliant blue, the gels were scanned and the distribution of radioactivity determined.*

residues per mole of protein. Another is that not all of the N protein molecules are phosphorylated. A third possibility is that there is another phosphoprotein which has an electrophoretic mobility similar to N. This other protein is either a different protein or a super-phosphorylated NS protein.

To investigate these possibilities a determination of the relative molar specific activities of the NS and N proteins was made. A 0.5 ml preparation of ^3H-uridine labelled SVCV (2.4×10^6 cpm/ml, 2.3 mg protein/ml) was extracted by phenol and 97% of the label recovered as RNA. By measuring the optical density and radioactivity, the RNA was found to have a specific activity of 4.7×10^7 cpm/mg RNA. It was calculated therefore that in the original virus preparation there was (per ml) 0.051 mg RNA per 2.3 mg protein, i.e. an RNA to protein ratio of 1 to 45. The relative amounts of the various viral polypeptides have been estimated by both comparative optical measurements of stained protein gels (see Fig.1), as well as from the ratio of ^3H-leucine labelled viral polypeptides resolved by electrophoresis in 8% polyacrylamide gels. Both methods have yielded similar results with the relative per cent distributions for the L : G : N : NS : M of the order of 3 : 23 : 48 : 4 : 22 respectively. On the basis that SVCV has a genome similar in size to that of VSV (i.e. 3.8×10^6, ref.7), and only one RNA molecule per virus particle, the number of N protein molecules can be calculated. Thus we calculate that the gram moles of N protein per gram moles of virus RNA is $3.8 \times 10^6 \times 45 \times 0.48 \div$ N molecular weight (45,000) i.e. some 1800 moles N per mole of RNA. This amount is equivalent to a mass ratio of about 21 N protein to 1 of RNA, which is in the same range as that obtained for VSV (1). We are unable directly to estimate the molecular weight of SVCV NS protein due to its variable electrophoretic mobility (Fig.1). However it is probably reasonable to assume that, like the NS protein of VSV Indiana which based on its mRNA size must be around 25,000 daltons, SVCV NS protein has a molecular weight of around 25,000 daltons. If we make this assumption then the calculated number of NS molecules per virus particle is $3.8 \times 10^6 \times 45 \times 0.04 \div$ 25,000 i.e. 270. Note that if the NS molecular weight is 27,000 or 50,000 there would be respectively some 230 or 135 NS molecules per virus particle. Despite the uncertainity of its molecular weight, it can be concluded that NS represents *no more than* 15% of the number of N protein molecules present in the virus particle (i.e. $270 \times 100 \div 1800$).

In a separate experiment with ^{32}P-labelled SVCV, a virus suspension (2.3×10^6 cpm of acid-insoluble radioactivity after SDS treatment) was found to have 65×10^3 cpm of N phosphoprotein and 50×10^3 cpm of NS phosphoprotein. The rest of the ^{32}P label was recovered as RNA. On the basis that there are of the order of 11,000 nucleotides per SVCV genome (i.e. similar to that of VSV Indiana, ref.7), these amounts of radioactivity per virus particle are equivalent to 311 phosphate molecules for the total viral N protein (i.e. $55 \times 10^3 \times 11{,}000 \div 2.3 \times 10^6$) and 240 molecules for the total NS protein. Since we have estimated that there are some 1800 molecules of N protein and 270 molecules of NS protein per virus particle these results indicate that whereas essentially all the NS molecules are phosphorylated, no more than one in six of the N molecules are phosphorylated.

From these results it appears either that only a subset of N protein molecules are phosphorylated and that the phosphorylated species have the same electrophoretic mobility as the non-phosphorylated N, or that another phosphorylated viral protein comigrates with N. To investigate this, we analysed the electrophoretic mobilities of ^{35}S-methionine and ^{32}P labelled SVCV proteins in large slab gels of 8% polyacrylamide containing 0.1 M phosphate buffer, pH 7.0. After autoradiography all SVCV proteins were evident on the radioautogram as faint bands

(^{35}S label), while the phosphoproteins were identified by well marked bands (Fig.2A). Comparison of the gel stained with Coomassie brilliant blue confirmed (as indicated in Fig.2A) the location of the major virion polypeptides of SVCV. It was observed that the slower migrating SVCV phosphoprotein did not comigrate with the N polypeptide but instead moved slightly ahead of the N band. The NS band identified by autoradiography exactly comigrated with the stained NS band. Although it was not possible completely to excise the slower moving phosphoprotein free from N, when the ratios of ^{35}S to ^{32}P in the two phosphoproteins were compared it was found that relative to their ^{35}S content, the NS protein had only 1.5 times more ^{32}P than N the slower migrating phosphoprotein. Longer electrophoreses or other gel systems will be needed to obtain a better separation of the N from the phosphoproteins. Also more exact ratios of ^{3}H-amino acid to ^{32}P will be needed to confirm that the slower migrating phosphoprotein is free from non-phosphorylated N protein. Experiments are in progress to determine by peptide maps if the slower phosphoprotein is a phosphorylated N polypeptide, a super-phosphorylated NS or another virion protein.

PROTEIN KINASE ACTIVITY OF SVCV

The presence of a virion protein kinase activity capable of phosphorylating certain viral polypeptides has been demonstrated for VSV, pike fry rhabdovirus and all the rhabdoviruses of the VSV subgroup (4, 5, 8, 9, 11). The origin of this protein kinase enzyme is unknown although specific activity measurements for VSV grown in different cell types suggested that the enzyme may be a host function which is specifically picked up by budding virus particles (4). From all the evidence obtained so far, it is clear that the preferred substrate for the protein kinase enzyme is the resident viral phosphoprotein. Evidence obtained in the laboratories of D.H.L. Bishop, J.F. Obijeski and the late F. Sokol (unpublished observations) indicated that the phosphorylation obtained with the protein kinase is not a replacement phosphorylation of pre-existing NS phosphate groups, but rather addition phosphorylation to a maximum extent by 5 h. of *in vitro* phosphorylation of on average about one phosphate group per molecule of NS.

The protein kinase activity of SVCV has been investigated for virus grown in fathead minnow cells (FHM) and BHK-21 cells. With virus grown in both cell types the preferred substrates of the endogenous protein kinase activity were the two SVCV viral phosphoproteins (Fig.3). The protein kinase temperature optimum for SVCV grown in FHM cells is between 25 and 30°C whereas the temperature optimum for the BHK grown virus is

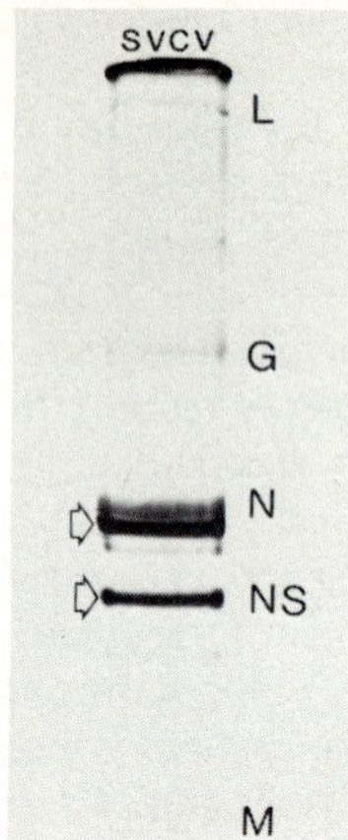

Fig.2. *High resolution gel electrophoresis of SVCV phosphoproteins: A preparation of* ^{35}S*-methionine and* ^{32}P*-phosphate labelled SVCV was extracted for proteins (see text) and resolved by electrophoresis on a 30 cm 8% polyacrylamide slab gel at pH 7.0. After autoradiography the positions of the major viral polypeptides (L, G, N and M) were identified as faint bands (as indicated by the adjacent letters), while the two phosphoproteins (arrows) were identified as well pronounced bands. Staining the gel after autoradiography confirmed the positions of the major viral polypeptides (L, G, N, NS and M). The two unassigned faint bands which migrated respectively further than N phosphoprotein, or NS protein, did not show up as stained polypeptides and presumably are contaminant phosphoproteins of unknown origin. They were not studied further.*

between 30 and 35°C. For virus prepared from either cell type the two phosphoproteins were phosphorylated at either low (25°C, Fig.3A) or high temperature (37°C, Fig.3B). Since we have not resolved the protein kinase reaction products on long slab gels we cannot say which of the two phosphoproteins are more efficiently phosphorylated although if the phosphoprotein which comigrates with N is really an N phosphoprotein, we would have to conclude that despite the larger number of viral N molecules, it is a less efficient substrate than the NS protein. However the validity of this statement will depend on whether the phosphoprotein which migrates like N is really a phosphorylated N or an alternate protein.

THE PHOSPHOPROTEIN OF RABIES VIRUS

The fact that the phosphoproteins of SVCV could be differentiated from the major nucleocapsid protein of the virus by

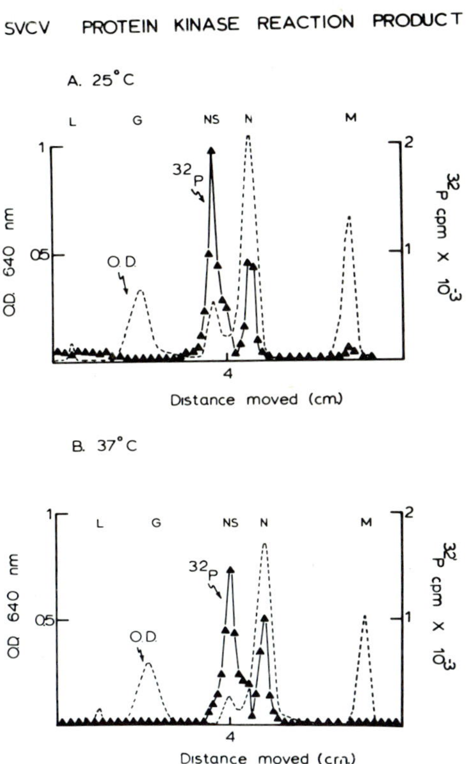

Fig.3. *Protein kinase reaction products of SVCV. Protein kinase reaction products of SVCV grown in BHK cells (11) labelled by ^{32}P-γ-ATP, and incubated for 2h at (A) 25°C or (B) 37°C, followed by repetitive 5% TCA precipitation, were dissolved in 1% SDS, 1% β-mercaptoethanol, and 0.01 M phosphate buffer pH 7.0, and resolved by discontinuous gel electrophoresis at pH 8.9 (see Fig.1). After staining, the distribution of the ^{32}P was determined.*

electrophoresis in long slab gels has encouraged us to determine whether the phosphoprotein of rabies can be distinguished from the N protein. In calculations made to determine the relative phosphorylation of the rabies N protein we have determined that less than 1 in 10 of the N protein molecules are phosphorylated. A 1 ml preparation of ^{3}H-uridine labelled rabies ERA (1.2 mg protein) was extracted from RNA and, from the RNA specific activity (5.32×10^7 cpm per mg RNA), the original RNA to protein ratio was determined to be 1 to 58. The ratio of the relative amounts of the major viral polypeptides was determined to be 4 : 35 : 20 : 15 : 16 for the viral L : G : N : M1 : M2 respectively, as judged by both stained and ^3H-leucine labelled virus resolved by 8% polyacrylamide gel electrophoresis. By RNA gel electrophoresis rabies ERA viral RNA is indistinguishable in size from that of VSV Indiana (D.H.L. Bishop, personal communication). Therefore we calculate that per gram mole of rabies virus there is $58 \times 3.8 \times 10^6 \times 0.2$ gm of N protein (i.e. 44×10^6 gm). Since rabies N protein is reported to have a molecular weight of 62,000 we calculate that there are some 710 molecules of N protein per RNA (i.e. $44 \times 10^6 \div 62,000$). In another experiment involving ^{32}P-labelled rabies ERA, a virus suspension with 1.4×10^6 cpm of ^{32}P RNA yielded 9×10^3 cpm of phosphoprotein. Assuming rabies, like VSV, has some 11,000 nucleotides (7), from the mononucleotide specific activity ($1.4 \times 10^6 \div 11,000$), it can be calculated that there are 70 phosphates resident in the total viral phosphoproteins. Thus if rabies N protein is a phosphoprotein, no more than one in ten is phosphorylated. Preliminary evidence indicates that the rabies phosphoprotein can be resolved from the bulk of the N protein by slab gel electrophoresis similar to that seen for SVCV in Fig.2. Again whether the rabies phosphoprotein is a phosphorylated N or another viral polypeptide will have to await eventual peptide analyses.

ACKNOWLEDGEMENTS

This investigation was supported by Public Health Service grant AI 13402 from the National Institute of Allergy and Infectious Diseases. We thank Reginald Anderson and Gloria Robeson for excellent technical assistance.

REFERENCES

1. Bishop, D.H.L. and Roy, P. (1972). *J. Virol.* 10, 234.
2. Bishop, D.H.L. and Smith, M. (1977). *In* "The Molecular Biology of Animal Viruses (D. Nayak, ed.) p.167. Dekker, New York.
3. Emerson, S.U. and Yu, Y.-H. (1975). *J. Virol.* 15, 1348.
4. Imblum, R.L. and Wagner, R.R. (1974). *J. Virol.* 13, 113.
5. Moyer, S.A. and Summers, D.F. (1974). *J. Virol.* 13, 455.
6. Obijeski, J.F., Marchendo, A.P., Bishop, D.H.L., Cann, B.W. and Murphy, F.A. (1974). *J. Gen. Virol.* 22, 21.
7. Repik, P. and Bishop, D.H.L. (1973). *J. Virol.* 12, 969.
8. Roy, P., Clark, H.F., Madore, H.P. and Bishop, D.H.L. (1975). *J. Virol.* 15, 338.
9. Sokol, F. and Clark, H.F. (1973). *J. Virol.* 52, 246.

10. Sokol, F., Clark, H.F., Wiktor, T.J., McFalls, M.L., Bishop, D.H.L. and Obijeski, J.F. (1974). *J. Gen. Virol.* 24, 433.
11. Strand, M. and August, J.T. (1971). *Nature (London) New Biol.* 233, 137.
12. Wagner, R.R., Prevec, L., Brown, F., Summers, D.F., Sokol, F. and MacLeod, R. (1972). *J. Virol.* 10, 1228.

MOLECULAR ANALYSES OF INFLUENZA C VIRUS

H. MEIER-EWERT, R.W. COMPANS,
D.H.L. BISHOP and G. HERRLER

Institut für Medizinische Mikrobiologie der Techn. Universitat München, West Germany,

and

The Department of Microbiology, University of Alabama Medical Center, Birmingham, Alabama 35294 U.S.A.

Since the time when the designation "Influenza C" was assigned to a group of viruses which cause mild upper respiratory tract infections in humans (5, 12) there has been some uncertainty as to whether these viruses should be classified together with other Orthomyxoviruses. In contrast to types A and B influenza viruses, our knowledge concerning the structural components or the process of replication of influenza C virus is fragmentary. The main reasons for this lack of information are the difficulty of propagating this virus in cell cultures and also the limited efforts devoted to its investigation.

Differences in the receptor on the membrane of red blood cells, as well as the receptor-destroying enzyme, between type C and other influenza viruses were reported in early studies (6). In a detailed investigation, Kendal (7) obtained convincing evidence that the receptor-destroying enzyme of influenza C virus is not the α-type neuraminidase that liberates sialic acid from a number of well-characterized substrates. In addition influenza C virus has morphological characteristics not seen with the A and B influenza viruses (1, 4).

We report in this study new findings concerning the structural components of influenza C virions, including evidence for the content of sialic acid residues in the viral envelope.

MORPHOLOGY OF THE VIRIONS

The overall appearance of purified influenza C virions is much more pleomorphic than a standard preparation of influenza A_0 WSN virus. The roughly spherical particles have diameters ranging from 75 to 180 nm, including a layer of surface projections of $\sim$10 nm in length. Long filamentous forms of virus particles are often found in preparations grown in fibroblast cell cultures. The width of these filaments is similar to the more spherical virions, but they frequently display a length up to 7 μm. The surface projections of spherical as well as filamentous virus particles are often arranged in a regular hexagonal structure (1, 4). These characteristic arrangements of the spikes have been observed with six different strains of influenza C virus.

VIRUS RNA

Analyses of RNA extracted from preparations of the purified JHB/1/66 strain of influenza C indicates that the genome consists of multiple segments of RNA (4). When ^{3}H-uridine labelled virus RNA was resolved by electrophoresis in a 2% polyacrylamide gel (2), six distinct genome RNA species were resolved. Coelectrophoresis with ^{32}P-labelled A/WSN viral RNA demonstrated that the overall pattern, as well as most of the individual peaks of influenza C viral RNA, could be clearly distinguished. Using values obtained for A/WSN viral RNA as standards, the molecular weights of the six RNA genome segments of influenza C virions have been estimated to be 1.25×10^6, 1.20×10^6, 0.98×10^6, 0.85×10^6, 0.50×10^6 and 0.37×10^6 respectively. Accordingly, the minimal sum molecular weight for the viral RNA is 5.15×10^5 (4).

VIRUS POLYPEPTIDES

Polyacrylamide gel electrophoresis of the proteins of influenza C virus grown in embryonated eggs and labelled with ^{3}H-leucine gave three major polypeptide peaks with a distinct shoulder on the trailing side of the fast migrating component. Isolation of the ribonucleoproteins from disrupted virus on a glycerol gradient indicated that the intermediate of the three polypeptide peaks was the nucleoprotein subunit NP (4). In order to obtain information on the glycosylated polypeptides, the C/Taylor/47 strain has been grown in eggs in the presence of ^{3}H-glucosamine and ^{14}C-amino acids (Fig. 1A). The glucosamine label was clearly associated with the largest viral polypeptide. In addition glucosamine label was found in the region of the nucleocapsid protein and on the trailing side of the fast-migrating peak. Based on coelectrophoresis with A/WSN virion polypeptides, the sizes of the three glycoproteins were estimated to be 88,000 daltons for the largest (gp88), 65,000 daltons for the intermediate (gp65) and 30,000 for the smallest glycoprotein (gp30). The sizes of the ribonucleoprotein subunit (NP) and the fastest migrating polypeptide, designated M, were estimated as 66,000 and 26,000 respectively. This pattern of virion polypeptides was found for a number of influenza C strains isolated in different parts of the world, including C/JHB/1/66; C/JHB/1/67; C/JHB/1/654; C/Yamagata/1/64; C/Taylor/1233/47 and C/Ann Arbor 1/50.

In contrast to the results obtained for egg grown virus, when influenza C is grown in chick embryo cells, a different pattern of ^{3}H-glucosamine labelling is obtained. Only gp88 was identified for ^{3}H-glucosamine labelled virus (Fig. 1B), although when ^{3}H-leucine was employed, gp88, NP and M were clearly evident (data not shown). This situation is reminiscent of similar findings with influenza A viruses, where the hemagglutinin is present only in its uncleaved state if the virus is grown in cell cultures without the enzyme necessary for cleavage into HA_1 and HA_2 (8, 10). Since HA_1 and HA_2 are linked by disulfide bonds (9, 11) it was of interest to determine whether any differences could be detected in the glycoproteins under reducing as opposed to non-reducing conditions. The patterns obtained for the glycoproteins from a single preparation of influenza C/Taylor treated with and without mercaptoethanol prior to electrophoresis were therefore compared (Fig. 1C, D). The results indicated that gp65 and gp30 are linked by disulfide bonds, with the resulting complex having an electrophoretic mobility similar to gp88. These results indicate therefore that the glycoproteins gp65 and gp30 of influenza C virions may be analogous to the HA_1 and HA_2 glycoproteins of influenza A and B viruses.

THE PRESENCE OF SIALIC ACID IN INFLUENZA C VIRUS PREPARATIONS

Enveloped viruses which lack neuraminidase generally contain sialic acid in their envelope. This may serve as an *in vitro* receptor for the hemagglutinin of influenza A viruses, leading to extensive clumping of the particles and causing hemagglutination inhibition (3). Since influenza C has been reported to lack neuraminidase (7), a preparation of influenza C/Taylor virus was added to an excess of influenza A/WSN virus and examined by electron microscopy (Fig. 2). By comparison to electron micrographs of the A/WSN preparation alone (Fig. 2A), the mixed virus preparations clearly contained aggregates consisting of the characteristic influenza C virus particles (see Fig. 2B, C) with A/WSN particles sticking to their surface.

The question of whether influenza C can cause inhibition of hemagglutination by influenza A was examined by first incubating chicken red blood cells with influenza C virions to destroy the C-receptor. This did not affect the hemagglutination titer for untreated A/WSN virus (Table 1, upper half). With treated cells, it was observed that influenza C virions caused a marked inhibition of hemagglutination by influenza A virus (Table 1,

3H dpm x10-2
gp88 NP gp65
gp30 M
14C dpm x10-2
A
45
36
27
18
9
20
16
12
8
4
3H dpm x10-2
B
gp88
10
8
6
4
2
3H dpm x10-2
C
115
gp88
gp65
gp30
9
3
3H dpm x10-2
D
30
gp88
118
6

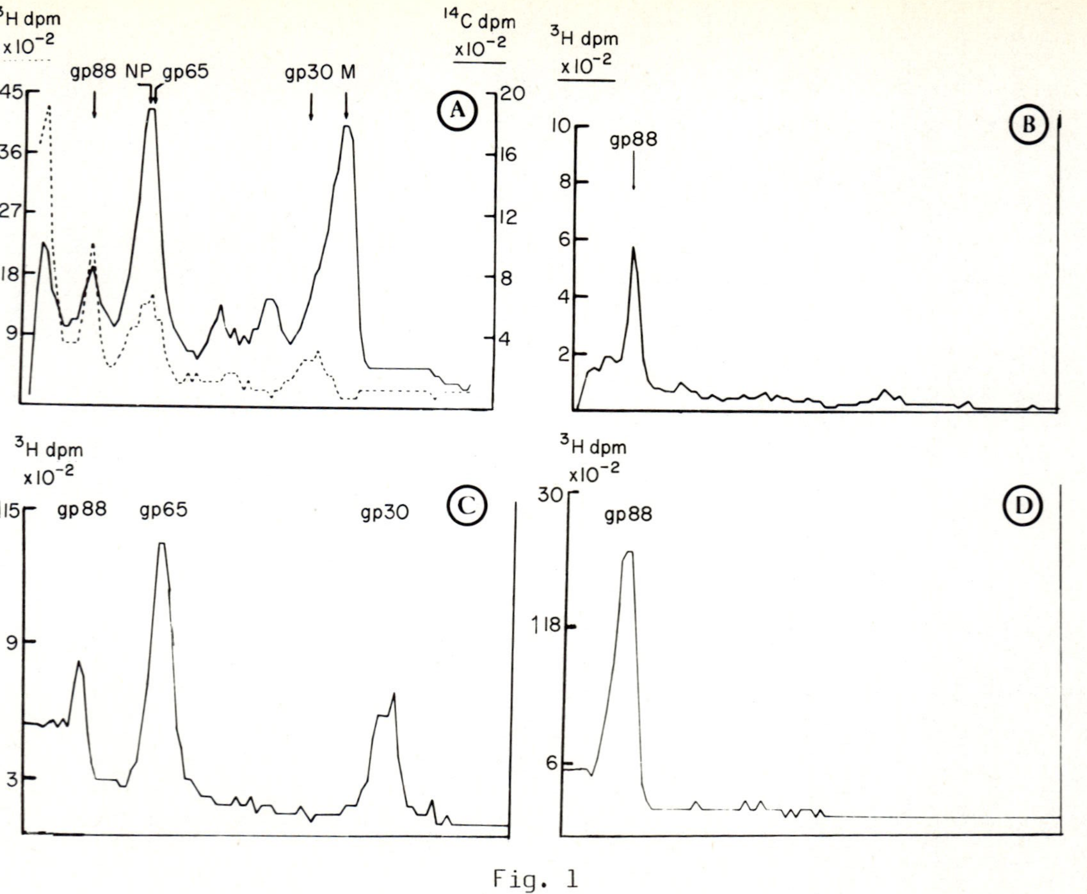

Fig. 1

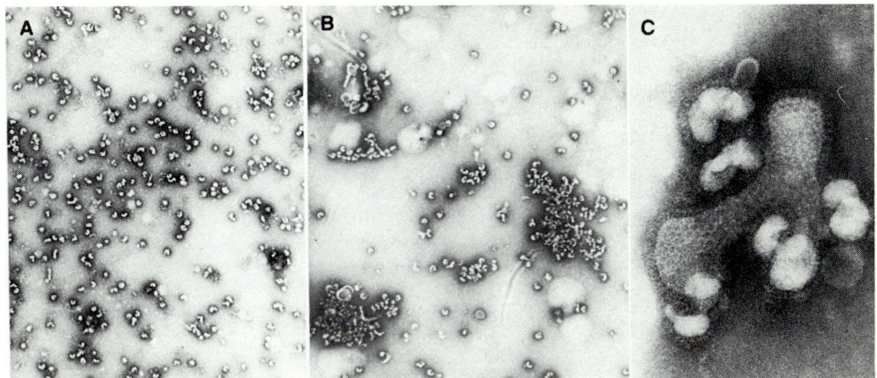

Fig. 2. *(A) Influenza A/WSN virions distributed on an electron microscope grid. (B) The same preparation immediately after mixing with influenza C/Taylor virions showing extensive aggregation. (c) Virions of A/WSN attached to a pleomorphic influenza C/Taylor virus particle which shows the characteristic hexagonal pattern of the surface spikes. Magnification: A and B, x7000. C, X120,000.*

lower half). If the neuraminic acid residues in the membrane of influenza C virus are responsible for the hemagglutination inhibition, then it should be possible to abolish this affect by treating the influenza C virion with neuraminidase. The results obtained (Table 1, lower half) showed that after neuraminidase treatment the hemagglutination-inhibition titer was drastically reduced, thus providing additional evidence for the presence of sialic acid in influenza C virus preparations.

Fig. 1. *(A) polyacrylamide gel electrophoresis of polypeptides of influenza C/Taylor virions grown in embryonated eggs and labeled with ^{3}H-glucosamine and ^{14}C-amino acids. The major polypeptides and glycoproteins that have been clearly identified are indicated by arrows and designated as described in the text. (B) polypeptides of influenza C/JHB/1/66 virions obtained from CEF cell cultures labeled with ^{3}H-glucosamine, showing label associated with the major glycoprotein gp88. (c) ^{3}H-glucosamine-labeled glycoproteins of egg-grown influenza C/Taylor; mercaptoethanol was added to the sample prior to the electrophoresis. (D) the same preparation without mercaptoethanol.*

TABLE 1

Evidence for neuramininic acid in influenza virions

	HA titer, control RBC	HA titer, RBC pretreated with influenza C[a]
Influenza type C	2048	<4
Influenza type A, WSN	4096	4096

	HI titer, RBC pretreated with influenza C[b]
Influenza type C	512
Neuraminidase-treated Influenza type C[c]	16

a) A 1% suspension of chicken red blood cells was incubated at 37°C for 90 min with 512 HA units/ml of influenza C/JHB/1/66. The RBC were pelleted and resuspended to a final concentration of 1% in phosphate-buffered saline.

b) 8 HAU/25 μl of purified influenza A/WSN were added to each well of the twofold dilution row of the purified preparation of influenza C/JHB/1/66. After incubating for 30 min at 4°C, 25 μl of influenza C pretreated 1% RBC were added to each well. Agglutination patterns were read 45 min thereafter.

c) Influenza C/JHB/1/66 in 0.01 M phosphate-buffer was incubated for 2h at 37°C with neuraminidase from Behringwerke/Germany at a concentration of 25 units/ml. The HI test was carried out in the cold.

SUMMARY AND CONCLUSIONS

The results presented here indicate that influenza C has a number of similarities to influenza A and B viruses: a multisegmented genome, as well as the number and molecular weight of the major structural polypeptides. However, there are also certain important differences such as the nature of the receptor-destroying enzyme and the presence of sialic acid on the viral envelope. Biological parameters such as host range and optimal growth temperatures also are different. The significance of these dissimilarities between influenza C and other orthomyxoviruses can only be assessed when more information is available about the infection process and the role of the various virus components (e.g. the virus receptor-destroying enzyme) is elucidated.

ACKNOWLEDGEMENTS

This research was supported by Grants AI 12680 and AI 13402 from the National Institute of Allergy and Infectious Diseases, U.S.P.H.S., PCM 76-09711 from the National Science Foundation, and NATO Research Grant 1184. We thank Mr. John Fitzpatrick for excellent technical assistance.

REFERENCES

1. Apostolov, K., Flewett, T.H. and Kendal, A.P. (1970). *In* "The Biology of Large RNA Viruses" (Barry, R.D. and Mahy, B.W.J., eds). Academic Press, New York, p.3.
2. Bishop, D.H.L., Obijeski, J.F. and Simpson, R.W. (1971). *J. Virol.* 8, 74.
3. Compans, R.W. (1974). *J. Virol.* 14, 1307.
4. Compans, R.W., Bishop, D.H.L. and Meier-Ewert, H. (1977). *J. Virol.* 21, 658.
5. Francis, T., Quilligan, J.J. and Minuse, E. (1950). *Science* 112, 495.
6. Hirst, G.K. (1950). *J. Exp. Med.* 91, 177.
7. Kendal, A.P. (1975). *Virology* 65, 87.
8. Klenk, H.-D., Rott, R., Orlich, M. and Blodorn, J. (1975). *Virology* 68, 426.
9. Laver, W.G. (1971). *Virology* 45, 275.
10. Lazarowitz, S.G., Compans, R.W. and Choppin, P.W. (1973). *Virology* 52, 199.
11. Stanley, P.M. and Haslam, E.A. (1971). *Virology* 46, 746.
12. Taylor, R.W. (1951). *Arch. Gesamte Virusforsch.* 4, 485.

POLYPEPTIDES OF MEASLES VIRUS

David L.J. Tyrrell* and Erling Norrby

Department of Virology, Karolinska Institutet, School of Medicine, S.B.L., Stockholm 1, Sweden.

Permanent address: Division of Infectious Diseases, Department of Medicine, University of Alberta, Edmonton, Alberta, Canada.

In 1973 Hall and Martin described six polypeptides of measles virus. The molecular weights were 75K, 69K, 60K, 53K, 51K and 45.7K. The 69K and 53K polypeptides were glycosylated completing a pattern similar to that of other paramyxoviruses (13). Concurrent with the above study Waters and Bussell (19) described six major polypeptides of measles with molecular weights of 78K, 71K, 61K, 53K, 45K and 37K in SDS-PAGE. Further studies indicated only the 78K polypeptide was glycosylated (2). These two studies differed in some molecular weight determinations, but most significant was the difference in the glycosylation patterns. In an attempt to clarify these discrepancies we analyzed the polypeptides of two strains of measles virus in SDS-PAGE. Both whole and partially degraded virus were studied under various labelling conditions to define the structural-functional relationships of measles polypeptides. In parallel with our studies, Mountcastle and Choppin (14) have analyzed four strains of measles virus. They have found six major polypeptides with molecular weights of 80K, 70K, 61K, 55K, 42K and 37K. Only the 80K polypeptide was glycosylated.

METHODS

The growth, purification and labelling of the measles virus with ^{35}S-methionine, ^{3}H-fucose and ^{3}H-glucosamine have recently been described (17). The protein sample preparation and SDS-PAGE system were similar to the method of Laemmli (7). We substituted Tris-H_3PO_4 for Tris-HCl in the buffer for the spacer gel. In protein samples containing small amounts of solubilized proteins, insulin was used for co-precipitation. In these samples ethanol was omitted from the washing procedure. The gel apparatus was similar to that described by Studier (16). The gels were impregnated with PPO for autofluorography using the method of Bonner and Laskey (1). The gels were dried using the method described by Maizel (11) prior to exposure to X-ray film for 72 hr at -70°.

RESULTS AND DISCUSSION

Slab gels of LEC and Edmonston strains of measles virus grown in either Vero or green monkey kidney cells gave the same polypeptide pattern. Six major polypeptides with molecular weights of 79K, 72K, 60K, 43K, 40K and 36K are seen (Fig.1). A number of minor polypeptides bands are also

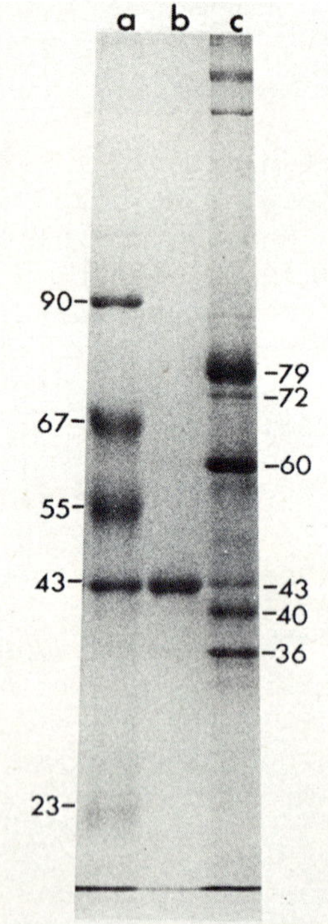

Fig.1. *Coomassie Brilliant Blue stained 10% acrylamide slab gel. a) Standard proteins: phosphorylase (90K), bovine serum albumin (67K), kappa chains (55K), actin (43K) and lambda chains (23K). b) actin. c) purified measles virions.*

present. However, most of these polypeptides comigrate with polypeptides seen in gels of uninfected Vero cells or appear in the autofluorograms of measles virus grown on Vero cells prelabelled with ^{35}S-methionine. On this basis these bands are considered to be cellular contaminants and will not be discussed.

The autofluorograms of ^{35}S-methionine labelled virus and virus grown on ^{35}S-methionine prelabelled cells are compared in Fig.2.The autofluorogram of labelled virus shows that

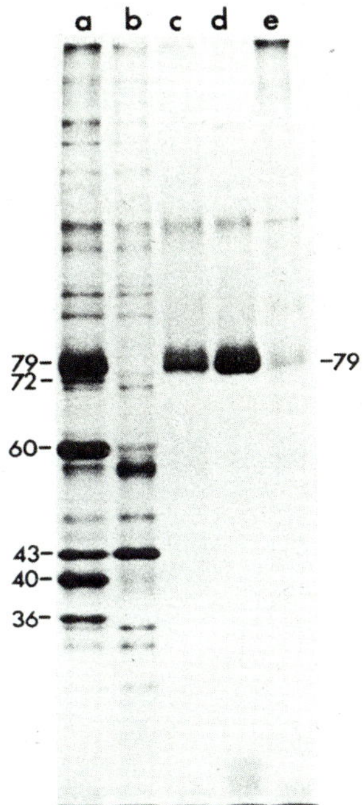

Fig.2. *Autofluorogram of 10% acrylamide slab gel. a) ^{35}S-methionine labelled virus. b) virus grown on ^{35}S-methionine prelabelled Vero cells. c) ^{3}H-glucosamine labelled. d) ^{3}H-fucose labelled. e) externally labelled glycoproteins.*

all six major polypeptides are heavily labelled. In contrast, only the 43K polypeptide is heavily labelled in virus grown on prelabelled cells. This polypeptide band comigrates with cellular actin. Actin is present in the polypeptide analysis of other enveloped viruses (3, 18), Sendai virus (8) and measles virus (14). Actin is sensitive to cleavage by trypsin (9). The actin band remains unaltered when purified measles virions are treated with trypsin or nonionic detergent (see below). On this basis we believe the actin identified in purified measles virions is probably a structural component inside the virion rather than a surface contaminant.

When measles virus is grown in medium containing ^{3}H-fucose or ^{3}H-glucosamine the 79K polypeptide is the only measles protein labelled (Fig.2). We have also used the method of Luukkonen *et al.* (10) to externally label the glycoproteins of purified virus. The 79K polypeptide is the only polypeptide of measles labelled by this method (Fig.2).

Treatment of measles virus with trypsin removes the surface projections and prevents the induction of hemagglutination-inhibiting (HI) antibodies (15). We treated purified virions with 0.01% trypsin for 2 hr at 37°. The solubilized proteins and trypsin were separated from the sedimentable particles on discontinuous cesium chloride gradients (17). Comparison of untreated and trypsin-treated virions shows that the 79K polypeptide is removed by trypsin treatment (Fig.3A). The other

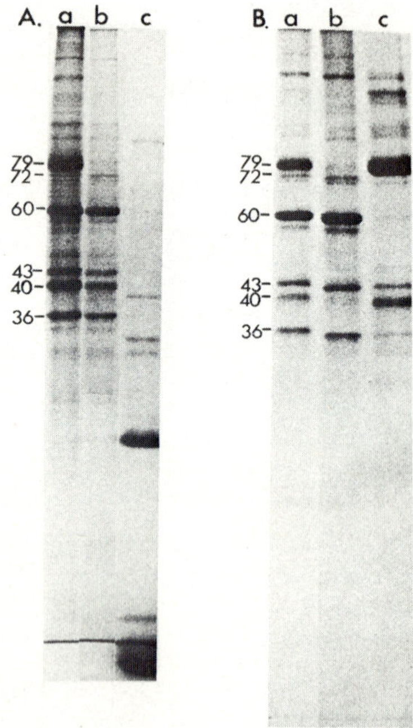

Fig.3. *Coomassie Brilliant Blue stained 15% acrylamide slab gel. A. Trypsin treatment. a) untreated virus. b) sedimentable particle after trypsin. c) solubilized proteins after trypsin. B. Cutscum treatment. a) untreated virus. b) sedimentable particle after Cutscum. c) solubilized proteins after Cutscum.*

measles polypeptides remain unchanged after trypsin treatment. When considered with the previous morphological and serological data the results in SDS-PAGE confirm that the 79K glycosylated polypeptide is the hemagglutinin antigen (HA).

Although trypsin treatment reduces the ability of measles virus to induce HI antibodies, it enhances the ability of the virus to induce hemolysin-inhibiting HLI) antibodies (15). We treated purified measles virions with 0.1% Cutscum+ at 22° for 15 min and separated the solubilized proteins from sedimentable particles. Fig.3B shows that the 79K and 40K polypeptides are removed by this treatment. The 40K polypeptide remains with the viral particle after trypsin treatment, but is removed by Cutscum treatment indicating that it is an envelope protein. Since surface proteins are usually glycosylated, it is surprising that the 40K polypeptide is a nonglycosylated surface protein. We believe the 40K polypeptide is the protein component responsible for cell fusion and hemolysin activities. Graves *et al.* (4) found that if cell fusion was inhibited in measles infected Vero cells, two measles specific glycoproteins, 80K and 62K, were evident in SDS-PAGE of infected cells. In the same study a 40K polypeptide appeared in pulse-chase experiments. Possibly the protein associated with cell fusion and hemolysin function is transported to the surface of the infected cell as the 62K glycosylated protein. On maturation of the virus, this surface protein could be cleaved to the 40K nonglycosylated protein. This would explain both the pulse-chase experiments and the fact that the only purified measles virion, unlike other paramyxoviruses, has only one glycosylated surface protein.

The 60K polypeptide is the most abundant protein and has previously been identified as the nucleocapsid protein (2, 5, 14). The 36K polypeptide has previously been identified as the membrane (M) protein (2, 14).

Polypeptides of approximately 53K and 70K have been identified in the previous studies of measles virus. We identified a polypeptide of 72K, but its function remains unclear. Hall and Martin (5) identified two glycoproteins at 69K and 53K in measles. Both glycoproteins were readily removed by bromelain or Tween 20 treatment (6). However both Bussel *et al.* (2) and Mountcastle and Choppin, (14) found only a small amount of nonglycosylated protein at 53K. We could not detect a measles polypeptide in the 53K region in our studies.

In summary, our analysis of purified measles virion in SDS-PAGE shows six major measles polypeptides. The 79K and 40K polypeptides are envelope proteins. The 79K polypeptide

---

+Registered trade mark Fischer Scientific Co.

is the HA antigen and is the only glycosylated polypeptide. The 40K polypeptide is probably associated with cell fusion and hemolysin functions. The 60K and 36K polypeptides have previously been identified as the NP and M proteins. The 72K polypeptide is an internal protein, but its role has not been determined. The 43K polypeptide appears to be cellular actin incorporated into the virus during assembly.

ACKNOWLEDGEMENTS

This research was supported by a grant from the Swedish Medical Research Council (Project No. B18-16X-00116). David Tyrrell held a Centennial Fellowship from the Medical Research Council of Canada. The authors thank Drs W.E. Mountcastle and P.W. Choppin for providing a copy of their manuscript prior to its appearance in press. We thank Anita Bergström for skilled technical assistance.

REFERENCES

1. Bonner, W.M. and Laskey, R.A. (1974). *Eur. J. Biochem.* 46, 83.
2. Bussell, R.H., Waters, D.J., Seals, M.K. and Robinson, W.S. (1974). *Med. Microbiol. Immunol.* 160, 105.
3. Fleissner, E. and Tress, E. (1973). *J. Virol.* 11, 250.
4. Graves, M.C., Silver, S.M. and Choppin, P.W. (1977). *Abstr. Am. Soc. Microbiol.*, 283.
5. Hall, W.W. and Martin, S.J. (1973). *J. Gen. Virol.* 19, 175.
6. Hall, W.W. and Martin, S.J. (1974). *J. Gen. Virol.* 22, 363.
7. Laemmli, U.K. (1970). *Nature* 227, 680.
8. Lamb, R.A., Mahy, B.W.J. and Choppin, P.W. (1976). *Virology* 69, 116.
9. Lazarides, E. and Lindberg, U. (1974). *Pro. Nat. Acad. Sci. USA* 71, 4742.
10. Luukkonen, A., Gahmberg, C.G. and Renkonen, O. (1977). *Virology* 76, 55.
11. Maizel, T.V. (1971). *In* "Methods in Virology" (K. Maramorosch and H. Koprowski, eds.), Academic Press, New York.
12. Mountcastle, W.E., Compans, R.W., Caliguiri, L.A. and Choppin, P.W. (1970). *J. Virol.* 6, 677.
13. Mountcastle, W.E., Compans, R.W. and Choppin, P.W. (1971). *J. Virol.* 7, 47.
14. Mountcastle, W.E. and Choppin, P.W. (1977). *Virology* 78, 463.
15. Norrby, E. and Gollmar, Y. (1975). *Infection and Immunity* 11, 231.
16. Studier, F.W. (1973). *J. Mol. Biol.* 79, 237.
17. Tyrrell, D.L.J. and Norrby, E. (1977). *Virology* (submitted).

18. Wang, E., Wolf, B.A., Lamb, R.A., Choppin, P.W. and Goldberg, A.R. (1976). *In* "Cold Spring Harbor Conference on Cell Motility". Cold Spring Harbor Laboratory 3, 589.
19. Waters, D.T. and Bussell, R.H. (1973). *Virology* 55, 554.

15

BIOCHEMICAL COMPARISON OF MEASLES AND SUBACUTE SCLEROSING PANENCEPHALITIS (SSPE) VIRUSES

W.W. HALL, W.R. KIESSLING, V. ter MEULEN

Institute for Virology and Immunobiology, University of Würzburg, Versbacher Landstrasse 7, 8700 Würzburg, W. Germany.

The pathogenicity of subacute sclerosing panencephalitis (SSPE) is still not understood despite the fact that a measles-like virus (referred to as SSPE virus) has been demonstrated in SSPE brain material and in some cases isolated from brain tissue cultures (12). The epidemiological observations of the rural distribution of this disease and the rarity of its occurrence cannot be linked directly to the ubiquitous measles virus infection. Therefore both a genetic predisposition and an immunologic defect have been postulated to explain this disease. However, no laboratory evidence has supported either possibility. The failure to demonstrate a host factor responsible for this disease has turned the interest to studies on the infectious agent associated with this neurological illness.

Biochemical studies have recently shown that SSPE viruses differ from measles virus in their genomic RNA (10). Moreover, the results indicated that the SSPE strains may have additional information over that contained in the measles virus. In order to characterize the observed differences further, and to localise them, SSPE and measles viruses were compared in two further ways. The total antigenicity of the viruses was studied by an indirect radioimmunoassay and the patterns of messenger RNA synthesis in infected cells were analysed. The results obtained demonstrate that the viruses are antigenically distinct and that the differences are probably associated with the virus membrane protein.

ANTIGENIC COMPARISON OF SSPE AND MEASLES VIRUSES

It is likely that any differences in the genetic information of SSPE and measles virus should ultimately be expressed in the comparative antigenic content of these viruses. Therefore SSPE, measles and canine distemper viruses grown in Vero cells were purified and used to immunize rabbits. After three consecutive antigen applications at 10 day intervals the immune sera were submitted to standard serological assay. In the neutralization test (NT) a high antibody activity was present against the homologous virus. Since several studies have shown that no significant differences were detectable between SSPE and measles virus strains in biological assays such as neutralization, hemagglutination inhibition and hemolysin inhibition tests the three viruses and their homologous antisera were submitted to a binding-inhibition test based on an indirect radioimmunoassay. In this assay an appropriate dilution of a specific antiviral serum is mixed with different concentrations of an inhibiting antigen and incubated at 37°C for one hour. After the incubation period an aliquot of this mixture is added to microtiter wells, coated with a specific viral antigen and control antigen. The amount of antiviral antibody which has not reacted with the inhibitor is determined by the indirect radioimmunoassay (IRIA) (17). Fig.1 summarizes the results obtained for measles virus (Edmonston), SSPE (Lec) and Canine Distemper viruses. In the Edmonston system the Edmonston virus bound at a high concentration all anti-Edmonston antibodies, whereas the SSPE (Lec) bound approximately 95% and CDV inhibited only to 30%. In the SSPE Lec system only Lec virus was capable of removing completely all anti-SSPE antibodies. The Edmonston virus inhibited up to 90% despite an antigen excess. It is of interest to note that canine distemper virus reacted with anti-Lec antibodies to a higher degree than with antimeasles antibodies. In the canine-distemper system the homologous virus only eliminated at higher concentration anti-canine distemper virus antibodies whereas the Lec virus inhibited up to 50%, and Edmonston virus only up to 25%.

The degree of inhibition of these viruses and one other measles and SSPE virus strain in the homologous or heterologous antigen-antibody systems is summarized in Table 1. The results show the high level of relatedness of these viruses. It is obvious that measles and SSPE virus do have small distinct antigenic differences which prevent these viruses giving 100% inhibition in the heterologous system. The results also provide some evidence that the SSPE antigens are capable of stronger binding to the measles antisera than the binding of measles antigens to SSPE sera. However, whether the differences are

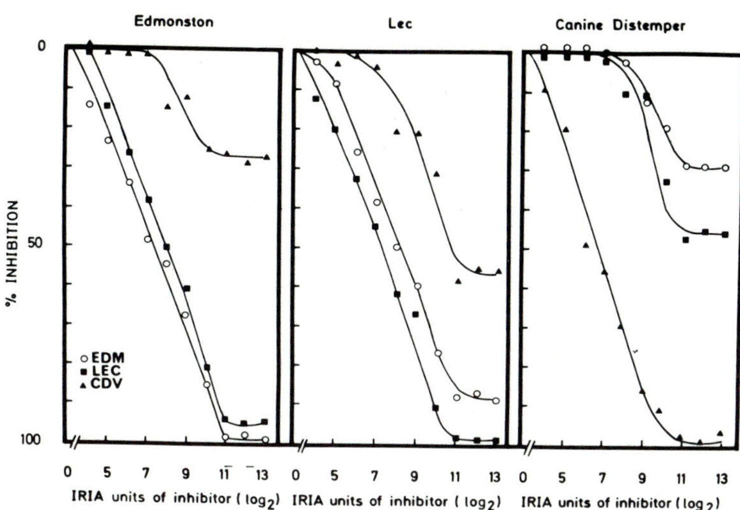

Fig.1. *Antigenic comparison of measles, SSPE and canine distemper viruses using a binding inhibition test. Details are as described by Yung et al. (1977) and are given in the text.*

significant remains to be established. In addition it is noteworthy that SSPE virus USC seems to be antigenically different from the Lec strain whereas the measles virus strain Woodfolk reacts like the Edmonston strain. This result provides preliminary evidence that although SSPE viruses are antigenically distinct from measles virus, it may also be possible that individual SSPE viruses are antigenically distinct. However, the greatest antigenic differences between SSPE and measles virus can be seen in the canine distemper system. Despite a relative high standard deviation the two SSPE viruses are antigenically more related to canine distemper virus than the two measles virus strains. This finding is in accordance with serological studies carried out using neutralization tests. Sato *et al.* (14) found that SSPE serum contained proportionally higher levels of neutralizing antibodies to canine distemper virus than sera isolated after a regular measles infection.

The observed immunological differences raise the question where the antigenic differences are located in the virus particle. So far standard serological assays using viral antigens with biological activities failed to reveal antigenic differences between SSPE and measles viruses. It is therefore suggested that the differences are probably located in a viral protein which is biologically inactive.

TABLE 1

Comparative Antigenicity of Measles and SSPE Viruses
Summary of Binding Inhibition Tests

Anti-Serum	Measles Viruses EDM	Measles Viruses WOODF	SSPE Viruses LEC	SSPE Viruses USC	Canine Distemper Virus
EDM	99.2 ± 1.3 n = 5	99.2 ± 1.3 n = 5	95.0 ± 5.2 n = 5	91.5 ± 4.8 n = 6	27.6 ± 10.0 n = 9
LEC	88.2 ± 4.3 n = 5	92.4 ± 5.6 n = 5	99.4 ± 1.3 n = 5	89.7 ± 7.8 n = 8	56.5 ± 7.3 n = 7
CDV	29.0 ± 14.9 n = 3	31.7 ± 19.5 n = 4	45.6 ± 19.6 n = 3	34.6 ± 15.1 n = 3	98.8 ± 1.1 n = 5

Table 1. *Summary of the comparative antigenicity of measles, SSPE and canine distemper viruses. n = number of individual experiments.*

COMPARISON OF mRNA SYNTHESIS IN INFECTED CELLS
GEL ELECTROPHORETIC ANALYSIS OF MEASLES VIRUS 12s to 36s mRNAs

Measles virus mRNAs (12s to 36s) isolated after sucrose gradient sedimentation (11) were subjected to electrophoresis on 2.5% polyacrylamide gels under denaturing conditions (97% formamide and 6 M Urea) (16). A densitometer tracing of a typical fluorogram obtained is shown in Fig.2a. It can be

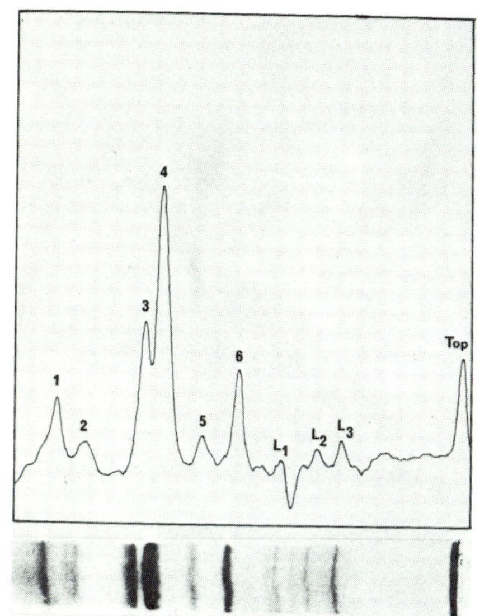

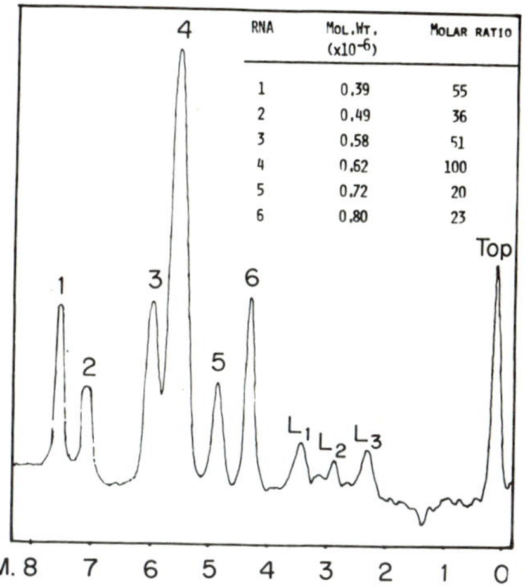

Fig.2. *Gel electrophoresis of measles virus 12s to 36s mRNAs on 2.5% polyacrylamide gels in the presence of formamide and urea. (a) total 12s to 36s mRNAs, (b) poly (A) rich mRNAs selected by oligo d(T) cellulose chromatography. Shown are densitometer scans of prepared fluorograms. Mol. wts. of the RNAs were calculated relative to 16s, 18s and 23s ribosomal RNAs. In the case of the 28s mRNA the relationship between log. mol. wt. and migration was no longer linear. Molar ratios were normalised to 100 units of RNA 4.*

seen that six major RNAs can be clearly resolved having molecular weights of (1) 0.39×10^6, (2) 0.49×10^6, (3) 0.58×10^6, (4) 0.62×10^6, (5) 0.72×10^6, and (6) 0.80×10^6 daltons. In addition, at least three minor bands (designated L_1, L_2 and L_3) were usually present and these had molecular weights in the range of approximately 2×10^6 daltons (Fig.2b). At the present time it is assumed that these are probably aggregation products but one may contain or represent a large mRNA species since measles virus contains and synthesizes a unique large protein (mol. wt. 200,000) in infected cells (see chapter 17; Graves, Silver and Choppin, personal communication). Electrophoresis of poly(A)-rich RNA as selected by oligo d(T) cellulose chromatography also shows that the six major RNAs and the L_1, L_2 and L_3 species are also present (Fig.2b). In a large number of analyses it could be seen that although the RNAs are synthesized in unequal amounts, the relative quantity of each individual RNA remains approximately the same throughout the virus growth cycle. The only exceptions are RNA 1 which tends to have a slightly decreased rate of synthesis at the later stages of infection and RNA 5 , which in some experiments showed a slight increase. In contrast, however, when input multiplicity is varied it can be shown that the relative amounts of the RNAs will differ considerably from infection to infection without apparently following any definite pattern (data not shown). The reasons for these variations are not known but they have also been observed with the synthesis of measles virus polypeptides in infected cells (Rima and Martin, personal communication). Attempts to correlate the individual mRNAs with the polypeptides in the purified virus have met with mixed success. The electrophoretic patterns of the virus polypeptides obtained on 7.5% polyacrylamide gels are shown in Fig.3. In both viruses 6 major polypeptides were found and their mol. wts. and chemical composition are shown in Table 2. The patterns obtained are quite different from those previously reported by one of us (7-9) but are similar to those reported by other groups (5, 13, chapter 17). The reasons for the differences are at present still not known. On the bases of only mol. wts. we can correlate only 4 of the mRNAs with virus polypeptides (Table 2). An additional problem is polypeptide 5, which has a mol. wt. of 42,000 and may be host cell actin (13). However, workers at the Rockefeller University (Graves, Silver and Choppin) have shown that a second glycoprotein (fusion protein F_0) is synthesized in measles virus-infected cells and that this is cleaved to form two products, one unglycosylated which has a mol. wt. of 40,000 (F_1) and a smaller glycosylated peptide of mol. wt. 12 to 15,000 (F_2). The uncleaved product migrates in polyacrylamide gels in the 62,000 dalton range, and we feel therefore that one could probably assign RNA 3 as the mRNA for the fusion glycoprotein. We have not attempted to correlate mRNA 2 to protein 5 at the present time since it has not been established if the latter is unique or represents a degradation product of one of the other polypeptides. The possibility also remains that this mRNA could also be a template for an intracellular nonstructural protein. However, it must be firmly stated that an exact correlation of

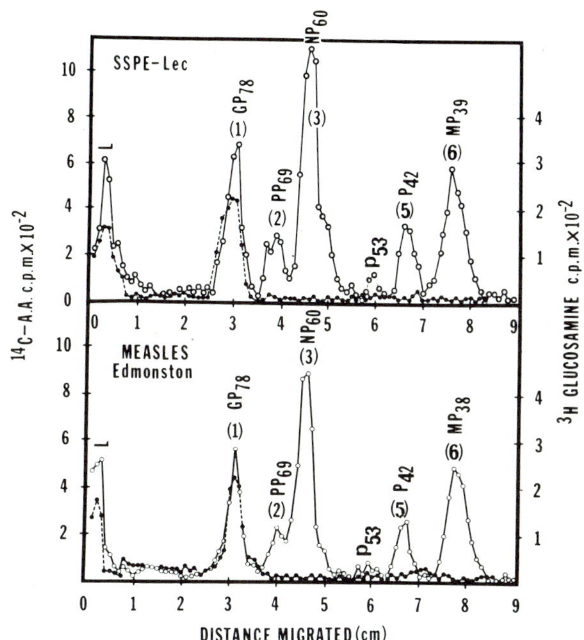

Fig.3. *Gel electrophoresis analyses on 7.5% SDS polyacrylamide gels of the polypeptide composition of measles and SSPE viruses.* ●——● ^{3}H*-glucosamine, cpm;* o——o ^{14}C*-amino acid, cpm. Migration is from left to right.*

any of the species is very tentative and will not be known until individual *in vitro* translation experiments are carried out.

MODE OF TRANSCRIPTION OF MEASLES AND SSPE VIRUSES

Recent studies with VSV and Sendai viruses have indicated that the genomes of these viruses may be organized in the form of a single transcriptional unit (1-3, 6). It has been proposed that a single promotor site is located at or near to the 3' end, that the entire complementary RNAs are synthesized as one complete molecule and that these may be subsequently cleaved to yield the individual mRNA species.

We have also attempted to determine if such a mechanism exists during the transcription of measles and SSPE viruses *in vivo*. In this report only experimental details of the measles virus experiments will be shown but identical results have been obtained for the Lec strain of SSPE virus. The methods used

TABLE 2

Molecular Weights of Measles Virus mRNAs and Virus Polypeptides

RNA (x 10^{-6})*		Polypeptide	
1: 0.39		6: 37,000	Membrane
2: 0.49	← →	5: 42.000	Unknown
3: 0.58		4: 52,000	
4: 0.62		3: 60,000	Nucleocapsid
5: 0.72		2: 69,000	Phosphoprotein
6: 0.80		1: 78,000	Glycoprotein

*Corrected for Poly(A) content.

Table 2. *Correlation of virus mRNAs and polypeptides. Mol. wts. of the virus polypeptides were calculated relative to those of Sendai virus.*

and the mathematical theory involved are those recently described by Glazier *et al*. (6) for Sendai virus. These workers have shown that the rate of inactivation of transcription relative to that of replication by UV irradiation may be used to decide whether a single or multiple promoter sites are involved in RNA transcription. Measles virus-infected Vero cells were irradiated with increasing quantities of UV light, and the intracellular RNAs were isolated and analysed by sucrose gradient sedimentation. By calculating the total ^3H-uridine incorporated into the 12 to 36s RNAs (transcription) compared to the 46 to 54s RNAs (replication) it was possible to measure their relative inactivation rates. The values obtained from four individual experiments showed that transcription was inactivated at approximately 0.5 the rate of replication, a value close to what would be expected for a single promoter model and the synthesis of 7 mRNA species (6).

Gel electrophoretic analyses of the individual transcription products after increasing doses of UV irradiation provided further evidence for a single promotor site and results obtained are shown in Fig.4 (a to h). It can be seen that the UV sensitivity of the individual transcripts did not always bear a direct relationship to their mol. wts. The most dramatic example is RNA 4 which was strongly resistant to irradiation even at levels which had reduced the total transcription by approx.

99% (600 and 1200 ergs/mm^2; Fig.4 (g,h)). Attempts were then made to calculate the individual inactivation rates of each RNA. However, several difficulties were encountered. The large species (L_1, L_2 and L_3) were found to be almost completely inactivated even after minimal doses of irradiation. A second problem was to load a sufficient or the same amount of radioactivity onto each gel slot. This proved to be difficult in the higher irradiation dose experiments, where large amounts of RNA had to be used and this tended to increase aggregation or "sticking" on top of the gel. It was also not possible in some cases to completely resolve RNA 3 from RNA 4 due to their closely similar migration. We feel that in some cases this may have led to a lower estimation of the inactivation rate than really exists for RNA 3. Fig.5 therefore shows the inactivation rates of the individual mRNAs (with the exception of the L's) within as much accuracy as the system would allow. Calculation of UV irradiation values required to inactivate each mRNA to a level of 37% showed that these were not inversely proportional to their mol. wts. (4, 6) as would be expected in the case of a multiple promotor system. We therefore propose that measles transcription, like that of VSV and Sendai virus, proceeds *via* a single promotor site and the gene order of the RNAs is as follows

3' - 4-(1/3) - 2 - (5/6) - L - 5'

This order presumes the existence of an L mRNA, and that its almost immediate inactivation indicates a position at one end of the genome.

Recently Villareal *et al.*(16) showed that the relative molar amounts of VSV mRNAs and proteins in infected cells corresponded to the gene order derived from UV inactivation experiments. A consideration of the relative molar ratios of the measles virus mRNAs (Fig.2b) shows that this may also be the case in the present system, and indicates that a polar gradient of transcriptional efficiency may occur. At the present time we have preferred not to make any firm decision on the relative location of RNAs 1 and 3 and 5 and 6 until more detailed experimental methods can be employed. Assigning tentative polypeptides to the individual RNAs would give a gene order

$$3' - (NP_{60}) - (MP_{38}/F_o) - (\underset{2}{RNA}) - (PP_{69}/GP_{78}) - (L) --- 5'$$

Comparison with the gene order of other paramyxoviruses show approx. the same sequence, i.e., 3' - NC - F_o - M ... which may indicate some common origin in paramyxovirus evolution.

COMPARISON OF mRNAs SYNTHESIZED IN MEASLES AND SSPE INFECTED CELLS

Fig.6 shows densitometer tracings of the fluorograms of the

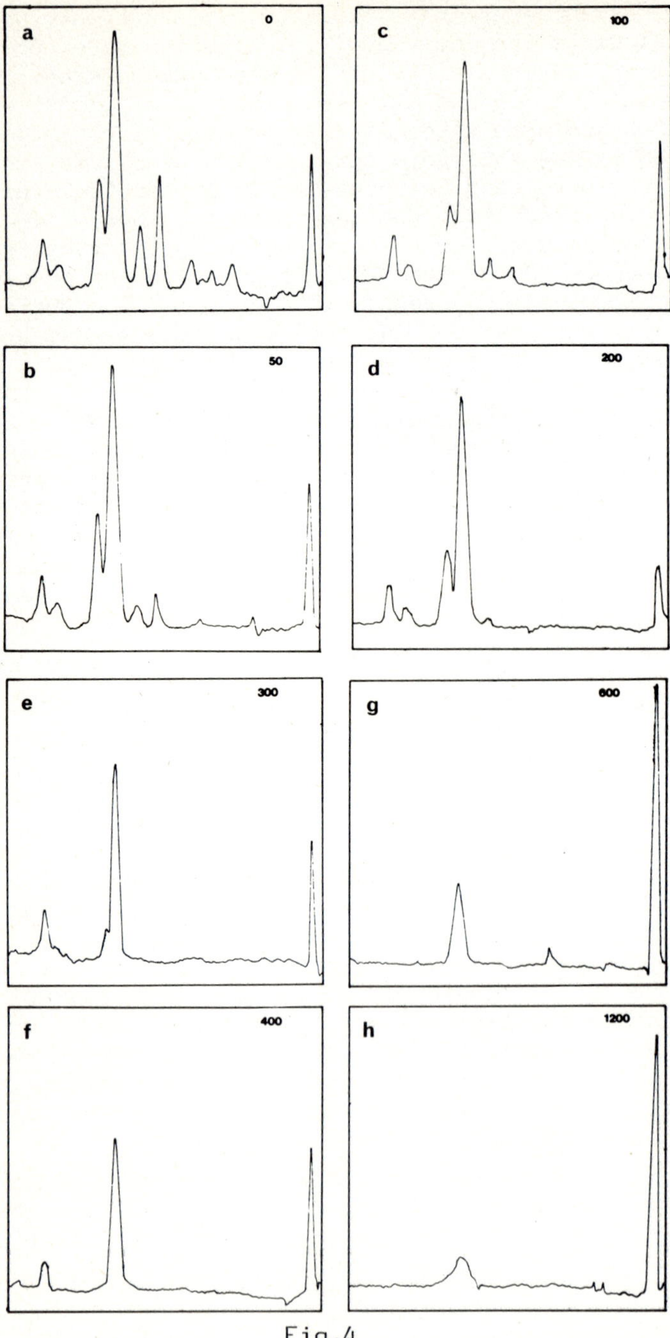

Fig.4

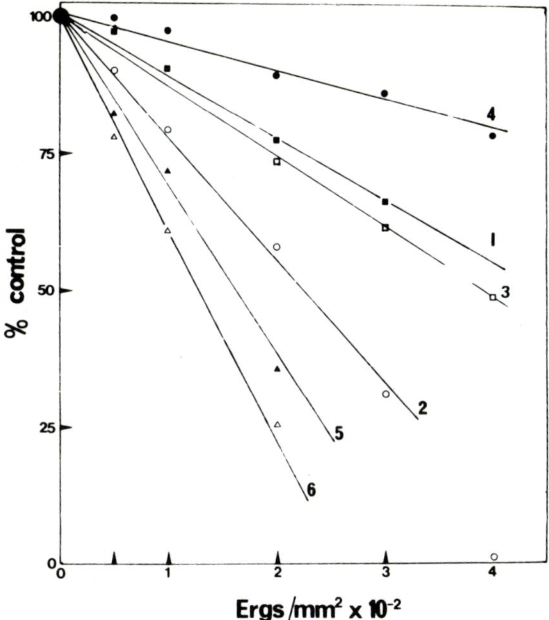

Fig.5. *Inactivation of the individual transcripts of measles virus mRNAs by UV irradiation. Each plot represents the % cpm in each RNA band relative to an unirradiated sample as in Fig. 4.*

total 12s to 36s RNAs synthesized in Vero cells infected with two measles viruses (Edmonston and Woodfolk) and two SSPE viruses (Lec and USC). The similarity in the migration of the six major RNAs is striking, although RNAs 3 and 4 are not always completely resolved. The one minor but important difference observed was that the smallest RNA (RNA 1) of the SSPE viruses was seen to migrate slower than the corresponding RNAs of the two measles viruses. The significance and relevance of this difference would not seem to be important if it were not for the fact that the differences expected between the viruses are very small and probably occur in only one part of the genome. However, more important have been the early ob-

---

Fig.4. *Inactivation of individual mRNA species by UV irradiation. Doses in ergs/mm^2 are shown and RNAs were examined by gel electrophoresis. Shown are densitometer tracings of prepared fluorograms. (a) 0, (b) 50, (c) 100, (d) 200, (e) 300, (f) 400, (g) 600, (h) 1200 ergs/mm^2.*

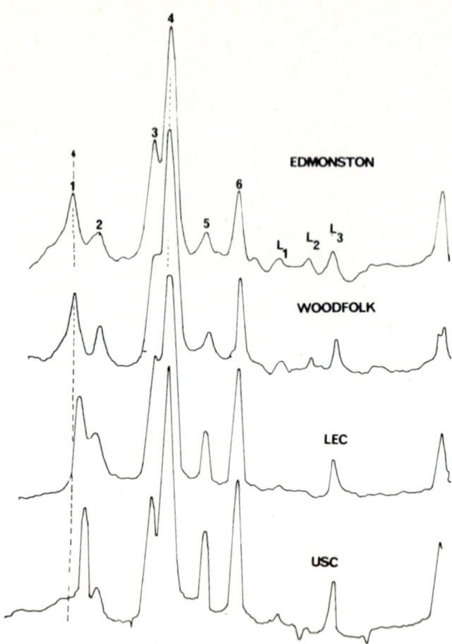

Fig.6. *Densitometer tracings of fluorograms of measles virus (Edmonston and Woodfolk) and SSPE (Lec and USC) 12s to 36s mRNAs.*

servations of Schluederberg *et al.* (15) and their confirmation and extension by Wechsler *et al.* (chapter 17) that the M protein of SSPE viruses is larger (migrates more slowly) than the M protein of measles virus in polyacrylamide gels. Since the M protein of measles virus is the smallest major protein of the virus (mol. wt. 37 - 38,000) and RNA 1 is the smallest mRNA species (corrected mol. wt. 390,000) we feel that this RNA may be the messenger species for the M protein. This being the case it would appear that the difference between the viruses is real and they lie in that part of the genome coding for the M protein.

CONCLUSION AND SUMMARY

The present report has described experiments which have attempted to increase our understanding of the biochemistry of measles virus and to use the information obtained to attempt to decide whether any differences exist between the viruses of measles and SSPE.

Using a competition-hybridization technique we have previously shown that there are small differences in genome homology between the viruses. However, the results tended to fall "one-way", that is, it appeared that the SSPE viruses were not only different but also could have some additional information over and above that contained in measles virus. In order to determine if this is the case and to establish the exact location of the differences we compared the total antigenicity of the viruses using a solid-phase radioimmunoassay. These experiments again provided similar results in that small differences in antigenicity (5 to 10%) did exist and these are probably associated with a biologically nonactive protein. Attempts were made to look in detail at the patterns of mRNA synthesis in infected cells. This was based on the fact that if a difference in the genetic content of the viruses really existed then it may be possible to see this in at least one of the transcription products. The mRNAs of measles and SSPE viruses sediment at 12s to 36s with a major peak at 18s in aqueous sucrose gradients. Like the vast majority of mRNAs in nature they contain poly(A) sequences at their 3' end (11).

Transcription of both of the viruses apparently proceeds *via* a single promotor site and the tentative gene order has been shown to be the same for each virus. However, the one difference observed was that the smallest mRNA, probably the message coding for the virus membrane (M) protein migrated slower in the SSPE virus systems tested. These findings are consistent with observations by other workers that the M protein of SSPE viruses is larger than its counterpart in measles and we feel that this is where the difference lies between the two systems. In an accompanying report (chapter 77) we shall provide evidence that in fact the membrane proteins of measles and SSPE viruses are antigenically distinct.

The obvious questions that follow are how has this difference occurred and what is its significance when one considers the etiology of SSPE itself. One possible answer is that the SSPE viruses have arisen from measles virus as a result of a modification or mutation in that part of the genome coding for the membrane protein. Whether this mutation occurs in the host or whether the host is infected by both wild type and mutant virus cannot be established definitely at present. However, it is interesting to consider the implication of what could happen during such a mutation process. It is possible that during the mutation the production of a non-functioning or defective M protein would occur and this could not maintain its normal function as a controller of virus assembly. This in turn could result in both the initiation and maintenance of a "non-productive" or persistent infection in that no progeny virus could be assembled. Moreover, the inability and general failure (~ 80%) to isolate SSPE viruses from brain tissue despite the fact that the cells are fully infected and contain large quantities of virus "antigen" could be explained. Such a defect could provide a simple explanation for measles virus persistence without the need to involve more complicated mechanisms, for example the involvement of defective or sub-genomic RNAs.

ACKNOWLEDGEMENT

This paper is supported by the Deutsche Forschungsgemeinschaft Schwerpunkt für "Multiple Sklerose und verwandte Erkrankungen" Az.: Me 270/16.

REFERENCES

1. Abraham, G. and Banerjee, A.K. (1976). *Proc. Nat. Acad. Sci. USA*, 73, 1504.
2. Ball, L.A. (1977). *J. Virol.* 21, 411.
3. Ball, L.A. and White, C.N. (1976). *Proc. Nat. Acad. Sci. USA*, 73, 442.
4. Brautigram, A.R. and Säuerbier, W. (1974). *J. Virol.* 13, 1110.
5. Bussell, R.H., Waters, D.J., Seals, M.K. and Robinson, W.S. (1974). *Med. Microbiol.* 160, 105.
6. Glazier, K., Raghow, R. and Kingsbury, D.W. (1977). *J. Virol.* 21, 863.
7. Hall, W.W. and Martin, S.J. (1973). *J. Gen. Virol.* 19, 175.
8. Hall, W.W. and Martin, S.J. (1974a). *J. Gen. Virol.* 22, 363.
9. Hall, W.W. and Martin, S.J. (1974b). *Med. Microbiol. Immunol.* 160, 143.
10. Hall, W.W. and ter Meulen, V. (1976). *Nature* 264, 474.
11. Hall, W.W. and ter Meulen, V. (1977). *J. Gen. Virol.* 35, 497.
12. Meulen, V. ter, Katz, M. and Müller, D. (1972). *Curr. Top. Microbiol. Immunol.* 57, 1.
13. Mountcastle, W.E. and Choppin, P.W. (1977). *Virology* 78, 463.
14. Sato, T.A., Yamanouchi, K. and Shishido, A. (1973). *Archiv. für ges. Virusforschung* 42, 36.
15. Schluederberg, A., Chavanich, S., Lipman, M.B. and Carter, C. (1974). *Biochem. Biophys. Res. Commun.* 58, 647.
16. Villareal, L.P., Breindl, M. and Holland, J.J. (1976). *Biochemistry* 15, 1663.
17. Yung, L.L., Loh, W. and ter Meulen, V. (1977). *Med. Microbiol. Immunol.* 163, 111.

SUBACUTE SCLEROSING PANENCEPHALITIS: AN ABORTIVE INFECTION BY A MEASLES-LIKE VIRUS

LARRY ERON, JUDY A. SPRAGUE, PAUL ALBRECHT, RUTH C. DUNLAP, JOHN T. HICKS and GURMIT S. AULAKH

Department of Health, Education and Welfare, Public Health Service, Food and Drug Administration, Bureau of Biologics, Division of Virology, Bethesda, Maryland, 20014, U.S.A.

Subacute sclerosing panencephalitis (SSPE) is a rare disease of children that emerges 5 to 7 years after a child has had an uneventful measles infection. The disease starts insidiously with behavioural disorders and terminates fatally with progressive stupor, dementia, and motor disorders. Patients with SSPE have high titers of antibody to measles in their cerebrospinal fluid, and measles antigen has been demonstrated in their brains by immunofluorescence (4, 21). Furthermore, measles-like viruses have been isolated from their brains (12, 18).

Since SSPE is a rare event with a prolonged latent period following a measles infection, it is likely that additional virus and/or host factors determine its occurrence. We undertook an examination of the properties of an SSPE isolate, IP-3, which continued to be cell-associated in tissue culture and which produced a chronic encephalitis in monkeys that appeared histologically and clinically similar to SSPE in man (2).

THE VIRION PARTICLE

Vero cells infected either by a vaccine-attenuated strain of measles virus, or by IP-3, an SSPE isolate, were harvested when at least 90% of the cells showed a cytopathic effect. Titers of infectious virus were determined by plaque titration of cell homogenates (cell-associated virus) and of culture medium (released virus). While measles produced 100 infectious particles per cell, 50% to 90% of which remained cell-associated, IP-3 formed 4 logs less of infectious virus, essentially all of which was cell-associated.

The buoyant densities and electron microscopic appearance

of these virus particles were then examined. By ultracentrifugation in density gradients of sucrose, we determined that the measles infectious particle banded at 1.23 g/cm^3, as reported previously (7), but the IP-3 infectious particle was slightly less dense (1.22 g/cm^3). The virion particles of measles possessed the typical paramyxovirus projections of the cell-surface thought to be hemagglutinin (Fig. 1a). However, the IP-3 particles seem to be devoid of these structures (Fig. 1b), except for rare particles (Fig. 1c).

THE STRUCTURAL PROTEINS

The structural proteins of measles virus were analyzed by SDS-polyacrylamide gel electrophoresis. Measles possessed four major and four minor polypeptides (Fig. 2). The molecular weights of these proteins were calculated by coelectrophoresis with known standards and shown to be 88,000 daltons for the glycoprotein (G), 80,000 daltons for VP_3, 69,000 daltons for the phosphoprotein (P), 58,000 daltons for the nucleocapsid (NC), 55,000 daltons for VP_6, 52,000 daltons for VP_7, 46,000 daltons for VP_8, and 38,000 for the matrix protein (M).

In contrast to measles, IP-3-infected cells contained almost entirely nucleocapsid (Fig. 2). A second polypeptide (VP'_1)

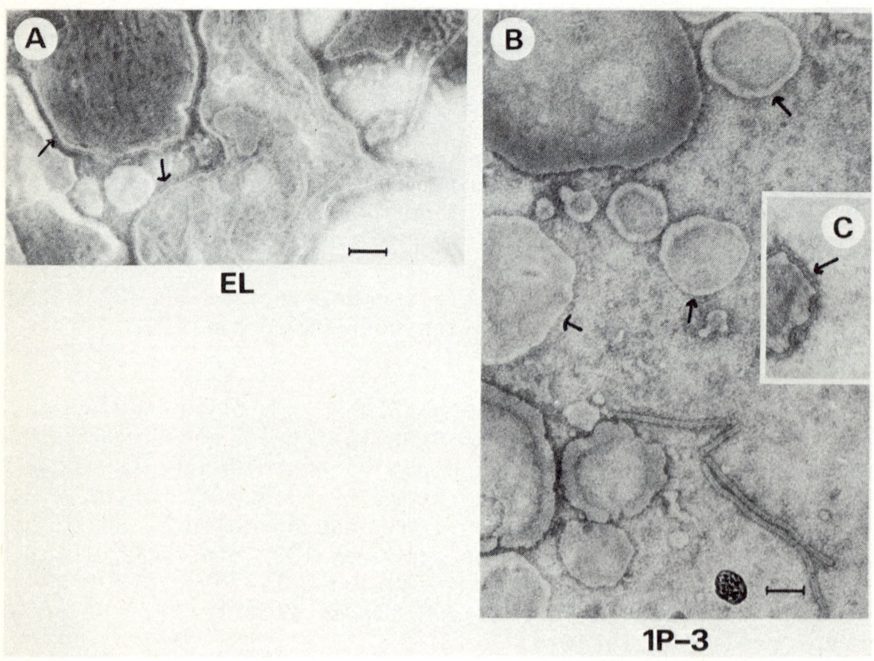

Fig.1

was visible and migrated more rapidly than the measles glycoprotein (G) on SDS-polyacrylamide gel electrophoresis. In addition, the IP-3 nucleocapsid (NC') appeared to migrate slightly more rapidly than measles NC.

THE NUCLEOCAPSIDS OF IP-3 AND MEASLES

To compare the native nucleocapsids of measles and IP-3, we analyzed their antigenic relatedness and size. Native nucleocapsids of these two viruses were purified and incubated with antisera prepared against measles nucleocapsid and virion particles. Immunoprecipitates were analyzed by SDS-polyacrylamide gel electrophoresis (Fig. 3). Both IP-3 and measles nucleocapsids were precipitated by measles antisera.

Size determination of the native nucleocapsids were made by rate-zonal sedimentation of nucleocapsids through sucrose density gradients. The S values of measles and IP-3 nucleocapsids were both 200S (Fig. 4). The attenuated measles virus nucleoprotein contained a second smaller peak at about 140S. There was no IP-3 nucleocapsid sedimenting at 110S, as reported for defective measles particles (8, 14).

In addition, electron microscopic measurements of the nucleocapsids were performed, and confirmed the size similarity. The

---

Fig. 1. *Electron micrographs of measles and IP-3 virion particles. A. Purified measles particles. The attenuated strain of measles virus was derived from a commercial vaccine (Eli Lilly & Co., Indianapolis, Indiana) and passaged twice in Vero cell cultures. Nucleocapsid strands appear inside the virion. Hemagglutinin spikes are visible on the periphery of particles (arrows). B. Purified IP-3 particles. This SSPE strain IP-3 was isolated by Dr. T. Burnstein from the brain biopsy of a 14-year-old child (3). It received 22 cell culture passes followed by one passage in rhesus monkey brain (2). Since its isolation from human brain the virus maintained its cell-restricted, non-productive character of growth in cell cultures. It spreads by cell fusion, gradually involving the whole cell sheet. Infectious seeds of the IP-3 virus were prepared in suckling hamster brain; they titered 10^6 pfu/ml of a 10% brain homogenate. Virions have no hemagglutinin spikes on the outer membrane (arrows). C. Rare IP-3 particle with surface projections (arrow). Bars = 100 nm. Negatively stained material was prepared by allowing a drop of viral suspension to remain on a formvar-carbon-coated grid 5 min. Excess fluid was removed with filter paper and the deposited material was stained with 1% phosphotungstic acid (PTA), pH 7.0, for 1 min. The excess stain was removed and the dried preparations were examined in an RCA EMU-3G electron microscope.*

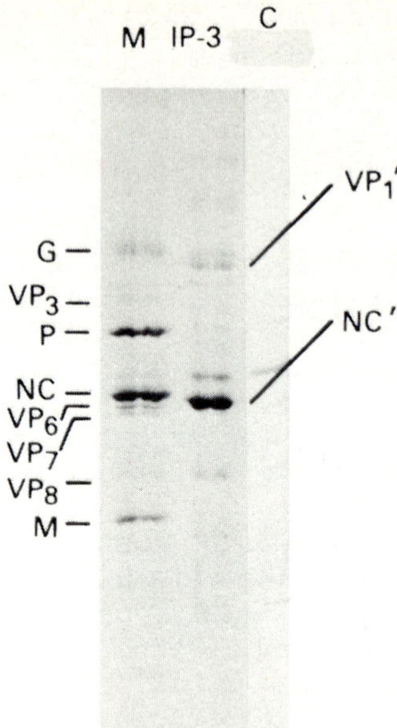

Fig. 2. *SDS-polyacrylamide gel electrophoresis of polypeptides isolated from measles (M) and IP-3-infected cells, as well as from uninfected cells (C). Vero cell monolayers were infected at a multiplicity of 1.0 for the attenuated strain and 0.01 for the SSPE strain. Both viruses caused almost complete CPE 2 and 3 days after infection, respectively, the latter virus due to a rapidly spreading process of cell fusion.*

Cultures infected with the attenuated virus were labeled 24 hr after infection; cultures infected with the IP-3 strain were labeled 48 hr after infection with 5 μCi/ml of ^{35}S methionine. Cultures infected with the attenuated virus were incubated at 36°C, cultures infected with IP-3 strain were incubated at 33°C-34°C.

Extracellular virus was concentrated by centrifugation at 100,000 x g for 1 hr or by polyethylene glycol precipitation (7). Following centrifugation, precipitates were resuspended in 0.01 volume of 0.01 M tris pH 7.4, 0.1 M NaCl, 1 mM EDTA (TNE buffer) and sonicated briefly by a Bronson probe sonicator at intermediate power settings.

Cell-associated virus was harvested by washing the cell sheet with PBS pH 7.4, swelling the cells in RSB buffer at 4°C for 20 min followed by 5 strokes in a Dounce homogenizer. The nuclei were removed by centrifugation at 600 x g for 5 min.

Purification was performed by centrifugation of concentrated virus material at 250,000 x g for 40 min on a discontinuous gradient of 20% and 65% (w/v) sucrose. The 20% - 65% sucrose inter-

majority of both IP-3 and measles nucleocapsids measured 1.2μ, despite a small degree of fragmentation (Fig. 5). In addition, there appeared to be a much smaller peak of measles nucleocapsid, measuring 0.2 to 0.3 μ.

Despite antigenic and size similarities between the native nucleocapsids of IP-3 and measles, disruption of the protein into its polypeptide subunits by SDS-polyacrylamide gel electrophoresis suggested that there might be some differences in primary structure. The IP-3 polypeptide subunit of the nucleocapsid appeared to be slightly smaller (56,000 daltons) than the measles counterpart (58,000 daltons) in Fig. 2. This difference was confirmed by tryptic peptide fingerprint analysis. Purified nucleocapsids were digested with trypsin and subjected to chromatographic analysis in two dimensions (Fig. 6). The polypeptides composing the two nucleocapsids possessed six methionine-containing tryptic peptides that chromatographed identically and two that chromatographed differently.

DISCUSSION

Several SSPE isolates have been compared to conventional measles viruses as to protein and nucleic acid composition. A small difference was noted between the membrane proteins (20), and there appeared to be differences in the nucleic acid composition (24) suggestive in one study (10) of the presence of sequences of another as yet unknown virus.

The SSPE isolate that we examined, IP-3, differs from the previously studied isolates in that it continues to be cell-associated after long-term passage in tissue culture, and it produces a subacute encephalitis in monkeys, resembling SSPE in humans (2). Measles and IP-3 virion particles differ in three important ways. First the infectious particles have different buoyant densities (1.23 versus 1.22 g/cm^3, respectively). Second, the large majority of IP-3 particles lack hemagglutinin on EM observation. Third, the IP-3-infected cells produce mainly nucleocapsid and perhaps one other protein detectable by gel

---

face was diluted 1:2 (v/v) with TNE buffer and centrifuged to equilibrium (135,000 x g for 16 hr) in a continuous gradient of 20% (w/v) sucrose in TNE buffer to 66% (w/v) sucrose in deuterium oxide.

SDS-polyacrylamide gels were composed of 10% acrylamide and 0.266% bisacrylamide. Electrophoresis was for 90 min at 200 volts and 30 ma (15).

Molecular weights were determined by running in parallel (^{35}S) methionine-labeled adenovirus proteins, as well as unlabeled pancreatic DNAse and bovine serum albumin. The gel was autoradiographed for one day.

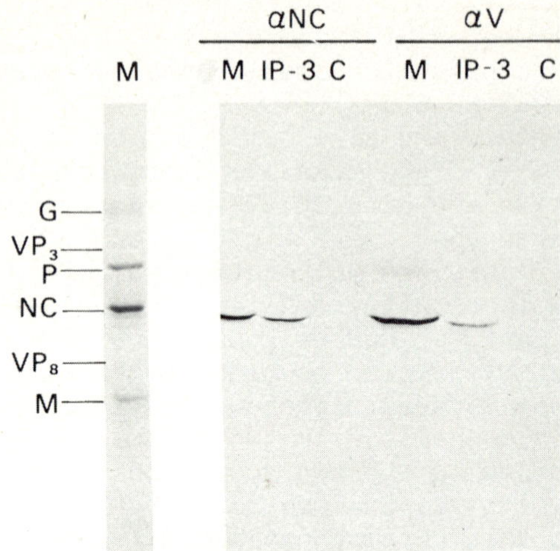

Fig. 3. *SDS-polyacrylamide gel electrophoresis of measles (M) and IP-3 polypeptides after immunoprecipitation. Uninfected (C) and infected cell extracts were precipitated with antisera directed against either measles virion (αV) or measles nucleocapsid (αNC). Immunoprecipitation of viral proteins was performed by mixing purified viruses and nucleocapsids (10^7 pfu of measles and 10^3 pfu of IP-3; 10^4 counts/min of nucleocapsid of both measles and IP-3) with 10 μl of antisera prepared against the whole measles virion particle and measles nucleocapsid. Antiserum against the virion particle was obtained from monkeys after natural measles infection and had a hemagglutination-inhibition titer of 1:512. The antiserum against measles nucleocapsid was prepared in rabbits by Dr. David J. Waters, and had a complement fixation titer of 1:256. The antigens and antisera in 0.1 ml of phosphate-buffered saline were incubated in 2% Triton X-100 for 1 hr at 37°C and either 10 μl goat anti-human IgG (anti-virion serum) or goat anti-rabbit IgG (anti-nucleocapsid serum) was added. Following a further incubation for 1 hr at 37°C, the mixtures were left overnight at 4°C and processed as described previously (5). The measles marker (left lane) was prepared as in Fig. 1. The gel was autoradiographed for 3 days.*

electrophoresis. On the other hand, the measles virions contain four major polypeptides - a nucleocapsid linked by disulfide bonds to a phosphoprotein, which is surrounded by an envelope consisting of glycoprotein-hemagglutinin and matrix protein, by analogy with other paramyxoviruses (6, 16, 19, 22).

Although the bulk of IP-3 structural proteins synthesized appears to be nucleocapsid, three observations suggest that limited quantities of envelope proteins are synthesized. First

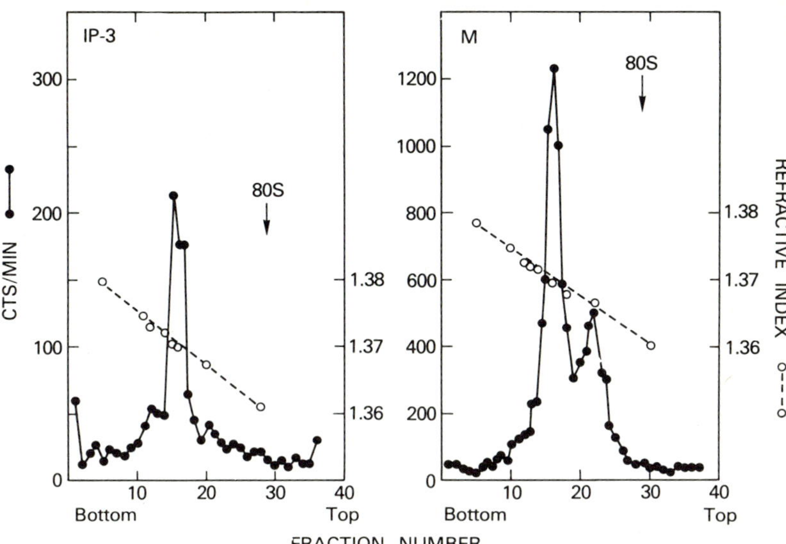

Fig. 4. *Rate-zonal sedimentation of measles and IP-3 nucleocapsids. Purified nucleocapsids (see legend to Fig. 2) were sedimented as described previously (8, 14). Fractions were collected and counted in Aquasol (New England Nuclear Corp.). Refractive indices of each fraction were determined in a Beckman refractometer. The position of the 80S ribosome marker is indicated by the arrow.*

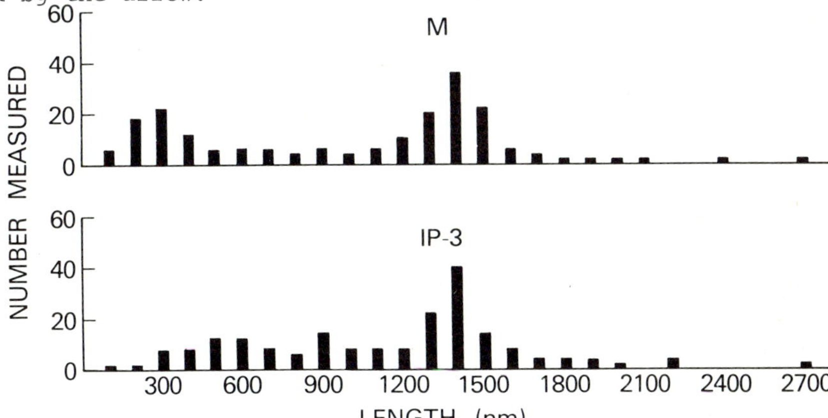

Fig. 5. *Histogram of length measurement of measles and IP-3 nucleocapsids. The nucleocapsids were obtained from scraped infected cell sheets by osmotic shock in water. Nuclei were removed by centrifugation (600 x g for 5 min). Supernatants were spread on grids, stained with 1% PTA and photographed by EM. All nucleocapsids in each micrograph whose complete lengths could be visualized were measured (185 strands per sample).*

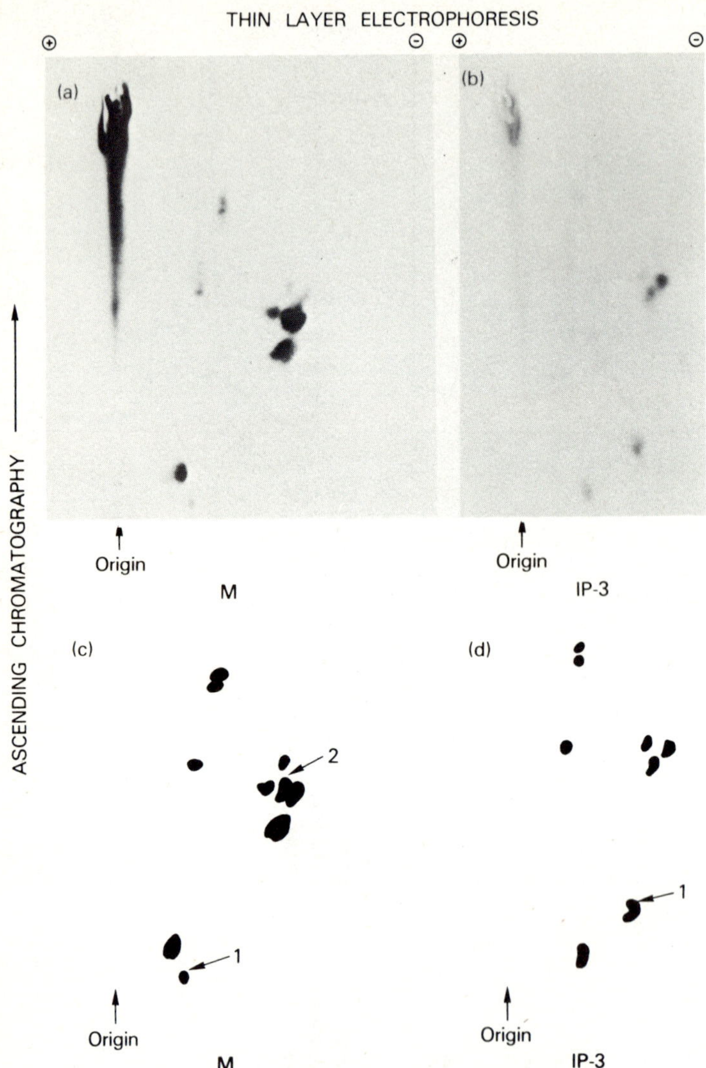

Fig. 6. *Two-dimensional thin layer chromatograms of tryptic peptides of measles (a) and IP-3 (b) nucleocapsids purified as in Fig. 4. Nucleocapsid was purified as described above, prior to digestion with trypsin. Samples of each tryptic digest (20,000 counts/min) were applied to the origin, chromatographed in one direction (bottom to top) and electrophoresed in the other (from cathode on the left to the anode on the right) as described previously (5). The chromatograms were autoradiographed for 3 days.*

IP-3 spreads from cell to cell by the process of cell fusion, suggesting the presence of a fusion factor. Second, rare IP-3 virion particles possess surface projections suggestive of hemagglutinin. Third, although hemagglutinating activity with infected cell homogenates was not yet demonstrated, large areas of the syncytia show hemadsorption of monkey RBC which can be blocked with measles-positive serum (Albrecht, unpublished results). It is possible that one or several of these activities are represented by VP'_1. Unlike measles, the amount of IP-3 nucleocapsid synthesized far exceeds that of envelope proteins. This difference between measles virus and IP-3 protein synthesis may explain the different *in vitro* growth characteristics (3) and pathogenesis (2). The imbalance or lack of certain viral envelope proteins incorporated into the cell surface, may prevent the nucleocapsid from proceeding to the cell surface for the normal maturation process.

The IP-3 and measles nucleocapsids cross-react antigenically as shown by immunoprecipitation of IP-3 nucleocapsid with antisera directed against measles components. The two nucleocapsids appear to be similar in size as shown by rate zonal sedimentation (200S) and electron microscopic measurements (1.2 μ). The size of the IP-3 nucleocapsid is clearly different from the value (110S) reported for defective particles of measles (8, 14). The EM measurements are similar to those reported elsewhere (O.O. Agbede, K.Y. Huang, and M. Reich, submitted to *Science*) but show less fragmentation. Furthermore, a second distinct size species of nucleocapsid was noted in the attenuated measles virus strain that appeared to be 10% to 20% of the normal size (17, 23). It is possible that the smaller species represents nucleocapsid of defective measles particles.

The polypeptide subunits composing the native IP-3 and measles nucleocapsids are of different sizes (56,000 daltons versus 58,000 daltons, respectively) and amino acid composition (nonidentical tryptic peptide fingerprints). Whether the alteration in the IP-3 nucleocapsid polypeptide occurred as a mutation prior to the onset of the SSPE syndrome, or whether it was generated subsequently during the passage of the virus in tissue culture remains to be determined. In the former instance, it is conceivable that the mutation could be important in the pathogenesis of SSPE.

It has been proposed that persistent virus infection such as SSPE is caused by defective viruses (13). At first glance, the IP-3 particles do not satisfy the established criteria of defective particles, i.e. that they possess the same structural proteins as wild-type particles, but only a fragment of the wild-type genome. Measurements of nucleocapsid size reported here as well as our preliminary results on the genetic composition of IP-3 indicate that the genetic information for the envelope proteins may be present in the IP-3 genome but remains unexpressed in the Vero cell infection, except at a very low rate. This does not exclude the possibility that the IP-3 genome has a portion of its measles information deleted and replaced with non-measles nucleic acid.

This replicative system may represent one of the mechanisms whereby IP-3 escapes immune surveillance by the host. In expressing mainly nucleocapsid information, the IP-3 genome may persist without synthesizing the quantities of glycoproteins necessary for altering the histocompatibility antigens on the cell surface (11). The mechanism by which the virus information may be selectively expressed in an abortive infection is under investigation and may be applicable to other slow virus infections.

ACKNOWLEDGEMENTS

We thank Dr. David J. Waters for his gift of antiserum directed against measles nucleocapsid, Heiner Westphal for ^{35}S-methione-labeled adenovirus proteins, and Dr. James Ramsey for 80S ribosomes. We are indebted to Mrs. Dorothy Smith and Mrs. Virginia Sheaffer for typing this manuscript.

REFERENCES

1. Albrecht, P. and Schumacher, H.P. (1972). *Archiv. fur die gesamte Virus-forschung* 36, 23.
2. Albrecht, P., Burnstein, T., Klutch, M.J., Hicks, J.T. and Ennis, F.A. (1977). *Science* 195, 64.
3. Burnstein, T., Jacobsen, L.B., Zeman, W. and Chen, T.T. (1974). *Infect. Immun.* 10, 1378.
4. Connolly, J.H., Allen, I.V., Hurvitz, L.J. and Millar, J.H. (1967). *Lancet* 1, 542.
5. Eron, L.J., Callahan, R. and Westphal, H. (1974). *J. Biol. Chem.* 249, 6331.
6. Graves, M.C., Silver, S.M. and Choppin, P.W. (1977). *Abstr. Ann. Meeting Am. Soc. Microbiol.* p.283.
7. Hall, W.W. and Martin, S.J. (1973). *J. Gen. Virol.* 19, 175.
8. Hall, W.W. and Martin, S.J. (1974a). *Med. Microbiol. Immunol.* 160, 155.
9. Hall, W.W. and Martin, S.J. (1974b). *J. Gen. Virol.* 22, 365.
10. Hall, W.W. and ter Meulen, V. (1976). *Nature (London)* 264, 474.
11. Hecht, T.T. and Summers, D.F. (1976). *J. Virol.* 19, 833.
12. Horta-Barbosa, L., Fucillo, D.A., Zeman, W. and Sever, J.L. (1969). *Nature (London)* 211, 974.
13. Huang, A.S. and Baltimore, D. (1970). *Nature (London)* 226, 325.
14. Kiley, M.-P., Gray, R.H. and Payne, F.E. (1974). *J. Virol.* 13, 721.

15. Maizel, J.V. Jr. (1971). *In* "Methods in Virology", Vol. V, (Maramorosch, K. and Kopowski, H., eds), Academic Press, New York, p.180.
16. Mountcastle, W.E. and Choppin, P.W. (1977). *Virology* 78, 463.
17. Norrby, E., and Hammarskjold, B. (1972). *Microbios* 5, 17.
18. Payne, F.E., Baublis, J.V. and Itabushi, H.H. (1969). *N. Eng. J. Med.* 281, 585.
19. Scheid, A. and Choppin, P.W. (1974). *Virology* 57, 476.
20. Schluederberg, A., Chavanich, S., Lipman, M.B. and Carter, C. (1974). *Biochem. Biophys. Res. Commun.* 58, 647.
21. Sever, J.L., Krebs, H., Ley, A., Barbosa, L.H. and Rubinstein, D. (1974). *J. Am. Med. Assoc.* 228, 604.
22. Waters, D.J. and Bussell, R.H. (1973). *Virology* 55, 554.
23. Waters, D.J., Hersh, R.T. and Bussell, R.H. (1972). *Virology* 48, 278.
24. Yeh, J. (1973). *J. Virol.* 12, 962.

A COMPARISON OF THE INTRACELLULAR POLYPEPTIDES OF MEASLES AND SUBACUTE SCLEROSING PANENCEPHALITIS VIRUS

S.L. WECHSLER, KATHRYN C. STALLCUP and B.N. FIELDS

Department of Microbiology and Molecular Genetics, Harvard Medical School, Boston, Massachusetts 02115.

and

Department of Medicine, Division of Infectious Diseases, Peter Bent Brigham Hospital, Boston, Massachusetts 02115.

Subacute sclerosing panencephalitis (SSPE) is a rare, slowly progressive neurological disease that occurs between the ages of four and twenty in individuals who have had uncomplicated measles, or less commonly, have received measles vaccine (27). Several lines of evidence have linked measles virus to SSPE. High titers of an antibody to measles virus are found in the serum and spinal fluid of patients with SSPE (3). Intracellular inclusions, similar to those found in measles infected cells, are present in the brain cells of SSPE patients (1). In addition, co-cultivation of SSPE brain cells with cells permissive for measles virus growth has yielded viruses that are similar in physical and serological properties to measles (15, 22).

The question of major interest is how measles virus, normally the agent of an acute, self-limiting disease, can persist and result in the delayed disease SSPE. Several theories have been advanced to explain this central issue. These theories attribute persistence to 1) an abnormal host immune response to measles virus, 2) a second viral agent that co-infects with measles virus, or 3) a mutation in the parental measles virus that results in altered pathogenicity (8, 16). The viral mutation theory of persistence is supported by the findings that certain temperature sensitive mutants of measles (12) produce delayed rather than acute disease in newborn animals. Furthermore, RNA-RNA hybridization studies have shown a 10% difference between the genomes of SSPE and measles virus (11). There has also been a single report comparing the polypeptides of measles and SSPE virions in which an aberrant migration of the M polypeptide was noted (26). Unfortunately no polypeptide profiles were presented and better documentation of the findings has not been published.

As a part of a study aimed at understanding the pathogenesis of SSPE we have compared the polypeptides of the Edmonston strain of measles (referred to in this report as wild type, or wt) and five different isolates of SSPE virus. This report describes such an analysis with a particular emphasis on the polypeptides synthesized in infected cells.

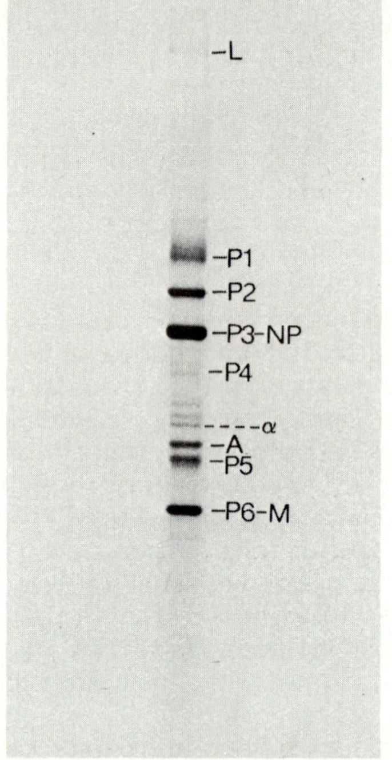

Fig.1. *Polyacrylamide gel electrophoresis of ^{35}S-methionine labeled polypeptides from purified measles virions. CV-1 cell monolayers in roller bottles (10^8 cells/bottle) were infected with the Edmonston strain of measles virus in 5 ml of medium at an MOI of 0.01. Following 2 hr adsorption at 37°C, each bottle was fed with 50 mls of Eagle's MEM supplemented with 10% fetal calf serum. 20 hr post-infection the medium was removed and replaced with 10 ml of methionine-free Eagle's MEM containing 0.5 mCi of ^{35}S-methionine. 48 hr post-infection, the medium was clarified of cell debris (1000 g for 30 mins) and virus was pelleted at 25,000 rpm in an SW27 rotor for 2 hr. The pellet was suspended in 1 ml of 1 x SSC, layered over a 15-50% (w/v) linear potassium tartrate gradient and centrifuged for 4 hr at 25,000 rpm in an SW27.1 rotor. The visible virus*

MEASLES VIRION POLYPEPTIDES

Measles virus, labeled with ^{35}S-methionine, was purified from infected CV-1 cells. Analysis of the purified virion polypeptides by SDS-polyacrylamide gel electrophoresis (SDS-PAGE), revealed 6 or 7 major, and several minor polypeptides (Fig.1). There is one glycoprotein (P1 or G) (21), and there are two phosphoproteins (P2 and P3) (2). The most abundant polypeptide, P3, is the nucleocapsid protein (NP) (9, 21, 31) while the smallest polypeptide, P6, is the nonglycosylated membrane protein, M (2, 21). One of the virion polypeptides, A, migrates at the same position as cellular actin and probably is not a virus specified protein (18, 30). Little is known about the largest polypeptide, L, or the polypeptides designated P4 and P5. Several minor bands between NP and A were detected in our virus preparations. The amounts of these minor components varied greatly in different preparations and they may represent cellular contamination or cleavage products of other virus polypeptides. Of these minor components only one has been unequivocally detected in extracts of infected cells. In order to distinguish it from the others we have designated this minor band α.

The estimated molecular weights of the polypeptides found in purified measles virions are L, 200,000; P1 (G), 80,000; P2, 70,000; P3 (NP), 60,000; P4, 55,000; α, 46,000; A, 43,000; P5, 41,000; and P6 (M), 37,000. The virus polypeptide pattern shown in Fig.1 is similar to the patterns reported by others (2, 10), and it is virtually identical to the pattern recently reported by Mountcastle and Choppin (21).

---

band (density = 1.23 mg/cc) was collected, diluted with 1 x SSC, pelleted and rebanded. The virus band from the second gradient was collected, pelleted and suspended in sample gel buffer containing 5% mercaptoethanol and 2% SDS. SDS polyacrylamide gel electrophoresis was carried out in a 10% polyacrylamide slab gel at a constant current of 45 mA for 2.5 hr. Gels were fixed in 50% methanol, 7% acetic acid for 1 hr, dried under vacuum and exposed for autoradiography using Kodak NS2T safety screen x-ray film.

Nomenclature of viral polypeptides: L, large molecular weight polypeptide; P1(G), glycoprotein; P2, nucleocapsid-associated phosphoprotein; P3-NP, major nucleocapsid protein (phosphorylated); P4, minor component; α, minor component; A, cellular actin; P5, variable component; P6-M nonglycosylated membrane protein.

INTRACELLULAR MEASLES VIRUS SPECIFIED POLYPEPTIDES

The analysis of intracellular virus polypeptides is complicated by the fact that measles virus does not efficiently shut off host cell protein synthesis. Thus the cellular background presents a major problem in the detection of virus polypeptides. At low MOI (<1 PFU/cell) cytopathic effects generally appear later than 36 hrs post infection (p.i.) and are characterized by widely spaced plaque-like cell destruction. When intracellular virus polypeptides are examined under these conditions, the cellular polypeptides are very prominent and thus even very late p.i. virus polypeptides are not readily detected over the cellular background. Attempts to use actinomycin D to lower the cell background have also proved unsatisfactory. The use of high MOI, on the other hand, results in the production of relatively large quantities of virus polypeptides which are easily detected, even against the background of cellular polypeptides.

The measles virus polypeptides that can routinely be detected in extracts of CV-1 cells infected at high MOI (50 PFU/cell) are P1 (G), NS1 and NS2 (2 nonstructural polypeptides with estimated molecular weights of 74,000 and 72,000 daltons), P2, NP, and M (Fig.2). These polypeptides, with the exception of the 2 nonstructural polypeptides (NS1 and NS2), correspond to polypeptides seen in purified virions. Three additional virion polypeptides, L, P4 and P5 are usually not well resolved in extracts of infected cells. Furthermore, when cells are fractionated into nuclear and cytoplasmic fractions, the P1 (G), NS1 and NS2 polypeptides are absent from the nuclear fraction (Fig.3).

SSPE POLYPEPTIDES

In order to compare SSPE virus with wt measles, cells were infected at high MOI with either the Edmonston strain of measles or with one of five different SSPE strains (Hallé, Mc Clellan, Munn, Fisher, or Mantooth). At 20 hr post infection, the cells were labeled with ^{35}S-methionine for 2 hr and cell fractions were prepared (Fig.3). The major nuclear and cytoplasmic polypeptides previously noted with measles virus are seen in cells infected with all the SSPE strains except Munn. While the general patterns of SSPE and measles polypeptides are similar, a close examination of Fig.3 reveals a slight difference in at least one virus polypeptide, P2. In four of the SSPE strains, this nucleocapsid-associated phosphoprotein has a more rapid electrophoretic migration than that of the wild type measles P2 polypeptide. Identical migrational differences can be seen on both the nuclear and cytoplasmic

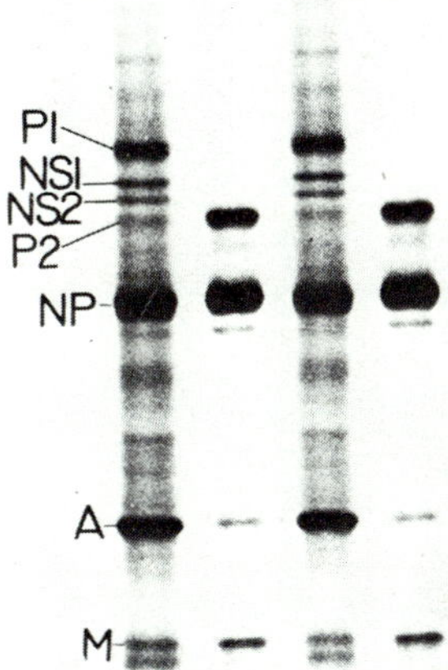

Fig.2. *Cytoplasmic and nuclear polypeptide synthesis in cells infected with measles virus. Confluent monolayers of CV-1 cells in Falcon multicluster wells (Falcon #3008) (10^5 cells/well) were infected with wt measles at an input multiplicity of infection (MOI) of 50 pfu/cell. After 2 hr adsorption at 33°C, the infected monolayers were fed with 1 ml of IMEMZO (International Biological Laboratories) supplemented with 10% fetal calf serum and then incubated at 33°C. At 24 hr post-infection, cells were washed and preincubated for 1 hr in methionine free medium, then labeled with ^{35}S-methionine in 0.5 methionine-free Eagles MEM containing 100 µCi ^{35}S-methionine (specific activity >400 mCi/mMole. After 2 hr the medium was removed and the monolayers rinsed twice with ice cold TMN (0.01M tris, 0.0015M$MgCl_2$, 0.14 M NaCl, pH 7.2). The cells were then collected with a glass rod and suspended in 1 ml TMN. Cell suspensions were made 0.5% with respect to NP40 and incubated on ice for 30 mins. The samples were then centrifuged at 1,000g for 1 min. The supernatant represents the cytoplasmic fraction and the pellet the nuclear fraction. The nuclei were suspended in 1 ml of TMN and 10 volumes of -20°C acetone were added to both fractions. Protein was precipitated overnight at -20°C and then pelleted at 1,500g for 2 hr. The acetone was poured off and the pellets allowed to air dry for 24 hr at room temperature. The pellets*

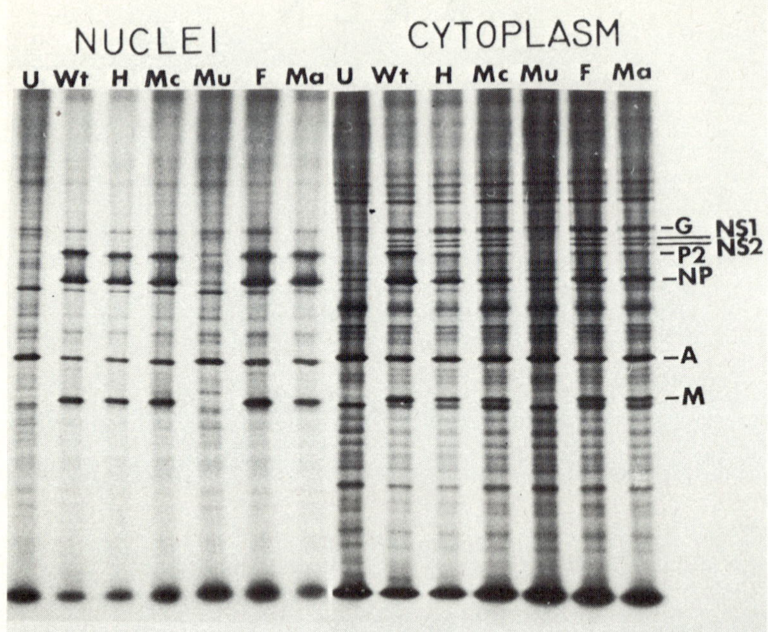

Fig.3. *Measles and SSPE virus-specified intracellular polypeptides. Confluent monolayers of CV-1 cells were infected with either wt measles or one of five strains of SSPE at an MOI of 50. 20 hr post-infection at 37°C, uninfected and infected cells were labeled with ^{35}S-methionine for 2 hr and processed for electrophoresis and autoradiography as described in the legends to Figs.1 and 2. Electrophoresis was for 2 hr at 45mA. U, uninfected; Wt, Edmonston wild type measles; SSPE strains: H, Hallé; Mc, McClellan; F, Fisher; Ma, Mantooth; Mu, Munn.*

fractions. Although it is not clearly seen in Fig.3, there is also the suggestion that differences in migration exist in the M proteins.

To determine the reliability of these findings, electrophoresis was performed (a) under various conditions that increase the resolution of the P2 and M regions and (b) on samples which

---

were then suspended in gel sample buffer. Electrophoresis and autoradiography is as described in the legend to Fig.1. C, infected cytoplasm; N, infected nuclei (duplicate samples of C and N are displayed).

were labeled at later times to determine whether polypeptides synthesized by the Munn strain (which produces cytopathic effects more slowly than wild type measles and the other four SSPE strains) could be detected. When electrophoresis was carried out for periods of time longer than the 2 hours described in Fig.3, the variations in electrophoretic mobility of P2 and M were better resolved. (Fig.4 and 5). In addition,

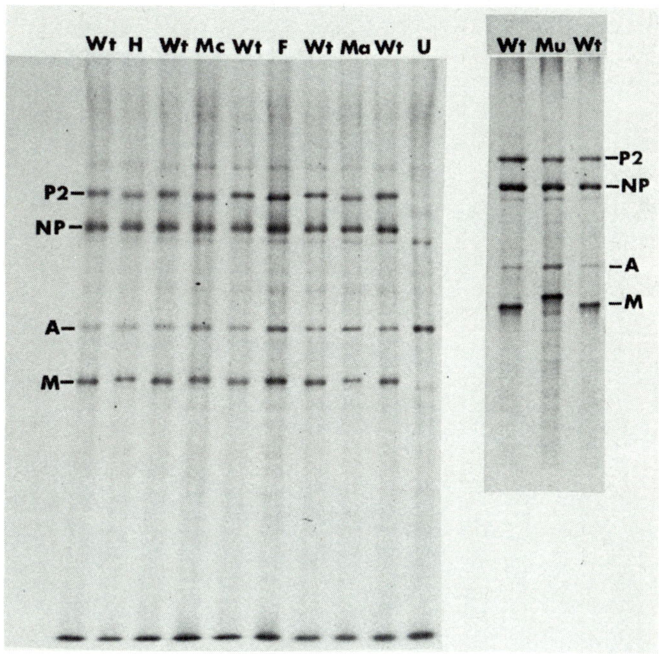

Fig.4. *Polypeptide synthesis in cytoplasm of cells infected with measles and SSPE virus. Experimental protocol is as described in the legend to Fig.3, except that a second monolayer infected with the SSPE strain Munn was labeled at 27 instead of 20 hr post infection (far right). Electrophoresis was for 2.5 hr at 45mA. U, uninfected, Wt, wt measles; SSPE strains, H, Hallé; Mc, McClellan; Mu, Munn; F, Fisher; Ma, Mantooth.*

when cells were infected with the Munn strain and labeled at 27 hrs post infection (instead of 20 hrs), virus specific polypeptides could be easily detected. The cytoplasmic fractions of cells infected with SSPE (Fig.4) reveal: 1) significantly faster migration of polypeptide P2 in all SSPE strains except Munn, 2) slightly slower migration of M in the same four strains and 3) a much slower migration of the M polypeptide of the Munn strain. These differences in the rates of migration of the M polypeptide are more dramatically illustrated in the nuclear extracts which were subjected to electrophoresis for an even longer period of time. (Fig.5). The striking abnormality of the M polypeptide in the Munn strain is easily seen, as are the smaller variations in the M polypeptides of the other strains. In these nuclear extracts, the four SSPE strains (Hallé, McClellan, Fisher, and Mantooth) also exhibit the aberrantly migrating P2 polypeptides as noted previously, while the P2 polypeptide from the fifth SSPE strain, Munn, does not differ significantly from the P2 polypeptide of wt measles (Fig.5).

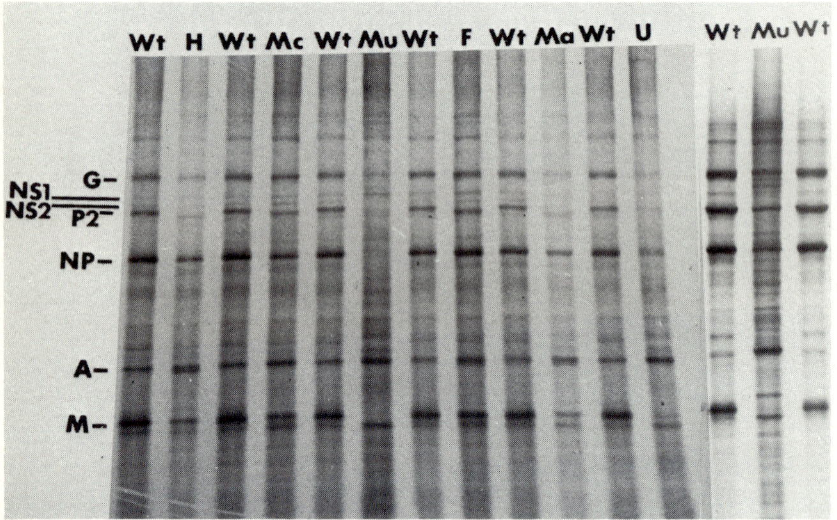

Fig.5. *Polypeptide synthesis in nuclei of cells infected with measles and SSPE virus. Experimental protocol is as described in the legend to Fig.1, except that SSPE Munn infected cells were labeled at 27 hr instead of 20 hours. Electrophoresis was for 3 hr at 45mA. U, uninfected; Wt, wt measles; SSPE strains, H, Hallé; Mc, McClellan; Mu, Munn; F, Fisher; Ma, Mantooth.*

DISCUSSION

The results described in this report have shown that the use of high multiplicities of infection allows the detection of measles and SSPE virus-specified polypeptides over the background of cellular polypeptides. Furthermore, the P2, NP and M virus polypeptides can be found in the nuclei as well as the cytoplasm of infected cells.

A comparison of SSPE and measles polypeptides demonstrates that most of the intracellular polypeptides of measles and SSPE viruses have identical electrophoretic mobilities. However, there are small but significant differences in the electrophoretic migrations of the P2 and M polypeptides of SSPE compared to the respective wt measles polypeptides (i.e. in all 5 SSPE strains, the M polypeptides migrated more slowly than the wt M polypeptide, while in 4 of the 5 SSPE strains, the P2 polypeptides migrated faster than the wt P2 polypeptide.) The precise biochemical alterations causing the aberrant electrophoretic migrations of P2 and M are not known. In addition to the measles strain described in this report, we have recently obtained a second strain of Edmonston measles (R. Rustigian, personal communication). This strain has been maintained by passage through Hela and Vero cells for more than 15 years. The polypeptides of the latter strain corresponded to those of the measles strain described here in all of the major polypeptides except P2. Furthermore, Mountcastle and Choppin, (21), in a recent study of four different measles strains, were also unable to detect migrational variants of the M polypeptide. Thus, while the P2 polypeptides of different measles strains may vary, no variants have been found in the M polypeptides from 6 different measles strains. Therefore, it is highly unlikely that the M polypeptide differences of the SSPE isolates described in this report reflect normal strain variations. Perhaps the most probable explanation for such altered polypeptide mobilities is mutation in the genome region coding for the M polypeptide. While single site, revertible, mutations can cause electrophoretic migrational variation of individual polypeptides similar in magnitude to those exhibited by M (4, 5, 32), it is not clear if this is the case here or if more complex explanations (multiple mutations or recombination) are in fact responsible for these changes.

Two independent lines of investigation support the potential significance of these findings: 1) Hall and ter Meulen (see chapter 15) have detected apparent alterations of the M polypeptide mRNA in SSPE isolates and 2) we have found that two cell lines persistently infected with measles virus (24, 25), have aberrant M polypeptides similar to those described above. These facts, together with our findings that all five SSPE isolates contain aberrantly migrating M polypeptides, support the hypothesis that a mutation in the M protein gene is associated with, and could account for, the persistence of measles virus in the disease SSPE.

While there is still no proof that an alteration of the M polypeptide leads to persistence and the disease SSPE, it is interesting that studies of other negative-strand viruses (influenza and VSV) assign to the M protein the function of associating the nucleocapsid with the membrane, a prerequisite of the budding process (20, 28, 29, 33). It has long been postulated that perturbation of the assembly process is involved in persistent infections (for review see 23) and in most systems the coalescence of viral components at the membrane is considered the penultimate step of infection (13, 14). Electron microscopic observations of several persistent systems have in fact shown accumulations of defective buds and cytoplasmic aggregates of nucleocapsids that remain unassociated with the cell membrane (6, 7). These findings could all be easily explained as the result of an altered M protein which is unable to recognize the nucleocapsid structure.

Oldstone has presented the theory that loss, or modulation, of surface antigens from infected cells allows maintenance of persistent infection. Antigen antibody complex is more easily lost from cells persistently infected with measles virus compared to acutely infected cells (17). While no physical link between the G and M polypeptides has been demonstrated in measles, evidence from VSV shows that ts mutants of the M polypeptide have an increased amount of soluble glycoprotein in the culture media (19). Thus another possible role of the M polypeptide could be in anchoring the glycoprotein in the membrane of infected cells. An alteration of the M polypeptide, such as those seen in the SSPE strains, could explain increased antigen loss.

While it is easy to envision a critical role for the M protein in altering the assembly of measles virus, the finding of an aberrant migration of this protein in several strains of SSPE does not prove that the altered polypeptide is the primary lesion. Hopefully further studies focusing on the biochemical changes responsible for these variations as well as the impact of such altered M proteins on various parameters of virus assembly and growth will shed further light on the pathogenesis of SSPE.

ACKNOWLEDGEMENTS

The expert technical assistance of K.B. Byers is gratefully acknowledged. This work was supported by research grant no. R.G. 991-A-3 from the National Multiple Sclerosis Society.

REFERENCES

1. Bouteille, M., Fontaine, C., Vedrenne, C. and Delarue, J. (1965). *Rev. Neurol.* 113, 454.
2. Bussell, R.H., Waters, D.J. and Seals, M.K. (1974). *Med. Microbiol. Immunol.* 160, 105.
3. Connolly, J.H., Allen, I.V., Hurwitz, L.J. and Millar, J.H.D.

(1967). *Lancet* 542.

4. Cross, R.K. and Fields, B.N. (1976a). *J. Virol.* 19, 174.
5. Cross, R.K. and Fields, B.N. (1976b). *Virology* 74, 345.
6. Dubois-Dalcq, M., Barbosa, L.H., Hamilton, R. and Sever, J.L. (1974). *Lab. Invest.* 30, 241.
7. Dubois-Dalcq, M., Reese, T.S. and Fuccillo, D. (1976). *J. Virol.* 19, 579.
8. Fields, B.N., (1972). *N. Eng. J. Med.* 287, 1026.
9. Hall, W.W. and Martin, S.J. (1973). *J. Gen. Virol.* 19, 175.
10. Hall, W.W. and Martin, S.J. (1974). *Med. Microbiol. Immunol.* 160, 143.
11. Hall, W.W., and ter Meulen, V. (1976). *Nature* 264, 474.
12. Haspel, M.V. and Rapp, F. (1975). *Science* 187, 450.
13. Hay, A.J. (1974). *Virol.* 60, 398.
14. Holland, J.J. and Kiehn, E.D. (1970). *Science* 167, 202.
15. Horta-Barbosa, L., Fuccillo, D.A., Zeman, W. and Sever, J.L. (1969). *Nature* 221, 974.
16. Johnson, R.T. (1970). *J. Inf. Dis.* 1213, 227.
17. Joseph, B.S. and Oldstone, M.B.A. (1975). *J. Exp. Med.* 142, 864.
18. Lamb, R.A., Mahy, B.W.J. and Choppin, P.W. (1976). *Virology* 69, 116.
19. Little, S.P. and Huang, A.S. (1977). *Virology,* 81, 37.
20. McSharry, J.J., Compano, R.W. and Choppin, P.W. (1971). *J. Virol.* 8, 722.
21. Mountcastle, W.E., and Choppin, P.W. (1977). *Virology* 78, 463.
22. Payne, F.E., Baublis, V.V. and Itabaski, H.H. (1969). *N. Eng. J. Med.* 281, 595.
23. Rima, B.K. and Martin, S.J. (1976). *Med. Microbiol. Immunol.* 162, 89.
24. Rustigian, R.J.(1966a). *Bacteriology* 92, 1792
25. Rustigian, R.J.(1966b). *Bacteriology* 92, 1805.
26. Schluederberg, A., Chauanich, S., Lipman, M.B., and Carter, C. (1974). *Biochem. Biophys. Res. Comm.* 58, 647.
27. Sever, J.L. and Zeman, W. (1968). *Eds. Conference on measles virus and subacute sclerosing panencephalitis. Neurology (Minneap.) Part 2.* 18, 1.
28. Shimizu, K. and Ishida, N. (1975). *Virology* 67, 427.
29. Wagner, R.R., Emerson, S.U. Imblum, R.L. and Kelley, J.M. (1975). *In* "Negative Strand Viruses", vol.I, p.l. (Mahy, B.W.J. and Barry, R.D., eds.) Academic Press, London.
30. Wang, E., Wolf, B.A., Lamb, R.A., Choppin, P.W. and Goldberg, A.R. (1976). *In* "Cell Motility", Book B. Cold Spring Harbor. Vol. 3, 589.
31. Waters, P.J. and Bussell, R.H. (1973). *Virology* 55, 554.
32. Wunner, W.H. and Pringle, C.R. (1974). *J. Gen. Virol.* 23,

97.
33. Yoshida, T., Nagai, Y., Yoshoo, S., Maeno, K., Matsumoto, T. and Hoshino, M. (1976). *Virology* 71, 143.

STUDIES ON THE STRUCTURE AND FUNCTIONS OF PARAMYXOVIRUS GLYCOPROTEINS

A. SCHEID, M.C. GRAVES, S.M. SILVER and P.W. CHOPPIN

The Rockefeller University,
New York, New York 10021, U.S.A.

The glycoproteins which form the spikes on the surface of influenza and parainfluenza viruses are involved in similar interactions between the virion and the host cell; however in the two types of viruses, the specific functions are differently apportioned between the spike glycoproteins (Fig.1).

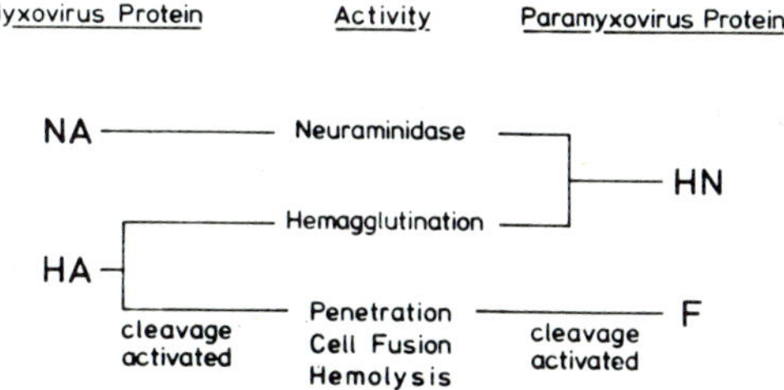

Fig.1. *The envelope glycoproteins of myxo- and paramyxoviruses and their biological activities, illustrating functional differences and homologies among the proteins of the different viruses.*

In influenza A and B viruses, receptor binding and neuraminidase are associated with different glycoprotein spikes (HA and NA) (11), whereas in parainfluenza viruses one glycoprotein (HN) carries both of these activities (33). Evidence for the involvement of the other parainfluenza virus glycoprotein (F) in hemolysis, cell fusion, and in the initiation of infection was first reported at this meeting in 1973 (7, 29); the expression of these biological activities requires a host-dependent proteolytic cleavage of a precursor (F_0) (8, 27). A correlation between expression of activity and proteolytic cleavage has also been found with influenza viruses, in which cleavage of the HA protein (13) enhances the infectivity of the virion (10, 12). The available evidence suggests that the HA protein of influenza and the F protein of parainfluenza viruses are analogous with regard to their role in the initiation of infection, and it appears likely that with both viruses this process involves fusion between virion and cell membranes.

The requirement for cleavage of the F protein has several biologically important implications regarding the interactions between virus and host. Spread of virus in a tissue requires that infectious virus, with cleaved F protein, be released from the host cell. There is evidence which suggests that proteolytic activation of the precursor glycoprotein F_0 has to occur during the maturation of the virus in the host cell, i.e., a target cell cannot activate adsorbed non-infectious virions even if it possesses a protease capable of cleaving the F_0 protein during synthesis of virus. Furthermore, there has to be a match between the specificity of the proteases available in the host and the susceptibility to cleavage of the F protein. The isolation of mutants of Sendai virus with altered susceptibility to specific proteases permitted the demonstration that this mechanism of proteolytic activation can determine tissue specificity of parainfluenza viruses (30, 31). On the basis of these findings with Sendai virus, of the precursor product relationship of a glycoprotein in NDV infected cells (9, 26) and of the protein in avirulent strains of NDV which could be interpreted to be the inactive precursor F_0 (14, 34), it was suggested that differences in virulence could be due to strain dependent variation in the susceptibility of the F_0 protein to host proteases, (30). Extensive studies on the naturally occurring NDV strains have clearly demonstrated a correlation between virulence and cleavage of the F_0 proteins, and in addition in two strains of a precursor of the HN protein (18, 19). This concept is further supported by the isolation of a mutant of an avirulent NDV strain in which increased virulence correlates with cleavability of the F_0 protein (see Chapter 61).

This report summarizes structural studies on the HN and F protein of parainfluenza viruses which are in progress in our laboratory. Present interests include the mode of organization of the polypeptides to form the spikes and the interactions between protein and lipid which are involved in the anchoring of the spike in the virus membrane. In addition, although a role of the F protein in the fusion of virus and cell membranes is clearly indicated, the mechanism by which it participates in or brings about membrane fusion remains to be elucidated.

Among the alternatives is the possibility that the F protein has some unknown enzymatic function, or that it interacts in some other, direct manner with the protein or lipid of the target membrane. The experimental techniques, and some of the findings reported here, have been described in detail elsewhere (4, 32).

THE PROTEIN WITH HEMAGGLUTINATING AND NEURAMINIDASE ACTIVITIES

The HN glycoprotein of parainfluenza viruses possesses the neuraminic acid specific binding site responsible for hemagglutination and neuraminidase activity, and the available evidence is compatible with the view that both functions are associated with the same active site (28). Like other spike glycoproteins of enveloped viruses, the HN glycoprotein is anchored in the virus envelope, and this is thought to result from the interaction of a hydrophobic portion of the protein, located at the base of the spike, with the lipid of the viral membrane. This has been inferred not only from the arrangement of the spike on the virion but also from the solubility properties of the spike glycoproteins. The spike glycoproteins are solubilized by non-ionic detergents, and removal of detergent causes the spikes to aggregate by their bases into rosette-like structures.

The HN glycoprotein of SV5 has been dissected to yield the hydrophobic portion responsible for anchoring the spike, and the portion of the molecule that carries the active site of the neuraminidase. These experiments were done with highly purified HN glycoprotein, isolated from virions by Triton X-100 fractionation (33) and affinity chromatography on fetuin-Sepharose (28). The isolated HN protein was exposed to immobilized chymotrypsin, and the products were analyzed by electrophoresis on SDS-polyacrylamide gels (Fig.2). With increasing duration of treatment, the $\sim$64,000 dalton HN polypeptide chain is cleaved to yield a fragment of $\sim$ 59,000 daltons. The smaller cleavage product is not resolved on this gel; however, it is evident from Fig.2 that the portion removed was of uniform size, as no intermediate size classes between the native 64,000 and 59,000 were detected.

The cleavage products have been separated by gel filtration and characterized. In contrast to the native HN protein, the isolated larger fragment, designated HNc1, is soluble in aqueous media in the absence of detergent, suggesting that the portion of the molecule responsible for aggregation of the intact protein has been lost with removal of the smaller peptide, HNc2. HNc1 possesses neuraminidase activity, and except for a 10-30% loss in the course of chymotrypsin treatment, all the activity can be isolated with the HNc1 fragment. Even though HNc1 contains the active site of the neuraminidase, it does not agglut-

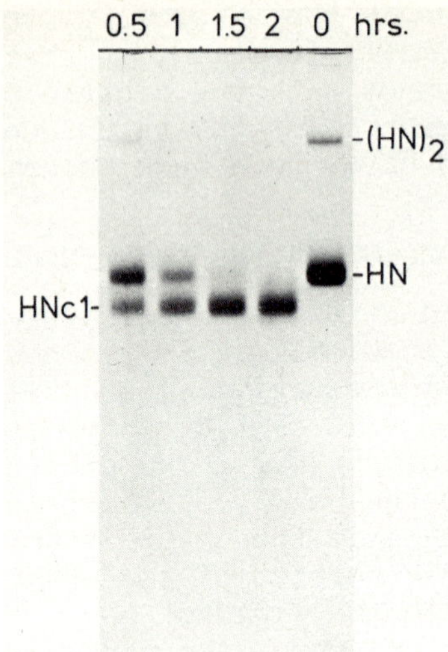

Fig.2. *Cleavage of the HN glycoprotein of SV5 by chymotrypsin in vitro to yield a watersoluble fragment, HNcl. Isolated HN incubated for the times indicated with chymotrypsin, covalently-linked to Sepharose, in the presence of Tris-chloride, pH 8.0, 50 mM; $CaCl_2$, 10 mM, and Triton X-100, 0.1%. Aliquots were precipitated and analyzed by SDS-polyacrylamide gel electrophoresis upon reduction by dithiothreitol, 10 mM. Staining with Coomassie blue.*

inate red cells, and this is compatible with the fact that it is univalent and that hemagglutinating structures must be polyvalent. The receptor binding sites do however seem to be present on HNcl, because HNcl can completely block the hemagglutination inhibiting activity of rabbit anti-HN antibodies.

In the native HN protein, two equal polypeptides are covalently linked by disulfide bond(s). We have found that in SDS-polyacrylamide gel analysis without reducing agents, the proteins of SV5 and Sendai virus display a molecular weight approximately 2-fold higher than the individual polypeptide chains, and this is also true for NDV (16). With the isolated HNcl polypeptide, a similar situation has been found; the reduced polypeptide has an apparent M.W. of ~59,000, and omission of reducing agent results in a molecular weight of ~120,000, indicating that interchain disulfide bonds are located on the watersoluble, larger fragment of HN (HNcl).

In addition to the covalent bonds between polypeptide chains, non-covalent interactions exist between the hydrophobic bases of HN. One observation that supports this is the partial dimerization of HN polypeptides even in the presence of reducing agent and SDS (cf. Fig.2; ref.33). Such dimers are not found with HNc1, and this suggests that dimerization of the reduced and denatured HN molecule is due to interaction between the hydrophobic peptide HNc2. Recent studies with the erythrocyte membrane glycoprotein glycophorin (3) have shown that non-covalent dimer formation between hydrophobic glycopolypeptides can occur in the presence of SDS.

A non-covalent interaction between the hydrophobic portions of HN is furthermore indicated by the sedimentation characteristics of this protein in sucrose gradients containing the non-ionic detergent Triton X-100. The native HN protein sediments with an S value of 8.6, whereas the HNc1 fragment sediments at 7 S. Upon reduction by DTT, the HNc1 sediments at 5 S, indicating that there are intermolecular disulfide bonds that are accessible to reduction under non-dissociating conditions. However such a decrease in sedimentation velocity on reduction is not observed with the 8.6 S native HN protein which possesses the hydrophobic base, suggesting that the dimers of HN are stable, in spite of the reduction of S-S bonds, and that the dimers are held together by hydrophobic interaction between the bases.

The site involved in cleavage into HNc1 and HNc2 appears to be a small region which is sensitive to several proteases, because treatment with trypsin or thermolysin also gives rise to large water soluble fragments which contain the active neuraminidase site and which are very similar in size to the chymotryptic fragment HNc1. The hydrophobic fragment is responsible for anchoring the protein in the membrane and for rosette-formation by the isolated protein. HN polypeptide chains form dimers by disulfide linkage(s) in the watersoluble distal position of the spike, as well as by non-covalent interaction at the hydrophobic base of the spike. N-terminal analyses on the fragments have led to the conclusion that HN and HNc1 contain blocked N-termini, and this suggests that HN is oriented with its carboxyterminus at the hydrophobic membrane-associated end of the molecule, (Fig.3).

THE POLYPEPTIDE CHAINS OF THE F PROTEIN

Proteolytic cleavage of the precursor glycoprotein F_0 (in Sendai virus $\sim$ 65,000 M.W.) yields a major glycopolypeptide of $\sim$53,000 M.W., now designated F_1. In the early studies (8, 27) only this cleavage product was identified, although other cleavage product(s) were thought to be present in the heterogeneous small molecular weight material found near the dye front of the 7.5% standard polyacrylamide gels used for polypeptide analysis. More recently, a smaller polypeptide, F_2, has been identified

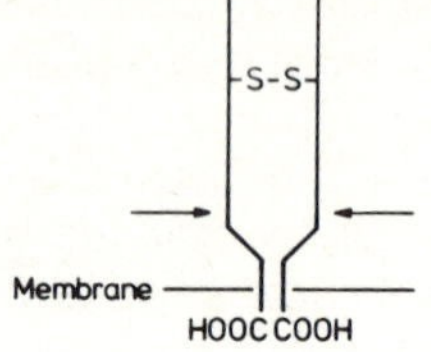

Fig.3. *Diagrammatic representation of the structural features of the HN proteins of the paramyxovirus SV5, as discussed in the text.*

which represents the other fragment of the cleaved F_0, and which remains covalently linked to F_1 by disulfide bonds (32).

The identification of F_2 was accomplished by isolating the disulfide-linked complex, $F_{1,2}$, under non-reducing conditions by preparative electrophoresis on SDS-polyacrylamide gels, followed by reduction and analysis on gels that are suitable for resolution of small peptides, i.e., containing 20% acrylamide. With Sendai virus, the apparent molecular weight of F_2 is ~ 13,000 daltons. A separate F_2 peptide is present only in the cleaved F protein and is not found associated with the isolated F_0 precursor, and this has established that it indeed arises from cleavage of F_0 and that it represents part of the precursor molecule. By labeling with ^3H-glucosamine it was established that F_2 is a glycopeptide, and that its relative glucosamine content is approximately three times that of the F_1 glycopolypeptide.

The F glycoprotein of other parainfluenza viruses, SV5 and NDV, also contains an F_2 glycopolypeptide of a similar molecular weight (~15,000 for SV5, and ~10,000 for NDV) which is disulfide-linked to the previously identified larger glycopolypeptide, now designated F_1. An F_0 precursor of the F_1 and F_2 chains has been found in cells infected with NDV (18) and SV5 (25). These findings with three different parainfluenza viruses suggest that it is a general property of the biologically active F protein that they contain two disulfide-linked polypeptide chains which are derived from an F_0 glycopolypeptide precursor.

In its native state, the F glycoprotein forms a spike with a morphological appearance which is distinct from the HN spike (33). Like the HN spike, this protein would be expected to contain a hydrophobic domain which accounts for its anchoring in the lipid of the virus envelope. Attempts to degrade this protein under controlled conditions, i.e., leaving the major portion intact and removing only the hydrophobic region, have yielded several peptides, and it has not yet been possible to use this approach for determining the polarity of the protein as described above for HN. However, end-group determinations on the F_0 precursor and the F_1 and F_2 polypeptide chains isolated from Sendai virus consistently failed to detect a free-amino terminal amino acid on F_2 and F_0, whereas a free phenylalanine was found on F_1 (Fig.4). This suggests that F_2 and F_0 share the same aminoterminus, and that F_2 is derived from the N-terminal region of F_0, whereas the N-terminus on F_1 is generated in the proteolytic cleavage of F_0. The analogy to other membrane glycoproteins whose C-termini are associated with the membrane (35, 36), including the HN glycoprotein of the same viruses (v.s.), would suggest that F_1 possesses the attachment site for the membrane, and that F_2 forms the distal end of the glycoprotein spike (Fig.4). This orientation is compatible with the finding that F_2 is rich in carbohydrate.

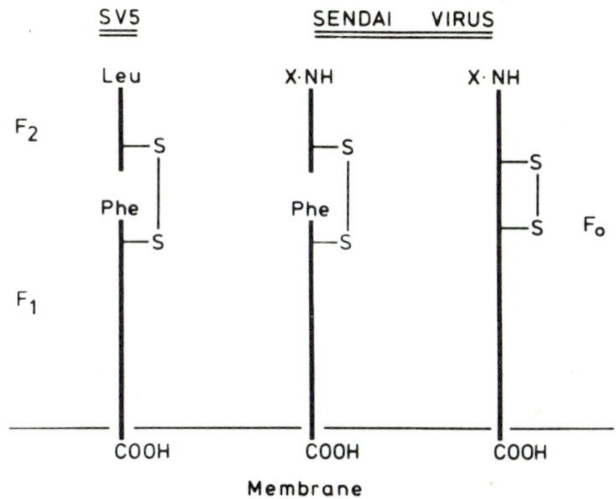

Fig.4. *Diagrammatic representation of the structural features of the F protein of paramyxoviruses, as discussed in the text.*

End-group determinations on F_1 polypeptides from three different viruses, SV5, NDV, and Sendai, revealed the same residue, i.e. phenylalanine. Further residues were determined by Edman degradation and end-group analysis of the newly exposed N-termini by the procedure of Weiner *et al.*(38). As shown in Fig.5, there clearly is similarity between these sequences, with phenylalanine in position 1, glycine in 3 and isoleucine in 6 shared by the F_1 polypeptides from all three viruses, and residues in position 4 and 5 shared by two out of three. Furthermore, the few differences that are found can be explained by point mutations, and wherever there is a change in an amino acid, the replacement is another uncharged residue. This striking similarity suggests that the N-terminal regions of the F_1 polypeptides are homologous and demonstrates that the proteolytic cleavage of the F_0 precursors has occurred at identical sites in the three viruses. This is remarkable in view of the fact that proteases of different hosts were involved in the proteolysis of the precursors, i.e., the proteases of the allantoic membrane of the chick embryo with NDV and Sendai virus, and of the MDBK cell with SV5.

N-Terminal Sequences of F_1

Sendai	H - Phe - Phe - Gly - Ala - Val - Ile -
SV5	H - Phe - Ala - Gly - Val - Val - Ile -
NDV	H - Phe - Ile - Gly - Ala - Ile - Ile -

Fig.5 *N-terminal sequences of the F glycopolypeptides of three paramyxoviruses.*

These findings suggest strongly that cleavage must occur at a sharply defined site for the cleaved F protein to be biologically active. We have begun to further investigate this point by analysis of protease activation mutants of Sendai virus (31) which are characterized by altered susceptibility to activating proteases. In these mutants, the molecular basis for altered susceptibility to specific proteases could lie either in an amino acid substitution which causes a change in tertiary structure of the molecule with a different stretch of the polypeptide exposes to proteases, or in a substitution which alters the susceptibility of a peptide bond in a specific position. The results to date suggest that the latter is the case. Pa-cl, a mutant which is cleaved and activated by chymotrypsin and elastase, but not by trypsin, was isolated with uncleaved F_0, and treated *in vitro* with elastase. The N-terminus of the isolated F_1 polypeptide was found to be identical to that of wild type virus cleaved *in vivo*, i.e. phe-phe-gly-. Thus in this mutant the change in susceptibility appears to have been brought about by amino acid substitution in the C-terminal position of F_2 by an exchange from a trypsin sensitive lysine or arginine to an amino acid that forms an elastate-sensitive peptide bond. This finding further emphasizes the apparent requirement for cleavage of F_0 at a defined site of F_0. Studies with other mutants will show whether there indeed is as little latitude for activating proteolysis of F_0 as the present data suggest.

THE POLYPEPTIDES OF MEASLES VIRUS

Measles virus closely resembles paramyxoviruses in its major structural features and its ability to cause hemolysis and cell fusion, however the lack of neuraminidase and failure to bind to neuraminic acid-containing receptors (21, 37) distinguish it from other paramyxoviruses and led to its classification in the morbillivirus subgroup of paramyxoviruses (2). The identification of measles virus polypeptides and the assignment of functions to them has been hampered by difficulties in obtaining virus and isolated polypeptides in sufficient quantity and purity for analysis. Furthermore, there appeared to be some discrepancy in the findings obtained. There was an analogy to other paramyxoviruses in that hemagglutination and hemolysis could be separately inactivated and had distinct antigenicities (23, 24), suggesting that measles virus possessed two envelope glycoproteins. However with the exception of the early report of Hall and Martin (5) studies in several laboratories identified only one glycoprotein with a molecular weight of ~80,000 on standard polyacrylamide gels (1, 15, 17). The evidence indicates that this glycoprotein is the hemagglutinin (see chapter 14) and we suggest that it be designated H.

The identity of a second measles virus glycoprotein was suggested by experiments in which a second glycosylated protein could be detected in polyacrylamide gel analysis of measles virus proteins in the absence of reducing agents (6, 17). On the basis of this and the finding that the F proteins of other paramyxoviruses contain two disulfide-linked glycopolypeptide chains, F_1 and F_2 (32) it was suggested that in measles virus, cleavage of an F_0 precursor could occur in such a way that none of the carbohydrate was present on the larger (F_1) fragment (M.W. ~40,000), but was present on a smaller fragment (F_2) which was not resolved on standard polyacrylamide gels (17). Recent studies on polypeptide synthesis in measles virus-infected cells have supported this concept, and in addition clarified some other features of measles virus proteins (4). Fig.6 shows the previously identified measles virus polypeptides, and in addition a large polypeptide (L) which is present in small amount in the virion (lane V), as well as in the infected cells (lane I). In infected cells, all the measles virus polypeptides could be detected with a 30 min pulse, with the notable exception of the 40,000 M.W. component (F_1). However, polypeptide F_1 could be detected in infected cells in pulse-chase experiments (Fig.6B), suggesting that its rate of appearance is limited by the processing of a precursor. Such a precursor (F_0) has been found in infected cells by pulse-labeling with ^3H-glucosamine, and a subsequent chase led to the appearance of a carbohydrate-labeled peptide with an apparent molecular weight of ~15,000, i.e., F_2. Thus, the processing of the F protein of measles virus appears to resemble

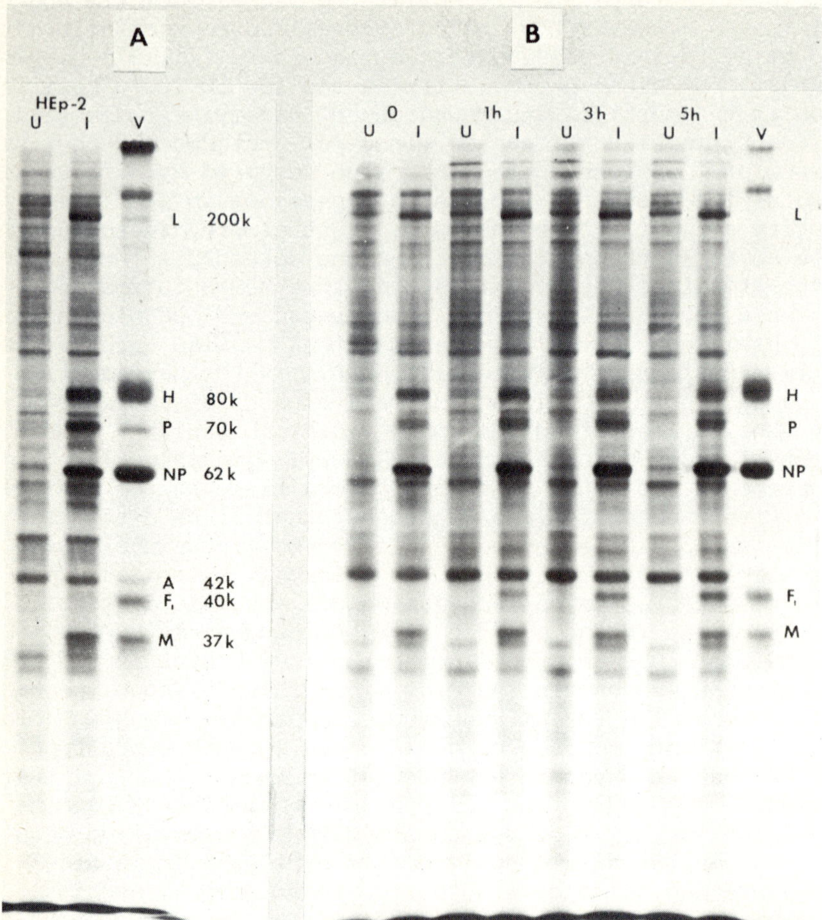

Fig.6. *Polyacrylamide gel electrophoresis of measles virus polypeptides in virions and virus-infected cells. A. HEp-2 cells were labeled for 30 min at 20 hr post infection and cell lysates were prepared and subjected to electrophoresis and fluorography. I = infected cells; U = uninfected cells; V = virions. Estimated molecular weights are given in 1,000's (K). B. Pulse chase experiment in measles-virus infected CV-1 cells. Cells were labeled for 30 min with ^{3}H-leucine at 24 hr post infection, washed, and chased with unlabeled medium. At times indicated cells were prepared for electrophoresis and fluorography.*

that of other paramyxoviruses, and the active protein seems similarly to be constituted of two disulfide-linked polypeptides, only the smaller of which contains a detectable amount of carbohydrate.

These studies of the synthesis of measles virus polypeptides in infected cells have been facilitated by an inhibitor of measles virus-induced cell fusion, carbobenzoxy-D-phenylalanyl-L-phenylalanyl-nitro-L-arginine (SV4814) (20, 22). This permitted following viral polypeptide synthesis for longer time periods without cell death, and also showed that although cell fusion is the major cause of cell death early after infection, infected cells ultimately die in the absence of fusion, indicating that measles virus possesses another mechanism for inducing cytopathological changes. The amino acids in this tripeptide inhibitor of cell fusion are similar to those at the N-terminal region of the F_1 polypeptides of other paramyxoviruses (see Fig.5), and this raises the possibility that this similarity is responsible for the inhibition of fusion by the compound.

CONCLUSIONS

The paramyxovirus spike with hemagglutinating and neuraminidase activities is composed of two identical, disulfide-linked glycopolypeptides (HN). Chymotrypsin treatment of this spike yields a water soluble protein with neuraminidase activity, which contains the larger polypeptide fragment (HNcl) as a disulfide-linked dimer. The other cleavage product is lipophilic and contains that part of HN which anchors the spike in the virus membrane. Failure to detect an N-terminal amino acid on HN or HNcl suggests that the hydrophobic tail of HN contains the carboxy-terminus. The other glycoprotein (F), which is required for virus-induced cell fusion, hemolysis, and the initiation of infection, consists of two disulfide-linked glycopolypeptides (F_1 and F_2) which are derived from a precursor (F_0), by proteolytic cleavage. The larger glycopolypeptide (F_1) has an estimated molecular weight of 48,000-54,000, and the smaller (F_2) ~ 10,000-16,000 depending on the virus. Although both polypeptides are glycosylated, F_2 contains more carbohydrate per unit protein than F_1. No free N-terminus was detected on F_0 or F_2 of Sendai virus, whereas N-terminal phenylalanine was found on F_1. This suggests that the order of the F_0 polypeptide is X-NH-F_2--F_1--COOH. Analysis of the N-terminal sequences of the F_1 polypeptides isolated from three viruses showed that these proteins are homologous and that identical N-termini are exposed by proteolytic activation of the F_0 precursors. Studies of the synthesis of measles virus polypeptides in cells protected from virus-induced fusion by the drug SV-4814, have revealed the following polypeptides: L, MW ~ 200,000; H, 80,000; P, 70,000; NP, 62,000; F_0, 62,000; F_1, 40,000; M, 37,000; and F_2 15,000.

H and F_0 are glycosylated, and the evidence suggests that F_1 and F_2 are derived from F_0 with only F_2 containing carbohydrate.

ACKNOWLEDGEMENTS

This work was supported by a grant from the Kroc Foundation and by Research Grants CA18213 from the National Cancer Institute and AI-05600 from the National Institute of Allergy and Infectious Diseases, DHEW, and PCM 76-09993 from the National Science Foundation.

REFERENCES

1. Bussell, R.H., Waters, D.J., Seals, M.K. and Robinson, W.S. (1974). *Med. Microbiol. Immunol.* 160, 105.
2. Fenner, F. (1976). *Intervirology* 7, 59.
3. Furthmayr, H. and Marchesi, V.T. (1976). *Biochemistry* 15, 1137.
4. Graves, M.C., Silver, S.M. and Choppin, P.W. (1977). *Abstr. Amer. Soc. Microbiol.*, p. 283.
5. Hall, W.W. and Martin, S.J. (1973). *J. Gen. Virol.* 19, 175.
6. Hardwick, J.M. and Bussell, R.H. (1976). *Abstr. Amer. Soc. Microbiol.*, p.232.
7. Homma, M. (1975). *In* "Negative Strand Viruses" (B.W.J. Mahy and R.D. Barry, eds.). p.685. Academic Press, London.
8. Homma, M. and Ohuchi, M. (1973). *J. Virol.*12, 1457.
9. Kaplan, J. and Bratt, M.A. (1973). *Abstr. Amer. Soc. Microbiol.*, p.243.
10. Klenk, H.-D., Rott, R., Orlich, M. and Blödorn, J. (1975). *Virology* 68, 426.
11. Laver, W.G. and Valentine, R.C. (1969). *Virology* 38, 105.
12. Lazarowitz, S.G. and Choppin, P.W. (1975). *Virology* 68, 440.
13. Lazarowitz, S.G., Compans, R.W. and Choppin, P.W. (1971). *Virology* 46, 830.
14. Lomniczi, B., Meager, A. and Burke, D.C. (1971). *J. Gen. Virol.* 13, 111.
15. Miller, C.A. and Fields, B.N. (1975). *Abstr. Amer. Soc. Microbiol.* p.245.
16. Moore, N., Cheyne, I.M. and Burke, D.C. (1975). *In* "Negative Strand Viruses". (B.W.J. Mahy and R.D. Barry, eds.), p.51, Academic Press, London.
17. Mountcastle, W.E. and Choppin, P.W. (1977). *Virology* 78, 463.
18. Nagai, Y., Ogura, H. and Klenk, H.-D. (1976). *Virology* 69, 523.
19. Nagai, Y., Klenk, H.-D. and Rott, R. (1976). *Virology* 72, 494.

20. Nicolaides, E., DeWald, H., Westland, R., Lipnick, M. and Posler, J. (1968). *J. Med. Chem.* 11, 74.
21. Norrby, E. (1962). *Arch. Gesamte Virusforsch.* 12, 164.
22. Norrby, E. (1971). *Virology* 44, 599.
23. Norrby, E. and Gollmar, Y. (1972). *Infect. Immun.* 6, 240.
24. Norrby, E. and Gollmar, Y. (1975). *Infect. Immun.* 11, 231.
25. Peluso, R.W., Lamb, R.A. and Choppin, P.W. (1977). *J. Virol.* 23, 177.
26. Samson, A.C.R. and Fox, C.F. (1973). *J. Virol.* 12, 579.
27. Scheid, A. and Choppin, P.W. (1974a). *Virology* 57, 457.
28. Scheid, A. and Choppin, P.W. (1974b). *Virology* 62, 125.
29. Scheid, A. and Choppin, P.W. (1975a). *In* "Negative Strand Viruses" (B.W.J. Mahy and R.D. Barry, eds.)p.177. Academic Press, London.
30. Scheid, A. and Choppin, P.W. (1975b). *In* "Proteases and Biological Control". (E. Reich, D.B. Rifkin, and E. Shaw, eds.), p.645. Cold Spring Harbor Laboratory.
31. Scheid, A. and Choppin, P.W. (1976). *Virology* 69, 265.
32. Scheid, A. and Choppin, P.W. (1977). *Virology* 80, 54.
33. Scheid, A., Caliguiri, L.A., Compans, R.W. and Choppin, P.W. (1972). *Virology* 50, 640.
34. Shapiro, S.C. and Bratt, M.A. (1971). *Proc. Soc. Exptl. Biol. Med.* 136, 834.
35. Skehel, J.J. and Waterfield, M.D. (1975). *Proc. Nat. Acad. Sci. USA* 72, 93.
36. Tomita, M. and Marchesi, V.T. (1975). *Proc. Nat. Acad. Sci. USA* 72, 2964.
37. Waterson, A.P. (1965). *Arch. Gesamte Virusforsch.* 16, 57.
38. Weiner, A.M., Platt, T. and Weber, K. (1972). *J. Biol. Chem.* 247, 3242.

SV5 AND SENDAI VIRUS POLYPEPTIDES: ASPECTS OF COMPOSITION, SYNTHESIS AND PHOSPHORYLATION

R.A. LAMB, R.W. PELUSO and P.W. CHOPPIN

The Rockefeller University, New York, N.Y. 10021, U.S.A.

The paramyxoviruses Sendai virus and simian virus 5 (SV5) contain proteins with similar biological functions. These are the nucleocapsid protein (NP), a glycosylated protein associated with hemagglutinating and neuraminidase activity (HN), a glycosylated protein associated with cell fusion and hemolysis activity (F), and a membrane protein (M) (2). Two internal proteins L and P are associated with the nucleocapsid and are probably involved in RNA transcriptase activity (1, 11). In addition, there is a 42,000 mol. wt. polypeptide (A), suggested previously to be cellular actin (9, 18), and in Sendai virus a minor component, polypeptide 5 (mol. wt. 54,000), is also found. On Sendai virions the F glycoprotein is present either as an inactive precursor F_o, or as a cleaved and active form consisting of two disulfide-linked chains F_1 and F_2 (15, 16). However, on SV5 virions only the active form $F_{1,2}$ has been found (14, 16).

In Sendai virus-infected cells, in addition to the virion polypeptides L, P, HN, F_0, NP, and M, two polypeptides, B and C, with no virion counterparts have been identified (9). This present report describes further the synthesis of Sendai virus polypeptides in infected cells and compares it with SV5 polypeptide synthesis.

THE IDENTIFICATION OF B AS A PHOSPHORYLATED FORM OF M

Studies of the subcellular distribution of Sendai virus-specific polypeptides showed that B and C were unstable in disrupted cells, but C was found associated with ribosome-containing fractions if the experiments were performed in the presence of protease inhibitors (7). On storage of disrupted infected cells at 0°C, polypeptide B decreased in amount with a concomitant increase in polypeptide M. However, in 5 min pulse-labels of infected cells with ^{35}S-methionine followed by a chase in the absence of radioisotope, polypeptide M was converted to polypeptide B. These results suggested that a post-translational modification of M was occurring in infected cells, with reversion of B to M after disruption of cells. The possible phosphorylation of polypeptides was investigated, and as shown in Fig.1A, when Sendai virus-infected CEF cells were labeled with $^{32}PO_4$ and analyzed on 13% and 8% polyacrylamide gels, polypeptides P, NP, and B were found to be phosphorylated. However, on gels containing urea, polypeptide B co-migrated with M and little phosphorus label co-migrated with NP, but new bands NP_{P1} (co-migrating with F_o) and, in lesser amount, NP_{P2}, were observed migrating slower than NP. Examination of the polypeptides of purified virions grown in the presence of $^{32}PO_4$ on 13% gels showed that L, P, HN, and NP were phosphorylated, and also that in the B region of the gel a phosphopolypeptide was detected, although no such polypeptide was observed in ^{35}S-methionine-labeled virus. When gels containing urea were used no $^{32}PO_4$ was associated with NP, but it now migrated predominantly as NP_{P2}, and to a lesser extent NP_{P1}. Thus, two new phosphopolypeptide bands have been found, NP_{P1} and NP_{P2}, in virions and infected cells, with NP_{P1} being predominant in infected cells and NP_{P2} in virions.

Peptide mapping by limited proteolysis and analysis on SDS gels of B, M, and the B/M band (from a urea gel) with *Staphylococcus aureus* V8 protease, papain and chymotrypsin showed that they contained similar peptide fragments (Fig.1B), and this was confirmed by tryptic peptide mapping using high voltage electrophoresis and chromatography. Peptide mapping of NP, NP_{P1} (Fig.1B), and NP_{P2} by limited proteolysis showed that NP, NP_{P1}, and NP_{P2} had the same peptide fragments except that NP_{P1} contained in addition the peptides of F_o, with which it co-migrates on gels. That polypeptide B was a phosphorylated form of M was further shown by *in vitro* conversion of M to B by a virion-associated protein kinase and, conversely, by conversion of B to M through the loss of phosphate in cell lysis.

Very little B was found in virions, although in infected cells B and M were found in similar proportions. It is not known whether M is selectively incorporated into virions or whether phosphate is removed from B during the maturation process. This raises several possible biological roles for phosphorylation of the membrane protein. Phosphorylation and dephosphorylation of proteins associated with the erythrocyte membrane resulting in charge and conformational changes of the adjacent lipoproteins have been found (4). Thus, phosphorylation and dephosphorylation of M could influence either the mechanism by which the glycoproteins associate with M at the plasma membrane or the presumed recognition mechanism by which the nucleocapsid associates with M at the plasma membrane.

Although phosphorylated forms of NP (NP_{P1} and NP_{P2}) have been found in virions and infected cells, unphosphorylated NP is the major species in both cases. Phosphorylated NP may be involved in transcription of Sendai virus RNA as both L and P are also phosphorylated and are found in Sendai virus transcriptive complexes (5, 11, 17). Alternatively, phosphorylated NP may bind only to plus-strand 50 S RNA, providing a mechanism by which the replicase could recognize the correct RNA strand and then produce negative-strand genome RNA. Another possibility is that as phosphorylated forms of NP are present in virions in much smaller quantity than non-phosphorylated NP and if phosphorylated NP bound only to plus-strand RNA, the additional charge would be a means by which positive-strand RNA nucleocapsids are largely excluded from the virion.

PEPTIDE MAPPING OF THE POLYPEPTIDES IN SENDAI VIRIONS AND INFECTED CELLS

Peptide mapping of all the virion polypeptides and the Sendai virus polypeptides synthesized in infected cells was done to investigate the identity of polypeptide C found only in infected cells. Limited proteolysis with *S. aureus* V8 protease (Fig.2A) and chymotrypsin revealed the following: 1. Polypeptide C has no virion counterpart; 2. P, HN, F_o, NP, and M are unique polypeptides; 3. The peptide composition of the virion polypeptides and their cell counterparts are the same. Minor differences, due to varying degrees of glycosylation, were observed between the glycoproteins obtained from MDBK cell-grown virus and infected CEF cells, but not when the polypeptides were obtained from MDBK cells; 4. Virion polypeptide 5 (mol. wt. 54,000) has most of the same peptides found in NP (Fig.2B) and is therefore probably a cleavage product of NP and not a primary gene product; 5. Polypeptide L from both infected cells and the virion has the same peptides (Fig.2B) and L is a unique species; 6. Polypeptide A and purified MDBK cell actin (Fig.2B) have identical peptide maps confirming that cellular actin is found in Sendai virus particles.

From these results the available evidence would suggest that polypeptide C is a non-structural polypeptide, but as described below an increase in synthesis of polypeptides that are presumably host-coded has been observed in SV5 and Sendai virus-infected cells. Therefore until a virus-specific mRNA coding for C has been identified and translated *in vitro*, the possi-

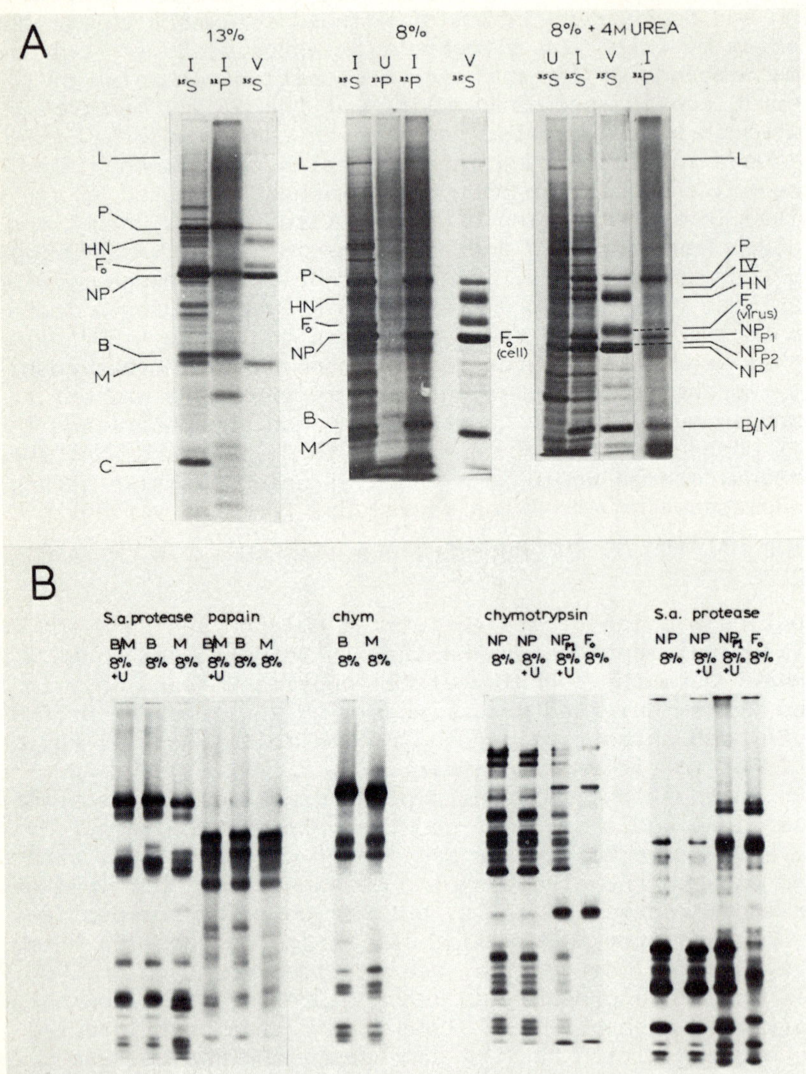

Fig.1A. *Phosphorylation of Sendai virus-infected cell polypeptides. At 14 hr p.i. Sendai virus-infected CEF cells were labeled for 30 min with either ^{35}S-methionine or $^{32}PO_4$, and the cell lysates were analyzed on 13%, 8%, or 8% + 4M urea gels. ^{35}S = ^{35}S-methionine-labeled; ^{32}P = $^{32}PO_4$-labeled; V = MDBK-grown marker virions; I = infected cell lysate; U = uninfected cell lysate. The conditions of cell growth, virus*

bility that it is a host-cell protein induced by Sendai virus infection cannot be eliminated.

COMPARISON OF SV5 AND SENDAI VIRUS POLYPEPTIDES IN INFECTED CELLS

In SV5-infected cells the virion polypeptides L, HN, NP, 5, and M were detected above the host-cell background after a short pulse-label (Fig.3A). However, no polypeptides comprising the smaller glycoprotein (F) were identified. Instead, a glycosylated precursor F_0 (mol. wt. ~66,000) was observed which, in pulse-chase experiments, was cleaved to yield polypeptides F_1 and F_2 (12). In addition to the structural polypeptides, up to five other polypeptides (I-V) were observed in SV5-infected cells, depending on the host-cell type. The available evidence suggests that polypeptides I-IV are host-cell polypeptides whose synthesis may be enhanced upon infection, as polypeptides of similar electrophoretic mobility are found in uninfected cells and polypeptides I-IV found in SV5-infected CEF cells are also found in Sendai-infected CEF cells (Fig.3A). A densitometer scan of an autoradiograph of this region of a SV5-infected cell showing the increase in synthesis of these polypeptides is shown in Fig.3B. Polypeptide IV has been found in all cell types examined, but the detection of polypeptides I-III depends on the cell type. Polypeptide V (mol. wt. ~ 24,000) has been observed in all cell types examined, has no counterpart in uninfected cells, and thus may represent a non-structural polypeptide analogous to the presumed non-structural polypeptide C (mol. wt. ~22,000) synthesized in Sendai virus-infected cells (Fig.3A).

High salt concentrations were used to selectively decrease the synthesis of host-cell proteins, and although the synthesis of most cellular proteins was greatly reduced relative to SV5 proteins, the synthesis of polypeptides I, III, and IV in CEF cells, was much less inhibited than that of the other host-cell polypeptides. This suggests that these polypeptides may be translated relatively more efficiently in infected cells than

---

purification, labeling of cells and polyacrylamide gel electrophoresis have been described previously (6, 8, 9).

Fig.1B. *Peptide mapping by limited proteolysis of polypeptides B, M, NP, and NP_{P1}. Polypeptides from Sendai-infected cells were removed from either an 8% gel (8%) or an 8% gel containing 4 M urea (8% + U), and digested with Staphylococcus aureus V8 protease, chymotrypsin, or papain as indicated. The conditions for peptide mapping were those of Cleveland et al. (3) modified as described previously (8).*

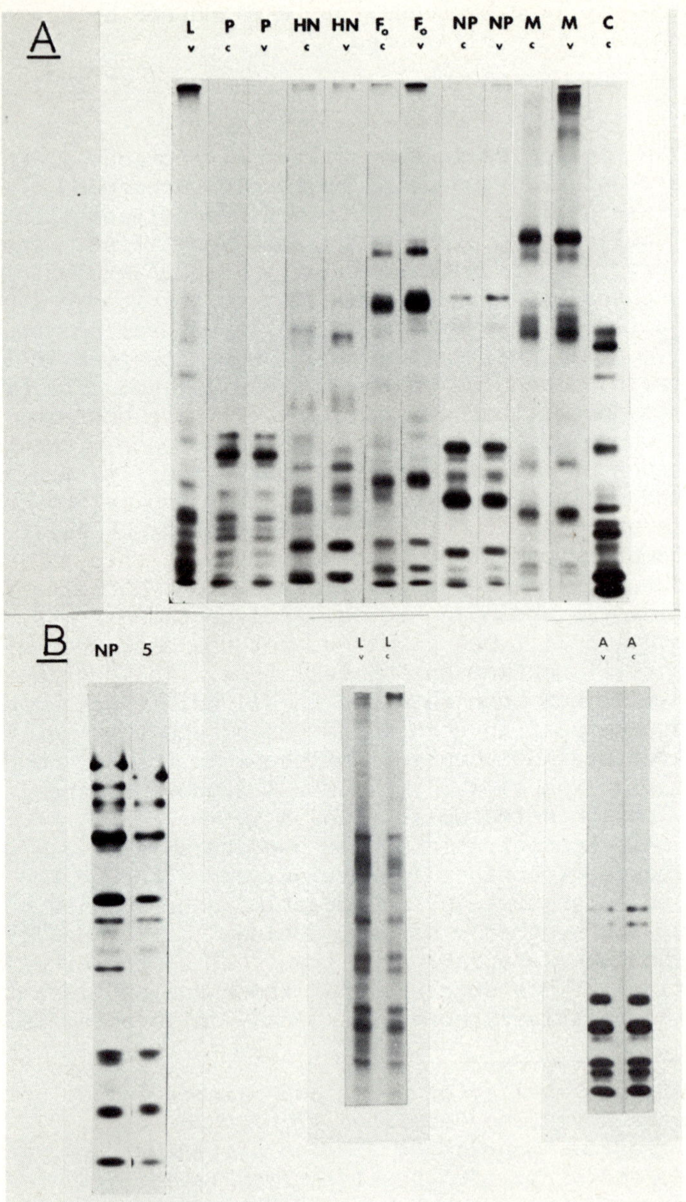

Fig.2A. *Peptide mapping of Sendai virion and infected cell polypeptides by limited proteolysis with S. aureus V8 protease. Virion polypeptides were obtained from ^{35}S-methionine-labeled virus grown in MDBK cells, and infected cell polypeptides, from*

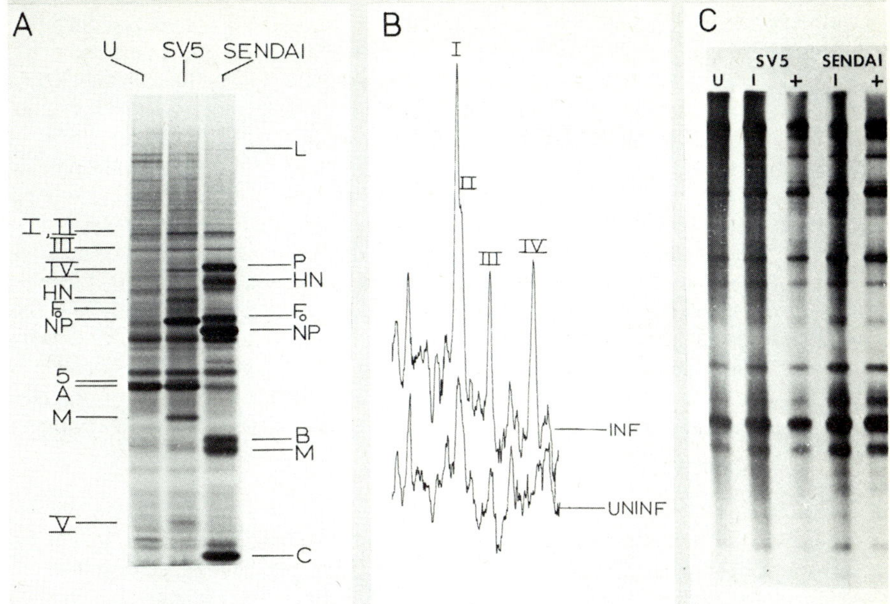

Fig.3A. *Comparison of polypeptides synthesized in SV5- and Sendai virus-infected CEF cells.*
Fig.3B. *Densitometer tracing of an autoradiogram of polypeptides I-IV in uninfected and SV5-infected CEF cells.*
Fig.3C. *Peptide mapping by limited proteolysis with chymotrypsin of polypeptide III from uninfected, SV5-, and Sendai-infected CEF cells. U = Uninfected; I = infected; + = treated with 150 mM NaCl prior to labeling.*

---

^{35}S-methionine-labeled CEF cells. Polypeptides L, F_O, and NP were obtained from 8% gels of both infected cell lysates and purified virions, and polypeptides P, HN, M, B, and C, from 13% gels. V = polypeptide from purified virions; C = polypeptides from infected CEF cells. Peptide mapping conditions were as described in Fig.1B.
Fig.2B. *Peptide mapping by limited proteolysis of Sendai virus polypeptides L, NP, 5, and A. ^{35}S-methionine-labeled virion polypeptides NP, 5, and A were obtained from a 13% gel, and L from virions and infected MDBK cells was obtained from 8% gels. Actin was purified from ^{35}S-methionine-labeled MDBK cells by affinity chromatography as described by Lazarides and Lindberg (10). NP, 5, and A were digested with S. aureus V8 protease, and L, with chymotrypsin, as described in Fig.1B.*

in uninfected cells. Polypeptide III from SV5- and Sendai-infected CEF cells, synthesized in the absence or presence of high salt, and the polypeptide of identical electrophoretic mobility from uninfected cells have been peptide mapped (Fig. 3C) and shown to contain the same peptides, confirming that this polypeptide is a host-cell polypeptide whose synthesis is enhanced after infection. It is not known whether polypeptides I-IV have any role in viral infection, e.g., as a component of the replicase enzyme, or whether the enhanced synthesis relative to other cell proteins is caused by nonspecific perturbations of cellular metabolism during infection. Induction of some host-cell polypeptides by amino acid analogues has been observed (see chapters 23 and 39) and also by glucose starvation (13), and therefore the mechanism and role of these induced host-cell polypeptides by non-transforming RNA viruses merits further investigation.

SUMMARY

Polypeptide B found in Sendai virus-infected cells has been shown to be a phosphorylated form of the membrane protein. Phosphorylated forms of NP have been found which can be separated from unphosphorylated NP on acrylamide gels. Peptide mapping of the virion and infected-cell polypeptides has shown that polypeptide C, found only in infected cells, is unique as are L, P, HN, F_0, NP, and M. The precursor (F_0) to F_1 and F_2 has been identified in SV5-infected cells. Sendai virus and SV5 infection causes enhanced synthesis of some host-cell polypeptides.

ACKNOWLEDGEMENTS

This research was supported by Research Grants AI-05600 from the National Institute of Allergy and Infectious Diseases, PCM76-09993 from the National Science Foundation, and by CA-09256 from the National Cancer Institute, under which R.W.P. is a Predoctoral Trainee.

REFERENCES

1. Buetti, E., and Choppin, P.W. (1977). *Virology* 82, 493.
2. Choppin, P.W., and Compans, R.W. (1975). *In* "Comprehensive Virology" (H. Fraenkel-Conrat and R.R. Wagner, eds.), Vol. 4, p.95. Plenum Press, New York.
3. Cleveland, D.W., Fischer, S.G. Kirschner, M.W., and Laemmli, U.K. (1977). *J. Biol. Chem.* 252, 1102.
4. Gazitt, Y., Ohad, I., and Loytes, A. (1976). *Biochem. Biophys. Acta* 436, 1.
5. Lamb, R.A., Ph.D. Thesis, University of Cambridge, England.
6. Lamb, R.A., and Choppin, P.W. (1976). *Virology* 74, 504.

7. Lamb, R.A., and Choppin, P.W. (1977a). *Virology* 81, 371.
8. Lamb, R.A., and Choppin, P.W. (1977b). *Virology* 81, 382.
9. Lamb, R.A., Mahy, B.W.J., and Choppin, P.W. (1976). *Virology* 69, 116.
10. Lazarides, E., and Lindberg, U. (1974). *Proc. Nat. Acad. Sci. USA*, 71, 4742.
11. Marx, P.A., Portner, A., and Kingsbury, D.W. (1974). *J. Virol.* 13, 107.
12. Peluso, R.W., Lamb, R.A., and Choppin, P.W. (1977). *J. Virol.* 23, 177.
13. Pouysségur, J., Shiu, R.P.C., and Pastan, I. (1977). *Cell* 11, 941.
14. Scheid, A., Caliguiri, L.A., Compans, R.W., and Choppin, P.W. (1972). *Virology* 50, 640.
15. Scheid, A., and Choppin, P.W. (1974). *Virology* 57, 475.
16. Scheid, A., and Choppin, P.W. (1977). *Virology* 80, 54.
17. Stone, H.O., Kingsbury, D.W., and Darlington, R.W. (1972). *J. Virol.* 10, 1037.
18. Wang, E., Wolf, B.A., Lamb, R.A., Choppin, P.W., and Goldberg, A.R. (1976). *In* "Cell motility" (R. Goldman, T. Pollard, and J. Rosenbaum, eds.), p.589, Book A. Cold Spring Harbor.

STUDIES ON THE PROTEINS AND CARBOHYDRATES OF INFLUENZA VIRUS

K. NAKAMURA, M.W. SHAW, D.G. PRITCHARD and R.W. COMPANS

Department of Microbiology,
University of Alabama Medical Center,
Birmingham, Alabama 35294, USA.

We have been interested in determining the detailed structure of the influenza virion, and the function of its various molecular components. A large amount of data has been obtained in recent years on the polypeptide and nucleic acid components of the virion, but comparatively little information is available concerning the viral carbohydrates. The hemagglutinin and neuraminidase of influenza virions undergo two types of secondary modifications, glycosylation and sulphation. Whether these modifications are required for the assembly of these glycoproteins into virions and their subsequent biological activities is not clearly understood. We have undertaken studies of the glycosylated and sulphated components of influenza virions, the sites of secondary modification in the infected cell, and the effects of inhibitors on these processes.

At the meeting on The Biology of Large RNA Viruses in 1969, we described the appearance of cytoplasmic electron-dense inclusions in influenza virus-infected cells (5). Although a large amount of information on influenza virus structure and replication has been obtained since that time, the nature of these inclusions and their possible significance in virus replication have not been established. We have recently developed a procedure for the isolation of these inclusions in pure form, and determined their polypeptide composition.

THE CARBOHYDRATE COMPONENTS OF INFLUENZA VIRIONS

In addition to the carbohydrates which are covalently linked to viral glycoproteins, host cell-derived mucopolysaccharides have been detected in purified virus preparations by labeling

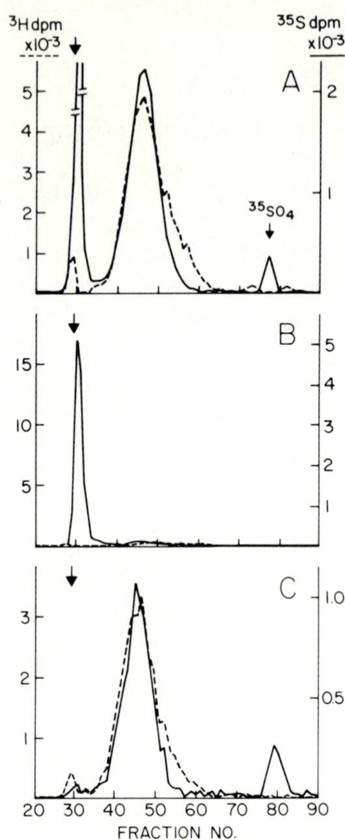

Fig.1. *Analysis of pronase digests of influenza virions labeled with ^{3}H-glucosamine (3 µCi/ml) and $^{35}SO_4$ (20 µCi/ml) by gel filtration on Biogel P-6. The conditions for pronase digestion and gel filtration were essentially those described by Sefton (17). A. Pronase digest of intact virions. B. Pronase digest of pellet fraction obtained after treatment of virions with 10% Triton X100 and centrifugation at 100,000 x g for 30 min. C. Pronase digest of supernatant from sample described in B.*

with $^{35}SO_4$ (4). To further characterize the carbohydrate components, purified virions grown in MDBK cells labeled with ^3H-glucosamine and $^{35}SO_4$ were digested with pronase and the resulting components analyzed by gel filtration on Biogel P6 (Fig. 1). The elution profile in Fig.1A shows 3 peaks of $^{35}SO_4$ label. The first peak, around fraction 30, is close to the void volume, and contains little or no ^{3}H-glucosamine label. The second peak coincides with a major peak of ^{3}H-glucosamine label, whereas the minor peak around fraction 77 corresponds to the elution position of free $^{35}SO_4$. This peak was not detected when pronase digests of purified viral glycoproteins were analyzed (data not shown).

Fractionation of virions by treatment with Triton X100 followed by high speed centrifugation results in solubilization of the viral glycoproteins, while the sulphated mucopolysaccharides are pelleted along with the nonglycosylated viral proteins (4). As seen in Fig.1B, the first peak of $^{35}SO_4$ was observed only in pronase digests of the Triton pellet, thus identifying it as the mucopolysaccharide component. The Triton supernatant, when analyzed after pronase digestion, contains the second peak of $^{35}SO_4$ label and almost all of the ^{3}H label, identifying this fraction as the glycopeptides which contain oligosaccharides of the viral glycoproteins (Fig.1C).

As seen in Fig.1, the oligosaccharides of influenza virus glycoproteins consist of two or more size classes. The major peak of $^{35}SO_4$ coincides with the largest peak of ^{3}H-glucosamine (peak I) but there are smaller oligosaccharides with a lower ^{35}S: ^{3}H ratio (peak II). Preliminary compositional analyses indicate that peak II contains mannose-rich oligosaccharide components; the mannose: galactose ratio was 2.6 for peak II as compared with 0.3 for peak I. In further studies of the oligosaccharide components (manuscript in preparation) we have observed that the HA_1 glycoprotein contains both glycopeptides I and II whereas HA_2 contains peak I only. The glycopeptides of virions grown in MDCK, BHK21-F, and chick embryo fibroblast cells also consisted of two size classes, with the smaller oligosaccharides rich in mannose. However, the sizes of the oligosaccharide components varied with the cell type, as expected from previous observations indicating that the electrophoretic mobility of the glycoproteins depends on the cell type (6, 8, 16).

CELLULAR SITE OF GLYCOPROTEIN SULPHATION

The site of incorporation of $^{35}SO_4$ into viral polypeptides was studied in MDBK cells infected with the WSN strain of influenza virus (14). In plasma membranes isolated by a modification of the procedure of Warren and coworkers (18), three polypeptides, HA, NP, and M were resolved, and $^{35}SO_4$ was demonstrated in the HA glycoprotein. Similarly, $^{35}SO_4$ incorporation was observed in HA in both smooth and rough cytoplasmic membranes isolated by the procedure of Caliguiri and Tamm (2), and no incorporation was observed in nonglycosylated proteins. Since HA polypeptides appear to be synthesized in association with rough endoplasmic reticulum (3, 9, 11), these results indicate that at least partial sulphation of HA is already completed in this cellular compartment. However, when infected cells were doubly labeled with $^{35}SO_4$ and ^{3}H-leucine, the ratio of $^{35}S/^{3}H$ was not uniform in cellular fractions; it was highest in virions and decreased progressively in plasma membranes, smooth membranes, and rough membranes. These results suggest that further sulphate incorporation into HA may occur after migration of HA to smooth and plasma membranes. Significant incorporation of $^{35}SO_4$ into HA was detected in the presence of 100μg/ml of cycloheximide, which completely inhibited the synthesis of viral polypeptides. Sulphate continued to be incorporated into preformed HA for 30-60 minutes after cessation of protein synthesis. Since HA has been observed to migrate rapidly to smooth membranes after synthesis, these results also suggest that sulphation of HA continues in smooth as well as rough membranes.

As was observed for HA, the incorporation of $^{35}SO_4$ into mucopolysaccharides was also detected in rough and smooth cytoplasmic membrane fractions as well as plasma membranes. However, sulphation of mucopolysaccharides was markedly inhibited by influenza virus infection. In whole infected cells, sulphation of mucopolysaccharides was observed at a rate of less than 10% of control by 9 hr post-infection. The inhibition was more pronounced in rough membranes than in other subcellular fractions.

We attempted to analyze the process of sulphation *in vitro* using various cell fractions and $^{35}SO_4$ or 3'-phosphoadenosine-5'-phosphosulphate as sulphate donors. It was observed that smooth as well as rough membranes possessed sulphotransferase activity, whereas a 100,000 xg supernatant fraction of cell extracts was required for activation of $^{35}SO_4$. The product of *in vitro* sulphation was observed to be sulphated mucopolysaccharide in every cell fraction tested; no sulphation of HA was detected under these conditions.

EFFECTS OF INHIBITORS ON SECONDARY MODIFICATIONS OF INFLUENZA VIRUS GLYCOPROTEINS

We have compared the effects of three inhibitors, glucosamine, 2-deoxy-D-glucose (2-DG) and tunicamycin (TM) on glycosylation and sulphation of influenza virus glycoproteins (Table 1). The electrophoretic mobility of the HA progressively increased with increasing concentration of glucosamine accompanied by a decrease in the incorporation of ^{3}H-fucose and $^{35}SO_4^{-2}$ into the glycoprotein. However, significant amounts of these labels were detected in association with HA even when cells were treated with glucosamine concentrations as high as

TABLE 1

Effects of Inhibitors on Secondary Modifications of Influenza Virus Glycoproteins

Compound	Glycosylation of HA	Sulphation of HA	Molecular Weight of HA_0 Polypeptide
Glucosamine (40mM)	partial	extensive	69,000
Deoxyglucose (10mM)	partial*	not detected	67,000
Tunicamycin (0.3μM)	not detected	not detected	63,000

*2-deoxyglucose is incorporated into HA

40mM. In this case, the peak of ^3H-fucose label associated with HA was distinct from the peak detected with labeled amino acid precursors, indicating that HA glycoproteins produced in the presence of glucosamine are heterogeneous with respect to the extent of glycosylation. Further, it was found that the peak of ^{35}S label associated with HA coincided with that of ^3H-fucose label rather than HA_0.

When cells were treated with increasing concentrations of 2-DG, the incorporation of $^{35}SO_4^{-2}$ into HA decreased in parallel with that of ^3H-glucosamine, and no incorporation of either label was detected in cells treated with 10 mM 2-DG.

TM completely inhibited the incorporation of ^3H-glucosamine into viral glycoproteins at a concentration as low as 0.29 μM. Although unglycosylated hemagglutinin (HA_0) was not detected in whole cells treated with TM, it was clearly resolved in isolated smooth membrane fractions. No $^{35}SO_4^{-2}$ label was detected in the HA_0 protein.

These results further support the conclusion that sulphate groups are linked to sugar components of HA glycoproteins. In addition, the presence of HA_0 in smooth membranes of TM-treated cells suggests that neither glycosylation nor sulphation is essential for membrane association or migration of HA from rough to smooth membranes. It also appears that such modifications of HA are not required for integration of the protein into plasma membranes since HA_0 was detected in plasma membranes isolated from both 2-DG and TM-treated cells.

INFLUENZA VIRUS-INDUCED CYTOPLASMIC INCLUSIONS

We wished to isolate the paracrystalline inclusions observed in influenza virus-infected cells in order to determine their composition and possible significance in virus replication. Initially we attempted to localize the inclusions by examination of subcellular fractions by thin section electron microscopy. It was observed that the crude nuclear pellet obtained after Dounce homogenization was enriched in inclusions, which could be solubilized by treatment with 1% Triton X100. Subsequent centrifugation in a discontinuous sucrose-metrizamide gradient yielded a fraction highly enriched in inclusions. A second approach for purification involved extensive Dounce homogenization, extraction of the resulting cytoplasmic extract with Genesolv D (13), and centrifugation of the aqueous phase in a discontinuous sucrose gradient. Essentially pure preparation of inclusions were obtained by this procedure. As shown in Fig.2A, the isolated inclusions negatively stained with uranyl acetate were free from adherant ribosomes or other cellular contaminants. They show a major periodicity of ~8nm in some images, and a periodicity of ~4nm was also frequently observed.

The polypeptides of isolated inclusions labeled with ^3H-leucine were analyzed by SDS-polyacrylamide coelectrophoresis (Figs.2B and C). A single radioactively labeled polypeptide was detected, and it was identified as the nonstructural (NS) protein of influenza virus by coelectrophoresis with marker proteins. Further, a single band was observed in the position of NS in stained gels of the polypeptides of isolated inclusions (not shown). These results indicate that influenza virus-induced cytoplasmic inclusions contain a single virus-coded polypeptide, the NS polypeptide. Studies to determine whether any nucleic acid is also present in the inclusions are currently in progress.

DISCUSSION

The oligosaccharide components of influenza virus glycoproteins are unusual because they lack sialic acid, but otherwise they show some similarities in size and composition to oligosaccharides of other viral glycoproteins (1, 7, 10). The molecular weights of the MDBK-grown influenza viral glycopeptides were estimated by cochromatography with glycopeptides obtained from Sindbis virus grown in chick embryo fibroblasts (1), and a value of ~ 3000 daltons was obtained for peak I and 2000-2300 daltons for peak II. The estimate of 11,500-12,000 daltons for the carbohydrate content of HA (12, 15, 19), together with data indicating that HA_1 contains about twice as much

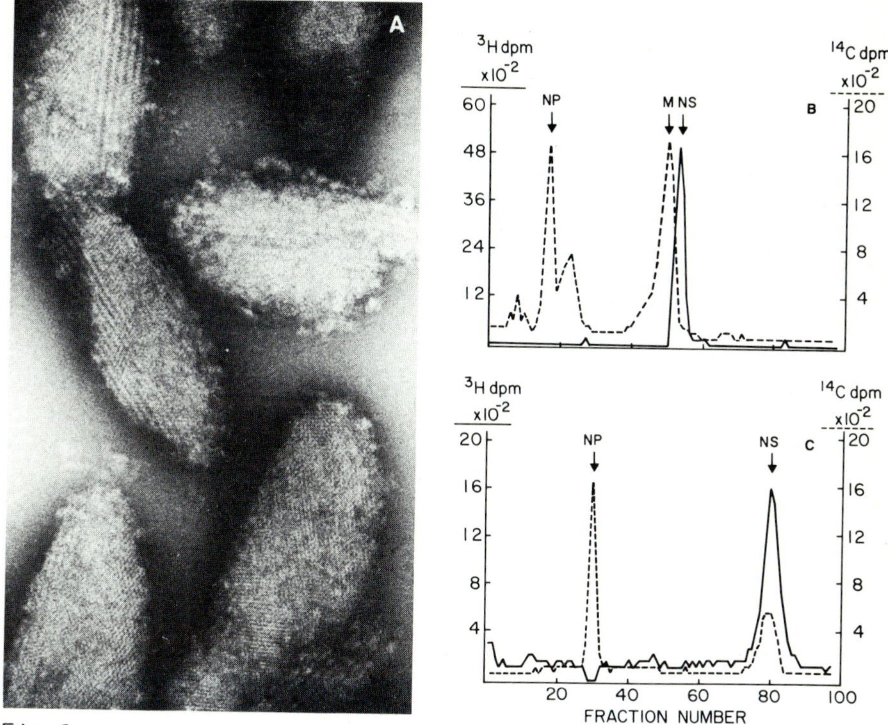

Fig.2. *A. Purified inclusions from influenza virus-infected cells, negatively stained with uranyl acetate. Magnification: X 100,000.*

B. Analysis of the polypeptides of ^{3}H-leucine-labeled purified inclusions by coelectrophoresis with ^{14}C-amino acid-labeled polypeptides of influenza (WSN) virions.

C. The same preparation shown in B, analyzed by co-electrophoresis with ^{14}C-labeled marker polypeptides of infected BHK21-F cells harvested at 23 hr post-infection. At this time NP and NS proteins are the only labeled proteins present in cells in significant amounts, even though cells were labeled during the entire growth cycle.

label (galactose or fucose) in peak I as does HA_2, indicate that HA_2 contains a single type I oligosaccharide, whereas HA_1 contains two type I oligosaccharides plus one type II oligosaccharide.

The isolation of the NS protein in pure form in paracrystalline inclusions from influenza virus-infected cells opens new possibilities for further studies of the structure and function of this protein. Yields of over 1 mg of pure NS protein may be obtained from batches of 10^9 infected BHK21-F cells. Further, the fact that the protein undergoes self-association into crystalline arrays suggests that it should be feasible to obtain larger crystals suitable for crystallographic analysis. The availability of large amounts of pure NS protein, which has not been subjected to denaturing conditions, should also facilitate studies of its possible enzymatic or regulatory functions in virus replication.

ACKNOWLEDGEMENT

This research was supported by Grant No. AI 12680 from The National Institute of Allergy and Infectious Diseases, USPHS.

REFERENCES

1. Burge, B.W., and Strauss, J.H. (1970). *J. Mol. Biol.* 47, 449.
2. Caliguiri, L.A., and Tamm, I. (1970). *Virology* 42, 100.
3. Compans, R.W., (1973). *Virology* 51, 56.
4. Compans, R.W., and Pinter, A. (1975). *Virology* 66, 151.
5. Compans, R.W., Dimmock, N.J. and Meier-Ewert, H. (1970). *In* "The Biolgoy of Large RNA Viruses"(R.D. Barry and B.W.J. Mahy, eds.). Academic Press New York, p.87.
6. Compans, R.W., Klenk, H.-D., Caliguiri, L.A., and Choppin, P.W. (1970). *Virology* 42, 880.
7. Etchison, J.R. and Holland, J.J. (1974). *Proc. Nat. Acad. Sci. USA* 71, 4011.
8. Haslam, E.A., Hampson, A.W., Radiskevics, I., and White, D.O. (1970). *Virology* 42, 566.
9. Hay, A.J. (1974). *Virology* 60, 398.
10. Keegstra, K., Sefton, B. and Burke, D. (1975). *J. Virol.* 16, 613.
11. Klenk, H.-D., Wollert, W., Rott, T., and Scholtissek, C. (1974). *Virology* 57, 28.
12. Laver, W.G. (1971). *Virology* 45, 275.
13. Nagayama, A., and Dales, S. (1970). *Proc. Nat. Acad. Sci. USA* 66, 464.
14. Nakamura, K., and Compans, R.W. (1977). *Virology* 79, 381.
15. Nakamura, K., and Compans, R.W. (1977). Submitted for publication.
16. Schulze, I.T. (1970). *Virology* 42, 890.
17. Sefton, B.M. (1976). *J. Virol.* 17, 85.
18. Warren, L., Glick, M.C., and Nass, M.K. (1966). *J. Cell Physiol.* 68, 269.
19. White, D.O. (1974). *Curr. Top. Microbiol. Immunol.* 63, 1.

STRUCTURAL STUDIES ON THE INFLUENZA VIRUS INTERNAL MEMBRANE PROTEIN

BETTY H. ROBERTSON, AJIT S. BHOWN, RICHARD W. COMPANS, and J. CLAUDE BENNETT

Department of Microbiology and Division of Clinical Immunology and Rheumatology, University of Alabama Medical Center, University of Alabama in Birmingham, Birmingham, Alabama, U.S.A.

INTRODUCTION

The nonglycosylated membrane (M) proteins found within negative-stranded viruses are thought to play an important role in viral structure and assembly. There are indications that the M protein may be of primary importance in determining the rigidity of the viral membrane (8, 12). In addition, due to its location within the virion, the M protein molecule also provides a system for the study of polypeptides associated with the cytoplasmic surface of cellular membranes. The present studies are directed toward elucidating the molecular basis for these functions by examining whether specific segments of the molecule contain a primary structure which might mediate such interactions.

COMPARISON OF ISOLATION PROCEDURES FOR THE M PROTEIN

We have used the WSN strain of influenza virus (H_0N_1) to investigate the primary structure of the internal membrane protein. The virus was grown in 2 liter roller bottle cultures of MDBK cells which yield ~ 1 mg of purified virions per culture and purified by polyethylene glycol precipitation and banding in a continuous potassium tartrate gradient (2, 3, 9). For purification of M protein, we compared three different isolation methods. These methods all involved obtaining an initial fraction of molecules enriched in M protein followed by gel filtration on a 6% agarose column in 5M guanidine. The theoretical yield of the M protein should be 40-50% of the virion protein; the actual amount recovered depended upon the isolation procedure.

The methods employed were as follows:

1. Protease digestion of intact virions selectively removes the external glycoproteins leaving a spikeless particle which contains the intact ribonucleoprotein and M protein within the lipid bilayer (3, 15). The resulting spikeless particles, after purification, were suspended in 8M guanidine and incubated at 60°C for 3 hr. Any remaining aggregated material was removed by centrifugation at 2000 rpm for 20 min and the soluble sample applied to a BioGel A5M column in 5M guanidine to further separate the nucleoprotein and M protein.
2. Treatment of purified virions with a final concentration of 10% triton for 10 minutes in a low salt buffer disrupts the lipid bilayer and solubilizes the glycoproteins (4, 14). The M protein and the nucleoprotein were pelleted at 100,000 x g for 1 hour, solubilized in 8M guanidine as described above, and fractionated on a BioGel A5M column to separate the proteins.
3. Disruption of the purified virions with a final concentration of 0.2% ammonium deoxycholate solubilizes the glycoproteins and ribonucleoprotein but results in flocculation of the M protein from solution (11). The flocculated M protein was pelleted by centrifugation at 10,000 x g and the pellet resuspended in 8M guanidine as described above. The sample was then applied to a Sepharose CL6B column in 5M guanidine to remove any residual contaminating polypeptides.

Purity of the M protein isolated by these methods was determined by SDS-PAGE and amino acid analysis. A comparison of the amino acid composition of the M protein isolated by the three methods is shown in Table 1. There is good agreement between the analysis of the M protein obtained by the different isolation procedures. The compositional analysis indicates that there is no cysteine found within the M protein molecule, while glutamic acid, alanine, and leucine account for approximately one-third of the amino acid residues of the polypeptide. The values obtained correlate well with previously published data on the M protein of influenza A viruses isolated by other methods (7, 10).

The manipulations involved in obtaining spikeless particles resulted in recovery of only 50% of the starting material, while the triton fractionation and ammonium deoxycholate treatment both resulted in approximately 100% recovery of the M protein. The latter method was routinely used because minor amounts of contaminating polypeptides were present after the detergent treatment, thus ensuring that a pure preparation of M protein was subsequently obtained.

TABLE 1

Amino Acid Composition of Purified M Protein

AMINO ACID	TRITON/A5M	DOC/CL6B	CHYMO/A5M
ASX	20.13*	20.38	18.56
THR	14.62	13.87	14.17
SER	13.44	14.15	15.47
GLX	26.32	27.44	30.30
PRO	10.07	11.37	10.81
GLY	15.98	16.88	16.47
ALA	21.76	20.73	22.71
CYS	0	0	0
VAL	14.82	14.65	16.20
MET	9.34	8.32	10.23
ILE	10.70	10.92	8.60
LEU	22.18	21.98	21.57
TYR	5.72	5.08	4.84
PHE	7.12	7.10	5.45
HIS	5.83	5.82	6.20
LYS	12.90	13.23	12.73
ARG	14.74	14.34	13.49

*Expressed residues/mole based on molecular weight of 25,000

CNBr PEPTIDES OF THE M PROTEIN

Attempts to obtain sequence data indicated that the amino terminal residue of the M protein is blocked, which prevents any direct approach for sequence determination from the NH_2-terminus. Therefore, the purified protein was treated with CNBr to yield peptides containing carboxy-terminal methionine derivatives and free αNH_2-terminal residues, to enable further structural analysis. The CNBr peptides, labeled with ^3H-leucine, were fractionated according to molecular weight on a Bio-Gel P6 column (Fig.1). The first three peaks were not completely resolved, while the peptides contained within the last

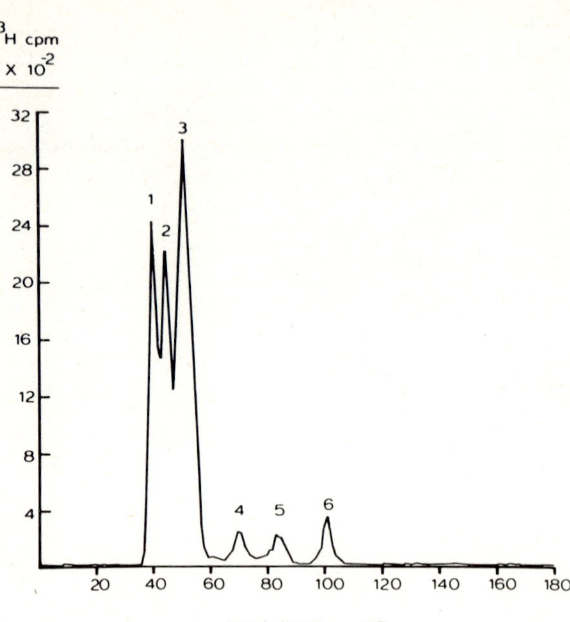

Fig.1. *Gel filtration separation of ^{3}H-leucine labeled M protein CNBr peptides on BioGel P6 (200-400 mesh) in 50% formic acid; 0.7 x 90 cm column; 0.35 ml fractions.*

three peaks were well separated and appeared to be chromatographically homogeneous. The approximate molecular weights of the peptides within the three initial peaks estimated from amino acid composition were 6000, 4000, and 2200 respectively. Thin layer silica gel chromatographic separation in butanol-pyridine-acetic acid-water (90:60:18:72) further resolved the peptides contained within the first three peaks. The first and second peaks contained one component, and the third peak a mixture of three or four peptides. The total number of peptides obtained correlated well with the expected number based on methionine residues determined by amino acid analysis.

The peptides eluting in peaks four to six contained a single leucine residue per peptide, and a C-terminal methionine derivative, homoserine, as seen from their amino acid composition (Table 2a). As expected, integral values were obtained for all the amino acids of Peak 4 and 5; the values for Peak 6 on the other hand gave other than integral values for a number of the amino acid residues. Preliminary results suggest that this peak may contain two peptides, one of which lacks leucine.

SEQUENCE ANALYSIS OF CNBr PEPTIDES

The three smaller CNBr peptides range in molecular weight from about 2000 to 1500 daltons which limits the approach to determination of primary sequence by the classical Edman degradation (5). Peptides in this molecular weight range are not easily sequenced using the automated liquid phase sequencer (6) due to losses during various wash cycles. In preliminary studies with Peak 4, a peptide containing two arginine residues which would facilitate adherence to the glass cup, there was visible loss of peptide material from the cup after the first cycle, and no further results were obtained. Therefore, a new and more sensitive approach to sequencing was employed. The individual peptides were coupled to an insoluble support via the C-terminal homoserine derivative, and sequenced with an LKB solid phase sequencer, using radioactive ^{35}S-phenylisothiocyanate as the coupling reagent for the Edman degradation (1). The residues were separated by thin layer chromatography (16) and detected by autoradiography. The sequences obtained to date for the last three peptides are shown in Table 2b. The amino terminal residues, depicted in parentheses, were determined by reaction with dansyl chloride, since all free amino groups are blocked with nonradioactive methylisothiocyanate during attachment of the peptides to the solid phase resin and are not detected by autoradiography. We have obtained sequence information on one-half of the residues in Peak 4 while only 4 residues within the Peak 5 peptide remain to be placed. The amino acids detected by sequencing correlate well with the residues determined by analysis.

DISCUSSION

Further information on the primary structure of the M protein from influenza virus may be obtained by analysis of other CNBr peptides. Preliminary results indicate that the N-terminal peptide is contained within the first P6 peak, while the C-terminal peptide is in the third P6 peak. To order the peptides and further define the overall structure, larger peptide fragments would be useful. A possible approach to obtain these may be limited tryptic digestion (13). Identification of limited sequences in these peptides will enable the CNBr peptides to be ordered within the molecule.

The correlation of structural data with the M protein functions requires identification of the functional regions of the molecule. The use of crosslinking reagents may enable identification of the regions of M involved in interaction with other

TABLE 2

A. Amino Acid Composition of Low MW CNBr Peptides from Influenza Virus M Protein

	Peak 4	Peak 5	Peak 6
ASX	1.97*		1.31
THR	2.26	1.81	
SER	0.67	1.22	1.52
GLX	1.99	1.53	3.30
PRO	0.90		
GLY			1.93
ALA	1.39	3.62	3.49
CYS			
VAL	1.02	0.92	
MET			
ILE	1.00		1.10
LEU	1.00	1.00	1.00
TYR			0.61
PHE			0.59
HIS	0.90		
LYS		1.01	
ARG	1.93		1.74
HSR	0.77	0.92	1.35
	16-17	11-13	17-19

*Data calculated on the basis of one leucine residue per peptide.

B. Partial Sequence of CNBr Peptides from Influenza Virus M Protein

Peak 4: (VAL)- ? -THR-THR-ASN-PRO-LEU-ILE-(ASX,SER,2GLX, ALA,HIS,2ARG)HSR
(position 1 = (VAL); position 5 = ASN)

Peak 5: -LEU-ALA-SER-THR-THR-ALA-GLN-ALA-(GLX,ALA, VAL,LYS)HSR

Peak 6: (GLY)/(ALA)-LEU-ILE-TYR-ASN/GLN-(ASX,2GLX,2SER,GLY,3ALA, PHE,2ARG)HSR

proteins. In addition, reassociation studies of the CNBr peptides with lipid vesicles may indicate whether parts of the molecule are involved in binding to the viral envelope. Such studies are now underway in our laboratory.

SUMMARY

The CNBr peptides from the M protein of influenza virus have been fractionated into six peaks based on molecular weight. The first three peaks are not completely resolved by gel filtration, and thin layer chromatography indicates that the third peak contains three or four peptides. The three smallest CNBr peptides appear homogeneous by gel filtration; their composition has been determined and a partial sequence obtained on each peptide.

ACKNOWLEDGEMENTS

This research was supported by Grants No. PCM 76-09711 from the National Science Foundation, AI 12680 from the National Institute of Allergy and Infectious Diseases, USPHS, and AM 03555 from the National Institute of Health. We thank Vivian Brown, Arthur Weissinger, and Patricia H. Springfield for excellent technical assistance.

REFERENCES

1. Bridgen, J., (1976). *Biochemistry* 15, 3600.
2. Choppin, P.W., (1969). *Virology* 39, 130.
3. Compans, R.W., Klenk, H.-D., Caliguiri, L.A., and Choppin, P.W., (1970). *Virology* 42, 880.
4. Compans, R.W. and Pinter, A., (1975). *Virology* 66, 151.
5. Edman, P., (1950). *Acta Chemica Scand.* 4, 283.
6. Edman, P., and Begg, G. (1967). *Europ. J. Biochem.* 1, 80.
7. Gregoriades, A., (1973). *Virology* 54, 369.
8. Landsberger, F.R., and Compans, R.W. (1976). *Biochemistry* 15, 2356.
9. Landsberger, F.R., Compans, R.W., Choppin, P.W., and Lenard, J.L. (1971). *Proc. Nat. Acad. Sci. USA.* 68, 2579.
10. Laver, W.G., and Baker, N., (1972). *J. Gen. Virol.* 17, 61.
11. Laver, W.G., and Webster, R.G. (1976). *Virology* 69, 511.
12. Lenard, J.L., Tsai, D., Compans, R.W., and Landsberger, F.R. (1976). *Virology* 71, 389.
13. Oxford, J.S., and Schild, G.C., (1977). *Virology* 74, 394.
14. Scheid, A., Caliguiri, L.A., Compans, R.W., and Choppin, P.W., (1972). *Virology* 50, 640.
15. Schulze, I.T., (1970). *Virology* 42, 890.
16. Summers, M.R., Smythers, G.W., and Oroszlan, (1973). *Anal. Biochem.* 53, 624.

INFLUENZA VIRUS SPECIFIC COMPONENTS IN INFECTED CELLS: MEMBRANE AND NON-STRUCTURAL PROTEINS

ANASTASIA GREGORIADES

Department of Virology, The Public Health Research Institute of The City of New York, Inc., New York, New York 10016, U.S.A.

Influenza virus induces the synthesis of eight known proteins in the infected cell and these are primary gene products specified by eight individual RNA segments (11, 19, 22). All except one of these are structural proteins of the virus. Within the infected cell the glycosylated haemagglutinin and neuraminidase are found exclusively in the cytoplasm, while the nucleoprotein (NP), the three P proteins, and a non-structural (NS) protein are also present in nuclei (12, 16, 25). The reasons for the presence of these in the nucleus is not clear although it has been implied that transcription occurs there (26). The membrane (M) protein, which constitutes the largest part of the virus (2, 23) has also been reported to be associated with the nucleus and has been detected either by immunofluorescence of infected cells (18) or by analysis of isolated nuclei on polyacrylamide gels (6, 7, 9).

The NS component is of particular interest and was detected by immunofluorescence in nucleoli and diffusely in the nucleoplasm and cytoplasm of infected cells (3) and, subsequently, by analysis of infected cells on polyacrylamide gels (16, 24). Large amounts of this protein are found in infected cell nuclei (12, 13, 16) and associated with ribosomes and polysomes (1, 20). Because the protein is not packaged into virus and is made in large amounts, approximating that of M (11, 15) it frequently crystallizes out. In chick embryo fibroblasts (CEF), for example, these crystals begin to appear around 5 hr after infection and are present in large amount by 20 hr in both nuclei and cytoplasm (17). Nevertheless, despite its considerable abundance, the function of NS in the replication of the virus is, at the moment, unknown although a number of functions have been suggested for it (27).

M and NS proteins have some characteristics in common and this has complicated their analyses. These proteins are between 23,000 and 25,000 daltons and are generally not resolved on standard phosphate buffered polyacrylamide gels. They have been resolved by gradient gels (21) or discontinuous systems using higher gel concentrations (11, 15). Aside from the similarity in molecular weights, both proteins are extractable by acidic chloroform-methanol (7). This procedure was used initially to detect the presence of hydrophobic "proteolipid" type molecules in influenza virus. M was the only protein extractable from virus of the WSN strain. A chloroform-methanol extractable component of about 25,000 daltons was also present in whole infected cells, isolated nuclei and polysomes of CEF and this was entirely

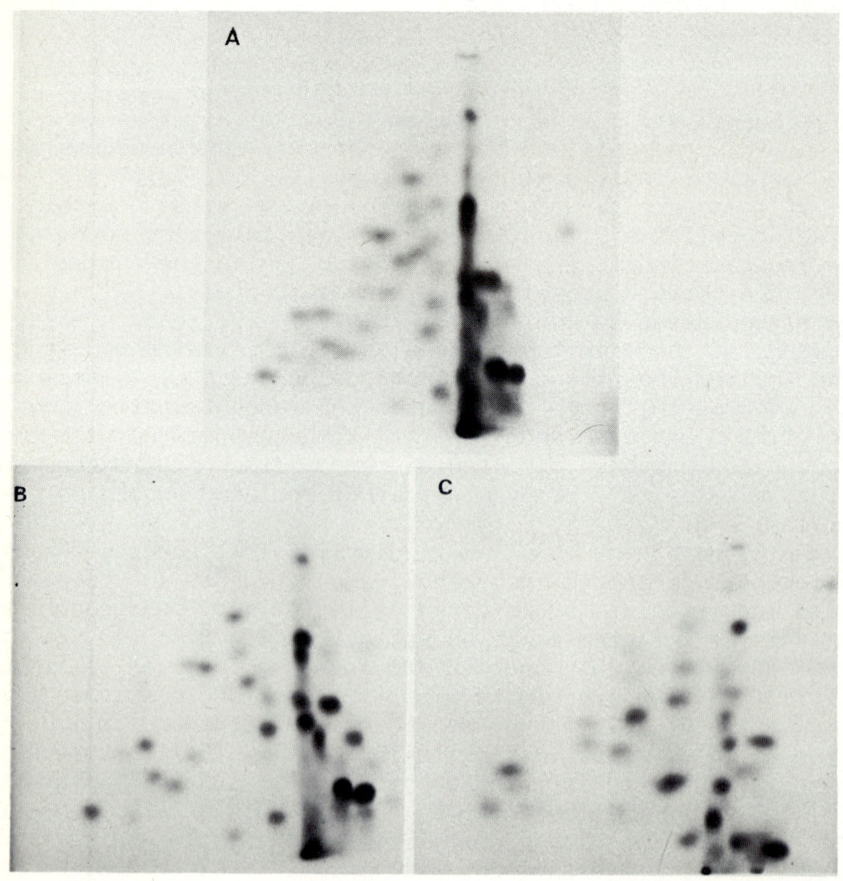

Fig.1.

extractable into this solvent.

The approximately 25,000-dalton material extracted into acidic CM was recently reinvestigated (7). CEF were infected for varying hours in the presence of ^{14}C-amino acids, and nuclei and cytoplasm were isolated and extracted with solvent as previously described (6). As a rich source of NS, nucleoli were isolated according to the procedure of Krug and Soiero (13). All fractions which would normally have been discarded in the preparation of nucleoli were saved, and the protein precipitated with ethanol and labeled nucleoplasm. The 25,000-dalton material was then isolated either by extraction into acidic chloroform-methanol or by 6 M guanidine hydrochloride column chromatography on biogel 5 M (4). Figure 1a shows the tryptic digest pattern observed on silica thin layer plates of the approximately 25,000-dalton fraction extracted into chloroform-methanol from cytoplasm of infected cells. Figure 1b is the M extracted by acidic CM from virus, and Fig. 1c is the approximately 25,000-dalton NS material isolated from a nucleolar fraction by column chromatography with 6 M guanidine hydrochloride. If M of 1b and NS of 1c are superimposed, the pattern of 1a is produced. Apparently, both proteins are soluble in acidic chloroform-methanol. This explains the quantitative removal of the approximately 25,000-dalton protein fraction from infected cells observed previously (8). It is also apparent from the chromatograms, that a large amount of overlapping occurs between M and NS peptides and these are resolved only by the chromatography in the second dimension. For this reason, it was difficult to resolve the simultaneous extraction of NS along with M in chloroform-methanol. As previously pointed out, both proteins are quantitatively extracted from nuclei by acidic CM, but NS is not extracted in good yields from nucleolar preparations (7). The same is true of NS protein extracted from crystals that are isolated from the cytoplasm of infected cells (17). The low yields of NS are probably due to a decrease in solubility observed with some proteins when they are concentrated through purification. This may explain why NS is entirely extracted from whole nuclei or cytoplasm but not completely from nucleolar preparations or NS crystals.

Fig. 1. *Two dimensional tryptic digest analysis of the 25,000 dalton protein extracted into acidic chloroform-methanol from CEF infected for 7 hours. (a) acidic CM extract of cytoplasm; (b) M extracted by acidic CM from virus; (c) nucleolar component isolated by 6 M guanidine hydrochloride column chromatography on agarose. The proteins were isolated, oxidized and processed as described (7).*

Previous reports indicated that NS could be detected in infected cells at 2 hr after infection and in increasing amounts thereafter (9, 15, 24). In addition, some reports suggested that NS was completely sedimentable at 200,000g for 30 minutes (1) while others indicated that it was still present in the soluble fraction of the cytoplasm after 200,000g for 60 minutes (9). This difference may be due to the high cellular background present in some cell lines, which could mask the presence of this protein.

The synthesis and distribution of NS in CEF were therefore investigated. CEF were infected and incubated in the presence of ^{35}S-methionine (75 μCi/monolayer) and harvested from 1 to 6 hr after infection. Nuclei were prepared as described previously (5) and a 100,000g pellet and supernatant were prepared by sedimenting the cytoplasm at 33,000 rpm in a SW-56 rotor for 1 hr. The proteins in the soluble fraction were precipitated with 5 volumes cold ethanol and all fractions were then solubilized with 5% SDS and 1% mercaptoethanol and analyzed on 7-15% polyacrylamide gel and 4% spacer (14). The results of these experiments indicated that in CEF, NS is detected at 1 hr after infection and increases with time. It is also detected in nuclei at 1 hr after infection. In the cytoplasm, approximately 50% is still present in the soluble fraction after 100,000g for 1 hr, except very late in infection (17 hr) when all of it is sedimentable (data not shown). Results also showed the association of M with nuclei for possible reasons which have been discussed elsewhere (7). M is entirely sedimentable at 100,000g as previously reported (9).

The NS present in the 100,000g cytoplasmic supernatant was further investigated by overlaying this material onto 4-40% sucrose gradients in 0.15 M NaCl, 10^{-3} M $MgCl_2$, 0.01 M Tris-HCl, pH 8.0 and sedimenting at 48,000 rpm in a SW-56 rotor for 20 hr at 4 C. Fractions of 4-5 drops were collected and aliquots monitored for radioactivity. SDS and mercaptoethanol were then added to the fractions and these were analyzed by electrophoresis on 12% discontinuous polyacrylamide gels (14). Figure 2 shows the sedimentation profile of the 100,000g supernatant on 4-40% sucrose and the (+) sign indicates those areas of the gradient where NS was present. NS was found to sediment at approximately 2.8 S, consistent with a protein of about 23,000 daltons. This indicates that NS present in the soluble fraction of the cell is not bound to any subviral or host component. This protein thus appears to be different from the nonstructural σ2A induced by reovirus which is never found free in the supernatant but is bound to single-stranded viral RNA (10). The influenza virus NS protein present in the soluble fraction of the cytoplasm should thus prove to be a good source of NS

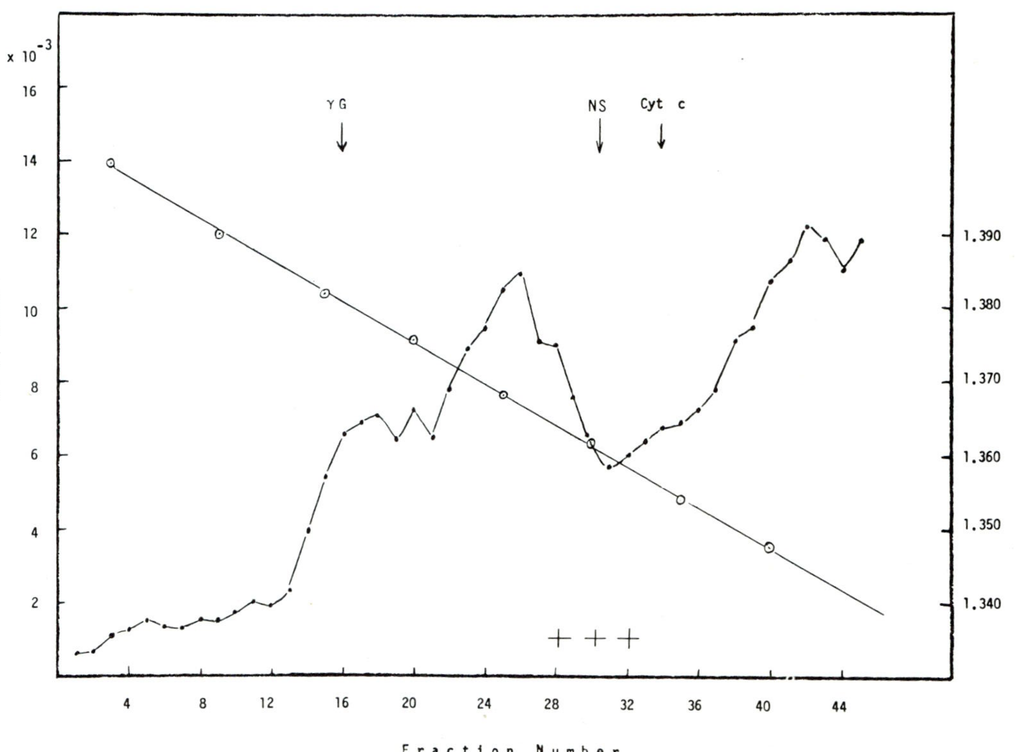

Fig.2. *Sedimentation of NS on sucrose gradients. CEF were harvested at 6 hr after infection and the cytoplasm was sedimented at 100,000g for 1 hr. The supernatant containing soluble NS was removed and sedimented on a 4-40% sucrose gradient for 20 hr at 48,000 rpm in a SW-56 rotor at 4°.*

for purposes of studying its function in the replication of the virus.

ACKNOWLEDGEMENT

This study was supported in part by Grant No. AI-12567 from the National Institute of Allergy and Infectious Diseases, U.S. Public Health Service.

REFERENCES

1. Compans, R.W. (1973). *Virology* 51, 56.
2. Compans, R.W., Klenk, H.D., Caliguiri, L.A. and Choppin, P.W. (1970). *Virology* 42, 880.
3. Dimmock, N.J. (1969). *Virology* 39, 224.
4. Fish, W.W., Mann, K.G. and Tanford, C. (1969). *J. Biol. Chem.* 244, 4989.
5. Gregoriades, A. (1970). *Virology* 42, 905.
6. Gregoriades, A. (1973). *Virology* 54, 369.
7. Gregoriades, A. (1977). *Virology* 79, 449.
8. Gregoriades, A. and Hirst, G.K. (1975). *In* "Negative Strand Viruses", Vol. 2 (Mahy, B.W.J. and Barry, R.D., eds). Academic Press, London, p.595.
9. Hay, A.J. and Skehel, J.J. (1975). *In* "Negative Strand Viruses", Vol. 2 (Mahy, B.W.J. and Barry, R.D., eds). Academic Press, London, p.635.
10. Huismans, H. and Joklik, W.K. (1976). *Virology* 70, 411.
11. Inglis, S.C., Carroll, A.R., Lamb, R.A. and Mahy, B.W.J. (1976). *Virology* 74, 489.
12. Krug, R.M. and Etkind, P. (1973). *Virology* 56, 334.
13. Krug, R.M. and Soeiro, R. (1975). *Virology* 64, 378.
14. Laemmli, U.K. (1970). *Nature, Lond.* 227, 680.
15. Lamb, R.A. and Choppin, P.W. (1976). *Virology* 74, 504.
16. Lazarowitz, S.G., Compans, R.W. and Choppin, P.W. (1971). *Virology* 46, 830.
17. Morrongiello, M.P., Beveridge, T. and Dales, S. (1977). *Intervirology* (in press).
18. Oxford, J.S. and Schild, G.C. (1975). *In* "Negative Strand Viruses", Vol. 2 (Mahy, B.W.J. and Barry, R.D., eds). Academic Press, London, p.611.
19. Palese, P. (1977). *Cell* 10, 1.
20. Pons, M.W. (1972). *Virology* 47, 823.
21. Ritchey, M.B. and Palese, P. (1976). *J. Virol.* 18, 738.
22. Scholtissek, C., Harms, E., Rohde, W.A., Orlich, M. and Rott, R. (1976). *Virology* 74, 332.
23. Schulze, I.T. (1970). *Virology* 42, 890.
24. Skehel, J.J. 91972). *Virology* 49, 23.
25. Taylor, J.M., Hampson, A.W., Layton, J.E. and White, D.O. (1970). *Virology* 42, 744.

26. Taylor, J.M., Illmensee, R., Litwin, S., Herring, L., Broni, B. and Krug, R.M. (1977). *J. Virol.* 21, 530.
27. White, D.O. (1974). *In* "Current Topics in Microbiology and Immunology", Vol. 63, Springer-Verlag, Berlin-Heidelberg-New York, p.1.

EFFECT OF THE HOST CELL ON EARLY AND LATE SYNTHESIS OF INFLUENZA VIRUS POLYPEPTIDES

R.A. LAMB and P.W. CHOPPIN

*The Rockefeller University,
New York, New York 10021, USA.*

Influenza virus has been reported to contain eight distinct pieces of RNA (3, 19, 22, 23), and in infected cells polypeptides P_1, P_2, P_3, HA, NA, NP, M, and NS have been detected (8, 11, 17, 26, 28). A genetic map of the genes coding for these proteins has been established for several strains (9, 26, 27). In influenza virus-infected cells the viral genome RNA species are transcribed by the virion polymerase into complementary RNA (cRNA) (2) and cRNA is the only virus-specific RNA found on infected cell polysomes (5, 7). This cRNA has mRNA activity, synthesizing influenza virus-specific proteins in *in vitro* systems (4, 6, 9, 25, 30).

Evidence for control of the synthesis of influenza virus polypeptides in infected cells has been found. In fowl plague virus-infected cells, polypeptides P_2, NP, and NS were detected before the others, with the synthesis of HA, NA, and M occurring later in infection. The rate of synthesis of M increased during infection, whereas the synthesis of NS declined after 4 hr post-infection (p.i.) (29). When cycloheximide was used to restrict RNA synthesis to that produced by primary transcription and then the cycloheximide removed and protein synthesis examined, only P_2, NP, and NS were detected immediately and the other polypeptides were synthesized later (29). In CEF cells infected with the WSN strain of influenza virus, the synthesis of M could be detected at early times in infection, and the synthesis of NS did not decline late in infection (11). In experiments designed to examine the polypeptides synthesized from mRNA's produced by primary transcription, similar to those described above, all the polypeptides with the possible exception of NA were found, indicating that the viral polypeptides could be translated from mRNA's produced by primary transcription. That control of synthesis occurred was evident from the increasing rates of synthesis of the M polypeptide relative to the other polypeptides after long periods of primary transcription and a relatively small amount of polypeptide HA synthesized immediately after removal of cycloheximide.

The control of influenza virus polypeptide synthesis appears to be largely at the transcriptional level, but control of translation must also occur. Pons (24) showed that all eight mRNA species were found on polysomes, very early after infection, and Glass *et al.* (7) showed that although 90% of the genome is represented on polysomes about 50% of the genome was represented in much greater quantities than the other half. Bosch, Hay and Skehel (chapter 47) have found that all the RNA species are synthesized during primary transcription, but that later in infection the abundance of each polypeptide closely reflects that of its corresponding mRNA. The differences in early protein synthesis observed between fowl plague virus and WSN may reflect either variations in virus strain or the host cell.

INFLUENZA VIRUS POLYPEPTIDE SYNTHESIS IN VARIOUS HOST CELL TYPES

That the host cell could affect the synthesis of influenza virus polypeptides was suggested by the finding that although in CEF cells the amounts of M and NS synthesized after removal of cycloheximide blocks were roughly equivalent, in MDBK and CHO-S cells the amount of NS was considerably greater than that of M (11, 14). Therefore, the synthesis of influenza virus polypeptides in CEF, L-929, BHK21-F, CHO-S, HeLa, MDBK, and CV-1 cells was examined using the MDBK cell-grown WSN strain as inoculum. In BHK21-F, L-929 (Fig.1A), and CEF cells the synthesis of polypeptide M was detected at ½ hr p.i. and in amount only slightly less than that of NS. In these cells, no amplification in the rate of synthesis of the three P polypeptides was detected between ½ hr and 6 hr p.i. In contrast, in MDBK, CV-1 (Fig.1B), HeLa, and CHO-S cells the synthesis of M was greatly reduced at early times compared with that of NS. Amplification of the three P polypeptides occurred between 1½ and 2½ hr p.i. in these cells. Thus, the host cell affects the level of synthesis of individual virus polypeptides.

EARLY PROTEIN SYNTHESIS IN THE PRESENCE OF AMINO ACID ANALOGUES

Previous data using cycloheximide to block protein synthesis suggested that the mRNA species for M and HA, as well as the other viral polypeptides, were transcribed in the absence of protein synthesis (11). To investigate whether functional early proteins were necessary for translation of the "late" proteins (HA, NA, M), the amino acid analogues, p-fluorophenylalanine (pFPA) and azetidine-2-carboxylic acid (ACA), were used to restrict protein synthesis to that directed by mRNA species synthesized by primary transcription. In the presence of amino acid analogues, any new transcriptase proteins synthesized are likely to be non-functional. CEF cells were treated with pFPA, or pFPA and ACA together, in phenylalanine-free Eagle's, infected with influenza virus, and maintained in amino-acid analogues. As shown in Fig.2A, polypeptides P_1, P_2, P_3, NP, M, and NS were readily detected in the presence of amino acid analogues, and, in addition, a small amount of HA was observed in the pFPA-treated cells. That the use of amino acid analogues to restrict protein and RNA synthesis to that directed by the input virion transcriptase was effective was demonstrated by the finding that no amplification of polypeptides occurred between 1 and 5 hr p.i. as compared to the untreated control samples. Similar results were obtained using CHO-S cells, except that the amount of M synthesized was greatly reduced compared to that of NS, again emphasizing variations between host-cell types. From these results it seems that "early" protein synthesis is not required for the translation of M, and probably also HA, in infected cells.

An induction in synthesis of three host-cell polypeptides (one migrating between P_1 and P_2, the second just ahead of HA, and the third co-migrating with NS) in the presence of pFPA has been observed (Fig.2A, compare pFPA 1 hr uninfected with control 1 hr uninfected), and this induction can be inhibited by the addition of actinomycin D. This observation has also been made by L. Hightower (see chapter 39).

IRRADIATION OF CELLS WITH ULTRAVIOLET (U.V.) LIGHT BEFORE INFLUENZA VIRUS INFECTION

Irradiation of cells prior to infection with U.V. light inhibits influenza virus replication (1), and Mahy and coworkers (18) and Minor and Dimmock (20) have found that under these conditions the synthesis of the "late" proteins, and in particular M, is drastically reduced. As discussed in the above studies, the synthesis of M in infected CHO-S cells occurs later than that of NS, and U.V. irradiation of CHO-S cells prior to infection caused a reduction in synthesis of the M polypeptide. However, with CEF cells pre-irradiation with varying dosages of U.V. light before WSN infection caused a reduction in synthesis of all virus-specific polypeptides, but not a specific reduction in M synthesis (Fig.2B). This result again suggests that the host-cell influences the expression of influenza virus polypeptides. When high dosages

A
BHK
U 1/2 1 1 1/2 2 3 4 5 6
P1
P2 P3
HA
NP
M
NS
L
U 1/2 1 1 1/2 2 3 4 5 6
P1
P2, P3
HA
NP
M
NS
B
MDBK
U 1/2 1 1 1/2 2 3 4 5 6
P1
P2, P3
HA
NP
M
NS
CV-1
U 1/2 1 1 1/2 2 3 4 5 6
P1
P2, P3
HA
NP
M
NS

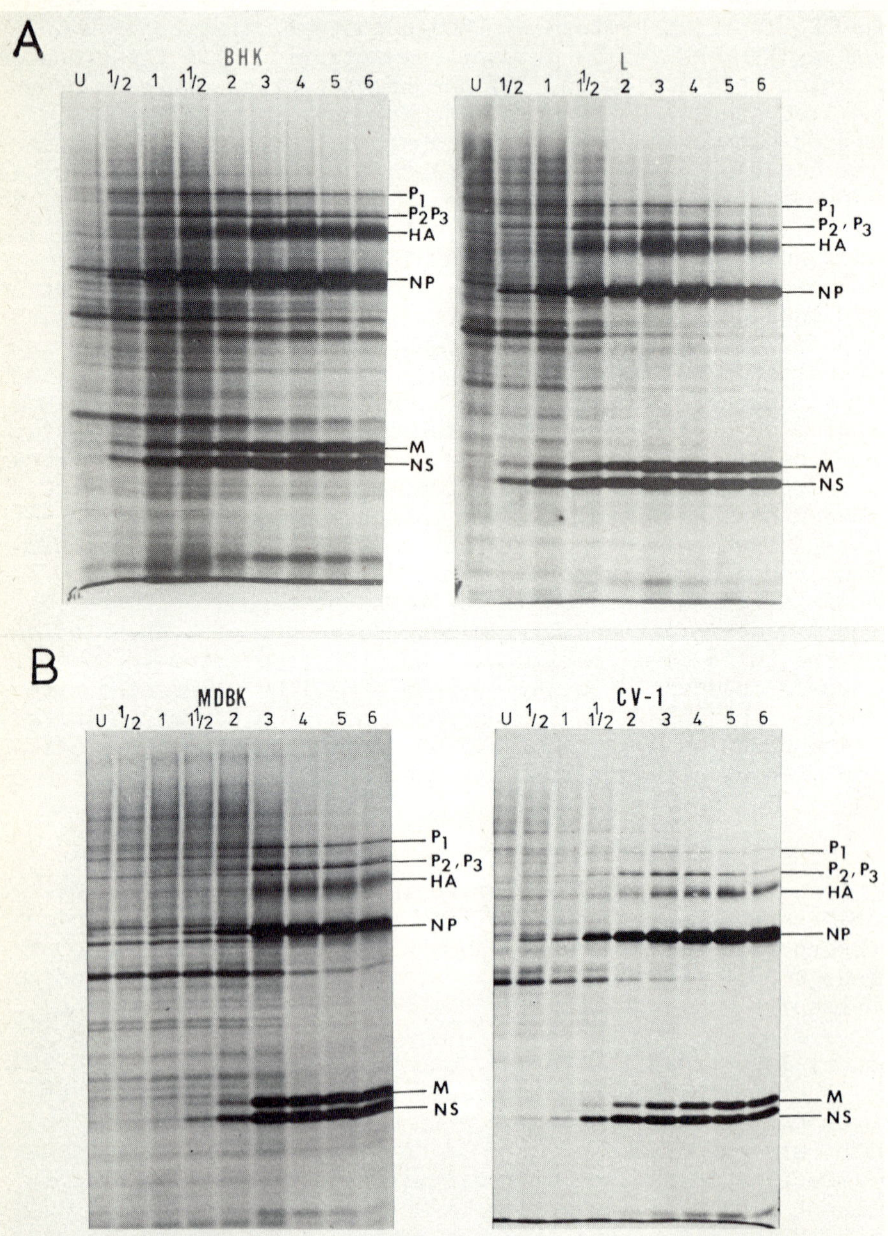

Fig.1

of U.V. light prior to infection were used, no amplification of polypeptide synthesis was observed from 1-5 hr p.i. and the level of polypeptide synthesis was very similar to that found after a cycloheximide block (11). This suggests that U.V. irradiation of cells before infection blocks the amplification mechanism after primary transcription.

PEPTIDE MAPPING OF THE "SMALL" POLYPEPTIDES SYNTHESIZED IN INFLUENZA VIRUS-INFECTED CELLS

In CEF cells infected with the WSN strain, four polypeptides (1-4) smaller than NS have been found with molecular weights of ~21,800, 19,000, 17,000 and 11,000, respectively (Fig.3A). The amount of these polypeptides found in infected cells varied considerably depending on the cell type. Occasionally a fifth small polypeptide (mol. wt. ~8,000) has been observed in BHK21-F and CHO-S cells. Tryptic peptide maps of polypeptides 1, 2, and 3 and a comparison with peptide maps of HA, NP, M, and NS, showed that these small polypeptides had many of the tryptic peptides of NS, indicating that they are not unique gene products, but are probably cleavage products (Fig.3B). Two of the large tryptic peptides of polypeptide 4 (mol. wt. ~11,000) were similar to those of M, but the remainder were different. Therefore the unique nature of this polypeptide, is open to further investigation. Polypeptide 4 described here probably corresponds to polypeptide reported 9 by Skehel (28). However, of polypeptides 1-4, the one which corresponds to NS_2, as described by Krug and Etkind (10), remains to be determined, because their NS_2 was found, like NS, in the nucleolus and therefore might be a fragment of NS, as polypeptides 1-3 appear to be. Although several strains of influenza virus, including WSN, have been reported to contain only 8 RNA species (21), fowl plague virus has been reported to contain 8 (19) or 10 RNA species (30, see chapter 47), and small polypeptide(s) have been synthesized by translation of infected cell RNA *in vitro* (30, see chapter 24). Therefore, the function and unique nature of the small polypeptide(s) required further investigation.

Fig.1. *The synthesis of polypeptides in influenza virus-infected BHK21-F and L-929 cells (A) and MDBK and CV-1 cells (B). Uninfected and cells infected with the WSN strain of influenza virus grown in MDBK cells were labeled with* ^{35}*S-methionine for 15 min at various times after infection. The cells were dissociated with lysis buffer (11), subjected to electrophoresis on 13% gels (12), and processed for autoradiography as described previously (15). U = uninfected.*

A
pFPA
pFPA / ACA
Control
1h 1h 3h 5h 1h 1h 2h 3h 5h 1h 1h 3h
U I I I U I I I I U I I
P1
P2 P3
HA
NP
M
NS
420

B
CONTROL
15sec UV
30 sec UV
60 sec UV
120 sec UV
150 sec UV
180 sec UV
P1
P2 P3
HA
NP
M
NS
P2,3

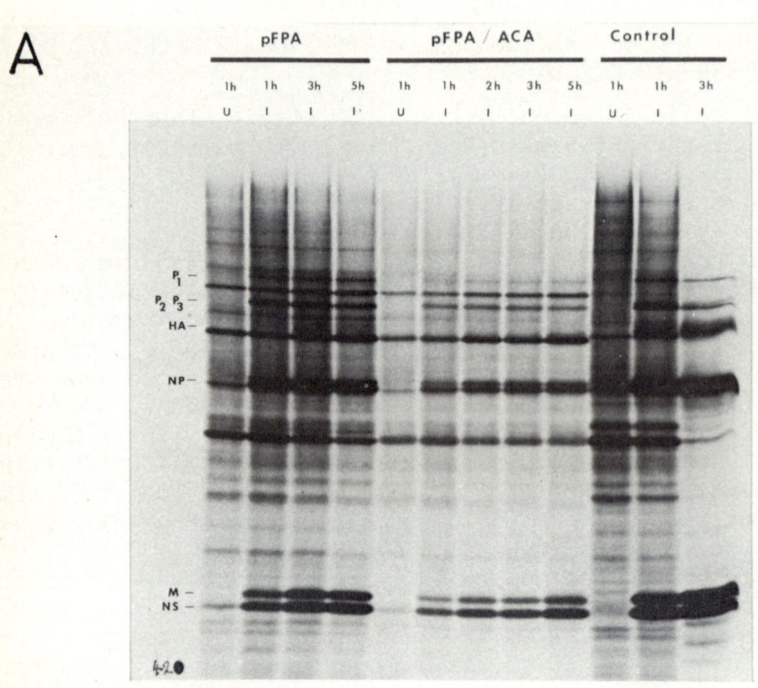

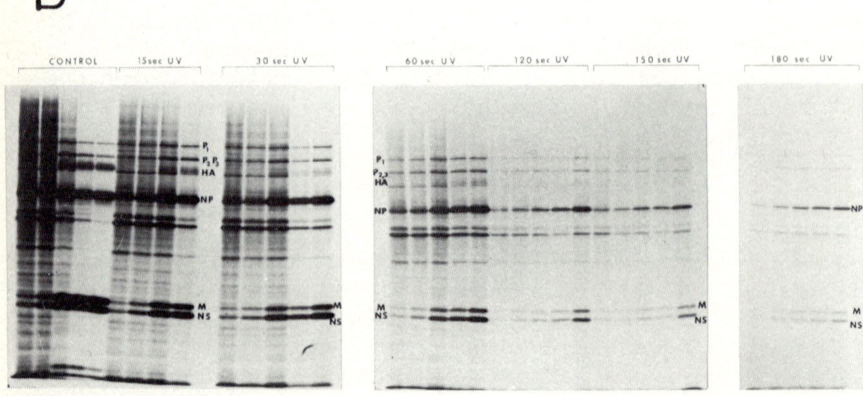

Fig.2

Fig.2A. *Influenza virus polypeptides synthesized in the presence of amino-acid analogues. Chick embryo fibroblast (CEF) cells were treated with either 3.1 mM p-fluorophenylalanine (pFPA) or 3.1 mM pFPA and 8 mM azetidine-2-carboxylic acid (ACA) in phenylalanine-free Eagle's medium 1 hr before infection. The cells were infected for 30 min in the presence of amino acid analogues and maintained in amino acid analogues after infection. At various times indicated, cells were labeled for 15 min in reinforced Eagle's medium deficient in phenylalanine and methionine and containing the amino acid analogues and ^{35}S-methionine, 20 μCi/ml. Samples were analyzed as described in Fig.1. U = uninfected; I = infected.*

Fig.2B. *Analysis of polypeptides synthesized in influenza virus-infected CEF cells which were pre-irradiated with varying dosages of ultraviolet (U.V.) light. CEF cells were washed twice with phosphate-buffered saline and irradiated for varying periods of time as indicated. The cells were inoculated for 30 min with the WSN strain of influenza virus and incubated in Eagle's medium for varying periods in the dark before labeling for 15 min with ^{35}S-methionine. Cells were dissociated and subjected to electrophoresis as described in Fig.1. For each dosage of U.V. light, infected cells were labeled at ½ hr, 1 hr, 2 hr, 3 hr, and 4 hr p.i., and these are shown in successive lanes of the gels from left to right for each group. (The 4 hr sample after 15 sec U.V. is not shown.)*

SUMMARY

Studies of influenza virus polypeptide synthesis in several different cell types has indicated that the host-cell influences the synthesis of individual viral polypeptides. Experiments using amino acid analogues have suggested that early protein synthesis is not required for late protein synthesis. Evidence has been obtained that in certain virus-cell systems, pre-irradiation of cells with U.V. light blocks amplification of virus polypeptide synthesis. Peptide mapping of four small polypeptides (mol. wt. ~11,000-22,000) has shown that three of these are fragments of NS.

ACKNOWLEDGEMENTS

We thank Richard Peluso for helpful discussion, Ann Duncan for excellent technical assistance, and Lawrence Hightower and Andrew Ball for valuable discussions concerning amino-acid analogues.

This research was supported by Research Grants AI-05600 from the National Institute of Allergy and Infectious Diseases and PCM76-09993 from the National Science Foundation.

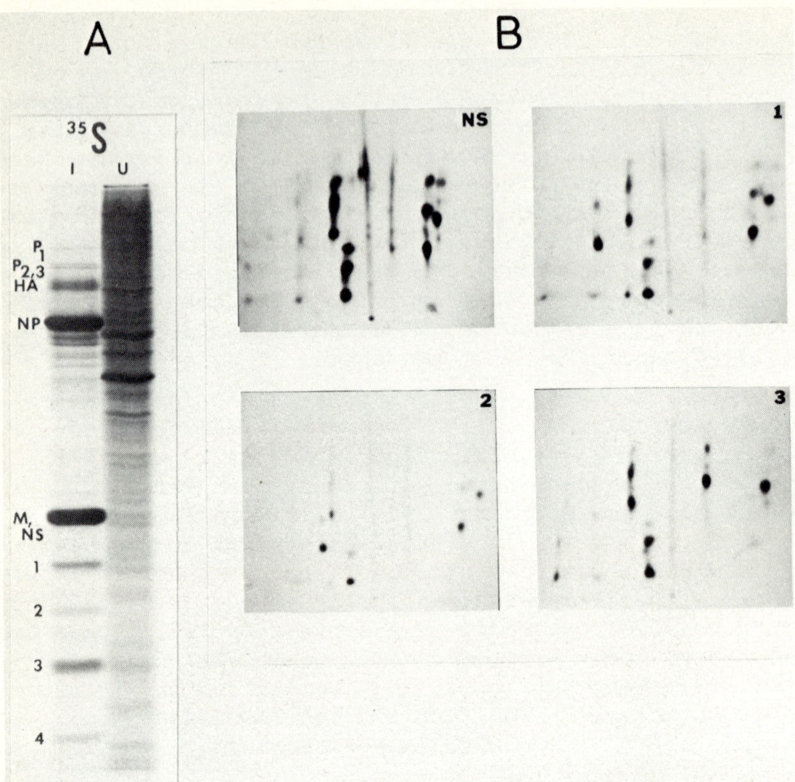

Fig.3A. *Examination of polypeptides synthesized in influenza virus-infected CEF cells on high percentage polyacrylamide gels. Uninfected and influenza virus-infected CEF cells were labeled with ^{35}S-methionine from 4-5 hr p.i. and the cells subjected to electrophoresis on a 15% gel (acrylamide/bis-acrylamide ratio 37.5:1) as described in Fig.1. I = infected; U = uninfected.*

Fig.3B. *Tryptic peptide maps of NS and the small polypeptides 1, 2, and 3. Influenza virus-infected CEF cells were labeled with ^{35}S-methionine (150 μCi/ml) from 4-6 hr p.i. After electrophoresis on a 13% gel to obtain NS, and a 15% gel to obtain polypeptides 1, 2, and 3, autoradiograms were prepared and used as templates to locate the polypeptides. The relevant bands were excised from the gels, rehydrated in NH_4HCO_3, and macerated before digestion with trypsin (20 μg) for 12 hr. The samples were desalted and subjected to chromatography and high voltage electrophoresis on thin-layer cellulose plates (13). Labeled peptides were detected by fluorography (16). To confirm that polypeptides 1, 2, and 3 had similar tryptic peptides to those of NS, mixing experiments were performed to establish* *that the major tryptic peptides co-migrated when subjected to chromatography and high voltage electrophoresis on the same plates (data not shown).*

REFERENCES

1. Barry, R.D. (1964). *Virology* 24, 263,
2. Bean, W.J. and Simpson, R.W. (1973). *Virology* 56, 646.
3. Bean, W.J. and Simpson, R.W. (1976). *J. Virol.* 18, 365.
4. Content, J. (1976). *J. Virol.* 18, 604.
5. Etkind, P.R., and Krug, R.M. (1974). *Virology* 62, 38.
6. Etkind, P.R., and Krug, R.M. (1975). *J. Virol.* 16, 1464.
7. Glass, S.E., McGeoch, D., and Barry, R.D. (1975). *J. Virol.* 16, 1435.
8. Inglis, S.C., Carroll, A.R., Lamb, R.A., and Mahy, B.W.J. (1976). *Virology* 74, 489.
9. Inglis, S.C., McGeoch, D., and Mahy, B.W.J. (1977). *Virology* 78, 522.
10. Krug, R.M., and Etkind, P.R. (1973). *Virology* 56, 334.
11. Lamb, R.A., and Choppin, P.W. (1976). *Virology* 74, 504.
12. Lamb, R.A., and Choppin, P.W. (1977a). *Virology* 81, 371
13. Lamb, R.A., and Choppin, P.W. (1977b). *Virology* 81, 382.
14. Lamb, R.A., and Choppin, P.W. (1977c). *J. Virol.* 23, 816.
15. Lamb, R.A., Mahy, B.W.J., and Choppin, P.W. (1976). *Virology* 69, 116.
16. Laskey, R.A., and Mills, A.D. (1975). *Eur. J. Biochem.* 56, 335.
17. Lazarowitz, S.G., Compans, R.W., and Choppin, P.W. (1971). *Virology* 46, 830.
18. Mahy, B.W.J., Carroll, A.R., Brownson, J.M.T., and Mc Geoch, D.J. (1977). *Virology* 83, 150.
19. McGeoch, D., Fellner, P., and Newton, C. (1976). *Proc. Nat. Acad. Sci. USA.,* 73, 3045.
20. Minor, P.D., and Dimmock, N.J. (1977). *Virology'* 78, 393.
21. Palese, P. (1977). *Cell* 10, 1.
22. Palese, P., and Schulman, J.L. (1976). *J. Virol.* 17, 876.
23. Pons, M.W. (1976). *Virology* 69, 789.
24. Pons, M.W. (1977). *Virology* 76, 855.
25. Ritchie, M.B., and Palese, P. (1976). *Virology* 72, 410.
26. Ritchie, M.B., Palese, P., and Schulman, J.L. (1976). *J. Virol.* 20, 307.
27. Scholtissek, C., Harms, E., Rohde, W., Orlich, M., and Rott, R. (1976). *Virology* 74, 332.
28. Skehel, J.J. (1972). *Virology* 49, 23.
29. Skehel, J.J. (1973). *Virology* 56, 394.
30. Stephenson, J.R., Hay, A.J., and Skehel, J.J. (1977). *J. Gen. Virol.* 36, 237.

CONTROL OF INFLUENZA VIRUS POLYPEPTIDE SYNTHESIS

STEPHEN C. INGLIS, G. CONTI* and B.W.J. MAHY

Division of Virology, Department of Pathology, University of Cambridge, Addenbrooke's Hospital, Hills Road, Cambridge

**Permanent address:*

Instituto di Microbiologia, Scuola di Medicina, Universita degli Studi di Parma, Parma 43100, Italy.

The synthesis of fowl plague virus (FPV) polypeptides in infected chick embryo fibroblasts (CEF) occurs in two distinct stages: an early stage in which synthesis of the matrix (M) polypeptide is low relative to the other major virus-induced polypeptides, and a late stage in which synthesis of M is amplified to become the most abundant of all (8).

The mechanism by which this temporal control of virus gene expression operates is not certain. Skehel (19) proposed that primary transcription, *i.e.* transcription of the infecting virus genome by the input virion polymerase, was restricted to the genes coding for the early proteins. This hypothesis was supported by Avery and Dimmock (1) who reported that only half of the FPV genome was represented as complementary sequences in CEF early in infection. However Glass *et. al.*,(6) and Barrett *et. al.*,(chapter 32) showed that RNA complementary to the entire FPV genome can be detected as early as 45 min after infection. These data suggested that primary transcription was not limited to the early protein genes, and further that the absence of synthesis of M polypeptide early in infection might be due to a translational rather than a transcriptional control mechanism.

We have investigated these possibilities by analysing the relative amounts of individual virus cRNAs in infected cells, as measured by translation of RNA from infected cells in a cell-free system from wheat germ. The methods employed for RNA extraction and cell-free synthesis, and the characterisation of the virus-specified *in vitro* translation products have been described elsewhere (9). These experiments were carried out using the Rostock strain of FPV to infect primary CEF cultures at a multiplicity of approximately 20 pfu/cell. Cell cultures and virus stocks were prepared as previously described (5). Procedures for ^{35}S-methionine labelling of cells, and polyacrylamide gel analysis were as detailed in Inglis *et. al.*(8).

COMPARISON OF *IN VITRO* AND *IN VIVO* TRANSLATION PRODUCTS THROUGHOUT INFECTION

Fig.1(a) shows a time course of polypeptide synthesis in FPV-infected CEF up to 8.5 hr after infection. Virus proteins, in particular NP and NS, are visible from 1.5 hr onwards, but the synthesis of M is not apparent until 2.5 hr from when it increases to become the most abundant of all the virus polypeptides. Fig.1(b) shows the polypeptides synthesised *in vitro* in response to RNA extracted from infected cells during the same experiment, and at the same times as shown in (a). RNA extracted at 1.5 hr post infection (p.i.) directs the synthesis of very little M polypeptide *in vitro* relative to NS. However, by 4.5 hr M is the most abundant *in vitro* product.

The pattern of virus polypeptide synthesis *in vitro* closely resembles that in the infected cell at each time after infection. This result is consistent with the possibility that virus gene expression is controlled at the level of transcription *i.e.* that the increase in synthesis of M in the infected cell at 2.5 hr is due to increased transcription of the mRNA for M. Nevertheless, if the autoradiogram shown in Fig.1 (a) is exposed for long enough, some synthesis of M can be detected in the translation product of RNA extracted at 1.5 hr p.i., indicating that at least some transcription of the gene for M protein has occurred.

VIRUS cRNA SYNTHESIS DURING PRIMARY TRANSCRIPTION

The synthesis of virus-specific cRNA was examined in cells infected in the presence of cycloheximide. Suppression of all protein synthesis should allow only primary transcription to occur; that is, only cRNA species which are transcribed from infecting vRNA by pre-existing RNA polymerase molecules should accumulate.

Fig.2 shows the *in vitro* translation products of cytoplasmic RNA extracted from untreated, and cycloheximide-treated infected cells at 1.5 hr and 6.5 hr p.i. These times were chosen to represent early and late stages during the infectious cycle. Each RNA directed the synthesis of all the virus polypeptides *in vitro*, with the possible exception of NA, whose polypeptide moiety we could not reliably detect in the cell-free product. However there were several differences in the relative amounts of their synthesis. As in the previous experiment, RNA extracted from untreated infected cells at 1.5 hr p.i. (track 2) directed the synthesis of very little M relative to NS_1 *in vitro*, but by 6.5 hr (track 4), M was the major cell-free product. The translation product of RNA extracted at either time from cycloheximide-treated infected cells (tracks 3 and 5) contained approximately equal amounts of M and NS_1. Furthermore the amounts of radioactivity in the other virus polypeptides were more equivalent using cycloheximide-treated infected cell RNA although the larger polypeptides were still less abundant than the smaller ones. Since smaller RNAs are translated more efficiently in the wheat germ system, probably because of the relatively high nuclease activity of wheat germ extracts (7) the larger virus cRNAs may be present in greater proportions than their translation products suggest.

These results indicate that the absence of synthesis of M polypeptide early in infection is not due to a failure of the virion polymerase to transcribe the genome segment coding for M. Furthermore, the data suggest that primary transcription could be equivalent (operate at the same rate) for each vRNA segment. The pattern of early polypeptide synthesis in the infected cell is therefore probably the result of selective amplification of transcription of the mRNAs for the early proteins, a process dependent on new, and presumably virus specified, protein synthesis. Amplification of the mRNA for M protein appears to occur later, and again seems to require new protein synthesis, since addition of cycloheximide to cells at 1.5 hr p.i. blocks any subsequent increase in the amount of M synthesised *in vitro* in response to infected cell RNA (S.C.I., unpublished results).

EFFECTS OF LOW CONCENTRATIONS OF ACTINOMYCIN D AND PREIRRADIATION OF HOST CELLS WITH ULTRA-VIOLET LIGHT ON cRNA SYNTHESIS IN INFECTED CELLS

It is well known that the multiplication of influenza virus can be blocked by inhibitors of cellular DNA function, such as actinomycin D (AMD), and by exposure of cells to ultraviolet (UV) light prior to infection (2, 3, 17). The mechanism of this inhibition is not understood, but it is possible that host cell DNA may provide some function necessary for virus RNA transcription or replication. This hypothesis gained support from the observation that α-amanitin, an inhibitor of cellular RNA polymerase II activity, also inhibits influenza replication (14, 18). The conclusion that some transcription of cellular DNA is necessary for virus replication has been considerably strengthened by the recent reports (10, 20) that α-amanitin does not inhibit multiplication of influenza virus in Chinese hamster ovary cells which possess an α-amanitin-resistant RNA

a
b
IN VIVO
0·5 1·5 2·5 3·5 4·5 5·5 6·5 7·5 8·5 U
IN VITRO
no RNA U 0·5 1·5 2·5 3·5 4·5 5·5 6·5 7·5 8·5
P1
P2
P3
HA
HA0
NP
M
NS1

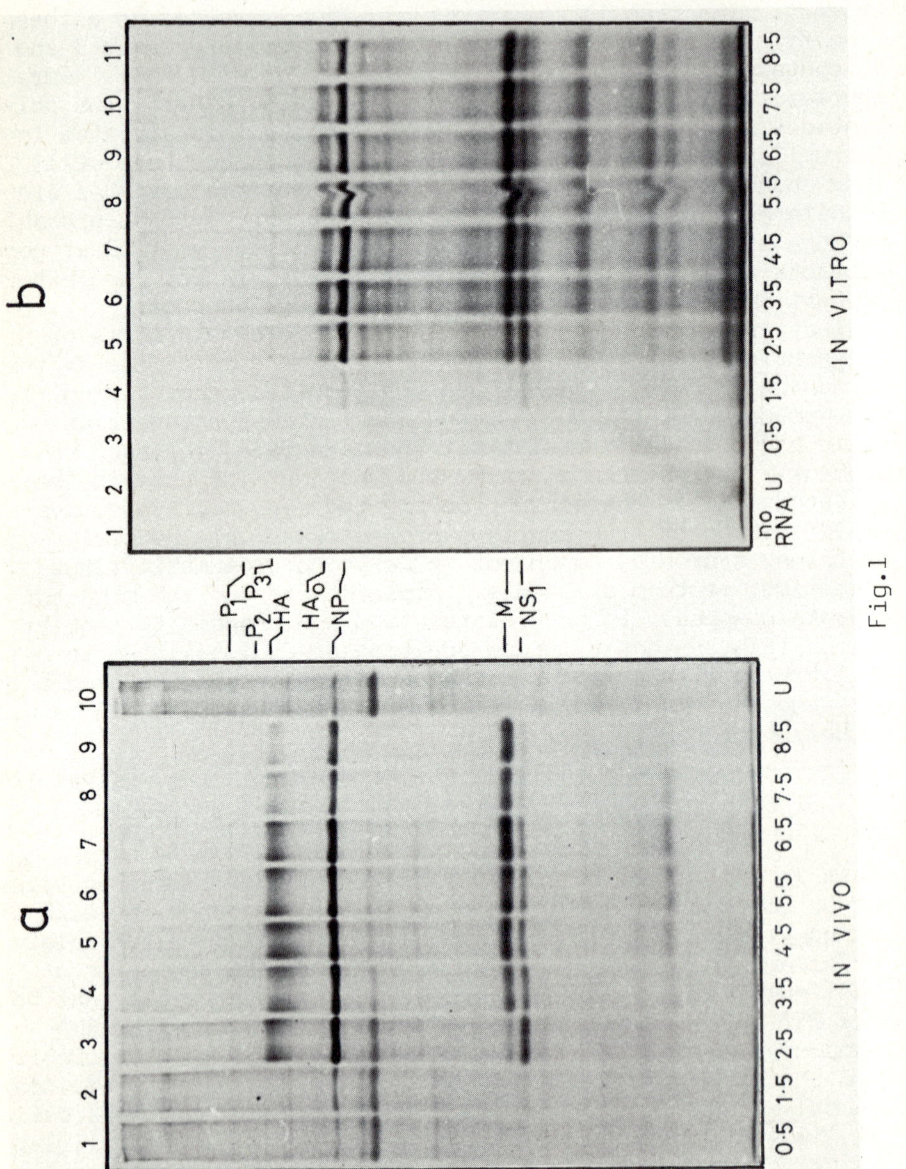

Fig.1

polymerase II activity (11).

Minor and Dimmock (15) postulated that two host cell DNA-dependent events were essential for virus replication, since they found that α-amanitin, and high concentrations of AMD, inhibited all virus protein synthesis, while low concentrations of AMD, and preirradiation of host cells with UV light, although still blocking production of infectious virus, allowed some virus protein synthesis. They reported that the latter treatments selectively inhibited production of the M, HA and NA polypeptides in the infected cell. A similar effect of AMD had been noted previously (19). Recently we showed that this effect of UV light could be abolished by incubating irradiated cells under conditions which allowed photoreactivation repair of the DNA (13).

We have confirmed that low concentrations of AMD and pre-irradiation of host cells with UV light preferentially inhibit synthesis of M polypeptide in FPV-infected cells. Fig.3(a) shows the polypeptides synthesised in FPV-infected CEF which were untreated (tracks 1 and 4), pre-irradiated with UV light (tracks 2 and 5), or treated with AMD (0.1 µg/ml) throughout infection (tracks 3 and 6). At 1.5 hr p.i. the pattern of virus polypeptide synthesis in untreated and treated cells was similar (tracks 1-3), but the increase in M synthesis observed in untreated cells at 6.5 hr p.i. (track 4) was inhibited in

---

Fig.1.

(a) Polyacrylamide gel (15%) electrophoresis of polypeptides synthesised in FPV-infected cells, and uninfected cells (U), during a 15 min pulse of ^{35}S-methionine (5 µCi/ml). Cells were labelled immediately before harvesting at the times after infection indicated below each track. Conditions of labelling, processing, and gel electrophoresis have been described elsewhere (8). Approximately equal amounts of cell protein were loaded in each slot, and labelled polypeptides were detected by autoradiography.

(b) Polyacrylamide gel (15%) electrophoresis of polypeptides synthesised by the wheat germ cell-free system in response to total cytoplasmic RNA from FPV-infected CEF. Numbers below each track represent the time (in hours) after infection at which cells were harvested. U - uninfected cell RNA; No RNA -endogenous activity of the wheat germ system. RNA extraction, cell-free protein synthesis, and analysis of in vitro products by gel electrophoresis were as previously described (9). Each RNA was added at 100 µg/ml, equal volumes of incubation mixtures were loaded in each slot, and ^{35}S-methionine labelled products were detected by autoradiography.

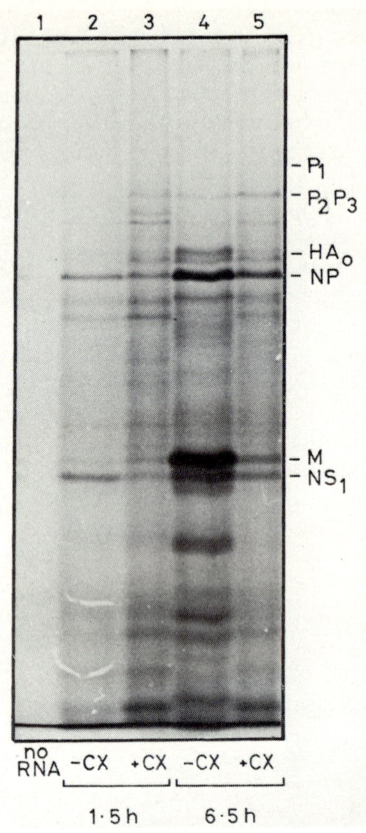

Fig.2. *Polyacrylamide gel (15%) electrophoresis of polypeptides synthesised by the wheat germ system without added RNA (1) and in response to total cytoplasmic RNA extracted from (2) untreated infected cells, 1.5 hr post infection (3) cycloheximide treated infected cells 1.5 hr p.i. (4) untreated infected cells, 6.5 hr p.i. (5) cycloheximide treated infected cells 6.5 hr p.i. Cells were treated with 100 μg/ml cycloheximide from 1 hr before virus addition continuously up to the time of harvest. Conditions for cell-free synthesis and analysis of the in vitro translation products were exactly as for Fig.1(b).*

UV and AMD treated cells (tracks 5 and 6). Other experiments (not shown) indicated that in UV and AMD treated cells synthesis of M polypeptide remained less than that of NS, up to 12 hr p.i., the latest time analysed. Since we assume that both AMD and UV-irradiation interfere with cellular DNA function, our results indicate that a host cell factor may be necessary for expression of M protein, in this virus/cell system at least. Such a factor could operate at the level of either transcription or translation of virus mRNA.

We examined these possibilities further, by analysis of the *in vitro* translation products of RNA extracted from infected cells which had been treated with these inhibitors.

The *in vitro* translation products of RNA extracted from infected cells which had been treated in exactly the same manner, during the same experiment, as presented in Fig.3(a), are shown in Fig.3(b). The cell-free products of UV-preirradiated and AMD-treated infected cell RNA extracted at 1.5 hr (tracks (2) and (3)) are similar to those of normal 1.5 hr RNA (track (1)). However, RNA extracted from UV- and AMD-treated infected cells at 6.5 hr p.i. (tracks (5) and (6)) directs the synthesis *in vitro* of far less M relative to NS than does RNA from normal infected cells at this time (track (4)). These results suggest that preirradiation of host cells with UV light, and low concentrations of AMD may selectively inhibit amplification of the mRNA for M protein, and therefore that a host factor may be involved in this process. Direct measurements by hybridization of the amounts of individual poly A - cRNA's present in UV-irradiated infected cells agree with this conclusion (13).

The results presented above are consistent with the hypothesis that FPV gene expression in CEF cells is controlled at the level of transcription of messenger RNA. Nevertheless they do not by themselves rule out the possibility of translational control. The *in vitro* translation experiments were carried out using cytoplasmic RNA, and so would not detect cRNA which might be sequestered in the nucleus. However other results from our laboratory (chapter 32) indicate that the relative proportions of different species of virus cRNA were similar in nucleus and cytoplasm throughout infection, so this possibility seems unlikely. It is also possible that the mRNA for M protein could require some modification to allow its translation in the cell-free system as well as in the infected cell, although one obvious modification which might affect translation, namely 'capping' of the virus mRNA (12) can apparently be carried out by the wheat germ extract itself (16).

CONCLUSIONS

The *in vitro* translation products of FPV-infected CEF cell RNA extracted at various times after infection closely resemble the polypeptides synthesised in the infected cell. RNA from cells infected in the presence of cycloheximide directs the synthesis of all virus polypeptides normally detectable in the

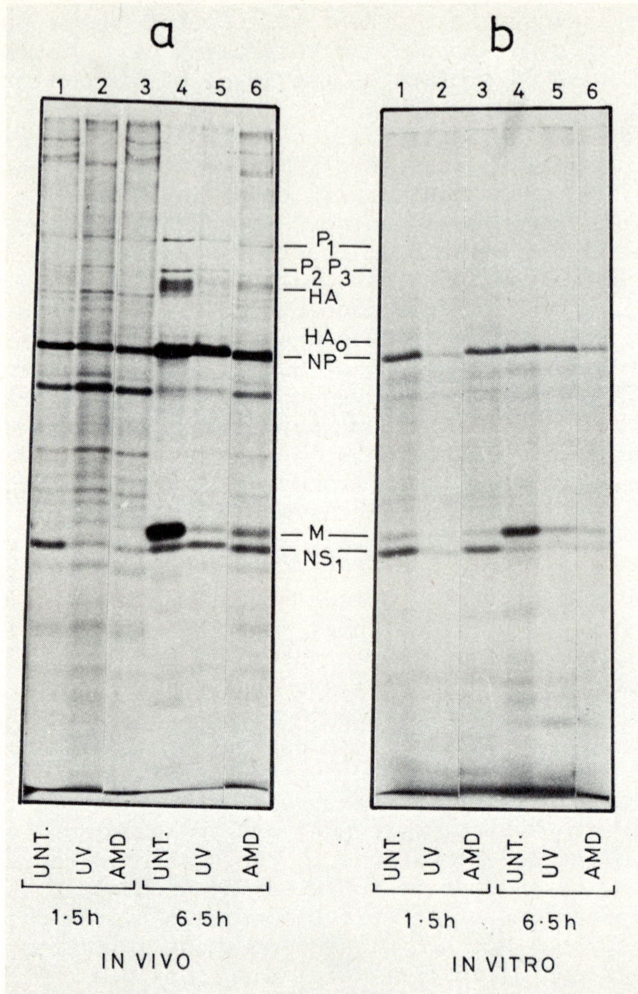

Fig.3.

(a) Polyacrylamide gel (15%) electrophoresis of polypeptides synthesised in FPV-infected CEF during a 15 min pulse of ^{35}S-methionine at 1.5 hr (tracks 1-3) and 6.5 hr (tracks 4-6) after infection. Infected cells were treated as follows. Tracks 1 and 4 - untreated, tracks 2 and 5 - irradiated with UV light (360 ergs/mm^2) for 3 min prior to infection; tracks 3 and 6 - AMD (0.1 μg/ml) present continuously from 1 hr before infection up to the time of harvest. The procedure for UV-irradiation has been detailed elsewhere (13). Labelling of polypeptides and gel analysis were performed as in Fig.1(a).

cell-free product, in amounts which suggest that primary transcription could be equivalent for each vRNA segment. The patterns of 'early' and 'late' polypeptide synthesis in infected cells may therefore result from selective amplification first of mRNA for the 'early' proteins, and subsequently of mRNA for the late proteins, both of which events require new protein synthesis. Pre-irradiation of host cells with UV light prior to infection, or low concentrations of AMD inhibit the transition from 'early' to 'late' virus protein synthesis, and appear to act by blocking amplification of the mRNA for M protein. A cellular factor may therefore be involved in this process.

ACKNOWLEDGEMENTS

This work was supported by a grant from the Medical Research Council. S.C.I. was in receipt of an MRC research studentship. G.C. acknowledges support from grant No.77.00.309.84 given by CNR 'Progetto finalizzato virus', Rome, Italy.

REFERENCES

1. Avery, R.J. and Dimmock, N.J. (1975). *Virology* 64, 409.
2. Barry, R.D. (1964). *Virology* 24, 563.
3. Barry, R.D., Ives, D.R. and Cruickshank, J.G. (1962). *Nature, Lond.* 194, 1139.
4. Bean, W.J. Jnr. and Simpson, R.W. (1973). *Virology* 56, 646.
5. Borland, R. and Mahy, B.W.J. (1968). *J. Virol.* 2, 33.
6. Glass, S.E., McGeoch, D. and Barry, R.D. (1975). *J. Virol.* 16, 1435.
7. Hunter, A.R., Farrell, P.J., Jackson, R.J. and Hunt, T. (1977). *Eur. J. Biochem.* 75, 149.
8. Inglis, S.C., Carroll, A.R., Lamb, R.A. and Mahy, B.W.J. (1976). *Virology* 74, 489.
9. Inglis, S.C., McGeoch, D. and Mahy, B.W.J. (1977). *Virology* 78, 522.
10. Lamb, R.A. and Choppin, P.W. (1977). *J. Virol.* 23, 816.

(b) Polyacrylamide gel (15%) electrophoresis of polypeptides synthesised by the wheat germ system in response to RNA extracted from infected cells at 1.5 hr (tracks 1-3) and 6.5 hr (tracks 4-6) after infection. Cell-free products in each track were of RNAs extracted from infected cells which had been treated in exactly the same way as those shown in the corresponding tracks in (a). Each RNA was added at 100 μg/ml, and ^{35}*S-methionine-labelled cell-free products were prepared and processed for gel electrophoresis as for Fig.1(b).*

11. Lobban, P.E., Siminovitch, L. and Ingles, C.J. (1976). *Cell* 8, 65.
12. Krug, R.M., Morgan, M.A. and Shatkin, A.J. (1976). *J. Virol.* 20, 45.
13. Mahy, B.W.J., Carroll, A.R., Brownson, J.M.T. and McGeoch, D.J. (1977). *Virology* 83, 150.
14. Mahy, B.W.J., Hastie, N.D. and Armstrong, S.J. (1972). *Proc. Natl. Acad. Sci. USA* 69, 1421.
15. Minor, P.D. and Dimmock, N.J. (1975). *Virology* 67, 114.
16. Muthukrishnan, S., Both, G.W., Furuichi, Y. and Shatkin, A.J. (1975). *Nature, Lond.* 255, 33.
17. Rott, R., Saber, S. and Scholtissek, C. (1965). *Nature, Lond.* 205, 1187.
18. Rott, R. and Scholtissek, C. (1970). *Nature, Lond.* 228, 56.
19. Skehel, J.J. (1973). *Virology* 56, 394.
20. Spooner, L. and Barry, R.D. (1977). *Nature, Lond.* 268, 650.

MECHANISM OF RNA SYNTHESIS *IN VITRO* BY VESICULAR STOMATITIS VIRUS

A.K. BANERJEE, R.J. COLONNO, D. TESTA and M.T. FRANZE-FERNANDEZ*

Roche Institute of Molecular Biology, Nutley, New Jersey, 07110, U.S.A.

**On leave from Departamento de Quimica Biologica, Facultad de Farmacia y Bioquimica, Universidad de Buenos Aires, Argentina.*

Vesicular stomatitis virus (VSV), a prototype of negative strand rhabdoviruses, contains a linear single-stranded genome RNA of molecular weight 4×10^6 which is packaged within a characteristic bullet-shaped particle containing five structural proteins designated as L, G, M, NS and N (32). A virion-associated RNA polymerase (6) transcribes the genome RNA *in vitro* into five monocistronic mRNA species (23) which code for the five virus structural proteins (12). The RNA synthesizing activity is carried out by the ribonucleoprotein (RNP) core particles containing only the N, NS and L proteins using the core-associated template genome RNA (11, 29, 30). Upon dissociation of the RNP core, it was shown that both the L and the NS proteins and the intact N protein-RNA complex are required for RNA synthesis (17, 24). In addition to the transcriptase activity, the RNP-core particles contain several enzyme activities which modify VSV mRNAs *in vitro*. These include (a) a guanyltransferase which guanylates the 5'-termini of the VSV mRNAs into a blocked structure as G(5')ppp(5')Ap... (3); (b) methyltransferases which methylate the 5'-blocked structure as $m^7G(5')ppp(5')A^m_p$... (4); and (c) a poly(A) polymerase which polyadenylates the newly synthesized mRNAs at the 3'-termini *in vitro* (9, 31), by a transcription dependent mechanism (1). The precise role of these activities in the purified virions in relation to the mechanism of RNA synthesis *in vitro* is still obscure.

The intriguing problem in the VSV system is that the synthesis of a complete complement of the genome RNA has not been detected *in vitro* although it has been found in VSV infected cells (21). Since a 42S + strand is the required intermediate in the replication of the VSV genome RNA, it seems that a regulatory mechanism must be operative *in vivo* which governs the transcription of the genome RNA into the complete complement and the five mRNA species.

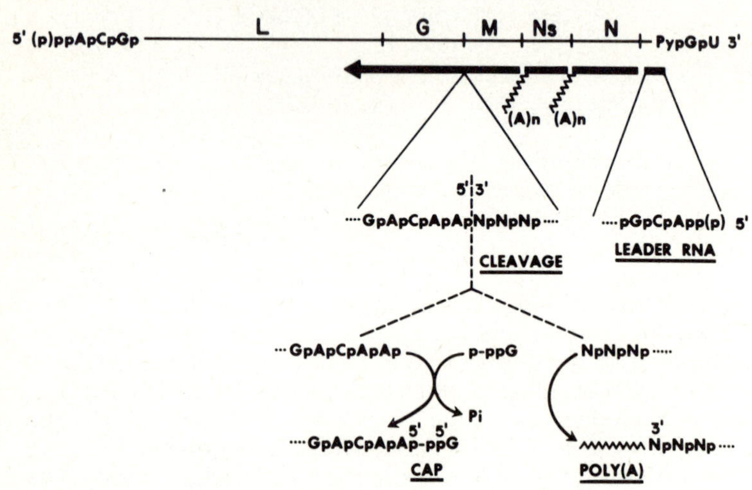

Fig. 1. *Model for the biosynthesis of VSV mRNAs.*

A MODEL FOR THE BIOSYNTHESIS OF VSV mRNA SPECIES

Recently we proposed a model for the possible mechanism by which the VSV mRNAs are generated *in vitro* (7). According to the model (Fig. 1) the virion-associated RNA polymerase initiates transcription at the 3'-terminus of the genome RNA with the synthesis of a small leader RNA molecule followed by the sequential synthesis of the VSV mRNA species in the order 3'-N-NS-M-G-L-5'. The biosynthesis of the various mRNA species possibly involves nucleolytic cleavage of a precursor RNA at specific sites as shown in the figure. This model is based on the following observations: (a) A discrete RNA molecule (leader RNA) 60-70 nucleotides long with a 5'-terminal sequence ppACG... complementary to the 3'-terminal sequence of VSV genome RNA is released during transcription *in vitro* (13). (b) Exposure of VSV particles to ultraviolet radiation resulted in a polar effect on the synthesis of individual mRNA species. The calculated target size of each gene included the sum of the molecular weights of that gene and its 3'-proximal genes (2, 5). (c) The 5'-termini of the VSV mRNA species are blocked, having the sequence GpppAACAG (25). Since only the α and β phosphates of GTP are incorporated into the blocked structure, the biosynthesis of this structure may involve cleavage of a larger RNA.

THE LEADER RNA

Product RNA was synthesized *in vitro* by VSV using α-^{32}P CTP as the labeled substrate. Following extraction by phenol-SDS and Sephadex G-50 chromatography, the poly(A)-containing RNAs were removed by binding to oligo (dT)-cellulose columns. The remaining unbound material was analyzed by electrophoresis on a 20% polyacrylamide slab gel. As shown in Fig. 2 (left panel), the autoradiogram demonstrates the synthesis of a small discrete

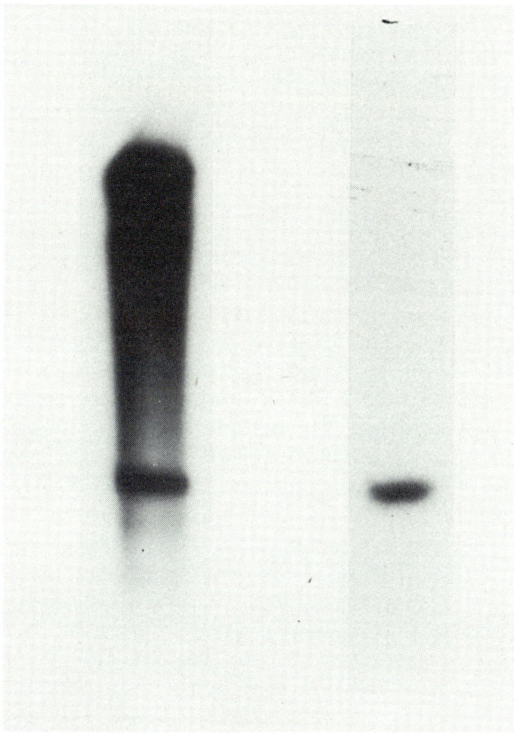

Fig. 2. *Polyacrylamide gel electrophoresis of VSV_{IND} leader RNA. RNA was synthesized in vitro in the presence of α-^{32}P CTP by VSV (Indiana serotype) virions (13). The resulting product RNA was purified by extraction with phenol-SDS, Sephadex G-50 chromatography and oligo(dT) cellulose chromatography. The ^{32}P CMP-labeled product RNA not binding to oligo(dT) was analyzed by electrophoresis on a 20% polyacrylamide slab-gel containing 8 M urea (13). The resulting autoradiogram shows total oligo(dT) unbound product RNA (left) and leader RNA recovered from the gel and re-electrophoresed (right).*

RNA species which migrates midway into the gel. The large patch of radioactivity near the origin represents the larger incomplete VSV mRNA species lacking poly(A). The leader RNA was recovered from the gel and reanalyzed by electrophoresis to show the integrity of purified leader RNA (Fig. 2, right panel). The leader RNA contains approximately 48% adenosine residues and the unblocked, polyphosphorylated 5'-termini ppACG (13). These characteristics are quite in contrast to the VSV mRNA species which contain blocked 5'-termini and 3'-terminal poly(A). A series of experiments designed to characterize the VSV leader RNA (14) have led to the following conclusions: (a) The leader RNA is transcribed from the 3'-end of the genome RNA since it can form an RNase-resistant hybrid with the 3'-end of the ^3H borohydride labeled genome RNA after annealing. (b) The leader RNA is coded for only once on the genome RNA. (c) The leader RNA is the first RNA product to be synthesized *in vitro* followed by the mRNA species. (d) The leader RNA hybridizes with the genome RNA of a defective interfering particle (DI) obtained from the heat resistant (HR) strain of VSV which contains only the 3'-terminal half of the wild-type genome RNA (15). Since only these DI particles have subsequently been shown to possess transcriptase activity synthesizing mRNAs coding for N, NS, M and G (15), it was suggestive that the 3'-terminal region of the VSV genome RNA is the recognition site for initiation of RNA synthesis by the virion-associated transcriptase. From the above results it was concluded that the leader RNA probably represents the initiated lead-in RNA segment which is removed during formation of VSV mRNAs by a possible processing mechanism.

In order to test whether the synthesis of the leader RNA is unique for other negative strand RNA viruses containing unsegmented genome RNA, we selected VSV New Jersey which is serologically distinct from VSV Indiana and a paramyxovirus, Newcastle disease virus. As shown in Fig. 3, both viruses were able to synthesize discrete RNA species which, as shown in the autoradiogram, migrate very similar to the leader RNA synthesized by VSV Indiana (Fig. 2). Table 1 summarizes the characteristics of the leader RNAs isolated from the above three viruses. They have very similar sizes, 5'-terminal sequences, and in the case of SV Indiana and New Jersey serotypes, the base compositions are also very similar. Nevertheless, the hybridization experiments clearly indicate that the respective leader RNAs are virtually unique and contain very little, if any, sequence homology. Assuming the VSV_{NJ} leader RNA like VSV_{IND} leader RNA is complementary to the 3'-terminal portion of the genome RNA, the above results indicate that VSV_{NJ} genome RNA should have 3'-terminal triplet ..CpGpU similar to VSV_{IND} genome RNA (10). It is interesting to note that the same 3'-terminal sequence is also present

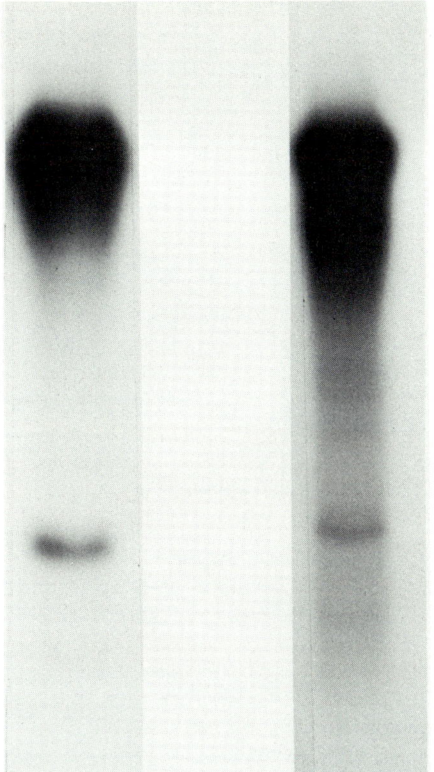

Fig. 3. *Polyacrylamide gel electrophoretic analysis of VSV_{NJ} and NDV (dT) unbound product RNAs. VSV_{NJ} (Ogden strain) and NDV (Beaudette C strain) were used to synthesize ^{32}P CMP-labeled product RNA in vitro as previously described (13, 16). The product RNA not binding to oligo(dT) was isolated and examined on 20% polyacrylamide gels as in Fig. 2. An autoradiogram of the gel shows the migration of VSV_{NJ} (left) and NDV (right) leader RNAs.*

in three widely dissimilar VSV DI particle genome RNAs (20). From the above results, it seems that the synthesis of leader RNAs may be a general strategy for the initiation of RNA synthesis by the negative strand viruses with an unsegmented genome RNA.

MODIFICATIONS AT THE 5'-TERMINI OF VSV mRNAs

(a) Capping

Although, as shown in Fig. 1, the processing model for the biosynthesis of VSV mRNAs *in vitro* has been favoured, a model involving "stop-start" by the polymerase has not been ruled out.

TABLE 1

Characterizations of Leader RNAs

Leader RNAs	Chain Length	5'-Terminal Sequence	Base Composition Mole Percent				Hybridization to the Genomes of		
			A	C	G	U	VSV_{IND}	VSV_{NJ}	NDV
VSV_{IND}	68[1]	(p)ppACG	48	20	11	21	+	-	-
VSV_{NJ}	72[1]	(p)ppACG	42	16	19	23	-	+	-
NDV	~70[2]	(p)pp$\overset{A}{G}$C	(not determined)				-	-	+

[1]Calculated from percentage of termini released following RNase T2 digestion.
[2]Estimated from migration in gel.

By this process, the RNA polymerase may initiate transcription at the 3'-terminus to synthesize the leader RNA and the same enzyme would re-initiate transcription of the N-gene, terminate and initiate transcription of the adjacent gene and so on. In this situation, the modification of the 5'-termini of the VSV mRNAs occurs by a discrete reaction involving GTP and the presumptive 5'-polyphosphorylated VSV mRNAs to form the blocked structure. Since only the α-phosphate of the terminal ATP is incorporated into the 5'-blocked structure of the VSV mRNAs, the above reaction should involve removal of β and γ phosphates of initiated VSV mRNAs during the blocking reaction. In this respect, the mechanism of capping in the VSV system seems to be quite different from the reovirus and vaccinia systems (19, 22). In the latter systems, the blocking with GTP occurs on the 5'-termini of the initiated mRNA molecules with α and β phosphates of the initiating base and the α phosphate of GTP being incorporated into the blocked structure. Moreover the capping reaction in the VSV system is not inhibited by the addition of P_i or PP_i (Abraham and Banerjee, unpublished observations). Pyrophosphate has been shown to inhibit the corresponding reactions in the vaccinia and reovirus systems.

Nucleoside triphosphate analogues such as GMP-P(NH)P have been used to elucidate the mechanism of capping reaction in the reovirus and the vaccinia systems (18, 22). Since the β, γ bond of this imido analogue of GTP is resistant to the action of phosphatases, the reovirus mRNA products synthesized *in vitro* using this analogue result in the 5'-terminal structure p (NH)ppG-C... (18). Likewise, in the vaccinia system using a similar analogue of GTP, GMP-P (CH_2)P, it was shown that only the methylated blocked structure of the form m^7Gp/ppAp... was synthesized and not m^7Gpp/pGp^m... . Thus, the cleavage of the β, γ bond of GTP is essential for the normal capping reaction in the vaccinia and the reovirus systems. We have used the β, γ imido analogue of GTP in the VSV system in order to examine the capping reaction further. Preliminary results (Testa and Banerjee, unpublished observations) indicate that GMP-P(NH)P is capable of supporting RNA synthesis at a level of 15-20% that of the control reaction with GTP. The synthesized RNA is methylated when S-adenosine-l-methionine (Sadomet) is included in the reaction mixture. The 5'-terminal analysis of the mRNAs showed the presence of a blocked and methylated structure of the form m^7GpppAp^m... . The synthesis of a normal cap using the GTP analogue demonstrates that there must be other enzymatic activities in the purified VSV which can convert GMP-P(NH)P to GTP in quantities sufficient to support the capping reactions. The above results clearly indicate that the capping reactions in the VSV system are quite distinct from the vaccinia and the reovirus systems and may involve other enzymatic steps yet to be identified.

(b) Methylation

In the presence of Sadomet, the 5'-terminal blocked structure of the newly synthesized VSV mRNAs are methylated as m^7GpppAp^m... (4). Similar methylase activities were also observed in several other animal viruses (28). The methylation process in VSV, like the capping reaction, was again quite distinct from the corresponding process in two other well-studied viral systems, i.e., vaccinia and reovirus. In the VSV system *in vitro* concentrations of the methyl donor specify the number and location of the methyl groups transferred to the capped 5'-termini of VSV mRNAs. Limited concentrations of SAM result in a single methylation of the penultimate base in the 2'-hydroxyl position, i.e. GpppAp^m... (*Cap B*) whereas, saturating concentrations of Sadomet methylate the blocking guanosine residue at the 7 position resulting in the dimethylated cap, m^7GpppAp^m... (*Cap A*). As shown in Fig. 4A, at a suboptimal concentration of Sadomet (2.2 μM), the nuclease P_1 and bacterial alkaline phosphatase resistant 5'-terminal portion of the mRNAs yielded two radioactive peaks (*Cap A* and *Cap B*) upon electrophoresis. Further analysis of *Cap A* by treatment with nucleotide pyrophosphatase and bacterial alkaline phosphatase confirmed it to be dimethylated cap (Fig.4B) while by a similar analysis *Cap B* was found to be monomethylated cap (Fig. 4C). VSV purified from different host cells including hamster, mouse and human contain both methyltransferase activities (Fig. 5). Thus, in the VSV system the sequence of events leading to the fully methylated 5'-terminal structure of VSV mRNAs *in vitro* can be depicted as:

$$GTP + ATP + UTP + CTP \rightarrow GpppAACAGp\ldots + PP_i$$

$$GpppAp\ldots + Sadomet\ (\leq 0.1\ \mu M) \rightarrow GpppA^{m}p\ldots + SAH$$

$$GpppA^{m}p\ldots + Sadomet\ (\geq 10\ \mu M) \rightarrow m^{7}GpppA^{m}p\ldots + SAH$$

This sequence of methylation is different from the vaccinia and reovirus systems where 7-methylation of guanosine precedes 2'-O-methylation of the penultimate base (19, 22), although both these viruses were purified from human and mouse cells similar to those used for VSV.

CONCLUSIONS

The mechanism of *in vitro* synthesis and modification of VSV RNAs involves the following events which have been fairly well under-

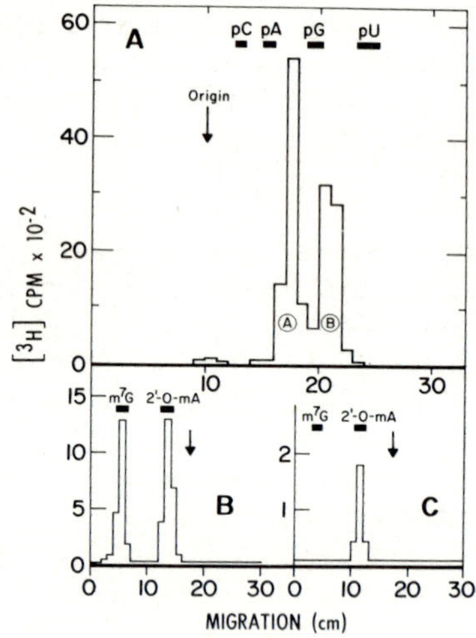

Fig. 4. Paper electrophoretic analyses of enzymatic digestions of ^{3}H methyl-labeled VSV mRNA synthesized in vitro. VSV mRNA was synthesized in vitro in the presence of ^{3}H Sadomet (2.2 μM) as the labeled substrate (26). The purified mRNA was digested with penicillium nuclease, followed by bacterial alkaline phosphatase and analyzed directly by paper electrophoresis (A) (3, 4). After electrophoresis, the radioactivity was eluted from the paper, further digested by nucleotide pyrophosphatase followed by bacterial alkaline phosphatase and reanalyzed by electrophoresis (B and C). Marker nucleotides and nucleosides were included in each sample, and their positions after electrophoresis are shown. (B) The material from panel A designated as region A or Cap A. (c) The material from panel A designated as region B or Cap B.

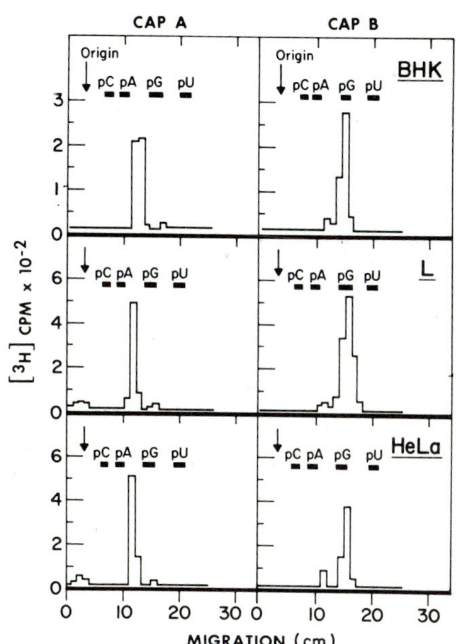

Fig. 5. *Paper electrophoretic analyses of Caps A and B synthesized in vitro by VSV purified from different cell lines. VSV was purified from BHK, L, and HeLa cells (8) and mRNA was synthesized in vitro in the presence of ^3H Sadomet at final concentrations of 13.3 μM (Cap A) or 0.037 μM (Cap B). The purified mRNA was digested with penicillium nuclease, followed by bacterial alkaline phosphatase and analyzed by paper electrophoresis with marker nucleotides.*

stood. (A) The virion-associated RNA polymerase initiates transcription at a single site at the 3'-end of the genome RNA with the synthesis of a leader RNA. The 5'-terminus of the leader RNA is unmodified and contains the sequence (p)ppACG... . The unique size of the leader RNA indicates that there is a specific termination event following the leader RNA synthesis. Since adenosine comprises nearly half of the leader RNA sequence, the 3'-terminal region of the VSV genome RNA presumably contains a high content of U residues and this portion of the genome serves as the recognition site for the virion-associated RNA polymerase. Whether L and NS proteins are both located at this site of the genome is still unknown. The New Jersey serotype of VSV and a paramyxovirus NDV, both synthesize distinct leader RNAs which may indicate a common strategy for the initiation of RNA synthesis by the negative strand RNA viruses with unsegmented genomes. (B) Following the leader RNA synthesis, there is a sequential synthesis of VSV mRNA species in the order 3' N-NS-M-G-L 5' (2, 5). Whether the leader RNA is cleaved off and mRNA species are formed by a processing mechanism awaits solid documentation. This mechanism of RNA synthesis is closely analogous to the early mRNA synthesis by bacteriophage T7 where the mRNAs are synthesized by cleavage with concomitant release of initiated lead-in RNA sequence or leader RNA (27). Alternatively, in the VSV system, the synthesis of the leader RNA may be terminated and then followed by sequential initiations of individual mRNA species. Since VSV mRNAs containing only the capped 5'-termini are observed, it is still ambiguous as to whether the mRNA molecules are initiated or cleaved products. (C) VSV mRNAs are modified at both ends. The 5'-termini are blocked by GTP in the form G(5')ppp(5')Ap... and poly (A) is added onto the 3'-termini. The precise mechanism of both the processes are still unclear. Moreover, in the presence of Sadomet two methyltransferase activities are discernible which methylate the blocked structure as $m^7G(5')ppp(5')A^mp$... . Although similar modifications at the 5'-termini of vaccinia and reoviruses are observed, the mechanism operative in the VSV system seems to be different and more intriguing. The proteins, either virus-specific or host, responsible for all the *in vitro* steps during VSV RNA initiation, modification, cleavage (if any), termination, etc. have not been characterized. Thus, VSV provides an excellent system to study the RNA-protein interaction leading to synthesis and modifications to RNA.

REFERENCES

1. Abraham, G. and Banerjee, A.K. (1976). *Virology* 71, 230.
2. Abraham, G. and Banerjee, A.K. (1976). *Proc. Nat. Acad. Sci. USA* 73, 1504.
3. Abraham, G., Rhodes, D.P. and Banerjee, A.K. (1975). *Nature (London)* 255, 37.
4. Abraham, G., Rhodes, D.P. and Banerjee, A.K. (1975). *Cell* 5, 51.
5. Ball, L.A. and White, C.N. (1976). *Proc. Nat. Acad. Sci. USA* 73, 442.
6. Baltimore, D., Huang, A.S. and Stampfer, M. (1970). *Proc. Nat. Acad. Sci. USA* 66, 572.
7. Banerjee, A.K., Abraham, G. and Colonno, R.J. (1977). *J. Gen. Virol.* 34, 1.
8. Banerjee, A.K., Moyer, S.A. and Rhodes, D.P. (1974). *Viro-*

logy 61, 547.
9. Banerjee, A.K. and Rhodes, D.P. (1973). *Proc. Nat. Acad. Sci. USA* 70, 3566.
10. Banerjee, A.K. and Rhodes, D.P. (1976). *Biochem. Biophys. Res. Commun.* 68, 1387.
11. Bishop, D.H.L. and Roy, P. (1972). *J. Virol.* 10, 234.
12. Both, G.W., Moyer, S.A. and Banerjee, A.K. (1975). *Proc. Nat. Acad. Sci. USA* 72, 274.
13. Colonno, R.J. and Banerjee, A.K. (1976). *Cell* 8, 197.
14. Colonno, R.J. and Banerjee, A.K. (1977). *Virology* 77, 262.
15. Colonno, R.J., Lazzarrini, R., Keene, J. and Banerjee, A.K. (1977). *Proc. Nat. Acad. Sci. USA* 74, 1884.
16. Colonno, R.J. and Stone, H.O. (1975). *Proc. Nat. Acad. Sci. USA* 72, 2611.
17. Emerson, S.U. and Yu, Y.H. (1975). *J. Virol.* 15, 1348.
18. Furuichi, Y. and Shatkin, A.J. (1977). *Virology* 77, 566.
19. Furuichi, Y., Muthukrishnan, S., Tomasz, J. and Shatkin, A. J. (1976). *J. Biol. Chem.* 251, 5043.
20. Keene, J.D., Rosenberg, M. and Lazzarrini, R.A. (1977). *Proc. Nat. Acad. Sci. USA* 74, 1353.
21. Morrison, T.G., Stampfer, M., Lodish, H.F. and Baltimore, D. (1975). *In* "Negative Strand Viruses, Vol. 1" (Mahy, B.W.J. and Barry, R.D. eds), Academic Press, New York, p.293.
22. Moss, B., Gershowitz, A., Wei, C.M. and Boone, R. (1976). *Virology* 72, 341.
23. Moyer, S.A. and Banerjee, A.K. (1975). *Cell* 4, 37.
24. Naito, S. and Ishihama, A. (1976). *J. Biol. Chem.* 251, 4307.
25. Rhodes, D.P. and Banerjee, A.K. (1976). *J. Virol.* 17, 33.
26. Rhodes, D.P., Moyer, S.A. and Banerjee, A.K. (1974). *Cell* 3, 327.
27. Rosenberg, M., Kramer, R. and Steitz, J.A. (1974). *J. Mol. Biol.* 89, 777.
28. Shatkin, A.J. (1976). *Cell* 9, 645.
29. Szilagyi, J.F. and Uryvayev, L. (1973). *J. Virol.* 11, 279.
30. Toneguzzo, R. and Ghosh, H.P. (1975). *FEBS Letters* 50, 369.
31. Villarreal, L.P. and Holland, J.J. (1973). *Nature New Biol.* 246, 17.
32. Wagner, R.R. (1975). *In* "Comprehensive Virology, Vol. 4" (Fraenkel-Conrat, H. and Wagner, R.R. eds.), Plenum Press, New York, p.l.

GENOME ORGANIZATION OF VESICULAR STOMATITIS VIRUS: MAPPING ts G41 AND THE DEFECTIVE INTERFERING T PARTICLE

GORDON J. FREEMAN, DONALD D. RAO, and ALICE S. HUANG

Department of Microbiology and Molecular Genetics, Harvard Medical School, Boston, Massachusetts 02115, U.S.A.

Many temperature-sensitive (ts), host-range and deletion mutants of vesicular stomatitis virus (VSV) have been isolated (4, 20, 25). Although their phenotypes have been extensively characterized and some gene assignments have been made (10, 23), definitive assignments and physical fine-mapping remain to be done.

This paper presents two approaches to the problem of mapping of VSV mutants. The first approach, heteroduplex formation between the virion RNA of ts G41 and the mRNA of its revertant, led to the physical location of the ts lesion contained in the mutant ts G41. The second approach, RNA fingerprinting, led to the identification and location of extragenomic sequences in the RNA of the DI-T particle.

LOCATION OF THE LESION IN ts G41

Duplex formation

Recently it has been shown that all five of the VSV mRNAs can be separated on gels as duplexes formed with virion RNA (6, 26). Figure 1 demonstrates this approach. Duplexes containing the L, G, N, NS, and M mRNAs were separated on a polyacrylamide gel (slot a). Subsequent isolation of the duplex molecules and re-electrophoresis yields relatively pure species of each of the double-strands (slots b-f).

Heteroduplexes with RNAs of ts G41

If the RNAs of a ts mutant and its wild-type were annealed with each other, the ts lesion may be identified by the formation of an altered duplex molecule. Such altered heteroduplexes

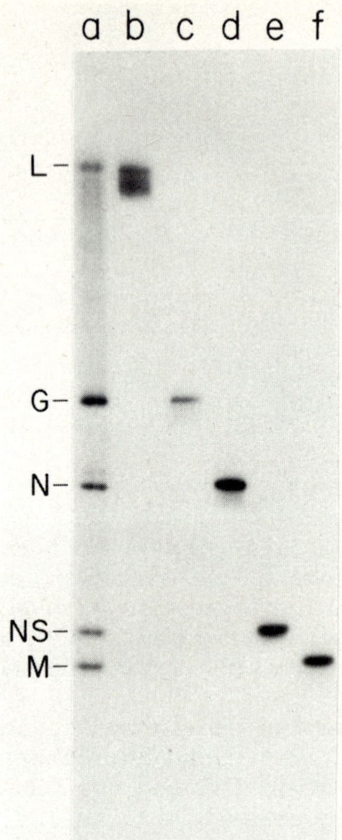

Fig. 1. *Purification of homoduplexes formed between virion RNA and mRNAs of vesicular stomatitis virus. Virus was grown and labeled with* ^{32}P *as described previously (6). All of the procedures for extraction, RNA-RNA hybridization and separation of the RNAs were as described (6), except that the hybridization was done at 72°, for 4 hr in 0.3M NaCl, 0.03M Na citrate, pH 6.5. Using an autoradiograph of the wet gel as a template, the regions containing labeled duplex RNA were excised and the RNA electrophoresed out of the gel (6). Contaminating polyacrylamide was largely removed by binding the double-stranded RNA to a CF-11 column in 25% ethanol in STE buffer (0.1M NaCl, 0.05M Tris, pH 6.85, 1 mM EDTA), washed with 20% ethanol and eluted with 0% ethanol in STE buffer (5). (a) Ribonuclease-resistant duplexes of* ^{32}P*-labeled virion RNA and unlabeled VSV mRNA; (b) re-electrophoresis of the L duplex; (c) re-electrophoresis of the G duplex; (d) re-electrophoresis of the N duplex; (e) re-electrophoresis of the NS duplex; and (f) re-electrophoresis of the M duplex.*

have been observed for SV40 DNA where a single base-mismatch is detected by S1 nuclease (27). Also, a heteroduplex formed between the RNAs of a ts mutant of reovirus and its wild-type can be identified by its aberrant migration rate on gels (13).

In order to test this approach for VSV ts mutants, RNAs were prepared from ts G41 and from its spontaneous ts^+ revertants. Homoduplexes formed between ^{32}P-labeled virion RNA and unlabeled mRNA made from ts G41-infected cells resembled duplexes formed between the RNAs of wild-type virus. The ts G41 duplexes labeled M, NS, N and G were readily visualized (Fig. 2, slot b). The two distinct bands near the top of the gel probably represent the L duplex or its breakdown product (Fig. 2, slot b). This separation of the L duplex was occasionally seen with RNAs from wild-type infections. Heteroduplexes formed between ^{32}P-labeled ts G41 virion RNA and unlabeled mRNA from cells infected with a revertant appeared identical to the homoduplexes (Fig. 2, slot e). However, ribonuclease digesting the hybridized RNA at lower salt concentrations than the standard one of 0.3M diminished the L, G, and N duplexes and produced a new duplex X (Fig. 2, slots d and f). The homoduplex controls digested with ribonucleases at the identical lower salt concentrations indicate that the L and G homoduplexes were also susceptible to digestions, but that the N homoduplex was more resistant than the N heteroduplex. Moreover, no X duplex was generated among the homoduplexes under the same conditions which generated it with heteroduplexes (Fig. 2, slots a and c).

This suggests that the X duplex resulted from a cleavage within the N duplex formed between RNAs from ts G41 and its revertant. Oligonucleotide fingerprints of the X and N duplexes are in progress. The portion of the N duplex not found in the X duplex may have been present near the bottom of the gel. Another revertant of ts G41 with similar results has been obtained (Freeman, unpublished results). Therefore, the temperature-sensitive lesion of the group IV mutant ts G41 is assigned to the N gene.

The assignment of the lesion in ts G41 to the N gene is in agreement with the findings that the N protein is unstable when synthesized at the nonpermissive temperature (14, 19). Also, lesions in the L cistron and the G cistron are most likely in the complementation groups I and V, respectively (12, 16).

Applicability of the approach of duplex formation

The generality of this approach for gene assignment and fine-mapping remains to be established. The logic of mapping mutations from the site of base mismatch of mutant and revertant is complicated by the mechanism of reversion. While the base change(s) leading to reversion may occur at the site of the original

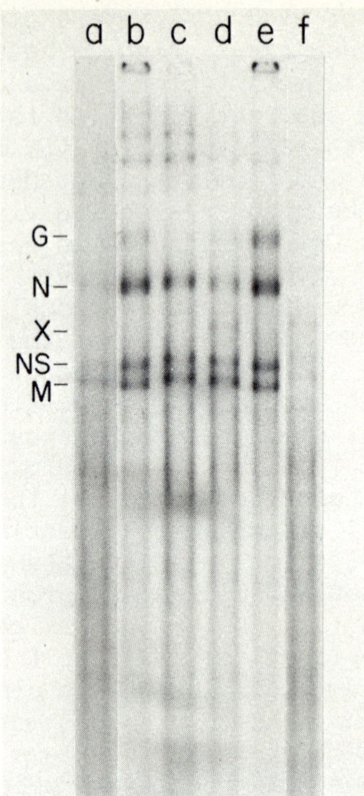

Fig. 2. *Mapping of ts G41 by heteroduplex formation. Ts G41, obtained from Dr. Craig Pringle, was cloned 5 successive times by plaque isolation on Chinese hamster ovary cells (28). Revertants, forming plaques at 38°, were picked and isolated immediately upon cloning of the mutant. Duplexes were formed between ^{32}P-labeled ts G41 virion RNA and unlabeled mRNA from ts G41- or ts^+ revertant-infected cells. The duplexes were digested with ribonucleases at 37° in varying concentrations of SSC (SSC = 0.15M NaCl, 0.015M NA citrate, pH 7.0) and separated by electrophoresis (6). The slots contain homoduplexes of ts G41 digested in (a) 0.5 x SSC, (b) 2 x SSC, and (c) 1 x SSC, or heteroduplexes of ts G41 digested in (d) 1 x SSC, (e) 2 x SSC, and (f) 0.5 x SSC.*

mutation or in the same coding triplet, the change might be elsewhere within the same cistron or even in another cistron. Reversion of a temperature-sensitive phenotype by means of a compensating change in a different phenotype or by means of a compensating change in a different protein has been shown in reovirus (24). In addition, at the site of a base mismatch, one of three degrees of mismatch may be present: (1) a strong degree of mismatch such as an A-G pairing, (2) an intermediate degree such as an A-C or G-U pairing, or (3) a weak mismatch such as a C-U pairing. Ribonucleases may not recognize all of these degrees of mismatch, so not all mutant-revertant heteroduplexes may be mappable. Because of these considerations, several mutants in a complementation group should be mapped in order to confidently assign a coding function.

FINGERPRINTING OF DI RNA

Comparison with standard RNA

DI-T particles of VSV were originally described by Huang *et al.* (9) and its RNA was found to anneal only to the largest mRNA (18). To test whether or not the genomes of DI-T particles were simple deletion mutants of wild-type genomes, both standard and DI-T particles were labeled with ^{32}P, their RNAs extracted and digested with ribonuclease T1, and the oligonucleotides separated on two-dimensional gels (3). Figure 3 shows that the two RNAs were indeed related to each other with many of the oligonucleotide spots found in DI RNA included in standard RNA. In addition, these fingerprints indicate that DI-T RNA contained fewer unique spots and was, as expected, about 1/3 as complex

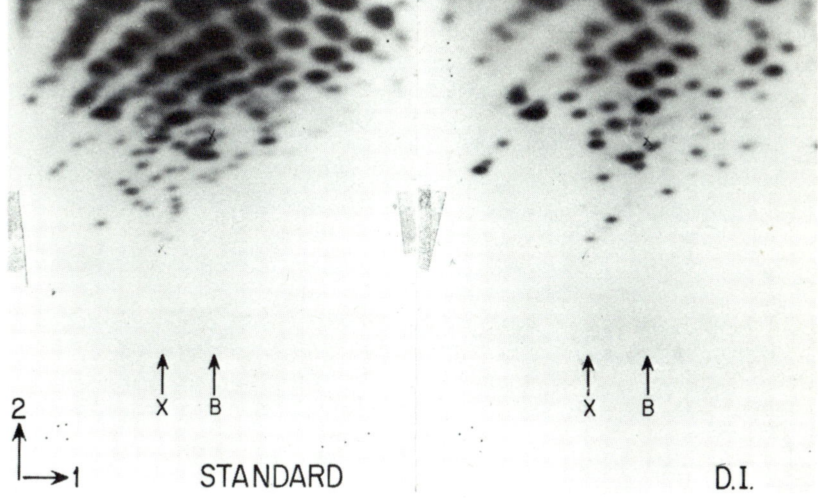

Fig. 3. *Comparison of oligonucleotides from the genomes of standard B particles and DI-T particles. The San Juan strain of the Indiana serotype of vesicular stomatitis virus and DI-T particles (9) were labeled with ^{32}P and purified by rate zonal centrifugation. The RNAs were phenol extracted and repurified on sucrose gradients prior to digestion with ribonuclease T1. The oligonucleotides were separated by two-dimensional acrylamide gel electrophoresis (3).*

as standard RNA (11).

There were, however, two unique spots that were found exclusively in DI-T RNA (Fig. 3). These two oligonucleotides have been numbered 3 and 6 on Figure 4c which is a similar fingerprint of DI-T RNA.

Evidence in support of terminal complementary sequences

When DI-T RNA was self annealed, 9% of the RNA was hybridized. Sizing of the annealed fragment after ribonuclease digestion indicates that hybridization was not due to the annealing of full size plus and minus strands (Rao, unpublished observations). Similar self-annealing of 3% to 80% has been reported for a variety of VSV DI RNAs (17, 21), as well as for Sendai DI RNAs (15).

Because DI RNAs have been shown to form panhandle-containing circular structures and because VSV standard and DI RNAs probably all contain some terminal complementary sequences (1, 2, 7, 22), a model has been proposed for the generation of VSV DI genomes (8). This model generates DI genomes with covalently linked subgenomic and extragenomic sequences, by the formation of a hairpin during the synthesis of the negative-stranded DI genome. Specific predictions of the model are: (1) the extragenomic sequences must be identical to sequences complementary to the DI genome and therefore, hybridizeable to genome RNA; and (2) some of the extragenomic sequences should be found only at or near the 3' end of the DI RNA.

Fig. 4. *Enrichment for oligonucleotides of DI-T RNA by borate-column selection and by self-hybridization. DI-T RNA was prepared as described for Figure 3. For the borate selection, 5 x 10^5 cpm of ^{32}P-labeled RNA was mixed with 40 μg of yeast RNA and digested with ribonuclease T1 at an enzyme to substrate ratio of 1:250 in 0.01M $MgCl_2$, 0.01M Tris-HCl pH 7.5 buffer, 0° for 30 min. The partially digested RNA was diluted 20-fold with loading buffer (0.05M morpholin-HCl, pH 8.5; 0.1M $MgCl_2$, 1 mM β-mercaptoethanol) and selected by boric acid gel column (Aldrich, particle size 0.1-0.4). The selected RNA was then digested to completion with ribonuclease T1. For self-hybridization, 1 x 10^6 cpm of RNA was suspended in 0.4M NaCl, 0.01M Tris-HCl, 0.01M EDTA pH 7.4 and hybridized at 70° for 6 hr. RNA was then digested with RNase T1 in the same buffer at 37° for 30 min. Double-stranded RNA was selected by CF-11 column as indicated for Figure 1, boiled and then digested with RNase T1 to completion. The oligonucleotides were separated as described for Figure 3. (a) borate-selected oligonucleotides of DI-T RNA; (b) self-hybridized oligonucleotides of DI-T RNA; (c) total oligonucleotides of DI-T RNA.*

a
X
B
b
c
37
36
35
34
33
32
31
30
29
28
27
26
25
24
23
22
21
20
19
18
17
16
15
14
13
12
11
10
9
8
7
6
5
4
3
2
1
39
38
2
1

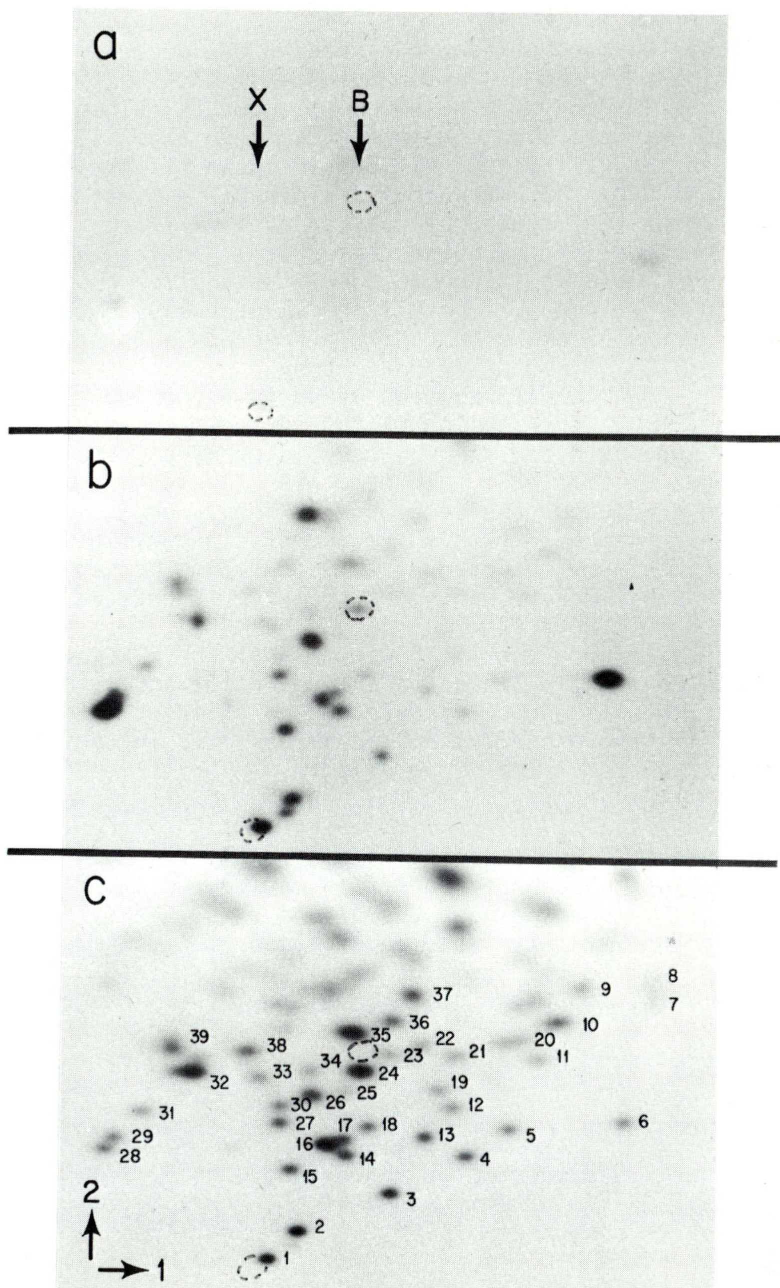

Fig. 4

Enrichment of an extragenomic oligonucleotide from DI-T RNA

In order to further test the predictions of our model, DI RNA was partially digested by ribonuclease T1 and the fragments containing the 3' ends were selected on a borate column. The selected fragments were then completely digested with T1 and fingerprinted as before. Figure 4, panel c shows all the oligonucleotides of DI RNA: the extragenomic spots are numbered 6 and 3. Panel a shows that spot 6 as well as spots 1 (very faint) and 28 were enriched by the borate selection. Therefore, the extragenomic sequences contained in spot 6 were close enough to the 3' end to be selected. Spots 1 and 28 probably represent sequences at the 3' end of the DI RNA that are identical to sequences at the 3' end of standard RNA.

Panel b shows the enrichment of spots 6 and 3 as a result of self-annealing. ^{32}P-labeled DI RNA was self-annealed, single-stranded RNA was hydrolyzed by ribonuclease, and the remaining double-strands completely digested with T1 and then finger-printed. As predicted, the sequences at the 3' end (spots 1, 28, and 6) were enriched by this procedure. In addition, other oligonucleotides were selected by this annealing procedure; these were presumably due to sequences from the 5' end, as well as additional sequences from the 3' end.

Such preliminary results indicate that there are subgenomic sequences covalently linked to extragenomic sequences and that these extragenomic sequences are indeed at or near the 3' end of DI RNA. Proof that the 5' end is involved in the self-annealing is currently being obtained.

CONCLUSION AND SUMMARY

Assignment of the lesion in the group IV VSV mutant, ts G41, to the N cistron by heteroduplex formation and the finding of extragenomic sequences near the 3' end of DI RNA by RNA fingerprinting are examples of the application of biochemical molecular techniques to the study of genome organization of VSV and their mutants. Continued efforts along these lines should shed light on the biological significance of such genomes and the consequences of reorganization between genomic and extragenomic segments.

ACKNOWLEDGEMENTS

This work is supported by research grants AI 10100 from the U.S.P.H.S. and VC-63 from the American Cancer Society. Gordon J. Freeman is supported by a U.S.P.H.S. Institutional Research Service Award CA 09031 and Alice S. Huang is a U.S.P.H.S. Career Development awardee. We thank Drs. Warren Jelinek and John Coffin for advice and help with the RNA fingerprinting. We thank Ms. Trudy Lanman and Priscilla Walker for excellent technical support.

REFERENCES

1. Banerjee, A. and Rhodes, D.P. (1976). *Biochem. Biophys. Res. Commun.* 68, 1387.
2. Colonno, R.J. and Banerjee, A.K. (1976). *Cell* 8, 197.
3. DeWachter, R. and Fiers, W. (1972). *Anal. Biochem.* 49, 184.
4. Flamand, A. and Pringle, C.R. (1971). *J. Gen. Virol.* 11, 81.
5. Franklin, R.M. (1966). *Proc. Natl. Acad. Sci. U.S.A.* 55, 1504.
6. Freeman, G.J., Rose, J.K., Clinton, G.M. and Huang, A.S. (1977). *J. Virol.* 21, 1094.
7. Hefti, E. and Bishop, D.H.L. (1975). *J. Virol.* 15, 90.
8. Huang, A.S. (1977). *Bacteriol. Rev.* 41, 811.
9. Huang, A.S., Greenawalt, J.W. and Wagner, R.R. (1966). *Virology* 30, 161.
10. Huang, A.S. and Baltimore, D. (1977). *In* "Comprehensive Virology, Vol. 10" (Fraenkel-Conrat, H. and Wagner, R.R. eds), Plenum Publishing Corp., New York
11. Huang, A.S. and Wagner, R.R. (1966). *J. Mol. Biol.* 22, 381.
12. Hunt, D.M., Emerson, S.U. and Wagner, R.R. (1976). *J. Virol.* 18, 596.
13. Ito, Y. and Joklik, H.K. (1972). *Virology* 50, 202.
14. Knipe, D., Lodish, H. and Baltimore, D. (1977). *J. Virol.* 21, 1140.
15. Kolakofsky, D. (1976). *Cell* 8, 547.
16. Lafay, F. (1969). *Compte rendu hebdomadaire des séances de l'Academie des Sciences, Paris,* 268, 2385.
17. Lazzarini, R.A., Weber, G.H., Johnson, L.D. and Stamminger, G.M. (1975). *J. Mol. Biol.* 97, 289.
18. Leamnson, R.N. and Reichmann, M.E. (1974). *J. Mol. Biol.* 85, 551.
19. Little, S.P. and Huang, A.S. (1977). *Virology* 81, 37.
20. Obijeski, J.F. and Simpson, R.S. (1974). *Virology* 57, 369.
21. Perrault, J. (1976). *Virology* 70, 360.
22. Perrault J. and Leavitt, R.W. (1978). *J. Gen. Virol.* 38, 35.
23. Pringle, C.R. (1975). *Immunol.* 69, 85.
24. Ramig, R.F., White, R.M. and Fields, B.N. (1977). *Science* 195, 406.
25. Reichmann, M.E., Pringle, C.R. and Follett, E.A.C. (1971). *J. Virol.* 8, 154.
26. Rhodes, D.P., Abraham, G., Colonno, R., Jelinek, W. and Banerjee, A. (1977). *J. Virol*. 21, 1105.
27. Shenk, T.E., Rhodes, C., Rigby, P.W.J. and Berg, P. (1974). *Cold Spring Harbor Symp. Quant. Biol.* 39, 61.

28. Stampfer, M., Baltimore, D. and Huang, A.S. (1971). *J. Virol.* 7, 409.

27

EVOLUTION OF RHABDOVIRUS GENOMES

J.P. CLEWLEY and D.H.L. BISHOP

Department of Microbiology, The Medical Center, University of Alabama in Birmingham, Birmingham, Alabama 35294, USA.

Rhabdoviruses have been obtained from many species of animals, insects and plants. At least forty virus isolates have been categorized into twenty-one serological groups with no evidence of any serological relationship between members of one group and members of another (5). Two of the groups which have been studied in detail both biochemically and serologically are the rabies group (rabies, Lagos bat virus, Mokola, kotonkan, Obodhiang, Bolivar and Duvenhage viruses), and the vesicular stomatitis virus group (VSV Indiana, VSV New Jersey, Cocal, VSV Brazil (Alagoas), Chandipura, Isofahan and Piry viruses).

By certain serological criteria, members of the VSV group of viruses are interrelated to each other although the exact antigenic determinants which are shared are not known. With intact virus preparations and sera raised against infectious virus, cross neutralization of infectivity has been demonstrted in reciprocal tests involving the sera and viruses of VSV Indiana (C), Cocal and VSV Brazil, although little cross neutralization of infectivity was obtained between these viruses and VSV New Jersey (M), Chandipura or Piry viruses (6). With dissociated virus preparations and the respective antisera substantial, reciprocal, cross neutralization has been reported for VSV Indiana (C), Cocal, VSV Brazil and VSV New Jersey (M), suggesting that their infectious nucleocapsids possess common antigenic determinants (6). Complement fixation studies with these viruses confirm the presence of common antigenic determinants for members of the VSV subgroup (6, 29).

Studies on defective particle interference (6), RNA hybridization (23, 27), and transcription of heterologous templates by VSV transcriptase (3) confirm that certain VSV group members are more closely related to each other than to other members of the group. Complementation between temperature sensitive (*ts*) mutants of Cocal and VSV Indiana (C) has been reported, although none was observed between *ts* mutants of VSV Indiana (C) and VSV New Jersey (M), or between VSV New Jersey (M) and Cocal virus (20, 21). The ability of wild type VSV New Jersey (Concan) to rescue only the *in vivo* primary transcription capability of the VSV Indiana (C) *ts* mutants G I-114 at nonpermissive temperatures has been demonstrated (22). All these observations indicate that although there are some antigenic and functional determinants which unite the various members of the VSV group, many of the viruses are biochemically and serologically distinguishable.

In this communication we document RNA sequence differences not only between several of the members of the VSV group but also between various isolates of the same virus serotype. The stability of the prototype VSV Indiana genome to consecutive cloning or passage in various cell lines in different laboratories, is documented. We have been able to induce some 50 nucleotide changes in the prototype strain of VSV Indiana by consecutive mutagenesis; these changes mimic to some extent the evolution which occurs in Nature.

OLIGONUCLEOTIDE FINGERPRINTS OF VARIOUS VSV INDIANA ISOLATES

When ^{32}P-labeled single-stranded RNA is digested by ribonuclease T_1 a series of oligonucleotides are generated which can be partially separated by a two-dimensional gel electrophoresis procedure to produce a characteristic fingerprint for the largest oligonucleotide species (13). The application of this technique to four independent isolates of VSV Indiana (Fig.1) has provided evidence that each isolate is related to, but distinguishable from the prototype strain (7). As far as we can determine, the French groups of P. Printz and A. Flamand as well as the American groups of D.F. Summers, R.A. Lazzarini, R.R. Simpson and ourselves, work with the prototype strain of VSV Indiana (Fig.1A). This strain, which is carried by the American Type Culture Collection (Fig.2B), was originally obtained by W.E. Cotton from diseased cattle at Richmond Indiana (10). The Canadian groups of A.F. Holloway, T. Nakai and L. Prevec work with a heat resistant (HR) derivative of the prototype strain. The English groups of F. Brown, C.R. Pringle as well as the U.S. group of M.E. Reichmann and associates, work with C strain of VSV Indiana (Fig.1C). This strain was obtained by A.H. Frank from diseased equine tissue during an outbreak of VSV in Colorado (28). R.R. Wagner, S.U. Emerson,

A.S. Huang, D. Baltimore and associates work with the San Juan isolate of VSV Indiana (Fig.1B). This isolate was obtained in 1956 by L.O. Mott and associates from vesicular tongue epithelium obtained from a diseased steer on the San Juan Indian reservation near Espanola, New Mexico. A fourth isolate of VSV (Fig.1D) has been obtained more recently by C.H. Calisher from *Aedes* mosquitoes collected at Rancho de Abiquin, in 1966 (31).

The oligonucleotide fingerprints are highly reproducible between duplicate analyses of the same RNA preparation or between various preparations of the same virus strain grown in different cell lines. After 36 consecutive clonings of the prototype strain in BHK-21 cell monolayers we have not detected any changes in the large oligonucleotides which are resolved by the 2-dimensional gel electrophoresis procedure. Although we do not know how many of the large oligonucleotides of the prototype strain represent multiple species, a nomenclature system for them has been proposed (Fig.2A, ref.7). This nomenclature system was arrived at by comparing the autoradiograms of samples run for different times in the first or second dimension to produce slightly better resolutions of some of the groups of oligonucleotides seen in Fig.1. Based on the different intensities of the various spots and the relative radioactivity in the oligonucleotide gel plugs, it is probable that some of the species indicated in Fig.2A represent multiple oligonucleotides. If, at some later date, this is shown to be true by sequence analyses then we suggest that they be assigned letters (e.g. 24A, 24B, etc). By computing the total radioactivity present in the 71 oligonucleotides identified in Fig.2A and relating this to the amount of ^{32}P RNA initially digested, we have determined that these 71 oligonucleotides represent approximately 10% of the VSV genome.

The fingerprint of VSV Indiana, San Juan, evidently differs from that of the prototype strain (Fig.1B). Eleven large oligonucleotides (#1, 2, 3, 16, 26, 37, 57, 58, 62, 63 and 69) are missing (indicated by * in Fig.1B). Eleven new oligonucleotides are present (indicated by arrows in Fig.1B). The oligonucleotide fingerprint of VSV Indiana C (Fig.1C) lacks six or seven large oligonucleotides by comparison to the prototype strain (#9, 42, 46, 57, 59 and 61, and possibly 51). This strain has six new oligonucleotides which are not present in the fingerprint of the prototype strain. The VSV Indiana, New Mexico strain lacks 11 oligonucleotides by comparison to the prototype strain (#7, 10, 15, 16, 24, 25, 49, 57, 58, 63 and 65), while 8 new oligonucleotides are present (Fig.1D). The HR derivative of VSV Indiana lacks oligonucleotide #17 of the prototype virus, but is identical in every other respect

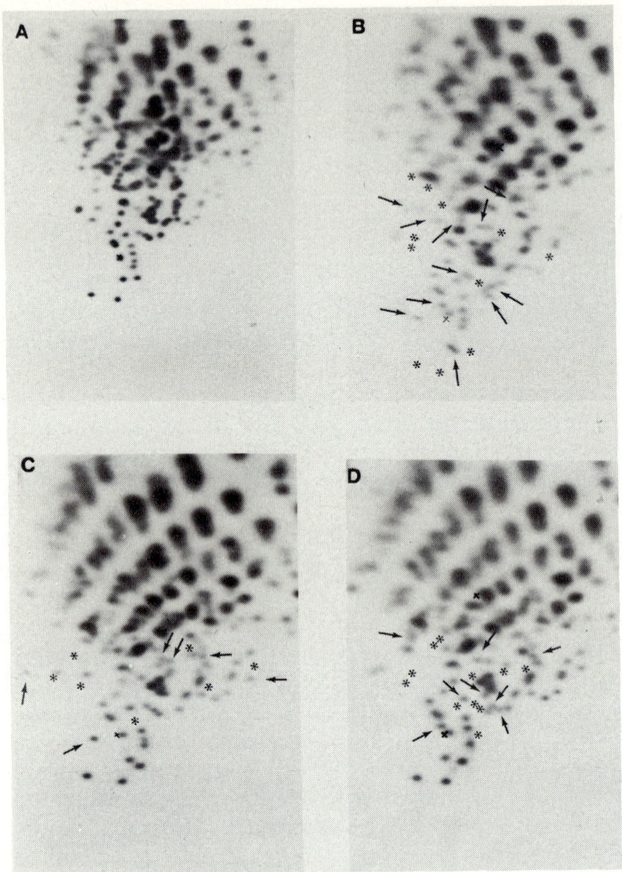

Fig.1. *The oligonucleotide fingerprints of various VSV Indiana isolates. Ribonuclease T_1 digests of 2-4x10^6 counts/min of purified ^{32}P-labelled viral RNA were prepared and resolved by two dimensional gel electrophoresis (7, 8, 13). The first dimension was from left to right and the second from bottom to top. The positions of the xylene cyanol FF and bromophenol blue dye markers are indicated by x. The oligonucleotide positions were determined by autoradiography. In (A) the fingerprint of VSV Indiana prototype, ts 0 I-5, is given. This fingerprint is identical to that of the parental French prototype strain as well as the ATCC VSV Indiana stock (Fig.2B), the wild type VSV Indiana used in the laboratories of D.H.L. Bishop (see Fig.5A), or C.Y. Kang, D.F. Summers, or L. Prevec. In (B) the oligonucleotide fingerpring of VSV Indiana, San Juan isolate is given. This virus was obtained from, and is used by R.R. Wagner; it is also used by S.U. Emerson, A.S. Huang and*

(C.Y. Kang, personal communication).

It is of interest to note that for the most part the various collections of VSV Indiana *ts* mutants were derived from different VSV isolates. In the case of the Orsay, Glasgow, and Massachusetts collections, the mutants were derived from the prototype strain, VSV Indiana C, and VSV Indiana, San Juan (respectively). The Winnipeg collection came from the HR derivative of the prototype strain.

OLIGONUCLEOTIDE FINGERPRINTS OF VSV INDIANA DEFECTIVE INTERFERING PARTICLES

What determines the origin or induction of defective interfering T particles of VSV (9) is not known. Different types of T particles have been shown to contain RNA species of different sizes. In some cases T particle preparations have been shown to contain complementary RNA sequences. In other cases no complementary RNA sequences have been detected (16, 18, 24, 25). In addition, some T particle preparations appear to be unique whereas others contain particles and corresponding RNA species of various sizes (4, 12, 14, 17, 19, 24, 27, 30).

The technique of oligonucleotide fingerprinting has allowed us to define the RNA sequences of particular T particle preparations in terms which can be directly compared with other T particles produced by the same virus strain. T particle preparations made in different cell lines and in different laboratories can also be compared by this technique, provided that the same initial virus isolate is used. Additional information which can be gleaned from such analyses includes the possibility of identifying oligonucleotides which are not present in the digest of B particle RNA, as in the case where positive strand RNA sequences are present.

Most T RNA species are less than half the size of the viral genome, except for those obtained from a large T particle (which was originally isolated from the HR mutant of VSV Indiana). These small RNA species anneal to the 28 *s* viral mRNA species (26, 30). By assuming that the small T RNA species come from contiguous sequences of the B genome, it should be possible to begin to order certain oligonucleotides by comparing the oligonucleotide fingerprints of the different size classes of overlapping T RNA sequences.

D. Baltimore. In (C) the oligonucleotide fingerprint of VSV Indiana C ts G I-114 is given. This fingerprint is identical to that obtained from the VSV Indiana C ts G I-II mutant. The parental VSV Indiana C is used by F. Brown, C.R. Pringle and M.E.Reichmann. In (D) the fingerprint of the VSV Indiana New Mexico strain is given.

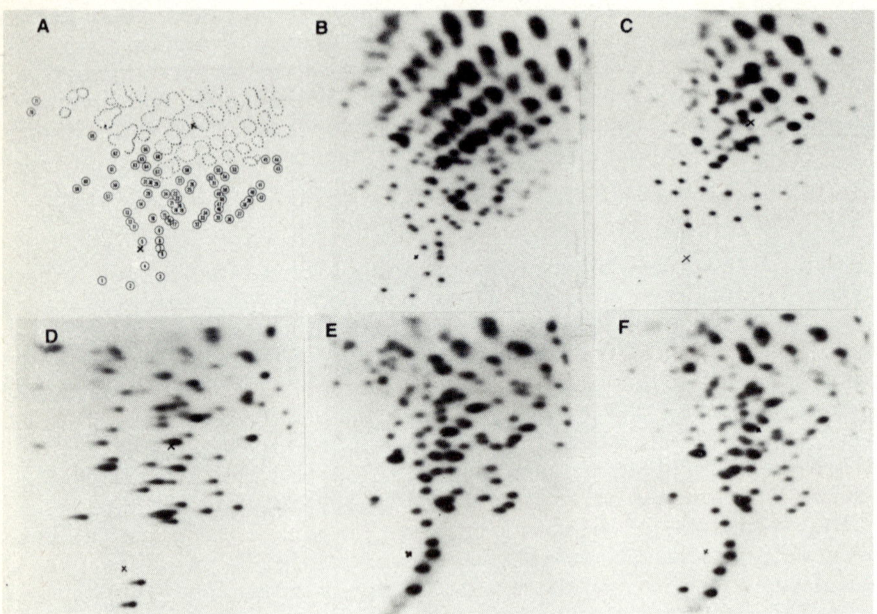

Fig.2. *A nomenclature system for certain VSV Indiana large oligonucleotides and the oligonucleotides of 4 defective interfering (dI) particles of VSV Indiana, prototype strain. In (A) seventy one large oligonucleotides have been identified in the fingerprint of the prototype strain of VSV Indiana. In (B) the digest of VSV Indiana prototype strain obtained from the ATCC is given. In (C) the fingerprint of the large dI obtained from the HR derivative of the prototype strain (19) is shown. In (D), (E) and (F) the fingerprints of the smaller dI obtained by C.Y. Kang from VSV Indiana prototype strain grown in rat cells, and two larger dI particles obtained in hamster cells are given.*

Three small T particle preparations have been prepared for us by C.Y. Kang from the prototype strain of VSV Indiana grown in either rat or hamster cells. From each of these T particle preparations RNA has been extracted and fingerprinted. The results shown in Fig.2D, E, F, indicate that each species of T RNA contains certain particular oligonucleotides, all of which are derived from the B particle RNA. The oligonucleotides present in the smallest T particle RNA include 14 of the 71 oligonucleotides identified in Fig.2A (Nos.2, 4, 18, 21, 22, 25, 38, 47, 50, 54, 56, 60, 64 and 70). The next larger size class of T RNA contains, in addition to those 14 oligonucleotides, 11 additional large nucleotides (Fig.2E; Nos.6, 8, 10, 14, 30 (or 31), 36, 37, 44, 64, 67 and possibly 59). The third largest size class of small T particle has, in addition to all of the nucleotides present in the smaller 2 species, four other nucleotides (Fig.2F; Nos.1, 24, 51 (or 52), and 53). It is clear from these results that the three T particle RNA preparations have overlapping sequences. None of the RNA species contain oligonucleotides which might be attributed to positive strand RNA.

In addition to the three T particles generated from the prototype strain of VSV, the RNA from the large T particle preparation obtained from the HR derivative of the prototype strain, has been fingerprinted (Fig.2C). As with the other T particle RNA species that we have analysed, all of the oligonucleotides present in the large T preparation appear to have originated from the B particle sequence, although few correspond to the nucleotides found in the smaller T RNA species. This result agrees well with the hybridization data obtained by M.E. Reichmann,R.A. Lazzarini and associates that the larger T RNA preparation corresponds to a different part of the genome of VSV Indiana.

It is of interest to note that five separate preparations of T particles obtained from five different laboratory strains of the prototype virus have given us oligonucleotide fingerprints identical to that shown in Fig.2E. This suggests that the derivation of this T particle is a common event although why this should be so is not known.

THE OLIGONUCLEOTIDE FINGERPRINTS OF VSV NEW JERSEY STRAINS

We have not been able to obtain the original VSV New Jersey isolated by W.E. Cotton (11). Five other strains have been cloned and their RNA species fingerprinted (VSV New Jersey, Concan, Guatemala, Ogden, Hazelhurst and M). The fingerprints of the Guatemala (Fig.3A), Concan and Ogden strains, although similar to each other, differ by either a few or several oligonucleotides (7). The fingerprints of the M (Fig.3B) and Hazelhurst RNA digests are quite different to those of the Concan, Ogden and Guatemala viruses (7). These results are consistent with a recent report of limited RNA sequence homology between the Ogden and M strains of VSV New Jersey as determined by hybridization (27).

THE OLIGONUCLEOTIDE FINGERPRINT OF COCAL VIRUS

Although Cocal virus is antigenically, and by RNA homology studies, related to VSV Indiana (15, 23), its RNA fingerprint is quite distinct to that of the prototype VSV Indiana strain (7).

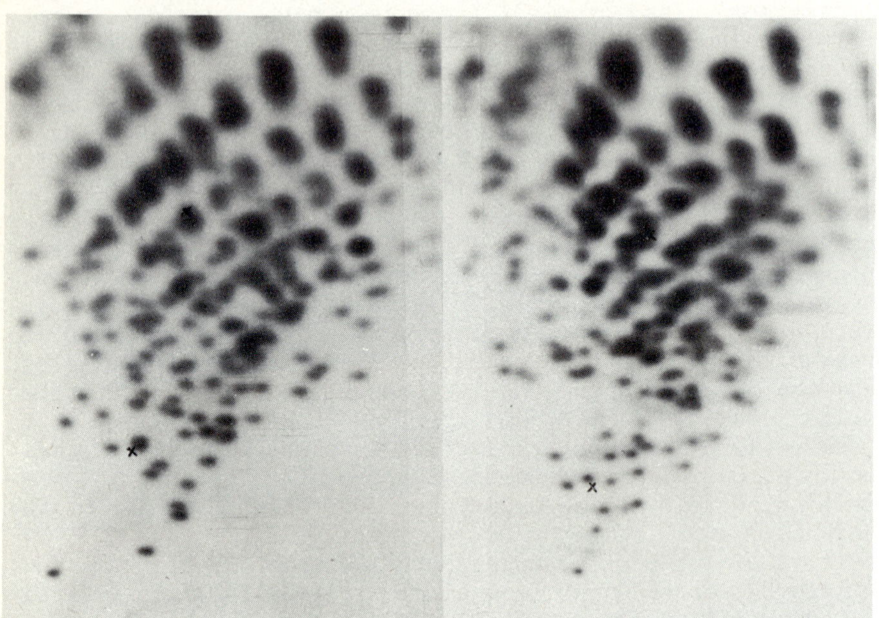

Fig.3. *The oligonucleotide fingerprints of VSV New Jersey strains. In (A) the fingerprint of VSV New Jersey, Guatemala, and in (B) the fingerprint of VSV New Jersey, M are given.*

THE OLIGONUCLEOTIDE FINGERPRINTS OF TWO CHANDIPURA STRAINS

In view of the foregoing analyses it was unlikely that any homology would be observed between the fingerprints of Chandipura and other VSV subgroup members. This was substantiated by the analyses of two Chandipura strains, one isolated from a human serum sample obtained in Nagpur, Maharashtra, India (2; Fig.4A) and the other from a hedgehog obtained by the Ibadan virus laboratory in Nigeria (1; Fig.4B). Compared to each other, the fingerprints of the two Chandipura strains are quite distinct.

IN VITRO EVOLUTION OF THE NUCLEOTIDE SEQUENCE OF THE PROTOTYPE STRAIN OF VSV INDIANA BY HIGH LEVEL MUTAGENESIS

We have used consecutive high level mutagenesis to study the evolution of the genome of the VSV Indiana prototype strain. In twenty-three consecutive passages of VSV in the presence of 400 µg 5-fluorouracil per ml of Eagle's medium, with cloning of the virus in the absence of the mutagen at each passage, five of the 71 large oligonucleotides identified in Fig.2A were

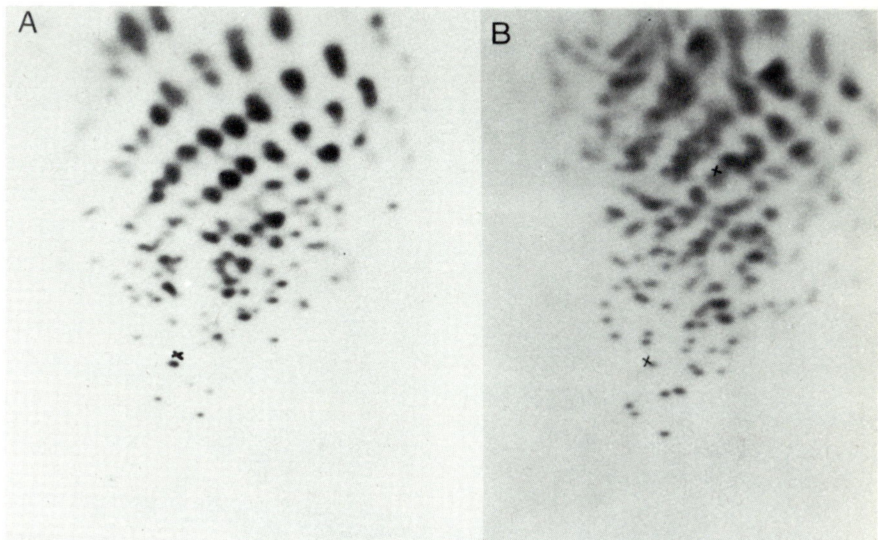

Fig.4. *The oligonucleotide fingerprints of Chandipura strains. In (A) the fingerprint of Chandipura (Nagpur) and in (B) the fingerprint of Chandipura (Ibadan) are given.*

sequentially and inheritably lost (indicated by * in Fig.5). By the ninth passage the oligonucleotides no.71 and 64 had disappeared while by the seventeenth passage oligonucleotide no.57 was lost. Between the twenty-first and twenty-third passage the oligonucleotides no.3 and no.8 disappeared. It is noteworthy that the two new oligonucleotides appeared in lateral positions to nucleotides which were lost (see arrows in Fig.5), suggesting that the nucleotide transitions which occurred in the original oligonucleotides (i.e. no.71 and no.8) did not involve G residues (e.g. C to U). Sequence analyses will be needed to confirm these suggestions. For the oligonucleotide no.3 (and possibly no.57 and no.64) presumably nucleotide transitions occurred in residues which resulted in a transition from U, C or A to a G residue. Since the 71 large oligonucleotides identified in Fig.2a represent some 10% of the viral genome, the five oligonucleotide losses probably correspond to some 50 overall nucleotide changes in the genome. The efficiency of the mutagenesis employing 5-fluorouracil is being compared in similar experiments to that involving 5-azacytidine and nitrosoguanidine.

DISCUSSION

Oligonucleotide fingerprinting of ribonuclease T_1 digests

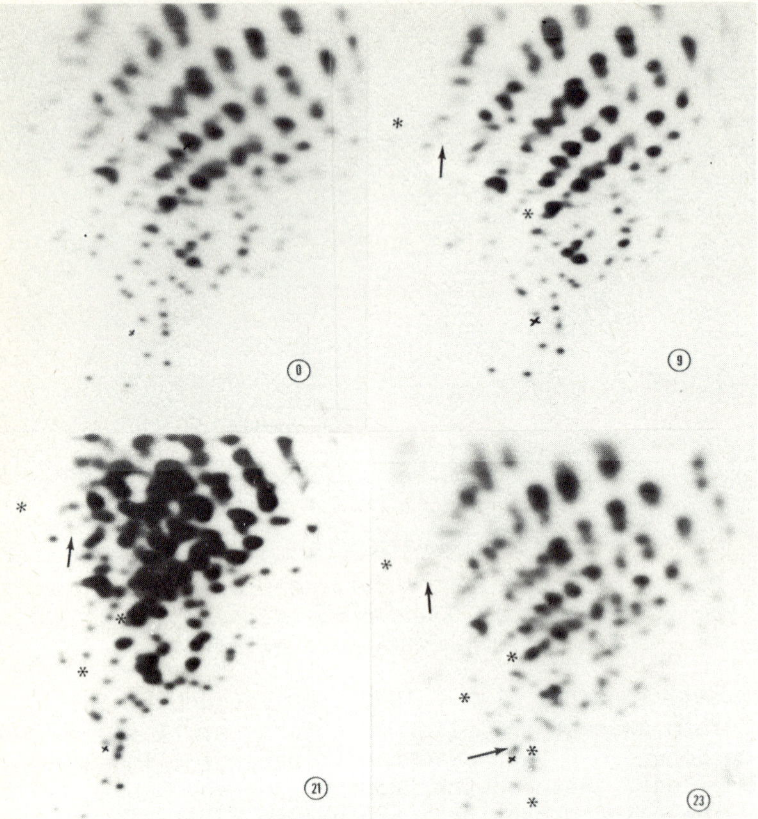

Fig.5. *Consecutive high level mutagenesis of VSV Indiana. In (A) the fingerprint of the VSV Indiana prototype strain used by D.H.L. Bishop is given. Its fingerprint is identical to that of the ATCC (Fig.2B) and French (Fig.1A) strains. Twenty-three consecutive mutageneses were made by growing the virus in the presence of 400 μg 5-fluorouracil per ml and cloning the progeny virus between each mutagenesis. The viral RNA fingerprints of virus from the ninth (B), twenty-first (C) and twenty-third (D) passage are given.*

of suitably labelled viral RNA species provides an extremely sensitive tool for differentiating the genomes of related and unrelated species of virus. Using this technique we have been able to show that different serotypes of the VSV group of rhabdoviruses are easily distinguished. In addition each of four different isolates of VSV Indiana have distinguishable fingerprints. Likewise each of five different isolates of VSV

New Jersey and two isolates of Chandipura virus are distinguishable (7). The evolution of these viruses in each of their various ecological niches can be mimicked in the laboratory by high level mutagenesis. After twenty-three consecutive passages in the presence of 400 µg of 5-fluorouracil per ml, we have been able to induce some 50 or so inheritable nucleotide changes. These are equivalent to an alteration of some 0.05 % of the viral genome. We have initiated studies with other mutagens (5-azacytidine and nitrosoguanidine) in order to determine if they induce greater rates of nucleotide substitution than 5-fluorouracil. The *in vitro* mutagenesis was initiated in order to reinvestigate the question of recombination for rhabdoviruses, and before we knew that the genome of VSV Indiana C was distinguishable from that of the VSV Indiana, prototype strain. Our idea was to so change the genome of VSV Indiana that any wild type recombinant generated from *ts* mutants of the original and derived strain could be shown by fingerprinting to have some (but not all) of the oligonucleotides of one parent and some (but not all) of the oligonucleotides of the other. Since the fingerprints of revertants should be identical to the fingerprint of one or the other parent, an unambiguous proof is available to distinguish a revertant from a recombinant. Now that we have obtained evidence that VSV Indiana C is distinguishable from VSV Indiana prototype we are in a position to test for recombination using the Glasgow and Orsay collections of *ts* mutants. Such experiments are in progress.

We have fingerprinted four different dI particles of VSV Indiana. Three small dI species come from that part of the genome which codes for the L protein (i.e. the 5' end). One larger dI comes from somewhere at or near the 3' end of the genome. We do not yet know if the three small dI species all correspond to the actual 5' end sequence of the B particle genome or whether the large dI corresponds to the actual 3' end of the B genome. Experiments are also in progress to investigate these questions as well as to obtain a more definitive ordering of the VSV genome.

ACKNOWLEDGEMENTS

This investigation was supported by Public Health Service grant AI 13402 from the National Institute of Allergy and Infectious Diseases. We would like to thank Dr. C.Y. Kang for providing the defective interfering particle preparations of VSV Indiana. Also we would like to thank those investigators who supplied us with the virus strains which were used in these studies, and information concerning the origin of VSV strains they possess. We thank Reginald Anderson and Gloria Robeson for excellent technical assistance.

REFERENCES

1. Berge, T.O. (1975). *International catalogue of arboviruses.* U.S.D.H.E.W. Publ. CDC 75-8301.
2. Bhatt, P.N. and Rodrigues, F.M. (1967). *Indian J. Med. Res.* 55, 1295.
3. Bishop, D.H.L., Emerson, S.U. and Flamand, A. (1974). *J. Virol.* 14, 139.
4. Bishop, D.H.L. and Roy, P. (1971). *J. Mol. Biol.* 58, 799.
5. Bishop, D.H.L. and Smith, M.S. (1977). *In* "The Molecular Biology of Animal Viruses" (Nayak, D., ed.) p.169.
6. Cartwright, B. and Brown, F. (1972). *J. Gen. Virol.* 16, 391.
7. Clewley, J.P., Bishop, D.H.L., Kang, C.-Y., Coffin, J., Schnitzlein, W.M., Reichmann, M.E. and Shope, R.E. (1977). *J. Virol.* 23, 152.
8. Clewley, J.P., Gentsch, J. and Bishop. D.H.L. (1977). *J. Virol.* 22, 459.
9. Cooper, P.D. and Bellett, A.J.D. (1959). *J. Gen. Microbiol.* 21, 485.
10. Cotton, W.E. (1927). *Amer. Vet. Med. Assoc. J.* 23, 168.
11. Cotton, W.E. (1927). *Vet. Med.* 22, 169.
12. Crick, J. and Brown, F. (1972). *J. Gen. Virol.* 18, 79.
13. De Watcher, R. and Fiers, W. (1972). *Anal. Biochem.* 49, 184.
14. Huang, A.S. and Wagner, R.R. (1966). *Virology* 30, 173.
15. Jonkers, A.H., Shope, R.E. Aitken, T.H.G. and Spence, L. (1964). *Amer. J. Vet. Res.* 25, 236.
16. Lazzarini, R.A., Weber, G.H., Johnson, L.D. and Stamminger, G.M. (1975). *J. Mol. Biol.* 97, 298.
17. Leamnson, R.N. and Reichmann, M.E. (1974). *J. Mol. Biol.* 85, 551.
18. Perrault, J. (1976). *Virology* 70, 360.
19. Petric, M. and Prevec, L. (1970). *Virology* 41, 615.
20. Pringle, C.R., Duncan, I.B. and Stevenson, M. (1971). *J. Virol.* 8, 836.
21. Pringle, C.R. and Wunner, W.H. (1973). *J. Gen. Virol.* 12, 677.
22. Repik, P., Flamand, A. and Bishop, D.H.L. (1976). *J. Virol.* 20, 157.
23. Repik, P., Flamand, A., Clark, H.F., Obijeski, J.F., Roy, P. and Bishop, D.H.L. (1974). *J. Virol.* 13, 250.
24. Roy, P. and Bishop, D.H.L. (1972). *J. Virol.* 9, 946.
25. Roy, P., Repik, P., Hefti, E. and Bishop, D.H.L. (1973). *J. Virol.* 11, 915.
26. Schnitzlein, W.M. and Reichmann, M.E. (1976). *J. Mol. Biol.*

101, 307.
27. Schnitzlein, W.M. and Reichmann, M.E. (1977). *Virology* 80, 275.
28. Shahan, M.S., Frank, A.H. and Mott, L.O. (1946). *J. Amer. Vet. Med. Assoc.* 108, 5.
29. Shope, R.E. (1975). *In* "The Natural History of Rabies", Vol.1, (Baer, G. ed.) p.141, Academic Press, New York.
30. Stamminger, G. and Lazzarini, R.A. (1974). *Cell* 3, 85.
31. Sudia, W.D., Fields, B.N. and Calisher, C.H. (1967). *J. Epid.* 86, 598.

A POSSIBLE HOST FACTOR REQUIREMENT FOR THE TRANSCRIPTION OF VESICULAR STOMATITIS VIRUS RNA

J.F. SZILÁGYI and C.R. PRINGLE

Medical Research Council Virology Unit,
Institute of Virology, University of Glasgow,
Glasgow, G11 5JR, Scotland.

CONVENTIONAL AND TEMPERATURE-DEPENDENT HOST RANGE MUTANTS

Three types of conditional lethal mutant have been isolated from wild type stock of VSV New Jersey after mutagenisation by 5-fluorouracil (7): (a) conventional temperature-sensitive (*ts*) mutants which are unable to form plaques at 39° on monolayers of BHK or secondary chicken embryo (CE) cells, (b) conventional host range (*hr* CE) mutants, which are able to form plaques in BHK cells both at 31° and 39°, but not in CE cells at either temperature, (c) temperature-dependent host range (*td* CE) mutants which form plaques both in BHK and CE cells at 31° but only on BHK cells at 39°. The *td* CE mutants were isolated comparatively frequently and four (*td* CE 1, 2, 3 and 4) have been studied in some detail. Attempts to place these mutants into complementation groups have not been successful so far, and their relationship to the six complementation groups of *ts* mutants is unknown.

A revertant clone (*td* CE/R1) was isolated from one of the *td* CE mutants (*td* CE 3). This mutant forms plaques in CE cells both at 31° and 39°, but it is distinguishable from wild-type virus by the morphology of its plaque at 39° (7).

IN VITRO RNA SYNTHESIS

To determine whether the virion-associated RNA transcriptase activity was affected by the mutation in the *td* CE mutants, the activity of this enzyme was assayed *in vitro* at 31° and 39° and the results were compared with those obtained for the wild-type virus (7).

Wild-type VSV New Jersey synthesized RNA linearly at 31° for about 60 min after which the rate of synthesis was reduced and a plateau reached by about 3 hr. The initial rate of RNA synthesis at 39° was approximately half of the initial rate at 31° and the duration was shorter, reaching a plateau after 60 to 90 min of incubation. The amount of RNA synthesized during 3 hr of incubation was approximately 20 to 30% of that synthesized at 31°.

The *td* CE mutants fell into two categories. Two (*td* CE 1 and *td* CE 2) synthesized RNA at both temperatures. The other two mutants (*td* CE 3 and *td* CE 4) synthesized RNA at 31°, but at 39° the total RNA synthesized by these mutants was not more than 2 to 4% of that synthesized at 31° during 3 hr of incubation. Thus, the mutation appears to affect the transcriptase activity of at least some of the *td* CE mutants.

IN VITRO RNA SYNTHESIS BY THE MUTANT *td* CE 3

Mutant *td* CE 3 has been studied in some detail using a transcribing nucleoprotein complex (TNP) prepared from purified virions (7, 8). The TNP contained polypeptides L, N and NS complexed to the virion RNA (9).

The amount of RNA synthesized after 120 min of incubation at 39° by the TNP of *td* CE 3 was approximately 4% of that produced at 31°, whereas the corresponding values for wild-type and revertant (*td* CE/R1) TNPs were 30% and 37% respectively (Fig.1).

Thus, in *td* CE /R1 the reversion of the *td* CE 3 phenotype is accompanied by the restoration of wild-type enzyme activity. This implies that the original *td* CE 3 mutation involves the virion transcriptase.

Irrespective of whether the mutant *td* CE 3 was grown in BHK or CE cells, only a very slight RNA synthesis was observed *in vitro* at 39°, therefore the cell in which the mutant was propagated had no influence on the heat sensitivity of the transcriptase (7).

In temperature-shift experiments it has been shown that the *in vitro* transcriptase activity of the mutant *td* CE 3 was immediately and fully restored by transfer from 39° to 31° (7). Thus inactivation of the transcriptase activity at 39° was due to the inhibition of enzyme function (presumably involving a configurational change in the mutated polypeptide) and not due to irreversible thermal inactivation as had been observed with some of the group I conventional temperature-sensitive mutants of VSV Indiana (1-3, 6, 7).

DISSOCIATION AND RECONSTITUTION OF THE TNPs AND THEIR EFFECT ON *IN VITRO* RNA SYNTHESIS BY *td* CE 3

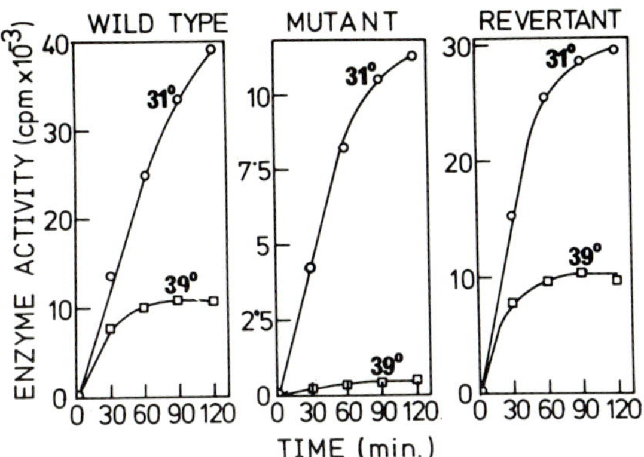

Fig.1. *In vitro synthesis by TNP preparations at 31° and 39°*

TNPs were prepared from VSV New Jersey (wild-type), the temperature-dependent host range mutant td CE 3 (mutant) and the revertant td CE/R1 isolated from td CE 3 (revertant). The preparation of the TNP and the in vitro assay conditions have been described (8). RNA synthesis at 31° (0) and at 39° (□).

Treatment of TNP with high concentrations of LiCl in the presence of 0.1% digitonin releases polypeptides that can be separated from the residual TNP by centrifugation through a glycerol gradient (4, 5, 8). For the following experiment TNP prepared from wild-type, mutant *td* CE 3 and the revertant *td* CE/R1 were treated in two different ways (8). In one of the treatments the concentration of LiCl was kept relatively low (0.9 M) to obtain a supernatant fraction that contained only polypeptide L. In the second treatment the concentration of LiCl was higher (2.0 M) to remove polypeptide L from the TNP and therefore to obtain a pellet fraction which contained almost exclusively polypeptides N and NS (Fig.2).

The 0.9 M LiCl supernatant and the 2.0 M LiCl pellet fractions were used in reconstitution experiments (Fig.3). The supernatant and the pellet fractions alone had no detectable transcriptase activity with the exception of the revertant pellet which retained a small amount of transcriptase activity. RNA synthesis at 31° was restored to a considerable extent when any of the three pellets was reconstituted with any of the three supernatants. The amount of RNA synthesized after 120 min of incubation depended on the supernatant and was related to the original activity of the TNP from which the supernatant was obtained.

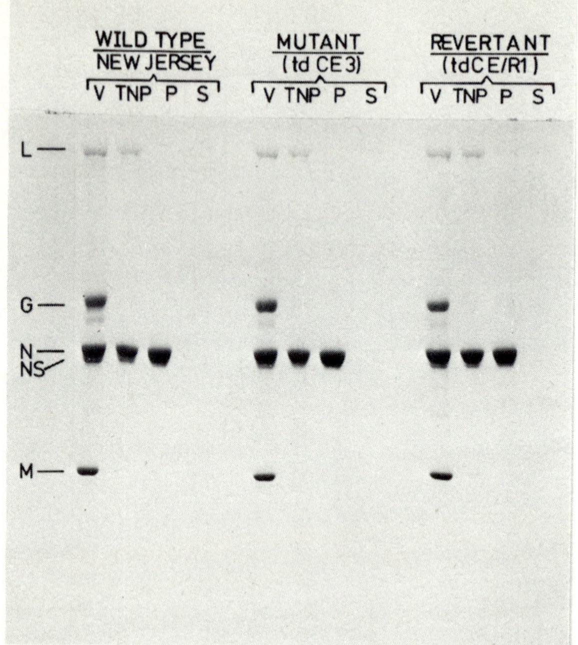

Fig.2. *Polyacrylamide gel electrophoresis of the polypeptides of wild-type New Jersey mutant td CE3 and revertant td CE/R1.*

For the electrophoresis we used either purified virions (V), transcribing nucleoprotein complexes (TNP), pellet fractions obtained by treating the TNPs with 2.0 M LiCl (P), or supernatant fractions obtained by treating the TNPs with 0.9 M LiCl (S). The polypeptides are large (L), glyco (G), nucleo (N), the so-called non-structural (NS) and the matrix (M). The method of fractionation of TNP and electrophoresis have been described (8).

Reconstitution of wild-type and revertant supernatants with any of the three pellets resulted in RNA synthesis at 39°. In contrast, when the mutant supernatant was recombined with any of the three pellets there was only very slight RNA synthesis at 39°.

Since the temperature-sensitive polypeptide appears to be located in the supernatant these results suggest that neither polypeptide in the pellet fraction (polypeptides N and NS) is affected by the mutation and thus implicate polypeptide L. Polyacrylamide gel electrophoresis showed that the 0.9 M LiCl supernatant contained hardly any viral polypeptide apart from polypeptide L making it more likely that this is the polypeptide

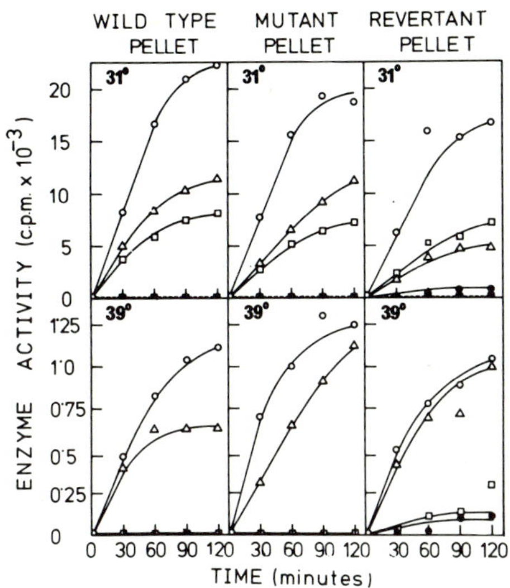

Fig.3. *In vitro RNA synthesis at 31° and 39° after reconstitution of 2.0 M LiCl pellets with 0.9 M LiCl supernatants.*

The method of dissociation and reconstitution of the TNPs has been described (8). Pellets of VSV New Jersey (wild-type pellet), td CE 3 (mutant pellet) and td CE/R1 (revertant pellet) were reconstituted with the supernatants of wild-type VSV (O), mutant td CE 3 (□) and revertant td CE/R1 (△). RNA synthesis by supernatant fractions alone (■ ---) and the pellet fractions alone (●) are also indicated. The upper row shows in vitro RNA synthesis at 31° and the lower row at 39°.

which determines the temperature sensitive phenotype of the *td* CE 3 mutant.

THE INFECTIVITY OF WILD-TYPE, MUTANT AND REVERTANT TNP

The TNP of VSV is infectious when assayed under conditions which facilitate its absorption (9). Therefore BHK and CE cells were infected at 31° and 39° with TNP of wild-type, mutant *td* CE 3 and revertant, to determine whether the mutant phenotype was affected by removal of the virion envelope. Plaque formation was obtained with all three TNP preparations following infection of BHK cells at 31° and 39°, and with wild-type and revertant TNP following infection of CE cells at 31° and 39°. The TNP of mutant *td* CE 3, however, failed to produce plaques on CE cell monolayers at 39°. Thus the host dependent phenotype did not depend on presence of the virion envelope, which suggests that the restriction of *td* CE 3 in CE cells at 39° probably does not involve an interaction with a receptor on the surface of the host cell.

DISCUSSION

Temperature-dependent host range mutants (*td* CE) of VSV New Jersey form plaques both at 31° and 39° on BHK cells but only at 31° on CE cells.

In vitro assay of the virion RNA transcriptase at 31° and 39° showed that the *td* CE mutants fall into two classes: those that synthesized RNA at 39° similarly to wild-type virus, and those that did not. One of the mutants of the latter category, *td* CE 3, was studied in some detail and it was shown that the inhibition of the transcriptase activity was fully and instantaneously reversible. Dissociation and reconstitution experiments using transcribing nucleoprotein complex (TNP) obtained from wild-type, *td* CE 3 and revertant viruses indicated that the polypeptide affected by the *td* CE 3 mutation is the L polypeptide.

Since the transcriptase of the mutant *td* CE 3 appears to be heat sensitive we postulate that virus development in BHK cells at 39° is due to the presence of a host factor which, in some way, helps the transcriptase to overcome its heat sensitivity. Since reconstitution experiments strongly indicate that polypeptide L was the polypeptide which was made heat sensitive by the *td* CE 3 mutation the interaction of the host factor with this polypeptide is a possibility. This host factor is presumably present in BHK cells, but missing from CE cells (or present in lower concentration). If such a factor does exist it is presumably a cytoplasmic component since mutant *td* CE 3 can grow in enucleated BS-C-1 cells (unpublished data). We examined the effect of cell extracts of BHK and CE cells on *in vitro* transcription by *td* CE 3 but the heat-sensitivity of the transcriptase was unaltered in the presence of these extracts.

Since no complementation has so far been achieved with *td* CE 3 we cannot be certain that the *td* CE phenotype represents a single mutational event. However, infection with TNP of *td* CE 3 indicated that the restriction of this mutant does not involve an interaction with a cell surface receptor which is a further indication that the *td* CE 3 mutant is probably the result of a single mutational event.

REFERENCES

1. Hunt, D.M., Emerson, S.U. and Wagner, R.R. (1976). *J. Virol.*

18, 596.
2. Pringle, C.R. (1975). *Curr. Top. Microbiol. Immunol.* 69, 85.
3. Repik, P., Flamand, A. and Bishop, D.H.L. (1976). *J. Virol.* 20, 157.
4. Szilágyi, J.F. (1975a). *In "In vitro* transcription and translation of viral genomes". (Haenni, A-L, and Beaud, G. eds.) INSERM, Paris. p.129.
5. Szilágyi, J.F. (1975b). *In* "Negative Strand Viruses". (Mahy, B.W.J. and Barry, R.D., eds.) Academic Press, p.421.
6. Szilágyi, J.F. and Pringle, C.R. (1972). *J. Mol. Biol.* 71, 281.
7. Szilágyi, J.F. and Pringle, C.R. (1976). *J. Virol.* 16, 927.
8. Szilágyi, J.F., Pringle, C.R. and MacPherson, T.M. (1977). *J. Virol.* 22, 381.
9. Szilágyi, J.F. and Uryvayev, L. (1973). *J. Virol.* 11, 279.

THE BINDING OF L AND NS PROTEINS TO THE RIBONUCLEOCAPSID OF VESICULAR STOMATITIS VIRUS

MARGARET G. MELLON and SUZANNE U. EMERSON

*Department of Microbiology,
University of Virginia School of Medicine,
Charlottesville, Virginia 22901, USA.*

Vesicular stomatitis (VS) virus, a negative strand rhabdovirus, is composed of a membrane envelope, 5 proteins and a single strand of RNA. The nucleocapsid core of the virus contains the RNA, the major structural protein N and two minor proteins L and NS. The two remaining proteins G and M are associated with the membrane (3).

VS virions can be dissociated with high-salt detergent solutions and the RNA tightly bound to N protein can be pelleted from solution leaving G, M, L and NS proteins in the supernatant (4). The N protein-coated RNA, but not naked viral RNA, constitutes the template for the endogenous RNA transcriptase carried by the virus (4). Transcriptase activity can be reconstituted by recombining the solubilized viral proteins and the template (4). Of the 4 proteins in the solubilized protein fraction, the two membrane proteins G and M are not required for transcriptase activity. However, experiments with separated L and NS proteins have shown that both of the minor nucleocapsid proteins are essential for transcription and therefore that one or both of these molecules constitute the transcriptase enzyme (6). In order to eventually elucidate the details of the transcriptase reaction we have initiated a study of the binding of the L and NS proteins to the purified template.

THE BINDING ASSAY

VS virus, New Jersey serotype, was grown in BHK cells and harvested by standard procedures (7). ^{35}S-methionine labelled virus was dissociated by mixing 1 volume of virus with 1 volume of 2 x high salt solubilizer (1.44 M NaCl, 18.7% glycerol, 1.2×10^{-3} M DTT, 3.74% triton X-100 and 10^{-2} M tris pH 7.4). After pelleting the template at 125,000 x g for 2½ hrs, the supernatant fraction was used directly as source of L and NS.

The template was prepared from VS virus New Jersey serotype labelled with ^{3}H-leucine. The virus was diluted to 43 ml with 10^{-2} M Tris (pH 7.4) and mixed with 43 ml of 2 x high salt solubilizer. The template was pelleted onto a renografin pad, then further purified by passage through an agarose column and finally pelleted onto 100% glycerol. Template prepared in this way is free of both L and NS proteins (7, unpublished observation, M. Mellon).

The template was recombined with the high-salt solubilized fraction under conditions suitable for transcriptase function, except that one of the 4 required nucleotides (UTP) was omitted to prevent transcription (see Legend Fig.1). After incubation at 31°C for 0.5 hr to allow binding, the entire reaction mix (0.4 ml) was layered on a 15-35% glycerol gradient and centrifuged at 125,000 x g for 1 hr. The complex of ^{3}H-labeled template and any bound ^{35}S-labeled supernatant proteins sedimented to the middle of the gradient, while unbound ^{35}S-proteins remained at the top. Aggregated material pelleted through the gradients. The ^{35}S-labeled supernatant proteins do not sediment to the middle of the gradient in the absence of template. The peak fractions from the middle of the gradient containing cosedimenting ^{3}H and ^{35}S were pooled and the proteins were analyzed by SDS polyacrylamide gel electrophoresis. Gels from two representative binding reactions are shown in Fig.1. Although the supernatant fractions added to the binding reaction contained large amounts of G and M proteins, essentially only L and NS proteins rebound to the template under these conditions.

SATURATION OF THE TEMPLATE WITH L AND NS PROTEINS

If the L and NS proteins are bound to specific sites and not merely aggregating nonspecifically on the template, it should be possible to saturate the template with L and NS proteins. For these experiments a series of reaction tubes were set up with increasing concentrations of the supernatant proteins relative to the concentration of template (Fig.2). As the initial relative concentration of the total ^{35}S-labeled supernatant proteins to ^{3}H-labeled template in the binding reactions increases, the amount of both L and NS proteins bound to the template increases linearly and then levels off. These data indicate that there are specific sites on the template which can be saturated with L and NS proteins.

BINDING OF SEPARATED L AND NS PROTEINS

Further experiments were done in which L and NS proteins were added separately to the template. L and NS proteins were separated from each other by a series of passages through phosphocellulose and DEAE columns. The NS fraction contained 95% NS protein when analyzed by SDS polyacrylamide gel electrophoresis; the L fraction contained both L and M proteins but since the M protein does not rebind to template no further purification was necessary. The L and NS fractions were enzymatically free of cross contamination in that when combined separately with template neither of the fractions could support RNA transcription but high levels of transcription were observed when both fractions were added together to template.

The L and NS fractions were added to binding reaction mixtures under the same conditions as described for the unfractionated ^{35}S-labeled supernatant proteins. After separation from the unbound proteins on glycerol gradients, the template-bound ^{35}S-labeled proteins were analyzed by SDS polyacrylamide gel electrophoresis. As can be seen in Fig.3, when the NS fraction is added alone, NS protein binds to the template (Fig. 3b). However, when the L fraction is added alone to the template, no L protein is bound (Fig.3c). When both fractions are added together, both L and NS proteins are bound (Fig.3a). Thus the binding of L protein is dependent on the presence of NS protein. The binding of NS protein does not appear to depend on L protein in that similar amounts of NS protein are bound in the presence and absence of L protein.

NUCLEOTIDE, Mg^{++} AND TEMPERATURE REQUIREMENTS

Since L and NS proteins bind to the template under the assay conditions for the transcriptase reactions (5), the initial conditions used for these experiments were those suitable for transcription. It was of interest however to determine to what extent these conditions were actually required for binding. Concentrations of template and solubilized viral proteins were chosen so that the template was saturated with L and NS proteins. The protocol described in legend for Fig.1 was followed except that all nucleotides and Mg^{++} were omitted from one reaction mixture. When compared to standard reaction mixtures supplemented with ATP, GTP, CTP and Mg^{++}, the amounts of template bound L and NS proteins were unchanged, indicating that nucleotides and Mg^{++} are not required to bind L and NS proteins to the template (data not shown).

Binding reactions, again under saturating conditions, were also set up in which instead of incubating at 31°C for 0.5 hr, the mixtures were left on ice. The gradients were loaded in the cold room and centrifuged in a precooled rotor at 4°C. When analyzed by SDS polyacrylamide gel electrophoresis, the

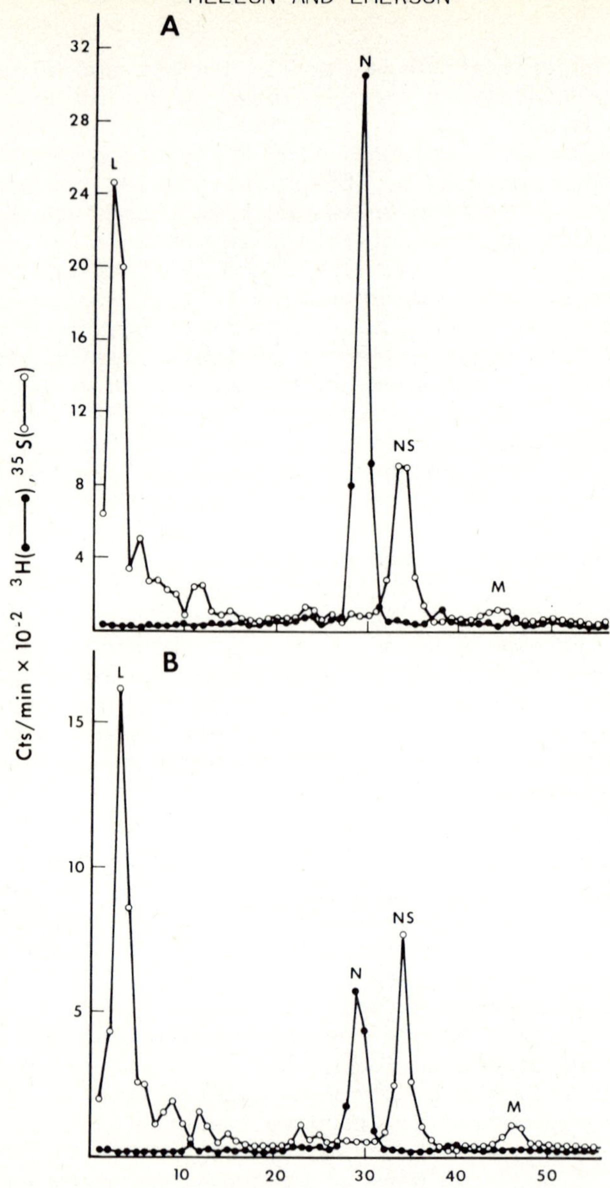

Fig.1. *SDS-polyacrylamide gel electrophoresis of rebound L and NS proteins.*

High salt-solubilized VS-New Jersey viral proteins labeled with ^{35}S were obtained as a supernatant fraction and used directly as a source of L and NS proteins (see text).

Template was prepared from ^{3}H-labeled VSV New Jersey virus (see text).

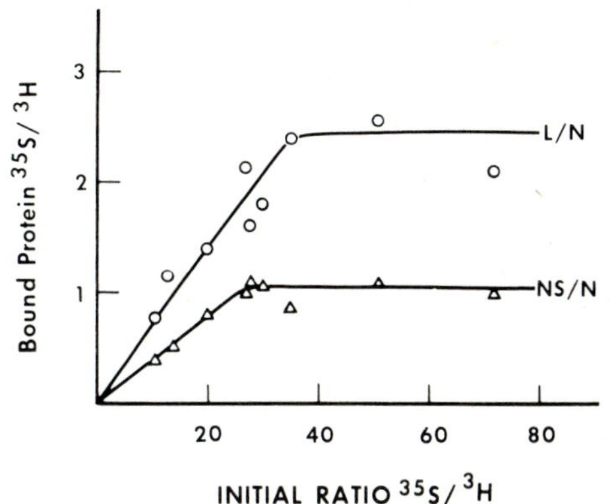

Fig.2. *Saturation of template with L and NS proteins.*

A series of reaction mixtures (0.4 ml total volume) were set up as described in Fig.1 containing equal volumes of the ^{35}S-labeled supernatant proteins and decreasing (80-10 µl) volumes of ^{3}H-labeled template. (The concentration of supernatant proteins relative to the concentration of template increases as the absolute amount of template decreases in the series.) After separation of bound from unbound ^{35}S-labeled-supernatant proteins on glycerol gradients, the pooled gradient fractions containing the rebound L and NS proteins were analyzed on SDS polyacrylamide gels (see Fig.1). The counts of ^{35}S in the L or NS proteins were normalized to the counts of ^{3}H-N-protein on each gel. These values were plotted against the initial ratio of ^{35}S-labeled supernatant proteins/^{3}H-labeled template as determined by counting small aliquots of each reaction mixture before application to the glycerol gradients.

Fig.1 (Contd.).

Either 70 µl (A) or 10 µl(B) of template were added to binding reaction mixtures containing 80 µl of the ^{35}S-labeled supernatant proteins and 0.2 ml of a prereaction mix (1.4 x 10^{-3} M ATP, CTP and GTP, 8 x 10^{-3} M Mg acetate, 1.2 x 10^{-3} M DTT and 2.5 x 10^{-2} M Tris pH 7.4). Sufficient 0.1 M Tris (pH 7.4) was included in each mix to bring the total volume to 0.4 ml. After a 30 min incubation at 31°C, the entire reaction mix (0.4 ml) was layered on a 15-35% glycerol gradient containing 0.144 M NaCl, 2% triton, 4 x 10^{-3} M Mg acetate and 2.5 x 10^{-2} M tris pH 7.4 and centrifuged at 125,000 x g for 1 hr at 4°C. Aliquots of the gradient fractions were counted to determine the peak of cosedimenting ^{35}S and ^{3}H. The peak gradient fractions were pooled and the bound proteins analyzed on 7½% SDS-phosphate polyacrylamide gels.

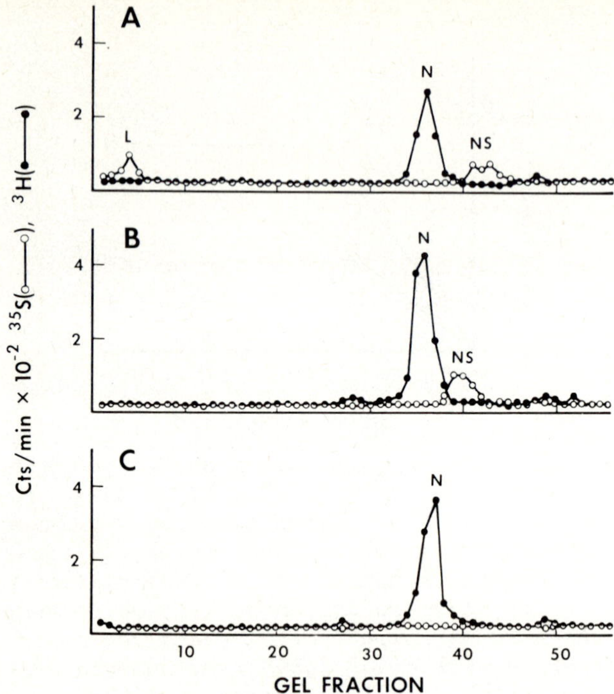

Fig.3. *Rebinding of separated L and NS proteins.*

^{35}S-labeled viral proteins solubilized from VS-New Jersey virus with high salt solubilizer were mixed with phosphocellulose and dialyzed overnight against 0.1 M NaCl in glycerol column wash (25% glycerol, 0.2% triton X-100 and 2 x 10^{-4} M DTT in 0.05 M Tris pH 7.2). The phosphocellulose and ^{3}H-labeled supernatant protein mixture was transferred to a 12 ml plastic syringe, drained and washed with 0.1 M NaCl in glycerol column wash. The flow-through solutions were saved and used later. The phosphocellulose column was then eluted with 0.5 M NaCl in glycerol column wash and the peak fractions of ^{35}S were called the L fraction. The reserved flow-through from the phosphocellulose column was applied to a DEAE-cellulose column and the NS protein was eluted with 0.4 M NaCl in glycerol column wash. The NS fraction used in this experiment was rechromatographed on phosphocellulose and DEAE-cellulose columns in order to completely remove L protein.

The L and NS fractions obtained from the columns were added either alone or together to 25 µl of ^{3}H-VS-New Jersey template in binding reaction mixtures similar to those described in Fig.1. Final concentrations of NaCl, ATP, GTP, CTP and Mg^{++} were identical in each tube. The bound L and NS proteins were separated from unbound proteins on glycerol gradients and reaction mixtures left on ice showed 60-70% of the L and NS proteins bound to template as in controls incubated at 31°C. Thus, although no transcription occurs at 4°C (unpublished observations, S.U. Emerson), substantial binding of L and NS proteins is observed at low temperatures (data not shown).

NEW JERSEY L AND NS PROTEINS BIND TO INDIANA TEMPLATE

A preliminary experiment was also done to determine if the L and NS proteins prepared from VS virus-New Jersey serotype could bind to template prepared from VS virus-Indiana serotype. Equal concentrations of solubilized viral proteins were added to reaction mixtures containing either Indiana or New Jersey-template. The total amounts of N protein in the two template preparations were similar (6.7 μg New Jersey, 7.7 μg Indiana) and the concentration of New Jersey-solubilized viral proteins were known to be saturating for the New Jersey-template. About 35% of the L and NS proteins bound to the homologous New Jersey-template were bound to the heterologous Indiana-template. Although saturation data are needed to establish the specificity of the heterologous binding, it is interesting that Indiana-template can bind New Jersey-L and NS proteins. The limited antigenic crossreactivity of the N proteins of New Jersey and Indiana serotypes may explain the heterologous binding (2). However it has also been shown that solubilized viral proteins of New Jersey serotype virus are incapable of restoring transcription when combined *in vitro* with template prepared from Indiana serotype virus (1).

DISCUSSION

These experiments represent a preliminary attempt to define the binding relationships among the L and NS proteins and the template under transcription conditions. The data show that NS protein can bind to template in the absence of L, and that the presence of NS protein facilitates or is necessary for the binding of L protein. Naito and Ishihama (8) have purified the transcriptase of VS-New Jersey virus by different methods and have shown that fractions containing N and L proteins bound to each other in a 1:1 stoichiometry are maximally active in a transcriptase assay. Their experiments confirm the previous finding of Emerson and Yu (4) that the NS protein is required for transcription. The NS protein requirement for transcription may reflect the L protein dependence on NS protein for

---

electrophoresed on SDS-acrylamide gels (see Fig.1). A) NS and L fractions added together B) NS fraction added alone C) L fraction added alone.

binding to the template. The NS protein may act to stabilize L protein or may serve as a bridge to bind L protein indirectly to the template. On the other hand the binding of NS protein may modify the template, thereby exposing an L protein binding site. Further experiments are needed to define these binding relationships exactly.

ACKNOWLEDGEMENTS

This work was supported by Public Health Service grant AI-11722 from the National Institute of Allergy and Infectious Diseases. M.G.M. is a Postdoctoral Fellow of the National Institute of Allergy and Infectious Diseases (R32-AI05292). S.U.E. is the recipient of Public Health Service Research Career Development Award AI-0013 from the National Institute of Allergy and Infectious Diseases.

REFERENCES

1. Bishop, D.H., Emerson, S.U., and Flamand, A. (1974). *J. Virol.* 14, 139.
2. Cartwright, B., and Brown, F. (1972). *J. Gen. Virol.* 16, 391.
3. Emerson, S.U. (1976). *Curr. Top, Microbiol. Immunol.* 73, 1.
4. Emerson, S.U. and Wagner, R.R. (1972). *J. Virol.* 10, 297.
5. Emerson, S.U. and Wagner, R.R. (1973). *J. Virol.* 12, 1325.
6. Emerson, S.U. and Yu, Y.H. (1975). *J. Virol.* 15, 1348.
7. Hunt, M.D., Emerson, S.U. and Wagner, R.R. (1976). *J. Virol.* 18, 596.
8. Naito, S. and Ishihama, A. (1976). *J. Biol. Chem.* 251, 4307.

INFLUENZA VIRAL RNA TRANSCRIPTION *IN VIVO* AND *IN VITRO*

R.M. KRUG, S.J. PLOTCH AND J. TOMASZ*

Memorial Sloan-Kettering Cancer Center, New York, N.Y., U.S.A.

**The Institute of Biophysics Szeged, Hungary.*

It has now been firmly established that influenza virus is a negative-strand virus: complementary RNA (cRNA), the product of transcription, is the viral messenger RNA (mRNA, (4, 6, 7, 9, 17, 19), and the virion contains a transcriptase (3, 15, 22). We have been investigating whether the virion-associated transcriptase has the capability to synthesize cRNA *in vitro* which is identical to the cRNA synthesized in the infected cell (16). A major goal of our investigation is to gain an insight into the host nuclear function required for influenza viral RNA transcription. In the infected cell, viral RNA transcription is inhibited by actinomycin D and α-amanitin (1, 17, 20, 21, 24), whereas *in vitro* transcription catalyzed by the virion transcriptase is not inhibited by these drugs (3, 15, 22).

Many of the characteristics of the *in vivo* cRNA which serves as mRNA are known. We have shown that this cRNA contains poly A, 140-170 nucleotides in length, internal 6-methyl adenosine residues and is capped and methylated at the 5' end (6, 7, 10, 16). Like the virion RNA (vRNA) from which it is transcribed, *in vivo* cRNA is segmented (8). The *in vivo* cRNA segments migrate more slowly than the corresponding vRNA segments during gel electrophoresis. This is due to the poly A sequences in cRNA not found in vRNA. After enzymatic removal of the poly A, the cRNA and vRNA segments appeared to comigrate when analyzed on cylindrical gels, indicating that they are approximately the same size (8).

In our previous studies of the virion transcriptase, we have shown that this enzyme is capable of synthesizing cRNA which is similar in several characteristics to *in vivo* cRNA (16). In the presence of the proper divalent cation, Mg^{++}, and a specific primer, ApG or GpG, the virion transcriptase synthesizes cRNA which is large and contains covalently-linked poly A. Though the *in vitro* cRNA migrates in the same molecular weight range as vRNA during gel electrophoresis, it is much more heterogeneous, so that discrete *in vitro* cRNA segments corresponding to the vRNA segments cannot be discerned.

GEL PATTERN OF DEADENYLATED *IN VITRO* cRNA

To determine whether the heterogeneity of *in vitro* cRNA is due to its large, heterogeneous poly A sequences, these sequences were removed enzymatically with RNase H (23) in the presence of poly dT. The deadenylated *in vitro* cRNA was coelectrophoresed with deadenylated *in vivo* cRNA on cylindrical gels containing 6M urea (Fig. 1). After removal of its poly A sequences, *in vitro* cRNA distributes into discrete segments. Both *in vitro* and *in vivo* cRNA are resolved into six peaks, which are designated L, M1, M2, M3, S1, and S2. We will show later that the L peak is actually composed of three closely-spaced segments. As analyzed on cylindrical gels, the *in vitro* and *in vivo* cRNA segments comigrate, indicating that they are approximately the same size. Both the ApG- and GpG-primed products contain all the cRNA segments, indicating that each segment can initiate with either A or G.

The *in vitro* and *in vivo* cRNA's differ in the relative proportion of the different segments which they contain. In both *in vitro* and *in vivo* cRNA, the smallest segment, S2, is present in the largest molar amount, about three times the amount of the M2 segment. The two cRNA's, however, differ in the representation of the M3 and S1 segments: in *in vitro* cRNA these segments are found in about 2.5 times the amount found in *in vivo* cRNA. Also, the L and M1 segments are present in larger amounts in *in vitro* cRNA. Similar ratios of the different segments were found in the *in vitro* product whether ApG or GpG was used as primer. The *in vivo* cRNA was obtained from cells treated with cycloheximide at 2.5 hr after infection and labeled with ^3H-adenosine for a subsequent 3 hr (7). Presumably, this *in vivo* cRNA represents the pattern of transcription at 2.5 hr.

UNPRIMED *IN VITRO* cRNA

With Mg^{++} as divalent cation, the synthesis of large, poly A-containing (poly A (+)) cRNA occurs in the presence of primer ApG or GpG (16). With Mn^{++}, however, only small, poly A(-) RNA is synthesized whether primer is present or not. It was therefore of interest to determine whether in the presence of Mg^{++} large, poly A (+) cRNA could be synthesized in the absence of a primer. With Mg^{++}, the amount of RNA synthesized in the absence of primer is only 1-2% of that made in the presence of a primer (16). About half of this product was found to bind to oligo dT cellulose and to contain poly A. After enzymatic removal of its poly A, the unprimed product distributes into the same six peaks as the primed product when analyzed on cylindrical gels (see Fig. 1). The only significant difference between

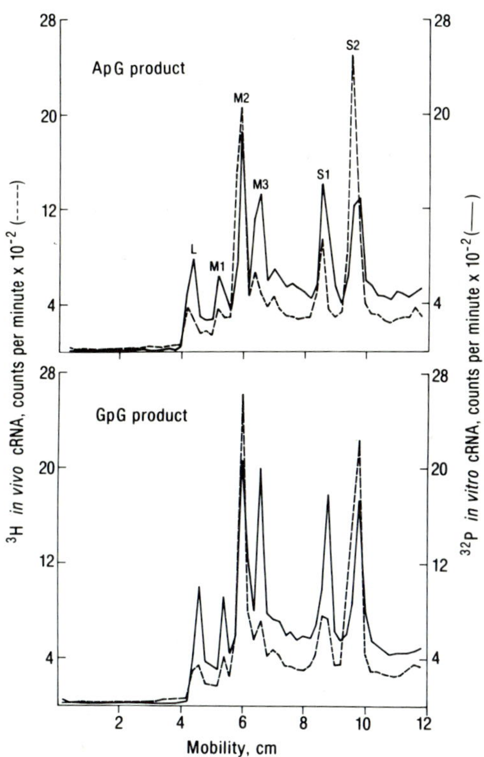

Fig. 1. *Gel electrophoresis of ^{32}P-labeled, deadenylated in vitro cRNA and [^{3}H] adenosine-labeled, deadenylated in vivo cRNA. The in vitro cRNA was synthesized with either ApG (top panel) or GpG (bottom panel) at 0.4mM as primer (16); Mg^{++} at 5 mM was the divalent cation, and the labeled precursor was α ^{32}P GTP at 0.3 mCi/μmole. In vivo cRNA was purified as described previously (7). The in vitro and in vivo cRNA was hybridized to poly (dT) and the poly A was removed by digestion with RNase H (8). After heating and fast-cooling, the cRNA's were coelectrophoresed on 12 cm cylindrical, 2.1% acrylamide-0.7% agarose gels containing 6M urea, using Peacock Dingman buffer.*

this unprimed product and the primed product is that about 100-fold less of each peak is made in the absence of the primer. These results indicate that the virion transcriptase in the presence of Mg^{++} has a real, but very limited, capacity to initiate cRNA chains in the absence of a primer. Once initiated, elongagion and poly A addition proceed normally.

COMPARISON OF THE SEGMENTS OF vRNA, *IN VITRO* cRNA, and *IN VIVO* cRNA BY SLAB GEL ELECTROPHORESIS

To compare the mobilities of the vRNA and cRNA segments, gel electrophoresis was carried out in slabs rather than in cylinders and the gels were analyzed by autoradiography. Fig. 2 (left panel) shows a comparison of the gel pattern of vRNA (v) and deadenylated *in vitro* cRNA (c). The eight known segments of vRNA (12, 14, 18) are resolved into seven bands. Based on its autoradiographic intensity relative to the other bands, the slowest-moving band of vRNA can be presumed to contain two unresolved segments. *In vitro* cRNA is also resolved into seven bands, including two large-size (L) bands. Each of the *in vitro* cRNA segments migrates slightly faster than the corresponding segment of vRNA.

To confirm this difference in mobility, a mixture of vRNA and *in vitro* cRNA was run in the same lane (Fig. 2 (right panel), lane 4). As compared to the pattern of vRNA (lane 1) or *in vitro* cRNA (lane 2), a doublet at every position is observed, thus

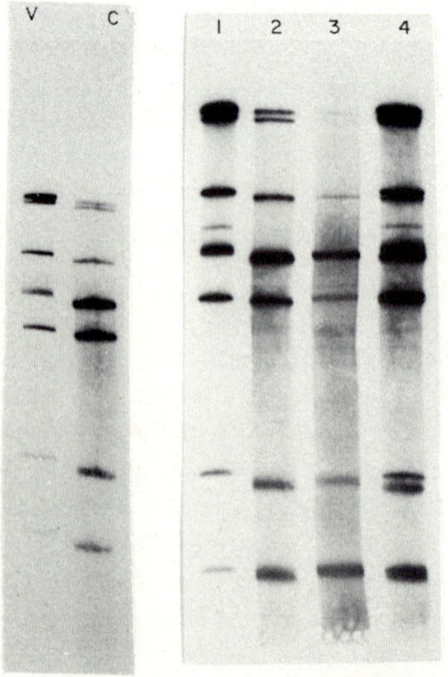

Fig.2.

verifying the migration difference between the vRNA and *in vitro* cRNA segments. These results suggest that the cRNA segments are slightly smaller than the corresponding vRNA segments. Lane 3 is the gel pattern of *in vivo* cRNA. The segments of *in vivo* cRNA migrate just between the vRNA and *in vitro* cRNA segments.

SLAB GEL ELECTROPHORESIS OF DOUBLE-STRANDED RNA

To establish that each of the vRNA segments is transcribed into cRNA *in vitro*, labeled deadenylated *in vitro* cRNA was hybridized to excess unlabeled vRNA and the double-strands formed were either electrophoresed directly (lane 1, Fig. 3) or were treated with RNase T2 prior to electrophoresis (lanes 2a and 2b). In both cases, the two small-size and three-medium size hybrids are clearly resolved. With the large-size hybrids, resolution is better in the RNase T2-treated sample (lane 2b): three distinct bands are seen. Thus all eight vRNA segments are transcribed into cRNA *in vitro*. It is also evident that the deadenylated *in vitro* cRNA:vRNA hybrids treated with RNase T2 migrate slightly faster than the untreated hybrids. This reduction in size probably results from the hydrolysis of a short sequence of vRNA in each hybrid which is not in double-stranded form because the complementary sequence is missing from the *in vitro* cRNA segment in each hybrid (see Conclusions).

ASSAYS FOR CAPPING AND METHYLATING ENZYMES IN THE VIRION

We have now described the characteristics of *in vivo* and *in vitro* cRNA which are similar. However, these two cRNA's differ in one important respect: *in vivo* cRNA is capped and methylated at its 5' end, whereas neither capping nor methylation of the *in vitro* transcripts has been detected. To assay for capping and methylating enzymes, we have utilized as primers the derivatives of ApG, GpG, and GpC containing one, two, or three phosphates at their 5' end. Fig. 4 shows the results of a represen-

---

Fig. 2. *Autoradiogram of a slab gel electrophoresis of vRNA, in vitro cRNA, and in vivo cRNA. Left panel: ^{32}P-labeled vRNA (v) and ^{32}P-labeled, deadenylated in vitro cRNA (ApG-primed) (c). Right panel: lane 1- ^{32}P-labeled vRNA; lane 2- ^{32}P-labeled, deadenylated in vitro cRNA (ApG-primed); lane 3- ^{3}H-adenosine labeled, deadenylated in vivo cRNA; lane 4- mixture of vRNA and deadenylated in vivo cRNA. Electrophoresis was carried out on 14 x 16 x 0.15 cm slab, 2.8% acrylamide gels containing 6 M urea. The slab gel shown in the right panel was treated with PPO-DMSO (2). The extra band seen between two of the middle-size vRNA segments is 18S ribosomal RNA, which is found in variable amounts in purified preparations of influenza virus (5, 10).*

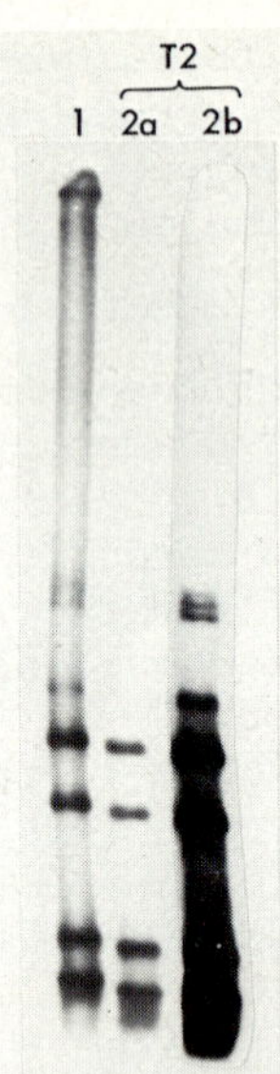

Fig.3. *Slab gel electrophoresis of double-stranded RNA. Lane 1: double-stranded RNA formed by hybridizing ^{32}P-labeled, deadenylated in vitro cRNA (ApG-primed) with excess unlabeled vRNA for 24 hr at 37° in 55% formamide (10). Lane 2a: as in the left lane, except that the hybrids were treated with RNase T2 (20 U/ml). Lane 2b: the material in lane 2a after longer exposure to the X-ray film. Electrophoresis was carried out in 14 x 16 x 0.3 cm slab, 3% acrylamide gels.*

tative experiment, using ppApG as primer. For the labeling of the product cRNA, we used α-^{32}P-GTP and ^{3}H-SAM. The RNA was digested with RNase T2 and the digest was chromatographed on DEAE-Sephadex in urea. The major doublet peak at -2 charge represents the mononucleotides. If any cap structure were present, we would have expected a peak of approximately 700 counts per min of ^{32}P and ^{3}H at -4.5 to -5.5 charge. No such peak is seen, indicating the absence of a cap structure.

These results are meaningful only if it can be shown that the 5' diphosphate of the ppApG primer is preserved in the product and thus is a suitable substrate for capping and methylating enzymes. If the diphosphate were preserved, then we should be able to cap and methylate the product cRNA with vaccinia virus capping and methylating enzymes (13). To ensure that these latter enzymes would be acting on the product cRNA, we designed our experiments such that any cap structure formed on the cRNA by these enzymes would contain radioactivity derived from a

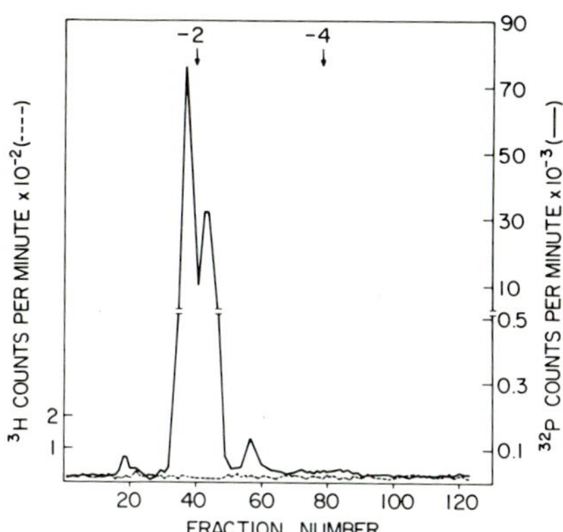

Fig. 4. *DEAE-Sephadex chromatography of the RNase T2 digest of a ppApG-primed cRNA product synthesized with* $\alpha^{32}P$ *GTP (12 mCi/μmole) as labeled precursor.*

precursor used in the *in vitro* transcription reaction. Thus, if initiation of *in vitro* cRNA synthesis occurs exactly at the 3' end of vRNA, then the beginning sequence of cRNA primed with ppApG should be ppApGpC, as the 3' terminus of vRNA is $GpCpU_{OH}$ (see Chapter 5). Consequently, the phosphate between the G and C at the 5' end of cRNA should be derived from $\alpha^{32}P$-labeled CTP. After capping and methylation by the vaccinia virus enzymes, the resulting cap structure, $m^7GpppA^mpG\overset{*}{p}C$ should be hydrolyzed by RNase T2 to form $m^7GpppA^mpG\overset{*}{p}$, which should have a charge of -5. The presence of methyl residues in such a cap fragment was monitored by employing ^3H-SAM during the capping and methylating reaction with the vaccinia virus enzymes.

As shown in Fig. 5 (top panel) DEAE-Sephadex chromatography of the RNase T2 digest of such a cRNA product reveals the presence of a ^3H-labeled methylated cap at -5 charge which contains ^{32}P derived from $\alpha^{32}P$ CTP. The amount of ^{32}P radioactivity in the -5 peak is consistent with one cap structure per RNA chain of about 1600 nucleotides. As this is approximately the average size of the *in vitro* cRNA product, this means that most, if not all, of the 5' diphosphate derived from the ppApG primer is preserved in the product. In contrast to the results obtained with the $\alpha^{32}P$ CTP precursor, when the precursor is $\alpha^{32}P$ GTP (bottom panel) little or no ^{32}P is found in the ^3H-labeled methylated cap at -5, thus confirming that the initial 5' sequence of the *in vitro* cRNA is ppApGpC. Both these DEAE-Sephadex analyses also show the presence of peaks of ^{32}P, but not ^3H, in the -3 and -4 charge region. We are presently characterizing these species. Despite the presence of these species, it is clear

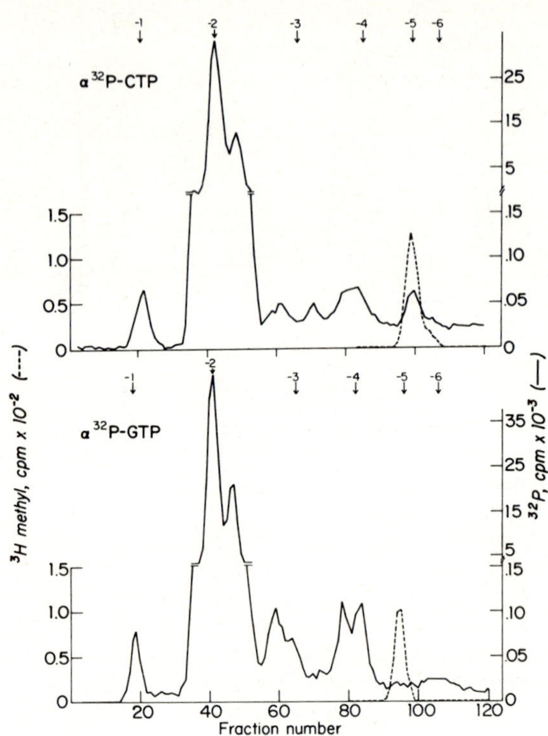

Fig. 5. *DEAE-Sephadex chromatography of the RNase T2 digest of a ppApG-primed cRNA product, which was subjected to capping and methylation by vaccinia virus capping and methylating enzymes. Top panel:* $\alpha^{32}P$ *CTP (16 mCi/μmole) as labeled precursor in the ppApG-primed reaction; bottom panel:* $\alpha^{32}P$ *GTP (18 mCi/μmole) as labeled precursor. The poly A(+) cRNA was capped and methylated with the vaccinia virus enzymes (13) using* ^{3}H*-SAM (12 mCi/μmole) as labeled precursor.*

that the cap structure incorporated onto the cRNA product made with $\alpha^{32}P$ CTP contains ^{32}P, which is not the case with the product made with $\alpha^{32}P$ GTP.

These results strongly suggest that the virion does not contain enzymes which cap and methylate at the initiated 5' terminus of the cRNA chain. Further, these results indicate that *in vitro*, transcription with ApG probably initiates exactly at the 3' end of vRNA.

PRIMING BY XpApG TRINUCLEOTIDES

As another approach to determine whether *in vitro* transcription initiates exactly at the 3' end of vRNA, we examined the ability of trinucleotides of the form XpApG to serve as primers (Table 1). If the trinucleotide acted as primer by hydrogen-bonding

TABLE 1

STIMULATION OF COMPLEMENTARY RNA SYNTHESIS *IN VITRO* BY XpApG TRINUCLEOTIDES

Oligonucleotide	pmol of GMP incorporated
None	0.5
ApG	23.7
ApApG	25.6
CpApG	18.0
UpApG	22.0

Reactions were for 1 hour at 30°C with $\alpha^{32}P$ GTP (0.3 mCi/μmole) as labeled precursor and 5 mM Mg^{++}. The trinucleoside diphosphates or ApG were added at 0.22 mM. Optimum synthesis with ApG is at 0.40 mM (16).

to the vRNA at some unique internal trinucleotide sequence, one of the XpApG trinucleotides would be expected to be a much more effective primer than the others. However, all three of the trinucleotides tested have similar priming activity, suggesting that the 5'-terminal base X is not involved in hydrogen-bonding to the vRNA template and that ApG acts therefore by hydrogen-bonding to the 3' terminal bases of vRNA.

CONCLUSIONS

Our results indicate that in the presence of Mg^{++} the influenza virion transcriptase transcribes each of the vRNA segments into cRNA segments of almost the same size. Only a very small amount of these cRNA segments is made in the presence of Mg^{++} alone, and a specific primer, ApG or GpG, causes about a 100-fold increase in the synthesis of the segments. This almost total dependence on the addition of a primer suggests that a primer RNA is also required for the effective transcription of vRNA into cRNA in the infected cell. We have already postulated that the synthesis of such a primer RNA may be the actinomycin D- and α-amanitin-sensitive host function known to be required for *in vivo* viral transcription (16, 24).

Our gel electrophoretic analyses of single-stranded *in vitro* cRNA and vRNA and of the *in vitro* cRNA:vRNA hybrids suggest that the *in vitro* cRNA segments are incomplete transcripts of the corresponding vRNA segments. One possible explanation for incomplete transcription is that initiation *in vitro* with ApG occurs internally and not at the 3' end of vRNA. Our results, however, strongly suggest that *in vitro* transcription with ApG as primer initiates exactly at the 3' end of vRNA. The initial 5' sequence of ApG-primed cRNA is ApGpC, which is complementary to the $GpCpU_{OH}$ 3' terminus of vRNA. Also, XpApG trinucleotides are effective primers regardless of the identity of their 5'-terminal base X. If further evidence verifies that *in vitro* transcription with ApG initiates exactly at the 3' end of vRNA, then the apparent smaller size of *in vitro* cRNA would result from termination of transcription before the 5' end of vRNA is reached. Initiation of each cRNA segment *in vitro* by GpG can best be explained by the ability of GpG to hydrogen bond to the penultimate GpC sequence at the 3' end of vRNA. It has been shown that when a GG dinucleotide is bound to a polynucleotide, G-to-G interactions between the dinucleotide and the polynucleotide are at least as strong as the Watson-Crick G-to-C interaction (11).

The *in vivo* cRNA segments migrate just between the vRNA and *in vitro* cRNA segments. It is not known whether this means that *in vivo* cRNA contains additional nucleotide sequence(s) not found in *in vitro* cRNA. The slight difference in mobility could reflect in part a secondary structure difference between *in vivo* and *in vitro* cRNA due to the 5'-terminal cap structure, which is present in *in vivo* cRNA and absent from *in vitro* cRNA. Using *in vitro* cRNA primed with ppApG, we have recently found that the addition of a 5'-methylated cap structure with the vaccinia virus capping and methylating enzymes causes a slight decrease in the mobility of the capped cRNA segments compared to the uncapped segments during gel electrophoresis (Plotch and Krug, unpublished experiments). Further evidence is clearly required to establish whether *in vivo* cRNA contains additional nucleotide sequence(s) not found in *in vitro* cRNA.

Unlike *in vivo* cRNA, the *in vitro* cRNA is not capped or methylated even when 5' di- or tri-phosphorylated derivatives of ApG or GpG are used as primers. As with all negative data, it is conceivable that we have not detected capping or methylation because we have not yet employed the proper conditions. Nonetheless, our data strongly suggests that the virion does not contain capping and methylating enzymes and that consequently host enzymes carry out these steps. This would be consistent with the presence of internal 6-methyl adenosine residues in *in vivo* cRNA (10).

ACKNOWLEDGEMENTS

We thank Barbara B. Broni for expert technical assistance.

This work was supported by Public Health Service grants CA 17085, CA 08748 and AI 11772 from the National Institutes of Health.

REFERENCES

1. Bean, W.J. Jr. and Simpson, R.W. (1976). *Virology* 56, 646.
2. Bonner, W.M. and Laskey, R.A. (1974). *Eur. J. Biochem.* 46, 83.
3. Chow, N.L. and Simpson, R.W. (1971). *Proc. Nat. Acad. Sci. USA* 68, 752.
4. Content, J. (1976). *J. Virol.* 18, 604.
5. Cox, N.J. and Barry, R.D. (1976). *Virology* 69, 304.
6. Etkind, P.R. and Krug, R.M. (1974). *Virology* 62, 38.
7. Etkind, P.R. and Krug, R.M. (1975). *J. Virol.* 16, 1464.
8. Etkind, P.R., Buchhagen, D.L., Herz, C., Broni, B.B. and Krug, R.M. (1977). *J. Virol.* 22, 346.
9. Glass, S.E., McGeoch, D. and Barry, R.D. (1975). *J. Virol.* 16, 1435.
10. Krug, R.M., Morgan, M.M. and Shatkin, A.J. (1976). *J. Virol.* 20, 45.
11. Lewis, J.B., Brass, L. Jr. and Doty, P. (1975). *Biochemistry* 14, 3164.
12. McGeoch, D., Fellner, P. and Newton, C. (1976). *Proc. Nat. Acad. Sci. USA* 73, 3045.
13. Moss, B. (1977). *Biochem. Biophys. Res. Commun.* 74, 374.
14. Palese, P. and Schulman, J.L. (1976). *Proc. Nat. Acad. Sci. USA* 73, 2142.
15. Penhoet, E., Miller, H., Doyle, M. and Blatti, S. (1971). *Proc. Nat. Acad. Sci. USA* 68, 1369.
16. Plotch, S.J. and Krug, R.M. (1977). *J. Virol.* 21, 24.
17. Pons, M.W. (1973). *Virology* 51, 120.
18. Pons, M.W. (1976). *Virology* 69, 789.
19. Ritchey, M.B. and Palese, P. (1976). *Virology* 72, 410.
20. Rott, R. and Scholtissek, C. (1970). *Nature* 228, 56.
21. Scholtissek, C. and Rott, R. (1970). *Virology* 40, 989.
22. Skehel, J.J. (1971). *Virology* 45, 793.
23. Stavrianopoulos, J.G. and Chargaff, E. (1973). *Proc. Nat. Acad. Sci. USA* 70, 1959.
24. Taylor, J.M., Illmensee, R., Litwin, S., Herring, L., Broni, B. and Krug, R.M. (1977). *J. Virol.* 21, 530.

STRUCTURALLY AND FUNCTIONALLY DISTINCT INFLUENZA VIRUS COMPLEMENTARY RNAs

A.J. HAY, J.J. SKEHEL, G. ABRAHAM
*J.C. SMITH and *P. FELLNER

Division of Virology,
National Institute for Medical Research,
Mill Hill, London NW7.

**Searle Research Laboratories,*
High Wycombe, Bucks.

The genomes of influenza viruses, which contain at least 8 single-stranded RNAs with molecular weights between 10^5 and 10^6 (6-9, 11, 13), are transcribed during infection into molecules complementary in sequence (3, 6, 12). Since virus-specific complementary cRNA (cRNA) has to serve two functions, as messenger RNA (4, 5, 10, 14), and as template RNA in genome replication, the question arises whether or not the same RNA molecules perform both functions. From the results of analyses of transcription in influenza virus-infected cells (6) it appears that they do not. In those studies it was observed that during replication two types of cRNA are produced which can be distinguished on the basis of a number of different properties. The cRNAs which are present in the polysomes of infected cells as mRNAs are polyadenylated. Their synthesis is controlled throughout infection both with respect to the amount of each transcript produced and to the time at which each is produced in maximal amount. On the other hand, unpolyadenylated cRNAs, which are not associated with polysomes, are produced in similar amounts during infection and it is suggested that these molecules serve as templates during the replication of the virus genome.

In addition to these differences, comparison of the electrophoretic mobilities of the double-stranded RNAs formed between virus RNA and either type of cRNA showed that the double-stranded molecules formed with the polyadenylated cRNAs following nuclease S_1 treatment migrated faster than the corresponding molecules formed with unpolyadenylated cRNAs (Fig.1) suggesting that the polyadenylated transcripts are shorter. The differences in electrophoretic mobility were most marked for the smaller RNAs and corresponded to differences in molecular weights of approximately 20-30,000, or 30-40 bases for the single-stranded cRNAs.

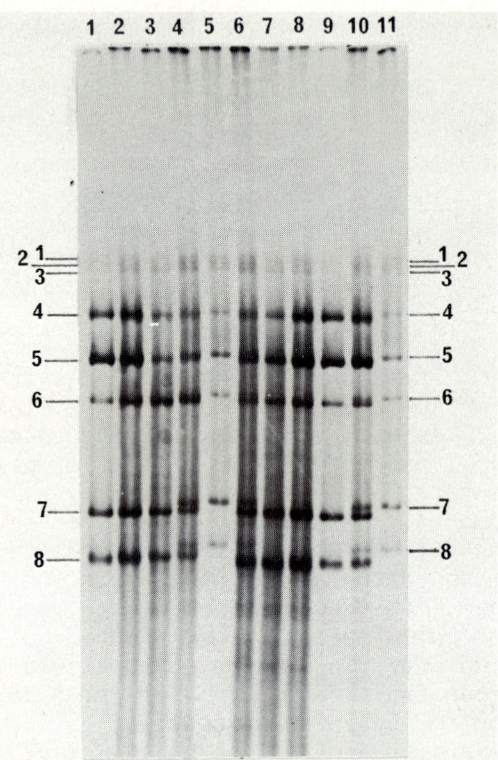

Fig.1. *Comparison of the double-stranded RNAs formed between virus RNA and cRNAs synthesised 'in vivo' or 'in vitro'.*

RNA extracted from chick cells, incubated in medium containing ^{3}H-uridine (100 μc/ml) between 1.5 and 2.5 hr after infection with fowl plague virus, was hybridized with an excess of unlabelled virus RNA in 63% dimethyl sulphoxide at 37° for 20 hr and the polyadenylated RNA and unpolyadenylated double-stranded RNA were separated, as described previously (6). ^{3}H-labelled cRNA was prepared 'in vitro' using either the virion-associated transcriptase or virus-specific ribonucleoprotein isolated from fowl plague virus infected cells (Hay, in preparation), and hybridized in the presence of added unlabelled virus RNA. The RNA samples in 0.1 m NaCl, 10 mM sodium acetate pH 4.5, 0.5 mM $ZnSO_4$ were incubated at 37° for 4 hr with nuclease S_1 (10-20 units/μg RNA) and the double-stranded RNAs were

To investigate these differences further, comparisons were made of the ribonuclease T_1 oligonucleotide fingerprints either of the corresponding cRNAs or of ^{32}P virus RNA before and after hybridization to cRNA and subsequent nuclease S_1 digestion. The results of these experiments indicated that whereas the unpolyadenylated cRNAs contain sequences complementary to all the genome sequence represented by the unique ribonuclease T_1-oligonucleotides, the polyadenylated cRNAs lack sequences complementary to one of the unique T_1-oligonucleotides, as illustrated for RNAs 7 and 8 in Fig.2. Furthermore the oligonucleotide missing in the case of these two RNAs is known from sequence analyses (see chapter 5) to be located close to the 5' end of the molecule indicating that the polyadenylated transcripts are incomplete at their 3' ends. This appears to be the case for all the mRNA molecules since in contrast to the double-stranded RNAs containing unpolyadenylated cRNA, the 5' terminal nucleotide of the virus RNA component of hybrid molecules containing polyadenylated cRNA is sensitive to nuclease S_1 digestion (Fig.3). Demonstration that the unpolyadenylated cRNAs are in fact complete genome transcripts will require further data with respect to the 3' terminus of virus RNAs and these experiments are still in progress. To obtain more detailed information concerning the nature and extent of these untranscribed regions, the nucleotide sequence at the 5' ends of the virus RNAs were investigated.

The method used in the sequence analysis was that developed by Donis-Keller *et al.* (2), which involves limited digestion of 5' terminally labelled virus RNA and subsequent polyacrylamide gel electrophoresis of the partial digestion products. This is exemplified by the results shown in Fig.4 obtained using virus RNA 7. Analysis of mild alkaline digestion products indicates the complete spectrum of oligonucleotides while the partial products produced by ribonuclease T_1 and U_2 digestion indicate the positions of G and A residues,

---

analysed either separately or as mixtures by electrophoresis on a 4% polyacrylamide slab gel at 4 volts/cm for 22 hr and detected by fluorography. The direction of migration was from top to bottom of the figure. The samples analysed contained double-stranded RNAs formed with polyadenylated cRNA (lanes 1 and 9); virion-associated transcriptase product alone (lane 3) or + polyadenylated cRNA (lane 2) or + unpolyadenylated cRNA (lane 4); unpolyadenylated cRNA (lanes 5 and 11); cell-associated transcriptase product alone (lane 7) or + unpolyadenylated cRNA (lane 6) or + polyadenylated cRNA (lane 8); polyadenylated cRNA + unpolyadenylated cRNA (lane 10).

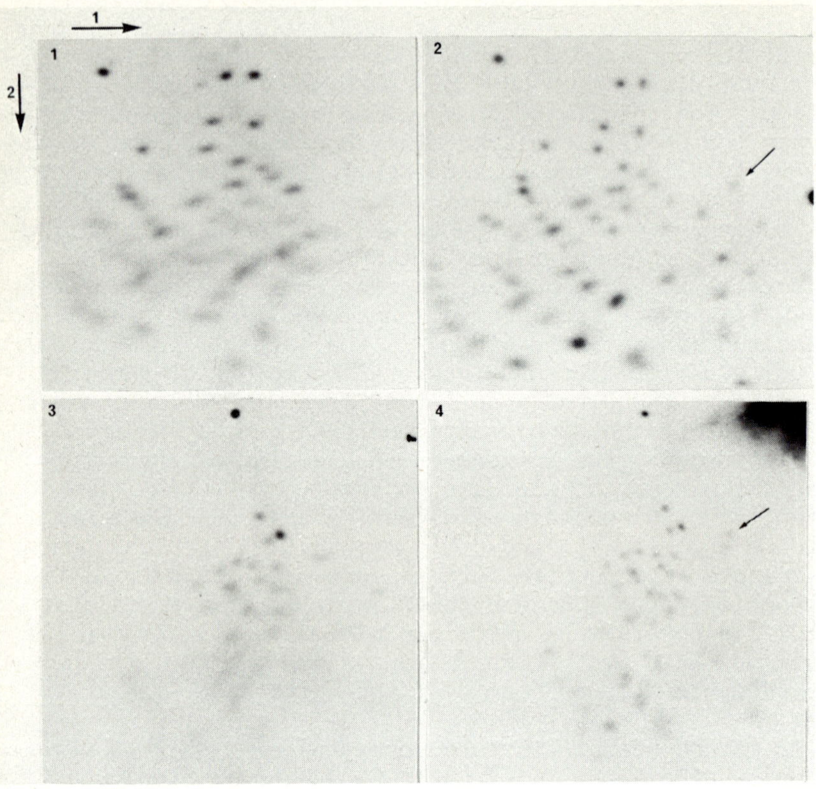

Fig.2. *Ribonuclease T_1 oligonucleotide fingerprints of ^{32}P virus RNAs 7 and 8 in double-stranded molecules formed with unpolyadenylated and polyadenylated cRNAs.*

^{32}P virus RNA was hybridized under conditions of cRNA excess to unlabelled RNA extracted from chick cells 3 hr after infection with fowl plague virus and the polyadenylated hybrid molecules and unpolyadenylated double-stranded RNAs were isolated by oligo (dT) cellulose chromatography and LiCl fractionation as described previously (6). Following digestion with nuclease S_1 (see Fig.1) the double-stranded RNAs were separated by electrophoresis on 4% polyacrylamide gels. The individual RNAs were eluted from the gel, denatured and digested with ribonuclease T_1 and the resulting oligonucleotides were analysed by 2-dimensional polyacrylamide gel electrophoresis (1). Migration in the first dimension was from left to right and in the second dimension from top to bottom. The fingerprints are of ^{32}P virus RNAs 7 (1 and 2) and 8 (3 and 4) present in double-stranded RNAs obtained from polyadenylated (1, 3) or unpolyadenylated (2, 4) hybrids after S_1 nuclease

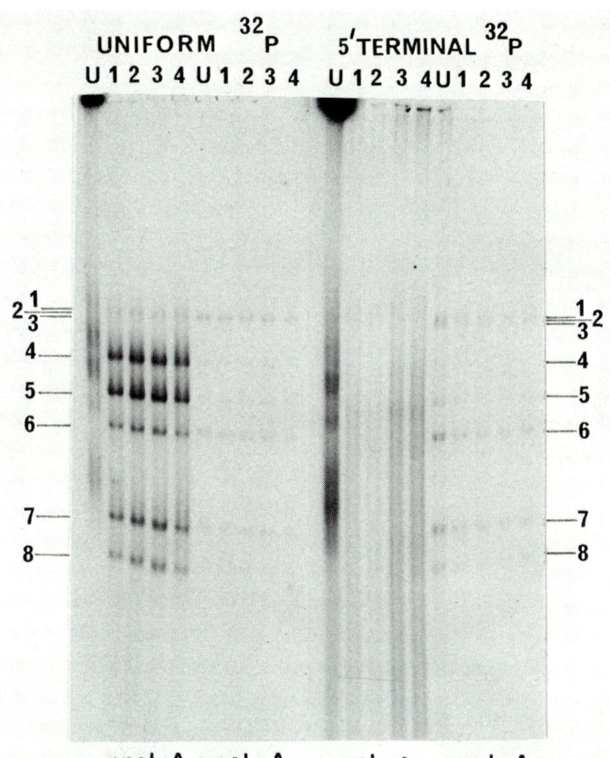

Fig.3. *Analysis of double-stranded RNAs formed between 5' terminal ^{32}P labelled virus RNA and poluadenulated or unpolyadenylated cRNAs*

Uniformly ^{32}P-labelled virus RNA and 5' terminally ^{32}P-labelled virus RNA (see Fig.4) were separately hybridized with unlabelled RNA extracted from cells 3 hr after infection with fowl plague virus. The polyadenylated and unpolyadenylated hybrid molecules were isolated, dissolved in 0.3 M NaCl, 10 mM sodium acetate pH 4.5, 0.5 mM $ZnSO_4$ and divided into 5 aliquots. 3 aliquots were incubated at 37° with nuclease S_1 (10 units/µg RNA) for 0.5 hr (1), 1 hr (2) or 2 hrs (3) respectively. To a fourth aliquot (4) 5 mM $MgCl_2$ was added and the sample incubated with nuclease S_1 for 1 hr. The fifth aliquot (U) was not incubated with nuclease S_1. The radioactive double-stranded RNAs were analysed by electrophoresis on a 4% polyacrylamide gel at 5 volts/cm for 17 hr and detected by autoradiography.

digestion. The arrows indicate the observed differences between the two fingerprints of each RNA.

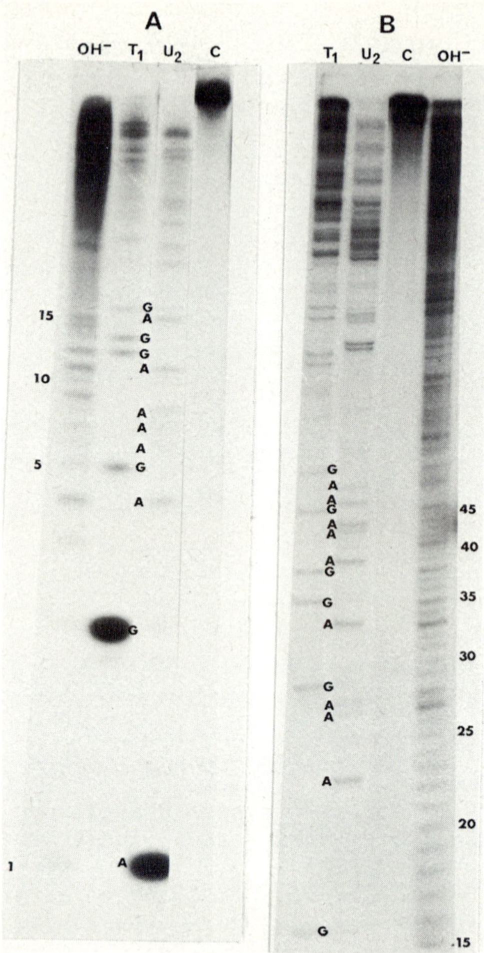

Fig.4. *The products of partial hydrolyses of 5'-end-labelled RNA 7 of fowl plague virus.*

Preparations of FPV RNA were dephosphorylated using bacterial alkaline phosphatase (Worthington), re-extracted with phenol and phosphorylated using γ-^{32}*P-ATP (Amersham) and polynucleotide kinase (PL biochemicals) as described by Donis-Keller et al. (2). Labelled RNAs were purified by polyacrylamide gel electrophoresis (6) and the partial hydrolysis procedures for sequence analysis described by Donis-Keller et al. (2) were followed exactly. The 5'-end-labelled RNA 7 samples for the nuclease digestions were dissolved in 20 μl aliquots of sodium citrate buffer (0.02 M, pH 5.0) containing urea, 6.7 M; EDTA, 0.001 M; tRNA, 10 μg; and either 0.005 units* respectively. At this stage C and U residues are not distinguished and the positions of pyrimidines are indicated by the lack of A or G residues at these positions and are represented by blanks in the results shown in Tables 1 and 2.

Table 1 shows the sequence of the first 36 nucleotides from the 5' end of virus RNAs 7 and 8 obtained using these procedures compared with partial sequence data obtained using conventional RNA sequence techniques (see chapter 5). These results, which suggest that the 'untranscribed' ribonuclease T_1-oligonucleotide in each case starts at nucleotide 17, indicate that the minimum size of the region not represented in mRNA is 17 nucleotides. Preliminary results of other experiments using this sequencing procedure to examine the ability of polyadenylated cRNA to protect the virus RNA against ribonuclease T_1 or U_2 digestion have also indicated that sequences complementary to the first 16 nucleotides are not present in mRNAs. Further experiments aimed at determining the exact size of the untranscribed regions and their sequences in the different RNAs are still in progress. Comparisons of the nucleotide sequences at the 5' ends of the genome RNAs of different strains of influenza A viruses have indicated that there is a marked degree of homology in the sequences of the first 22 nucleotides (Tables 1 and 2). In particular no differences were detected in nucleotides 1-13 which may therefore be identical in the different RNAs. The complement of this sequence which is present only in the unpolyadenylated cRNAs, the presumed template for genome replication, may possible function as a recognition site for the replicase.

The production of incomplete transcripts could be the result either of premature termination of transcription or of degradative processing of complete transcripts. The information presently available which indicates that both types of cRNA can be transcribed by the same enzyme and that, unlike the synthesis of mRNAs, the synthesis of complete unpolyadeny-

---

RNase T_1 or 1 unit RNase U_2 (Sankyo Ltd.). They were incubated together with the control sample lacking nuclease at 50° for 15 min. An additional aliquot of labelled RNA was dissolved in 20 µl of sodium bicarbonate buffer 0.05 M, pH 9.0 containing 0.001 M EDTA and 5 µg tRNA and incubated at 90° for 15 min. Following incubation, urea was added to the hydrolysate to a final concentration of 5 M and this together with the nuclease digestion products was applied directly to the polyacrylamide gel. The figure shows autoradiograms of polyacrylamide gels (containing 20% acrylamide) used in the analyses which were electrophoresed at 20 volts/cm for 8 hr (A) and 24 hr (B).

TABLE I

5' terminal nucleotide sequence of influenza virus RNAs 7 and 8

NUCLEOTIDE	1	2	3	4	5	6	7	8	9	10	11	12	13	14	15	16	17	18	19	20	21	22	23	24	25	26	27	28	29	30	31	32	33	34	35	36
RNA 7																																				
	A	G	-	A	G	A	A	A	-	-	A	G	G	-	A	G	-	-	-	-	-	-	A	-	-	-	A	A	G	-	-	-	-	A	-	G
	*A	G												*U	A	G	U	U	U	U	U	U	A	C	(U,	C)	<u>C</u>	A	G	C	(U,	U,	C)	(A	U)	G
RNA 8																																				
	A	G	-	A	G	A	A	A	-	-	A	G	G	G	-	G	-	-	-	-	-	-	A	-	-	A	A	-	-	A	A	A	-	-	A	G
	*A	G															*U	U	U	U	U	U	(A	U	U,	_	A	U	C)	(A	A	A	U)	<u>A</u>	A	G

*Sequence data obtained from Smith *et al*. (chapter 5). The large brackets indicate the ribonuclease T_1 oligonucleotides which are not transcribed into mRNA (see Fig.2). The order of the sequences in brackets was previously undetermined and the remaining ambiguities are indicated by commas. The differences between the two sets of sequence data, which are underlined, may be due to the different strains of fowl plague virus used.

TABLE II

5' terminal nucleotide sequence of influenza virus RNAs

NUCLEOTIDE		1	2	3	4	5	6	7	8	9	10	11	12	13	14	15	16	17	18	19	20	21	23	23
fowl plague virus RNA	*1-3	A	G	-	A	G	A	A	A	-	-	A	G	G	-	A	G	-	-	-	-	-	-	A
	4	A	G	-	A	G	A	A	A	-	-	A	G	G	G	A	G	-	-	-	-	-	-	-
	5	A	G	-	A	G	A	A	A	-	-	A	G	G	G	-	A	-	-	-	-	-	-	-
	6	A	G	-	A	G	A	A	A	-	-	A	G	G	A	G	A	-	-	-	-	-	-	-
	7	A	G	-	A	G	A	A	A	-	-	A	G	G	-	A	G	-	-	-	-	-	-	A
	8	A	G	-	A	G	A	A	A	-	-	A	G	G	G	-	G	-	-	-	-	-	-	A
	9	A	G	-	A	G	A	A	A	-	-	A	G	G	-	A	G	-	-	-	-	-	-	A
	10	A	G	-	A	G	A	A	A	-	-	A	G	G	-	A	G	-	-	-	-	-	-	A
X-31 virus RNA	*1-3	A	G	-	A	G	A	A	A	-	-	A	G	G	-	A	G	-	-	-	-	-	-	A
	4	A	G	-	A	G	A	A	A	-	-	A	G	G	G	-	G	-	-	-	-	-	-	A
	*5+6	A	G	-	A	G	A	A	A	-	-	A	G	G	G/A	-	A	-	-	-	-	-	-	A
	7	A	G	-	A	G	A	A	A	-	-	A	G	G	-	A	G	-	-	-	-	-	-	A
	8	A	G	-	A	G	A	A	A	-	-	A	G	G	G	A	G	-	-	-	-	-	-	A

*analysed as mixtures

lated transcripts is uniquely dependent upon continued virus-specific protein synthesis ((6); unpublished results) is consistent with both mechanisms. The required virus-specific protein(s) may alter the action of the transcriptase in such a way that it no longer recognises a termination signal or they might affect post-transcriptional modification either by binding to the 3' end of the complete transcript possibly as an initial stage in RNA replication or by inhibiting the processing enzymes. At present it is not possible to distinguish between these alternatives. However, since transcription of the virus genome *in vitro*, either by the virion-associated transcriptase or by viral ribonucleoprotein isolated from infected cells, produces unpolyadenylated transcripts equivalent in size to the polyadenylated transcripts synthesised *in vivo* (Fig.1), the former alternative appears more likely.

The available data is also insufficient to allow identification of the sequences involved in termination of transcription or processing. It is however of interest to note that both RNAs 7 and 8 contain a sequence of 6 consecutive U residues between positions 17 and 22 (Table I) and the other RNAs probably also have a similar, if not identical, sequence (Table II). Since the sequences from this position at least up to nucleotide 36 of RNAs 7 and 8 show little similarity it may be speculated that the sequence 17-22 contains the termination signal and may also act as an initiation signal for the addition of poly A.

ACKNOWLEDGEMENTS

We thank Helen Donis-Keller and Allan Maxam for communicating their analytical procedures prior to publication. A.J.H., G.A. and J.J.S. thank Bernard Precious and David Stevens for excellent assistance. J.C.S. and P.F. thank Dr N.H. Carey for valuable discussions and Dr A.J. Hale for provision of research facilities.

REFERENCES

1. DeWachter, R. and Fiers, W. (1972). *Anal. Biochem.* 49, 184.
2. Donis-Keller, H., Maxam, A.M. and Gilbert, W. (1977). *Nucleic Acid Res.* 4, 2527.
3. Etkind, P.R., Buchhagen, D.L., Hertz, C., Broni, B.B. and Krug, R.M. (1977). *J. Virol.* 22, 346.
4. Etkind, P.R. and Krug, R.M. (1974). *Virology* 62, 38.
5. Glass, S.E., McGeoch, D. and Barry, R.D. (1975). *J. Virol.* 16, 1435.
6. Hay, A.J., Lomniczi, B., Bellamy, A.R. and Skehel, J.J. (1977). *Virology* 83, 337.
7. McGeoch, D., Fellner, P. and Newton, C. (1976). *Proc. Nat. Acad. Sci. USA* 73, 3045.

8. Palese, P. (1977). *Cell* 10, 1.
9. Palese, P.and Schulman, J.L. (1976). *J. Virol.* 17, 876.
10. Pons, M.W. (1972). *Virology* 47, 823.
11. Pons, M.W. (1976) *Virology* 69, 789.
12. Pons, M.W. (1977). *Virology* 76, 855.
13. Scholtissek, C., Harms, E., Rohde, W., Orlick, N. and Rott, R. (1976). *Virology* 74, 332.
14. Stephenson, J.R., Hay, A.J. and Skehel, J.J. (1977). *J. Gen. Virol.* 36, 237.

STUDIES ON THE SYNTHESIS OF cRNA AND vRNA IN CELLS INFECTED WITH INFLUENZA VIRUS

T. BARRETT, J.M. BROWNSON, A.J. WOLSTENHOLME AND B.W.J. MAHY

Division of Virology, Department of Pathology, University of Cambridge, Laboratories Block, Addenbrooke's Hospital, Hills Road, Cambridge, England.

Early after infection of a cell by influenza virus, the virion RNA (vRNA) is transcribed into a complementary form (cRNA) by the virion-associated RNA transcriptase (9). The cRNA serves both as messenger for virus-specific protein synthesis and as template for production of new vRNA molecules. The details of these events are not yet understood.

Production of cRNA in infected cells has been measured by following the conversion of ^{32}P-labelled virion template RNA into ribonuclease-resistant complexes formed by hybridization with newly synthesised complementary RNA strands present in total RNA extracted at different times after infection with radioactively-labelled influenza virus (1). Under these conditions formation of cRNA was not detected until 60 min after infection, and so it was suggested that a host cell function needed to be expressed during the first 40-60 min before RNA transcription could be initiated. This method, however, gives an underestimate of both the time of appearance and the extent of transcription since only a small percentage of the infecting virus inoculum has been shown to be transcriptionally active (8).

Using ^{32}P vRNA as a probe for cRNA, Glass *et al* (4) could detect cRNA on polysomes by 45 min post infection and this cRNA was complementary to more than 80% of the virus genome indicating that all segments were probably transcribed by this time. The data, however, suggested that up to 2.25 hr post infection all cRNA sequences might not be represented in equimolar amounts. In agreement with this observation, it has been reported that some virus-specific proteins (P, NP, NS) are synthesised early in infection in greater amounts relative to others (M, HA and NA) (5, 10). This could reflect either unequal transcription of some virus genes early in infection or, perhaps, sequestering of certain transcripts in the nucleus, making them unavailable for translation. Using radioactive probes for both vRNA and cRNA we have studied in more detail the time course of appearance of both vRNA and cRNA in the infected cell. In addition, using radioactive probes for each individual gene segment, we have determined the relative proportions of the various cRNA's present at different times post-infection.

APPEARANCE OF POLYADENYLATED cRNA IN THE INFECTED CELL

In all cases only polyadenylated cRNA, selected on oligo dT columns, was studied. The amount of cRNA needed to saturate a constant small amount of labelled vRNA probe was determined. An amount of cRNA needed to give 25-30% hybridization with this amount of vRNA was then hybridized to saturation with increasing amounts of vRNA. Double reciprocal plots of the data were used to estimate the amount of cRNA present; extrapolation of the straight line reciprocal plots to an infinite number of input counts was used to obtain the maximum number of labelled vRNA counts bound to the cRNA at saturation. The results (Table I) are expressed as counts per minute vRNA bound per μg of polyadenylated cRNA, adjusted for losses incurred during preparation of nuclear and cytoplasmic RNA's (2). The concentration of cRNA was higher in the nucleus than in the cytoplasm at early times, reached a maximum at 2 hr post-infection and then declined. cRNA concentration rose steadily in the cytoplasm up to 2.5 hr post infection then increased dramatically by 3 hr. Since there is a much greater amount of cell RNA in the cytoplasm than in the nucleus, these results expressed as concentrations do not accurately reflect the total amounts of cRNA present in the two cell fractions. An estimate of the numbers of copies of cRNA was made from the relative amounts of polyadenylated RNA in terms of cell equivalents present in each preparation (Table I). The absolute number of gene copies was greater in the nucleus than in the cytoplasm at 0.5 hr post-infection and roughly equivalent in the two cell fractions up to 1.5 hr post-infection. From 2 hr post-infection the amount of cRNA declined in the nucleus but continued to rise in the cytoplasm at least up to 3 hr post-infection. Broadly similar results to these were obtained by Taylor *et al* (11) using the WSN strain of influenza virus.

ACCUMULATION OF POLYADENYLATED cRNA TO INDIVIDUAL VIRUS RNA SEGMENTS AT DIFFERENT TIMES POST-INFECTION

Radioactive probes for individual cRNA genome segments were prepared by separating I^{125} vRNA on polyacrylamide gels (7): segments 1 to 3 were not resolved, and were treated as one band. The bands were cut out and RNA eluted from the gel (7). When re-electrophoresed on polyacrylamide gels the individual bands ran in the expected order of mobility and there was little evidence of cross-contamination. The intrinsic ribonuclease resistance of the various RNA species varied from 0.8% to 2%.

For each vRNA band, the amount of nuclear and cytoplasmic polyadenylated cRNA hybridized at saturation was determined, and expressed relative to the molecular weight of the band (7). At all times post-infection cRNA hybridizing to band 1 (segments 1-3) was represented in the lowest relative amount. Accordingly, for each time studied, the amount of cRNA hybridizing to band 1 was given a molar value of 1.0, and the molar amounts of the other segments expressed relative to band 1.

At early times after infection, the amount of transcription was inversely related to the molecular weight of the segment, i.e. the smaller segments were transcribed to a greater extent (see Fig.1). Subsequently there was a relatively greater increase in the proportion of NS and NP gene transcripts reaching a maximum at 2.5 hr. NA and HA accumulated slowly over the period studied while cRNA to M rose steadily and at 4.5 hr post-infection was almost as abundant as that to NS. The accumulation of individual cRNA segments in the nucleus was not very different from that in the cytoplasm and there was no evidence for compartmentalisation of particular segments in the nucleus (Fig.1). These results suggest that some form of transcriptional control of polyadenylated cRNA production occurs but it does not rule out the possibility of additional control at the translational level.

ACCUMULATION OF vRNA AT VARIOUS TIMES POST-INFECTION

A ^{3}H-cDNA copy of total influenza (fowl plague) virus RNA was prepared using purified reverse transcriptase and oligo dG_{12-18} as primer. The cDNA, when hybridized in excess, protected the template vRNA from digestion by ribonuclease to 100%, indicating that all vRNA segments were copied. The cDNA did not hybridize to nuclear or cytoplasmic RNA from uninfected chick embryo fibroblast cells. To measure the amount of vRNA and cRNA in RNA from infected cells, samples were hybridized in vast excess to ^{3}H-cDNA or ^{125}I-vRNA and the degree of hybridization at which equilibrium was attained was measured, thus providing the ratio of vRNA:cRNA in the sample. These hybridizations were then repeated, but with a known amount of unlabelled vRNA included in the reaction

TABLE I

Accumulation of polyadenylated cRNA in chick embryo fibroblast cells infected with influenza (fowl plague) virus

TIME	^{32}P vRNA cpm hybridised per µg A^+ cRNA		No. of gene copies	
	Nucleus	Cytoplasm	Nucleus	Cytoplasm
0.5	4,511	548	1.4	0.82
1.0	9,125	2,231	3.0	3.6
1.5	25,244	20,759	29	37
2.0	105,479	158,068	113	297
2.5	86,582	185,534	78	373
3.0	17,572	931,262	27	2009

Legend to Table I. RNA was extracted from nuclei and cytoplasm of primary chick embryo fibroblasts (CEF) monolayers infected with fowl plague virus at a multiplicity of 50 pfu/cell. Cells were harvested into ice-cold saline. The cell pellets were resuspended in RSB (0.1M NaCl, 0.0015M $MgCl_2$, 0.01M Tris HCl, pH 7.4), kept in ice for 15 min and homogenised in a Dounce homogeniser. Crude nuclei were pelleted at 1000 x g for 1 min. The supernatant (cytoplasm) was kept and the nuclei resuspended in RSB. Deoxycholate was added to 0.2% w/v and NP40 to 1% v/v. The nuclei were vortexed for 1 min and pelleted at 1000 x g for 1 min. The supernatant was pooled with the previous cytoplasmic supernatant and centrifuged at 7000 x g for 10 min to remove mitochondria. The nuclei were washed three times with 0.32M sucrose, 1 mM $MgCl_2$ and the final nuclear pellet resuspended in RSB. Equal volumes of pronase (1 mg/ml) in 0.05M NaCl, 0.01M EDTA, 0.5% SDS, 0.1M Tris HCl, pH 7.5) were added to both the nuclear and cytoplasmic fractions which were then incubated at 37° for 1 hr. The nucleic acid was then extracted with equal volumes of chloroform-isoamyl alcohol (25:1) for 5 min at 37°. The aqueous phase was reextracted with chloroform-isoamylalcohol, and the final aqueous phase precipitated with ethanol at -20°.

The nucleic acid fractions were resuspended in 10 x RSB and incubated with DNase I (50 µg/ml final concentration) for 2 hr at 28°. The nucleic acid was then purified by pronase treatment and chloroform-isoamyl alcohol extraction as before. RNA

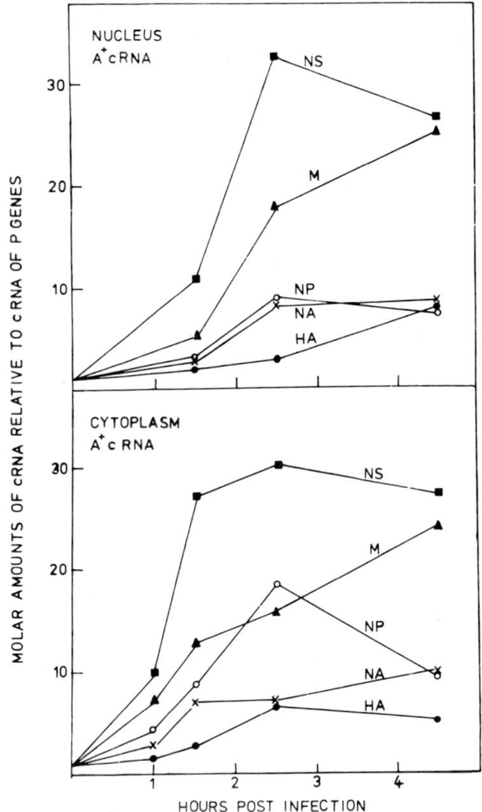

Fig.1. *CEF cell monolayers grown in roller bottles were infected with fowl plague virus and RNA was extracted from the nuclear and cytoplasmic fractions as described in the legend to Table I.*

The RNA samples were hybridised with radioactively labelled viral RNA fragments as described by Mahy et al (6). For these experiments vRNA was labelled with I^{125} as described by Commerford (3). Specific activities of 10^6-10^7 cpm/μg were obtained by this method. Individual labelled RNA fragments were obtained by separation of the labelled vRNA on 2.2% polyacrylamide-agarose gel and the RNA was extracted from the gel as described by McGeoch et al (7).

samples were dissolved in 1 mM EDTA, 10 mM Tris HCl pH 7.4, made 90% with DMSO, heated to 45° for 20 min, and ethanol-precipitated. Polyadenylated RNA was selected on oligo d(T) columns. RNA samples were hybridised with ^{32}P-labelled vRNA (1-10 x 10^6 cpm/μg) as described in reference (5).

mixture. The degree to which this amount of vRNA altered the equilibrium values for the hybridization was used to quantitate the amounts of both vRNA and cRNA in infected cells (see legend to Table II). Significant vRNA accumulation did not occur in either the nucleus or the cytoplasm up to 1.5 hr post-infection. In the nucleus the amount of vRNA increased until 2.5 hr post-infection and then declined. In the cytoplasm vRNA accumulated steadily at least up to 4.0 hr post-infection (see Table II).

TABLE II

Accumulation of vRNA in chick embryo fibroblasts infected with FPV

Time post-infection	0	0.5	1	1.5	2	2.5	3	4
Normal cells								
Nucleus	23	23	26	37	61	189	108	ND
Cytoplasm	42	44	49	146	256	350	827	1427
Total	65	67	75	183	317	539	935	-

Legend to Table II. RNA was extracted from the nucleus and cytoplasm of cells at the stated times post-infection as described in the legend to Table I, and annealed separately to ^{125}I-vRNA or ^{3}H-cDNA at 70° in 2 x SSC at an RNA concentration of 1 mg/ml (cytoplasmic) or 100 µg/ml (nuclear). Samples were removed at 2, 10, 30, 100, 300, 1000 and 3000 min and the extent of annealing was determined by digestion with either RNase A (50 µg/ml) + RNase T_1 (50 units/ml for ^{125}I-vRNA or S_1 Nuclease for ^{3}H-cDNA. These reactions were repeated with the addition of a known amount of non-radioactive vRNA. In one of the initial hybridizations equilibrium was reached at less than 100%, and the position of this equilibrium gives the ratio of vRNA/cRNA in that sample. The addition of unlabelled vRNA altered the equilibrium and from this change the amount of cRNA and vRNA present was calculated as follows:- If in the first hybridization $\frac{V}{C} = x$ and the addition of 2 µg vRNA alters the equilibrium so that $\frac{V + z}{C} = y$, then $C = \frac{z}{y-x}$ µg and $V = x\ C$ µg. If the approximate RNA contents of cytoplasm and nucleus are known, the number of gene copies present can be calculated from the formula:- $\frac{VR}{10^{-11}C}$ where

V = amount of virus-specific RNA (μg) in the hybridization mixture
R = RNA content per nucleus or cytoplasm (μg)
C = total amount of RNA in the hybridization mixture (μg)
10^{-11} = the weight of the influenza virus genome (μg)

CONCLUSIONS

Polyadenylated cRNA can be detected as early as 0.5 hr post-infection in both the nucleus and cytoplasm. The cRNA accumulates at approximately the same rate in both the nucleus and cytoplasm up to 1.5 hr post-infection. The accumulation of cRNA begins to decline in the nucleus at 2 hr post-infection, but continues in the cytoplasm. There is evidence for some transcriptional control of polyadenylated cRNA for individual genome segments in virus infected cells. However, there is no evidence for compartmentalisation of individual cRNA segments in the nucleus or cytoplasm.

vRNA begins to accumulate in both the nucleus and the cytoplasm at 1.5 hr post-infection and increases in the cytoplasm at least up to 4.0 hr post-infection. After an initial rise in the nucleus there is a decline in vRNA accumulation in the nucleus after 2.5 hr post-infection.

ACKNOWLEDGEMENT

We are grateful to Miss Nurit Kitron for excellent technical assistance. This work was supported by a grant from the Medical Research Council.

REFERENCES

1. Bean, J.W. and Simpson, R.W. (1973). *Virology* 56, 646.
2. Brownson, J.M. Ph.D. Thesis, University of Cambridge, 1977.
3. Commerford, S.L. (1971). *Biochemistry* 10, 1993.
4. Glass, S.E., McGeoch, D. and Barry, R.D. (1975). *J. Virology* 16, 1435.
5. Inglis, S.C., Carroll, A.R., Lamb, R.A. and Mahy, B.W.J. (1976). *Virology* 74, 489.
6. Mahy, B.W.J., Carroll, A.R., Brownson, J.M. and McGeoch, D.J. (1977). *Virology* 83, 150.
7. McGeoch, D.J., Fellner, P. and Newton, C. (1976). *Proc. Nat. Acad. Sci. USA* 73, 3045.
8. Repik, P., Flamand, A. and Bishop, D.H.L. (1974). *J. Virol.* 14, 1169.
9. Simpson, R.W. and Bean, W.J. (1975). *In* "The Influenza Viruses and Influenza", (E.D. Kilbourne, ed.), p.125. Academic Press, New York.

10. Skehel, J.J. (1973). *Virology* 56, 394.
11. Taylor, J.M., Illmensee, R., Litwin, S., Herring, L., Broni, B. and Krug, R.M. (1977). *J. Virol.* 21, 530.

TRANSCRIPTION AND REPLICATION OF THE INFLUENZA VIRUS GENOME EARLY AFTER INFECTION

G.E. MARK, J.M. TAYLOR, L. HERRING, B. BRONI and R.M. KRUG

The Institute for Cancer Research, Fox Chase Cancer Center, Philadelphia, Pa. 19111, and Memorial Sloan-Kettering Cancer Center, New York, N.Y. 10021, U.S.A.

Recently we have described the use of specific radioactive probes to study the transcription and replication of the influenza virus genome (16). Probes were made for the influenza genome RNA (vRNA) and its complement (cRNA): ^{32}P-labeled cDNA, synthesized with the avian sarcoma virus reverse transcriptase, and ^{125}I-vRNA, respectively. Measurements were made of the kinetics of annealing of these probes in the presence of an excess of unlabeled RNA extracted from canine kidney cells infected with the WSN strain of influenza virus. From these data it was possible to deduce the average number of molecules of vRNA and cRNA in the nucleus and cytoplasm as a function of time after infection. The times examined were 1.75, 2.75 and 3.75 hr after infection. These data showed an initial increase in vRNA and cRNA sequences in the nucleus but later the majority of both these sequences was found in the cytoplasm. The present study was undertaken to look more closely at the amount and cellular distribution of cRNA and vRNA at earlier times in the absence and presence of cycloheximide added at the time of infection. The transcription of vRNA and cRNA that occurs in the presence of such a cycloheximide treatment is operationally defined as primary transcription (1).

TIME COURSE OF APPEARANCE OF cRNA AND vRNA

For the following studies replica roller bottle cultures of canine kidney cells were infected with WSN strain influenza virus at a multiplicity of about 50 pfu per cell. Adsorption was for 1 hr at 4°C. Zero time of infection corresponds to when the virus inoculum was replaced by medium which had been warmed to 37°C. Cultures were removed at 0, 0.5, 0.75, 1, 1.25, 1.5, 2 and 2.5 hr after infection. The cells were fractionated into nucleus and cytoplasm and the RNAs extracted and examined for their content of cRNA and vRNA using the radioactive probes, as previously described (16). The results are summarized in Fig. 1a and 1b for cRNA and vRNA respectively.

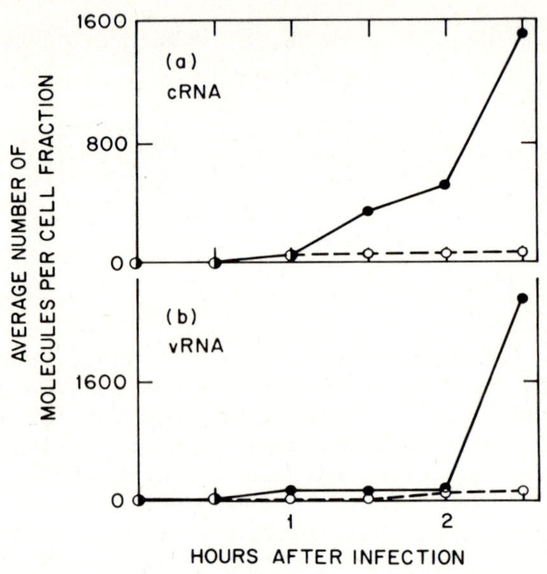

Fig. 1. *Time course of appearance of cRNA and vRNA in infected cells. The quantitation of cRNA and vRNA in nuclear (O--O) and cytoplasmic (●—●) fractions from infected cultures was carried out as previously described (16).*

At 0 hr the amount of cRNA is insignificant, as expected, and remains so until 1 hr (Fig. 1a). From 1.5 - 2.5 hr the amount of cRNA increases significantly. At these times the distribution of this cRNA is such that more than 85% is detected in the cytoplasm. Our previous data using times out to 3.75 hr after infection also detected the majority of cRNA in the cytoplasm.

Approximately 60 molecules of vRNA are detected in the infected cells at 0 hr (Fig. 1b). This presumably represents the inoculum vRNA. By 2 hr the number of molecules of vRNA has increased four-fold but, nevertheless, at this time there is still a three-fold excess of cRNA to vRNA. Between 2 and 2.5 hr the vRNA accumulates faster than the cRNA and achieves a two-fold excess. At all the times we have studied, the distribution of vRNA, and also of cRNA, is such that the majority is detected in the cytoplasm. An obvious interpretation is that vRNA and cRNA are predominantly synthesized in the cytoplasm. Alternatively, they may be synthesized in the nucleus and rapidly transported to the cytoplasm.

PRIMARY TRANSCRIPTION OF vRNA INTO cRNA

Theoretically primary transcription would be that early transcription of the infecting vRNA into cRNA using the inoculum transcriptase. The operational definition is that transcription which occurs when cycloheximide (100 μg/ml) is added at the time of infection (1). We have previously quantitated primary transcription at 2.5 hr after infection. The present study was undertaken to look at primary transcription at earlier times.

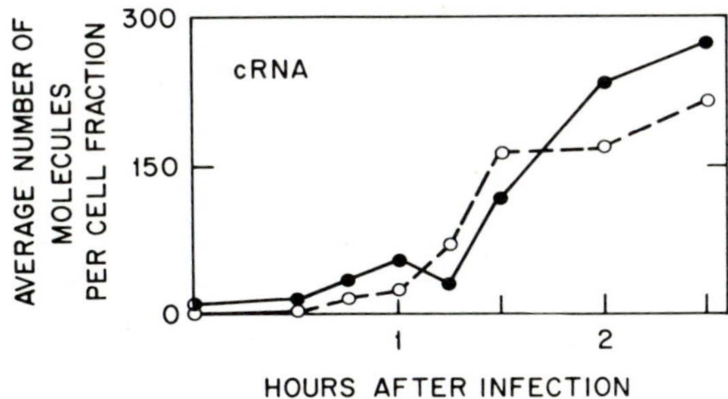

Fig. 2. *Time course of appearance of cRNA in infected cells treated at 0 hr with cycloheximide. The quantitation of cRNA in nuclear (O--O) and cytoplasmic (●—●) fractions from infected cultures was carried out as previously described (16).*

The time course of appearance of cRNA in the nucleus and cytoplasm of cycloheximide-treated cells is shown in Fig. 2. From 1 to 2.5 hr the amount of cRNA detected increases linearly up to a maximum of about 500 molecules, a value approximately one-third of that obtained in the absence of cycloheximide (Fig. 1a). At all the times analyzed the cRNA is approximately equally distributed between nuclear and cytoplasmic fractions. This distribution is in contrast to untreated cultures where the majority of cRNA is detected in the cytoplasm (Fig. 1a). One interpretation of these data is that cRNA is transcribed in the nucleus and that cycloheximide treatment interferes with a subsequent migration from nucleus to cytoplasm. From our previous studies we expect that at times beyond 2.5 hr the intracellular distribution of cRNA in cells treated with cycloheximide will change so that the majority (greater than 80%) of cRNA will be detected in the cytoplasm. In fact, in some cases such a change in intracellular distribution has occurred by 2.5 hr (16).

As expected, cycloheximide effectively inhibits the synthesis of vRNA. The inhibition, however, is not absolute. A three-fold increase in vRNA is seen by 2.5 hr (data not shown), but this is in contrast to the 45-fold increase observed in the absence of cycloheximide (Fig. 1b). Thus it can be presumed that the cRNA synthesized in cycloheximide-treated cells at times up to 2.5 hr is enriched for transcripts of the inoculum vRNA.

THE POLY (A) CONTENT OF PRIMARY TRANSCRIPTS

Clearly, the addition of poly (A) is the terminal event in the synthesis of a cRNA molecule. If poly (A) were present on the transcripts synthesized in the presence of cycloheximide, this would be consistent with these transcripts being of normal length. To test this, the RNA was extracted from the nucleus and cytoplasm of cycloheximide-treated cells at 2.5 hr. The poly (A)-containing cRNA was separated from the poly (A)-deficient cRNA by oligo (dT)-cellulose chromatography (8). These fractions were assayed for their content of cRNA with the radioactive probes. Of the cRNA in the nucleus and cytoplasm, 52 and 73%, respectively, was poly (A)-containing. Similar results were obtained with RNA extracted from cells treated with cycloheximide from the time of virus addition rather than from 0 hr.

Our previous studies have confirmed those of others which show that actinomycin added after infection specifically blocks the synthesis of cRNA (11, 14, 16). This is an inhibition of amplified transcription. When actinomycin is added with cycloheximide at 0 hr we can see the effect of actinomycin on primary transcription. It blocks the appearance of cRNA in the cytoplasm, but, nevertheless, a significant amount of cRNA is still detected in the nucleus (16). In parallel with the experiments described in the previous paragraph, we have analyzed for poly (A) content the cRNA species found at 2.5 hr in the nucleus of cells treated at 0 hr (or at the earlier time of virus addition) with both actinomycin and cycloheximide. Relative to cells treated with cycloheximide alone, the absolute amount of cRNA is depressed two-fold. About 95% of the cRNA is detected in the nucleus, as in the previous studies and of this, 82-92% contains poly (A). One interpretation of these data is that the residual cRNA synthesis in cells treated by both cycloheximide and actinomycin occurs primarily in the nucleus.

PRIMARY TRANSCRIPTION OF INDIVIDUAL GENOME SEGMENTS

An assumption of the method we have used to quantitate vRNA and cRNA in infected cells is that transcription is uniform. That is, for each time after infection, it is assumed that all genome segments, vRNA or cRNA, are present in relatively equimolar amounts. Studies of cytoplasmic poly (A)-containing cRNA and also studies on the translation of virus proteins suggest that at least in the cytoplasm, the cRNA species for individual segments are not present in equimolar amounts (3, 4, 6, 9, 15). The following experiment was undertaken to determine whether all the genome segments are represented in the cRNA transcribed in cycloheximide treated cells, and if so, whether they are present in equimolar amounts. Radioactive probes, ^{32}P-cDNA and ^{125}I-vRNA, were prepared for each of the eight genome segments. First, 250 μg of ^{3}H-labeled vRNA was subjected to electrophoresis on a polyacrylamide slab gel. As shown in Fig. 3, segments 4, 5, 6, 7 and 8 were resolved. Segments 1, 2 and 3 were not adequately resolved and were treated as a single pool. Radioactive probes were made against this pool and against segments 4, 5, 6, 7 and 8.

The probes to individual genome segments were tested for their kinetics of annealing to RNA extracted from the nucleus and cytoplasm of cells at 1.8 hr after infection, that had been treated with cycloheximide from 0 hr. These data were analyzed to

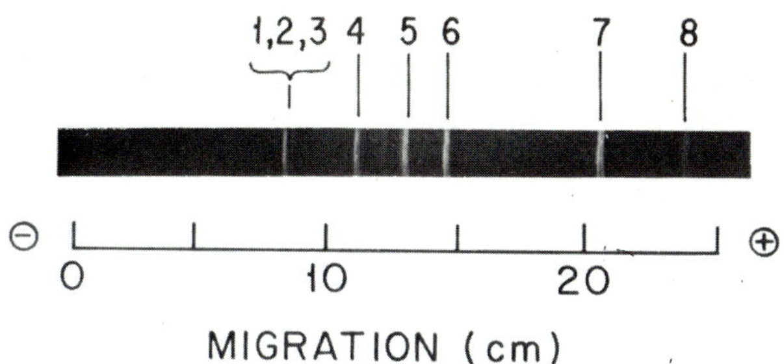

Fig. 3. *Preparative acrylamide slab gel electrophoresis of influenza genome segments. Virion RNA (250 μg) was applied to a 2.8% polyacrylamide slab gel (36 x 14 x 0.3 cm) containing 6 M urea and the buffer of Peacock and Dingman (10). After electrophoresis for 44 hr at 140 V, the gel was stained with ethidium bromide and visualized with UV light, as shown by the photograph above. The double nature of the RNA bands is an artifact of the photography. The numbers refer to the eight genome segments. The additional band between 4 and 5 is 18S rRNA, which has been observed by others to occur in variable amounts in the RNA extracted from purified virus (2, 5, 7). Genome segments 1-3 were not resolved and so were treated in the subsequent analysis as a single band. Thus, six bands were cut from the gel, crushed by passage through the needle of a syringe, and the RNA eluted into 0.5 M LiCl - 0.5% SDS - 0.01 M Tris, pH 7.4. After removal of particulate material by centrifugation, the RNA was collected by ethanol precipitation and purified by sedimentation for 26 hr at 40 K in the SW40 rotor using a gradient of 15-30% sucrose in: 0.1 M NaCl - 0.01 M Tris (pH 7.4) - 0.001 M EDTA.*

TABLE 1

Relative Molar Amounts of cRNA Segments Synthesized in the Nucleus as a Result of Primary Transcription

vRNA segment used to make radioactive probes	Designation of mRNA segment	Corresponding viral protein	Relative number of cRNA molecules per	
			nucleus	cytoplasm
1,2,3	L1,L2,L3	P_1,P_2,P_3	0.20	0.08
4	M1	HA	0.99	0.50
5	M2	NP	1.00	1.00
6	M3	NA	0.34	0.43
7	S1	M	0.78	0.43
8	S2	NS_1	1.54	2.66

The vRNA segments were isolated as described for Fig. 3. The designation of the mRNA segments is as used by Etkind *et al.* (3). The assignment of corresponding viral proteins is as described by Ritchey *et al.* (13) and Etkind *et al.* (3). The vRNA segments were used to make radioactive probes that were then tested for their kinetics of annealing to RNA isolated from cultures at 1.8 hr after infection, that had been treated from 0 hr with cycloheximide. The number of cRNA molecules corresponding to each RNA segment is shown relative to segment M2, which was determined to be 1160 and 820 molecules for nucleus and cytoplasm, respectively. The deduction of the number of molecules was carried out as previously described (16) with the modification that the $C_rt_{\frac{1}{2}}$ for each segment was taken to be proportional to the molecular weight of that segment.

determine the average number of molecules of vRNA and cRNA in the nucleus and cytoplasm for each genome segment. The relative amounts of cRNA for each segment detected in the nucleus and cytoplasm are summarized in Table 1. For both nucleus and cytoplasm the number of molecules of cRNA is not the same for each segment. In this and in a similar experiment we find there is less of the cRNA corresponding to P_1, P_2 and P_3 and maybe also of NA. There is significantly more of the cRNA corresponding to NS_1. Actually, the deduced relative amounts of each cRNA seem very similar to the results obtained by Pons (12) using a different assay to characterize that cRNA in the cytoplasmic polysomes at 20 min after infection.

CONCLUSIONS

In this manuscript we have presented data based on the use of molecular hybridization techniques to quantitate transcription and replication of the influenza virus genome. The sensitivity of the assay has allowed us to detect early transcription and replication in the infected cell.

With respect to the site(s) of synthesis of cRNA our studies detect the majority of cRNA in the cytoplasm (Fig. 1). If cycloheximide is added at 0 hr, the absolute amount of cRNA synthesis is depressed and approximately half of the residual cRNA species are detected in the nucleus. If both cycloheximide and actinomycin are added at 0 hr, cRNA synthesis is depressed even further, and those cRNA species transcribed are detected primarily in the nucleus. One interpretation is that the resistant cRNA synthesis occurs in the nucleus. It is possible that in untreated cultures cRNA is also synthesized primarily in the nucleus but migrates rapidly to the cytoplasm.

We have partially characterized that cRNA in the nucleus and cytoplasm of cycloheximide-treated cells. First, the majority of the molecules are poly (A)-containing. This is consistent with their being of full size, although direct tests have not yet been done. Secondly, the nuclear cRNA contains sequences representative of each genome segment. It was necessary to establish the latter because our assays using probes to the total genome are biased toward the detection of the large segments and are insensitive to the amount of small segments. For example, the three large P segments together account for as much as 54% of the total probe, whereas the smaller segments, M and NS_1, each represent only about 5% of the probe. Thus it is only with probes to individual segments that quantitation of the smaller segments can be obtained. Our initial studies have in fact demonstrated that at early times after infection primary transcription is not a uniform process.

ACKNOWLEDGEMENTS

We thank Alan Baseman for careful technical assistance. This work was supported by a grant PCM76-81525 from the National Science Foundation, grant AI-11772 from the National Institute of Allergy and Infectious Diseases, and by an appropriation from the Commonwealth of Pennsylvania.

REFERENCES

1. Bean, W.J. and Simpson, R.W. (1973). *Virology* 56, 646.
2. Cox, N.J. and Barry, R.D. (1976). *Virology* 69, 304.
3. Etkind, P.R., Buchhagen, D.G., Herz, C., Broni, B.B. and Krug, R.M. (1977). *J. Virol.* 22, 346.

4. Inglis, S.C., Carroll, A.R., Lamb, R.A. and Mahy, B.W.J. (1976). *Virology* 74, 489.
5. Krug, R.M., Morgan, M.M. and Shatkin, A.J. (1976). *J. Virol.* 20, 45.
6. Lamb, R.A. and Choppin, P.W. (1976). *Virology* 74, 504.
7. Lewandowski, L.J., Content, J. and Leppla, S.H. (1971). *J. Virol.* 8, 701.
8. McLaughlin, C.S., Warner, J.R., Edmonds, M., Nakazato, H. and Vaughan, M.H. (1973). *J. Biol. Chem.* 248, 1466.
9. Meier-Ewert, H. and Compans, R.W. (1974). *J. Virol.* 14, 1083.
10. Peacock, A.C. and Dingman, C.W. (1967). *Biochemistry* 6, 1818.
11. Pons, M.W. (1973). *Virology* 51, 120.
12. Pons, M.W. (1977). *Virology* 76, 855.
13. Ritchey, M.B., Palese, P. and Schulman, J.L. (1976). *J. Virol.* 20, 307.
14. Scholtissek, C. and Rott, R. (1970). *Virology* 40, 989.
15. Skehel, J.J. (1972). *Virology* 49, 23.
16. Taylor, J.M., Illmensee, R., Litwin, S., Herring, L. Broni, B. and Krug, R.M. (1977). *J. Virol.* 21, 530.

THE VIRION-ASSOCIATED AND CELL-ASSOCIATED *IN VITRO* ACTIVITIES OF THE INFLUENZA RNA POLYMERASE

MICHEL A. HORISBERGER and RETO BRAMBILLA

Research Department, Pharmaceuticals Division, CIBA-GEIGY Limited, Basle, Switzerland.

The RNA-dependent RNA polymerase specific for the influenza virus is found encapsidated into the virion (virion-associated polymerase) and is also found in influenza infected cells (cell-associated polymerase). It is associated with a ribonucleoprotein complex consisting of four viral polypeptides, NP, P1, P2 and P3 (12) and of 8 segments (separate genes) of viral RNA (14). The polymerase is firmly bound to the viral RNA template and does not respond to added RNA (4, 18). It has a requirement for ribonucleoside triphosphates and functions best at pH 8. It is absolutely dependent on the presence of a divalent cation but this requirement is different depending on the origin of the polymerase. The cell-associated polymerase responds equally well to Mn^{2+} and Mg^{2+} (10, 11), in contrast to the virion-associated polymerase which shows a marked preference for Mn^{2+} (5, 10, 15). The virion-associated polymerase gives a good response to Mg^{2+} only if additional factors are present: the specific dinucleoside monophosphates ApG and GpG (16) or cellular components such as ribosomes or RNA (9). The identity of the two activities was therefore questioned. Several parameters have been measured to solve this problem; they are the size of the RNA transcripts, their methylation and polyadenylation, their base composition, and the K_m of each ribonucleoside triphosphate under the various assay conditions.

SEDIMENTATION PROPERTIES OF THE RNA TRANSCRIPTS

The RNA transcripts synthesized by both polymerases in the presence of Mn^{2+} are clearly of small size and most probably do not represent complete copies of the template (Fig.1a and c). Mg^{2+} is the proper ion for the assay of the cell-associated polymerase since the transcripts display a similar size to the viral genome; in this case ApG (0.05 mM) stimulates the activity about 3-fold without influencing the size of the RNA transcripts (Fig.1d).

The virion-associated polymerase is poorly activated by Mg^{2+}, and the assay must be done in the presence of additional factors which stimulate the synthesis of RNA transcripts. ApG (0.25 mM) with Mg^{2+} gives transcripts of small size, as does Mn^{2+} alone (Fig.1a). Ribosomes (equivalent amount of protein to the virus) with Mg^{2+} direct the synthesis of long transcripts (Fig.1a and b) which code for viral proteins (6). When ribosomes or ApG are added separately they stimulate the polymerase activity more than 15 times; if they are added together they stimulate the RNA polymerase activity more than 50 times, and the RNA transcripts are of large size. These results indicate that dinucleoside monophosphates and ribosomes act synergistically and that they have different mechanisms of action on the polymerase; the ribosomes represent the necessary factor for the synthesis of large RNA transcripts.

BASE COMPOSITION

The radioactive RNA transcripts have been first tested for their ability to bind to polyU-Sepharose, (13). The RNA transcripts of the virion-associated polymerase do not bind, which indicates no polyadenylation of the RNA. This result is confirmed by analysing the base composition of the total RNA synthesized *in vitro*, which is that of the plus strand (Table I). The cell-associated polymerase directs the synthesis of polyadenylated and non-polyadenylated RNA transcripts, but in different proportions, depending on the divalent ion used in the assay. In the presence of Mg^{2+} (with or without ApG), 50% of the RNA is polyadenylated, but in the presence of Mn^{2+}, only 15% adenylation is found. The RNA lacking polyA synthesized in the presence of either cation is plus-stranded, and after correction for the presence of polyA (22%), the RNA bound to polyU-Sepharose also proves to be plus-stranded (Table I). Methylation with S-adenosyl-^{3}H-methionine under conditions which are sensitive enough to detect the methylation of one base per RNA transcript gives only negative results for both polymerases, and addition of crude methylases to the system does not result in methylation.

THE K_m VALUES

The K_m values are in general the lowest for CTP and GTP, and the highest for ATP (Table II). The lowest K_m values for the whole series of ribonucleoside triphosphates are obtained with 2 mM Mg^{2+} and ribosomes with the virion-associated polymerase, and with 2 mM Mg^{2+} with the cell-associated polymerase. It is interesting to note that under these conditions the polymerase directs the synthesis of the largest RNA transcripts (Fig.1). ApG increases the K_m of some triphosphates, and the use of Mn^{2+}ions in the assay considerably increases the K_m values for all the triphosphates.

DISCUSSION

Based on the following results, we conclude that the virion- and cell-associated polymerase activites are identical. (a) The proper divalent ion is Mg^{2+} for both polymerase assays, because the replacement of Mn^{2+} by Mg^{2+} led to a decrease of the K_m values of the triphosphates. We find also that both polymerases synthesize small RNA transcripts in the presence of Mn^{2+}, in confirmation of other reports (3, 8, 16). (b) The conditions for the synthesis of large RNA transcripts are very similar for both polymerases: Mg^{2+} and ribosomes for the virion-associated polymerase, and Mg^{2+} alone for the cell-associated polymerase. Because of the extraction procedure (11), the cell-associated polymerase preparation already contains ribosomes and ribosomal subunits active in the stimulation of the virion-associated polymerase (9). (c) Both polymerase activities are stimulated by ApG three to five fold in assay conditions described under (b). The high concentration of ApG or GpG required for the activation of the virion-associated activity precludes all physiological relevance of the dinucleoside monophosphates. (d) Both polymerases direct the synthesis of plus-stranded viral RNA as shown by base composition; however, the method is not sensitive enough to exclude completely the synthesis of a small proportion of negative-stranded RNA. We observe no polyadenylation of the RNA transcripts with the virion-associated polymerase and we suggest that the influenza virus RNA polymerase and the polyA polymerase are two distinct enzymes, the latter not being incorporated into the virion. (e) Both polymerases show the same K_m values for the ribonucleoside triphosphates under assay conditions where they synthesize full RNA transcripts.

Some of our results do not completely agree with some published reports. For instance, Skehel (19) observed considerable activity of the virion-associated polymerase in the presence of Mg^{2+}; others have found synthesis of polyA with the virion-associated polymerase (8, 16). These apparently contradictory results could be explained if, under certain conditions, the virion could incorporate some cellular components such as a polyA polymerase or ribosomes. For example, it is

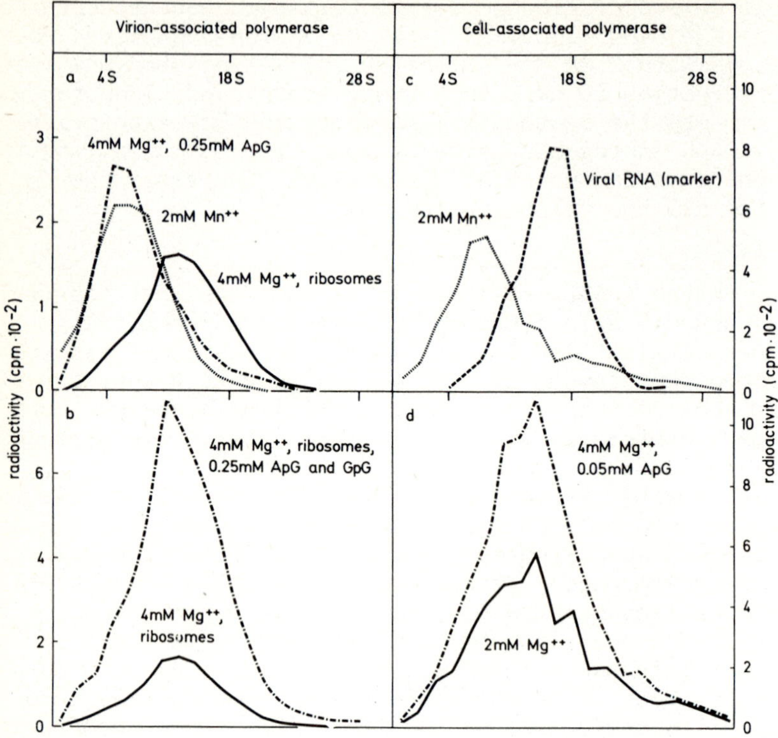

Fig.1. *Sedimentation properties of the RNA synthesized in vitro.*

Methods for the growth and purification of the NWS influenza virus (HON1), the disruption of the virions, and the preparation of ribosomes have been published (9). The purification of the cell-associated polymerase was essentially as published (11) with the following modification: The S-15 extract was adjusted to 0.4% Triton X-100, 10 mM EDTA and 0.5 M KCl before centrifugation. Fractions containing the polymerase activity were pooled to give the cell-associated polymerase preparation. The transcriptase assays were performed as published (9) with two modifications: the specific activity of ^{3}H-CTP was increased to 100 μCi/μmole in some experiments, and KCl was replaced by CH_3COOK which is less inhibitory at high concentrations. The concentrations of Mg^{2+} and Mn^{2+} indicated on the figure gave the maximal rate of synthesis in a particular assay; protein concentrations of virus, ribosomes, and cell-associated polymerase were 200 to 400 μg/ml in each assay. The radioactive RNA synthesized in vitro was phenol extracted at pH 9 and its sedimentation properties were examined on SDS-

already known that rRNA is present in influenza virions (7). The presence of such cellular components in the influenza virion could explain the above mentioned discrepancies.

REFERENCES

1. Ada, G.L., and Perry, B.T. (1955). *Nature* 175, 854.
2. Bishop, D.H.L., Obijeski, J.F., and Simpson, R.W. (1971a). *J. Virol.* 8, 66.
3. Bishop, D.H.L., Obijeski, J.F., and Simpson, R.W. (1971b). *J. Virol.* 8, 74.
4. Bishop, D.H.L., Roy, P., Bean, W.J., and Simpson, R. (1972). *J. Virol.* 10, 689.
5. Chow, N.L., and Simpson, R.W. (1971). *Proc. Nat. Acad. Sci. USA* 68, 752.
6. Content, J., Dewit, L., and Horisberger, M. (1977). *J. Virol.* 22, 247.
7. Content, J., Duesberg, P.H. (1971). *J. Mol. Biol.* 62, 273.
8. Ghendon, Y., and Blagoveshienskaya, O. (1975). *Virology* 68, 330.
9. Horisberger, M.A. (1976). *FEBS Letters* 63, 134.
10. Horisberger, M.A., and Guskey, L.E. (1974). *J. Virol.* 13, 230.
11. Horisberger, M.A., and Schulze, C. (1974). *Arch. ges. Virusforsch.* 46, 148.
12. Inglis, S.C., Carroll, A.R., Lamb, R.A., and Mahy, B.W.J. (1976). *Virology* 74, 489.
13. Lindberg, U., and Persson, T. (1972). *Eur. J. Biochem.* 31, 246.
14. Palese, P. (1977). *Cell* 10, 1.
15. Penhoet, E., Miller, H., Doyle, M., and Blatti, S. (1971). *Proc. Nat. Acad. Sci. USA* 68, 1369.
16. Plotch, S.J., and Krug, R.M. (1977). *J. Virol.* 21, 24.
17. Rogg, H., Brambilla, R., Keith, G., and Staehelin, M. (1976). *Nucl. Acids Res.* 3, 285.
18. Scholtissek, C., and Becht, H. (1969). *J. Gen. Virol.* 4, 125.
19. Skehel, J.J. (1971). *Virology* 45, 793.

---

sucrose gradients after denaturation of the RNA. ^{32}P-labelled virus RNA was obtained from purified virus grown in primary cultures in the presence of ^{32}P-orthophosphate.

TABLE I

Base composition of the RNA synthesized *in vitro*

Polymerase:	cell-associated				virion-associated	WSE (HON1)	
PolyU-Sepharose:	bound RNA		unbound RNA		total RNA	(Ada and Perry 1955)	
Assay conditions: / Base	2mM Mn^{2+}	2mM Mg^{2+}	2mM Mn^{2+}	2mM Mg^{2+}	4mM Mg^{2+} 0.25mM ApG + GpG + ribosomes		Base
	%	%	%	%	%	%	
A	46.5	48.6	35.7	34.2	33.4	33.3	U
C	15.2	13.9	20.1	18.9	20.3	20.1	G
G	17.4	17.4	19.8	20.6	23.2	24.1	C
U	20.9	20.1	24.4	26.3	23.1	22.5	A

The transcriptase assay was done with four tritium-labelled triphosphates whose specific activity was determined after removing contaminants by thin layer chromatography with isobutyric acid - 0.5 M ammonium hydroxide - 0.1 M EDTA (100:6:1.6, by vol.). The RNA was extracted as in Fig.1, and further purified on Sephadex G-50. The RNA was digested to nucleosides (17) for 4 hrs at 37°; the nucleosides were separated by two dimensional thin layer chromatography; first direction: propanol - NH_3 conc.-H_2O (60:30:10, by vol.); second direction: isopropanol - HCl conc.-H_2O (102:26.4:21.6, by vol.).

TABLE II

K_m Values of the Triphosphates

RNA-dependent RNA polymerase	Assay conditions	K_m (μ mole) ATP	CTP	GTP	UTP
Virion-associated	2mM Mg^{2+} + ribosomes	84	3.6	9.6	23
	4mM Mg^{2+} + 0.25mM ApG	81	99	10.9	21.1
	4mM Mg^{2+} + 0.25mM ApG + ribosomes	84	15.8	9.9	24.4
Cell-associated	2mM Mg^{2+}	122	3.0	6.2	17.6
	4mM Mg^{2+} + 0.05mM ApG	366	6.9	9.9	30
	2mM Mn^{2+}	341	104	83	36
Virion-associated (Bishop *et al.*(2))	1mM Mn^{2+} + 8mM Mg^{2+}	430	48	100	42

The velocity of the RNA synthesis was determined for several limiting concentrations of a ^3H-labelled triphosphate, whereas the concentration of the three others was kept constant at saturation levels. The linear regression curve and the K_m values were determined by computer procedures from the reciprocals of the substrate concentration and reaction velocity.

CELL-FREE COUPLING OF INFLUENZA VIRUS TRANSCRIPTION AND TRANSLATION

J. CONTENT, L. DE WIT,
and M. HORISBERGER*

Department of Virology, Institut Pasteur du Brabant, Brussels 1040, Belgium

and

**Research Department, Pharmaceuticals Division, CIBA-GEIGY Ltd., CH 4002 Basel, Switzerland*

Influenza viruses are enveloped animal viruses whose genome is composed of a segmented single-stranded, probably entirely "negative stranded" RNA (9, 12, 15). Each of the gene segments (20) is transcribed into a monocistronic mRNA terminating with poly A at its 3' end and with methylated "cap" at its 5' end (17). Inside the virion each RNA segment is wrapped into a ribonucleoprotein (RNP) structure (8, 23) containing the viral protein NP and other internal peptides probably responsible for the viral RNA polymerase activity found in each of those RNP segments (3, 14, 25). In cells infected with influenza virus, it seems that at least two types of RNP are found: a) a parental-type RNP (-) with its associated polymerase activity, and b) a mRNA-containing RNP (+) associated with the polysomes (16, 22).

Some *in vivo* translation experiments indicate that the expression of the influenza genome is regulated, both with respect to time and to different gene products (18, 26). The molecular basis for this regulation is not fully understood. This problem could be approached and analysed in more detail if a cell-free system could be designed for the transcription and translation of influenza mRNA. Recently, we have shown that despite the low activity of influenza virion-associated polymerase it is possible to establish such a coupled cell-free system, yielding most of the virus structural proteins (10).

Here we extend these observations and show that detergent-disrupted purified influenza virion can generate all the non-glycosylated viral proteins including large amounts of P_1 and

P_2 and at least one non-structural protein in the nuclease-treated rabbit reticulocyte cell-free system (21). A cytoplasmic RNP fraction (30-85s) can be translated in the same cell-free system and fractionated into individual RNP's promoting the synthesis of distinct viral proteins. Finally we have compared the synthesis of viral peptides in the same coupled reticulocyte cell-free system directed with standard and defective interfering (von Magnus) virions.

COMPARISON OF THE TRANSLATION PATTERN OBTAINED WITH NWS AND THE FPV STRAINS OF INFLUENZA VIRUS IN THE COUPLED CELL-FREE SYSTEM

The activity of the coupled cell-free system for transcription and translation of influenza virus mRNA has been recently improved by using a higher concentration of mammalian tRNA (150 µg/ml) and by reducing the amount of purified disrupted virus from 1 mg/ml (FPV) to 0.1-0.2 mg/ml in the case of NWS. Amino acid incorporation in this case is usually prolonged for at least 3 hrs and the stimulation of protein synthesis after addition of the disrupted virus is 5-10 fold as compared to a 2 fold stimulation in our previous report (10). As shown in figure 1, P_1 and P_2 (representing probably P_2 and P_3, which are

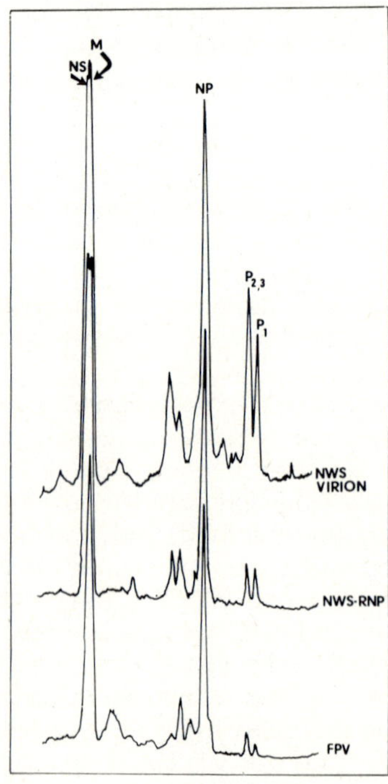

Fig.1.

not resolved by the gels used in this work) are readily detected. In the case of NWS, large amounts of P_1 and P_2 are produced, and NS_1, distinct from M, is detectable. All these proteins as well as NP have been characterised by tryptic peptide analysis in one dimension (high voltage paper electrophoresis) or two dimensions (thin layer electrophoresis and chromatogrraphy) and demonstrated to be indistinguishable from the authentic *in vivo* labelled viral proteins. In contrast with our previous report, HA and NA or their non-glycosylated counterparts have not been detected unambiguously. They may be masked by adjacent endogenous peptides or by the relative abundance of NP.

The abundant synthesis of P_1 and P_2 is certainly striking in the case of NWS and shows that all the RNA segments (except those corresponding to HA and NA) are translated into complete viral peptides. This is in contrast with the situation described for rhabdoviruses where a similar kind of coupled system never yields the L protein (1, 5, 24) probably because the L protein cistron is located at the distal 3' end of the long RNA genome (1).

Fig. 1. *Analysis of products synthesized in the coupled reaction programmed with different viral templates. The cell-free micrococcal nuclease preincubated rabbit reticulocyte lysate was prepared and used as described (10) except that the reaction was supplemented with 150 μg/ml of calf liver tRNA (Boehringer, Mannheim). The reactions were incubated for 2 hr at 31° with 6 x 10^8 cpm ^{35}S-methionine per ml, analysed by electrophoresing for 4-5 h at 80 volts on a 10% polyacrylamide slab gel and after fluorography the films were scanned at 540 nm (10). The letters NS, M, NP and P_1-P_3 represent the position of the authentic in vivo proteins obtained in a parallel lane with an ^{35}S-methionine labelled FPV or NWS infected cytoplasmic extract. The top tracing represents the products of an in vivo reaction programmed with 0.14 mg/ml of purified-disrupted NWS virion (see ref. 10). The bottom tracing represents the products of an in vitro reaction programmed with 1.0 mg/ml of purified-disrupted FPV. The middle tracing was obtained with 0.15 mg/ml of NWS-cytoplasmic RNP. This component was prepared as described previously (13) from HeLa cells 7.5 hr after infection with NWS virus. After lysis, an S 15 fraction was treated with 0.4% Triton X-100, 0.5 M KCl and 10 mM EDTA, sedimented on a 15-45% linear sucrose gradient for 180 minutes at 4° and 40,000 rpm in a SW 41 Beckman rotor. The fractions containing the polymerase activity were pooled and diluted with an equal volume of glycerol.*

TRANSLATION PATTERN OBTAINED WITH NWS VIRUS CYTOPLASMIC RIBONUCLEOPROTEINS IN THE COUPLED CELL-FREE SYSTEM

Large amounts of viral RNP's can be prepared and purified from the cytoplasm of influenza infected cells, and contain the virus cell-associated polymerase (7, 13). If such RNP-polymerase complexes are active in the coupled cell-free system, their products might reflect more closely the *in vivo* situation than those obtained from the virion-associated cell-free system. In contrast to the coupled system with the disrupted virion which was inhibited from 70 to 100% when the four ribonucleoside triphosphates were omitted, the cytoplasmic RNP-directed reaction was usually not inhibited more than 40-50%. Furthermore, 0.8 mM of the disodium disulfonate salt of bathocuproine, a selective inhibitor of influenza virus RNA polymerase (19), inhibits completely the coupled cell-free system programmed with NWS disrupted virions, but inhibits only from 20 to 30% the same system programmed with the cytoplasmic RNP's; the translational machinery is very little affected by this concentration of bathocuproine since the translation of globin or TMV mRNA is not inhibited more than 13%. These results suggest that the RNP preparation may contain two types of RNP's; the RNP (-) type would be transcribed and translated like the RNP's of the virion, and the RNP (+) type would be directly translated.

The cytoplasmic RNP's are resolved by sucrose gradient sedimentation into various segments. The distribution of the polymerase activity shows two peaks (Figure 2). In fractions 15

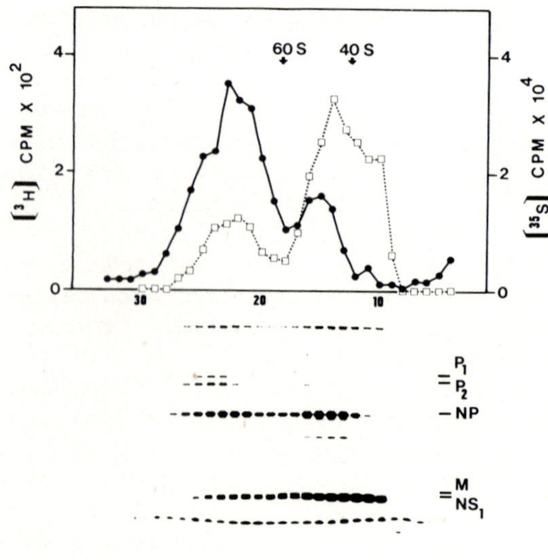

Fig.2.

and 22, respectively 0.12 and 0.25 μg viral RNA were synthesized per ml, during an incubation period of 20 minutes; this amount is sufficient for programming the coupled cell-free system (10). However, when each fraction was assayed in the coupled cell-free system, it became apparent that no strict correlation exists between the distribution of the transcriptase activity and the extent of protein synthesis (Figure 2). It can be seen that within each peak there is a certain correlation between the size of the RNP's and the size of the peptides programmed in the cell-free system. These results would be consistent if we postulate the existence of two types of RNP's: an RNP (-) type associated with the transcriptase, sedimenting between 40 and 85S and an RNP (+) type sedimenting between 30 and 55S without associated polymerase.

CELL-FREE PROTEIN SYNTHESIS WITH STANDARD AND DEFECTIVE INTERFERING (DI) VIRUS

DI influenza virions have been obtained from the NWS strain after 3 undiluted passages in calf kidney primary cultures. As shown in Table 1 the incomplete virus had a PFU/HA ratio decreased 20 fold as compared to the standard virus preparation. These particles usually retain a good virion-associated transcriptase activity (2). However, the pattern obtained after protein synthesis in the coupled cell-free system described above differs dramatically: the DI particles synthesize about 4 or 5 times less P_1 and P_2 proteins than the standard virion (Table 1). This observation certainly confirms the model of the von Magnus phenomenon proposed earlier (4, 6): under high multiplicity of infection, the synthesis of the largest viral RNA segments is preferentially reduced (6, 11) possibly because of an imbalance between the cellular overload in viral RNA-transcriptase complex

---

Fig. 2. *In vitro transcription and translation of cytoplasmic NWS-RNP segments in the reticulocyte cell-free system. NWS cytoplasmic RNP's were prepared as described in figure 1 except that the sedimentation was on a 15-45% linear sucrose gradient containing 0.2 M urea and 0.1% NP-40, for 16 hr at 22500 rpm in a SW 27 rotor. 200 μl of each fraction from the gradient was assayed for RNA polymerase activity (Mg^{2+} assay; 13), and 2 μl of each fraction was assayed for translation in the reticulocyte cell-free system (21) in the conditions described recently (10). The reaction volume was 15 μl. Hot TCA insoluble radioactivity was measured in 5 μl, and 2 μl were analysed by slab gel electrophoresis and fluorography as described in figure 1. Closed circles: RNA polymerase activity, open squares: protein synthesis.*

TABLE 1

Proteins obtained after transcription and translation of detergent-disrupted influenza virion in the reticulocyte cell-free system

		PFU/ml	HA	log PFU/HA	P_1**	P_2**	NP**	M + NS**
NWS	1*	3.3×10^6	64	3.71	6.8%	7 %	41.3%	44.4%
NWS	3	2.8×10^5	128	2.34	1.2%	1.6%	38.2%	58.9%

*Number of undiluted passages in calf kidney primary cultures.

**Purified virions were disrupted, incubated in the reticulocyte cell-free system at 0.1 mg/ml as indicated in fig. 1. The products of the reaction were analysed by polyacrylamide gel electrophoresis on a 10% slab gel that was processed for fluorography. In each lane the regions of the dry gel corresponding to the indicated viral proteins, identified from the fluorography, were excised and counted in toluene based scintillation fluid. The ^{35}S-methionine radioactivity in each region of the gel is expressed as a percentage from the "total counts" (P_1 + P_2 + NP + M + NS).

and the limited capacity of the host cell machinery to synthesize large molecular weight RNA's. Our results show directly that the phenomenon, which is initially host cell dependent (6), becomes later virion dependent and self-aggravating since the synthesis of those peptides (P_1, P_2, P_3) responsible for virion RNA transcription and replication (20) is preferentially reduced. However, we cannot conclude from the data if the DI particles differ only by their reduced content in the largest RNA segments or if the distribution of the viral transcriptase itself among the RNA segments is also modified.

CONCLUSION

Our results, obtained with two strains of influenza virus, show that purified, standard virions direct the synthesis of five structural and one non-structural viral peptides in an *in vitro* system for transcription and translation that has been described previously (10); the proportion of the viral peptides obtained *in vitro* seems compatible with the translation of primary RNA transcripts (18). With defective, interfering (von Magnus) virions, the *in vitro* synthesis of P_1 and P_2 is selectively decreased. The fact that glycoproteins are not detected after *in vitro* synthesis and that all the other structural and one non-structural viral proteins have been authenticated by tryptic peptide analysis (results not shown) demonstrates that the viral peptides generated in the coupled cell-free system have really been translated and that they do not result from amino acid exchange, for instance. A preparation of heterogeneous cytoplasmic RNP's has been isolated from the cytoplasm of NWS-infected cells, which is capable of programming the synthesis of the major viral proteins in the reticulocyte cell-free system. Some of the separated individual RNP segments code for distinct viral peptides. Since the sedimentation coefficient of these RNP's is distributed between about 30 and 85S and since their activity is only partly inhibited by the omission of the 4 nucleotide triphosphates or by a selective inhibitor of influenza virus transcriptase (the sodium disulphonate of bathocuproine) they may represent two biologically active forms of RNP's. The coexistence of RNP (+) and RNP (-) in the same cytoplasmic preparation has already been described (16). We suggest that the viral peptides synthesized in the cell-free system where the influenza polymerase has been severely inhibited, are coded by RNP's containing plus-stranded RNA. Consistent with this hypothesis is the evidence that viral mRNA is in the form of an mRNP or RNP (+) on the polysomes of influenza infected cells (22) and that eukaryotic messenger RNA's can be isolated from the cytoplasm or from purified polysomes as mRNP's; furthermore, these mRNP complexes can be translated efficiently in cell-free systems (see ref. 27).

ACKNOWLEDGEMENTS

Two of us (J.C. and L.D.) were supported by grants from the Belgium F.R.S.M., F.N.R.S., and the Fondation Rose et Jean Hoguet.

REFERENCES

1. Ball, L.A., and White, C.N. (1976). *Proc. Natn. Acad. Sci., U.S.A.* 73, 442.
2. Bean, W.J., and Simpson, R.W. (1976). *J. Virol.* 18, 365.
3. Bishop, D.H.L., Roy, P., Bean, W.J., Jr., and Simpson, R.W. (1972). *J. Virol.* 10, 689.
4. Blair, C.D., and Duesberg, P.H. (1970). *Ann. Review Microbiol.* 24, 539.
5. Breindl, M., and Holland, J.J. (1976). *Virology* 73, 106.
6. Choppin, P.W., and Pons, M.W. (1970). *Virology* 42, 603.

7. Compans, R.W., and Caliguiri, L.A. (1973). *J. Virol.* 11, 441.
8. Compans, R.W., Content, J., and Duesberg, P.H. (1972). *J. Virol.* 10, 795.
9. Content, J. (1976). *J. Virol.* 18, 604.
10. Content, J., De Wit, L., and Horisberger, M. (1977). *J. Virol.* 22, 247.
11. Duesberg, P.H. (1968). *Proc. Natn. Acad. Sci., U.S.A.* 59, 930.
12. Etkind, P.R., and Krug, R.M. (1975). *J. Virol.* 16, 1464.
13. Horisberger, M., and Schulze, C. (1974). *Arch. Gesamte Virusforsch.* 46, 148.
14. Inglis, S.C., Carroll, A.R., Lamb, R.A., and Mahy, B.W.J. (1976). *Virology* 74, 489.
15. Inglis, S.C., McGeoch, D.J., and Mahy, B.W.J. (1977). *Virology* 78, 522.
16. Krug, R.M. (1972). *Virology* 50, 103.
17. Krug, R.M., Morgan, M.A., and Shatkin, A.J. (1976). *J. Virol.* 20, 45.
18. Lamb, R.A., and Choppin, P.W. (1976). *Virology* 74, 504.
19. Oxford, J.S., and Perrin, D.D. (1974). *J. Gen. Virol.* 23, 59.
20. Palese, P. (1977). *Cell* 10, 1.
21. Pelham, H.R.B., and Jackson, R.J. (1976). *Eur. J. Biochem.* 67, 247.
22. Pons, M.W. (1975). *Virology* 67, 209.
23. Pons, M.W., Schulze, I.T., Hirst, G.K., and Hauser, R. (1969). *Virology* 39, 250.
24. Preston, C.M., and Szilagyi, J.F. (1977). *J. Virol.* 21, 1002.
25. Rochovansky, O.M. (1976). *Virology* 73, 327.
26. Skehel, J.J. (1973). *Virology* 56, 394.
27. Williamson, R. (1973). *FEBS Lett.* 37, 1.

TRANSCRIPTION *IN VITRO* OF THE SEGMENTED RNA GENOME OF UUKUNIEMI VIRUS

MARJUT RANKI and RALF F. PETTERSSON

Department of Virology, University of Helsinki, Haartmaninkatu 3, SF-00290, Helsinki, Finland

Uukuniemi virus is a tick-borne member of the newly defined large Bunyaviridae family of arthropod-borne viruses (11). These viruses are antigenically very diverse. The majority of the bunyaviruses fall into the Bunyamwera serologic supergroup, whereas the rest are scattered in a number of minor antigenically unrelated groups. Some bunyaviruses cause encephalitis or hemorrhagic fever in both man and animals, whereas Uukuniemi virus, the prototype of a distinct serologic group, apparently is nonpathogenic to man.

We have used Uukuniemi virus as a model system for studying the molecular biology of the bunyaviruses. The virus particle (about 95 nm in diameter) consists of an outer lipoprotein envelope and an inner ribonucleoprotein core (8). Two glycoproteins, G1 and G2, both with a molecular weight of about 65,000-75,000, form the highly organized spike layer (14). The core consists of three species of ribonucleoproteins, each containing one size of RNA and multiple copies of the N protein (6, 9) and probably a few copies of a minor large polypeptide species, L (molecular weight 180,000-200,000) (N. Guttman and R. Pettersson unpublished data). The ribonucleoproteins are circular as revealed by electron microscopy (9) and probably have a helical configuration inside the virion (15). Recently we have also found by electron microscopy that the deproteinized RNA species are circular when exposed to non-denaturing or mildly denaturing conditions. The circularization apparently results from base-pairing between inverted complementary sequences at the 3' and 5' ends of the molecules (5).

We have previously shown that purified Uukuniemi virions

contain an RNA-dependent-RNA polymerase activity (12), a finding which suggested that the genome of the bunyaviruses may be of negative polarity (1). Consistent with this concept is the lack of infectivity of virion RNA (Pettersson, unpublished data), and the absence of both a capped, methylated 5' end and a poly (A) tract at the 3' end, both characteristics of most eukaryotic messenger RNAs. The 5' end of each species of bunyavirus genome RNA is instead pppAp (7), a structure found in many negative-strand RNA viruses.

To characterize further the RNA-polymerizing activity present in Uukuniemi virions, we have isolated the three size classes of ribonucleoprotein particles and have used them individually as an enzyme-template source. Each ribonucleoprotein species possesses enzymatic activity, suggesting that primary transcription of each species can take place independently. The optimal conditions for *in vitro* enzymatic activity and the nature of the products synthesized have been characterized. To determine whether each of the three RNA species contains a unique RNA sequence, and therefore whether each codes for different proteins, we have compared the ^{32}P-labeled T_1-oligonucleotides derived from each RNA by two-dimensional gel electrophoresis (7). We conclude that the genome of Uukuniemi virus consists of three unique RNA segments with molecular weights of about 2.4×10^6 (L), 1.2×10^6 (M) and 0.5×10^6 (S), each containing information for different proteins.

OLIGONUCLEOTIDE MAPS OF THE VIRION RNA SPECIES

Three species of single-stranded RNA sedimenting at about 29S (L), 22S (M) and 17S are consistently found in plaque-purified Uukuniemi virions (6). To study whether these RNA species contain unique primary sequences or whether the sequences present in the two smaller M and S RNAs also are present in the L RNA, we compared the oligonucleotide fingerprints derived from each RNA. The RNA was labeled with ^{32}P-orthophosphate and fractionated on a sucrose gradient. Each RNA species was then digested with RNase T_1 to yield oligonucleotides terminating in a G residue. The oligonucleotides were fractionated by two-dimensional gel electrophoresis (3, 7). This method fractionates RNA in the first dimension mainly according to base composition and in the second dimension on the basis of chain length. Since the probability that long oligonucleotides (more than about 15 nucleotides) with identical length and base composition occur twice in the same molecule or any unrelated molecule is very small, the relationship between different RNAs can be determined by comparing the pattern of their unique T_1-oligonucleotides. If the L RNA contained M and S sequences, then all or some of the unique M and S oligonucleotides should also be present in the

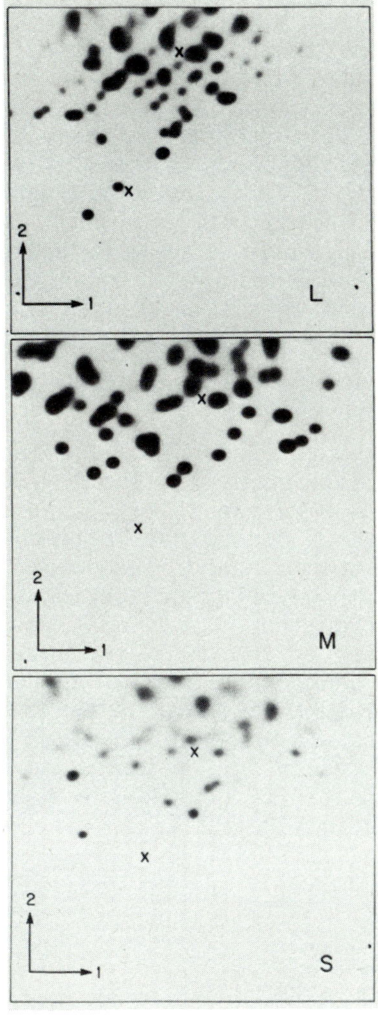

Fig. 1. *Two-dimensional gel electrophoresis of* ^{32}*P-labeled* T_1*-oligonucleotides from Uukuniemi virus L, M and S RNAs. Uukuniemi virus was grown in the presence of* ^{32}*P-orthophosphate, purified, and the L, M and S RNA species fractionated as described previously (7). About 1 x* 10^6 *cpm of each RNA species was digested with RNase* T_1 *and the resulting oligonucleotides fractionated by two-dimensional gel electrophoresis. The first dimension was on a 10% acrylamide gel in 6 M urea (pH 3.5) and the second dimension on a 22% acrylamide gel (pH 8.0) (2,3). The two x's indicate the positions of the xylene cyanol dye (lower) and bromophenol blue (upper).*

L RNA. Figure 1 shows autoradiograms of the lower parts of fingerprints obtained from L, M and S RNAs. By visual inspection, it is clear that the fingerprints are completely different. Detailed analysis of the relative position as well as the base composition of the oligonucleotides has revealed that 30 of the L spots are unrelated to any of the M spots, and 15 of the M spots are unrelated to the L spots. Similarly, at least 5 oligonucleotides present in the S fingerprint did not display any relationship to the L or M oligonucleotides. The faint spots in the S fingerprint represent oligonucleotides derived from contaminating L and M sequences. All three Uukuniemi virus RNA species thus contain a unique primary sequence with little, if any, overlapping.

TRANSCRIPTION *IN VITRO* OF THE UUKUNIEMI VIRUS RNAs

The Uukuniemi virus RNA polymerase activity was localized in the virion ribonucleoprotein particles which were isolated and used as enzyme-template complexes for *in vitro* transcription. The isolation of the RNPs resulted in a purified fraction which was essentially free of contaminating ribonuclease (see Fig. 4 and 5; analysis of template and product RNA after transcription *in vitro*), and which supported transcription for longer periods (5 hr) than unfractionated, disrupted virus (2 hr).

After purification of the RNPs, the optimal conditions for transcription were also changed (Fig. 2). Mn^{++} remained as the more active divalent cation, but the concentration required for optimal activity was lowered from 5 mM to 1 mM. Mg^{++}, which previously was completely ineffective, now supported transcription to 40% of the activity obtained with Mn^{++} alone. When

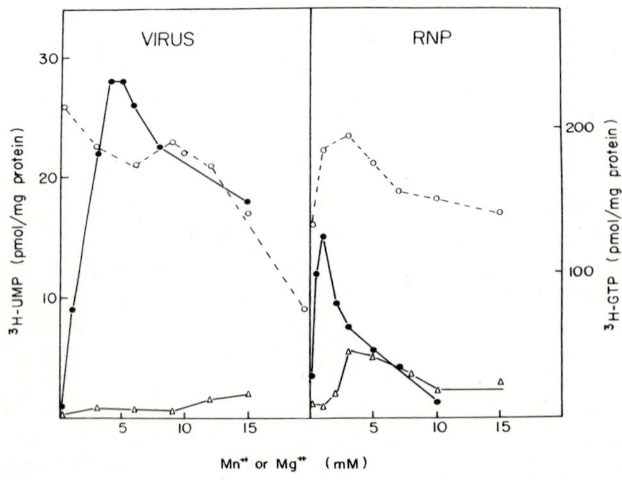

Fig.2.

Mg^{++} was added to Mn^{++}-containing reactions a further increase in activity of the RNP enzyme resulted, but the activity exhibited by unfractionated disrupted virus remained unchanged or was slightly inhibited.

Similar findings have been reported for one other negative-strand virus. As a result of RNP purification, the influenza virus transcriptase activity was catalyzed equally well by either Mn^{++} or Mg^{++} (13). Mg^{++} together with suitable dinucleoside monophosphate primers, however, was the only divalent cation capable of supporting faithful initiation of transcription, leading to complete transcription of the templates (10).

In the presence of Mn^{++} and Mg^{++}, the RNA produced by the Uukuniemi virus RNP-associated polymerase was of heterogenous size, ranging from around 1000 nucleotides to pieces smaller than 80 nucleotides (identical in mobility to the marker tRNA). In the presence of Mn^{++} alone, the average product size was definitely smaller. The amount of RNA produced in the presence of Mg^{++} alone was too small to allow detailed size analysis.

Because *in vitro* transcription of the Uukuniemi virus RNAs did not result in full-sized transcripts, it is difficult to evaluate the significance of the increased activity in the presence of the two divalent cations. The results are, however, in accordance with Mg^{++} being an important and necessary component of the transcriptase reactions.

In order to test the L, M and S RNPs separately for polymerase activity, the three size classes of RNPs, labeled with ^{32}P in their RNA, were separated on a sucrose gradient (Fig. 3). Each of the isolated RNP species was able to direct RNA synthesis *in vitro*, which implies that they all contained the polymerase and apparently were capable of independent primary tra-

---

Fig. 2. *Divalent cation dependence of the Uukuniemi virus RNA polymerase associated with unfractionated, disrupted virus or with the purified RNPs. The RNP fraction was isolated as follows: Uukuniemi virus, some of which was labeled with* ^{32}P*, was disrupted with Triton X-100, layered onto a discontinuous sucrose gradient (20% and 60% sucrose), and the RNPs centrifuged to the 60% sucrose cushion. The assays for polymerase activity contained in 50 µl: 50 mM Hepes pH 8.5, 50 mM NaCl, 10 mM 2-ME, 0.5 mM ATP, GTP, CTP and 85 pmol of* ^{3}H*-UTP or unlabeled UTP and 270 pmol of* ^{3}H*-GTP, and 25 µg of disrupted virus or the RNP derived from an identical amount of virus, which corresponded to 7µg of protein N. The amount of* ^{3}H*-UTP incorporated into acid precipitable form was determined after 1 hr incubation at 37° (12). Closed circles,* Mn^{++} *alone: open circles,* Mg^{++} *in the presence of 5 mM* Mn^{++} *(virus) or 1 mM* Mn^{++} *(RNP): open triangles,* Mg^{++} *alone.*

nscription. The transcriptional activities of the RNP species were calculated by knowing the specific activity of ^{3}H-UTP used in the *in vitro* reaction and the specific activities of the ^{32}P-labeled L, M and S RNAs, which were assumed to be identical (Table 1). A roughly equal number of UMP molecules were incorporated into RNA by each RNP species. Because the size distribution of the RNAs transcribed by the three different RNPs was

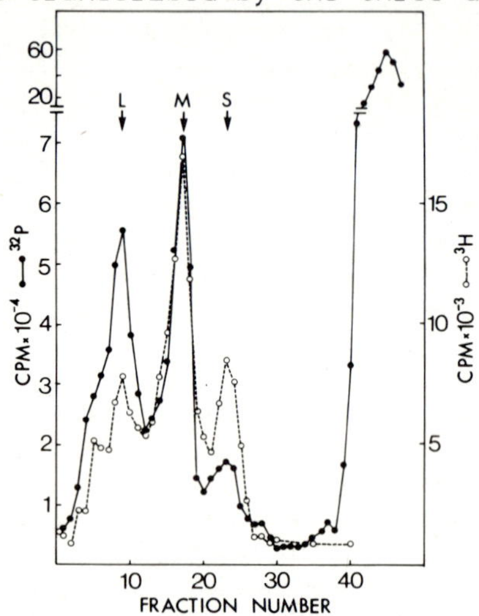

Fig. 3. *Separation of Uukuniemi virus RNPs. 1 mg of Uukuniemi virus, some of which was labeled with* ^{32}P, *was disrupted with Triton X-100 and the released RNPs separated on a linear 15-30% sucrose gradient in 10 mM Tris, pH 7.5, 100 mM NaCl, 0.1% Triton X-100. The gradient was layered over a 1 ml cushion of 60% sucrose, in order to collect any RNP aggregates. Centrifugation was in a SW41 rotor at 39,000 rpm for 2.5 hr at 4°. Fractions of 250 μl were collected and the* ^{32}P *radioactivity determined. A standard assay for polymerase activity (see Fig. 2) containing 1 mM* $MnCl_2$ *and 3 mM* $MgCl_2$ *was performed using 20 μl aliquots of fractions 1-40, and the* ^{3}H-*UTP incorporated into acid-precipitable form determined.*

very similar, it can be concluded that an equal number of product molecules was synthesized per template irrespective of the template length. This suggests also that the number of polymerase molecules associated with each of the three RNP species was very similar. The conclusion is, however, correct only if initiation of transcription *in vitro* occurs in an uninhibited manner and is not affected by factors that restrict transcription.

The viral-specific nature of the RNA produced *in vitro* by the three isolated RNP species and the extent of transcription of

TABLE 1

Transcriptional activity of the Uukuniemi virus RNPs

Sample	Template RNA (pmol)	^{3}H-UMP (pmol incorporated) into product RNA	per pmol template
L RNP	3.2	44	14
M RNP	7.6	86	11
S RNP	4.4	49	11

The 1 mg of virus used for the RNP preparation contained 30 μg of RNA, calculated according to the known chemical composition of the virus (Renkonen and Pettersson, unpublished results). The specific activities of the virus RNAs were assumed to be identical after a 48 hr labelling with ^{32}P during virus growth. The RNA molecular weights used in the calculation were 2×10^6 (L), 1×10^6 (M) and 0.5×10^6 (S) daltons. The values for ^{32}P-activity in the template RNAs and ^{3}H-UTP incorporated into product RNA by the RNPs have been obtained from the experiment described in Fig. 3.

the template RNA was further characterized. Product RNA, labeled with ^{3}H-UTP, and the template, prelabeled with ^{32}P, were isolated from *in vitro* transcription reactions directed by the L, M and S RNPs, respectively, and each of the isolated RNA mixtures was incubated under hybridizing conditions prior to analysis by sucrose gradients. The reactions directed by the M RNP (Fig. 4) and the S RNP (Fig. 5) species are shown. The results indicate that most of the template RNAs had remained intact during the *in vitro* incubation (3 hr, 37°), and thus the small sized product RNAs were not due to breakdown of the template RNA. In all cases the product RNA was hybridized to its template ($\geq$ 80% resistant to ribonuclease). The degree of double strandedness of the template RNAs varied: the L RNA was 2-4%, the M RNA 4-10%, and the S RNA 8-25% resistant to ribonuclease. This implied that a relatively larger proportion of the smallest template (S RNA) was transcribed than of the larger templates (L and M RNA). The sedimentation pattern of the S RNA after hybridization was more heterogeneous, reflecting the larger percentage of double-strandedness. The result is in accord with the size analysis on polyacrylamide gels which, as discussed previously, revealed that all the product RNAs were within a similar size range.

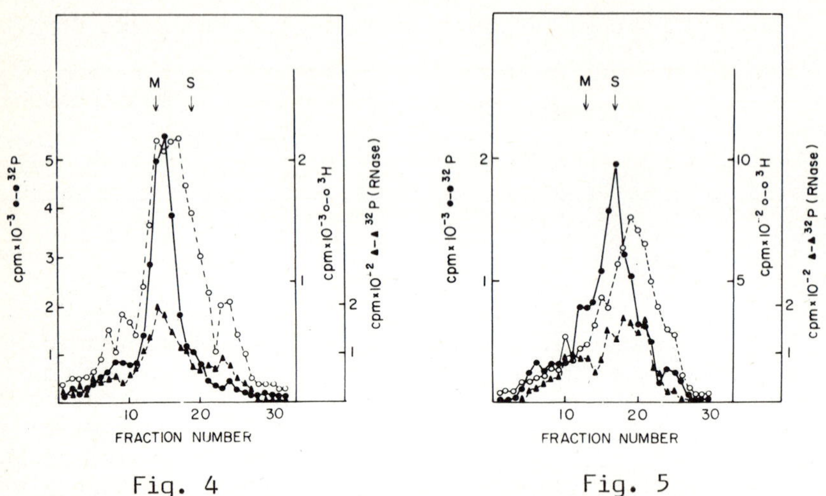

Fig. 4 Fig. 5

Fig. 4. *Gradient analysis of the template and product RNAs from the M RNP-directed in vitro transcription reaction. The M RNP, isolated in the experiment described in Fig. 3, was added to polymerase reactions containing 4 μM ^{3}H-UTP and 5 μM unlabelled UTP. All other conditions were as described under Fig. 2. After incubation for 3 hr at 37°, SDS was added and the reaction mixtures extracted with phenol. The RNA was purified over a cellulose column (4) and was precipitated with ethanol. Before analysis, the product RNA was hybridized to its particular template in 3 x SSC, 50% formamide, 0.5% SDS at 37° for 3 hr and then layered over a 15-30% sucrose gradient and centrifuged in the SW27 rotor for 17 hr at 25,000 rpm at 22°. 0.5 ml fractions were collected, 200 μl of which were left untreated and 200 μl were treated with pancreatic ribonuclease (10 μg/ml, 2 x SSC, 37°, 30 min). The acid precipitable radioactivities in each were determined.*

Fig. 5. *Gradient analysis of the template and product RNAs from the S RNP-directed in vitro transcription reaction. S RNP, isolated in the experiment described in Fig. 3, was used in an in vitro polymerase reaction. The RNAs were isolated after incubation, exposed to hybridization conditions and subsequently analyzed on a 15-30% sucrose gradient as in Fig. 4.*

That transcription of the Uukuniemi virus RNA species probably initiates at a single site can be concluded from the following observations: i) the number of complementary RNA molecules produced per template was constant and ii) the proportion of the template which became double stranded after hybridization to its product RNA did not exceed the maximum product RNA length.

The results presented here allow us to conclude that the three RNA species of Uukuniemi virus are composed of unique nucleotide sequences. Each of the RNA segments thus contains information for different polypeptides. The virion RNAs are transcribed into complementary RNAs which apparently function as mRNAs, by a polymerase present in each of the RNP species. How exactly the information of the three genome segments is transferred to at least four, previously identified, virus-specified proteins remains to be determined.

REFERENCES

1. Baltimore, D. (1971). *Bacteriol. Rev.* 35, 235.
2. Coffin, J.M. and Billeter, M.A. (1976). *J. Mol. Biol.* 100, 293.
3. DeWachter, R. and Fiers, W. (1972). *Anal. Biochem.* 49, 184.
4. Franklin, R.M. (1966). *Proc. Natl. Acad. Sci.* 55, 1504.
5. Hewlett, M.J., Pettersson, R.F. and Baltimore, D. (1977). *J. Virol.* 21, 1085.
6. Pettersson, R.F. and Kääriäinen, L. (1973). *Virology* 56, 608.
7. Pettersson, R.F., Hewlett, M.J., Baltimore, D. and Coffin, J.M. (1977). *Cell* 11, 51.
8. Pettersson, R.F., Kääriäinen, L., von Bonsdorff, C.-H. and Oker-Blom, N. (1971). *Virology* 46, 721.
9. Pettersson, R.F. and von Bonsdorff, C.-H. (1975). *J. Virol.* 15, 386.
10. Plotch, S.J. and Krug, R.M. (1977). *J. Virol.* 21, 24.
11. Porterfield, J.S., Casals, J., Chumakov, M.P., Gaidamovich, S.Y., Hannoun, C., Holmes, I.H., Horzinek, M.C., Mussgay, M., Oker-Blom, N. and Russell, P.K. (1976). *Intervirology* 6, 13.
12. Ranki, M. and Pettersson, R.F. (1975). *J. Virol.* 16, 1420.
13. Rochovansky, O.M. (1976). *Virology* 73, 327.
14. von Bonsdorff, C.-H. and Pettersson, R.F. (1975). *J. Virol.* 16, 1296.
15. von Bonsdorff, C.-H., Saikku, P. and Oker-Blom, N. (1969). *Virology* 39, 342.

TRANSCRIPTION, TRANSLATION, AND MAPPING OF THE GENES OF NEWCASTLE DISEASE VIRUS

L.A. BALL, P.L. COLLINS and L.E. HIGHTOWER

Microbiology Section,
Biological Sciences Group,
University of Connecticut,
Storrs, Connecticut 06268, USA.

Newcastle disease virus (NDV) is an avian paramyxovirus whose natural host includes the chicken (16). Its genome is a single negative strand of RNA of molecular weight about 5.4×10^6 (22), which contains the genes for the six known virus-specific proteins (18). Three of these proteins (HN, F, and M) are associated with the lipid envelope of the virus: HN and F are glycoproteins which constitute the surface spikes and M is a non-glycosylated matrix protein which probably underlies the viral membrane. The F protein is formed by proteolytic cleavage of a larger glycosylated polypeptide, F_0 (18). Together with the genome RNA, the other three proteins (L , NP, and 47K) constitute the nucleoprotein core of the virus particle. By analogy with rhabdoviruses (28), which share much of the molecular biology of paramyxoviruses, the L and 47K proteins may be components of the viral transcriptase (RNA nucleotidyltransferase, EC 2.7.7.6) whose template is the genome RNA complexed with NP. The aggregate molecular weight of the six known proteins accounts for more than 90% of the coding capacity of the virus.

In infected cells, the virus proteins are synthesized from mRNA species which are complementary to the genome (21, 4) and which appear to be monocistronic. Two major size classes of mRNA have been detected: a 35s class which contains the mRNA species for the L protein, and an 18s class which contains at least five mRNA species (6, 20, 33). In cell-free extracts of wheat-germ, these mRNAs direct the synthesis of polypeptides which resemble all of the authentic viral proteins by the criteria of their electrophoretic mobilities and/

or their tryptic peptide fingerprints (5, 27).

In vitro, low concentrations of non-ionic detergents activate the virion transcriptase, resulting in the synthesis of RNA which is complementary to the genome (19). This RNA resembles authentic messenger RNA in that it contains 3'-terminal poly(A) (32) and 5'-terminal methylated cap structures (8, 9). Here we describe its translation in cell-free extracts of primary chick cells, to yield polypeptides which correspond to at least four of the six viral proteins. Experiments designed to show whether NDV has a single or multiple sites for the initiation of transcription are also presented.

The discovery that another negative-strand RNA virus, vesicular stomatitis virus (VSV), has a single initiation site has far-reaching implications concerning the relationship between transcription and replication (2), so it was of interest to determine whether or not this is a general property of non-segmented negative strand viruses. The problem was approached by the technique of transcriptional mapping using UV irradiation (14, 15), which relies on UV-induced pyrimidine dimers in the transcription template to block the passage of RNA polymerase. The sensitivities to UV radiation of the expression of the different viral genes are proportional to the distances between the genes and their promoter site(s). In the case of multiple genes transcribed from a single initiation site, these data can provide an unambiguous transcriptional map, as we showed for VSV (1, 2). More recently, use of the same technique has shown that the genes of Sendai virus (12) and many of the late genes of adenovirus (13) also occupy a single transcriptional unit.

TRANSCRIPTION AND TRANSLATION *IN VITRO*

The transcriptase activity of purified NDV was about 0.3 μg RNA synthesized/mg virus protein/hr, a figure which agrees closely with the results of Huang *et al.* (19). However, addition of a cell-free extract of mouse L-cells or primary chick embryo cells to the reaction mixture stimulated transcription by 30 to 50-fold, and under these conditions RNA was synthesized at a rate of about 12 μg/mg viral protein/hr for 6-8 hours. A similar, though less dramatic, stimulation of VSV transcription by cell extracts has been observed (2), but we do not understand the mechanism of either stimulation at present. All of the *in vitro* RNA which was used in this work was made in the presence of a cell-free extract of mouse L-cells.

The products of NDV transcription *in vitro* were labeled by incorporation of [^{3}H]UTP and the poly(A)-containing species were subjected to electrophoresis on polyacrylamide-agarose slab gels in the presence of 6M urea (10). For comparison, the RNA species labeled during incubation of NDV-infected chick cells with [^{3}H]uridine in the presence of actinomycin D were electrophoresed in parallel. The results showed that the major RNA species in the *in vitro* product corresponded in mobility to each of the five major bands in the 18s size-class from infected cells (designated 1-5 in order of decreasing mobility, Fig.1). In addition, a small amount of 35s RNA (band 8) was usually synthesized *in vitro*. RNA bands 6 and 7 appear to be largely aggregates of the 18s RNA species (33), but they too were represented among the products made *in vitro*. Overall, the RNA made by detergent-activated NDV *in vitro* yielded a gel profile which was almost indistinguishable from that of the viral mRNA extracted from infected cells.

Attempts to couple NDV transcription with cell-free protein synthesis, as was achieved successfully with VSV (2), yielded at best a two-fold stimulation of amino acid incorporation. The spectrum of polypeptide products obtained was difficult to resolve from the endogenous products of the cell-free system. Transcription and translation were therefore executed separately. The poly(A)-containing fraction of the *in vitro* RNA product (which usually comprised about 60% of the total) was purified by affinity chromatography on a column of oligo-dT cellulose, and used to direct cell-free protein synthesis in extracts of primary chick embryo cells. By SDS-polyacrylamide gel electrophoresis, the protein products were compared with those resulting from cell-free translation of NDV mRNA extracted from infected cells, and with the NDV-specific proteins labeled *in vivo*. In all cases, chick embryo cells, which constitute a permissive host for NDV, were used so that the *in vitro* and *in vivo* systems were directly homologous.

Cell-free systems which were programmed by mRNA made *in vivo* or *in vitro*, synthesized polypeptides with mobilities corresponding to those of the viral proteins NP, F and M from infected cells (Fig.2). No products corresponding in size to the L, HN or F_0 proteins were detected, but polypeptides of apparent molecular weight 59,000 (P_{59}) and 67,000 (P_{67}) were directed by both mRNA preparations. In addition, 20-30% of the products of the cell-free systems migrated as a heterodisperse population of unresolved polypeptides with molecular weights below 55,000. Thus, unambiguous identification of a cell-free system product corresponding to the 47K protein from virus-infected cells was difficult.

The individual protein products of the cell-free system directed by NDV mRNA that had been transcribed *in vitro* were excised from polyacrylamide gels, digested with trypsin, and analysed by two-dimensional peptide mapping. The results

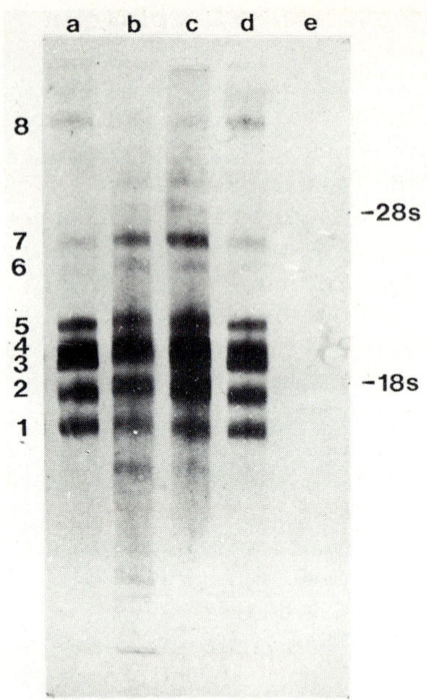

Fig.1. *Comparison of the NDV mRNA species synthesized in vivo and in vitro. (a) RNA extracted from infected cells; (b) poly(A)-containing RNA made in vitro; (c) total RNA made in vitro; (d) as (a); (e) RNA extracted from an equivalent number of uninfected cells. The NDV-specific bands are labeled 1-8 in order of decreasing mobility, and the positions of the 18s and 28s ribosomal RNAs are shown. In vitro RNA was synthesized in a reaction mixture which contained 30 mM Tris-HCl, pH 7.6; 33mM NH_4Cl; 7mM KCl; 50mM NaCl; 4mM Mg acetate; 1mM dithiothreitol; 0.2mM spermidine; 1mM ATP; 0.5 mM GTP and CTP; 0.1mM ^{3}H-UTP (0.98 Ci/mM); 10mM creatine phosphate; 80μg/ml NDV (strain AV, purified as described (17) and 30% (v/v) mouse L-cell extract, (11) to give a final concentration of about 1.5 mg of cellular protein per ml. Incubation was at 30° for 8 hr, after which the reaction was stopped by adding SDS to 1% and NaCl to 0.1M, and the RNA purified (5). Alternatively, the poly(A)-containing RNA was purified on a column of oligo-dT cellulose (3). In vivo RNA was labeled by incubating monolayers of uninfected or NDV-infected secondary chick embryo cells with ^{3}H-uridine (100μCi/ml; 40 Ci/mM) from 4 to 9 hr post-infection in the presence of 5μg actinomycin D per ml, added 3.25 hr post-infection. Cytoplasmic RNA was extracted from the cells (5). Aliquots of the RNA preparations were subjected to electrophoresis on a 2.5% polyacrylamide - 0.5% agarose slab gel (10, 26). Electrophoresis was at 120V for 10 hr at 4°, after which the gel was fixed and*

confirmed the identification of the cell-free system products which comigrated with NP , F and M although they showed that the latter species was contaminated with a fragment of NP (Fig.3). Moreover, they showed that P_{67} was related to HN, whereas P_{59} was related to NP. The 47,000 MW polypeptides made *in vivo* and *in vitro* also shared many tryptic peptides (not shown), but without a peptide map of the 47K protein from purified virions, their origin remains unclear.

On the basis of these results, the transcription and translation of the genes of NDV can be described as follows. The M protein gene was transcribed into an mRNA (probably band 1 based on the estimated molecular weights of the RNAs and proteins), which on translation yielded a polypeptide that comigrated with authentic M protein. The 47K protein gene was probably transcribed into band 2 mRNA but the evidence of its authentic translation is weak at present. The gene for NP was transcribed into an mRNA (probably band 3), which on translation yielded a polypeptide that comigrated with authentic NP. In addition, a fraction of the NP-related product migrated more slowly than mature NP from infected cells, and had an apparent molecular weight of 59,000. If this species is, indeed, a larger version of NP, then it would raise the possibility that NP is synthesized by precursor processing, as has been suggested before (30), and is discussed further by Hightower and Smith (see chapter 39). However, there are several alternative explanations for P_{59} which we cannot exclude at this stage.

The gene for F_0 was transcribed into an mRNA (probably band 4) which on translation yielded a polypeptide that comigrated with authentic F protein. Since cell-free systems are generally deficient in both precursor cleavage and glycosylation reactions (29), it seems likely that the cell-free system product was the non-glycosylated version of F_0. Its comigration with the authentic F protein was probably a coincidence which resulted from the carbohydrate portion of F and the extra peptide portion(s) of F_0 having similar molecular weights. The alternative explanation for their comigration is that the cell-free system executed both glycosylation and cleavage, but this seems improbable (cf.5). The gene for HN was transcribed into an mRNA (probably band 5) which on translation yielded P_{67}, a putative non-glycosylated precursor to HN. Finally, the gene for the L protein was probably transcribed into band 8 mRNA but in such small amounts that its translation was not detected Firm assignment of the proteins to individual mRNAs will have to wait for the results of translation of the separated species, and this approach may also shed light on the nature and origins of RNA bands 6 and 7.

stained. A fluorograph of the dried gel (24) is shown.

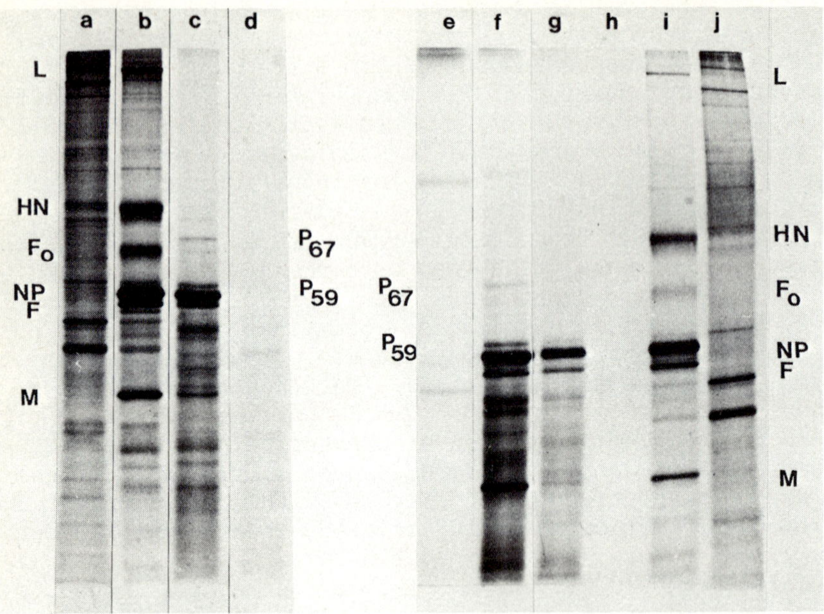

Fig.2. *Comparison of the NDV protein species synthesized in vitro and in vivo. Proteins labeled in: (a) uninfected chick embryo cells; (b) NDV-infected cells; (c) a cell-free extract of chick embryo cells directed by NDV mRNA transcribed in vitro; (d) a cell-free extract directed by RNA from a transcription reaction lacking NDV; (e) a cell-free extract directed by poly(A)-containing RNA extracted from uninfected chick embryo cells; (f) a cell-free extract directed by poly(A)-containing RNA extracted from NDV-infected chick embryo cells; (g) as (c); (h) a cell-free extract incubated in the absence of exogenous mRNA; (i) as (b); (j) as (a). Proteins made in vitro were labeled with [^{35}S]methionine during synthesis in reaction mixtures which contained 26 mM Hepes buffer, pH 7.6; 27 mM KCl; 3.0 mM Mg acetate; 0.2 mM spermidine; 1mM ATP; 0.5 mM GTP, CTP and UTP; 10 mM creatine phosphate; 80 µg creatine kinase per ml; 50 µM 19 unlabeled amino acids; 0.5 mCi/ml [^{35}S]methionine (Amersham-Searle; about 500 Ci/mM); 30% (v/v) cell-free extract of primary chick embryo cells (3) to give a final concentration of about 1.2 mg protein per ml; and the appropriate mRNA, prepared as described in the legend to Fig.1, at 0.5-3.5 µg/ml. Incubation was at 30° for two hours, after which the products were prepared for electrophoresis on polyacrylamide slab gels (23). Proteins labeled in vivo were prepared by incubating monolayers of uninfected or NDV-infected secondary chick embryo cells with [^{35}S]methionine (50µCi/ml)*

TRANSCRIPTIONAL MAPPING BY UV IRRADIATION

The technique of transcriptional mapping by UV irradiation depends on measuring the expression of individual genes from irradiated templates. Since undamaged genome RNA can undergo replication to produce progeny genomes capable of directing transcription and translation, it is essential that the experiment be performed under conditions where replication is prohibited. This was achieved in three ways: first, we used UV-irradiated NDV to direct transcription *in vitro*, and assayed the mRNAs and their translation products by polyacrylamide gel electrophoresis. Secondly, NDV-infected cells were irradiated in mid-infection, treated with cycloheximide to block further genome replication, and the products of viral transcription from the intracellularly-irradiated templates were labeled with [^{3}H]uridine and analysed by polyacrylamide gel electrophoresis. Thirdly, we used UV-irradiated NDV to infect cells in the presence of the amino acid analogue, p-fluorophenylalanine, in order to block viral replication by making non-functional proteins. [^{35}S]methionine was used to pulselabel the viral polypeptides, all of which are presumed to result from translation of the products of primary transcription.

Results from experiments of this type can be analysed in two ways. If each gene has its own promotor site, its UV sensitivity will be directly proportional to its size. If, on the other hand, all the genes are transcribed from a single promoter site, their UV sensitivities will depend on their distances from that site. So a simple qualitative comparison of gene sizes and sensitivities can suggest the transcriptional strategy and map. A more rigorous, quantitative, analysis of the data can be achieved if the UV-sensitivities of individual gene products can be related to that of the entire genome, as measured by the sensitivity of viral infectivity, for example. Under these conditions, the sensitivities of individual genes can be expressed in terms of RNA molecular weight, as UV target-sizes, and compared with the values that are predicted by different models of the transcriptional strategy and map of the virus.

Transcriptional mapping in vitro

Purified NDV was exposed to various doses of UV radiation

---

from 6.0 to 6.5 hr post-infection. Samples (a)-(d) were electrophoresed on an 11.5% polyacrylamide slab gel for 16 hr at 15 mA. Samples (e)-(j) were electrophoresed on a 11.5% gel for 18 hr at 15 mA. Autoradiographs of the fixed dried gels are shown.

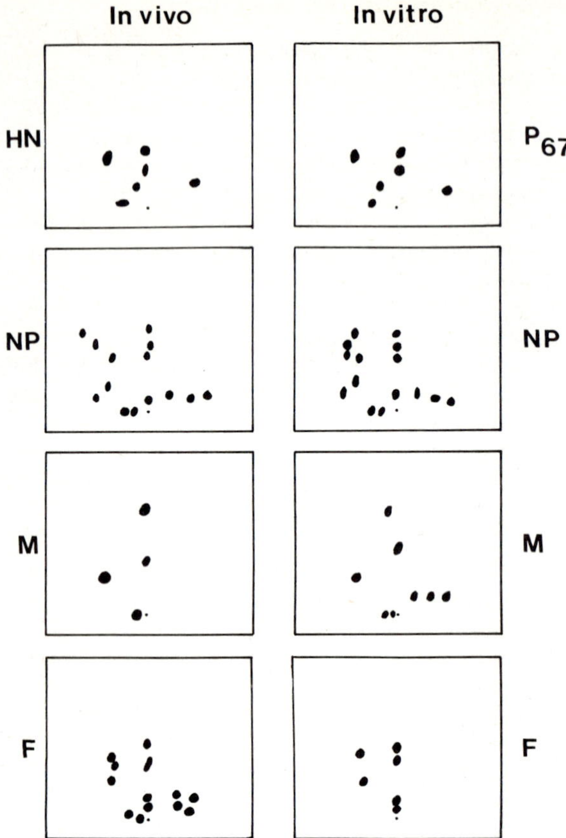

Fig.3. *Two-dimensional peptide maps of the proteins of NDV made in vivo and in vitro from mRNA made in vitro. NDV proteins, labeled with [^{35}S]methionine either in infected cells or in cell-free extracts of chick embryo cells directed by mRNA transcribed in vitro, were separated by polyacrylamide gel electrophoresis. The methods are described in the legends to Figs.1 and 2. The appropriate protein bands were excised from the dried gel using the autoradiograph as a guide, and digested with trypsin (18). The resulting peptides (about 10,000 cpm) were spotted onto 20 x 18 cm sheets of glass-fiber paper impregnated with silica gel (Gelman ITLC sheets, type SA) and resolved first by electrophoresis at pH 6.5 (horizontal dimension; buffer of pyridine:acetic acid:water (25:1:474); 300V for 3 hr), and then by ascending chromatography (vertical dimension; solvent of n-butanol:acetic acid:water (3:1:1)). The origin was in the middle of the bottom of each sheet; the cathode was to the right. Tracings of the autoradiographs of the thin-layer sheets are shown.*

as described previously for VSV (2), and used to direct transcription *in vitro* as detailed in the legend to Fig.1. The RNA products were subjected to chromatography on columns of oligo-dT cellulose and the purified poly(A)-containing fractions were analysed both by polyacrylamide gel electrophoresis and by translation in chick cell-free extracts. The relative UV-sensitivities of the RNA bands increased in the order: 3 < (4,2) < 1 < 5 < 8. The UV-sensitivities of the proteins for which they coded increased in the order: NP < F < M < P_{67} (not shown). The relationship between these two orders was consistent with the tentative assignments of proteins to mRNA bands discussed above. However, neither of these hierarchies corresponded to the order of increasing gene size, which was 1-8 in terms of their mRNAs, or M, 47K, (NP,F), HN, L in terms of their proteins, and this result suggested that the genes were not transcribed from separate promoter sites.

Transcriptional mapping in vivo by RNA analysis

The RNAs transcribed from viral genomes that had been irradiated in infected cells were labeled with [^3H]uridine and analysed by polyacrylamide gel electrophoresis. No aberrant RNA species appeared as a result of irradiation. Instead, the yields of the eight identifiable virus-specific RNA bands decreased progressively as a function of UV dose (Fig.4). If bands 6 and 7, which were probably RNA aggregates (33), were ignored, the relative UV sensitivities of the putative mRNAs increased in the order: 3< (4,2) < 1 < 5 < 8. This agreed with the order of sensitivities determined *in vitro*, and further suggested that the genes do not have six independent promoter sites.

On some occasions, the resolution achieved by the polyacrylamide gels was sufficient to reveal that band 2 was composed of two RNA species that showed different UV sensitivities (Fig.4, channels i, j and k). The component with lower mobility had a UV sensitivity similar to that of RNA band 3, whereas the component with higher mobility had a sensitivity similar to or slightly less than that of band 5. The identity of these components is unclear; one possibility is that the slower species is a fragment of RNA 3 (the putative NP mRNA) and the faster species is the mRNA for the 47K protein. Investigation of this must await translation of the separated RNA bands, but meanwhile it is clear that the apparent UV sensitivity of band 2 RNA should be interpreted with caution.

Quantitative analysis of the RNAs synthesized *in vivo* was impossible because no known target-size was available with which to compare the UV sensitivities. Loss of viral infectivity, measured after irradiation of purified virions, was

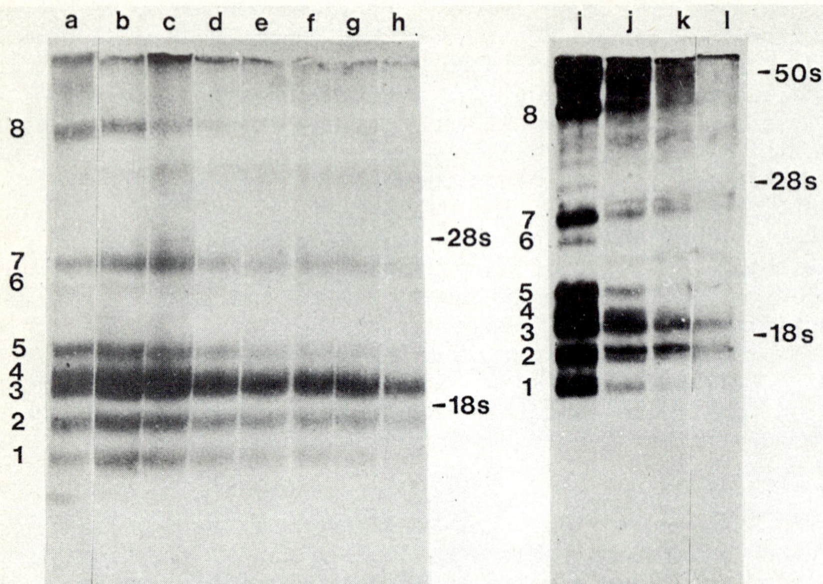

Fig.4. *Polyacrylamide gel electrophoresis of the viral RNA species labeled in NDV-infected cells that had been exposed to various doses of UV radiation. Samples (a)-(h) and (i)-(l) were from two separate experiments. The UV doses, in ergs mm*$^{-2}$*, were: (a) 0; (b) 32; (c) 65; (d) 97; (e) 130; (f) 162; (g) 195; (h) 227; (i) 0; (j) 130; (k) 260; (l) 390. Monolayers of secondary chick embryo cells were infected with NDV at a multiplicity of 10 pfu/cell, and 6 hr post-infection they were washed several times with PBS and exposed to the doses of UV radiation shown above (dose-rate = 6.5 ergs mm*$^{-2}$ *sec*$^{-1}$*). Following irradiation, medium containing actinomycin D (2μg/ml) and, in the case of samples (a)-(h), cycloheximide (50μg/ml) was added and the monolayers were incubated at 37° for 30 min. [*^{3}*H]uridine (100μCi/ml; 48 Ci/mM) was added and the incubation continued for a further 3 hr. The cytoplasmic RNA was extracted and subjected to electrophoresis on polyacrylamide slab gels at 115V for 11 hr as described in the legend to Fig.1. Fluorographs of the dried gels are shown.*

unacceptable as a standard, since the sensitivity of viral genomes changes dramatically after infection, as has been well documented (25). In experiments involving UV-irradiation of Sendai virus-infected cells, the target-size of 50s genome RNA was used, because no attempt was made to block replication (12). While this procedure provided an internal standard for quantitation, it left open the possibility of transcription from gēnome RNA synthesized after irradiation, which can seriously affect target-sizes obtained by this method. Although the results from Sendai virus do not appear to be distorted by replication, its effect on the apparent UV sensitivities of the genes of NDV can be seen from Fig.4, channels i-l, which show the results of performing the experiment in the absence of cycloheximide. Compare, for example, channels a and e (0 and 130 ergs mm^{-2} irradiation; RNAs labeled in the presence of cycloheximide), with channels i and j (0 and 130 ergs mm^{-2} irradiation; RNAs labeled in the absence of cycloheximide). We ascribe the much more dramatic effect of UV in the absence of cycloheximide to transcription from genome RNA synthesized after irradiation from templates which escaped UV damage. The presence of a prominent band of 50s genome RNA in channels i and j, which is absent from channels a and e (Fig.4), supports this interpretation.

Transcriptional mapping in vivo by protein analysis

The third method used for transcriptional mapping of NDV was to analyse the proteins directed by UV-irradiated virus during translation of primary transcripts *in vivo*. A similar approach with VSV showed that within the first hour of infection, replication and secondary transcription did not obscure the results (1). However, with NDV infection under the same conditions, viral protein synthesis was barely detectable. We therefore used the amino acid analogue, p-fluorophenylalanine (FPA) to block replication while permitting accumulation and translation of the products of primary transcription. Under these conditions, the virus proteins NP, F, and M were clearly labeled during a 30 min pulse given 5 hr post-infection, although, as expected for an infection restricted to primary transcription and translation, viral protein synthesis was much reduced compared with infection in the absence of FPA (Fig.5). We presume that this accounted for the low levels of the L, HN, and 47K proteins under these conditions.

In response to UV irradiation of the infecting virus, the synthesis of each of the viral proteins was reduced with characteristic single-hit kinetics, and in this case, quantitative comparisons with the target-size of viral infectivity, measured by plaque assay of the irradiated virus, were possible (Fig.6). The results showed that, in terms of RNA molecular weight, the UV target-sizes for expression of the NP, F and M protein genes were 0.8, 1.6 and 2.0 (all $\pm$ 0.3) x 10^6 daltons, respectively. Only for NP did the measured target-size approximate the estimated physical size of the gene (0.6 x 10^6 daltons). The target-sizes of the others exceeded their physical sizes by 3 to 5 times, results which were incompatible

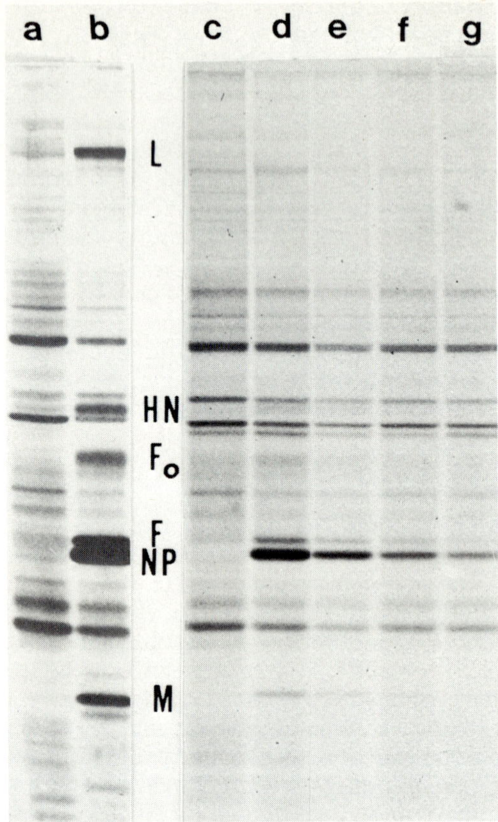

Fig.5. *Polyacrylamide gel electrophoresis of the proteins labeled in uninfected cells and in cells infected with NDV which had been exposed to various doses of UV radiation. Samples (a) and (b) were from cells maintained in the absence of D, L- p-fluorophenylalanine (FPA); samples (c)-(g) were from cells maintained in the presence of FPA, as described below. Proteins labeled in: (a) and (c), uninfected cells; (b) and (d), cells infected with unirradiated NDV; (e) - (g), cells infected with NDV that had been exposed to: (e) 195 ergs mm^{-2}; (f) 390 ergs mm^{-2}; (g) 780 ergs mm^{-2}. Mouse L-cells grown in suspension culture, were infected with irradiated or unirradiated NDV (m.o.i. = 100 pfu [unirradiated virus] per cell), in modified Eagle medium lacking (a and b) or containing (c-g) 0.75 mg FPA per ml. Five hr after infection, the cells were isotopically labeled with [^{35}S]methionine (15μCi/ml; 36 Ci/mM) for 30 min and the labeled proteins extracted and solubilized as described before (1). Aliquots were subjected to electrophoresis at 15 mA for 10 hr on a discontinuous 8.5% polyacryl-*

with the three genes being transcribed independently. On the other hand, the results were in agreement with the target-sizes which would result from the genes being transcribed from a single promoter site, in the order NP, F, M (which would give target-sizes of 0.6, 1.2 and 1.6 x 10^6 daltons, respectively). The small discrepancies between the predicted and measured target-sizes leave open the possibility that there is another (small) gene before or between NP, F and M, but we regard this as unlikely. Two alternative explanations for the discrepancies are: (1) a contribution to the target-sizes from the NDV leader sequence which has been discovered recently (A.K. Banerjee, personal communication), and which appears to be directly analogous to the VSV leader RNA (7) or (2) a non-random distribution of potential sites for UV hits.

The UV sensitivity of HN (measured *in vitro*) and of its putative mRNA (band 5) was greater than that of M and band 1 RNA, which suggests that HN is located distal to M in the transcriptional map. By a similar argument, the L protein gene, represented by band 8 mRNA, can be positioned distal to HN as the most sensitive gene detectable. Assuming that the transcriptional map is colinear with the physical map of the virus, the gene order of NDV can be represented: 3'-NP-F_0-M-HN-L-5'. The location of the gene for the 47K protein is unclear, but on the basis of three pieces of evidence, we would assign it to a region of the map distal to the M gene. First, it is a minor protein in infected cells (18), and for both VSV and Sendai virus which share the same transcriptional strategy, the amount of product from each gene reflects its position in the gene order (12, 31). Secondly, the UV-sensitive component of band 2 RNA which is the putative messenger for the 47K protein, mapped close to HN; and thirdly, the best candidate in Sendai virus for a protein analogous to 47K is the P protein, which maps adjacent to HN in the Sendai virus transcriptional map (12).

The transcriptional strategies and maps of three negative-strand viruses have now been determined. All have a single promoter site, presumably at the 3' end of the viral RNA, and maps which show extensive homology:

VSV: 3'-N-NS-M-G-L-5'
NDV: 3'-NP-F_0-M-(47K)-HN-L-5'
Sendai virus : 3'-NP-F_0-M-P-HN-L-5'

The possession of a single promoter site implies that tran-

---

amide slab gel in the presence of SDS (23). Under these conditions, the viral F protein migrated more slowly than NP; on 11.5% gels (Fig.2) their mobilities were reversed.

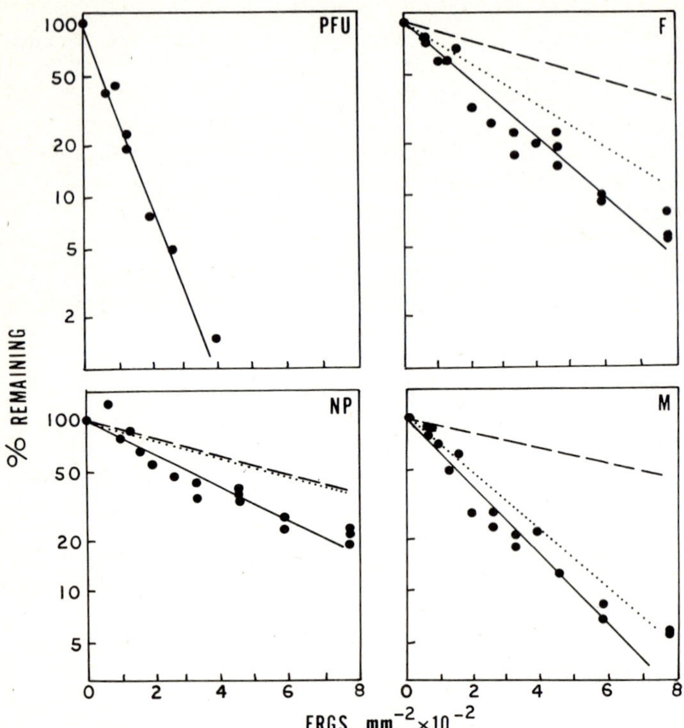

Fig.6. *Effect of UV irradiation of NDV on viral infectivity and the synthesis of proteins F, NP and M. Viral infectivity was measured by plaque assay of the irradiated virus on monolayers of secondary chick embryo cells. Viral proteins were labeled with [^{35}S]methionine for 30 min, 5 hr p.i., in cells infected with irradiated virus in the presence of 0.75 mg FPA per ml, as described in the legend to Fig.5. The proteins were subjected to electrophoresis and individual bands were quantitated by densitometry of the autoradiographs of polyacrylamide gels similar to those shown in Fig.5. In each case, the amount remaining is displayed on a semi-log scale as a percentage of the value for unirradiated NDV. In addition to the experimental points, two theoretical lines are shown for each protein. The dashed lines indicate the UV sensitivities which are predicted from the estimated molecular weights of the corresponding mRNAs assuming that each is transcribed from an independent initiation site. The dotted lines indicate the sensitivities which are predicted assuming that the genes are transcribed from a single initiation site, in the order NP, F, M.*

scription and replication may share a common process of positive strand synthesis. We postulate that mRNAs are generated from the transcribing complex by cleavage of the nascent strand, although other mechanisms can be imagined. In the absence of cleavage, the product would be an intact positive copy of the viral genome which is a necessary intermediate in replication. The homologies between the maps of these viruses may suggest that they were derived by evolution from a common ancestor. Alternatively, since the position of a gene in the genome of rhabdo- and paramyxoviruses is a major factor controlling the amount of the gene product which is made, the similar maps may reflect the similar replicative strategies of these classes of virus.

ACKNOWLEDGEMENTS

We thank Carol White and Mark Smith for invaluable help, and Drs Worth Clinkscales, Trudy Morrison and Michael Bratt for generous communication of their results before publication. This work was supported by a grant to Lawrence E. Hightower from the University of Connecticut Research Foundation and by Public Health Service grants HL 19490 and CA 14733. We benefited greatly from the use of a cell culture facility supported by the latter grant. Peter L. Collins was an N.S.F. graduate fellow.

REFERENCES

1. Ball, L.A. (1977). *J. Virol.* 21, 411.
2. Ball, L.A. and White C.N. (1976). *Proc. Nat. Acad. Sci. USA* 73, 442.
3. Ball, L.A. and White, C.N. (1978). *Virology*, 84 479.
4. Bratt, M.A. and Robinson, W.S. (1967). *J. Mol. Biol.* 23, 1.
5. Clinkscales, C.W., Bratt, M.A. and Morrison, T.G. (1977). *J. Virol.* 22, 97.
6. Collins, B.S. and Bratt, M.A. (1973). *Proc. Nat. Acad. Sci. USA* 70, 2544.
7. Colonno, R.J. and Banerjee, A.K. (1976). *Cell* 8, 197.
8. Colonno, R.J. and Stone, H.O. (1975). *Proc. Nat. Acad. Sci. USA* 72, 2611.
9. Colonno, R.J. and Stone, H.O. (1976). *J. Virol.* 19, 1080.
10. Floyd, R.N., Stone, M.P. and Joklik, W.K. (1974). *Anal. Biochem.* 59, 599.
11. Friedman, R.M., Metz, D.H., Esteban, R.M., Tovell, D.R., Ball, L.A. and Kerr, I.M. (1972). *J. Virol.* 10, 1184.
12. Glazier, K., Raghow, R. and Kingsbury, D.W. (1977). *J. Virol.* 21, 863.

13. Goldberg, S., Weber, J. and Darnell, J.E. Jr. (1977). *Cell* 10, 617.
14. Hackett, P.B. and Sauerbier, W. (1974). *Nature* 251, 639.
15. Hackett, P.B. and Sauerbier, W. (1975). *J. Mol. Biol.* 91, 235.
16. Hanson, R.P. (1964). *In* "Newcastle disease virus; an evolving pathogen". (Univ. of Wisconsin press, Madison and Milwaukee.
17. Hightower, L.E. and Bratt, M.A. (1974). *J. Virol.* 13, 788.
18. Hightower, L.E., Morrison, T.G. and Bratt, M.A. (1975). *J. Virol.* 16, 1599.
19. Huang, A.S., Baltimore, D. and Bratt, M.A. (1971). *J. Virol.* 7, 389.
20. Kaverin, N.V. and Varich, N.L. (1974). *J. Virol.* 13, 253.
21. Kingsbury, D.W. (1966). *J. Mol. Biol.* 18, 204.
22. Kolakovsky, D., Boy de la Tour, E. and Delius, H. (1974). *J. Virol.* 13, 261.
23. Laemmli, U.K. (1970). *Nature* 227, 680.
24. Laskey, R.A. and Mills, A.D. (1975). *Eur. J. Biochem.* 56, 335.
25. Luria, S.E. and Latarjet, R. (1947). *J. Bacteriol.* 53, 149.
26. McGeoch, D., Fellner, P. and Newton, C. (1976). *Proc. Nat. Acad. Sci. USA* 73, 3045.
27. Morrison, T.G., Weiss, S., Hightower, L.E., Spanier-Collins, B. and Bratt, M.A. (1975). *In* "*In vitro* transcription and translation of viral genomes". (A-L. Haenni and G. Beaud, eds.), p.281. INSERM, Paris.
28. Naito, S. and Ishihama, A. (1976). *J. Biol. Chem.* 251, 4307.
29. Roberts, B.E., Paterson, B.M. and Sperling, R. (1970). *Virol.* 59, 307.
30. Samson, A.C.R. and Fox, C.F. (1973). *J. Virol.* 12, 579.
31. Villareal, L., Briendel, M. and Holland, J.J. (1976). *Biochem.* 15, 1663.
32. Weiss, S.R. and Bratt, M.A. (1974). *J. Virol.* 13, 1220.
33. Weiss, S.R. and Bratt, M.A. (1976). *J. Virol.* 18, 316.

NEWCASTLE DISEASE VIRUS MESSENGER RNA'S

G.P. THOMAS*, R.D. BARRY*,
P. FELLNER** AND J. SMITH**

**Division of Virology, Department of Pathology, University of Cambridge, Cambridge, England*

***G.D. Searle & Co. Ltd., Research Division, P.O. Box 53, High Wycombe, Bucks., HP12 4HL.*

RNA from Newcastle Disease virus-infected cells, labelled in the presence of actinomycin D, was originally characterised on sucrose gradients, and three size classes were resolved that sedimented at 18-22S, 35S and 50S (4,5). It has since been shown that the 18-22S and 35S classes together contain sequences complementary to the entire genome (20). The greater resolution of polyacrylamide gel electrophoresis separated the 18-22S RNA into five major and two minor species that, although in variable amount, were also observed under denaturing conditions (7, 14, 24). RNA synthesized *in vitro* by the virion-associated polymerase was resolved into very similar patterns on gradients and on gels, including the two minor species (23). Estimates of the size of each RNA species agreed with those expected for messengers coding for the virus proteins. Direct evidence for messenger function of the 18-22S RNA of paramyxoviruses has been obtained from cell-free translation studies (6, 8, 15, 18).

The 18-22S RNA is complementary to 60% of the genome (4, 20), and the 35S RNA is complementary to the remaining 40%. The latter has been reported to direct the synthesis of the L protein (18). However, since the 18-22S RNA contains more than five species, and there are only six unique virus proteins (12), what function, if any, have the two minor species? To answer this question, we have examined the nature and coding capacity of the NDV-specific RNAs found in infected cells. We present data concerning the separation and identification of virus-specific mRNAs. This RNA is complementary to the genome RNA when annealed individually, is polyadenylated, and directs the synthesis of virus proteins when translated *in vitro* either as a total population or as individual species. NDV-infected cells contain at least six unique species, as shown by oligonucleotide mapping: these comprise the six mRNAs, and two other mRNA-like species, one of which contains some of the sequences of two of the unique RNAs. We suggest that these RNAs may be aberrant transcripts, whose existence would support the accumulating evidence for the sequential transcription of paramyxoviruses (10).

POLYACRYLAMIDE GEL ELECTROPHORESIS OF NDV-INFECTED CELL RNA

^{32}P-labelled NDV-specific RNA was obtained from chick cells infected with NDV (California). During virus adsorption and throughout infection, the cells were treated with 5 μg/ml actinomycin D and labelled from 0-6 hr p.i. with 10 mCi/ml ^{32}P orthophosphate. Following RNA extraction, separation was achieved on 2% acrylamide 0.6% agarose gels in the presence of 6M urea (11). The results are shown in Fig. 1. Complete separation of 8 RNAs was obtained on prolonged electrophoresis (Fig. 1B). The species are numbered in order of increasing chain length. Bands 1 to 7 correspond to those described previously

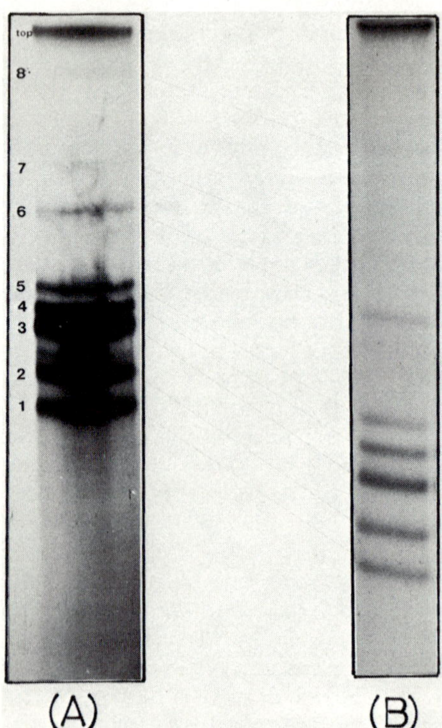

Fig.1.

(7, 24), and band 8 is probably the L protein in RNA.

The material at the origin is probably genome-sized RNA. The patterns shown are qualitatively identical to those described by others (7, 14, 24); identical results are obtained when RNA is labelled with either ^{32}P or ^{3}H-uridine.

We have estimated the chain lengths of species 1 to 5, relative to FPV RNA (17) coelectrophoresed under a wide variety of conditions. These are given in Table 1 (right-hand column);

TABLE I

The Proteins and Messenger RNAs of Newcastle Disease Virus

Protein	M.W. (daltons x 10^{-3})[a]	Minimum mRNA chain length (nucleotides)[b]	Estimated mRNA chain length (nucleotides)[c]
L	180 - 200	6000	-
HN_o	82	2700	-
HN	75	2500	2700
HN_o*	68	2270	
F_o	68	2270	
F	57	1900	2490
F_o*	56	1870	
NP	56	1870	2280
47k	47	1570	1640
M	41	1370	1410

(a) Values are based on migration in SDS-polyacrylamide gels, relative to appropriate markers, of the virion polypeptides and *in vitro* translation products obtained by others (12) and ourselves (Fig. 4 and unpublished).

(b) Calculated assuming an average molecular weight of 100 for an amino-acid, and that each is encoded by a triplet of nucleotides.

(c) Based on electrophoretic mobility relative to influenza genome RNA (17) determined on a large number of gels run under many different conditions (i.e. gel strength, presence and absence of agarose, concentration of urea, duration of electrophoresis).

Fig. 1. *Autoradiograms of ^{32}P-labelled NDV-infected cell RNA, separated on 2% acrylamide 0.6% agarose slab gels in the presence of 6M urea. Total cytoplasmic RNA was denatured in 90% DMSO for 30 min at 45°, brought to 6M urea and electrophoresed for (A) 18 hr at 180 volts, or (B) 24 hr at 180 volts.*

the probable coding capacity of each species is shown. Estimates were made on total NDV-specific RNA, and have not been corrected for poly(A) content, because it is our experience better resolution of the species is achieved when unfractionated RNA is electrophoresed than with either the poly(A)-containing or poly(A)-deficient populations alone (see 22). The chain lengths of species 6, 7 and 8 could not be readily determined as they lie outside the size range of FPV RNA, and calibration becomes nonlinear and unreliable. There is good agreement between both these estimates and those made on formamide gels (24), and with the sizes expected for messengers coding for the virus polypeptides.

INTEGRITY OF THE RNA SPECIES AFTER EXTRACTION FROM GELS

Individual species of RNA could be recovered intact from gels, although they were not necessarily pure. When re-run under the same conditions, only band 1 was uncontaminated by other species (Fig. 2). The other RNAs were obviously contaminated, and this was reflected in their oligonucleotide fingerprints. Contamination may have been due to incomplete separation (at least as far as species 4 and 5 are concerned) or to aggregation in the first gel, despite the presence of 6M urea and denaturation prior to electrophoresis. Only bands 1 to 5 were run in this gel, because insufficient radioactivity was recovered in bands 6, 7 and 8.

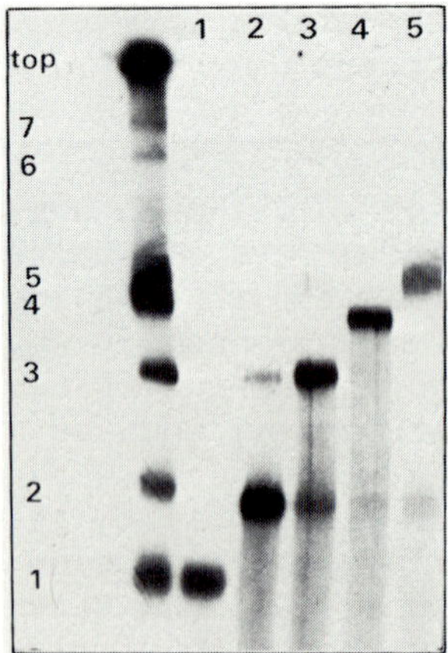

Fig. 2. *Bands 1 to 5, as numbered in Fig. 1(A), were extracted (13, 17). Aliquots were denatured by boiling for 30 secs in 0.01M Tris HCl (pH 7.5)- 90% formamide, loaded directly onto tracks 1 to 5, and electrophoresed in a 2.75% polyacrylamide gel containing 6M urea. Unfractionated infected cell RNA was treated as above and run in a parallel track.*

HOMOLOGY BETWEEN GENOME RNA AND VIRUS-SPECIFIC CELL RNA

Individual RNA bands were extracted from a preparative gel and annealed with an excess of virion RNA. It is clear from Table II that they were all virus-specific, and complementary to virion RNA. These results confirm those with unfractionated 18-22S and 35S RNAs, and suggest that they are the virus messenger RNAs(4)

The level of ribonuclease resistance following self-annealing for each RNA species was about that expected, if they contain a poly(A) tract of about 120 residues (23). Band 8 was a possible exception however. If this is the mRNA coding for the L protein, it would be about 6000 nucleotides, and should have had about 2% resistance after annealing. The observed value of

TABLE II

Homology Between NDV Genome RNA and RNA Species Extracted from Gels

RNA Species	% RNase resistance after self-annealing[1]	% RNase resistance after annealing[1] with 50S genome RNA[2]
1	6.4	90.1
2	4.7	95.4
3	4.9	95.0
4	3.7	85.1
5	4.8	116.7
6	4.7	86.5
7	2.8	102.7
8	6.2	94.2

(1) 10 µl aliquots of the individual ^{32}P-mRNA species (2-3 x 10^3 cpm) in 0.3M NaCl, 0.4M Tris HCl (pH 7.5), 0.002M EDTA and 50 µg/ml yeast tRNA, and a parallel series which contained, in addition, 0.5 µg 50S virion RNA, were boiled for 2 min in sealed capillaries, then transferred to 72° and annealed for 12 to 14 hr. The contents were diluted into 0.5 ml 2 x SSC. One half was immediately precipitated with 10% TCA after addition of carrier RNA, the other half was digested with ribonuclease A (50 µg/ml) and T1 (50 U/ml) for 30 min at 37°, precipitated with carrier, collected onto filters, washed and counted. Values are the average of duplicate determinations on two mRNA preparations.
(2) Prepared by sucrose gradient fractionation of RNA from partially purified egg-grown virus.

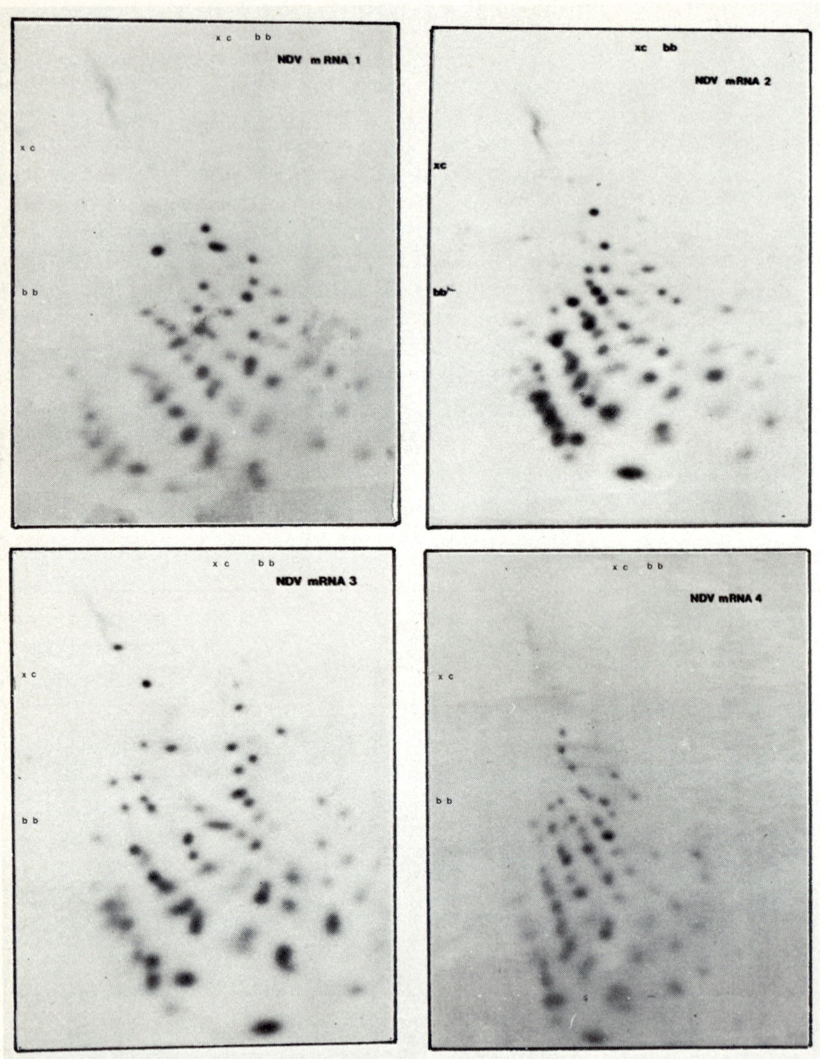

(A) RNA species 1 to 4

Fig. 3. *Two dimensional gel electrophoresis of ribonuclease T1 digests of NDV-specific RNAs, performed as according to deWachter and Fiers (9). The first dimension (10% polyacrylamide, 6M urea, pH 3.5) was from left to right, and the second dimension (20% polyacrylamide, pH 8.3) was from top to bottom. (XC and BB indicate the positions of the dye markers, xylene cyanol and bromophenol blue).*

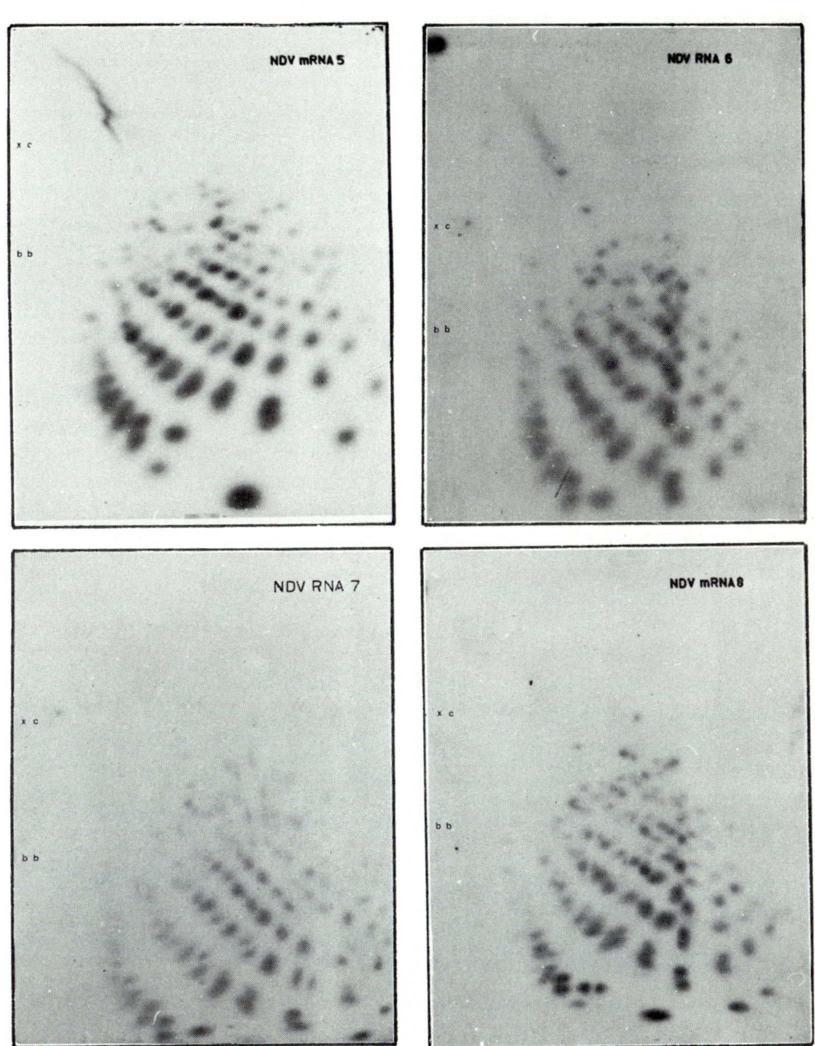

(B) RNA species 5 to 8

almost 7% may indicate that there were some intramolecular complementary sequences in band 8.

OLIGONUCLEOTIDE MAPPING OF THE VIRUS-SPECIFIC RNAs

We have prepared two-dimensional oligonucleotide maps (9) of the ribonuclease T1 digests of each of the virus-specific RNAs (Fig. 3). In this system, unique oligonucleotides migrate behind the bromophenol blue dye marker in the second dimension. It is apparent in Fig. 3 that there are two populations of oligonucleotides, those present in molar quantities and those in non-molar quantities. Since each RNA species is extracted intact (Fig. 2), the molar oligonucleotide spots observed in each case are characteristic for that particular RNA while the non-molar spots are the result of contamination or breakdown products. The fingerprints obtained for species 6, 7 and 8 are of low intensity because of the difficulty in obtaining sufficient radioactivity in these bands. Nevertheless, it appears that species 1, 2, 3, 4, 5 (and 8) probably contain unique base sequences. RNA 6 appears to contain some of the oligonucleotides derived from species 3 and 4, the possible significance of which will be discussed later.

Increased sequence complexity (as shown by increased numbers of molar characteristic oligonucleotides) accompanies increase in chain length, with the exception of species 4, which appeared less complex than species 3. This was observed for two different preparations of the mRNAs.

All eight species contain poly(A) tracts, and this is consistent with the finding that these RNAs bind to oligo-(dT)-cellulose and poly-U-sepharose (22, 23). The poly(A) tracts observed in species 6, 7 and 8 are not evident in these photographs.

The putative nucleocapsid (NP) protein mRNA, band 3, contains an oligonucleotide which migrated close to the end of the poly(A) tract. Composition analysis with pancreatic ribonuclease indicated that it contains predominantly adenylate residues. A possible explanation of the nature of this oligonucleotide is as follows. The major function of NP protein is to bind to RNA, a function that might be effected by the interaction of a short stretch of basic amino acids with the phosphate bonds of the RNA. The codons for lysine (AAA,AAG), (glutamine (CAA,CAG) and arginine (CAU,CAA,CAC and CAG), are relatively rich in adenylate and low in guanylate residues: if there were a sequence coding for three or so basic amino acids, it is possible that ribonuclease T1 digestion (G-specific) would give rise to an A-rich oligonucleotide, such as observed in RNA 3.

IN VITRO TRANSLATION PRODUCTS

Unfractionated RNA from NDV-infected cells directed the synthesis of polypeptides co-migrating with the M (41000 daltons), 47k (47000 daltons) and the NP (56000 daltons) proteins of purified virus and infected cells when added to the wheat germ *in vitro* protein synthesising system (Fig. 4). No synthesis of the L protein was observed, but two other polypeptides, HN_0^* (68000 daltons and F_0^* (56000 daltons) were apparent. NDV virions contain two glycoproteins, HN (74000 daltons) and F (56000 daltons), and in infected cells these arise from proteolytic cleavage of the precursor glycoproteins HN_0 (82000 daltons) and F_0 (57000 daltons, respectively (19). Because of the almost complete absence of both proteolytic and glycosylating activities in wheat germ extracts, we suggest that HN_0^* and F_0^* represent the uncleaved and unglycosylated precursors HN_0 and F_0, respectively. (The partial chymotryptic digests of HN_0^* and HN_0 are virtually indistinguishable: data not shown). These results agree with those recently obtained for 18-22S RNA from NDV-infected Chinese hamster ovary cells (6).

The translation of individual mRNA species is shown in tracks 1 to 4 of Fig. 4; only in track 1 was it possible to detect a translation product co-migrating with a virus protein. Band 1 mRNA directed the synthesis of a protein which co-migrated with the virus M protein, and this is perhaps the best evidence that these RNA species are really virus mRNAs (at least the smallest, band 1). Addition of other, separated mRNAs to the wheat germ system did not induce the synthesis of detectable virus proteins. A possible explanation for this failure could be that the relatively high levels of transfer RNA used as co-precipitant in the extraction procedure inhibited translation. We have found that high levels of non-translated RNA (e.g. tRNA and rRNA) are inhibitory in this system.

We are currently investigating both different cell-free systems and alternative gel extraction methods, by which we hope to identify the coding capacities of the remaining species.

DISCUSSION

NDV-infected cell RNA has been separated into eight virus-specific RNAs (Fig. 1). Oligonucleotide mapping indicates that species 1, 2, 3, 4, 5 (and 8) have unique base sequences, and we assume that they code for the M, F, 47k, NP, HN and L proteins, respectively, on the basis of size (Table I).

Recently it was shown that the mRNAs of Sendai virus are transcribed sequentially from a 3' promoter, as previously found for VSV (1, 2, 3, 10). The gene order determined by measurement

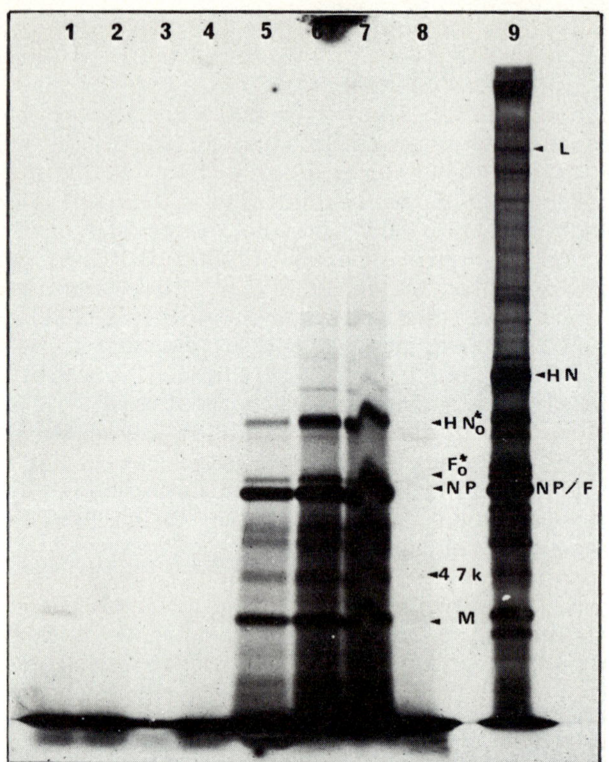

Fig. 4. *Autoradiogram of an SDS-polyacrylamide gel of* ^{35}*S-methionine labelled in vitro translation products. Poly(A)-containing infected cell RNA (purified on oligo-(dT)-cellulose) was mixed with* ^{32}*P-labelled marker RNA, and electrophoresed in a preparative gel, as described in Fig. 1. After extraction from the gel, the RNAs were dissolved in small volumes of water and incubated in the wheat germ extract as described (13). Tracks 1 to 4 show the translation products directed by bands 1, 2, 3+4, and 5, respectively (bands 3 and 4 did not separate sufficiently well to allow separate extraction, and were taken together). Tracks 5, 6 and 7 show the products directed by unfractionated infected cell RNA, added at 80, 160, and 240 μg/ml. Track 8 shows the polypeptides of NDV-infected CEF labelled for 15 min with* ^{35}*S-methionine at 6 hr p.i. Unlabelled purified virus was run in the track between tracks 8 and 9, and the virus proteins were located by staining the gel with Coomassie brilliant blue. All the samples were run in the same 8.5% SDS-polyacrylamide gel (16). The gel was exposed for 3 weeks in order to detect any products in tracks 1 to 4.*

of the ultraviolet inactivation target sizes of the mRNAs is 3'. NP.F.M.P.HN.L.5'. If NDV is also transcribed in this fashion, the equivalent gene order, obtained by substituting the NDV proteins into the positions of their Sendai virus counterparts, would be 3'. NP.F.M.47k,HN.L.5'.

RNA species 6 and 7 do not correspond to any virus specific polypeptide found in infected cells, but were complementary to the genome RNA (Table II). RNA 6 also appeared to contain some sequences of mRNAs 3 and 4 (Fig. 3). This may have been due to aggregation, however, under the conditions of electrophoresis (denaturation of RNA in 90% DMSO, which denatures RNA almost completely (21), followed by addition of urea to 6M and immediate electrophoresis in 6 M urea) aggregation should be minimised and the bands observed on the gels should represent single-stranded continuous RNA molecules. We have not yet been able to demonstrate that bands 6 and 7 are continuous single strands, but if it does represent aggregation it neither explains why RNA 6 contained only some of the oligonucleotides of mRNAs 3 and 4, nor does it tell us why there should be specific aggregation between mRNAs 3 and 4. We are now engaged in determining the nature of RNAs 6 and 7 to discover why they represent specific aggregates or are *bona fide* transcripts. Should it turn out that RNA 6 is a continuous single strand containing sequences of part of both mRNAs 3 and 4, its existence could support the putative gene order, which places gene 3 next to gene 4. It may be that aberrant initiation of transcription within gene 3 might continue into gene 4, perhaps out of phase and thereby avoiding the normal processing mechanisms, and at some special feature of structure or sequence in gene 4, terminate and polyadenylate to give rise to RNA 6. Similar aberrant transcription could generate RNA 7. The fact that these species also occur in *in vitro* transcripts might suggest that special features of either or both genome RNA and transcriptase could be capable of generating these (presumably untranslated) RNAs.

The detailed mechanism of mRNA synthesis of both the rhabdoviruses and the paramyxoviruses is not known; both the processing of a full-length positive strand or a 'stop-start' mechanism have not been ruled out. If these RNAs are aberrant transcripts and are generated as speculated above, this could prove to be a useful system in which to investigate the processes by which normal mRNAs are produced.

REFERENCES

1. Abraham, G. and Banerjee, A.K. (1976). *Proc. Nat. Acad. Sci. U.S.A.* 73, 1504.
2. Ball, L.A. and White, C.A. (1976). *Proc. Nat. Acad. Sci. U.S.A.* 73, 442.

3. Ball, L.A. (1977). *J. Virol.* 21, 411.
4. Bratt, M.A. and Robinson, W.S. (1967). *J. Mol. Biol.* 23, 1.
5. Blair, C.D. and Robinson, W.S. (1968). *Virology* 35, 537.
6. Clinkscales, C.W., Bratt, M.A. and Morrison, T.G. (1977). *J. Virol.* 22, 97.
7. Collins, B.S. and Bratt, M.A. (1973). *Proc. Nat. Acad. Sci. U.S.A.* 70, 2544.
8. Davies, J.W., Portner, A. and Kingsbury, D.W. (1976). *J. Gen. Virol.* 33, 117.
9. deWachter, R. and Fiers, W. (1972). *Anal. Biochem.* 49, 194.
10. Glazier, K., Raghow, R. and Kingsbury, D.W. (1977). *J. Virol.* 21, 863.
11. Floyd, R.W., Stone, M.P. and Joklik, W.K. (1974). *Anal. Biochem.* 59, 599.
12. Hightower, L.A., Morrison, T.G. and Bratt, M.A. (1976). *J. Virol.* 16, 1599.
13. Inglis, S.C., McGeoch, D.J. and Mahy, B.W.J. (1977). *Virology* 78, 522.
14. Kaverin, N.V. and Varich, N.L. (1974). *J. Virol.* 13, 253.
15. Kingsbury, D.W. (1973). *J. Virol.* 12, 1020.
16. Laemmli, U.K. (1970). *Nature (London)* 227, 680.
17. McGeoch, D.J., Fellner, P. and Newton, C. (1976). *Proc. Nat. Acad. Sci. U.S.A.* 73, 3045.
18. Morrison, T.G., Weiss, S.R., Hightower, L.A., Collins, B.S. and Bratt, M.A. (1975). *I.N.S.E.R.M.* 47, 281.
19. Nagai, Y., Klenk, H.-D. and Rott, R. (1976). *Virology* 72, 494.
20. Roux, L. and Kolakofsky, D. (1975). *J. Virol.* 16, 1426.
21. Strauss, J.H. Jr., Kelly, R.B. and Sinsheimer, R.L. (1968). *Biopolymers* 6, 793.
22. Varich, N.L., Lukashevich, I.S. and Kaverin, N.V. (1976). *J. Virol.* 18, 111.
23. Weiss, S.R. and Bratt, M.A. (1974). *J. Virol.* 13, 1220.
24. Weiss, S.R. and Bratt, M.A. (1976). *J. Virol.* 18, 316.

EFFECTS OF CANAVANINE ON PROTEIN METABOLISM IN NEWCASTLE DISEASE VIRUS-INFECTED AND UNINFECTED CHICKEN EMBRYO CELLS

LAWRENCE E. HIGHTOWER
and MARK D. SMITH

*Microbiology Section, Biological Sciences Group,
University of Connecticut, Storrs, Connecticut, 06286,
U.S.A.*

In a previous study, we determined relationships among the polypeptides of NDV by using both tryptic peptide and kinetic analyses (7). Our results suggested that there are at least six unique viral polypeptides. These include two glycosylated membrane proteins (HN, F_0), one nonglycosylated membrane protein (M), a nucleocapsid protein (NP), and at least two minor polypeptides having molecular weights of 220,000 daltons (L) and 47,000 daltons. Several fragments related to the nucleocapsid protein were found in both purified virions and infected cells. In addition, the structural glycoprotein F_1 proved to be derived from F_0 by proteolytic cleavage. Recently, the small product F_2 resulting from the cleavage of F_0 has been identified and shown to be linked by disulfide bonds to F_1 (17). In the present investigation, viral proteins were synthesized in the presence of the arginine analogue canavanine in an effort to inhibit proteolytic processing and facilitate the detection of short-lived precursors which might have been overlooked in our earlier studies. We will also describe several alterations in cellular protein metabolism effected by canavanine.

VIRAL PROTEIN SYNTHESIS IN THE PRESENCE OF CANAVANINE

Virus-infected cells were exposed briefly to culture medium containing ^{35}S-methionine and either arginine or canavanine during the steady-state of viral infection (6) as described in the legend of Fig. 1. Gel electropherograms of the proteins labeled in the presence of arginine (Fig. 1A, slot 1) show the usual profile of viral polypeptides dominating a much reduced background of host-cell proteins. The gel pattern of polypeptides from

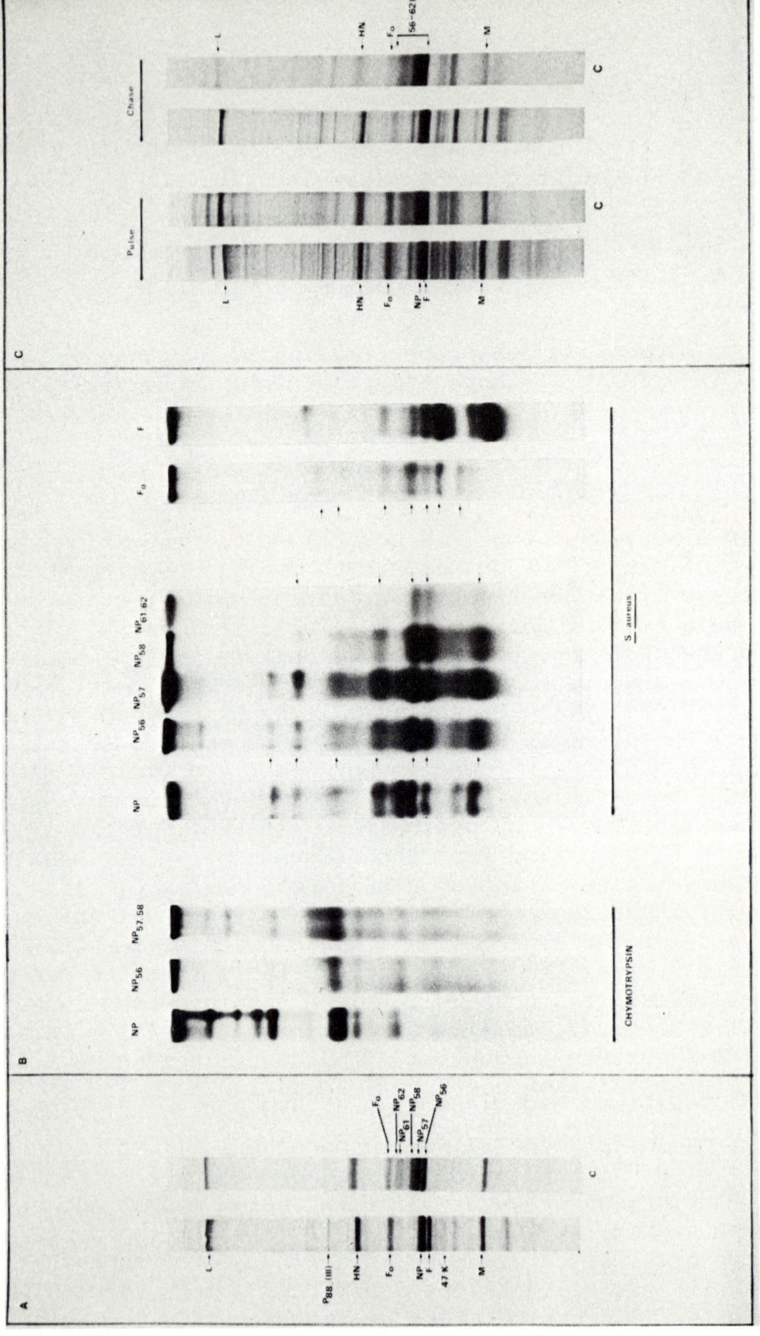

Fig. 1

Fig. 1. *Effects of canavanine on viral protein synthesis*
A. Electropherograms of viral proteins synthesized in either the presence or absence of canavanine. Confluent secondary cultures of chicken embryo cells were infected with strain AV (Australia-Victoria, 1932) of Newcastle disease virus (NDV) (moi=10) and incubated at 40°, 5% CO_2 *for 9 hr. Cultures were then washed with either MEM (Earles) supplemented with 2% dialyzed calf serum and containing either 0.6 millimolar L-arginine (channel 1) or L-canavanine (channel 2). After washing, the cultures were exposed to either normal or analog-substituted medium containing 5% normal concentrations of methionine and 15 μCi/ml* ^{35}S*-methionine for 30 min. The cultures were then solubilized in gel sample buffer (7) and analyzed by SDS-polyacrylamide slab gel electrophoresis on 10% gels (9). The gels were fixed, stained, dried, and exposed to Kodak XR-5 x-ray film for 24-48 hr. Gel slots corresponding to cultures labeled in the presence of canavanine are designated by a "c".*
B. Peptide maps of ^{35}S*-methionyl proteins. Analysis of peptides generated by limited proteolysis in the presence of sodium dodecyl sulfate on 15% polyacrylamide gels was carried out by the method of Cleveland et al. (4), with the modification that appropriate bands were excised from stained, dried gels and rehydrated in Laemmli gel sample buffer prior to exposure to protease. 50 μl aliquots of either chymotrypsin or Staphylococcus aureus protease solution at concentrations of 50 μg/ml were applied to each well of the gel containing the rehydrated gel slice. Electrophoresis was carried out at 50V, constant voltage for approximately 15 hrs. The gels were then stained and processed for fluorography (3). The* ^{35}S*-methionyl peptides were detected by exposing the processed gel to Kodak XR-5 film for about a week at -70°. Control experiments showed that canavanine substitution does not significantly alter the spectrum of peptides generated by chymotrypsin and S. aureus protease. Polypeptides NP,* F_0*, and F were synthesized in normal medium while the remaining radioactive proteins were synthesized in the presence of canavanine. The subscripts designate the molecular weight of the original band excised from gels. Arrows mark the positions of faint peptide bands on the fluorograms of* NP_{61-62} *and* F_0*.*
C. Electropherograms of pulse-chase experiment. Chick embryo cells were infected, labeled with ^{35}S*-methionine for thirty mimutes in the presence of either arginine or canavanine (C) at 6 hr post-infection, and analyzed by polyacrylamide gel electrophoresis as described in part A. The channels marked pulse show the protein patterns of cultures solubilized in gel sample buffer immediately following the labeling period, while the channels marked chase show the polypeptides obtained from cultures which were incubated for an additional 4 hr in medium containing arginine and a ten-fold excess of methionine.*

infected cultures labeled in the presence of canavanine revealed differential effects on viral protein accumulation (Fig. 1A, slot 2). The viral polypeptides L, HN, and M were least affected by analog-substitution. Although a precursor glycoprotein HN_0 having an apparent molecular weight of 82,000 was recently discovered in cells infected by avirulent strains of NDV (14), no glycopolypeptide with the expected characteristics of a precursor to HN accumulated in canavanine treated cultures infected by the virulent strain AV. All of the viral proteins except M exhibited slightly slower electrophoretic mobilities and small increases in band width on polyacrylamide gels. Similar alterations occurred in the behavior on gels of cellular proteins such as actin, tubulin, and LETS protein after canavanine-substitution (Fig. 3, slots 1,2). These changes in polypeptide patterns in both infected and uninfected cells could be quickly reversed by returning the culture to arginine-containing medium prior to the addition of radioactive label. They probably reflect slight changes in protein conformation or SDS binding-properties of the canavanyl (canavanine-substituted) polypeptides.

The most dramatic alterations involved the processing of F_0 and the accumulation of the nucleocapsid protein. No detectable polypeptide corresponding in electrophoretic mobility to authentic F accumulated in canavanine-treated cultures. Furthermore, peptide mapping of the major size classes of viral polypeptides synthesized in the presence of canavanine did not reveal an aberrant or intermediate form of F. Therefore, it is likely that the proteolytic processing of F_0 was blocked by the substitution of canavanine for arginine in the polypeptide chain. Cleavage of F_0 can be accomplished *in vitro* by trypsin (13) and the above data is consistent with the hypothesis that cleavage in cell culture is carried out by proteases with trypsin-like specificities.

In addition to F_0, the accumulation of NP was radically altered in canavanine-treated cultures. Instead of the single band at 56,000 daltons which is normally observed (Fig. 1A, slot 1), a triplet consisting of two major species having molecular weights of 56,000 daltons and 57,000 daltons and a partially resolved minor band at about 58,000 daltons accumulated (Fig. 1A, slot 2). Additional polypeptides covering a rather broad size range beginning at about 56,000 daltons and extending up to a faint doublet at 61-62,000 daltons were consistently detected as well. Varying the duration of canavanine treatment from five minutes to 60 minutes prior to addition of radioactive methionine did not alter the pattern of canavanyl polypeptides. Similar patterns were observed after radioisotopic labeling periods varying from two minutes to thirty minutes.

In order to establish the origin of the canavanyl polypeptides, we have compared their peptide maps with those of the known viral proteins by the polyacrylamide gel mapping procedure of Cleveland *et al.* (4). The peptide maps generated by either *S. aureus* protease or chymotrypsin of the 56K and 57K canavanyl proteins were very similar to the map of authentic NP (Fig. 1B). Furthermore, the maps generated by *S. aureus* protease of the 58K canavanyl protein and the 61-62K region were also similar to the map of NP. Peptide fingerprints of F_0 and F are included for comparison since both aberrant cleavage products and intermediates in the glycosylation of F_0 could accumulate in this region. However, gel analysis has shown that the 56-62K region is not contiguous with the 66K F_0 band but rather begins at the doublet centered at 61,000 daltons. Although we cannot completely rule out the presence of F-related material in the 56-62K regions, our mapping data suggests that NP-related polypeptides with molecular weights as high as 61-62K and extending down into the 56,000 dalton region accumulate in canavanine-treated cultures.

The behavior of the canavanyl polypeptides during an amino acid chase following a brief labeling period also supports our hypothesis that the majority of radioactive material in the 56-62K region is distinct from F_0 (Fig. 1C). The membrane-associated proteins HN, F_0, and M disappear relatively quickly during a chase while the amount of radioactive polypeptides in the 56-62K region declines only slightly. This observation is consistent with studies (see ref. 5) which suggest that abnormal hydrophobic proteins (which are likely to be membrane-associated) may be degraded faster than abnormal cytoplasmic proteins in animal cells. The recent finding (Chapter 37) that an NP-related polypeptide having a higher apparent molecular weight than NP extracted from virions or infected cells is synthesized in cell-free extracts directed by NDV mRNA is also consistent with the existence of a larger primary gene product. Samson and Fox (16) have also suggested the possibility of a precursor to NP in order to explain the kinetics of their radioisotopic pulse-chase studies.

The nucleocapsid protein of NDV, having a molecular weight of 56,000 daltons, is significantly smaller than those of the other well-characterized paramyxoviruses, Sendai virus (60,000), SV5 (61,000) (11) and of four strains of measles virus (60-62,000) (10). Our studies and those summarized above raise the possibility that the primary gene product for the nucleocapsid protein of NDV may also be as large as 61-62,000 daltons and may be rapidly processed to a 56,000 dalton species. The susceptibility of the nucleocapsid proteins of paramyxoviruses to proteolytic cleavage and attendant changes in nucleocapsid structure are well-documented (12). Such processing may have important implications for the maturation of virions, the cytopathology of infection, and may also play a role in determining templates for viral replication and transcription. Of course, other explanations exist for the appearance of the canavanyl polypeptides related to NP such as phosphorylation or alterations in initiation or termination of translation. The firm establishment of product-precursor relationships among the proteins in the 56-62K size range will require their separation on two-dimensional gels followed by more detailed peptide mapping and biochemical analyses.

CELLULAR PROTEIN SYNTHESIS IN THE PRESENCE OF CANAVANINE

During our studies of the effects of canavanine on viral protein metabolism, we observed striking alterations in cellular protein accumulation. If chicken embryo cell cultures were first exposed to medium in which canavanine was completely substituted for arginine, and then periodically pulsed with ^{35}S-methionine, cellular protein accumulation responded in three distinct ways (Fig. 2). The rate of accumulation of some cellular proteins such as actin was unperturbed, while the accumulation of other proteins such as tubulin and the LETS protein were markedly reduced. In contrast, the rate of accumulation of a relatively

---

Fig. 2. *Time course of enhancement and reversal. Confluent secondary cultures of chicken embryo cells were washed and incubated in either canavanine-substituted or normal MEM (Earles) supplemented with 2% dialyzed calf serum at 40°, 5% CO_2 (time zero). At 15 min intervals, control and experimental cultures were washed, labeled for 15 min with 15 μCi/ml ^{35}S-methionine in MEM having 5% normal concentration of methionine and supplemented with 2% dialyzed calf serum. Solubilized cultures were analyzed by polyacrylamide gel electrophoresis as described in Fig. 1A. At 3 hr after addition of analog, the remaining cultures were washed to remove canavanine and incubation was continued in normal medium (reversal). Incorporation of ^{35}S-methionine into acid-precipitable material was inhibited by 30-40% in analog-treated cultures after 3 hr. Incorporation immediately returned to control levels following reversal.*
Symbols: LETS, 230,000 dalton large external transformation-sensitive protein; Tb, 55,000 dalton tubulin; Ac, 43,000 dalton actin. $P_{71,72}$ is so designated because this band can sometimes be resolved into a doublet. The enhanced proteins labeled in the right-hand margin were sized relative to Bovine serum albumin, trypsin inhibitor, subunits of E. coli RNA polymerase (all from Boehringer Combithek Kit) and phosphorylase a.

LETS
Tb
Ac
P110
P88, (III)
P71, 72
P23
CAN
REV
0
0.5
1.0
1.5
2.0
2.5
3.0
3.75
4.25
4.75
5.25
5.75
6.25
C
TIME IN HOURS

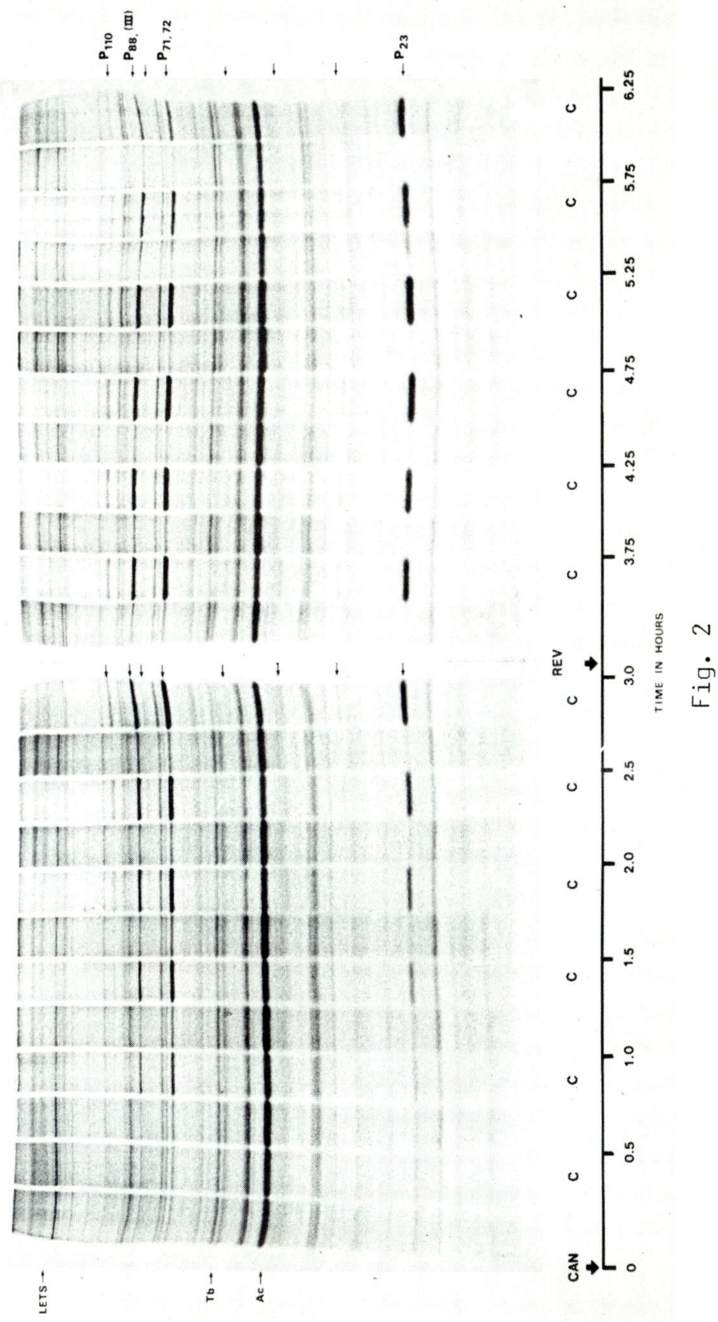

Fig. 2

small number of cellular proteins was dramatically increased. At least eight size classes of polypeptides which exhibit enhanced levels of accumulation have been identified (marked by arrows in the right margin of Fig. 2). Three of these showed particularly dramatic increases: P_{88} underwent a 6-fold increase in relative rate of accumulation when compared with polypeptides of similar mobility extracted from normal cultures, while $P_{71,72}$ (a double band) and P_{23} underwent 5-fold and 26-fold increases respectively. Thus in the course of about 3 hr these polypeptides achieved rates of accumulation comparable to that of cellular actin, the most abundant cellular protein. These responses of cellular protein metabolism to canavanine treatment were not simply due to arginine starvation since depriving culture cells of arginine for 3 hr was not sufficient to effect these changes. Furthermore, addition of normal levels of arginine to canavanine-substituted medium at the time of its addition to cell cultures completely antagonized the effects of the analogue on cellular protein metabolism. Therefore, canavanine, probably acting as an amino acid analogue and not an unrelated contaminant in the analogue preparation was responsible for these alterations. Kelley and Schlesinger (8) have observed similar effects of canavanine on protein metabolism in both cultured chick embryo cells and mammalian cell cultures.

Although we have not assigned functions to any of the enhanced proteins, several additional experiments have contributed information about their properties. For P_{88}, $P_{71,72}$, and P_{23} we have shown by peptide mapping that the canavanine-enhanced proteins are identical to proteins of similar electrophoretic mobility isolated from cultures maintained in normal medium. Therefore, canavanine treatment results in a modulation in the rate of accumulation of these proteins rather than the induction of new proteins. Peluso *et al.* (15) recently reported that a series of chicken embryo cellular proteins (designated I-IV) exhibit enhanced rates of accumulation following infection by either SV5 or Sendai virus. We have compared electrophoretic mobilities of the canavanine-enhanced proteins with Sendai-enhanced proteins provided to us by Dr. R.A. Lamb, Rockefeller University. The only proteins with identical mobilities were the canavanyl protein P_{88} and Sendai enhanced protein III, and preliminary peptide maps suggest that these proteins are the same. The P_{88}(III) protein is also resistant to inhibition of cellular protein synthesis during NDV infection (Clinkscales and Madansky, personal communication; see also Fig. 1A, slot 1). Thus, the accumulation of P_{88} can be enhanced both by canavanine treatment of uninfected cells and by paramyxoviral infection.

After accumulation of P_{88}, $P_{71,72}$, and P_{23} was enhanced by a 3 hr exposure to canavanine, the analogue could be removed without diminishing the high rates ofaccumulation (see Reversal, Fig. 2). Enhanced accumulation of P_{88} and $P_{71,72}$ continued for at least 3 hr while P_{23} accumulated at an elevated rate for at least 10 hr before each underwent an abrupt decline. Both actinomycin D and cordycepin effectively blocked the enhancement

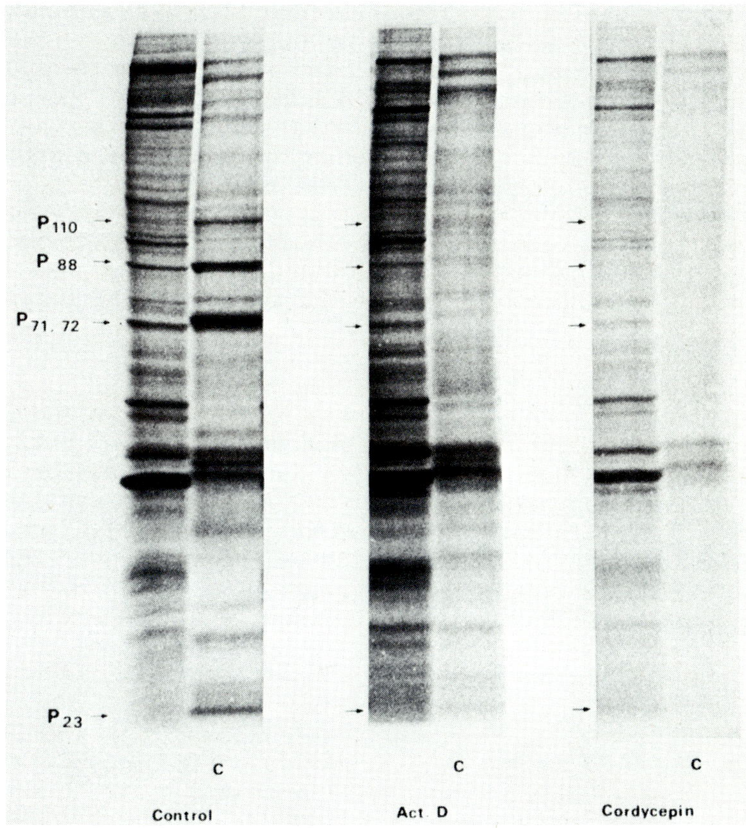

Fig. 3. *Antagonism of the canavanine-induced enhancement of cellular proteins by actinomycin D and cordycepin. Cell cultures were washed and incubated for 3 hr in either canavanine-substituted or normal culture medium. Cultures were labeled with ^{35}S-methionine for 30 min and analyzed by polyacrylamide gel electrophoresis as described in Fig. 1A. The channels marked control show the polypeptide pattern of cultures which did not receive drugs, while those marked actinomycin D received 2 μg/ml and those labeled cordycepin were exposed to 50 μg/ml of the drug added at time zero of incubation.*

(Fig. 3, and Kelley and Schlesinger, personal communication, for act. D). This result suggests that the enhancement requires gene expression and the subsequent appearance of newly synthesized mRNA in the cytoplasm.

Recently, Aksamit and Long (1) suggested that increased degradation of canavanine-containing regulatory proteins could explain their finding that canavanine induces endogenous xenotropic type C virus from transformed BALB/c cells. Canavanine-substituted proteins may also interfere with a regulatory mechanism in protein metabolism under our conditions. However, induction of RNA tumor viral proteins is an unlikely explanation for an enhancement of protein accumulation of the magnitude which we have observed. A simple, but not unique, hypothesis which is consistent with both the time course and drug antagonism of the enhancement is that the accumulation of canavanine-substituted proteins in the cell results in increased rates of transcription of genes normally expressed at much lower levels under conditions of cell cultures. The enhanced rates of protein accumulation would then reflect higher mRNA levels.

ACKNOWLEDGEMENTS

We gratefully acknowledge the aid of Pat Klitzke and Glenn Smith in the preparation of this manuscript; Andrew Ball, Charles Madansky, and Worth Clinkscales for valuable discussions; Robert A. Lamb for making unpublished manuscripts available and providing Sendai-infected cell samples; Philip Kelley and Milton J. Schlesinger for sharing with us the results of their analog experiments prior to publication; U.S. Public Health Service for grant HL 19490 and the University of Connecticut Research Foundation for support. This work benefited from the use of the Cell Culture Facility supported by Grant #CA14733.

REFERENCES

1. Aksamit, R.R. and Long, C.W. (1977). *Virology* 78, 567.
2. Ball, L.A., Collins, P.L. and Hightower, L.E., this volume.
3. Bonner, W.M. and Laskey, R.A. (1974). *Eur. J. Biochem.* 46, 83.
4. Cleveland, D.W., Fischer, S.G., Kirschner, M.W. and Laemmli, U.K. (1977). *J. Biol. Chem.* 252, 1102.
5. Goldberg, A.L. and St. John, A.C. (1976). *In* "Annual Review of Biochemistry, Volume 45" (Snell, E.E., ed), Annual Reviews, Inc., Palo Alto, Calif., pp.747-803.
6. Hightower, L.E. and Bratt, M.A. (1975). *J. Virol.* 15, 696.
7. Hightower, L.E., Morrison, T.G. and Bratt, M.A. (1975). *J. Virol.* 16, 1599.

8. Kelley, P.M., Leone, A. and Schlesinger, M.J. (1977). *Fed. Proc.* 36, 898.
9. Laemmli, U.K. (1970). *Nature (London)* 227, 680.
10. Mountcastle, W.E. and Choppin, P.W. (1977). *Virology* 78, 463.
11. Mountcastle, W.E., Compans, R.W., Caliguiri, L.A. and Choppin, P.W. (1970). *J. Virol.* 6, 677.
12. Mountcastle, W.E., Compans, R.W., Lackland, H. and Choppin, P.W. (1974). *J. Virol.* 14, 1253.
13. Nagai, Y. and Klenk, H.-D. (1977). *Virology* 77, 125.
14. Nagai, Y., Klenk, H.-D. and Rott, R. (1976). *Virology* 72, 494.
15. Peluso, R.W., Lamb, R.A. and Choppin, P.W. (1977). *J. Virol.* 23, 177.
16. Samson, A.C.R. and Fox, C.F. (1973). *J. Virol.* 12, 579.
17. Scheid, A. (1976). *In* "Animal Virology" Vol.IV (Huang, A.S. and Fox, C.F., eds), Academic Press, New York, pp.457-470.

STEPS IN THE ASSEMBLY OF SENDAI VIRUS NUCLEOCAPSIDS

D.W. KINGSBURY

Division of Virology, St. Jude Children's Research Hospital, P.O. Box 318, Memphis, Tennessee 38101, U.S.A.

In the infected cell, a newly made core (nucleocapsid) of a negative strand virus faces one of three alternative pathways. Two of these are metabolic: transcriptive and replicative RNA synthesis; a third pathway assembles cores into progeny virus particles (2, 3).

We know little about how cores are selected for each of these pathways. We do know that a virus-specific protein must be synthesized to establish and maintain viral RNA replication, but not transcription (7, 11). However, this protein, an important regulatory device if a single RNA polymerase species executes both kinds of RNA synthesis (7), has not been identified. Little consideration has been given to the third pathway, beyond the idea that nonglycosylated virus envelope proteins mediate the union of the nascent virus envelope and the core (3). This idea is well-supported, but it doesn't address a potentially important question: are the cores that enter budding virions selected randomly from the entire intracellular core population, or is there a special class of cores earmarked for export? In this brief summary of work that will be fully documented elsewhere, I pose that question and offer an answer.

THE KINETICS OF PROTEIN ADDITION TO CORES

Evidence suggesting the existence of a separate virion-designated core population in infections by the model paramyxovirus, Sendai virus, came unexpectedly from experiments directed at a more general problem. A study of the kinetics of assembly of the different polypeptide species into Sendai virions had shown that the major core polypeptide, NP, entered virions gradually, over several hours, whereas the next most abundant polypeptide, P, entered rapidly, its effective pool becoming exhausted in about an hour (10).

To learn if these differences reflected intracellular events in virus core assembly, the same pulse-chase protocol was applied to an examination of cores isolated from infected cells. This revealed a marked enhancement of the labeling pattern previously observed in virions. Immediately after a pulse of ^{35}S-methionine, core-associated polypeptide P was heavily labeled, several fold in excess of the labeling of core-associated NP (Fig. 1, Free Cores). However, the distribution of label in polypeptides from virions released by these cells favored P only slightly (Fig. 1, Virions), in agreement with previous results (10). After 60 min of chase, label in NP had gained on P in both cores and virions. Also, by this time, the incorporation of labeled glycopolypeptides HN and F_0 into virions could be seen between NP and P in the electropherogram (Fig. 1).

The intense labeling of P in the cores immediately after the pulse suggests that P and NP enter cores independently and sequentially, with P following NP. This conclusion is supported by the observation that P attached to preformed cores when RNA replication and core assembly were shut off in cells infected by a *ts* mutant of Sendai virus (8). However, the new data created a new difficulty: although the labeling patterns of both cores and virions favored P, they did not match; P was much more heavily labeled in the intracellular cores (Fig. 1). Thus, it became unlikely that the core population isolated in these experiments was a direct source of the cores in virions.

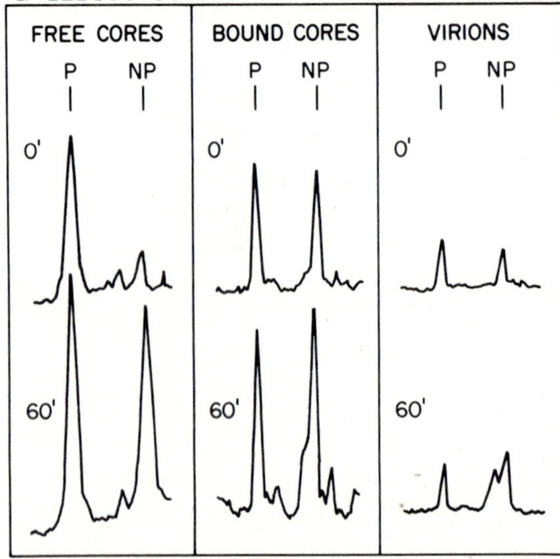

Fig.1.

IDENTIFICATION OF A SECOND CORE POPULATION

In earlier work, substantial amounts of Sendai virus core polypeptides had been seen in the "nuclear" fraction of a cell homogenate (13). But in these new experiments, a nonionic detergent was added to dissolve membranes, and it was expected that membrane-bound cores would no longer be found in the "nuclear" fraction. Surprisingly, despite the detergent, substantial core polypeptide label remained in that fraction (Fig. 1, Bound Cores). Also, about half of the total intracellular virus genome-sized RNA was there (Fig. 2). Electron microscopy confirmed that the "nuclear" fraction contained abundant Sendai virus cores. These were located outside of the nuclei, in tangled masses, possibly associated with microfilaments (Murti, G. and Kingsbury, D.W., unpublished data).

This new core fraction was designated "bound", in reference to its anomalous sedimentation behavior, compared to the "free" cores studied earlier. Kinetically, bound cores were distinguished by a decreased preference for labeling of polypeptide P, notably right after the pulse, but also after 60 min of chase (Fig. 1). Other differences were evident in the RNA (Fig. 2). Although abundant in 50S species, the bound core fraction contained scant RNA that sedimented near 18S, like virus messages; most of the candidate messages were in the supernatant fraction. Another important difference was the paucity of RNA species sedimenting between virus messenger and genomic RNA. Those species,

Fig. 1. *Incorporation of pulse-labeled polypeptides into Sendai virus core fractions and virions. Infection of chick embryo lung cells and labeling with ^{35}S-methionine have been described (10). Virions were recovered from the medium by high-speed centrifugation. Cells were homogenized in 0.01 M Tris-HCl, 0.01 M NaCl, 0.0015 M $MgCl_2$, 1% Triton X-100 (pH 7.4) with a Dounce homogenizer and centrifuged at 4° in a Sorvall SS-34 rotor for 10 min at 10,000 rpm. The pellet contained nuclei, mitochondria, and "bound" cores. The supernatant fluid was centrifuged at 10° in a Spinco SW 50.1 rotor for 4 hr at 47,000 rpm over two layers of sucrose solution. The bottom 1.5 ml had a density of 1.33 gm/cc; above it had been placed 1.5 ml of a 1.15 gm/cc solution (13). After centrifugation, the "free" core fraction was taken from the interface between the sucrose layers. The designated fractions were precipitated with trichloroacetic acid and a portion of each was subjected to electrophoresis and autoradiography (10). Presented are regions of densitometer scans from samples taken immediately after the 10 min labeling period (0') and after a 60 min chase with unlabeled methionine (60'). The direction of electrophoretic migration is from left to right.*

which we have characterized as transcriptive intermediates, complexes of nascent message and 50S RNA templates (9), were notably abundant in the supernatant fraction. These apparent differences in RNA synthetic activity have precedent: in tests of cell-free RNA-dependent RNA synthesis, cell fractions containing free cores had been shown to be the most active (6, 12).

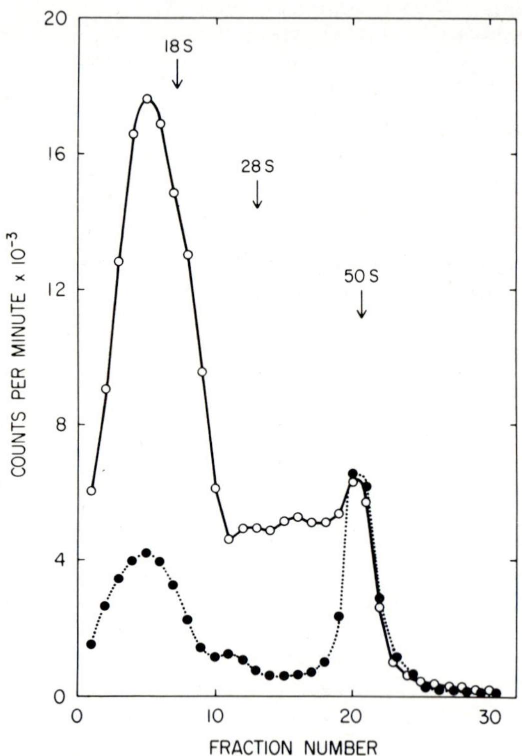

Fig. 2. *Sendai virus-specific RNA in fractions from infected cells. Infected chick embryo lung cells were labeled for 2 hr with ^{3}H-uridine in the presence of actinomycin D (9). The cell homogenate was centrifuged at 10,000 rpm (Fig. 1). Without further fractionation, the supernatant fluid was made 0.1 M in NaCl and 1% in SDS. The pellet was suspended in TEN buffer (0.01 M Tris-HCl, 0.1 M NaCl, 0.001 M EDTA, pH 7.4) and made 1% in SDS. Samples were heated at 56° for 2 min and centrifuged through 15 to 30% linear sucrose gradients in TEN containing 0.5% SDS, in a Spinco SW 27 rotor, 18,000 rpm, 20°, for 16 hr. Acid-insoluble radioactivity in 1 ml fractions was measured (9). Open circles: supernatant fluid;* **closed** *circles: pellet.*

THE ITINERARIES OF CORES

The independent entry of P and NP into free cores and the accentuation of P labeling just after the pulse of ^{35}S-methionine were interpreted above to suggest late addition of P to these cores. Assuming that a single "free" pool of P supplies all cores and that a similar situation obtains for NP, it follows that the relative proportions of labeled P and NP appearing in a core after a pulse of labeled amino acid provide a measure of the age of that core. For example, cores that are incompletely assembled when the pulse begins will acquire labeled NP, but finished cores can accept none. The rapid labeling of NP in the bound cores places this fraction close to the initial assembly event, the union of RNA and NP, whereas the slow entry of NP label into the free cores suggests that assembled cores enter this fraction from elsewhere. In contrast, P appears to enter both types of cores at about equal rates. Furthermore, the labeling pattern of bound core polypeptides closely resembles the labeling pattern in virions, whereas free cores do not, especially right after the pulse (Fig. 1). Considering that newly made core polypeptides enter Sendai virions without delay (10), bound cores are evidently better candidates than the free core population for the source of cores in virions.

These interpretations are summarized in Fig. 3. I hypothesize that newly synthesized virus genomic RNA is joined by NP to form a nascent "RNA-NP" complex. P protein is added in the

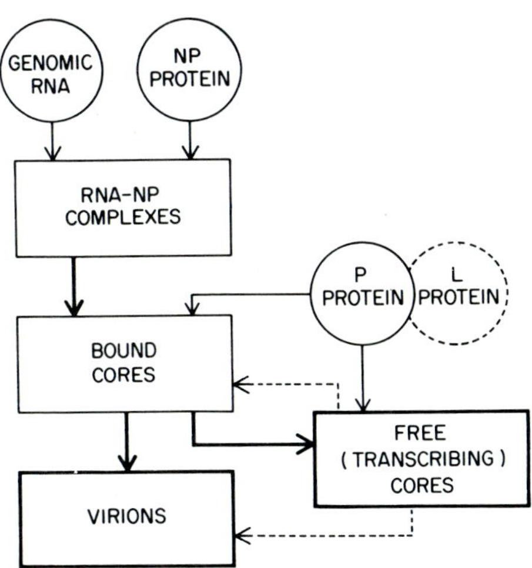

Fig. 3. *A model of Sendai virus core maturation and allocation. Details are discussed in the text.*

bound core compartment. This compartment, inactive in RNA synthesis, is viewed as the single source of cores for virions on the one hand and for RNA transcription on the other hand, in the free core compartment. Additional P joins the cores in the free core compartment.

Fig. 3 also includes some loose ends, like the unknown kinetic pathway of the sparse L protein (5), which was not measured in these experiments. The diagram suggests that L behaves like P, since both proteins are implicated in the function of RNA synthesis, whereas NP probably serves only as a structural element. Also suggested are the additional possibilities, neither indicated nor ruled out by any data, that some free cores exchange back with the bound core fraction and that a portion of free cores contributes to virion assembly. The compartment supporting virus RNA replication is not identified.

POSSIBLE CONTROL MECHANISMS

Assuming that the scheme outlined above is valid, what factors might determine the emergence of a core from the bound core compartment? Since newly made Sendai virus cores can enter virions without delay, bound cores may be situated in the cell periphery. Subsurface regions of the cell are especially rich in cytoskeletal elements (microfilaments) that might bind or entrap the bound cores (1), accounting for their unusually rapid sedimentation rate. A prominent cellular polypeptide that migrates like actin in gel electrophoresis is abundant in the bound core fraction (not shown) and in virions (5, 10). Since actin is a constituent of microfilaments, the possibility arises that newly made cores and microfilaments relate to each other in a meaningful way. Perhaps these cores are delivered close to nascent virus envelopes at the cell surface by the contractile action of microfilaments. Any cores that do not exit may then be transported into a deeper portion of the cell, where they begin RNA synthesis.

Another possibility is that compositional changes in cores determine their fate. Addition of more than a threshold amount of P (or L) might initiate RNA synthesis and prevent a core from entering a virion. Or post-assembly modifications of core constituents, like phosphorylation of a polypeptide, might signal transcription (4, Lamb and Choppin, *Virology, in press*). Such possibilities do not preempt topographical factors; both events could cooperate to shepherd cores into the appropriate pathways.

However, a difficulty encountered by any model that invokes a change in core composition as a device for switching on RNA synthesis is the fact that cores in virions are completely equipped to start making RNA on entry into a host cell. Virion cores are also capable of making RNA *in vitro*, which tends to rule out a need for a process provided by the host cell to switch on primary transcription. What, then, determines that a core should be fully endowed with the capability for RNA synthesis yet be inactive before its assembly into a virion? Such questions may yield to further investigation of the separable, metabolically distinct core fractions appearing in cells infected by Sendai virus.

ACKNOWLEDGEMENTS

The advice of Drs. K. Jones and J. Tannenbaum and the technical assistance of Mrs. D. Clift are appreciated. This work was supported by U.S. Public Health Service Research Grant AI 05343 and by ALSAC.

REFERENCES

1. Allison, A.C. and Davies, P. (1973). *In* "Advances in Cytopharmacology", Vol. 2 (Ceccarelli, B., Clementi, F. and Meldolesi, J., eds), Raven Press, New York, p.237.
2. Baltimore, D. (1971). *Bacteriol. Rev.* 35, 235.
3. Choppin, P. and Compans, R. (1975). *In* "Comprehensive Virology", Vol. 4, (Fraenkel-Conrat, H. and Wagner, R.R., eds), Plenum Press, New York, p.95.
4. Clinton, G.M. and Huang, A.S. (1977). *Abstracts 77th Ann. Meeting Am. Soc. Microbiol.* p.297.
5. Lamb, R.A., Mahy, B.W.J. and Choppin, P.W. (1976). *Virology* 69, 116.
6. Mahy, B.W.J., Hutchinson, J.E. and Barry, R.D. (1970). *J. Virol.* 5, 663.
7. Perlman, S.M. and Huang, A.S. (1973). *J. Virol.* 12, 1395.
8. Portner, A. (1977). *Virology* 77, 481.
9. Portner, A. and Kingsbury, D.W. (1972). *Virology* 47, 711.
10. Portner, A. and Kingsbury, D.W. (1976). *Virology* 73, 79.
11. Robinson, W.S. (1971). *Virology* 44, 494.
12. Stone, H.O., Portner, A. and Kingsbury, D.W. (1971). *J. Virol.* 8, 174.
13. Stone, H.O., Kingsbury, D.W. and Darlington, R.W. (1972). *J. Virol.* 10, 1037.

THE REPLICATION OF PICHINDE VIRUS - AN ARENAVIRUS

WAI-CHOI LEUNG

Department of Pathology,
McMaster University,
Hamilton, Ontario, Canada.

Members of the arenavirus group characteristically produce persistent infection in their natural hosts. The mode of persistent infection has been extensively documented for lymphocytic choriomeningitis virus (LCM), the prototype virus of the group. Arenaviruses can also readily establish persistence *in vitro*. Since the behavior of virus persistence *in vitro* closely mimics that of virus replication in persistently infected animals, studies of replication of the viruses *in vitro* may provide valuable information about the mechanism of persistence (9, 10).

Some members of the group occasionally infect man. The outcome of these infections vary from febrile illness to fatal disease. Although at present these diseases are limited to certain geographical locations (Lassa fever in Western Africa, Argentine hemorrhagic fever and Bolivian hemorrhagic fever in South America), increased international travel has increased the risk of transcontinental spread of the viruses. Therefore, this virus group, as a possible health threat, merits special attention.

Although the biology of the virus and the immune responses of the host to virus infection have been studied extensively in the case of LCM virus (5, 15), only limited progress on the molecular biology of the virus group has been made in the last decade. Biochemical characterization has been carried out mainly with Pichinde virus because it is relatively stable during laboratory handling and does not appear to be hazardous to laboratory personnel (1).

Like other arenaviruses, Pichinde virus is surrounded by an

envelope with spikes on its outer surface (6, 14). The genetic information of the virus consists of single-stranded RNA that upon extraction sediments as two segments, 31S and 22S (3). Ribosomal RNA can also be extracted from virions and is derived from host cell ribosomes within the virion (7). Ribonucleoprotein complexes containing mainly genomic RNA can be isolated from the virus (2, 16). Four major structural polypeptides have been identified in the virion by polyacrylamide gel electrophoresis and two of these are glycosylated (17). An RNA-dependent RNA polymerase has been found in purified virions (4).

This paper is concerned with the replication of Pichinde virus, dealing particularly with the likelihood that the virus RNA is negative-stranded and that the virion-associated ribosomes are not required for the replication of the virus.

VIRION-ASSOCIATED RNA POLYMERASE AND POLY-[U]POLYMERASE

RNA-dependent RNA polymerase activity is associated with purified Pichinde virus, but a less stringent requirement for guanosine triphosphate has been observed for this enzymatic activity (4). In view of the association of cellular components, especially ribosomes, with the virion, it is possible that other cellular enzymes might be present. Since it has been known that a polysome-associated poly-[U] polymerase also exhibited less stringent requirements for ribonucleoside triphosphates other than UTP (19, 20), the possibility that this enzymatic activity is also virion-associated was investigated. Table I shows that the RNA polymerizing activity of purified Pichinde virus which had been frozen for 4 days did not require the presence of all 4 ribonucleoside triphosphates for maximal activity. In fact, deletion of one or 3 of the unlabelled substrates did not affect significantly the RNA polymerizing activity. This property is very similar to that of the poly-[U] polymerase associated with polysomes extracted from BHK-21 cells, and is in marked contrast to the stringent requirement for all 4 ribonucleoside triphosphates shown by RNA-dependent RNA polymerases associated with VSV. Even freshly purified Pichinde virus did not exhibit the property of RNA-dependent RNA polymerase, the level of requirement for ribonucleoside triphosphates being somewhere between that of the VSV RNA polymerase and the polysomal poly-[U] polymerase. These results suggest the presence of multicomponents in the RNA synthesizing system associated with Pichinde virus; some might be more labile than the others.

Physical separation was used to search for different RNA polymerizing activities associated with Pichinde virus. The different ribonucleoprotein components in the virus were labeled

TABLE I

Effect of substrates on RNA polymerase activity

Reaction Condition	Enzyme Source: Vesicular Stomatitis Virus RNA Polymerase	BHK Polysome Poly-[U] Polymerase	Pichinde Virus (Fresh)	Pichinde Virus (Frozen at -20°C for 4 days)
Complete	100% (227.0)	100% (5.4)	100% (27.0)	100% (6.2)
-GTP	12%	82%	54%	92%
-CTP	10%	94%	37%	81%
-ATP	18%	86%	42%	104%
-GTP, CTP, ATP	7%	85%	40%	88%

The 100,000 x g post-mitochondrial pellet was prepared from BHK-21 cells. The pellet was resuspended in 0.01 M Tris, pH 8.0, sonicated for 3 minutes and provided the source of poly-[U] polymerase. The RNA polymerase assay was performed as described (4). Numbers in brackets indicate specific activity as picomoles ^{3}H-UTP incorporated per mg protein per 30 min at 30°C.

with ^3H-uridine. The purified virus was then treated with 0.5% Nonidet P-40 and 0.5 M KCl, and subsequently centrifuged in a 20-70% (W/V) sucrose gradient in 0.1M KCl. In Fig.1, upper graph, about 8% of the ^3H-UR cpm were located in a fast sedimenting component; about 15% were found in the 80S component and the rest consisted of a slowly sedimenting fraction at top of the gradient. Examination of the RNA species in the fast sedimenting structure indicated that they were mainly 31S and 22S RNA, although about 11-20% of the radiolabel was 28S and 18S RNA. These observations suggest that nucleocapsid-like structures containing the genomic RNA were released from the virion. However, the 28S and 18S RNA might be due to a small amount of aggregated ribosomes cosedimented with the nucleocapsid. On the other hand, only 28S and 18S RNA was extracted from the 80S components. When a nonlabeled virus preparation was similarly treated and each fraction was assayed for RNA polymerase activity, 2 major peaks of activities were found (Fig.1, lower graph). A fast sedimenting peak (peak 1) cosedimented with the nucleocapsid and slowly sedimenting peak (peak 2) cosedimented with the 80S ribosomes, although only about 5 to 12% of the total input polymerase activity was recovered in the sucrose gradient. The enzymatic activity in peak 1 exhibited a stringent requirement for all 4 ribonucleoside triphosphates for maximal activity (Table II). The considerable degree of ribosome cosedimenting with the nucleocapsid might contribute to the residual RNA polymerizing activity when one or more of the ribonucleoside triphosphates were deleted from the assay. On the other hand, peak 2 exhibited properties similar to the poly-[U] polymerase.

The *in vitro* products of the polymerase were characterized by nearest neighbor analysis. The transfer of α-^{32}P-UTP into adjacent ribonucleotides by peak 1 was evenly distributed among the four 2'(3')-ribonucleoside monophosphates, indicating the synthesis of heteropolymer. In case of the peak 2 activity, about 60% of the label was transferred to UMP, indicating that homopolymers of mainly UMP sequence were formed. The average chain length of the synthesized product was estimated by the ratio of ribonucleotide to ribonucleoside. Peak 1 enzyme synthesized RNA 57 to 228 nucleotides long while peak 2 activity synthesized RNA averaging 11 nucleotides in length. Reannealing experiments with virion RNA suggested that the ribonucleotides synthesized by peak 1 can hybridize extensively to virion RNA, but not to ribosomal RNA or t-RNA. The *in vitro* synthesis product of peak 2 did not hybridize with any of the above mentioned RNAs. The above data suggest that an RNA polymerizing activity associated with the fast sedimenting nucleocapsid is capable of synthesizing relatively long stretches of hetero-

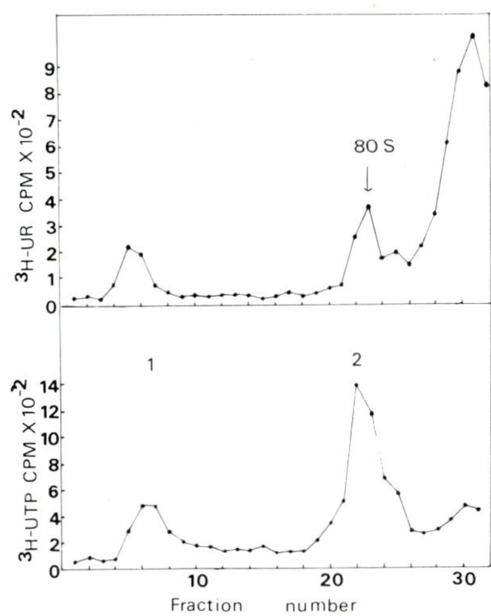

Fig.1. *Sedimentation of ribonucleoprotein complexes and RNA polymerizing activities released from Pichinde virion. Upper graph, Pichinde virus was grown in BHK-21 cell monolayers in the presence of 10 μCi/ml ^{3}H-uridine. Purified virus was treated with 0.5% Nonidet P-40, 0.5 M KCl for 10 min at 4°C and then centrifuged on a 20-70% (w/v) sucrose gradient containing 0.01 M Tris pH 7.5, 0.1 M KCl, 0.005 M $MgCl_2$ in an IEC SB-283 rotor for 1 hr at 100,000 x g. TCA precipitable radioactivity was calculated from each fraction. Lower graph, nonlabeled Pichinde virus was similarly treated as in the upper graph. Portions of each fraction were then assayed for RNA polymerizing activities as in Table I.*

polymeric RNA. The polarity of the synthesized RNA is complementary to the virion RNA, thus conforming with the known properties of RNA-dependent RNA polymerase. Also present in the virion is a poly-[U] polymerase which is associated with the ribosomes and synthesizes mainly short lengths of homopolymeric RNA. The poly-[U] polymerase might mask the activity of the RNA-dependent RNA polymerase and contribute to the difficulty of assaying the RNA polymerase in this system. Apart from its relatively low specific activity, the RNA-polymerase was comparatively unstable when compared to the poly-[U] polymerase. As indicated by Fig.2, the poly-[U] polymerase remained relatively active upon storage at -20°. However, less than 20% of the RNA polymerase activity remained upon storage at -20° for one day. Attempts to preserve the enzymatic activity by storage at varied temperatures (-50° to -90°) or in the presence of glycerol or bovine serum albumin have not been successful.

TABLE II

Effect of substrates on the RNA polymerizing activity of Pichinde virus

Reaction mixture	Peak 1	Peak 2
Complete	100% (1.2)	100% (2.7)
-GTP	34%	87%
-CTP	27%	96%
-ATP	39%	87%
-GTP, CTP, ATP	28%	92%

Fractions from the lower graph of Fig.1 were pooled: peak 1 included fractions 5 to 8, and peak 2 included fractions 21 to 25. The RNA polymerizing activities of peaks 1 and 2 were assayed according to Table I. The numbers in brackets indicate specific activity as picomoles of ^3H-UTP incorporated per 150 µl enzyme aliquot per hour at 30°.

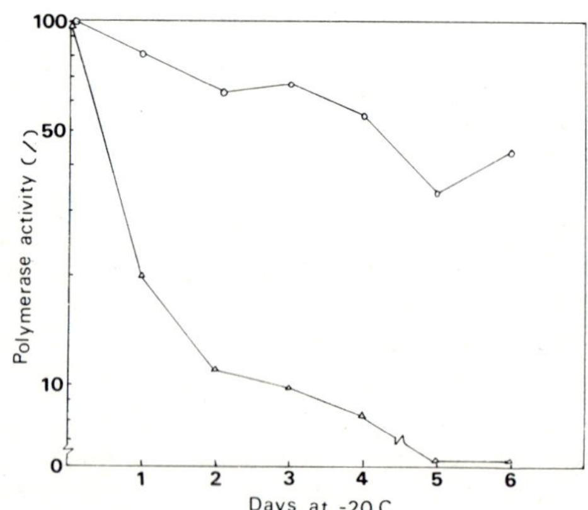

Fig.2. *Relative stability of the polymerizing activities from Pichinde virus. Fractions from Fig.1, lower graph, were pooled: peak 1 included fractions 5 to 8, peak 2 included fractions 21 to 25. Peak 1 and peak 2 were stored at -20° in 20% glycerol. At indicated time one ampoule of stored enzymes was removed and assayed for RNA polymerizing activity;* Δ-Δ-Δ, *peak 1 activity; 0-0-0, peak 2 activity.*

LACK OF MESSENGER FUNCTION FOR THE VIRION RNA

The presence of an RNA-dependent RNA polymerase in the Pichinde virion suggests that the polarity of the RNA is negative-stranded. The virion RNA of positive-stranded viruses serves as messenger RNA, whereas for negative-stranded viruses, the complementary RNA transcript of the virion RNA is the messenger. We therefore looked for messenger activity of the virion RNA. These studies included a search for features of messenger RNA, such as poly-[A] tracts, capping or methylation, as well as functional activities such as the ability to be translated *in vitro* into viral specific polypeptides (11). ^{3}H-uridine, ^{3}H-adenosine or ^{32}P-labeled Pichinde virion RNA was unable to bind to oligo-dT cellulose. The possibility of internal steric hindrance due to the presence of host ribosomal RNA in the Pichinde RNA preparation was considered. ^{32}P-labeled Pichinde RNA was digested with a mixture of pancreatic and T1 RNases, a procedure which retains only the poly-[A] sequence. The RNase digested product was still unable to bind to oligo-dT cellulose, indicating the absence of poly-[A] sequence in Pichinde RNA. Moreover, we were unable to detect any capped and methylated structures following digestion of about 10^{7} cpm of ^{32}P-labeled virion RNA with *Penicillium* P1 nuclease and nucleotidyl pyrophosphatase, whereas the same procedure produced capped and methylated structures from BHK-21 cell poly-[A] containing RNA.

The possibility that Pichinde virion RNA could be translated was investigated in the wheat germ and BHK-21 cell *in vitro* translation systems. No significant increase in the incorporation of ^{35}S-methionine into hot TCA insoluble cpm was observed (Table III). In wheat germ, a 2-3 fold increase in incorporation of radiolabel over background stimulated by virion RNA was similar to the increased incorporation observed when cellular ribosomal RNA was added to the system. Since virion RNA contained ribosomal RNA, increased incorporation could not be specifically attributed to viral genomic RNA. Analysis of the *in vitro* translation product by electrophoresis in a 7.5 to 15% gradient acrylamide gel did not reveal the presence of any of the 4 viral structural polypeptides (Fig.3). The above data strongly suggest the virion RNA did not possess any messenger function.

COMPLEMENTARITY BETWEEN VIRION RNA AND POLYSOMAL RNA FROM VIRUS INFECTED CELLS

Complementarity of the polysomal RNA from virus infected cells to virion RNA is a prediction for negative stranded viruses. Certain difficulties have been encountered, however,

TABLE III

In vitro translation of Pichinde virus RNA and polysomal RNA

Extract	Assay Condition		^{35}S-methionine incorporated cpm/0.01 ml/90 min
Wheat germ	-RNA		4228
	+ Pichinde RNA	2.0 μg	12205
	+ BHK cell ribosomal RNA	2.0 μg	11633
	+ Vesicular stomatitis virus m RNA	0.5 μg	211927
BHK-21	-RNA		110642
	+ Pichinde RNA	3.0 μg	121397
	+ Mock-infected BHK cell polysomal RNA	2.5 μg	364277
	+ Pichinde-infected BHK cell polysomal RNA	2.5 μg	356712

The BHK-21 cell free extract was prepared according to McDowell *et al.* (13), except that after preincubation at 30°C for 1 hr, it was dialyzed in 1 x HEPES buffer overnight at 4°C. The BHK-21 cell free extract was stored in liquid nitrogen until used. The Pichinde virus RNA and polysomal RNA from virus-infected cells were isolated as described (11). The reaction mixture for the *in vitro* protein synthesis system was as follows: BHK-21 cell free extracts, 0.6 ml/ml, 30 mM Hepes, pH 7.5, 86 mM KCl, 2.5 mM $MgAc_2$, 10 mM dithiothreitol, 1 mM ATP, 0.2 mM GTP, 11 mM creatine phosphate, 1 mg/ml creatine kinase, 0.2 mM each of 19 unlabeled amino acids, 165 μCi/ml ^{35}S-methionine (New England Nuclear) and exogenous RNA was added as indicated. The reaction mixture was incubated at 30°C for 1 hr. Twenty microliter aliquots were spotted on Whatman 3MM paper and hot TCA precipitate cpm were calculated.

in the radiolabeling of intracellular Pichinde virus-specific polysomal RNA. High doses of actinomycin-D which suppress cellular RNA synthesis also drastically inhibit the replication of Pichinde virus (18). In the presence of low doses of actinomycin-D, which inhibit ribosomal RNA synthesis, the virus-specific RNA constituted at most only about 1% of the total labeled RNA. We therefore used ^{32}P-labeled virion RNA as a probe. Virus was grown in cells in the presence of 0.05 µg/ml of actinomycin-D to suppress ribosomal RNA synthesis, and the RNA extracted from purified virions was further purified on a sucrose gradient to remove low molecular weight RNA. Analysis by gel electrophoresis indicated that about 70% of the ^{32}P-label was at the 31S and 22S genomic RNA region, while the rest was associated with the 28S and 18S ribosomal RNA. Reannealing experiments were performed with the ^{32}P-labeled virion RNA and polysomal RNA from virus-infected cells. About 60% of the ^{32}P-labeled probe was protected against digestion by RNase A and T1. A similar degree of hybridization was also obtained in hybridization between ^{32}P-labeled virion RNA and poly-[A] containing polysomal RNA prepared from virus-infected cells. The reason for partial complementarity is not known, but assuming that only the 31S and 22S RNA can reanneal with virus-specific polysomal RNA, it was calculated that the degree of reannealing was about 85%. The remaining nonhybridized fractions could be due to untranscribed regions on the virion RNA (11).

The possibility that the polysomal RNA used in the hybridization experiments was mRNA was examined by *in vitro* translation. When polysomal RNA from virus-infected cells was added to the BHK-21 *in vitro* translation system, it stimulated the synthesis of VP1, a 66000 dalton nucleocapsid polypeptide (Table III, Fig.3). Synthesis of the other 3 virus structural polypeptides was not observed. However, the inability to detect these structures on the *in vitro* synthesizing system may be related to the sensitivity of our system since *in vivo* labeling with radioactive amino acids of the 3 polypeptides was not successfully accomplished. Moreover, VP2 and VP3 are glycopeptides and glycosylation occurs poorly *in vitro*. Nevertheless, these studies indicated that mRNA for at least one virus-specific polypeptide was present in polysomes, and polysomal RNA was complementary to a major portion of the virion RNA.

VIRION-ASSOCIATED RIBOSOMES ARE NOT REQUIRED FOR THE REPLICATION OF PICHINDE VIRUS

Electron microscopic and biochemical studies established that arenaviruses contain ribosomal particles enclosed within the viral envelopes. Although the amount of ribosomal RNA is variable, it has been found to constitute up to 50% of the total RNA extracted from purified virion in some preparations. This unique structural feature of arenaviruses certainly raises

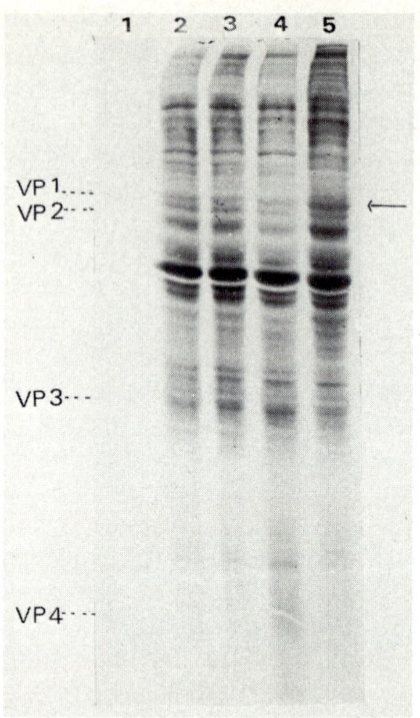

Fig.3. *Autoradiograph of polypeptides synthesized in vitro and analyzed by gradient slab gel electrophoresis. To a BHK-21 cell-free protein synthesizing system was added (5) polysomal RNA from BHK-21 cells infected with Pichinde virus, (4) polysomal RNA from mock-infected BHK-21 cells, (3) RNA from purified Pichinde virus and (2) no exogenous RNA. The in vitro protein synthesizing reaction was carried out as described in Table III. The in vitro synthesized product was then solubilized by 1% SDS, 0.1 M 2-mercaptoethanol and electrophoresed on a 7.5 to 15% gradient slab-polyacrylamide gel. Lane (1) contained Pichinde virus as a marker. After electrophoresis for 16 hr at 10°C, the gel was stained with Coomassie blue and destained. The slab gel was then dried and processed for autoradiography with Kodak RP x-mat film.*

the possibility that the virion-associated ribosomes are involved in the replication of the virus. We have investigated the growth characteristics of Pichinde virus grown in cells that are temperature-sensitive with respect to protein synthesis. The inability of these mutant cells to synthesize proteins at the nonpermissive temperature was attributed to altered stability of the 60S ribosomal subunit (8). Pichinde virus was allowed to replicate in the mutant cell at the permissive temperature. The efficiency of plating of the virus carrying the ts ribosome was similar to the control virus carrying the wt ribosome. If both viruses were grown in wild type cells for 48 hr at the permissive and nonpermissive temperatures respectively, the efficiency of replication for the virus carrying ts ribosomes was similar to the virus carrying the wt ribosomes. Since the virus-associated ts ribosomes would not be expected to function at the nonpermissive temperature, these observations suggest that virion-associated ribosomes are not required for the replication of Pichinde virus with respect to their translation function (12).

DISCUSSION

The existence of ribosomal particles enclosed within the envelopes of arenaviruses is a distinctive and central feature of the group. Our studies with virus grown in cells that possess temperature-sensitive ribosomes indicate that the ribosomal particles are not required for virus replication, at least in the initiation phase. The available data suggest that the ribosomes are carried adventitiously into the virus during virus assembly, and thus represent a cellular contaminant. However, because this process is consistently observed in the members of arenavirus, it is likely that the morphogenesis of the virus is intimately associated with the rough endoplasmic reticulum.

Although no functional role has yet been assigned to the virion-associated ribosomes, they complicate the experimental investigation of Pichinde virus. For example, the presence of polysome-associated poly-[U] polymerase greatly complicates the assay for the RNA-dependent polymerase. It is not known if this enzyme has any role in the synthesis or modification of virus-specific RNA. Other possible ribosome-associated enzymes, such as phosphodiesterase, might also complicate the detection of RNA-dependent RNA polymerase activity.

Initially the presence of ribosomes within the virion raised the speculation that virion RNA is positive-stranded. It was speculated that the virion-associated ribosomes facilitate the translation of virion RNA and could thus contribute to the relative ease with which the arenaviruses establish persistent infection. The demonstration of the presence of RNA-dependent RNA polymerase, the lack of messenger structure and function in virion RNA, the complementarity between virion RNA and polysomal RNA indicates that Pichinde virus RNA is negative-stranded. These findings contribute little to our understanding of the mechanisms for viral persistence which remain to be unraveled.

ACKNOWLEDGEMENTS

This work was supported by a grant from the Medical Research

Council of Canada. I am indebted to Dr William E. Rawls for critically reviewing this manuscript and to Maria Leung for excellent technical assistance. W.C.L. is a Medical Research Council Scholar.

REFERENCES

1. Buchmeier, M.J., Adam, E., and Rawls, W.E. (1974). *Infect. Immun.* 9, 821.
2. Buchmeier, M.J., Gee, S.R., and Rawls, W.E. (1977). *J. Virol.* 22, 175.
3. Carter, M.F., Biswal, N. and Rawls, W.E. (1973). *J. Virol.* 11, 61.
4. Carter, M.F. Biswal, N., and Rawls, W.E. (1974). *J. Virol.* 13, 577.
5. Cole, G.A., Golden, D.H. Monjan, A.A., and Nathanson, N. (1971). *Fed. Proc.* 30, 1831.
6. Dalton, A.J., Rowe, W.P., Smith, G.H., Wilsnack, R.E., and Pugh, W.E. (1968). *J. Virol.* 2, 1465.
7. Farber, F.E., and Rawls, W.E. (1975). *J. Gen. Virol.* 26, 23.
8. Haralson, M.A., and Roufa, D.J. (1975). *J. Biol. Chem.* 250, 8618.
9. Hotchin, J. (1971). *In* "Monographs in Virology", vol.3, p.96. S. Karger, Basel.
10. Lehmann-Grube, F. (1967). *Nature* 213, 770.
11. Leung, W.C., Ghosh, H.P., and Rawls, W.E. (1977). *J. Virol.* 22, 235.
12. Leung, W.C., and Rawls, W.E. (1977). *Virology* 81, 113.
13. McDowell, M.J., Joklik, W.K., Villa-Komaroff, L., and Lodish, H.F. (1972). *Proc. Nat. Acad. Sci. USA* 69, 2649.
14. Murphy, F.A., Webb, P.A., Johnson, K.M., Whitfield, S.G. and Chappell, W.A. (1970). *J. Virol.* 6, 507.
15. Oldstone, M.B.A., and Dixon, F.J. (1970). *J. Immunol.* 105, 829.
16. Pedersen, I.R., and Konigshofer, E.P. (1976). *J. Virol.* 20, 14.
17. Ramos, B.A., Courtney, R.J., and Rawls, W.E. (1972). *J. Virol.* 10, 661.
18. Rawls, W.E., Banerjee, S.N., McMillan, C.A., and Buchmeier, M.J. (1976). *J. Gen. Virol.* 33, 421.
19. Villareal, L.P., and Holland, J.J. (1974). *J. Virol.* 14, 441.
20. Wilkie, N.M., and Smellie, R.M.S. (1968). *Biochem. J.* 109, 229.

LUMBO VIRUS GENOME AND ITS REPLICATION IN BHK 21 CELLS

MICHÈLE BOULOY

Viral Ecology Unit, Institut Pasteur, Paris France

Lumbo virus is an arbovirus belonging to the California complex of the newly defined bunyaviridae family (16). The virus particles are roughly spherical (90 nm in diameter) surrounded by a lipid envelope and protruding spikes. Its multiplication is located in the cytoplasm and is not affected in the presence of actinomycin D (2).

STRUCTURAL COMPONENTS

Lumbo virus is composed of three major proteins, two glycoproteins G1 (mol. wt. 90,000 daltons) G2 (mol. wt. 33,000 daltons) and a small polypeptide N (mol. wt. 25,000 daltons) which is tightly associated with the RNA to form the ribonucleoprotein. A large minor polypeptide has been resolved in polyacrylamide slab gels (mol. wt. 140-150,000 daltons). Lumbo virus genome is a segmented, single-stranded RNA, for which three RNA species have been resolved. The large molecule (L RNA) has a sedimentation value of 32 S and a molecular weight of 3.0×10^6 daltons. The medium and the small size molecules (M RNA and S RNA) have respectively molecular weights of 2.0×10^6 and 0.5×10^6 daltons and sedimentation coefficients of 24 S and 16 S (2).

MORPHOLOGY

Lumbo virus RNAs exhibit some novel structural properties. Circular RNA molecules, falling into three size classes were observed by electron microscopy (19). These circles resist urea-formamide treatment or thermal denaturation (100 or 110°C) suggesting that they are probably covalently linked or base paired by self-annealing sequences, strongly resistant to severe denaturation. As a consequence of the circular RNA configuration, internal structures or ribonucleoproteins (RNP) also are circular (18). Purified RNPs obtained after treatment with a non-ionic detergent (Nonidet-P40) were examined by electron microscopy. They appear as circular coiled or supercoiled strands, about 8 to 9 nm wide, which probably have helical symmetry. Three size classes of nucleocapsids have been observed, with mean lengths of 0.7 µm, 0.5 µm and 0.3 µm. In linear sucrose gradients their sedimentations were estimated to be: 105 S (L RNP), 85 S (M RNP) and 45 S (S RNP) (4).

Three species of RNA have also been reported in other bunyaviruses such as Uukuniemi virus (13), Lacrosse virus (12), Snowshoe hare, Trivittatus, Melao virus (6, 10). Although circular nucleocapsids were also described in Uukuniemi virus (14) and Lacrosse virus (12), circular RNAs do not seem to be a general feature among bunyaviruses. Uukuniemi virus genome appeared as circles maintained by base pairing (11) whereas the absence of circular molecules was reported for Lacrosse virus RNA (7).

POLARITY OF THE GENOME

Another interesting finding is that Lumbo virus genome is a negative-strand RNA (3). This conclusion could be drawn from several findings. RNA extracted from purified virus is not infectious, it does not contain any poly A tract and a virion-associated RNA-dependent RNA polymerase has been detected. In addition, when viral RNA was tested in wheat germ extracts it did not stimulate any protein synthesis. The presence of a polymerase in Uukuniemi virus also suggests that it belongs to the group of minus strand RNA viruses (17).

The presence of three classes of RNAs (or RNPs) in Lumbo virus led us to investigate the relationship between the segments and their significance. L RNA may contain the full genetic information while M and S RNA are shorter molecules missing some sequences such as the defective molecules described for VSV. Alternatively, the genome may be segmented with each segment coding for distinct polypeptides.

The replication of Lumbo virus in BHK 21 cells has been studied. The data obtained suggest that each RNA segment replicates separately. Moreover messenger RNAs, complementary to the virion RNAs, were hybridized with each segment. These results provide direct evidence for the segmented nature of the genome.

RNA SYNTHESIS IN INFECTED CELLS

Short pulse experiments were performed to detect any possible precursor molecules involved in RNA synthesis. Lumbo virus infected BHK 21 cells were labeled with uridine in the presence of actinomycin D (2.5 µg/ml) at 4 hr after infection for 5, 10 or 30 min. Fig. 1A shows that after a 5 min radioactive exposure, most of the RNA molecules were resistant to RNase treatment and sedimented in a broad peak (20 S - 16 S). When the exposure times were increased (Fig. 1B and C), single-stranded RNAs could be detected. It was noticeable that 16 S, 24 S and 32 S RNA molecules appeared in this order, suggesting that the two smaller molecules (16 S and 24 S) are not cleavage products derived from the large species (L RNA).

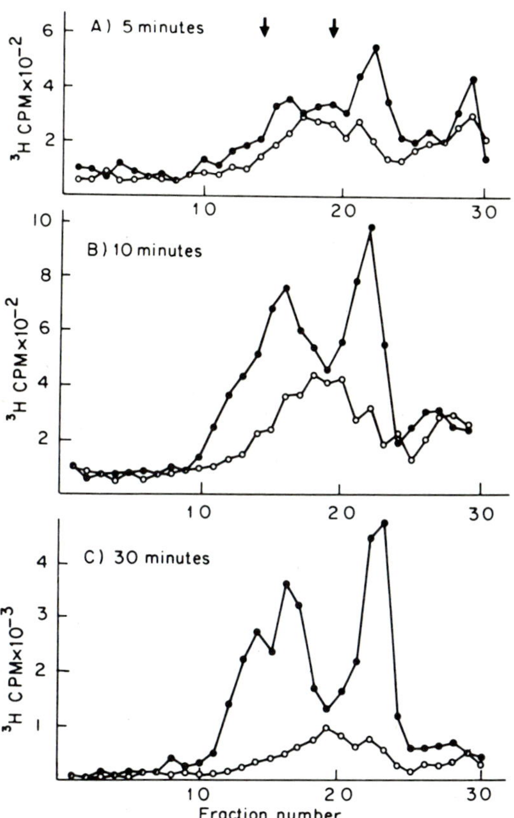

Fig. 1. *Lumbo virus RNAs in BHK21 cells. BHK21 cells infected with Lumbo virus at a m.o.i. of 10 pfu/cell were labeled with 10 µCi/ml of ^{3}H-uridine at 4 hr p.i. in the presence of actinomycin D (2.5 µg/ml). After a pulse of 5 min, 10 or 30 min, the cells were collected and cellular RNAs were extracted by SDS phenol. This material was centrifuged in a 15-30% sucrose gradient in STE for 14 hr at 24,000 rpm in a SW27 rotor. Each fraction was tested for RNase resistance (0-0).*

POLYMERASE ACTIVITY ASSOCIATED WITH EACH RIBONUCLEOPROTEIN

Cytoplasmic extracts were previously shown to contain RNPs sedimenting into three peaks. In each size class, two types of RNP were revealed by self annealing experiments, one containing the virion RNA and the other containing complementary RNA (4). It seems likely that these complementary RNPs are the templates for viral RNA progeny synthesis, implying that each segment replicates separately. Consequently, each class of RNP should contain its own polymerase activity. Lumbo virus ribonucleoproteins were obtained from cytoplasmic extracts collected at 8 hr p.i. and fractionated in a linear sucrose gradient. Each fraction was tested for polymerase activity under the conditions already described (3). Fig. 2 shows that the enzyme activity cosedimented with each size class of RNP (respectively 105 S, 85 S and 45 S). In control experiments with uninfected cells, no polymerase activity could be detected. The enzyme associated with cytoplasmic RNPs has the same properties (salt requirement) as the virion polymerase. Table 1 indicates that this enzyme requires the presence of both Mg^{++} and Mn^{++} cations and is strongly inhibited when high NaCl concentrations are added. The reaction was more rapid and efficient at 39°C than at lower temperatures.

These findings provide evidence that each RNP is active in RNA synthesis as transcriptive complex. Since viral and complementary RNPs were present in infected cells, transcriptive

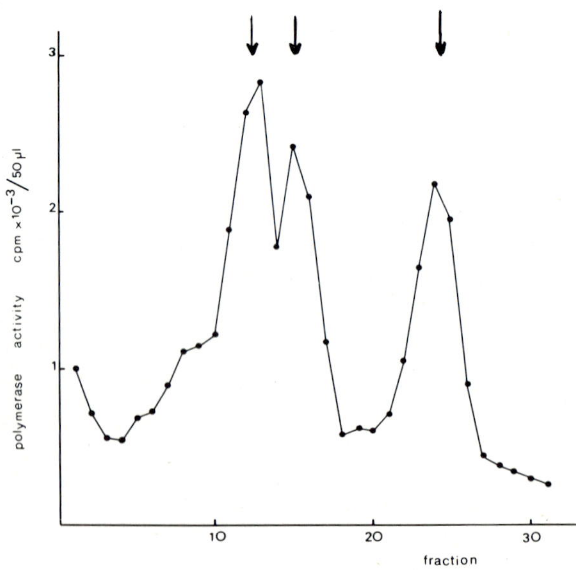

Fig.2.

TABLE 1

Comparison of Polymerase Activity Associated with Lumbo Virus and Viral Cytoplasmic RNPs

Reaction Mixture	% Activity	
	Virus	RNP
Complete	100	100
- Mn^{++}	2.1	5
5 mM $MgCl_2$	89.2	103
25 mM NaCl	95	97
200 mM NaCl	11.9	9.8

The reaction mixture was incubated for 3 hr at 39°C. Complete reaction mixture was the same as described in the legend of Fig. 2. Under these conditions RNA polymerase associated with virion and RNP incorporated respectively 800 cpm and 850 cpm of ^{3}H-UMP.

and replicative complexes would be detected simultaneously if the replicase has the same properties as the virion transcriptase.

ISOLATION OF CYTOPLASMIC STRUCTURES IN METRIZAMIDE GRADIENTS

When cytoplasmic extracts were fractionated in linear sucrose gradients, it has been reported that the polysomal fraction contained three classes of RNA equivalent in size to 32 S, 24 S and 16-14 S molecules (4). However, further analysis of 32 S and 24 S molecules revealed that the polysomes were contaminated by the larger RNPs. Since isopycnic centrifugation in metrizamide gradients has been shown to provide a convenient preparative

Fig. 2. *Polymerase activity associated with cytoplasmic RNPs. Lumbo virus-infected BHK21 cells were collected at 8 hr after infection. Cytoplasmic extracts were prepared in isotonic buffer containing 10 mM $MgCl_2$ and 0.5% NP40. The extract was centrifuged in a 15-30% sucrose gradient in isotonic buffer at 40,000 rpm for 3 hr in a SW41 rotor. Polymerase activity was assayed on a 50 µl aliquot of each fraction for 3 hr at 39°C. Reaction mixture contains ATP, CTP, GTP (0.5 mM each), ^{3}H-UTP (20 µCi/ml), 70 mM NaCl, 10 mM $MgCl_2$, 1.5 mM $MnCl_2$ and 10 mM 2-mercaptoethanol. The arrows indicate the position of L, M and S RNPs.*

method for isolation of cellular structures such as RNA - protein complexes i.e. polysomes and nucleoproteins (5, 8), we applied this new method to the separation of Lumbo virus polysomes and cytoplasmic complexes.

Non-infected and Lumbo-infected BHK21 cell extracts were centrifuged in metrizamide gradients. Polysomes from non-infected cells banded at a density of 1.275 - 1.280. Equilibrium sedimentation pattern of infected cytoplasmic extracts revealed the presence of three distinct peaks at 1.275, 1.225 and 1.205 (Fig. 3). Using viral RNP from Nonidet-P40 treated virions it was found that the structures sedimenting at 1.225 were RNPs. Three RNA species were isolated from these structures (Fig. 4B). Complexes banding at a density of 1.205 were analyzed. After isolation and glutaraldehyde fixation, they banded in CsCl gradients at a density of 1.44-1.49. Similar values were reported for poliovirus replicative complexes (1). The nature of the RNA was determined by sucrose gradient centrifugation after SDS-phenol extraction (Fig. 4C). Two species of molecule were observed, one sedimenting at 20 S and partially resistant to RNase

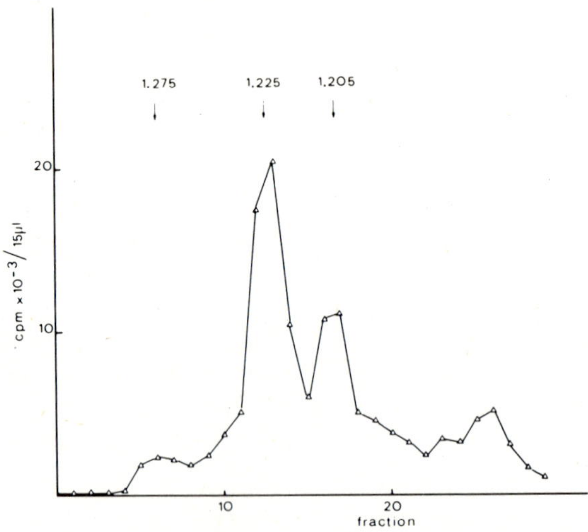

Fig. 3. *Cytoplasmic complexes:fractionation in metrizamide gradient. Lumbo virus-infected cells were labeled with uridine in the presence of actinomycin D for 2 hr at 3 hr p.i. Cytoplasmic extracts were prepared in isotonic buffer (rat liver supernatant and heparin 20 µg/ml). The material was centrifuged in a preformed metrizamide gradient (28-52%) for 14 hr at 38,000 rpm in a SW50 rotor. The density was determined from the refractive index.*

(35%) while the other sedimented at 16 S and was almost completely sensitive to RNase treatment. It seems likely that these cytoplasmic structures represent replicative and/or transcriptive complexes involved in RNA synthesis and include nascent single-stranded molecules still attached on their templates. After deproteinization by SDS-phenol, a large fraction was apparently released and sedimented at 16 S.

Polysomes isolated in the region d = 1.275 were deproteinized by SDS-phenol and mRNA were analyzed in sucrose gradients (Fig. 4A). They sedimented in two classes at 15 S and 12 S. Polyacrylamide gel electrophoresis revealed three RNA classes with mol. wt. of 0.6-0.7 x 10^6, 0.4 x 10^6 and 0.3 x 10^6 daltons. According to their size they might be monocistronic messengers encoding the polypeptide moieties of the three major structural proteins G1, G2 and N. It is obvious that these sequences do not contain the full genetic information since at least the messenger which codes for the minor large polypeptide is missing (or is present in low quantities so that it could not be resolved). Uridine-labeled mRNAs obtained in similar experiments were used in hybridization experiments with unlabeled, individual genome

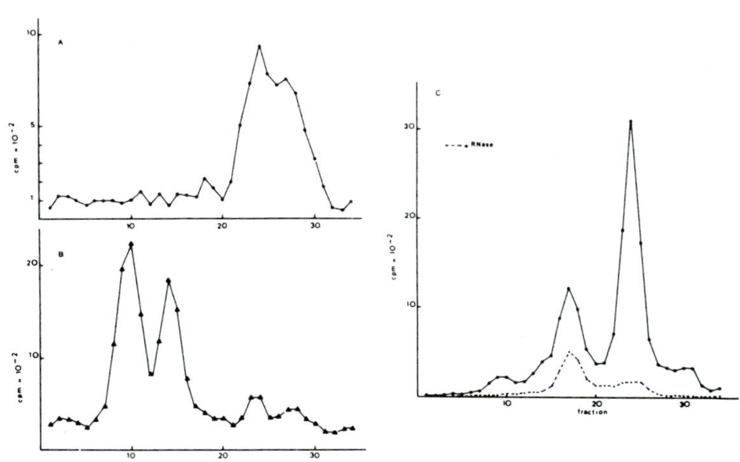

Fig. 4. *RNA composition of cytoplasmic complexes isolated in metrizamide gradient. RNA from material banding at density of 1.275 (A) 1.225 (B) or 1.205 (C) in metrizamide gradient (see Fig. 3) was extracted by SDS phenol and centrifuged on a 15-30% sucrose gradient for 15 hr at 30,000 rpm in a SW41 rotor. Each fraction of gradient "C" was tested for RNase resistance (100 μg/ml for 30 min at 37°C).*

segments to determine the relationships between the three virion RNAs. The results indicated that 64% and 71% of the labeled mRNAs became RNase resistant in annealing experiments respectively with L RNA and M RNA. In addition 96% of the messengers were annealed in the presence of a mixture of L and M RNAs. Therefore M RNA is not a deletion product of L RNA, since these molecules contain sequences encoding for distinct genetic information. From these experiments, we cannot ascertain whether L and M RNAs contain overlapping regions because they could be slightly cross-contaminated. When annealing experiments were performed using the small genome species as a probe, messengers became almost completely resistant to RNase treatment. This result was unexpected but could be explained by the fact that the S RNA could be contaminated by degradation products originated from L and M RNA: even in low amounts, they could be detected because of the sensitivity of the technique. Further experiments will be needed to determine whether S RNA contains unique genetic information or bears sequences already included in L and M RNAs.

DISCUSSION

The annealing technique has provided evidence for the segmented genome of Lumbo virus. Using the fingerprinting technique, similar conclusions were obtained for other bunyaviruses such as Uukuniemi virus (15), Lacrosse virus and Snowshoe hare virus (6). Other findings support this conclusion. First, a polymerase activity is associated with each RNP in the case of Lumbo virus and Uukuniemi virus (Ranki and Pettersson, personal communication). Secondly, three size classes of complementary RNA (or RNP) templates for viral RNA synthesis were observed in Lumbo-infected cells and this suggests that each segment replicates separately. Thirdly, genetic studies using ts mutants revealed a high frequency of recombination in cells co-infected by two mutants, suggesting that genome segment reassortment occurs (9, S. Ozden, personal communication).

ACKNOWLEDGEMENTS

I gratefully acknowledge M. Pierre Vialat for excellent technical assistance. This work was supported in part by D.R.M.E. Contract Number 73.34/192.

REFERENCES

1. Baltimore, D. and Huang, A.S. (1968). *Science* 162, 572.
2. Bouloy, M., Krams-Ozden, S., Horodniceanu, F. and Hannoun, C. (1973/74). *Intervirology* 2, 173.
3. Bouloy, M. and Hannoun, C. (1976a). *Virology* 69, 258.
4. Bouloy, M. and Hannoun, C. (1976b). *Virology* 71, 363.

5. Buckingham, M.E. and Gros, F. (1975). *FEBS Letters* 53, 355.
6. Clewley, J., Gentsch, J. and Bishop, D.H.L. (1977). *J. Virol.* 22, 459.
7. Dahlberg, J.E., Obijeski, J.F. and Korb, J. (1977). *J. Virol.* 22, 203.
8. Dissous, C., Lempereur, C., Verwaerde, C. and Krembel, J. (1976). *Eur. J. Biochem.* 64, 361.
9. Gentsch, J. and Bishop, D.H.L. (1976). *J. Virol.* 20, 351.
10. Gentsch, J., Bishop, D.H.L. and Obijeski, J.F. (1977). *J. Gen. Virol.* 34, 257.
11. Hewlett, M.J., Pettersson, R. and Baltimore, D. (1977). *J. Virol.* 21, 1085.
12. Obijeski, J., Bishop, D.H.L., Palmer, E. and Murphy, F. (1976). *J. Virol.* 20, 664.
13. Pettersson, R. and Kaariainen, L. (1973). *Virology* 56, 608.
14. Pettersson, R. and von Bonsdorff, C.H. (1975). *J. Virol.* 15, 386.
15. Pettersson, R., Hewlett, M., Baltimore, D. and Coffin, J. (1977). *Cell* 11, 51.
16. Porterfield, J., Casals, J., Chumakov, M.P., Yagaidamovich, S., Hannoun, C., Holmes, I., Horzinek, M., Mussgay, M. and Russell, P. (1973/74). *Intervirology* 2, 270.
17. Ranki, M. and Pettersson, R. (1975). *J. Virol.* 16, 1420.
18. Samso, A., Bouloy, M. and Hannoun, C. (1975). *C.R. Acad. Sc. Paris* 280, 779.
19. Samso, A., Bouloy, M. and Hannoun, C. (1976). *C.R. Acad. Sc. Paris* 282, 1653.

POLYADENYLATE (POLY A) SEQUENCES ASSOCIATED WITH RABIES VIRUS INTRACELLULAR RIBONUCLEIC ACID (RNA) SPECIES

A. ERMINE and P. ATANASIU

Rage Recherche Rhabdovirus Institut Pasteur, Paris, France

Rabies virus is a bullet shaped, membrane maturing virus classified as a rhabdovirus. The genome consists of a single negative stranded RNA with a molecular weight of approximately 4.6 x 10^6 (7, 9). The RNA is associated in the virion with a nucleoprotein (N) forming a nucleocapsid core. The viral envelope consists of two non-glycosylated membrane proteins (M_1 and M_2) and a glycoprotein (G) (8).

Rabies virus replicates entirely in the cytoplasm of susceptible cells. Recently it has been reported that prolonged treatment with actinomycin D or cytosine arabinoside has only a non-specific effect on rabies virus growth (4). It has been shown that virus-specific RNA extracted from infected BHK_{21} cells exhibits a heterogeneous profile on sucrose gradient analysis. 60% of this rabies virus-specific RNA is of small size (10-25S), is complementary to the minus strand RNA of the virion, and is thought to be the viral messenger RNA (mRNA) (3).

Polyadenylate sequences have been found in numerous RNA species with putative or verified messenger function and in RNA from several viruses in which the genome is its own messenger (2). Vesicular stomatitis virus (VSV), a rhabdovirus, has adenylate sequences associated with all virus messenger RNAs, whereas 40S template RNA from the virion contains no poly(A) (10). There is also evidence that mRNA of viruses like reovirus, TMV, and Lumbo, do not contain any poly(A) tracts (5, 6, 11). In this paper, we present experimental evidence in support of the hypothesis that rabies virus mRNA contains poly(A) tracts, and compare the amount of poly(A) associated with rabies virus mRNA with that of Lumbo virus and VSV.

RNA PREPARATION

Confluent monolayers of BHK_{21} cells were infected with either rabies or VSV and incubated. Actinomycin D was added to the culture medium half an hour before labelling with ^3H-uridine, then the cells were disrupted and cellular extracts were sedimented through equilibrium sucrose gradients. For rabies virus the gradient was divided into three regions: I, containing RNA molecules between 8S and 25S; II, composed of 25S to 35S molecules; III, a sharp peak of 39S. We previously reported (3) that: (a) RNAase resistance observed before annealing is high in region II. This finding suggests the presence of partially or totally double-stranded RNA molecules, likely to be trans-

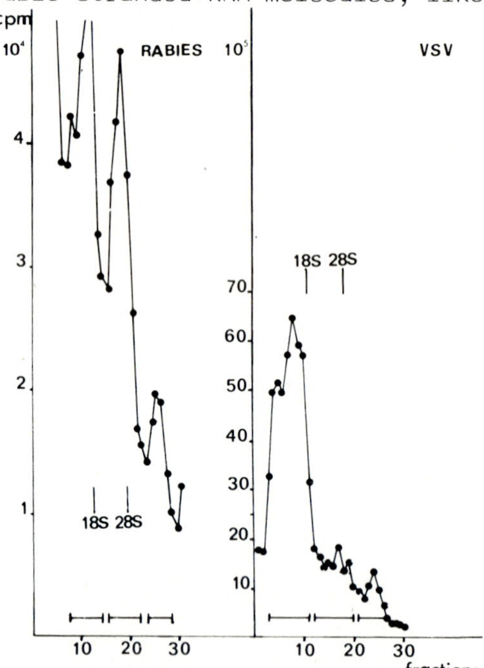

Fig. 1. *Virus RNA in BHK_{21} cells. Confluent monolayers of BHK_{21} cells were infected with virus (MOI:5) and incubated at 4° for 1 hr. All the cultures were then incubated at 37° with 4 ml MEM, 0.3% BSA in 5% CO_2. Actinomycin D (2 μg/ml) was added 0.5 hr before labelling with ^3H-uridine (1-3 hr post-infection for VSV, 8-12 hr post-infection for rabies). Then the cells were washed and disrupted in 1 ml of extraction buffer (1% NP_{40}, 1 mM EDTA, 50 μg/ml dextran sulphate, 10% RNase inhibitor, 1% SDS, 10% heparin). Cellular extracts were sedimented through an equilibrium sucrose gradient in the SW27 rotor (16 hr at 24000 rpm). After centrifugation the gradients were fractionated with an Isco gradient fractionator.*

criptive or replicative intermediates. (b) Annealing with an excess of cold virus RNA indicates that 65% of the labelled material is (+) molecules of viral origin. (c) Most of the (+) molecules are found in region I. They are postulated to be mRNA. (d) Region II is also composed of (+) single strands. It may contain the mRNA for the L protein. (e) Region III contains single-stranded (+) and (-) molecules (60% and 40%) of the same length as the genome. Fractions of each region were pooled and used directly for poly(A) detection.

For Lumbo virus (which contains a segmented RNA virus genome) it has been shown (1) that virus RNA extracted from infected cells sedimented as three species, 14S, 24S and 32S. A major part of mRNA is found in the 14S region, and double-stranded RNA in the 20-22S region. Genome RNA is found in all three regions.

AFFINITY CHROMATOGRAPHY ON POLY(U) SEPHAROSE

Samples were applied to the column and unbound fractions were obtained by washing with buffer 0.4M NaCl. The bound material was eluted with formamide.

Table 1 shows there is no difference in percent binding with non-infected cells or cells infected with Lumbo virus. With VSV and rabies virus binding is observed with regions I and II. The amount of binding was lower for rabies than for VSV.

ANALYSIS OF RABIES VIRUS RNA BOUND TO POLY(U) SEPHAROSE

In order to determine the nature of RNA bound to poly(U) sepharose, fractions of regions I and II obtained after elution

TABLE 1

Percentage of Poly(A) in Virus RNA

Fractions	% of VSV RNA bound on Poly(U)	% of Rabies RNA bound on Poly(U)	% of Lumbo RNA bound on Poly(U)	Uninfected Cells
I	75%	25%	5%	6%
II	55%	12%	7%	4%
III	23%	8%	5%	5%

Approximately 1 mg of poly(U) sepharose in 10 ml of buffer (NETS 0.1M NaCl 0.01M Tris (pH 7.4) 0.01M EDTA, 0.2% SDS) was used to prepare three small columns. The columns were washed with NETS buffer (0.4M NaCl). Samples (0.4M NaCl) were applied to the column at room temperature and the unbound fraction obtained by washing the column with 0.4M NaCl NETS buffer. The bound material was eluted by passage of 90% formamide elution buffer through the column.

with formamide were ethanol precipitated. The pellets were dissolved in the extraction buffer and centrifuged through a sucrose gradient.

The fraction of region I showed normal RNA profile, identical with the initial profile of region I. In contrast the fraction of region II exhibited a heterogeneous profile similar to the total virus RNA profile obtained from infected BHK_{21} cells.

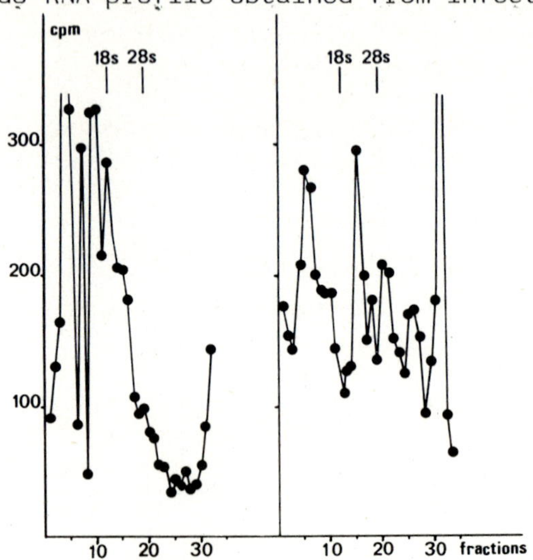

Fig. 2. *Composition of RNA bound to poly(U) sepharose. Fraction of regions I and II obtained after elution with formamide were ethanol precipitated (overnight, -20°). The pellets were dissolved in 1 ml of extraction buffer and centrifuged through an equilibrium sucrose gradient.*

CONCLUSIONS

Our results show that 25% of rabies virus-specific RNA binds to poly(U) sepharose in region I, and 12% in region II. Previous work (3) has shown that these two regions contain rabies mRNA. Thus we conclude that there are poly(A) tracts on rabies mRNA. In contrast there is no RNA binding in region III, which is known to contain RNA of genome length (60% of (+) molecules and 40% of (-) molecules).

With Lumbo virus, we could not observe any difference between binding to poly(U) sepharose of RNA extracted from infected cells or uninfected cells. These results suggest that Lumbo mRNA does not possess any poly(A) tracts.

In control experiments, we found 75% RNA binding for VSV-specific RNA in region I, which contrasts with the much lower percentage obtained with rabies virus. This low degree of binding is not due to RNA degradation because: (a) there is no loss of

TCA-insoluble radioactivity during RNA extraction and gradient centrifugation, (b) the ribosomal RNA profile shows no evidence of RNA degradation, (c) RNA profiles of region I are identical before and after binding to the poly(U) column, (d) there is no specific degradation of poly(A) tracts because in similar conditions a high percentage binding of VSV RNA is seen in agreement with published results (10), (e) inhibition of binding by proteins can be ruled out since a higher level of RNA binding is not obtained after pronase treatment (1 mg/ml 16 hr at 37°) and phenol-chloroform extraction.

From our results, the low level of binding can be accounted for by a low percentage of polyadenylation of rabies virus mRNA or by unusually short poly(A) tracts.

In region II we can observe (Fig. 2) that molecules bound to poly(U) sepharose have a very heterogeneous profile. This can be due to the presence of double-stranded RNA in this region (3), which is partially adsorbed to the poly(U) sepharose column. This double-stranded RNA can be denatured by formamide to RNA molecules of genome size and RNA molecules of mRNA size. But all the molecules in the region 8-25S cannot be accounted for by denaturation (with Lumbo virus there is not so much binding in the double-strand region). There is most probably, in region II, a small amount of mRNA (3) which is partially adenylated.

REFERENCES

1. Bouloy, M. *These de doctorat d'etat Universite Paris VII.*
2. Darnell, J.E., Jelinek, W.R. and Molloy, G.R. (1973). *Science* 181, 1215.
3. Ermine, A. and Flamand, A. (1977). *Ann. Microbiol. (Inst. Pasteur)* 128-A, 477.
4. Flamand, A., Pese, D. and Bussereau, F. (1977). *Virology* 78, 323.
5. Fraser, R.S.S. (1973). *Virology* 56, 379.
6. Siegel, A., Zaitnin, M. and Duda, C.T. (1973). *Virology* 53, 75.
7. Sokol, F. (1975). *In* "The Natural History of Rabies" (Baer, G.M. ed), Academic Press,Inc., N.Y., p.79.
8. Sokol, F. and Clark, H.F. (1973). *Virology* 52, 246.
9. Sokol, F., Shlumberger, H.D., Wiktor, T.J. and Koprowski, H. (1969). *Virology* 38, 651.
10. Soria, M. and Huang, A.S. (1973). *J. Mol. Biol.* 77, 449.
11. Ward, R., Shatkin, A.J., Banerjee, A.K. and La Fiandra, A. (1972). *J. Virol.* 9, 61.

PHENOTYPIC MIXING BETWEEN RABIES VIRUS AND VESICULAR STOMATITIS VIRUS

ANNE FLAMAND and FRANÇOISE BUSSEREAU

Bât.400, Université Paris-Sud,
91405 Orsay-Cedex, France.

To study the relationship between rabies virus and vesicular stomatitis virus (VSV), mixed infections between wild type or *ts* mutants of the VSV serotype Indiana and the CVS strain of rabies virus were performed. Because VSV will form plaques in conditions where rabies virus will not (the reverse situation does not exist), the study was limited to the VSV progeny. The possibility of complementation of *ts* mutants of VSV by rabies virus and the eventual appearance of phenotypically mixed particles of VSV was tested (for reviews see refs.2,4). In our system pseudotypes were defined as virions resistant to anti-VSV serum and containing a VSV genome.

MIXED INFECTION BETWEEN THE WILD TYPES OF RABIES VIRUS AND VSV

The replication cycle of rabies virus is much longer than that of VSV (30 hours and 6 hours respectively). Therefore in all experiments BHK cells were infected with the CVS strain of rabies before superinfection with VSV (for conditions see Fig.1). At the time of the superinfection all the cells were brightly fluorescent when treated with fluorescent antibodies directed against the rabies nucleocapsid N protein, so all the cells were infected by rabies virus. The results presented in Fig.1 show that virus was produced at 37, 38.5 and 39.6°, indicating that at a time where the cells contained rabies material in great amount, adsorption, penetration and multiplication of VSV can occur.

Neutralization tests were performed on the harvests, using anti-rabies or (and) anti-VSV serum. If the VSV inoculum or the harvest obtained from controls was treated with anti-V,

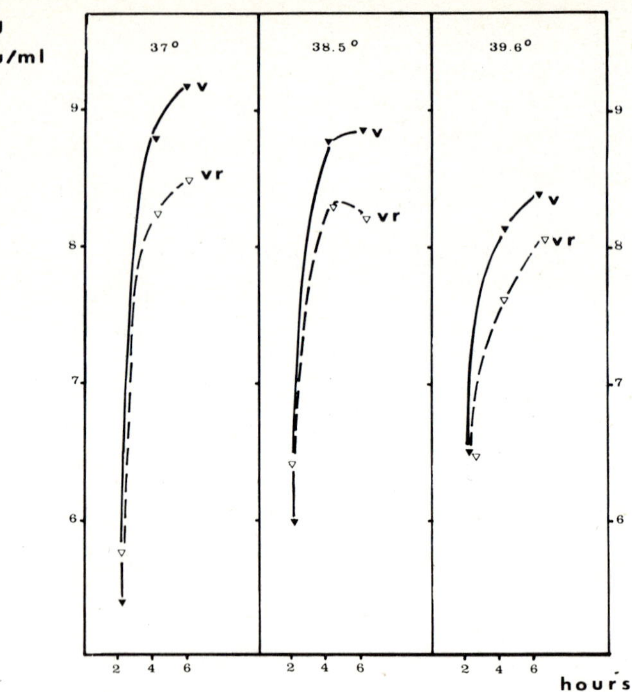

Fig.1. *Multiplication of VSV at different temperatures in BHK cells preinfected or not with rabies virus.*

Confluent monolayers of BHK cells (3 x 10^6 cells per plate) were infected with the CVS strain of rabies virus at a multiplicity of infection (moi) of 3 PFU/cell or mock infected. Cultures were incubated at 37° for 24 hours then superinfected with VSV, Indiana serotype, at a moi of 10 PFU/cell and incubated at 37°, 38.5° or 39.6°. The titrations were done under conditions in which only virions of VS genome form plaques. The yield was measured at 2, 4 and 6 hours after superinfection.
▼ *v = VSV (wild type) production in control cells*
▽ *v,r= viral production in cells preinfected with rabies virus.*

the titer was depressed by 5.5 log. Anti-R had no action. A mixture of both antisera, anti-V + R, gives the same degree of neutralization as anti-V alone; so no interference between both antisera was found.

When the harvests obtained from mixed infections were treated with anti-V, the inhibition was not as complete as with the control. The following percentages of survivors were obtained at 37°, 4.0, 0.3 and 0.1% (2,4 and 6 hours after superinfection respectively; at 38.5°: 1.5, 1.3 and 0.8% (2,4 and 6 hours after superinfection respectively; at 39.6°: 0.04, 0.03 and

0.05% (2, 4 and 6 hours after superinfection respectively). Moreover if the same harvests were treated with anti-V + R, those virions which were resistant to anti-V were totally inactivated to the same degree of neutralization as with VSV alone.

The fact that survivors to anti-V were found that were totally inactivated by anti-V + R was considered as evidence of the existence of pseudotypes. It seems that production of such virions was dependent on the temperature: more pseudotypes were found at 37° and 38.5° than at 39.6°. The proportion of pseudotypes was higher early after superinfection.

COMPLEMENTATION AND PHENOTYPIC MIXING BETWEEN *ts* MUTANTS OF VSV AND THE WILD TYPE OF RABIES VIRUS

Cells preinfected with the CVS strain of rabies virus were infected with VSV *ts* O 15, VSV *ts* O 111 23, and VSV *ts* O V 45, at the non-permissive temperature for the mutants of VSV (for conditions see Fig.2). These three *ts* mutants are affected respectively in the transcriptase, the M and the G proteins (3). As shown in Fig.2, the harvests were always lower in the presence of rabies virus than in the control cells. Therefore, there is no indication of complementation between rabies virus and VSV.

In vitro transcriptase activity has been demonstrated in rabies virus (Flamand, Delagneau and Bussereau unpublished data), therefore the absence of complementation between VSV *ts* O 1 5 and rabies virus suggests that rabies transcriptase cannot work efficiently with VSV genomes.

It has been shown that in cells infected with VSV *ts* O 111 23 and VSV *ts* O V 45 normal amounts of nucleocapsids were found at the non-permissive temperature (Combard, Martinet and Printz-Ané, personal communication). Absence of complementation between these mutants and rabies virus suggests that VSV nucleocapsids cannot be wrapped in rabies envelope to give infectious virions. Because phenotypic mixing was more efficient at lower temperatures, the experiments were repeated at 37° which is a semi-permissive temperature for both mutants, and the results are shown in fig.3. As expected, the residual growth of VSV *ts* O 111 23 and VSV *ts* O V 45 was higher at 37° than at 39.6°. In superinfected cultures, the yields were slightly below the level obtained with the control cultures; again there was no indication of complementation between the two viruses. The virus produced in mixed infection with VSV *ts* O 111 23 was absolutely insensitive to anti-R and completely inactivated by anti-V (or anti-V + R). Therefore one can assume that no pseudotypes were formed. The virus produced in mixed infection with *ts* O V 45 was insensitive to anti-R or

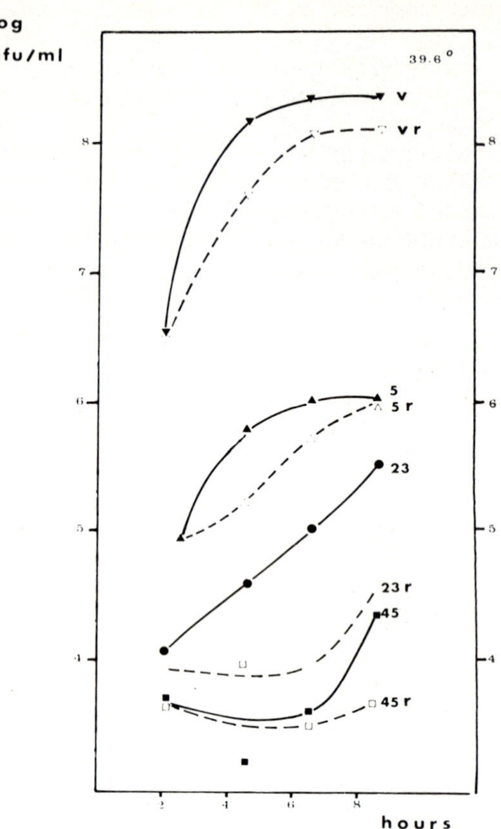

Fig.2. *Multiplication of VSV (wild type and ts mutants) in normal, and BHK cells preinfected with rabies virus at 39.6° (non-permissive temperature for ts mutants).*

Confluent monolayers of BHK cells (3 x 10^6 cells per plate) were infected with the CVS strain of rabies virus at a moi of 3 PFU/cell or mock infected. They were incubated at 37° for 24 hours, then superinfected with VSV, Indiana serotype, (wild type or ts mutants) at a moi of 10 PFU/cell and incubated at 39.6°. VSV production was measured at 2, 4 and 6 hours after superinfection. a) VSV production in control cells

▼ *v = VSV, wild type*
▲ *5 = VSV ts O 1 5*
● *23 = VSV ts O 111 23*
■ *45 = VSV ts O V 45*

b) Viral production in cells preinfected with rabies virus and superinfected with VSV.

anti-V but completely inactivated by anti-V + R. Moreover the infectious virus was much more fragile than the parental strain (which is already thermolabile, ref.1) being rapidly inactivated at room temperature. Our hypothesis is that the envelope of the virions produced in mixed infection with *ts* 0 V 45 included G proteins of both viruses. This would explain why they were resistant to one or the other antiserum but were neutralized by both of them. In fact it is quite possible that complementation could not be demonstrated because of the differential inactivation of normal and phenotypically mixed virions.

DISCUSSION

During mixed infection between rabies and VSV there is no exclusion of VSV by rabies virus. The harvest contained between 0.03 and 4% of pseudotypes, depending on temperature and length of further incubation.

Lack of complementation between rabies virus and VSV *ts* 0 1 5 indicates that rabies transcriptase is not efficient in transcribing VSV genome. Failure to detect complementation or phenotypic mixing between rabies virus and VSV *ts* 0 111 23 suggests that pseudotypes are not composed of VSV nucleocapsids coated in rabies envelopes.

Results with VSV *ts* 0 V 45 indicates that G protein of both viruses may be included in the envelope. The pseudotypes are extremely fragile which could explain why complementation was not detected.

At the moment it is seen that the only protein that could be exchanged between rabies virus and VSV is the G protein. Experiments are in progress to see if the N or Ns proteins may be provided by rabies virus.

ACKNOWLEDGEMENTS

We wish to thank Ph. Vigier for helpful discussions and J. Crick for reading the manuscript.

The excellent technical assistance of D. Pese is gratefully acknowledged. M. Sliwa typed the manuscript.

This research was supported by the "Centre National de la Recherche Scientifique" through the L.A. 40086 and the NATO grant 994.

▽ *v r = VSV, wild type*
△ *5 r = VSV ts 0 1 5*
○ *23 r = VSV ts 0 111 23*
□ *45 r = VSV ts 0 V 45*

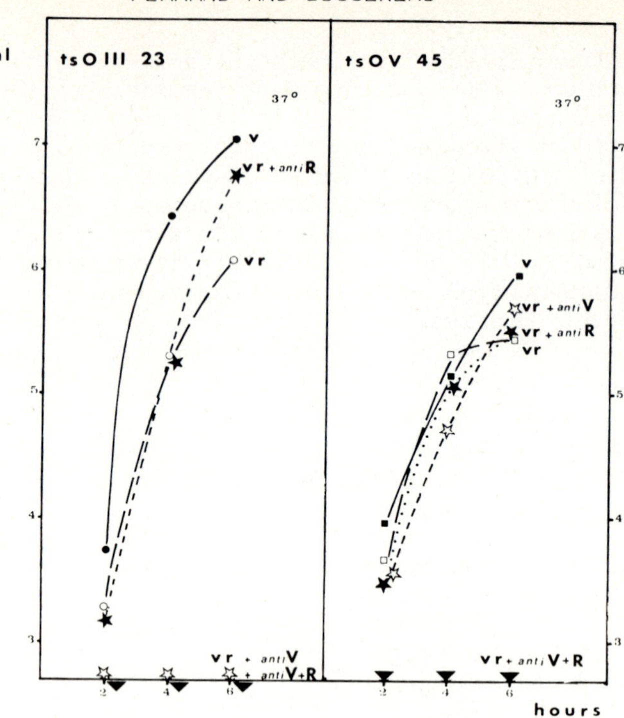

Fig.3. *Mixed infection between rabies virus and two ts mutants of VSV at 37° (semi-permissive temperature for ts mutants).*

Confluent monolayers of BHK cells (3 x 10^6 cells per plate) were infected with the CVS strain of rabies virus at a moi of 3 PFU/cell or mock infected. Cultures were incubated at 37° for 24 hours, then superinfected with VSV ts 0 111 23 or VSV ts 0 V 45 at a moi of 10 PFU/cell and incubated at 37°. Viral production in control (v) or mixed infections (v, r) was measured at 2, 4 and 6 hours after superinfection.

The harvest produced in mixed infection was also titrated in the presence of anti-VSV serum (v, r anti-V) or anti-rabies virus serum (v, r anti-R) or a mixture of both antisera (v, r anti-V + R).

REFERENCES

1. Deutsch, V. and Berkaloff, A. (1971). *Ann. Inst. Pasteur* 121, 101.
2. Dodds, J.A. and Hamilton, R.I. (1976). *Advances in virus research* 20, 33.
3. Wagner, R.R. (1975). *Comprehensive Virology* 4, 1.
4. Zavada, J. (1976). *Arch. Virol.* 50, 1.

EARLY AND LATE EVENTS IN THE REPLICATION CYCLE OF PARAMYXOVIRUSES

D.P. DURAND, J. BENEKE, J. BURDICK and M. SANBORN

Department of Bacteriology,
Iowa State University,
Ames, Iowa 50011, U.S.A.

Enveloped viruses, such as the paramyxoviruses Sendai and Newcastle disease virus (NDV), introduce virion components into the host cell membrane during the initial infection stage (at the time of viral envelope-cell membrane fusion) and just prior to final assembly when newly formed virion proteins or glycoproteins associate with the cell membrane (12). We will report the use of a radioimmunoassay procedure in our attempts to monitor these events. The enveloped viruses, especially the paramyxoviruses, are rather careless in their assembly process and produce heterogeneous particles which vary in size from 100 to 400 nm in diameter; their shape is spherical to ellipsoidal to filamentous (3, 5, 15). As a consequence, virus populations are produced with virions capable of attachment but not infectious; infectious but lacking fusion from without (FFWO) and hemolytic activity; or virions which are infectious, cause FFWO and are hemolytic. We will report our analysis of NDV virions separated into different populations by physical and temporal means.

DETECTION OF SENDAI VIRUS ANTIGENS IN INFECTED CELLS

In a productive infection (Sendai virus in CEL), parental virus components were depleted from the cell surface at a fairly rapid rate until approximately 8 hr into the cycle (Fig.1, upper panel). At this time, an increase in viral antigen at the surface was observed and probably represents parental virus material. Under the conditions of culture which we used, newly formed Sendai virus proteins are not observed until approximately 11-12 hr post infection which is represented in the second major peak seen for surface antigen. The coordinate reduction of surface antigen with the appearance of nucleocapsid antigen could reflect a burst of viral assembly and release. The following rise in both viral surface antigen and nucleocapsid antigen was expected and represents continued viral antigen production. The observed further depletion of surface antigen indicates a continued cyclic production of viral material, followed by assembly and release. This could be representative of a system presented by Kingsbury (9) which is a cyclic one, alternating between translation and replication, depending upon the amount of nucleocapsid available.

The non-productive BHK-21 cell line presented an entirely different picture for virus-host interaction. Since the labeled antibodies measured changes during the infectious cycle, it appeared that some virus antigen was produced though very inefficiently.

CHARACTERISTICS OF NDV VIRION POPULATIONS

The cholesterol to phospholipid ratio (C/P) has been an important determinant for several lipid studies involving viruses (2, 4, 10). Cholesterol in membranes has been suggested to be a stabilizing or condensing agent (13). This reported effect of cholesterol led to the speculation that the stability of NDV was, in part, linked to its cholesterol content. Cells reportedly have a C/P ratio of about 1.0 (11). Paramyxovirus populations have been separated into different size classes by differential and density gradient centrifugation, and the larger particles so derived have been shown to have greater hemolytic and FFWO ability (1, 8). It was implied that either the larger sized particles acted to enhance cell fusion by allowing for increased contact of the virus with the cell surface (8) or that the envelope of the larger particles was less stable (9). We looked at the different sized NDV virions to determine if any correlation existed between the virion C/P ratio and virion function. Table I presents the results of these determinations on NDV populations separated by density gradient centrifugation. The larger virions show a decreased C/P ratio and increased FFWO ability. If NDV buds through cholesterol-rich areas of the membrane (which appears to be the case since NDV has a higher C/P ratio than its host cell membrane), it is likely that virus produced late in the replication cycle would have available less cholesterol and so a lower C/P ratio than virus produced early in the cycle. In Table II we present the results obtained when progeny produced early and late in

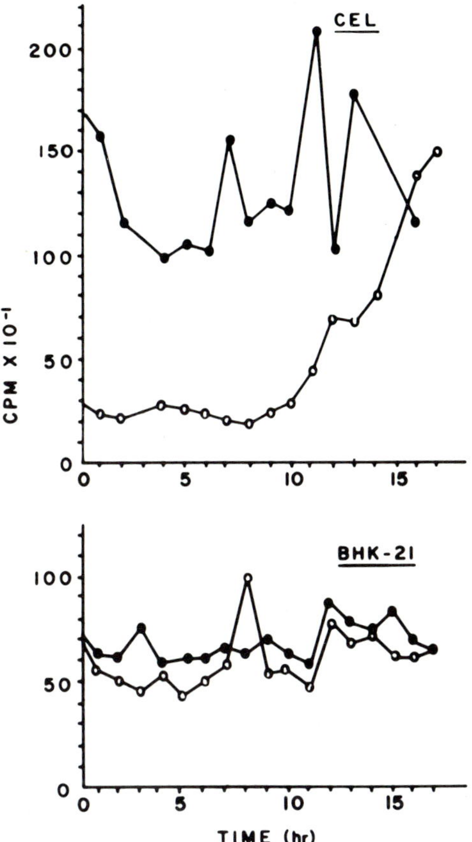

Fig.1. *Detection of Sendai virus antigens in chicken embryo lung cells (CEL) and baby hamster kidney cells (BHK-21). Virus was added to monolayers of CEL or BHK-21 at 4° for 10 min for attachment, unattached virus was removed by washing with BME and the cells incubated at 31° with BME. At hourly intervals cells were washed with PBS and fixed with either 2% paraformaldehyde for surface detection of antigen (7) or 70% ethanol for internal antigen detection (6). Antibody was raised in rabbits against whole Sendai virus or nucleocapsid which was isolated by the method of Scheid and Choppin (17). Antibody was tagged by the method of Rifkin et al. (14) as modified by Sanborn and Durand (16). Tritiated antibody was added for 45 min then excess antibody removed by washing 5 times with PBS. The cells were digested and the CPM of tritiated antibody attached to the cell surface (●——●) or internal nucleocapsid (o——o) determined.*

TABLE I

Characteristics of NDV virion populations separated by sucrose density gradients

Gradient[a] Fractions	Morphology[b]	HAU/pfu[c]	C/P[d]	FFWO[e]
Middle	relatively uniform, intact virions of ~ 150 nm diameter	439	1.04	1.16
Bottom	pleomorphic, fragile virions ~ 200-300 or greater in diameter	270	0.89	1.28

[a]Aqueous sucrose, 4 to 54%; 2 hr at 32,000 x g in a SW 27 rotor; 19, 2 ml fractions, middle fraction tubes 8-13, bottom fraction tubes 14-19.

[b]Negative stain with phosphotungstic acid.

[c]Ratio of hemagglutination units to plaque forming units.

[d]Ratio of cholesterol to phospholipid.

[e]Fusion from without = average number of nuclei/cell; MOI = 100 pfu/cell.

TABLE II

Characteristics of NDV virion populations collected at different times postinfection

Time post-infection[a]	Morphology[b]	HAU/pfu[c]	C/P[d]	FFWO[e]
6 hr	relatively uniform, intact virions of ~150 nm diameter	380	1.08	1.14
12 hr	pleomorphic, fragile virions, ~200-300 or greater in diameter	702	0.84	1.33

[a]Five units neuraminidase was added at 5 hr, medium was collected at 6 hr, the same treatment was repeated at 11 hr.

[b]Negative stain with phosphotungstic acid.

[c]Ratio of hemagglutination units to plaque forming units.

[d]Ratio of cholesterol to phospholipid

[e]Fusion from without = average number of nuclei/cell; MOI = 100 pfu/cell.

the replication cycle were analyzed. As predicted, the virions produced late in the cycle were larger and more fragile, with a lower C/P ratio and greater FFWO ability than those virions produced early in the cycle. In neither case could we show any significant differences in the protein profiles obtained by PAGE.

SUMMARY

Sendai virus replication in CEL cells was monitored by the use of tritiated antibody to detect the appearance of viral antigens at the infected cell surface. A cyclic response was observed with three major peaks appearing during the 17 hour test period. Generation of nucleocapsid protein seemed to stimulate a burst of virus assembly and release which was detected as a reduction of virus antigen at the cell surface.

When NDV populations were separated by centrifugation or collected at different times post infection, the larger particles were found to be more fragile, to have a lower C/P ratio, and to induce greater FFWO levels. During assembly by budding, the paramyxovirus envelope appears to be enriched in cholesterol; it therefore is possible that virus produced late in the cycle would have to bud through modified cell membranes depleted in cholesterol and the virion envelope would thus be less stable. Data presented support this suggestion.

REFERENCES

1. Apostolov, K. and Waterson, R.P. (1975) *In* "Negative Strand Viruses" (Mahy, B.W.J..and Barry, R.D., eds.) p.799. Academic Press.
2. Blair, C.D. and Brennan, P.J. (1972). *J. Virol.* 9, 813.
3. Blough, H.A. and Tiffany, J.M. (1975). *Curr. Topics Microbiol. Immunol.* 70, 1.
4. Choppin, P.W., Compans, R.W., Scheid, A., McSharry, J.J. and Lazarowitz, S.G. (1972). *In* "Membrane Research" C.F. Fox, ed., p.163. Academic Press.
5. Feller, U., Dougherty, R.M. and Distefano, H.S. (1969). *J. Virol.* 4, 753.
6. Girardi, A., Hampar, B., Hsu, K.C., Oroszlan, S., Hornberger, E., Kelloff, G. and Gilden, R.V. (1973). *J. Immunol.* 111, 152.
7. Gonatas, N.K., Stieber, A., Gonatas, J., Gambetti, P., Antonine, J.C. and Avrameas, S. (1974). *J. Histochem. Cytochem.* 22, 999.
8. Hosaka, Y. (1975). *In* "Negative Strand Viruses, (Mahy, B.W.J. and Barry, R.D., eds.), p.885. Academic Press.
9. Kingsbury, D.W. (1974). *Med. Microbiol. Immunol.* 160, 73.

10. Klenk, H.-D. (1973). *In* "Biological Membranes". (Chapman, D. and Wallach, O.F.M. eds.), p.145. Academic Press.
11. Klenk, H.-D. (1974). *Curr. Topics Microbiol. Immunol.* 68, 29.
12. Lenard, J. and Compans, R.W. (1974). *Biochem. Biophys. Acta.* 344, 51.
13. Papahadjopoulos, D., Cowden, M. and Kimelberg, H. (1973). *Biochem. Biophys. Acta.* 330, 8.
14. Rifkin, D.B., Compans, R.W. and Reich, E. (1972). *J. Biol. Chem.* 247, 6432.
15. Roman, J.M. and Simon, E.H. (1976). *Virology* 69, 287.
16. Sanborn, M.R. and Durand, D.P. (1974). *Infect. Immun.* 10, 445.
17. Scheid, A. and Choppin, P.W. (1973). *J. Virol.* 11, 263.

NEGATIVE STRAND VIRUSES IN ENUCLEATE CELLS

T.H. PENNINGTON and C.R. PRINGLE

Department of Virology and M.R.C., Virology Unit, Institute of Virology, University of Glasgow, Glasgow, Scotland.

Recent studies on the growth of RNA viruses in enucleate cells have shown that these viruses can be placed in four categories according to the degree of their dependence on the host cell nucleus (14). These categories comprise (i) viruses which multiply well in enucleate cells, examples including VSV and Semliki forest virus (3, 4, 16); (ii) viruses which multiply poorly in enucleate cells, examples including picornaviruses and reoviruses (4, 12); (iii) viruses which fail to produce infectious virions but which induce the synthesis of virus-specific gene products, an example being measles virus (5) and (iv) viruses which fail to induce the synthesis of virus-specific gene products in enucleate cells, exemplified by influenza virus (3, 7).

This paper reports the results of studies in virus growth and macromolecular synthesis in enucleate cells infected with Sendai virus and Bunyamwera virus, these viruses being chosen as typical representatives of the Paramyxoviridae and Bunyaviridae respectively. A continuous line of African green monkey kidney cells (BSC-1) was used in all enucleation experiments. Enucleate cells were prepared by treating cell monolayers with the mould metabolite cytochalasin B; the extruded nuclei were then sheared from the cells by the application of centrifugal force. This procedure results in the enucleation of almost all the cells in the monolayer, with minimal cell loss. The method described by Follett (2) was used. Cells were grown on 3 cm plastic petri dishes. When confluent, monolayers were inverted in centrifuge tubes containing Eagles medium with cytochalasin B (5µg/ml) at 37°. The tubes were then centrifuged at 37°

for 20 min at 9,000 r.p.m. (MSE 10 x 100 rotor, MSE 65 centrifuge) or 10,000 r.p.m. (Beckman type 21 rotor). Cells were then incubated in normal medium lacking cytochalasin for at least 1 hr at 37° before infection. Control monolayers which had been treated with cytochalasin B but not centrifuged were used as controls in all experiments.

The efficiency of enucleation was determined by counting at least 500 cells from six separate areas of a monolayer stained with Giemsa's stain. Monolayers with more than 3% nucleate cells remaining were discarded. The ability of enucleate cell monolayers to support the growth of VSV was used to test their viability (Table I).

TABLE I

Yield from cytochalasin-treated and enucleated cells as a percentage of the yield from cytochalasin-treated nucleate cells.

Time of enucleation (hours) before or after infection	Virus				
	VSV Cocal	Sendai	Bunyamwera	SSPE Measles	Influenza WSN
-1	28.3	0.3	0.3	0.3	0.2
+2	64.7	0.3	0.5	3.0	-
+4	93.8	-	0.9	16.0	20.0
+6	100.0	1.3	1.1	63.0	-

SENDAI VIRUS

Infection of BSC-1 cells with Sendai virus leads to the production of non-infectious virus particles; these particles can, however, be rendered infectious by trypsin treatment (Pennington, unpublished results), probably due to the cleavage of the inactive F spike to its active form (15). Accordingly, virus samples were assayed in the presence of trypsin.

Infection of enucleate cells with Sendai virus resulted in virus yields which were less than 1% of those obtained from cytochalasin treated nucleate cells (Table I). As enucleated monolayers usually contained between 2% and 3% nucleate cells the possibility could not be ruled out that the small yield of virus observed in these experiments originated from these residual nucleate cells. It appeared likely, therefore that enucleate cells did not support a productive infection with this virus.

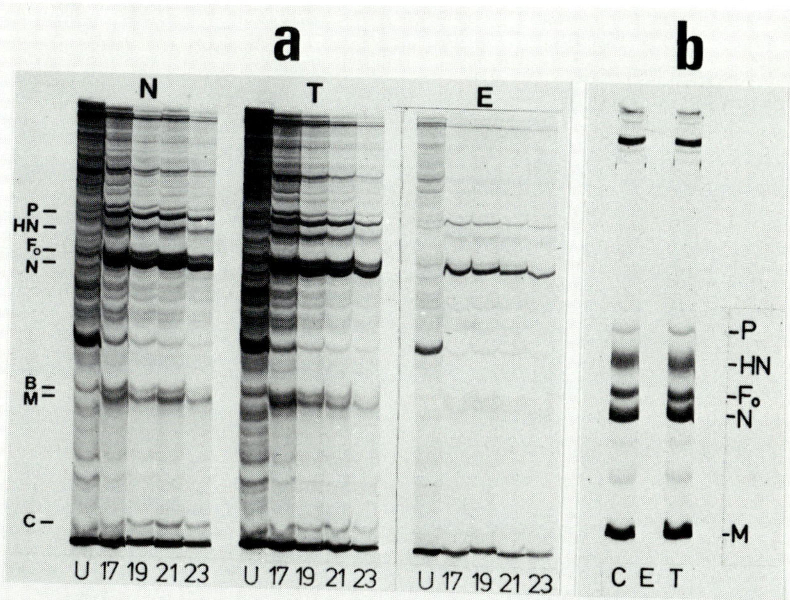

Fig.1. *Sendai virus polypeptide synthesis and particle production.*
(a) Virus-induced polypeptide synthesis in control and enucleate cells. Cells were labelled with ^{35}S-methionine for 30 min starting at the times indicated (hr post-infection, multiplicity of infection 20 p.f.u./cell). Samples were analyzed by SDS polyacrylamide gel electrophoresis and autoradiography using methods previously described (10). The autoradiogram of gels analyzing enucleate cell samples was exposed for twice as long as the other autoradiograms. N: control cells, no cytochalasin; T: control cells treated with cytochalasin but not centrifuged; E: enucleate cells; U: uninfected cells.
(b) Autoradiogram showing virus particle production by control and enucleate cells. Infected cells were labelled for 24 hr with ^{35}S-methionine and virus particles were partially purified as described in the text. Samples were analyzed by SDS polyacrylamide gel electrophoresis and autoradiography; C: infected control cells, no cytochalasin; E: infected enucleate cells; T: infected control cells treated with cytochalasin but not centrifuged.

Virus-induced polypeptide synthesis was examined in enucleate and nucleate cells; infected cells were pulse-labeled with ^{35}S-methionine at various times after infection. In autoradiograms of polyacrylamide gels analyzing control cell samples we observed (Fig.1) a pattern of virus-induced polypeptides similar to that described by Lamb *et al.* (8). Treatment of cells with cytochalasin had no effect on the number, mobility or time of appearance of these virus-induced polypeptides. Similarly, all virus-induced polypeptides were synthesized in enucleate cells, and they could not be distinguished either by mobility or time of appearance from those in normal or treated cells. The amount of virus-induced protein synthesized in enucleate cells was, however, much less than that synthesized in control cells; at 17 hr post infection in the experiment illustrated in Fig.1 the amount of polypeptide N synthesized in enucleate cells was 9% of that synthesized in cells treated with cytochalasin B.

Haemadsorption was tested using sheep erythrocytes; both infected enucleate and control cells showed a similar pattern of massive haemadsorption. These results confirm those of Cheyne and White (1) who demonstrated Sendai virus-induced haemadsorption in mechanically produced anucleate fragments prepared from radiation-induced giant cells.

In view of the apparent discrepancy between the lack of production of virus particles and the synthesis of virus polypeptides and haemagglutinin, the ability of enucleate cells to produce virus particles was assessed using biochemical techniques. Infected cells were labelled for 24 hr with ^{35}S-methionine and the virus particles released into the medium during this period were concentrated and partially purified by differential centrifugation with added purified Sendai virus as carrier.

Polyacrylamide gel analysis showed that only insignificant amounts of virus were produced by enucleate cells (Fig.1). It was therefore concluded that the failure of enucleate cells to support a productive infection with this virus could be explained, in part at least, by a defect in the assembly of virus particles. This conclusion received support from experiments in which infected cells were enucleated at different times after infection, as enucleation at 2 hr post-infection and 6 hr post-infection had as profound an inhibitory effect on virus yield as enucleation before infection (Table I).

BUNYAMWERA VIRUS

Assay of the yield of infectious virus particles released from enucleate BSC-1 cells infected with Bunyamwera virus showed that this virus, like Sendai virus, grew poorly, if at all, in such cells (Table I). The small yield of virus detected could easily be explained as originating from the nucleate cells remaining after cytochalasin treatment and centrifugation. Enucleation of cells at different times after infection also had severe inhibitory effects on virus growth (Table I). Virus-induced polypeptide synthesis was also investigated. Bunyamwera virus-induced polypeptides can be divided into two groups, early and late, according to their time of synthesis (11). Polypeptides N (Mol.Wt. 23,000) and L (Mol.Wt. 200,000) are synthesized early; they can be detected at 2 hr post-infection, their rates of synthesis reaching peaks simultaneously at 4 hr post-infection and then declining. The late polypeptides comprise the virion glycoproteins G1 (Mol. Wt. 128,000) and G2 (Mol. Wt. 31,000); their synthesis is first detected at 4 hr post-infection. The rate of synthesis of these polypeptides rapidly rises and then remains constant for several hours before gradually declining. Accordingly, virus-induced polypeptide synthesis in enucleate cells was investigated at various times after infection. These experiments showed that the three major virus-induced proteins N (early), G1 and G2 (late) were synthesized in enucleate cells (Fig.2), although the amounts made were much smaller than in control cells. In some experiments (Fig.2) only very small amounts of polypeptides G1 and G2 were synthesized in enucleate cells; in other experiments (Fig.2) appreciable amounts of these polypeptides were made, although the reduction of their synthesis in enucleate cells was still significantly greater than that observed with polypeptide N.

DISCUSSION

Sendai and Bunyamwera viruses resemble each other in that they are both unable to grow in cells enucleated by treatment with cytochalasin B followed by centrifugation. Infection of such cells with either virus, however, leads to the induction of virus-specific protein synthesis. In this system, the behaviour of these viruses resembles that of measles virus (5); they can be distinguished from this virus, however, by their differing response to enucleation at various times after infection (Table I). This type of experiment shows that enucleation only exerts a major inhibitory effect on measles virus when carried out before, or during the early stages of infection, whereas such an effect follows enucleation at later times during the Sendai or Bunyamwera virus growth cycles.

Comparison with other negative strand viruses (Table I) shows that Sendai and Bunyamwera viruses differ markedly from vesicular stomatitis virus, which grows well in enucleate cells, and from influenza virus, which requires a nucleus during the early stages of its growth cycle and which, in the absence of a nucleus, does not synthesize any virus polypeptides. The inhibitory effect of enucleation on the growth of Sendai and Bunyamwera viruses can only be explained in part by the smaller amounts of virus-induced proteins which are synthesized in such cells, and it is not clear why enucleation should exert such a

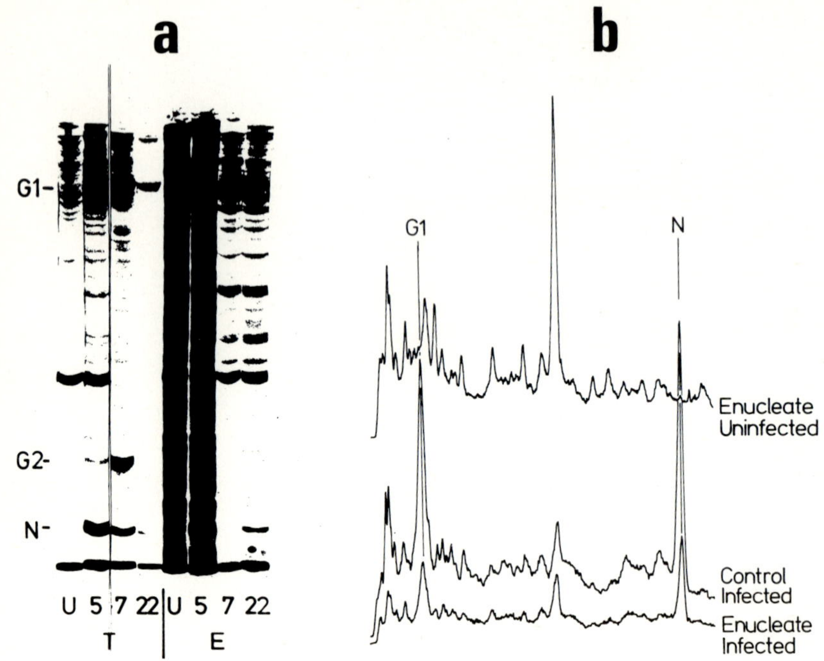

Fig.2. *Bunyamwera virus polypeptide synthesis in control and enucleate cells. Cells were infected with Bunyamwera virus (input multiplicity 40 p.f.u./cell) and were labelled with ^{35}S-methionine for 30 min starting at the times indicated (hr post-infection). Samples were analyzed by SDS polyacrylamide gel electrophoresis and autoradiography.*

(a) Autoradiogram. The two tracks on the left were exposed for 1/3 as long as the remaining tracks. U: uninfected cells. T: cells treated with cytochalasin but not centrifuged; E: enucleate cells.

(b) Virus-induced polypeptide synthesis in cells labelled at 19 hr post-infection: densitometer tracing of autoradiogram. In this experiment the synthesis of polypeptide G1 was more marked than in the experiment illustrated in Fig.2(a).

profound effect on the growth of these viruses. Sendai virus might be expected to replicate efficiently in such cells since it resembles vesicular stomatitis virus in being a negative strand virus with an unsegmented genome and a virion transcriptase. Likewise, although genetical and biochemical evidence indicates that Bunyaviruses, like myxoviruses, possess segmented genomes (6, 9, 13) they differ from myxoviruses in that they replicate in the presence of actinomycin D (13) and would also be expected to replicate in the absence of a nucleus.

Enucleate cells have a limited life span and Sendai and Bunyamwera viruses have relatively long growth cycles, and it could be argued that the inhibitory effect of enucleation might be due to the premature termination of virus growth occasioned by early cell death. This explanation is rendered unlikely by the demonstration that respiratory syncytial virus, a virus with an extended growth cycle, productively infects enucleate cells (4).

The results of experiments in which cells were enucleated at various times after infection show that the inhibitory effect of enucleation is not restricted to the early stages of growth of these viruses, and one possibility is that damage to cells occasioned by enucleation in some way interferes with virus assembly. Some support for this is provided by the experiments which showed that few Sendai virions were released from enucleate cells. Members of the Bunyaviridae mature by budding into vesicles in the region of the Golgi apparatus (13) and while the Golgi apparatus remains in the cell after nuclear extrusion it rapidly loses its structural integrity thereafter (17). Further experiments will be required to critically assess these possibilities.

REFERENCES

1. Cheyne, I.M. and White, D.O. (1969). *Aust. J. Exp. Biol. Med. Sci.* 47, 145.
2. Follett, E.A.C. (1974). *Exptl. Cell Res.* 84, 72.
3. Follett, E.A.C., Pringle, C.R., Wunner, W.H. and Skehel, J.J. (1974). *J. Virol.* 13, 394.
4. Follett, E.A.C., Pringle, C.R. and Pennington, T.H. (1975). *J. Gen. Virol.* 26, 183.
5. Follett, E.A.C., Pringle, C.R., Pennington, T.H. and Shirodaria, P. (1976). *J. Gen. Virol.* 32, 163.
6. Gentsch, J. and Bishop, D.H.L. (1976). *J. Virol.* 20, 351.
7. Kelly, D.C., Avery, R.J. and Dimmock, N.J. (1974). *J. Virol.* 13, 1155.
8. Lamb, R.A., Mahy, B.W.J. and Choppin, P. (1976). *Virology* 69, 116.
9. Objeski, J.F., Bishop, D.H.L., Palmer, E.L. and Murphy, F. (1976). *J. Virol.* 20, 664.
10. Pennington, T.H. (1974). *J. Gen. Virol.* 25, 433.
11. Pennington, T.H., Pringle, C.R. and MacCrae, M.A. (1977). *J. Virol.* 24, 397.
12. Pollack, R and Goldman, R.D. (1973). *Science* 179, 915.
13. Porterfield, J.S., Casals, J., Chumakov, M.P., Gaidamovich, S. Ya., Hannoun, C., Holmes, I.H., Horzinek, M.C., Mussgay, M., Oker-Blom, N., and Russell, P.K. (1975/76). *Inter-*

virology 6, 13.
14. Pringle, C.R. (1977). *Current Topics in Microbiology and Immunology*, 76, 49.
15. Scheid, A. and Choppin, P.W. (1974). *Virology* 57, 475.
16. Wiktor, T.J. and Koprowski, H. (1974). *J. Virol.* 14, 300.
17. Wise, G.E. and Prescott, D.M. (1973). *Exptl. Cell Res.* 81, 63.

RNA AND PROTEIN SYNTHESIS IN A PERMISSIVE AND AN ABORTIVE INFLUENZA VIRUS INFECTION

F.X. BOSCH*, A.J. HAY and J.J. SKEHEL

Division of Virology,
National Institute for Medical Research,
Mill Hill, London NW7.

**Present address:*
Institut für Virologie an der Veterinarmedizinischen Fakultat der Justus Liebig-Universitat Giessen, 63 Giessen, Frankfurter Strasse 87, Germany.

Many influenza virus infections of cells in culture do not result in the production of infectious virus. The results of analyses of polypeptide synthesis in a variety of cells infected with several different influenza viruses initially indicated that such abortive infections can be divided into two main groups: those in which virus particles are produced which contain uncleaved, inactive haemagglutinin and are therefore noninfectious (6, 9), and those in which virus-specified products are produced but assembly of virus particles does not occur. As an example of the latter, fowl plague virus (FPV) infection of L cells results in virus yields of less than 1 pfu/cell in contrast to permissive infections of chick embryo fibroblasts (CEF) in which yields of about 100 pfu/cell are obtained under equivalent experimental conditions. In an attempt to define the factors responsible for lack of virus particle formation these two systems have been compared and the results of analyses of virus-specific RNA and protein synthesis are presented here.

VIRUS-SPECIFIC RNA SYNTHESIS

The results of analyses of the synthesis of virus-specific RNA during the permissive infection of CEF by FPV have recently been reported (4) and the essential features may be summarized as follows: 1. The genome of the virus used in these studies contains 10 single-stranded RNAs which are all transcribed into corresponding molecules of complementary sequence. 2. The rates of production of the complementary RNAs (cRNA) and of virion RNA (vRNA) are maximal at around 2 hr after infection.

3. Two classes of cRNA molecules are produced: polyadenylated messenger RNAs which are incomplete transcripts and lack sequences complementary to the 5' ends of the virus RNAs (see chapter 31); and unpolyadenylated cRNAs which in contrast are complete genome transcripts, are not associated with polysomes in infected cells and probably function as templates in genome replication. 4. The syntheses of these two classes of cRNA are separately controlled. Thus, the synthesis of mRNA can proceed in the absence of protein synthesis, but the production of complete transcripts is dependent on the continued synthesis of virus-specific proteins (unpublished observations). Moreover, unpolyadenylated cRNAs are synthesized in similar amounts during infection, but the synthesis of individual messenger RNAs varies throughout infection both with respect to the amount of each mRNA produced and the time after infection at which it is synthesized in maximum amounts (Figs.1 and 2).

The results of similar analyses of RNA synthesis in FPV infected L cells have indicated that various aspects of the transcription and replication of virus RNA occur normally but that certain features of the control of mRNA synthesis are distinctly different from those observed in permissively infected CEF. In both studies virus-specific cRNAs and vRNAs were detected by annealing ^{3}H-labelled RNA extracted from infected cells either in the presence of an excess of unlabelled vRNA or unlabelled cRNA prepared from cells infected and incubated in medium containing cycloheximide. The relative amounts of vRNA and cRNA synthesized during infection were similar in both cell types although maximum synthesis occurred between 30 and 60 min later in L cells. The hybrid molecules containing labelled polyadenylated cRNAs and labelled unpolyadenylated cRNAs were separated by either oligo (dT) cellulose chromatography or by selective precipitation in 2 M lithium chloride, incubated with nuclease S_1 to remove any unhybridized, non-complementary sequences and then analyzed by electrophoresis in polyacrylamide gels.

The results of analyses of the polyadenylated messenger RNAs synthesized at different times during infection of both CEF and L cells are shown in Figs.1 and 2. All the virus-specific mRNAs were synthesized in L cells and as in CEF their syntheses varied independently during infection. However, the relative amounts of the various mRNAs synthesized in the two systems were different and the regulation of transcription, therefore, appears to vary with the cell type. In L cells mRNAs 6, 7 and 9 were produced in relatively small amounts throughout infection and mRNAs 1, 2 and 3, although synthesized in greater relative amounts than during the first hour of infection in CEF, were also more drastically reduced later in infection. Other features of the control of mRNA synthesis, however, such as the preferential synthesis of mRNA 8 at early times and mRNA 4 at later times during infection appeared to be similar in the two systems.

Analyses of RNA synthesis in cells infected in the presence of cycloheximide also indicated that primary transcription in L cells differed from that in CEF. Relatively larger amounts of mRNAs 1, 2 and 3 were produced and relatively smaller amounts of mRNA 7 were consistently observed - features also apparent in the early stages of a normal infection of these cells (figs. 1 and 2). As in the case of infected CEF, primary transcription remained unaltered in infected L cells incubated for at least 5 hr in medium containing cycloheximide.

Comparisons of the synthesis of unpolyadenylated cRNA in the two types of cell indicated that they were synthesized in similar amounts and in similar relative proportions (Fig.3).

VIRUS-SPECIFIC POLYPEPTIDE SYNTHESIS

Recent analyses of the virus-specific polypeptides synthesised in FPV infected CEF have indicated the synthesis of 10 distinct primary gene products (Fig.3), c.f. Inglis *et al.*, (5). Two additional virus-specific polypeptides which were distinguished by using certain polyacrylamide gel electrophoresis systems (10) were shown by tryptic peptide analyses to be indistinguishable from virus specific polypeptides MP and 8, respectively. With the possible exception of polypeptide 9, all virus-specific polypeptides are synthesised in infected L cells and the haemagglutinin cleavage products HA_1 and HA_2 are produced (Fig.4). However, compared to the controlled synthesis of these polypeptides in CEF (11, 12) their relative rates of synthesis in L cells are quite different (Fig.5). In particular, polypeptides P_1, P_2, P_3, NA (not shown) and 7 were all synthesised in reduced amounts relative to polypeptide 5 while polypeptide 10 was synthesised in greater amount. Furthermore the relative rates of synthesis of the various polypeptides remained fairly constant between 2 and 7 hr after infection in contrast to those in CEF.

The translation products of primary transcripts synthesised in both cell types were also examined. In these experiments cells were infected in the presence of cycloheximide for 5-6 hr and labelled immediately after reversal of the cycloheximide block. As shown in Fig.6 in both CEF and L cells, the synthesis of polypeptides P_1, P_2, P_3, NP, MP and 8 was detected and again the synthesis of MP was reduced in the latter system.

From these comparisons of the rates of production of messenger RNA and polypeptides in FPV infected CEF it is therefore apparent that the controlled production of messenger RNA is

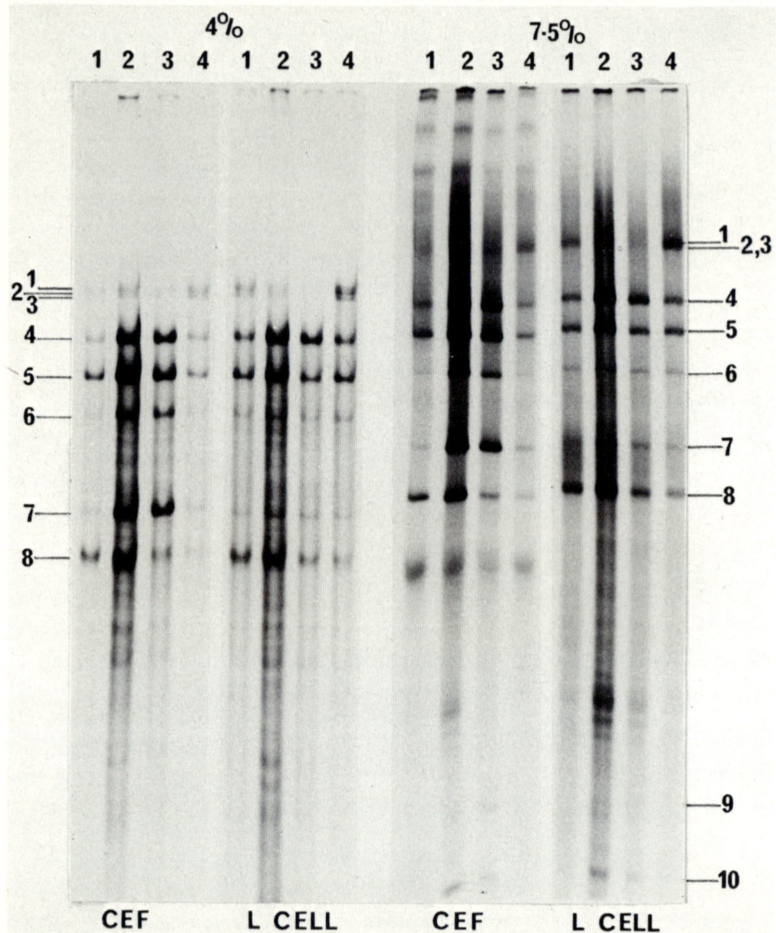

Fig.1. *The synthesis of polyadenylated cRNAs in fowl plague virus-infected CEF and L cells.*

Monolayers of the cells infected with FPV (100 pfu/cell) were incubated in medium containing ^{3}H-uridine (100 μc/ml for 60 min periods from 0 hr (1), 1½ hrs (2) or 3 hrs (3) after infection or from 0 hr (4) after infection in the presence of cycloheximide (100 μg/ml). The extracted RNA was denatured in 90% dimethyl sulphoxide at 45° in the presence of an excess of unlabelled virus RNA and annealed in 63% dimethyl sulphoxide at 37° for 20 hr and the polyadenylated and unpolyadenylated hybrid molecules isolated by oligo (dT) cellulose chromatography and LiCl precipitation, as described previously (4). The RNA in each fraction was incubated at 37° for 2 hr with nuclease S_1 (10-20 units/μg RNA) and the double-stranded RNAs were separated

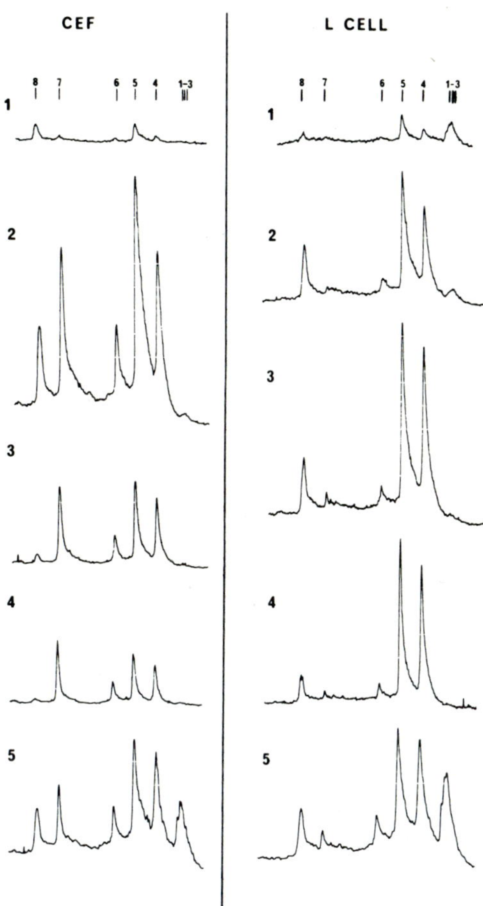

Fig.2. *Microdensitometer tracings of fluorography of polyadenylated cRNAs synthesised in CEF or L cells at different times after infection with FPV.*

The experimental details were as described in fig.1. Cells were labelled for 60 minute periods from 1. 0 hr, 2. 1½ hr, 3. 3 hr or 4.5 hr after infection with fowl plague virus or from 5.0 hr after infection in the presence of cycloheximide (100 µg/ml). Double-stranded RNAs derived from polyadenylated hybrid molecules by nuclease S_1 treatment were separated by electrophoresis on 4% polyacrylamide slab gels.

---

by electrophroesis either in 4% polyacrylamide slab gels at 4 volts/cm for 15 hr or in 7.5% polyacrylamide slab gels containing urea at 3 volts/cm for 40 hr and detected by fluorography.

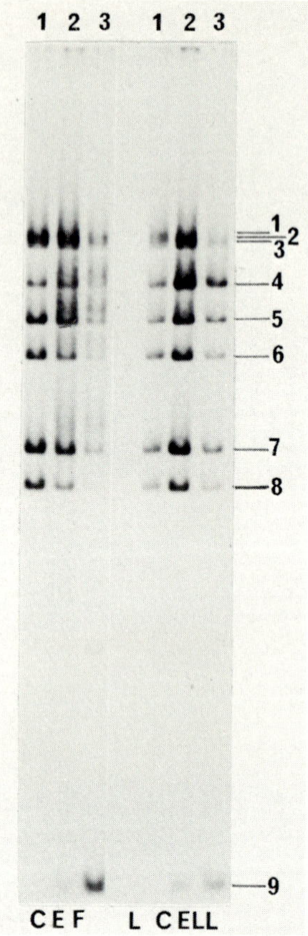

Fig.3. *The synthesis of unpolyadenylated cRNAs in FPV infected chick and L cells.*

The RNA samples were prepared and analysed as described in fig.1. Infected cells were labelled between 0-1 hr (1), 1½-2½ hr (2) and 3-4 hr after infection.

predominantly responsible for determining the amounts of the virus-specific polypeptides produced. This appears also to be the case in infected L cells and the reduced rates of synthesis of polypeptides P_1, P_2, P_3, NA, MP and 9 in these cells as compared to those in CEF can be correlated with the reduced synthesis of the corresponding mRNAs. This conclusion is in agreement with previous suggestions based on the results of experiments involving analyses of polypeptide synthesis in the presence of actinomycin D (12). However, the additional conclusion from that study that the initial transcription of the virus genome was selective is not supported by the results of

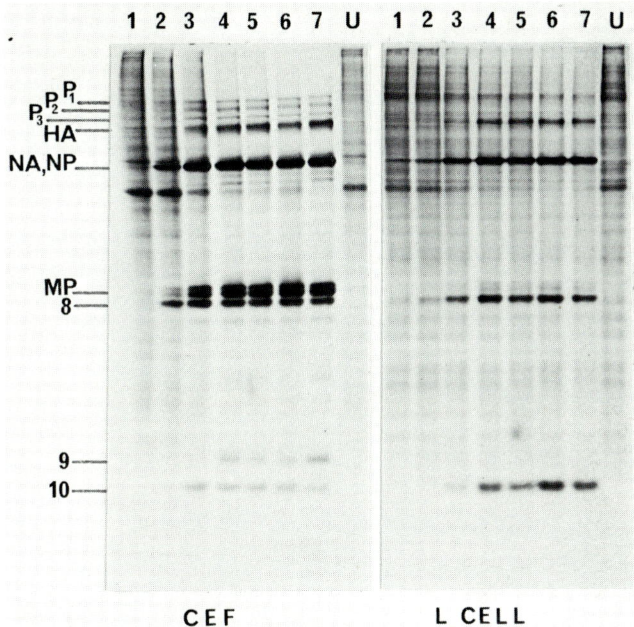

Fig.4. *The synthesis of virus-specific polypeptides in CEF and L cells.*

The infection procedure was as described in fig.1. At hourly intervals, as indicated, duplicate cultures were incubated in medium containing ^{35}S-methionine (5 μc/culture) for 10 min after which they were washed with saline and the cells were dissociated in SDS, urea and β-mercaptoethanol (1%, 8 M and 0.1% respectively) as described previously (11). The labelled polypeptide components were separated by electrophoresis on polyacrylamide gels containing 17% acrylamide and the buffer system described by Laemmli (7). Electrophoresis was at 5 volts/cm for 16 hr. The polypeptides are designated as previously (10). The uninfected control culture (U) was processed with the 4 hour samples.

the direct analyses of transcription reported here. Moreover, more sensitive analyses of polypeptide synthesis immediately after removal of cycloheximide (fig.6) and by Lamb and Choppin (8) also indicate that the majority of virus-specific polypeptides can be detected under these conditions. These results are, therefore, compatible with the above conclusion that, with the possible exceptions of the virus-specific glycoproteins, in both CEF and in L cells the extent of polypeptide synthesis in infected cells is a direct reflection of the intracellular population of virus-specific messenger RNA.

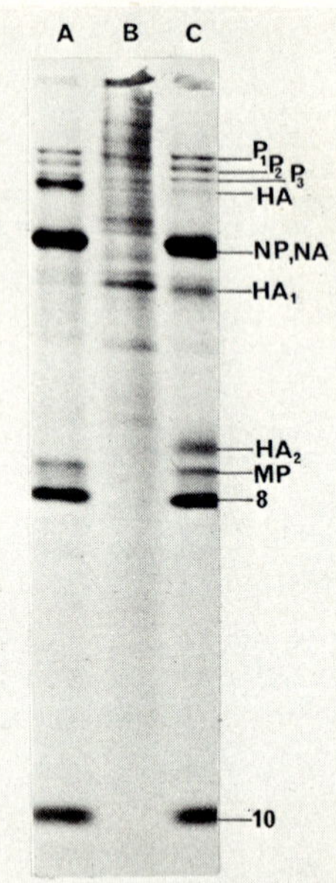

Fig.5. *The formation of the haemagglutinin polypeptide components in infected L cells.*

At 4 hours after infection duplicate cell cultures were A, incubated in ^{35}S-methionine for 10 min; and C, incubated in ^{35}S-methionine for 10 min and then in medium containing non-radioactive methionine for 1 hour. Uninfected control cells, B, were pulse labelled in parallel. All other experimental details were as for fig.4.

The information presently available does not allow an explanation for the differences between the transcription control in CEF and in L cells. The unpublished results of other experiments in which the transcription of the genome was determined in CEF infected and incubated in the presence of fluorophenylalanine have suggested that newly synthesized virus-specific proteins are involved in altering the pattern of both primary and secondary transcription. From the results presented here it seems likely that host cell components are also involved. Whether or not these are the same as those host factors which appear to be required for the initial expression of the virus genome in nucleate cells (2) and its sensitivity to antibiotics such as actinomycin D (1) remains to be determined.

The intracellular locations of the virus-specific polypeptides in L cells were also similar to those reported previously in CEF (3) particularly with regard to the virus-specific polypeptide composition of the ribonucleoprotein and membrane

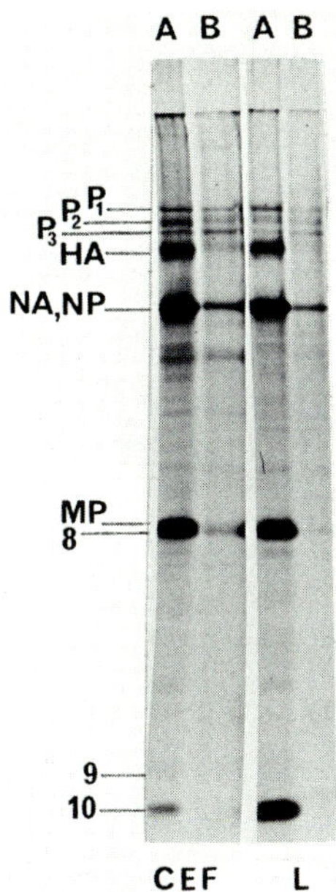

Fig.6. *The synthesis of virus-specific polypeptides in cells following incubation in cycloheximide.*

Monolayer cultures of cells were incubated in medium containing 100 μg/ml of cycloheximide for 1 hr before infection and then infected and incubated for a further 5 hr in the presence of cycloheximide. In parallel, duplicate cultures were infected as normally (fig.1). After 5 hr of incubation all cultures were washed in medium lacking cycloheximide and after 10 min incubation were pulsed with ^{35}S-methionine for a further 10 min. All other experimental details were as for fig.4 except that the polyacrylamide gel contained 15% acrylamide. For both cell types lane A contains the polypeptides synthesized 5 hr after a normal infection and lane B contains those synthesized immediately after removal of cycloheximide.

fractions of the infected cells. No defects were therefore apparent in the intracellular positioning of virion components for virus assembly and it appears most likely that in these cells the decreased level of infectious virus particle formation is due directly to the under production of specific virion proteins.

ACKNOWLEDGEMENTS

We thank D. Stevens, B. Precious and E. Fernandes for excellent assistance.

REFERENCES

1. Barry, R.D., Ives, D.R. and Cruickshank, J.G. (1962). *Nature* (Lond.), 194, 1139.
2. Follett, E.A.C., Pringle, C.R., Wunner, W.H. and Skehel, J.J. (1974). *J. Virol.* 13, 394.
3. Hay, A.J. and Skehel, J.J. (1975). *In* "Negative Strand Viruses" (B.W.J. Mahy and R.D. Barry, eds.), Vol.2, p. 635. Academic Press, London.
4. Hay, A.J., Lomniczi, B., Bellamy, A.R. and Skehel, J.J. (1977). *Virology* 83, 337.
5. Inglis, S.C., Carroll, A.R., Lamb, R.A. and Mahy, B.W.J. (1976). *Virology* 74, 489.
6. Klenk, H-D., Rott, R., Orlich, M. and Blodorn, J. (1975). *Virology* 68, 426.
7. Laemmli, U.K. (1970). *Nature* (Lond.), 227, 680.
8. Lamb, R.A. and Choppin, P.W. (1976). *Virology* 74, 504.
9. Lazarowitz, S.G. and Choppin, P.W. (1975). *Virology* 68, 440.
10. Lomniczi, B., Bosch, F.X., Hay, A.J. and Skehel, J.J. (1977). *J. Gen. Virol.* 35, 187.
11. Skehel, J.J. (1972). *Virology* 49, 23.
12. Skehel, J.J. (1973). *Virology* 56, 394.

MACROMOLECULAR SYNTHESES DURING ABORTIVE INFECTION OF KB CELLS BY INFLUENZA VIRUS

P. VALCAVI, G. CONTI and G.C. SCHITO

Istituto di Microbiologia,
Facolta' di Medicina E Chirurgia,
Universita' Degli Studi di Parma, Italy.

Complete expression of the influenza virus genome during replication depends on the continuous intervention of cellular functions in which, besides unidentified contributions, host DNA transcription seems to be involved (3, 11, 15, 18). Given these unique requirements there might exist host cells in which subtle modifications of the appropriate components, while not affecting the physiology of the cell, render these structures incapable of performing a correct role during the interaction with viral products. Such nonpermissive cells would more clearly define host-controlled steps in the influenza virus life cycle.

Although some data have already been gathered on abortive infection of several influenza virus strains in various cell lines (2, 6, 7, 8, 9, 13, 14, 20,21), a systematic analysis of host cell functions affecting virus development has not yet been undertaken. We have initiated an investigation along these lines employing fowl plague influenza A (FPV) as a model and report here the preliminary characterization of abortive infection of this virus in KB cells.

FPV, strain Rostock, obtained through the courtesy of R.D. Barry and B.W.J. Mahy, was grown in 11-day old fertile eggs and assayed by plaque formation on primary chick embryo (CE) cells. The results obtained were regarded in all further experiments as the permissive reference value. A range of cell lines was screened for ability to support FPV replication, including BHK-21, KB, EUE, Sirc, MA, Hep-2 and HeLa. Cultures were grown in MEM supplemented with 10% fetal calf serum, and monolayers formed in 6 cm plastic Petri dishes were infected with FPV at an input multiplicity of 20 pfu per cell.

Preliminary experiments demonstrated that all the cultures tested were able to adsorb and eclipse the virus but only BHK-21 and MA cells gave productive yields after 18 h at 37° of about 50 infectious particles per cell, similar to that observed with CE cells. All other cell lines yielded less than 3 infectious particles per cell and were considered to be defective in some function required for normal influenza virus replication. Of these, KB cells were selected for further examination.

To confirm and quantitate the lack of infectious particle production by KB cells, the technique of infected-cell complexes, as developed by Durand and Borland (5), was employed. After three hours adsorption, KB monolayers were trypsinized, washed once with PBS, suspended in growth medium and counted. Appropriate dilutions were made and aliquots were added to primary CE cell monolayers. Agar overlay medium was added; after incubation for 48-60 hours at 37°, plaques evidenced by neutral red were counted. As shown in fig.1 at no multiplicity of infection were KB cells able to reproduce FPV. Furthermore, when compared to the behaviour of permissive cells it is quite clear that infectivity of the parent virus is totally lost upon infection of KB cells.

While mature particles are not produced in KB cells, expression of some genetic potentialities of the virus chromosome is apparent upon further analysis of the abortive complexes. In fact, no significant differences in percentage of hemadsorbing cells was detected in comparison to the level found in permissive cells. Also, the amount of haemagglutinin (4) and neuraminidase (1) produced in KB cells and chick embryo fibroblasts was essentially comparable.

On the other hand analysis of the pattern of RNP synthesis in KB cells revealed a topographical alteration in the distribution of this antigen during abortive infection. To follow the production of the viral ribonucleoprotein KB cells were grown on glass coverslips in 6 cm dishes. After infection at a multiplicity of 20 pfu per cell coverslips were removed 2, 4, 6, 8 and 10 hours later, fixed and stained employing the indirect immunofluorescence technique. Specific antiserum to FPV S antigen was prepared in rabbits (19). Both in KB and CE cells, accumulation of S antigen was first detected in the nuclei 4 hours postinfection in agreement with previously reported data (7). However, while in CE cells the RNP migrated from the nucleus into the cytoplasm starting at about 6 hours postinfection (fig.2A), in KB cells the fluorescein-tagged material remained almost entirely confined to the nuclear region throughout the observation period (fig.2B).

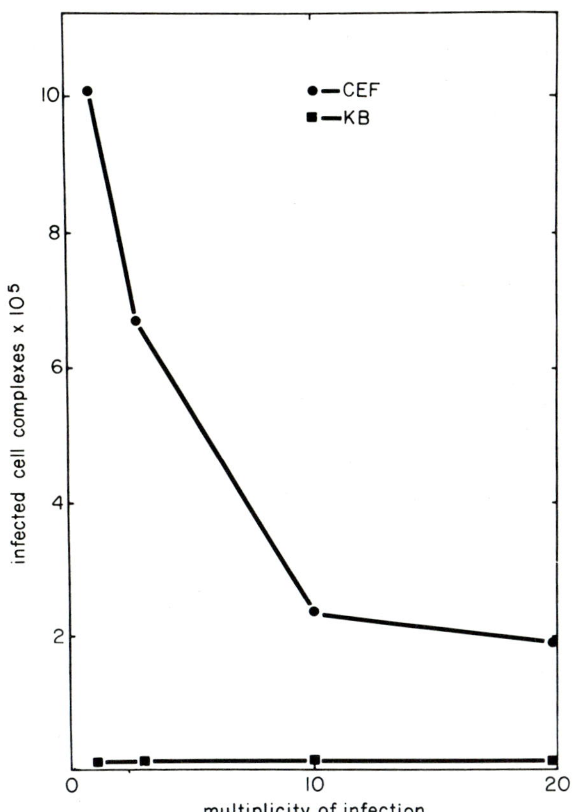

Fig.1. *Number of cells producing infectious virus in FPV infected KB and chick embryo cells at various multiplicity of infection.*

A more complete survey of the extent of FPV genome expression was obtained by analyzing the patterns of virus-specific polypeptide synthesis in KB as compared to permissive CE fibroblasts.

These experiments involved incubating infected monolayers in medium containing ^{35}S-labelled methionine, solubilizing the cells in SDS and separating the radioactive proteins by polyacrylamide slab gel electrophoresis under denaturing conditions (12). The radioactively labelled polypeptides were detected by autoradiography and a typical result is shown in Fig.3. Comparison of the bands appearing in permissive and abortive cells clearly indicates that in KB cells all of the known virus coded polypeptides are synthesized in the correct

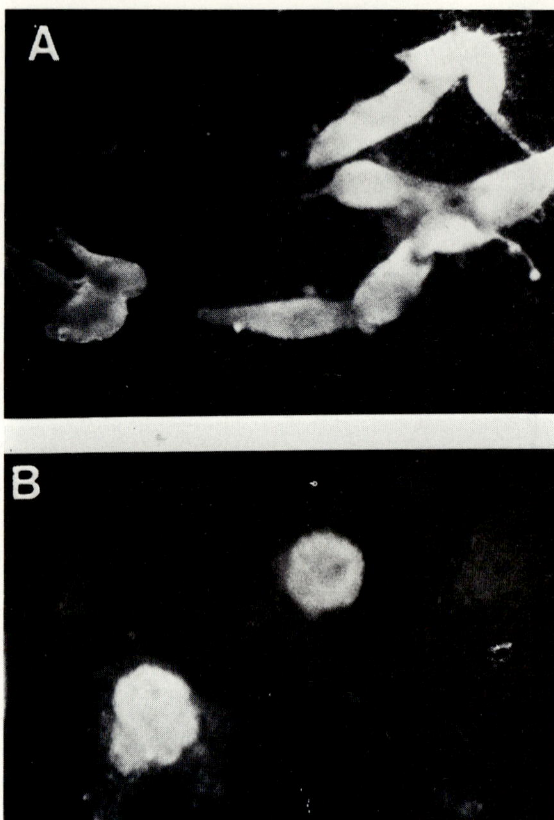

Fig.2. *Location of RNP antigen as evidenced by indirect immunofluorescence.*
A. Chick embryo fibroblasts stained at 8 hours p.i. showing both nuclear and cytoplasmic fluorescence.
B. KB cells stained at 8 hours p.i. showing exclusively nuclear fluorescence.

temporal sequence except for protein M which is totally absent. These results have been confirmed under a variety of experimental conditions, and seem to be a constant feature of the KB-FPV complexes.

Several differences are therefore apparent when the process of FPV infection in KB cells is compared to that in chick cells. While early viral proteins and most of the late polypeptides (12) are clearly synthesized in KB cells, the most striking feature of abortive infection in this host is the total lack of late M protein production. Similar results

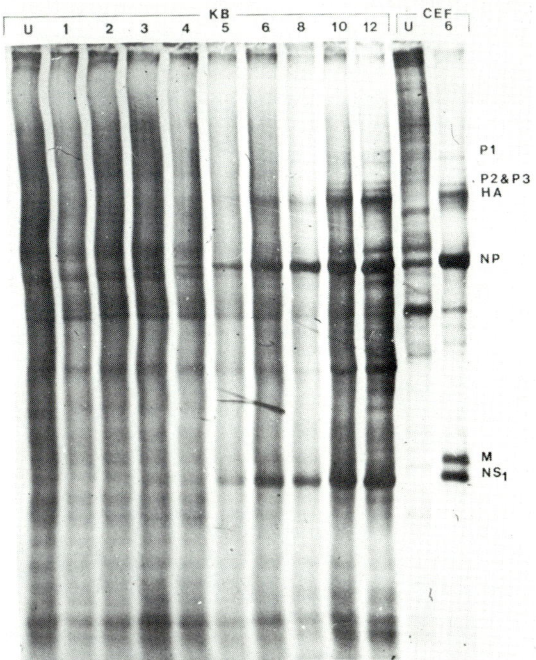

Fig.3. *Synthesis of polypeptides in FPV-infected KB cells and CE fibroblasts. Infected and uninfected cells were labelled with ^{35}S-methionine for 30 min at various times after infection. Whole cell lysates were subjected to polyacrylamide slab gel electrophoresis and processed for autoradiography. Uninfected cell lysates are labelled U. Numbers indicate the time when pulses of radioactive aminoacid were performed. Polypeptide separation was on 15% polyacrylamide gel and migration is from top to bottom.*

have been reported for cells treated with a variety of inhibitors of transcription, notably camptothecin (17) and UV light (16).

Although we are unable to provide at present data on the replication of viral RNA or the synthesis of complementary RNA in KB cells, extrapolation from what is known to occur in UV-treated hosts (16) suggests that in this non-permissive system a factor required for regulation of transcription of late viral function may be defective.

Whatever the origin of the defect, the absence of the major structural protein of the virus clearly explains the abortive nature of FPV infection in the KB host. While it is far from clear whether the inability to synthesize the M polypeptide is the only and primary failure occurring in KB cells during FPV gene expression, the abnormal finding of RNP accumulation in the nucleus is probably a secondary event related to the total absence of viral morphogenesis in these cells. Similar evidence has been presented in fact for a number of hosts incapable of producing normal yields of infectious particles (6, 7, 8, 11).

Replication of FPV is arrested in L cells because this host does not provide the appropriate functions necessary for virion RNA synthesis. In KB cells a defective contribution at a level of cRNA transcription or translation might be involved in generating the same phenomenon. Since the nature of the abortive step seems to differ from one virus-cell system to the other, further characterization of KB and other nonpermissive cells promises to be useful in dissecting the molecular basis of the host contribution to the replication of influenza virus.

ACKNOWLEDGEMENTS

This research was supported by CNR, Rome "Progetto Finalizzato Virus" grant no. 77.00309.84.

REFERENCES

1. Aminoff, D. (1961). *Biochem. J.* 31, 384.
2. Avery, R.J. (1975). *J. Virol.* 16, 311.
3. Barry, R.D. (1964). *Virology* 14, 398.
4. Davenport, F.M., Rott, R. and Schafer, W. (1960). *J. Exp. Med.* 112, 765.
5. Durand, D.P. and Borland, R. (1969). *Proc. Soc. Exptl. Biol. Med.* 130, 44.
6. Franklin, R.M. and Breitenfeld, P.M. (1959). *Virology* 8, 293.
7. Fraser, K.B. (1967). *J. Gen. Virol.* 1, 1.
8. Gandhi, S.S., Bell, H.B. and Burke, D.D. (1971). *J. Gen. Virol.* 13, 423.
9. Henle, G., Girardi, A. and Henle, W. (1955). *J. Exp. Med.* 101, 25.
10. Kelly, D.C. and Dimmock, N.J. (1974). *Virology* 61, 210.
11. Kelly, D.C., Avery, R.J. and Dimmock, N.J. (1974). *J. Virol.* 13, 1155.
12. Inglis, S.C., Carroll, A.R., Lamb, R.A. and Mahy, B.W.J. (1976). *Virology* 74, 489.
13. Lerner, R.A. and Hodge, L.D. (1969). *Proc. Nat. Acad. Sci. USA* 64, 544.
14. Low, I.E., Eaton, M.V. and Uretsky, S.B. (1972). *J. Immunol.* 89, 414.
15. Mahy, B.W.J., Hastie, N.D. and Armstrong, S.J. (1972).

Proc. Nat. Acad. Sci. USA 69, 1421.
16. Mahy, B.W.J., Carroll, A.R., Brownson, J.M.T. and McGeoch, D. (1977). *Virology* 83, 150.
17. Minor, P.D. and Dimmock, N.J. (1975). *Virology* 69, 336.
18. Rott, R., Saber, S. and Scholtissek, C. (1965). *Nature* 205, 1187.
19. Schild, G.C. and Pereira, H.G. (1969). *J. Gen. Virol.* 4, 355.
20. Sugiura, A. (1972). *Virology* 47, 517.
21. Wong, S.C. and Kilbourne, E.D. (1961). *J. Exp. Med.* 113, 95.

ASSEMBLY OF INFLUENZA VIRAL RIBONUCLEOPROTEIN STRUCTURES IN PERMISSIVE AND NON-PERMISSIVE CELLS

L.A. CALIGUIRI and H. GERSTEIN

Department of Microbiology and Immunology
Albany Medical College of Union University
Albany, New York, U.S.A.

The yield of infectious influenza virus is dependent on the host cell in which the virus is grown. The WSN strain of influenza virus A_0 undergoes a productive infection in the MDBK line of bovine kidney cells. These cells produce a high yield of infectious virus and are considered permissive for influenza virus (5). In contrast, HeLa cells produce very little infectious virus and are non-permissive for influenza virus (6). The host-dependent process which limits the production of virus in non-permissive cells is not known.

The results reported in this paper compare the formation of a subclass of viral RNPs in permissive and non-permissive cells inoculated with influenza A/WSN (H_0N_1) virus. Three cell lines were studied: MDBK cells which produce 200 to 700 pfu/cell; BHK-21F cells which produce 70 to 200 pfu/cell; and HeLa cells which produce 0.5 to 10 pfu/cell. Virus produced in these cell lines have similar infectivity to hemagglutination ratios ($\sim 10^4$) which indicate that the ratios of physical particles to infectious virus are also similar. The kinetics of formation and accumulation of a subclass of nuclear and cytoplasmic virus-specific RNPs were determined in the three cell lines, and differences in the turnover of viral RNPs interpreted in the light of host-dependent restriction of virus production.

SEPARATION OF SUBCLASSES OF VIRUS-SPECIFIC RNPs

We have recently developed a procedure to isolate a subclass of viral RNPs using renografin (methyl-glucamine salt of 3,5-diacetamide-2,4,6-triiodobenzoic acid) density gradients (3). The sedimentation of cytoplasmic extracts of infected and mock-infected MDBK cells in renografin density gradients are shown in

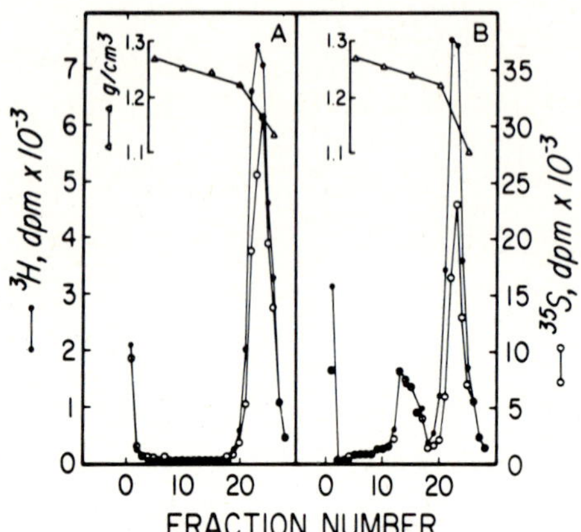

Fig. 1. *Isopycnic centrifugation of cytoplasmic extracts in renografin gradients. MDBK cells were mock-infected or infected with influenza A/WSN (H_0N_1) at a multiplicity of 50 pfu/cell. Input virus had previously been passaged 5 times in MDBK cells. ^{3}H-uridine (10 μCi/ml) and ^{35}S-methionine (10 μCi/ml) were added to the cultures 4 hr after infection and the cells were fractionated into cytoplasmic and nucleoplasmic fractions at 6 hr after infection as described previously (3). Cytoplasmic extracts were layered onto preformed 45-65% (w/v) renografin gradients in SET buffer (0.1M NaCl; 0.001M EDTA: 0.05M Tris, pH 7.4) and centrifuged at 30,000 rpm for 16 hr at 4° in a Beckman SW40 rotor. Fractions were collected and processed for radioactivity. The refractive index of selected fractions was determined and the buoyant density obtained using standard curves derived by weighing renografin samples of known concentration. Sedimentation in this and all subsequent gradients is to the left. (A) Mock-infected cells. (B) Influenza virus-infected cells.*

Figure 1. The conditions of infection, and other experimental details are provided in the legend to Figure 1. There is a peak of radioactivity near the top of the gradient in both the infected and mock-infected cytoplasmic extracts. This large, heterogeneous peak of material is not resolved into individual components under these conditions. Resedimentation of this material in a 25-50% renografin gradient reveals a broad peak in the range from 1.15 to 1.24 g/cm^3 with an average buoyant density of 1.21 g/cm^3 (3). The average buoyant density of 1.21 g/cm^3 will be used to refer to the RNPs present in this material. A second ribonucleoprotein peak which sediments in the range from 1.25 to 1.27 g/cm^3 is clearly separated in the samples from infected cells and it has an average buoyant density of 1.26 g/cm^3. The buoyant density of this peak of RNPs after glutaraldehyde-fixation and re-sedimentation in CsCl is 1.34 g/cm^3 (3) which is the same buoyant density in CsCl as that reported for glutaraldehyde-fixed RNPs from purified influenza virions and infected cells (8, 9). These results show that a species of virus-specific RNPs with an average buoyant density of 1.26 g/cm^3 can be isolated free of host cell RNPs in renografin density gradients.

Viral RNPs with a buoyant density of 1.26 g/cm^3 are also present in the nucleoplasm of infected cells (Fig. 2). Nucleoplasmic and cytoplasmic fractions of infected and mock-infected MDBK cells were centrifuged in 45-65% renografin gradients after a 2 hr labeling period with ^3H-uridine and ^{35}S-methionine beginning 4 hr post-infection. The only peak found in the mock-infected cells is near the top of the renografin gradient (Fig. 2A and 2B). Two peaks of radioactivity are present in the nu-

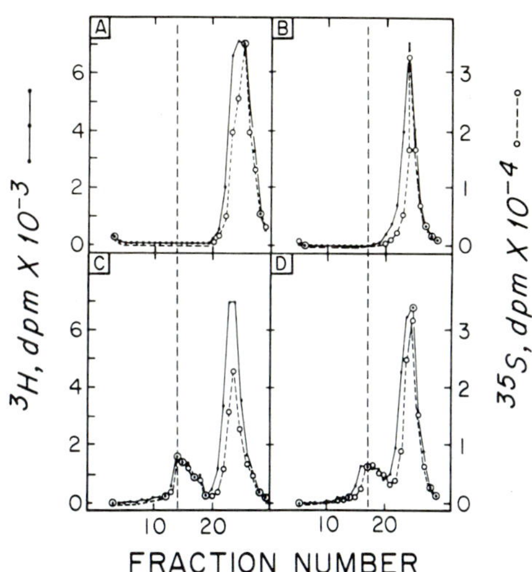

Fig. 2. *Isopycnic centrifugation of cytoplasmic and nucleoplasmic fractions in renografin gradients. Mock-infected or infected MDBK cells were labeled with ^3H-uridine and ^{35}S-methionine as described in Fig. 1. Nucleoplasmic and cytoplasmic fractions from each culture were subjected to isopycnic centrifugation in preformed 45-65% (w/v) renografin/SET gradients at 30,000 rpm for 16 hr at 4° in a Beckman SW40 rotor. Fractions were analyzed for radioactivity and buoyant density as described previously (3). The broken line indicates the position of a buoyant density of 1.27 g/cm^3 in the gradients. (A) Mock-infected cytoplasmic extract. (B) Mock-infected nucleoplasmic fraction. (c) Influenza virus-infected cytoplasmic extract. (D) Influenza virus-infected nucleoplasmic extract.*

cleoplasm and cytoplasm of infected cells, one of which has a buoyant density of 1.26 g/cm^3 as indicated by the vertical broken lines (Fig. 2C and 2D). Viral RNPs are not present in the nucleolar fraction of infected cells (3).

These results show that there are at least two subclasses of viral RNPs in the nucleoplasm and cytoplasm of infected cells. One subclass has an average buoyant density of 1.26 g/cm^3 and is clearly separated from host material. Another subclass cosediments with host cell RNPs at 1.21 g/cm^3 in 45-65% renografin gradients, and these viral RNPs can be resolved in discontinuous renografin gradients (Caliguiri, unpublished results). In addition to different buoyant densities in renografin, these two subclasses of virus-specific RNPs are further distinguished by differences in their time of appearance in the nucleoplasm and cytoplasm and by polypeptide composition. The 1.21 g/cm^3 RNPs appear first in the nucleoplasm within 2-4 hr after infection, whereas this subclass is not found in the cytoplasm until after 4 hr (3). The 1.26 g/cm^3 RNPs are present in both the nucleoplasm and cytoplasm within 2 hr after infection.

The 1.26 g/cm^3 RNPs are deficient in P polypeptides, whereas the 1.21 g/cm^3 RNPs are enriched in P polypeptides relative to NP (Fig. 3). Infected cells were labeled with ^3H-uridine and ^{35}S-methionine from 2 to 4 hr after infection. The RNP peaks at buoyant densities of 1.21 and 1.26 g/cm^3 from the cytoplasm and nucleoplasm were analyzed by polyacrylamide gel electrophoresis (1). The NP polypeptide is the major polypeptide present

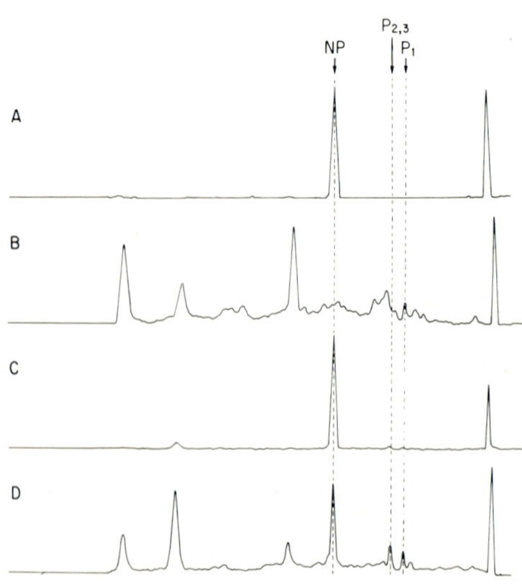

Fig.3.

in the 1.26 g/cm^3 RNPs from both the cytoplasm and nucleoplasm (Fig. 3A and 3C). There is a trace amount of the P polypeptides associated with the 1.26 g/cm^3 RNPs compared to that present in the 1.21 g/cm^3 peak (Fig. 3D). It should also be noted that P and NP polypeptides are absent in the 1.21 g/cm^3 material from the cytoplasm at this time (Fig. 3B). However, after 4 hr of infection, viral RNPs are in the 1.21 g/cm^3 peak from the cytoplasm (Fig. 4). The buoyant density of 1.26 g/cm^3 in renografin is consistent with an estimated RNA content of 10-20% which is the RNA content of virion RNPs (4, 8, 12). The similarity to virion RNPs is further supported by the sedimentation behavior of these RNPs in CsCl after glutaraldehyde treatment (3). Since the P polypeptides are minor components of viral RNPs (7), the lack of these polypeptides would not be expected to change the buoyant density of these structures.

These results show that two major subclasses of viral RNPs in the nucleus and cytoplasm of infected cells can be separated in preformed 45-65% renografin gradients. The 1.26 g/cm^3 subclass has an RNA content similar to virion RNPs and is deficient in P polypeptides relative to NP. The 1.21 g/cm^3 subclass has a higher RNA content than virion RNPs and is enriched in P polypeptides. The 1.26 g/cm^3 RNPs are resolved from host cell RNPs by this method, however the 1.21 g/cm^3 RNPs are not.

CUMULATIVE LABELING OF VIRAL RNPs

The functional role of the subclasses of virus-specific RNPs in virus replication is not clear. Virus-specific RNPs with a buoyant density of 1.26 g/cm^3 are found in the nucleus and cytoplasm of permissive and non-permissive host cells after infection. The cumulative labeling of the 1.26 g/cm^3 RNPs was determined in three cell lines. MDBK, BHK-21F, and HeLa cells were infected at a multiplicity of 20-50 pfu/cell with influenza virus derived from MDBK cells after five passages. ^3H-uridine

---

Fig. 3. *Polypeptides of RNPs from influenza virus-infected MDBK cells labeled from 2 to 4 hr after infection with ^{35}S-methionine (10 μCi/ml). The peak fractions of the 1.26 and 1.21 g/cm^3 regions of 45-65% (w/v) renografin/SET gradients after isopycnic centrifugation as described in Fig. 1 were dialyzed overnight against 0.05M Tris (pH 7.9) and subjected to electrophoresis in 10% polyacrylamide gel slabs and a densitometer tracing of an autoradiogram obtained (1). The broken lines indicate the positions of viral polypeptides obtained by co-electrophoresis of marker influenza virus. (A) 1.26 g/cm^3 peak from cytoplasmic extract. (B) 1.21 g/cm^3 peak from cytoplasmic extract. (C) 1.26 g/cm^3 peak from nucleoplasmic fraction. (D) 1.21 g/cm^3 peak from nucleoplasmic fraction.*

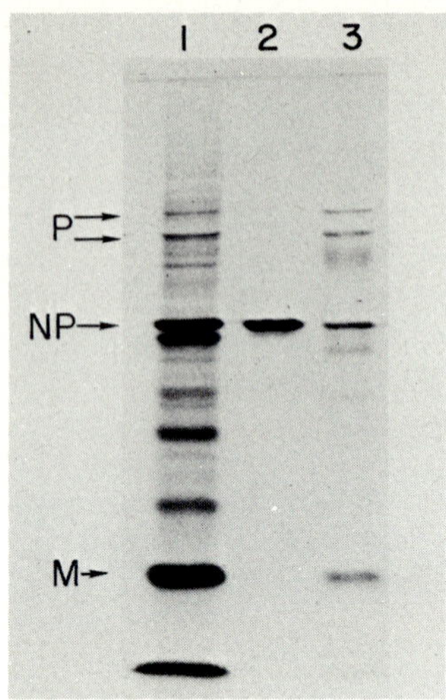

Fig. 4. *Polypeptides of RNPs from cytoplasmic extracts of influenza virus-infected MDBK cells labeled from 4 to 6 hr after infection with ^{35}S-methionine (10 μCi/ml). The peak fractions of the 1.26 g/cm^3 and 1.21 g/cm^3 regions of 45-65% (w/v) renografin/SET gradients were processed as described in Fig. 3, subjected to electrophoresis in 10% polyacrylamide gel slabs, and an autoradiogram of the gel developed as described previously (1). The arrows indicate the positions of viral polypeptides obtained by co-electrophoresis of marker influenza virus. Lane 1, influenza virus A/WSN (H_0N_1). Lane 2, 1.26 g/cm^3 peak from cytoplasmic extract. Lane 3, 1.21 g/cm^3 peak from cytoplasmic fraction.*

(15 μCi/ml) was added 1 hr after infection and cultures were harvested at 4 and 6 hr. Nucleoplasmic and cytoplasmic 1.26 g/cm^3 RNPs were isolated and the amount of radioactively-labeled RNA was determined (Table 1). The results were normalized to the uptake of ^3H-uridine in MDBK cells. This was necessary to compare the results in different cell lines since the uptake of uridine, the intracellular pools of precursors and the number of cells per culture are different for each cell line. At 4 hr after infection, the amount of radioactively-labeled RNA present in the nucleoplasmic RNPs in BHK-21F cells is higher than the other cell lines (Table 1). Similar results are found for the cumulative incorporation of RNA into cytoplasmic RNPs. At 4 hr, the amount of radioactively-labeled RNPs in the cytoplasm of BHK-21F cells is slightly higher than either MDBK or HeLa cells. However, by 6 hr the total amount of radioactively-labeled cytoplasmic RNPs in MDBK cells is 3-fold greater than BHK-21F and 5-fold greater than HeLa cells. These results show

that the highest amount of newly formed 1.26 g/cm^3 RNPs accumulate in MDBK cells, which also produce the highest yields of infectious virus.

TABLE 1

Cumulative Incorporation of ^3H-Uridine into 1.26 g/cm^3 RNPs in Three Cell Lines

Cell Line	Nucleoplasm		Cytoplasm	
	4 Hours	6 Hours	4 Hours	6 Hours
MDBK	4.7*	14.9	23.8	173.1
BHK-21F	15.7	14.2	28.8	57.3
HeLa	3.9	5.2	10.5	31.3

*Means of three experiments expressed as DPMx10^3 normalized to uptake of ^3H-uridine in MDBK cells.

These findings suggest that the amount of 1.26 g/cm^3 RNPs formed in host cells is directly correlated with the amount of infectious virus produced and that this subclass of viral RNPs may have a functional role in virus replication. The properties of the 1.26 g/cm^3 RNPs suggest several possibilities for the function of this subclass of RNPs in virus replication. These RNPs may act as precursors of virion RNPs which require incorporation of additional P polypeptides before incorporation into virions. Previous results suggest a difference in the rate of synthesis of P polypeptides and rate of incorporation into RNPs (2, 11). Alternatively, the 1.26 g/cm^3 RNPs may function as messenger RNPs in the synthesis of viral proteins or as template in the synthesis of viral RNA. Further experiments are required to decide among these possibilities.

RATE OF FORMATION OF 1.26 g/cm^3 VIRAL RNPs

The data from cumulative labeling experiments suggest a correlation between the amount of 1.26 g/cm^3 RNPs formed and the yield of infectious virus in permissive and non-permissive cells. It was important to determine the time course of appearance of the 1.26 g/cm^3 RNPs in infected MDBK, BHK-21F, and HeLa cells. The cells were infected as described above, however the cultures were labeled with ^3H-uridine (15 μCi/ml) for 1 hr at various times after infection. Viral RNPs with a buoyant density of 1.26 g/cm^3 were isolated from the nucleoplasm and cytoplasm and the

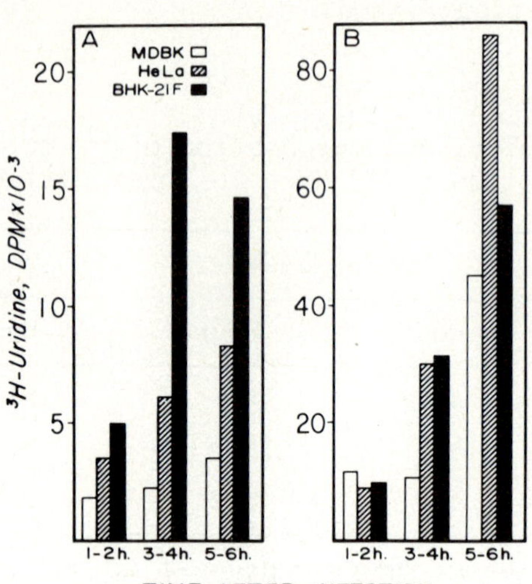

Fig. 5. *Rate of assembly of 1.26 g/cm^3 viral RNPs in three cell lines. MDBK, HeLa, and BHK-21F cells were infected with influenza virus A/WSN (H_0N_1) at a multiplicity of 20 to 50 pfu/cell and labeled for 1 hr with ^3H-uridine (15 μCi/ml) and ^{35}S-methionine (10 μCi/ml) at various times after infection. Cells were fractionated into cytoplasmic and nucleoplasmic fractions and subjected to isopycnic centrifugation in 45-65% (w/v) renografin/SET gradients as described in Fig. 1. The amount of radioactively-labeled RNA in the peak of 1.26 g/cm^3 RNPs was obtained and normalized to the uptake of ^3H-uridine in MDBK cells. Each value represents a mean of 4 experiments. (A) Nucleoplasmic 1.26 g/cm^3 RNPs. (B) Cytoplasmic 1.26 g/cm^3 RNPs.*

radioactivity of the peaks of RNPs was determined at each interval. The mean of 4 experiments for each cell line are represented in Figure 5. The standard errors are within 10% of each mean. The radioactivity in the three cell lines were normalized to the uptake of ^3H-uridine in MDBK cells.

Figure 5A shows that the rate of incorporation of RNA into *nucleoplasmic* RNPs is highest in the BHK-21F cells. In BHK-21F cells it reaches a maximum between 3 and 4 hr after infection and then declines slightly. The incorporation into nucleoplasmic RNPs in HeLa cells and MDBK cells follows a slightly different pattern than in BHK-21F cells with a continuous increase in incorporation during the 6 hr period of observation. The rate of incorporation in HeLa cells is somewhat higher than that found in MDBK cells (Fig. 5A).

The rate of incorporation of RNA into *cytoplasmic* RNPs is greater than that found for nucleoplasmic RNPs (Fig. 5B). In all three cell lines, there is an increase in the rate over the 6 hr period of observation. The highest rate of formation of cytoplasmic RNPs occurs in HeLa cells between 5 and 6 hr after infection. However, the rates of assembly of RNPs is higher in both HeLa and BHK-21F cells than in MDBK cells at all times studied except from 1 to 2 hr after infection. These results show that the 1.26 g/cm^3 subclass of viral RNPs are formed at a faster rate in the nucleus and cytoplasm of HeLa and BHK-21F cells than in MDBK cells. However, the yield of virus from HeLa or BHK-21F cells is much lower than that produced in MDBK cells.

Our results show a dichotomy between the relative rates of assembly of 1.26 g/cm^3 RNPs and the total amount of 1.26 g/cm^3 RNPs accumulated in the three cell lines. The amount of RNPs accumulated correlates directly with the yields of virus; the kinetic data do not. This paradox suggests that the 1.26 g/cm^3 RNPs are dynamic structures, and that the turnover of this subclass of RNPs is greater in non-permissive cells than in permissive cells. Several possible mechanisms can be proposed to explain this difference in turnover of viral RNPs. Viral RNPs formed in non-permissive cells are less stable than those formed in permissive cells. This instability may reflect a defect in the protein or RNA synthesis in non-permissive cells. The small, heterogeneous RNAs synthesized in HeLa cells (6) may represent products of defective transcription in these cells. Another possibility is that degradation of newly formed RNPs occurs. Degradation of viral RNPs may be an active process which requires a host cell factor and reflects the state of non-permissiveness. Either of these possibilities would lead to greater turnover of viral RNPs in non-permissive cells.

Host-dependent control of amplification of synthesis of viral proteins has been suggested as a possible mechanism for host restriction of virus replication (10). Amplification of viral proteins may be controlled at the level of transcription or translation. Our results suggest that rapid turnover of viral RNPs in non-permissive cells may limit amplification of protein synthesis either by reduction in the amount of messenger for translation, or by reduction of template for transcription. Further analysis of the function of the subclasses of viral RNPs that have been separated will provide a better understanding of their role in virus replication and a more precise definition of the host-dependent process which limits virus production in these cells.

ACKNOWLEDGEMENTS

This work was partly supported by U.S.P.H.S. Grant AI-11913

from the National Institute of Allergy and Infectious Diseases and by the Lillian Miller Memorial Grant VC-168 from the American Cancer Society.

We thank Marion Corpening and Gregory Lawrence for excellent technical assistance, Kathleen Benedetto for expert typing of the manuscript, and Dr. Lawrence Sturman for many helpful discussions and review of the manuscript.

REFERENCES

1. Beckman, L.D., Caliguiri, L.A. and Lilly, L.S. (1976). *Virology* 73, 216.
2. Caliguiri, L.A. and Compans, R.W. (1974). *J. Virol.* 14, 191.
3. Caliguiri, L.A. and Gerstein, H. (1977). *Virology* (submitted for publication).
4. Chan, R.T.L. and Scheffler, I.E. (1974). *J. Cell Biol.* 61, 780.
5. Choppin, P.W. (1969). *Virology* 39, 130.
6. Choppin, P.W. and Pons, M.W. (1970). *Virology* 42, 603.
7. Inglis, S.C., Carroll, A.R., Lamb, R.A. and Mahy, B.W.J. (1976). *Virology* 74, 489.
8. Krug, R.M. (1971). *Virology* 44, 125.
9. Krug, R.M. (1972). *Virology* 50, 103.
10. Lamb, R.A. and Choppin, P.W. (1976). *Virology* 74, 504.
11. Meier-Ewert, H. and Compans, R.W. (1974). *J. Virol.* 14, 1083
12. Pons, M.W., Schulze, I.T. and Hirst, G.K. (1969). *Virology* 39, 250.

THE REPLICATION OF ENVELOPED RNA VIRUSES IN CELLS WITH ALTERED MEMBRANE CARBOHYDRATES

S. SCHLESINGER, R. LEAVITT, C. GOTTLIEB and S. KORNFELD

Departments of Microbiology and Immunology, Medicine, and Biochemistry, Washington University School of Medicine, St. Louis, Mo. 63110, U.S.A.

Enveloped RNA viruses contain one or more kinds of glycoprotein arranged in spike-like structures on the outer surface of the virion. These same glycoproteins are found on the surface of infected cells. The formation of infectious virions requires specific interactions between the glycoproteins and internal viral components leading to budding of the virus particle from the cell plasma membrane. Our studies with these glycoproteins have focused on the role of the carbohydrate residues in the structure and replication of the enveloped RNA viruses - vesicular stomatitis virus (VSV) and Sindbis virus. In this communication we describe our results with VSV.

VSV has only a single glycoprotein (G). Each molecule of G contains two oligosaccharide chains linked to an asparagine residue in the polypeptide chain (12). The oligosaccharides contain a core composed of N-acetylglucosamine, mannose and fucose residues and outer branches with sialic acid, galactose and N-acetylglucosamine residues (1, 10). Etchison *et. al.* have proposed a structure for these oligosaccharides (2).

We have investigated the role of carbohydrates by two different means. One way has been to prevent glycosylation with a specific inhibitor. The inhibitor we have used is tunicamycin (TM) - a glucosamine containing antibiotic first described by Takatsuki and Tamura (19). TM inhibits the transfer of UDP-N-acetylglucosamine to the lipid carrier, dolichol pyrophosphate, and thus prevents the addition of the core oligosaccharide unit to asparagine residues in glycoproteins (18, 21). This type of approach had previously been carried out using D-glucosamine and analogues of glucose such as 2-deoxy-D-glucose to interfere with glycosylation of viral glycoproteins (14).

The second way has been to grow viruses in cell variants that are altered in the activity of one or more glycosyltransferases. These variants can provide a means for determining how specific defects in host cell glycosylation affect the structure and replication of an enveloped virus. It has been possible to isolate this type of variant by selecting for cells that will grow in the presence of lectins which inhibit normal cell growth (5, 11, 17). A lectin such as ricin binds to galactose on the cell surface and ultimately interacts with the ribosome to inhibit protein synthesis. Some of the resistant clones that have been isolated do not have galactose residues available on their surface and are unable to bind ricin.

THE EFFECT OF TUNICAMYCIN ON VSV

At concentrations of 0.5 μg per ml TM inhibits the replication of VSV in BHK cells more than 99% (9). Similar results have been obtained with other enveloped viruses including Newcastle disease virus (20), Sindbis virus (9) Semliki Forest virus (15), fowl plague virus and avian RNA tumor virus (15). The release of particles is inhibited in all cases but the latter. TM does not inhibit the replication of the non-enveloped virus, encephalomyocarditis virus (9).

We have analyzed the pattern of proteins synthesized in TM-treated BHK cells infected with VSV using sodium dodecyl sulphate-polyacrylamide gel electrophoresis. All of the virus-specific proteins are synthesized in the presence of TM but the G protein migrates faster than normal G protein (Fig.1) and is not labeled with ^{3}H-glucosamine (9). These findings support the conclusion that the initial steps in the synthesis of the oligosaccharide chains on the viral glycoprotein require the synthesis of the N-acetylglucosamine-lipid intermediate.

The G protein is synthesized on membrane bound polyribosomes and becomes inserted into the surface plasma membrane of the host cell (7). The localization of G on the outer cell surface can be demonstrated by iodination with lactoperoxidase (9), chymotrypsin treatment of intact cells (7) and indirect immunofluorescence of intact cells with VSV antiserum (Leavitt, Schlesinger and Kornfeld, in press). The same procedures show that non-glycosylated G fails to reach the outer surface of infected cells.

In order to determine if the inability of G to migrate to the cell surface is due to an intrinsic property of the nonglycosylated molecule we compared the solubility properties of the glycosylated and non-glycosylated forms of G. The glycosylated molecule was extracted from cellular membranes by 0.2% Triton-X100 whereas the non-glycosylated protein remained completely insoluble (Fig.1). Sodium hydroxide, lithium diiodosalicylate and IM potassium chloride also failed to solubilize non-glycosylated G. Guanidine hydrochloride gave variable results, but a combination of 6M guanidine hydrochloride and 0.2% Triton-X100 did lead to complete extraction.

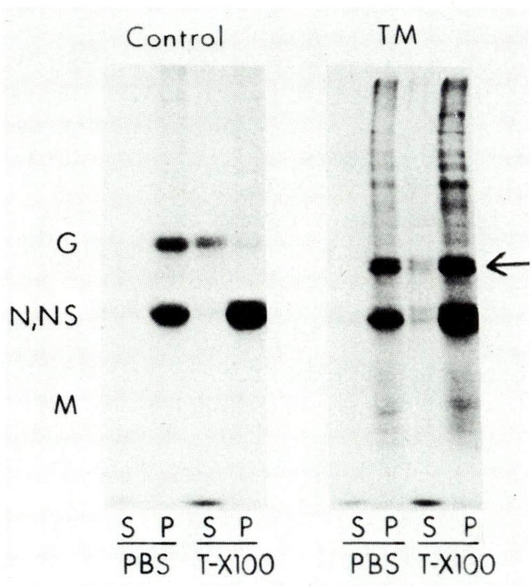

Fig.1. *Autoradiogram of SDS-polyacrylamide gels of ^{35}S-methionine labeled 100,000 x g particulate fraction from VSV infected BHK cells. Cells were infected with VSV and TM (0.5 μg/ml) was added at 1 hr postinfection. At 6 hr postinfection cells were labeled for 10 min with ^{35}S-methionine (10 μCi/ml for control infected cells, 16 μCi/ml for infected cells treated with TM). The monolayers were washed with cold phosphate buffered saline and a 100,000 x g particulate fraction was prepared. A portion of the particulate fraction (50 μg protein) was extracted with either phosphate buffered saline (PBS) or 0.2% Triton X-100 in PBS (T). After 15 min at 4°, the samples were centrifuged at 100,000 x g for 90 min to obtain the supernatant (S) and pellet (P) fractions. The proteins in the two fractions were precipitated with trichloroacetic acid and prepared for electrophoresis. The arrow indicates the nonglycosylated G protein.*

The insolubility of nonglycosylated G suggests that carbohydrates have a profound effect on the structure of this protein. In the absence of carbohydrates the protein may fold improperly or aggregate. An abnormal conformation could explain why the protein is unable to migrate to the cell surface.

REPLICATION OF VSV IN CELLS ALTERED IN GLYCOSYLTRANSFERASE ACTIVITY

We have been studying the replication of both VSV and Sindbis virus in three cell variants isolated by their ability to grow in the presence of the plant lectin, ricin. The biochemical properties of these variants are described in Table I. The variant 15B isolated from CHO cells has a decreased level of UDP-N-acetylglucosamine: glycoprotein N-acetylglucosaminyltransferase activity (3, 5). The cell membrane glycoproteins are deficient in N-acetylglucosamine, galactose and sialic acid. VSV appears to replicate normally in these cells. The yield of infectious virions and the plaque-forming units per mg of protein were essentially the same as those obtained with CHO cells (13). The G protein of VSV obtained from 15B cells migrates slightly faster than normal G on polyacrylamide gels (13). This alteration was seen more clearly when the size distribution of the glycopeptides was determined (Fig.2). The largest glycopeptide of VSV from CHO cells has an approximate molecular weight of 3100; the glycopeptides of VSV from 15B have a molecular weight of about 1800. Thus the oligosaccharides of the virion reflect the changes predicted by the enzymatic defect in the cell.

The glycopeptides of VSV obtained from L cells, clone 3 and clone 6 have also been analyzed (4). Those from clone 3 were slightly higher in molecular weight than those from L cells and those from clone 6 were smaller. Treatment of the glycopeptides of VSV from clone 3 with neuraminidase caused the expected shift to smaller molecular weight components indicating that the increase in size was due to sialic acid.

CONCLUSION

Our studies as well as those from several other laboratories provide evidence that the addition of carbohydrate to viral glycoproteins is essential for the formation of infectious virions. In the absence of any carbohydrate the glycoprotein of VSV is unable to migrate to the outer surface of the cell plasma membrane - a step that must precede virion assembly. The requirement for carbohydrate, however, does not extend to the outer sugars. Our results with cell variants show that these outer sugars on the oligosaccharide structure are not necessary for the formation and infectivity of VSV.

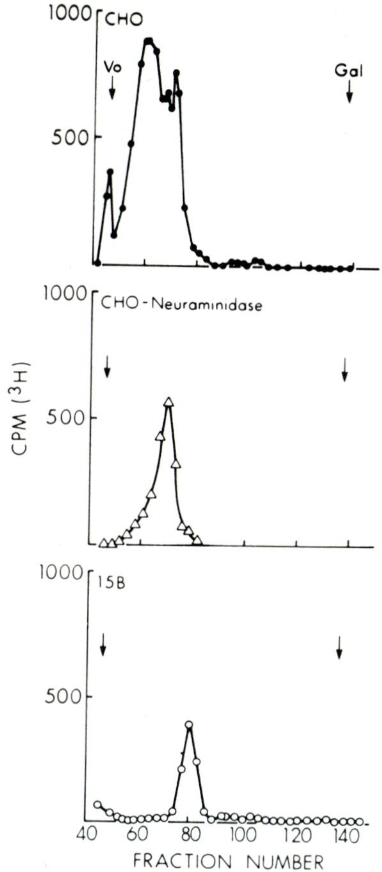

Fig.2. *Bio-gel P 6 chromatography of ^{3}H-glucosamine-labeled glycopeptides from VSV grown on CHO or 15B cells. After purification the virus samples were digested with pronase before chromatography (4). Fractions 50-75 from the analysis shown on the top panel were pooled, treated with neuraminidase and one-half of the sample was applied to the Bio-gel column (Δ --Δ).*

ACKNOWLEDGEMENTS

This work was supported by Public Health Service grants R01 Al 11377 (to S.S.) from the National Institute of Allergy and Infectious Diseases, R01 CA 80759 (to S.K.) from the National Cancer Institute and 5T05GM02016 (to R.L.) from the National Institute of General Medical Sciences.

REFERENCES

1. Etchison, J.R. and Holland, J.J. (1974). *Proc. Nat. Acad.*

TABLE I

Cell Variants Isolated by Resistance to Ricin

Variant	Parent	Defect	Reference
15B	CHO cells	decreased N-acetylglucosaminyl-transferase activity	(3, 5)
Cl 6	L cells	decreased N-acetylgluco-saminyltransferase and galactosyltransferase activities	(4)
Cl 3	L cells	increased sialyltransferase activity	(4)

Sci. USA 71, 4011.

2. Etchison, J.R., Robertson, J.S. and Summers, D.F. (1977). *Virology* 78, 375.
3. Gottlieb, C., Baenziger, J. and Kornfeld, S. (1975). *J. Biol. Chem.* 250, 3303.
4. Gottlieb, C. and Kornfeld, S. (1976). *J. Biol. Chem.* 251, 7761.
5. Gottlieb, C., Skinner, A.M. and Kornfeld, S. (1974). *Proc. Nat. Acad. Sci. USA* 71, 1078.
6. Kiely, M.L., McKnight, G.S., and Schimke, R.T. (1976). *J. Biol. Chem.* 251, 5490.
7. Knipe, D.M., Baltimore, D. and Lodish, H. (1977). *J. Virol.* 21, 1121.
8. Krag, S.S. and Robbins, P.W. (1977). *J. Biol. Chem.* 252, 2621.
9. Leavitt, R., Schlesinger, S. and Kornfeld, S. (1977). *J. Virology* 21, 375.
10. McSharry, J.J. and Wagner, R.R. (1972). *J. Virol.* 7, 412.
11. Meager, A., Ungkitchanukit, A., Nairn, R. and Hughes, R.C. (1975). *Nature* 257, 137.
12. Moyer, S.A., Tsang, J.M., Atkinson, P.H. and Summers, D.F. (1975). *J. Virol.* 18, 167.
13. Schlesinger, S., Gottlieb, C., Feil, P., Gelb, N. and Kornfeld, S. (1976). *J. Virol.* 17, 239.

14. Scholtissek, C. (1975). *Curr. Top. Microbiol. Immunol.* 70, 101.
15. Schwarz, R.T., Rohrschneider, J.M. and Schmidt, M.F.G. (1976). *J. Virolcgy* 21, 1121.
16. Sefton, B.M..(1977). *Cell* 10, 659.
17. Stanley, P., Caillibot, V. and Siminovitch, L. (1975). *Cell* 6, 121.
18. Takatsuki, A., Kohno, K. and Tamura, G. (1975). *Agric. Biol. Chem.* 39, 2089.
19. Takatsuki, A., and Tamura, G. (1971). *J. Antibiot.* 24, 224.
20. Takatsuki, A. and Tamura, G. (1971). *J. Antibiot*24, 785.
21. Tkacz, J.S. and Lampen, J.O. (1975). *Biochem. Biophys Res. Commun.* 65, 248.
22. Waechter, C.J. and Lennarz, W.J. (1976). *Ann. Rev. Biochem.* 45, 95.

HOST CELL FUNCTION DEPENDENT INDUCTION OF DEFECTIVE INTERFERING PARTICLES OF VESICULAR STOMATITIS VIRUS

C. YONG KANG, THOMAS GLIMP and RAE ALLEN

Department of Microbiology,
The University of Texas Health Science Center,
Southwestern Medical School,
Dallas, Texas, USA.

Most animal virus systems have been shown to contain not only the standard infectious virions but also defective virus particles which interfere with the replication of their related standard infectious viruses (5).

The existence of defective interfering (DI) particles in a preparation of an animal virus was described first for influenza virus (20). Huang and Baltimore (6) postulated a role for DI particles in host defense mechanisms concerned with overcoming viral infections. The physical and biological properties of DI particles have been reviewed previously (5). DI particles normally are generated during serial undiluted, high multiplicity passages of standard infectious virus. However, very little is known about the mechanism of induction of DI particles upon infection by standard virus particles. One of the major goals in DI particle mediated viral interference research is to understand this mechanism.

Vesicular stomatitis virus (VSV) provides an excellent model system to study the induction of DI particles and the mechanism of viral interference since the DI particles are readily separable from the standard virus particles. Cooper and Bellett (2) first described the autointerference phenomenon of VSV and Huang and Wagner (7) showed that truncated defective particles in preparations of VSV are responsible for homologous viral interference.

DI particles of VSV can be generated from a plaque purified clone of standard virus by serial undiluted passages of virus (19). Different sizes of DI particles were generated by independent clonal isolates of virus propagated in one cell type or the same clonal isolates propagated in different hosts (4, 17). It appears that the generation of DI particles of VSV is related in some way to the host cells in which the DI particles are generated.

To test the hypothesis that a host function may determine the generation of DI particles, we have investigated the effects of host cells on the generation of DI particles. This report describes studies on the rate of DI particle generation and type of DI particles generated in four different species of host cells, and on the induction of DI particles in cells pretreated with actinomycin D. We present direct evidence for host-cell function dependent induction of DI particles of VSV.

YIELD OF VSV IN FOUR DIFFERENT HOSTS DURING SERIAL UNDILUTED PASSAGES OF A CLONE OF STANDARD VIRION

To determine host effects on the generation of DI particles during serial undiluted high multiplicity passages of VSV, three independent plaque isolates of Indiana serotype of VSV (VSV_{IND}) were propagated in human (RSa), hamster (BHK_{21}), rat [R(B77)] and mouse (L_2) cell lines. The yield of standard infectious virions in each passage was measured by the plaque assay for VSV. Fig.1 demonstrates that three clones of VSV_{IND}, passaged independently in the four different cell lines, yielded different amounts of infectious B virion in different serial passages. RSa cells produced a tenfold reduction of infectious VSV at each passage and over a thousandfold reduction from the original titer of virus by the 4th passage. In contrast, L_2 cells produced a constant amount of virus up to the 6th passage and then showed a rapid decline in titer in the subsequent passages. The BHK_{21} cells and R(B77) cells behaved similarly. Five to six passages were required before the yield of virus was reduced by approximately 3 logs from the initial titer.

It is clear that the rate of reduction in production of the infectious viruses was determined by the host cell rather than by the virus since all three clones of virus showed the same pattern of reduction of virus titer in any given cell type.

To determine whether the reduction of titer after the serial undiluted passages of VSV_{IND} was the result of the generation of DI particles, the viruses from each passage were analyzed by sucrose gradient centrifugation. Fig.2 demonstrates that the production of DI particles was directly correlated with the titer of virus shown in Fig.1. The RSa cells produced detectable amounts of DI particles at the second undiluted passage. The DI particles produced in the RSa cells were rather heterogeneous in size. The BHK_{21} cells produced a detectable amount of DI particles at the third undiluted passage and we could detect at least two distinct sizes of DI particles produced in BHK_{21} cells. In contrast, R(B77) cells produced three different size classes of DI particles which were not detectable until the fourth passage. L_2 cells took 7 to 8 passages to produce detectable amounts of DI particles. There were three different sizes of DI particles produced in L_2 cells (Fig.2). It is clear from these data that various host cells generate different size classes of DI particles after a different number of passages. These results strongly suggest that a host function determines the type of DI particles produced since different host cells produced multiple size classes of DI particles even though a monoclonal isolate was used to initiate the infections. We have repeated these experiments using the same clone of virus in the same host cells and found the results are reproducible as long as we use the same clonal isolate of virus with the same cell type (data not shown).

We conclude from these data that the rate of DI particle generation and type of DI particles generated during serial undiluted high multiplicity passages are largely dependent upon the host cell.

DETERMINATION OF BIOLOGICAL ACTIVITIES OF DI PARTICLES GENERATED DURING UNDILUTED PASSAGES

When DI particles were mixedly infected with standard B virions the production of the standard B virion was decreased and the input DI particles were bred true (14).

To test the biological activities of DI particles generated by multiple passages, we infected cells with purified DI particles mixed with the first passage of VSV_{IND}, which did not give rise to detectable amounts of DI particles. We used DI particles from the 5th passage of VSV_{IND} which was grown in R(B77) cells as shown in Fig.2 and purified by sucrose gradient centrifugation. The inserted picture in Table I shows the types of DI particles produced in this passage and DI-1, -2 and -3 are designated as indicated.

Purified DI particles were mixed with infectious standard B virions and grown in four different hosts. Although all 3 DI particles exhibited their interfering capacity in all four host cells, the degree of interference in the four different hosts was varied (Table I). The infection of all three DI particles alone did not give rise to any detectable level of infectious virus production. We found that the two DI particles generated in BHK_{21} cells and the 3 DI particles generated in L_2 cells (Fig.2) were all defective in their replication and capable of interference. Thus, we conclude that all small truncated particles generated in the four different cells are defective interfering particles.

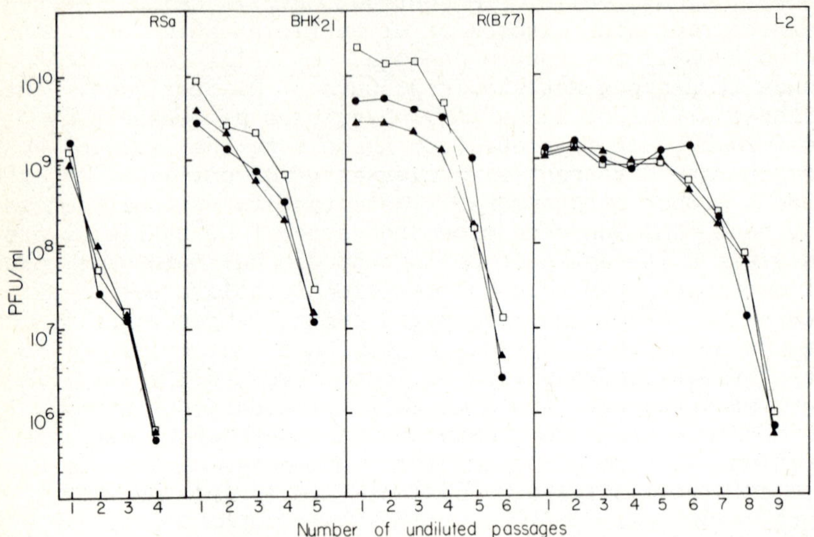

Fig.1. *Yield of standard B virions during successive undiluted passages of three clonal isolates of VSV_{IND} in four different hosts.*

Growing cultures of B77 strain of avian sarcoma virus transformed rat cells R(B77) were obtained from Dr Howard Temin, McArdle Laboratory, University of Wisconsin. Mouse L_2 cells were provided by Dr Kathryn Holmes, Uniformed Services University Medical School. Baby hamster kidney cell clone 21 (BHK_{21}) was purchased from the American Type Culture Collection. Baby hamster kidney cell clone 53 (BHK_{53}) was provided by Dr D.H.L. Bishop, University of Alabama, and the human cell line established by double infection with Rous sarcoma virus (RSV) and Simian virus 40 (SV40) which is designated as RSa was the kind gift of Dr T. Kuwata, Chiba University, Japan (12). We have cloned these cells at least once in our laboratory. None of these cells produce any detectable amounts of the endogenous viruses. The prototype strain of the VSV_{IND} used in this paper has been described previously (1, 9). Virus stocks of VSV_{IND} clones 1 (□), 2 (▲), and 3 (●) diverged at the first plaque isolation from the original stock virus and stocks of each clone were prepared after six additional independent plaque purifications followed by one low multiplicity passage using 0.01 PFU/cell in L_2 cells. Confluent monolayer cultures of approximately 2 x 10^7 cells per 100 mm culture dish were infected with 2 x 10^9 PFU of the stock virus for the first passage. After 45 min absorption at 37°, 7 ml of prewarmed

SERIAL UNDILUTED PASSAGES OF VSV_{IND} IN CELLS WITH AND WITHOUT ACTINOMYCIN D PRETREATMENT

The data presented in Figs.1 and 2 indicate strongly that the induction of DI particles during serial undiluted passages of VSV_{IND} depends on a host cell function. These data prompted us to investigate induction of DI particles in cells pretreated with actinomycin D, a known inhibitor for DNA-directed information flow (15) with no known effects on VSV replication (10). The stock virus was plaque purified and prepared in cells pretreated with actinomycin D and serial undiluted passages were carried out in cells with and without actinomycin D pretreatment. Fig.3 shows the titer of VSV_{IND} propagated in BHK_{21} cells with and without actinomycin D pretreatment. The untreated BHK_{21} cells yielded approximately the same amount of virus up to the 3rd passage and then showed a rapid reduction in virus titer during the subsequent passages. In contrast, when the same virus was propagated in BHK_{21} cells pretreated with actinomycin D, the titer of VSV in each passage up to the 12th passage remained constant. We carried out similar experiments using R(B77) cells to rule out the possibility that the phenomenon is dependent upon a particular cell type. There was no reduction in virus titer up to the 12th passage when the virus was propagated in R(B77) cells pretreated with actinomycin D, while it took 7 passages to give approximately a 3 log reduction from the original titer in the absence of actinomycin D (Fig.3).

To determine whether the reduction of titer after serial undiluted passages of VSV was the result of the generation of DI particles, the viruses from each passage were analyzed by sucrose gradient centrifugation. Fig.4 demonstrates that two different size classes of DI particles were produced in BHK_{21} cells when the virus was propagated in cells without actinomycin D pretreatment, whereas no detectable amounts of DI particles were produced from the cells pretreated with actinomycin D. At least three classes of DI particles were produced in untreated R(B77) cells and no DI particles were detectable when the virus was propagated in actinomycin D pretreated cells (Fig.4).

Dulbecco's MEM, supplemented with 5% fetal calf serum was added to each culture dish and incubated at 37° CO_2 incubation for 7 hr, when cell-free virus fluids were harvested and centrifuged at 600 x g for 15 min. Subsequent passages were obtained by adding 0.5 ml of undiluted previous passage virus to 2 x 10^7 cells/dish. At the end of each passage, the yield of infectious virus was measured by plaque assay on BHK_{53} cells. (□ clone No.1, △ clone No.2, and ● clone No.3).

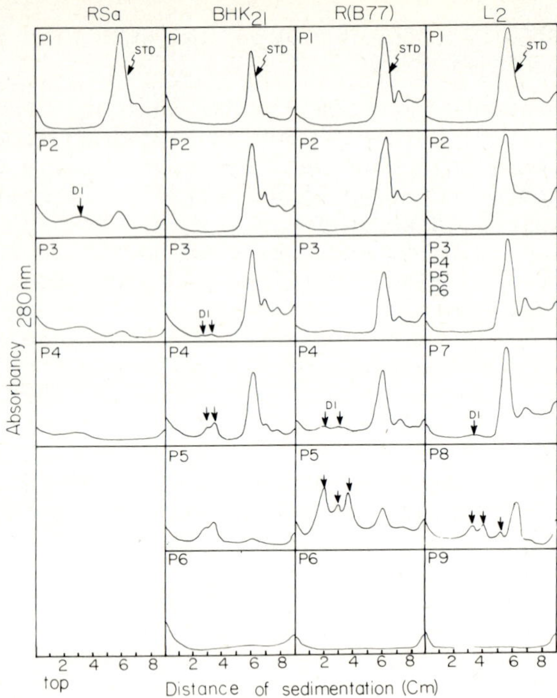

Fig.2. *Analysis of total virus particles from the continuous undiluted passages of clone 1 of VSV_{IND}*

Samples of lysate from four culture dishes of each passage illustrated in Fig.1 were centrifuged at 600 x g for 15 min to remove cellular debris. The virus particles were pelleted from the supernatant by centrifugation at 81,000 x g for 75 min in a Spinco SW27 rotor, resuspended in 0.3 ml of PBS and layered on a linear 5-30% sucrose gradient made in PBS. After centrifugation at 110,000 x g for 35 min at 5° in a Spinco SW41 rotor, the gradients were collected from the top using a Buchler Auto-Densi-Flow IIC. The presence and distribution of virus in the gradient was monitored continuously by passing the effluent through an LKB Uvicord II, with absorbance at 280 nm being continuously recorded on a Fisher Recordall 5000. The types of cells, passage numbers, and positions of the standard B virion (STD), and defective interfering particles (DI), are indicated in the figure. Electron microscopic examination of material in fractions sedimenting just in advance of the virus revealed only clumps of particles identical to B virions (11).

PRODUCTION OF DI PARTICLES IN CELLS WITH AND WITHOUT ACTINOMYCIN D PRETREATMENT USING STOCK VIRUS KNOWN TO CONTAIN DI PARTICLES

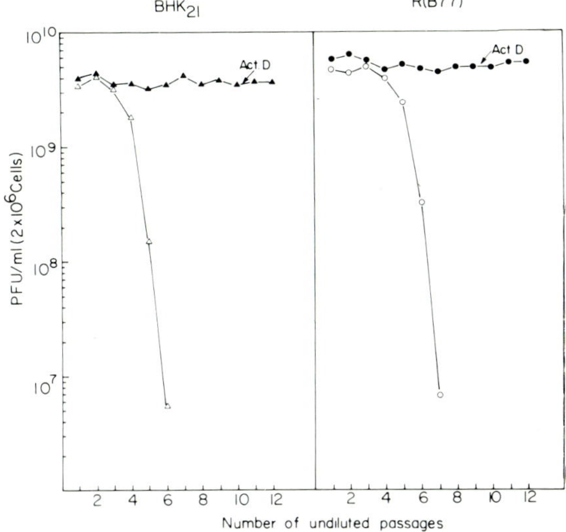

Fig.3. *Yield of standard B virions from cells with and without actinomycin D pretreatment during successive undiluted passages.*

Virus stock of VSV_{IND} was obtained from 5 consecutive plaque purifications, followed by 2 additional plaque purifications in cells pretreated with actinomycin D. Confluent monolayer cultures of either R(B77) cells or BHK_{21} cells in 100 mm culture dishes containing approximately 2 x 10^7 cells were treated with 1 µg/ml of actinomycin D for 24 hr. Actinomycin D pretreated monolayer cultures were infected with diluted virus to give approximately 50 plaques in each culture dish and overlayed with agar. Single, well-isolated plaques were picked and used to inoculate other cultures also pretreated with actinomycin D. The viruses from this second plaque isolation from either BHK_{21} cells or R(B77) cells were used to infect cells pretreated with 1 µg/ml actinomycin D at a multiplicity of infection of 0.1 PFU/cell. 100 mm culture dishes containing approximately 2 x 10^7 cells in monolayer were treated with 1 µg/ml of actinomycin D for 24 hr. A parallel control culture was not treated with actinomycin D. The monolayer cultures were infected with stock virus at approx. 100 PFU/cell for the first passage; for the subsequent passages, 0.5 ml/100 mm dish of undiluted virus from the previous passage was used. At the end of each 7 hr passage, the yield of infectious virus was measured by plaque assay on a BHK_{53} cell monolayer in 60 mm culture dish. Actinomycin D pretreated cells (Act. D) and the cell types are indicated. The number of cells in actinomycin D pretreated cells was less than control cultures at the time of infection. The amount of culture fluid was adjusted so that the yield of virus in 1 ml accounts for 2 x 10^6 cells.

TABLE I

Biological Activities of Three DI Particles Produced in R(B77) Cells

Inoculum	Titer of Virus, PFU/ml (% of control titer) Cell Lines for Interference Assay			
	R(B77)	BHK_{21}	L_2	RSa
Passage 1 VSV_{IND} alone	9.0×10^9(100)	2.0×10^9(100)	1.4×10^9(100)	4.8×10^8(100)
DI-1 alone	0	ND	0	ND
DI-1 + Passage 1 VSV_{IND}	2.0×10^7(0.22)	1.2×10^7(0.60)	2.2×10^7(1.59)	3.8×10^7(7.92)
DI-2 alone	0	ND	0	ND
DI-2 + Passage 1 VSV_{IND}	1.8×10^7(0.20)	1.4×10^7(0.70)	1.9×10^7(1.35)	1.9×10^7)3.96)
DI-3 alone	0	ND	0	ND
DI-3 + Passage 1 VSV_{IND}	1.9×10^7(0.21)	1.4×10^7(0.70)	1.6×10^7(1.14)	1.9×10^7(3.86)

TABLE I, (see facing page).

DI particles produced in the fifth passage of VSV_{IND} in R(B77) cells (Fig.2) were purified by sucrose gradient centrifugation as described in the legend to Fig.2. The three DI particles produced from R(B77) cells of DI particles are shown in the inserted photograph. 0.1 ml aliquots of each DI particle were mixed with 0.1 ml of passage 1 virus (which gave rise to no detectable DI particles) or with 0.1 ml of culture medium. These 0.2 ml mixtures were inoculated onto monolayers (2×10^7 cells) of 4 different cell lines. In addition, 0.1 ml of passage 1 virus mixed with 0.1 ml of culture medium was inoculated onto monolayers of the same cell types. After 45 minutes of absorption at 37° 10 ml of prewarmed DMEM supplemented with 5% fetal calf serum was added to each culture dish and incubated 7 hours at 37°. The virus containing culture fluid was harvested after the 7 hour growth period and the yield of infectious virus was determined by plaque assay on BHK_{53} cells. Inoculum used and cell type are indicated in the table.

We examined the growth of DI particles in cells pretreated with actinomycin D using virus stocks containing known DI particles to determine whether the production of DI particles is also dependent upon a cellular function. Two stock viruses containing DI particles prepared from cells without actinomycin D pretreatment, (that is, the 4th passage virus from BHK_{21} cells and the 5th passage virus from R(B77) cells shown in Fig.4) were propagated in cells pretreated with actinomycin D and in untreated cells. The data in Fig.5 demonstrate virtually identical patterns of DI particle production in cells with or without actinomycin D pretreatment. The actinomycin D pretreated cells produce DI particles and the standard infectious virions (Fig.5, BHK_{21} passage number 1) as effectively as untreated cells demonstrating that actinomycin D pre-treated cells are capable of supporting the growth of not only the standard infectious virions but also DI particles already present in a virus stock. Accordingly, we conclude that the induction of DI particles is host function dependent, but their replication does not require the same host function.

DISCUSSION

Although indirect evidence has suggested that host function(s) determine the generation of DI particles, we know very little about DI particle induction during serial undiluted passages of standard infectious virions. It has been postulated that DI particles arise as artifacts during viral preparation or by

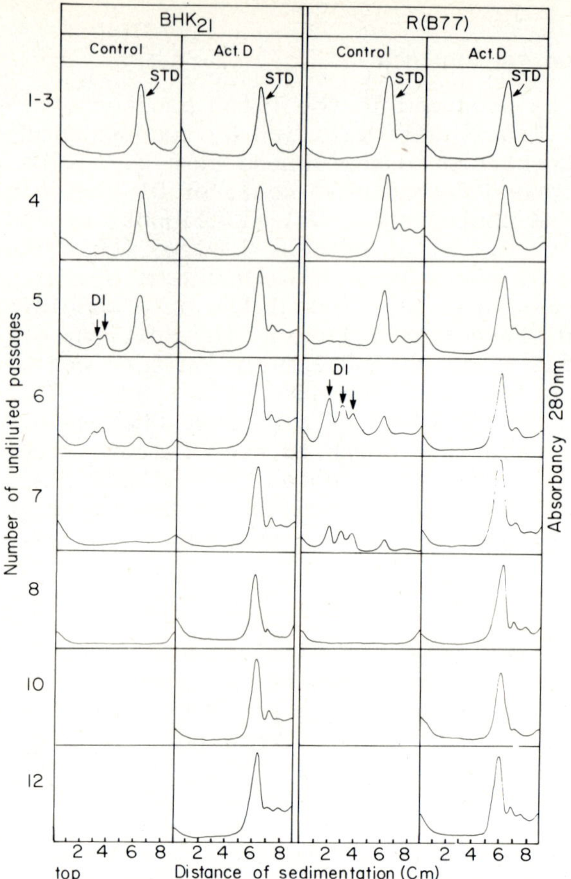

Fig.4. *Sucrose gradient analysis of total virus particles from successive passages.*

Samples of lysates from four culture dishes of each passage illustrated in Fig.3 were analyzed in sucrose gradients as described in the legend to Fig.2. Types of cells used and the numbers of passages are indicated.

mutational events during the replication of standard virions (3, 20). However, DI particles could not be generated by mechanical disruption or UV irradiation of the virus (4). We have demonstrated in this paper that the number of passages required for a visible amount of DI particle synthesis, at least in the case of VSV_{IND}, was largely dependent on the host cells. Three different virus clones passaged in four different hosts produced DI particles at the same number of passages in a given cell type. This observation suggests that the initial

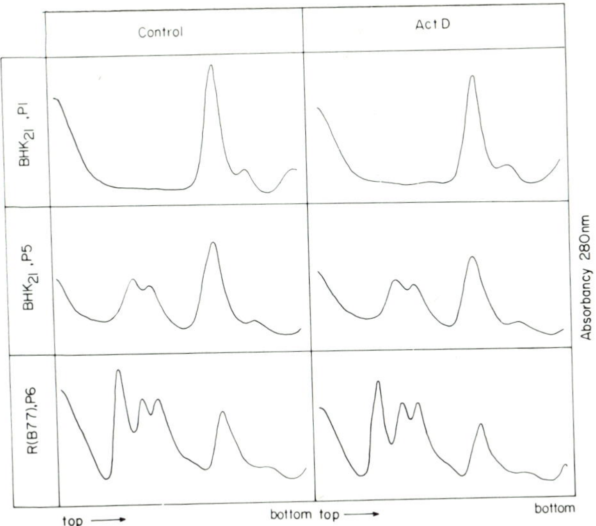

Fig.5. *Production of DI particles in cells pretreated with actinomycin D.*

Approximately 2 x 10^7 cells/100 mm culture dish were pretreated with 1 μg/ml of actinomycin D (Act. D) for 24 hr, or left untreated (Control), and infected with the viruses from passages 0 and 4 of the control BHK_{21} cells (Fig.4) and the virus from passage 5 of the control R(B77) cells (Fig.4). Virus was grown for 7 hr and analyzed in sucrose gradients as described in the legend to Fig.2.

induction and subsequent amplification of DI particles in various hosts is different. It appears that the rate of induction determines the overall production of DI particles since the yield of standard B virion during the first passage in these four different hosts did not vary significantly.

The patterns of DI particles generated during serial undiluted passages of cloned infectious B virions of VSV was also dependent in part on the host cell since a single clonal isolate generated different patterns of DI particles in different cells. Thus, the host plays an important role in controlling the generation of DI particles. We have observed that a given clonal isolate of B virions always generated the same pattern of DI particles in a clone of given cell type. This is consistent with the observations of Reichmann *et al.* (16) and Schincariol and Howatson (18) who generated particular size DI particles with particular temperature sensitive mutants of VSV, or heat-resistant strain of VSV_{IND} respectively. A part of our data is also consistent with the observation that the same clonal isolate generates different patterns of DI particles in different hosts during high multiplicity passages (4).

Defective particles were also interfering particles and those generated in one cell type were capable of interfering with the production of infectious particles in different hosts. However, the degree of interference in different hosts varied significantly as determined by standard B virion production. This observation is consistent with data previously reported (13). We have analyzed T_1 ribonuclease resistant oligonucleotide fingerprints of two DI particle RNAs from BHK_{21} cells and of three DI particle RNAs from R(B77) cells (11, and see also p.276). Surprisingly, five DI particle RNAs share a completely overlapping nucleotide sequence in spite of the fact that two are generated in BHK_{21} cells and three are generated in R(B77) cells. They represent a constant region of the infectious, negative-stranded B virion genome and no positive-stranded RNA sequences were detected. We do not understand yet how a deletion takes place during the replication of VSV.

The direct evidence for the host-cell function in the generation of DI particles was their absence when the virus was plaque purified and propagated in cells pretreated with actinomycin D. It is important to stress that the virus must be plaque purified in cells pretreated with actinomycin D in order to demonstrate the lack of induction of DI particles. We have performed the experiment with plaque purified stock virus prepared in cells without prior actinomycin D treatment and found that DI particles were generated at later passages even though the virus was serially passed in actinomycin D pretreated cells. Our data suggest that induction of DI particles by virus-specified premature termination of RNA replication, improper cleavage of mRNA transcripts or mutational events of the virus itself during the virus replication are unlikely. We cannot rule out the possibility that there is a virus-induced cellular function which may be responsible for induction of DI particles.

Our results define another possible host defense mechanism occurring at the cellular level which protects the host from a highly cytolytic virus infection. DI particles may play a critical role in host defense during the initial infection of virus, especially with fast growing cytolytic viruses. In such cases, the host is unable to respond quickly enough to combat the infecting virus early in infection with production of protecting levels of antiviral antibodies. Other possible host defense mechanisms, such as induction of interferon (8) or the generation of cytotoxic T cells (21), may not be as effective early in infection as DI particle-mediated viral interference. We have evidence that a host-cell function is critical for the induction of DI particle formation and suggest that the induction of these particles may not be a self-destructive viral mechanism, but rather a host-cell mechanism to protect the host from further spread of virus.

ACKNOWLEDGEMENTS

We are very grateful to Jan Reed for her excellent technical support. We thank Drs J. Bednarz-Prashad, B. Ozanne, and J. Shadduck for stimulating discussions and their detailed reviews of this manuscript. This investigation was supported by Public Health Service Grants CA 16479 and CA 20012 from the National Cancer Institute.

REFERENCES

1. Clewley, J.P., Bishop, D.H.L., Kang, C.Y., Coffin, J., Schnitzlein, W.M., Reichmann, M.E., and Shope, R.E. (1977). *J. Virol.* 23, 152.
2. Cooper, P.D. and Bellett, A.J. (1959). *J. Gen. Microbiol.* 21, 485.
3. Hackett, A.J. (1964). *Virology* 24, 51.
4. Holland, J.J., Villarreal, L.P., and Breindl, M. (1976). *J. Virol.* 17, 805.
5. Huang, A. (1973) *Ann. Rev. Microbiol.* 27, 101.
6. Huang, A.S. and Baltimore, D. (1970). *Nature* 226, 325.
7. Huang, A.S. and Wagner, R.R. (1966). *Virology* 30, 173.
8. Isaacs, A. and Lindenmann, J. (1957). *Proc. Roy. Soc. London* 147, 258.
9. Kang, C.Y. and Prevec, L. (1969). *J. Virol.* 3, 404.
10. Kang, C.Y. and Prevec, L. (1971). *Virology* 46, 678.
11. Kang, C.Y., Glimp, T., Clewley, J.P. and Bishop, D.H.L. (1977). *Virology* 84, 142.
12. Kuwata, T., Oda, T., Sekiya, S. and Morinaga, N. (1976). *J. Natl. Cancer Inst.* 56, 919.
13. Potter, K.N. and Stewart, R.B. (1976). *Can. J. Microbiol.* 22, 1458.
14. Prevec, L. and Kang, C.Y. (1970). *Nature* 228, 25.
15. Reich, E., Franklin, R.M., Shatkin, A.J. and Tatum, E.L. (1962). *Proc. Natl. Acad. Sci. USA* 48, 1238.
16. Reichmann, M.E., Pringle, C.R. and Follett, E.A.C. (1971). *J. Virol.* 8, 154.
17. Roy, P., Repik, P., Hefti, E., and Bishop, D.H.L. (1973). *J. Virol.* 11, 915.
18. Schincariol, A.L. and Howatson, A.F. (1970). *Virology* 42, 732.
19. Stampfer, M., Baltimore, D. and Huang, A.S. (1971). *J. Virol.* 7, 409.
20. Von Magnus, P. (1954). *Adv. Virus Res.* 2, 59.
21. Zinkernagel, R.M. and Doherty, P.C. (1974). *Nature* 248, 701.

CHARACTERIZATION OF DISTINCT VESICULAR STOMATITIS VIRUS, NEW JERSEY SEROTYPE, ISOLATES WITH RESPECT TO NUCLEIC ACID HOMOLOGIES, INTERFERENCE BY DI PARTICLES AND PROTEIN STRUCTURE

P.S. METZEL, W.M. SCHNITZLEIN and M.E. REICHMANN

Department of Microbiology,
University of Illinois,
Urbana, Illinois 61801, U.S.A.

Vesicular stomatitis virus (VSV) was originally classified into two major serotypes, New Jersey and Indiana, which exhibited little cross neutralization of activity in reciprocal serological tests (2). The latter serotype was further divided into four subgroups - Indiana, Argentina, Brazil and Cocal (2, 4). These classifications were presumably based on properties involving antigenic sites of the viral glycoprotein (7) and did not necessarily reflect relationships of other viral proteins or nucleic acid homology. While some of these classifications were partially confirmed by measuring the extent of heterotypic interference by defective interfering (DI) particles (3), reciprocal annealings between virus particle RNAs and mRNAs showed very little homology (10%) between the closely related Cocal and Indiana subgroups (15).

The recent mapping of DI particle RNAs utilized annealing experiments between viral mRNAs and particle RNAs of the Indiana subgroup from many different origins (9, 13, 16). The data demonstrated that all isolates of the Indiana subgroup were 90-100% homologous in the nucleotide sequences of their RNA. However, recent similar studies with the New Jersey serotype demonstrated a considerable lack of homology between two isolates of this group (17). Later work has shown that the New Jersey serotype isolates, used in most U.S. and Canadian laboratories, contained nucleotide sequences which were 80-100% homologous with the Ogden isolate RNA. However, the M (Glasgow) isolate was unique, with a possible 25-30% of its nucleotide sequences being in common with all the other isolates. This observation was of particular interest, since the currently available New Jersey serotype temperature sensitive mutants were derived from the M (Glasgow) isolate (12).

More recently, we have examined other New Jersey serotype isolates and found that the Hazelhurst virus resembled the M (Glasgow) virus in its RNA nucleotide sequences and showed a similar lack of homology with other North American isolates. Epidemiologically, the M (Glasgow) and Hazelhurst isolates originated from outbreaks in swine, whereas all the other viruses could be traced to epidemics in cattle and horses (1, 5, 6). Whether this distinction of host origin has any bearing on the nucleotide sequence relationships of these isolates can at present not be answered, and will require more comparative work of this nature.

Defective interfering (DI) particles, generated by the M (Glasgow) isolate were found to interfere equally efficiently with infections by the Ogden virus with which they shared a maximum equivalent sequence of 230 nucleotides (17). However, we now find that the HR (heat resistant isolate, Indiana) DI particle, which was previously reported to interfere equally well with Indiana and New Jersey serotype viral infections (11, 16), will only slightly interfere (10% as compared to homotypic interference) with infections by the M (Glasgow) viral isolate. Since this DI particle RNA contains the 3' half of the viral genome (9, 13, 16, 20) which might be important in the first step of the replicative cycle (14), the data suggest a possible difference in these replicase recognition sequences between the various New Jersey isolates.

The classification of the M (Glasgow) and Hazelhurst isolates as New Jersey serotype suggested that, in spite of the lack of nucleic acid homology with other members of this group, the antigenic sites in the G protein (7) were conserved. We have examined the primary structure of three major proteins (G, M and N) of the M (Glasgow) and Ogden isolates. A comparison of tryptic peptide fingerprints revealed major differences in the primary structure of the corresponding proteins of the two isolates. However, one dimensional electrophoresis of these peptides showed an unexpected similarity in the patterns, which suggested an extensive conservation of charged amino acids. This was also reflected in the equal number of tryptic peptides obtained from equivalent proteins of the two isolates.

CELLS AND VIRUSES

All isolates were grown in monolayers of BHK-21 C13 cells (5). The M (Glasgow) isolate was obtained from Dr. C.R. Pringle and originated in Dr. J.B. Brooksby's laboratory [New Jersey M (1)]. It was isolated from swine in an epidemic at

Kansas City, Missouri, in 1943 (6). The Hazelhurst isolate was given to us by Dr. R.A. Lazzarini and was originally obtained from the American Type Culture Collection (ATCC No.159). It was first isolated from epithelial tissue of swine at Hazelhurst, Georgia in 1952 (6). The previously described Prevec (P) isolate (17) was obtained from Dr. Prevec and can be traced back to a 1949 outbreak in Concan, Texas, where it was isolated from a diseased cow (6). It will be referred to as Concan (Prevec). The Ogden isolate was obtained from Dr. J.J. Holland, originated in Dr. F. Schaffer's laboratory, and was isolated from an afflicted cow at Ogden, Utah in 1949 (5). Two Indiana isolates (Glasgow, HR) have been described before (11, 16).

Virions and HR DI particles were grown and purified as described previously (16). Cytoplasmic and particle-bound RNA was isolated and annealed by procedures outlined elsewhere (16, 17).

COMPARISON OF THE NUCLEIC ACID HOMOLOGIES OF VARIOUS NEW JERSEY ISOLATES BY ANNEALING EXPERIMENTS

The degree of homology between the various isolates was determined by the reciprocal annealings of radioactively labeled viral mRNA species with an excess amount of unlabeled virion RNA. As a control, the mRNA species were also annealed with their respective virion RNAs. The extent of complementarity between the two classes of RNA species is expressed as the percentage of mRNA species resistant to pancreatic and T_1 ribonuclease digestion after annealing (Table I). The mRNA species were usually 90-100% complementary to their respective virion RNAs. The M (Glasgow) and Hazelhurst isolates had 80-100% homologous nucleotide sequences, whereas their RNAs were not more than 30% homologous to the RNAs of the other isolates. Likewise, the Ogden and Concan viral RNAs exhibited homologies of 90% or more, but the Concan RNA had only partial homology (approximately 30%) with the RNA of the M (Glasgow) isolate.

HETEROTYPIC INTERFERING ABILITY OF HR (Indiana) DI PARTICLES

Heterotypic interference by HR (Indiana) DI particles with New Jersey serotype virion infections was originally demonstrated with the Concan (Prevec) isolate (11) and later confirmed with the same isolate (16). In Fig.1, the interfering ability of this DI particle with infections by New Jersey M (Glasgow), New Jersey Concan (Prevec) and Indiana (Glasgow) are shown. The data is expressed as the log difference between the infectious output (plaque-forming units/ml) from virus-infected cells in the presence and absence of DI particles. As previously reported, the HR DI particles interfered to the

TABLE I

Annealing of New Jersey Viral mRNAs with Various Virion RNAs

Type of virion RNA	% Viral mRNA resistant to ribonuclease digestion[a]	
	Ogden mRNA	
	13-18S	30S
Ogden	99	94
Concan (Wagner)[b]	93	92
Hazelhurst	31	24
M (Glasgow)	25	19
None	1	5
	Concan (Prevec) mRNA	
	13-18S	30S
Concan (Prevec)	97	86
Ogden	91	80
M (Glasgow)	24	16
None	0	3
	M (Glasgow) mRNA	
	13-18S	30S
M (Glasgow)	94	88
Hazelhurst	82	77
Concan (Wagner)	33	33
None	1	14

[a]Excess unlabeled virion RNAs were annealed to $0.9\text{-}1.2 \times 10^3$ cpm of ^3H-labeled viral mRNA species (13-18S or 30S) and then treated with pancreatic and T_1 ribonuclease as described before (16, 17).

[b]This Concan isolate of New Jersey VSV was obtained from Dr. Wagner.

same extent with infections by the heterologous New Jersey Concan (Prevec) and the homotypic Indiana virions (Fig.1, solid and open circles, respectively). On the other hand, a weak interference (approximately 10% as compared to homotypic interference) with infections by New Jersey M (Glasgow) virions was observed (Fig.1, open triangles). Similarly, infections with Hazelhurst virions remained unaffected by low concentrations of HR DI particles, which caused a 90% reduction in homotypic virion production (R.A. Lazzarini, personal communication). It should be noted that, in contrast to HR (Indiana) DI particles, DI particles generated by New Jersey M (Glasgow) showed no difference in interfering ability with the M (Glasgow)

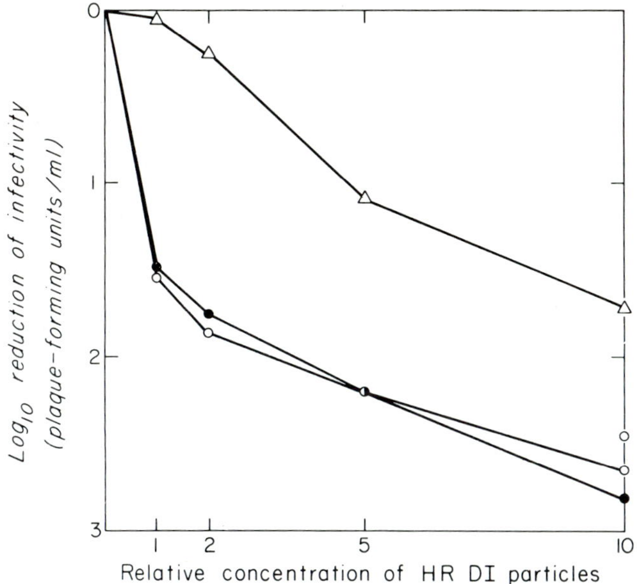

Fig.1. *Homotypic and heterotypic interfering ability of HR DI particles.*

Infectious output from BHK cells infected with virions in the presence or absence of HR DI particles was assayed as previously described (16, 17). Results are expressed as the log_{10} *reduction in Glasgow (Indiana) (—o—o—), Concan (Prevec isolate, New Jersey)(—●—●—) and M (Glasgow, New Jersey) (—Δ—Δ—) viral yields as a function of HR DI particle concentration in the respective inocula. In the absence of HR DI particles, the infectious yields were approximately* 8×10^8 *plaque-forming units/ml. The concentration of HR DI particles in the inocula is expressed in relative units, where 1 unit = 0.04 optical density units at 260 nm.*

virus and with the Ogden virus even though the DI particle RNA contained only a maximum sequence of 230 nucleotides in common with the latter virus (17).

CHARACTERIZATION OF G, M AND N PROTEINS OF ISOLATES WITH NON-HOMOLOGOUS RNA NUCLEOTIDE SEQUENCES

For purposes of comparison, proteins were isolated from purified M (Glasgow) and Ogden virions. The viral proteins were first radioiodinated as follows. Virions, suspended in 0.1 M phosphate buffer, pH 7.0, were disrupted in 1% SDS and subsequently iodinated with ^{125}I by the chloramine T reaction (19). The preparations were electrophoresed in 5% polyacrylamide gels for 4.0 hr at 8 mA/gel as described earlier (18). After electrophoresis, gels were frozen, sliced into 1mm sections and radioactive peaks determined by gamma counting. The chemically modified proteins separated well under these conditions (Fig.2), although some overlap of the G and N proteins occurred. For fingerprinting, only the non-overlapping peak fractions were eluted (Fractions 30-31 and 36, respectively). The profiles of proteins isolated from these fractions, as analyzed by 10% discontinuous gel electrophoresis (Fig.3A and 3B, respectively,) indicated that very little, if any, cross contamination had taken place.

For comparison of tryptic peptides, one or two gel slices from each of the protein peaks were pooled and washed in 20 ml of methanol-7% acetic acid (1:1) to remove the buffer and SDS. The slices were dried and overlayed with 1 ml of 1% ammonium carbonate containing 25 µg of trypsin TPCK. After digestion for 24 hr at 37°, the supernatant was removed with a pasteur pipette and dried by evaporation under a stream of N_2.

The peptides, resulting from enzymatic digestion of the radioiodinated viral proteins, were separated by both one-dimensional electrophoresis and two-dimensional "fingerprinting". After drying, the digested sample was suspended in 40 µl of distilled H_2O, spotted on a 1 cm line on Whatman No.3 MM chromatography paper (46 x 57 cm) and electrophoresed as described before (10). For "fingerprinting", the peptides were electrophoresed as described above and then chromatographed in an ascending system (10) for 18 hr. In order to locate the peptides, the chromatography paper was dried and subjected to autoradiography for 24-28 hr using Kodak No-screen Medical X-ray film. Autoradiographs of strips with the corresponding G, N and M proteins of the M (Glasgow) and Ogden isolates are shown in Fig.4A, B, and C, respectively. In spite of the lack of homology of the viral genomes, the tryptic peptides of the corresponding proteins appeared remarkably similar. The overall impression of similarity disappeared when the electrophoresed peptides were submitted to chromatography in the second dimension. The complete fingerprints shown in Fig.5A, B, and C reflected

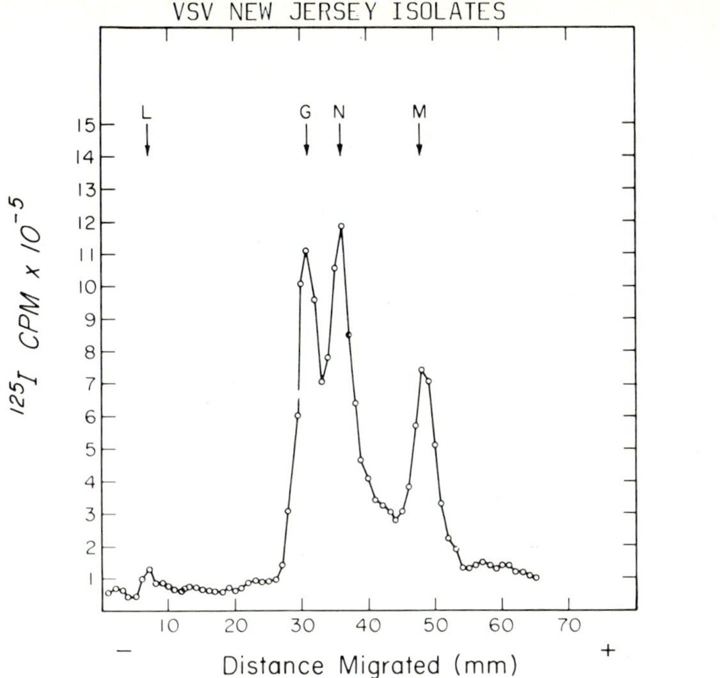

Fig.2. *Polyacrylamide gel electrophoresis of ^{125}I-labeled VSV (New Jersey) proteins*

Virus was disrupted in 1% SDS and extrinsically labeled with ^{125}I by the chloramine T catalyzed reaction (19). The reaction mixture was loaded directly onto a cylindrical (6 x 75 mm) 5% polyacrylamide gel and electrophoresed for 4 hr at 8 mA/gel. The arrows indicate the position of ^{3}H-amino-acid-labeled New Jersey viral proteins electrophoresed on a parallel gel. Fractions 30-31, 36 and 48-49 were used for further analysis of the G, N and M proteins, respectively.

considerable differences in the primary structure of the corresponding proteins, which was consistent with the lack of nucleic acid homology of the two isolates. The similarity in electrophoretic pattern and the equal number of tryptic peptides of corresponding proteins reflected a conservation of charged amino-acid residues including arginines and lysines which constitute the enzymatic cleavage site.

DISCUSSION

The Indiana serotype of VSV, particularly isolates of the Indiana subgroup, showed a high degree of homology in their RNA nucleotide sequences as measured by reciprocal annealing (9, 13, 16). In contrast, two isolates of the New Jersey serotype, although similar to each other, lacked extensive

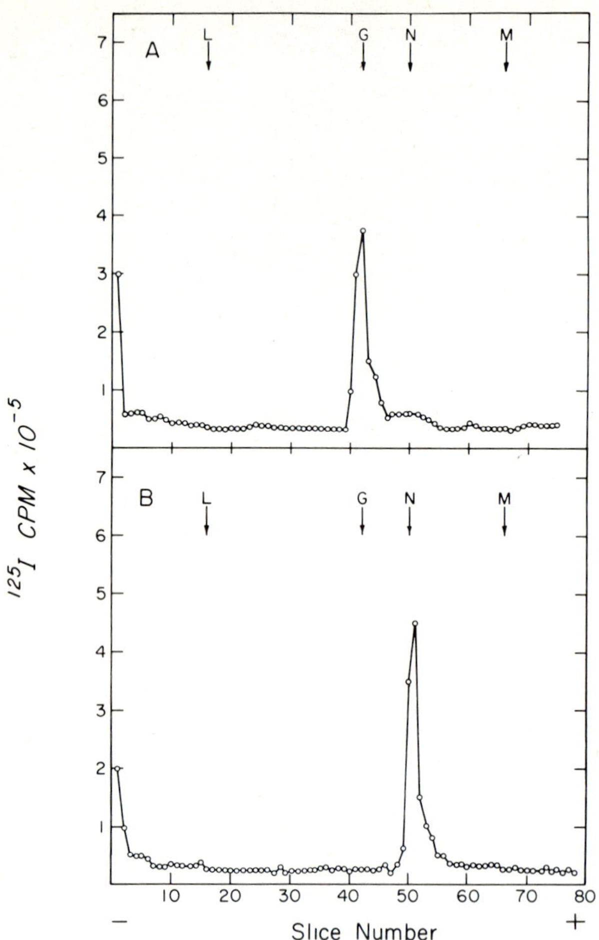

Fig.3. *Analysis of purified VSV (New Jersey) proteins by 10% discontinuous gel electrophoresis*

Viral proteins, isolated from a polyacrylamide gel (see Fig.2), were electrophoresed on 10% discontinuous gels. Electrophoresis was for 3.25 hr at 2.5 mA/gel. Arrows indicate the position of the ^{3}H-amino-acid-labeled New Jersey viral proteins electrophoresed on a parallel gel. A, G protein; B, N protein.

homology with other isolates of this serotype. Based on annealing data, New Jersey isolates seem to fall into two subgroups; one contains two members, M (Glasgow) and Hazelhurst, which were isolated from epidemics in swine; the other contained isolates of bovine origin. Since data on a large number of isolates is lacking, the significance of host difference has so far not been established.

The two New Jersey subgroups also differed in their susceptibility to heterotypic interference by HR (Indiana) DI particles, but not to homotypic interference by DI particles of New Jersey serotype origin. It has been shown previously that the RNA of the New Jersey DI particles mapped near the 5' end of the viral genome inside the L protein (30S mRNA) cistron (17). On the other hand, the HR DI particle RNA mapped in the 3' half of the viral genome and contained all other cistrons except that of the L protein (9, 13, 16). Since interference by DI particles probably takes place at the level of RNA replication, the results suggest that the viral RNA of the two New Jersey subgroups may differ in replicase recognition sites near the 3' end of their RNAs. The enzyme of the M (Glasgow) and Hazelhurst isolates may therefore not efficiently recognize the corresponding sequence of HR DI particle RNA, resulting in a partial abortive synthesis of the latter and in decreased interference. In contrast, the 5' terminal nucleotide sequences, (which are transcribed into 3' terminal replicase recognition sites on the positive RNA strand) of members of both subgroups are either identical or similar enough to secure replication of any DI particle RNA originating at this end.

Tryptic peptide fingerprints of the major viral proteins (G, N and M) reflected the lack of homologies of the two subgroups. The one dimensional electrophoretic pattern of equivalent proteins suggested a high degree of conservation of charged amino acid residues in the two virus subgroups. The conservation of arginines and lysines (sites of tryptic cleavage) was also reflected by the equal number of peptides in the digests of corresponding proteins of the two subgroups. The critical conservation of secondary and tertiary structures may be linked with the functional requirements of these proteins and may be dependent on the distribution and nature of charged amino acid residues in the polypeptide chains.

ACKNOWLEDGEMENTS

We thank Drs. Holland, Lazzarini, Pringle, Prevec and Wagner for their VSV isolates. The excellent technical assistance of Ms. P. Bay was appreciated. This work was supported in part by USPH grant AI 12070. One of the authors (PM) was the recipient of a NIH training grant (GM 941).

REFERENCES

1. Brooksby, J.B. (1948). *Proc. Soc. Exptl. Med.* 67, 254.
2. Cartwright, B. and Brown, F. (1972). *J. Gen. Virol.* 16, 391.

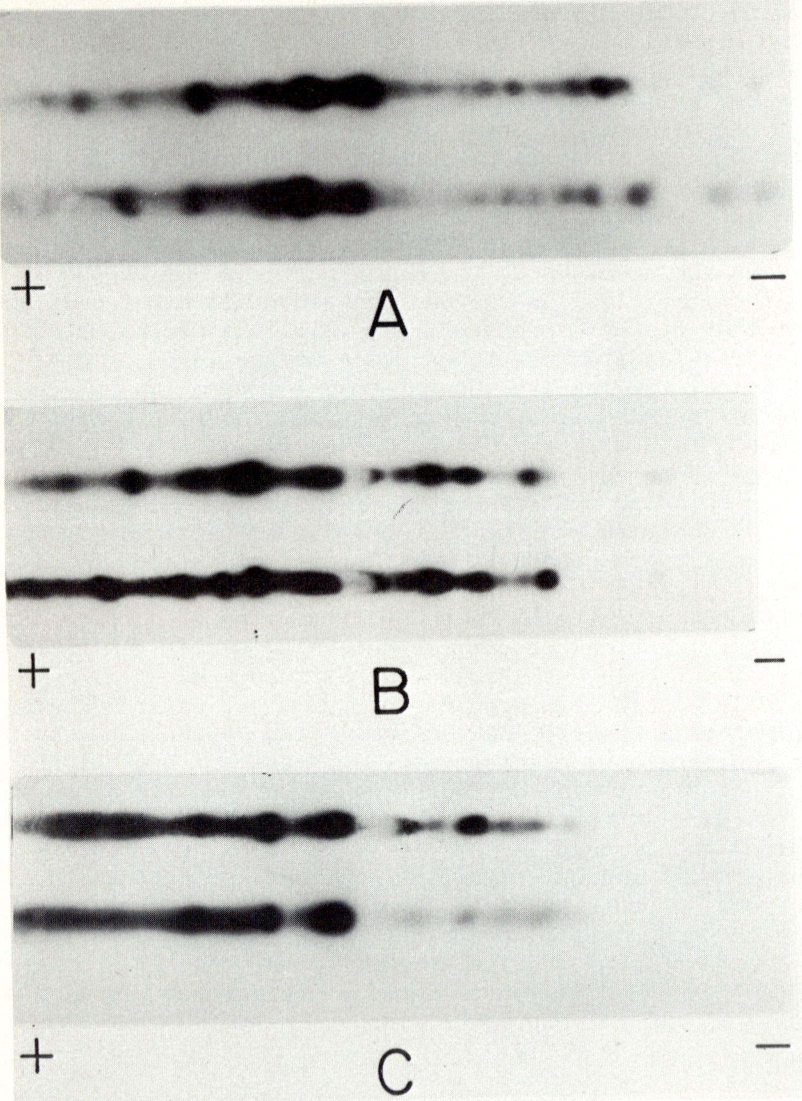

Fig.4. *One-dimensional electrophoresis of tryptic digestion from VSV (New Jersey) proteins.*

Radioiodinated proteins were purified by polyacrylamide gel electrophoresis and the protein-containing slices were digested in trypsin and then spotted on Whatman No.3 MM chromatography paper. Electrophoresis was for 2.5 hr at 2500 V in 1 M pyridine-acetate, pH 6.5. A, G protein; B, N protein; C, M protein. Upper track in each pair is derived from M (Glasgow) isolate, lower track is from Ogden isolate.

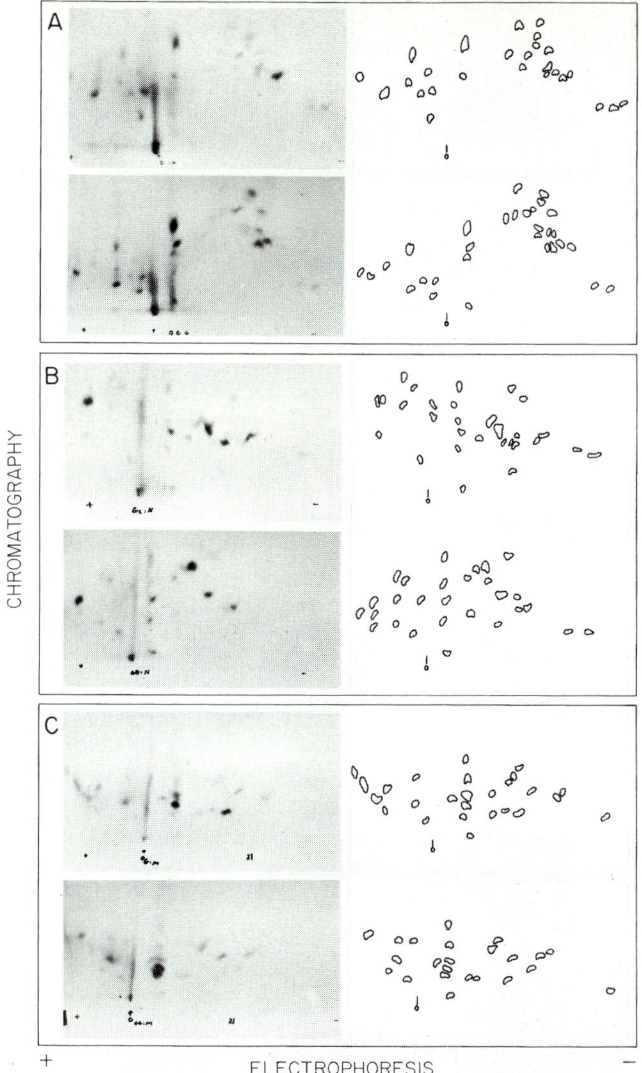

Fig.5. *Two-dimensional fingerprints of tryptic digestion products from VSV (New Jersey) proteins.*

Tryptic digests of the three major proteins of the M (Glasgow) and Ogden isolates were subjected to high voltage electrophoresis in the first dimension (see Fig.4) and ascending chromatography in the second dimension. A, G protein; B, N protein; C, M protein. The upper pattern in each box is derived from the M (Glasgow) isolate and the lower pattern is from the Ogden isolate. The figures on the right hand side are diagrammatic representations of the corresponding autoradiograms.

3. Crick, J. and Brown, F. (1972). *J. Gen. Virol.* 18, 79.
4. Federer, K.E., Burrows, R. and Brooksby, J.B. (1967). *Res. Vet. Sci.* 8, 103.
5. Hanson, R.P. (1952). *Bact. Rev.* 16, 179.
6. Holbrook, A.A., Geleta, J.N. and Patterson, W.C. (1956). *Proc. U.S. Livestock Sanit. Assoc.*,60th Ann. Meet. p.293.
7. Kelley, J.M., Emerson, S.U. and Wagner, R.R. (1972). *J. Virol.* 10, 1231.
8. Laemmli, U.K. (1970). *Nature, Lond.* 227, 680.
9. Leamnson, R.N. and Reichmann, M.E. (1974). *J. Mol. Biol.* 85, 551.
10. Lesnaw, J.A. and Reichmann, (1969). *Virology* 39, 738.
11. Prevec, L. and Kang, C.Y. (1970). *Nature, Lond.* 228, 25.
12. Pringle, C.R., Duncan, I.B. and Stevenson, M. (1971). *J. Virol.* 8, 836.
13. Reichmann, M.E. and Leamnson, R.N. (1974). *In* "Mechanisms of Virus Disease" (W.S. Robinson and C.F. Fox, eds.), p.101. W.A. Benjamin, Inc., Menlo Park, California.
14. Reichmann, M.E. and Schnitzlein, W.M. (1977). *In* "Microbiology-1977" (D. Schlessinger, ed.), p.439. American Society for Microbiology, Washington, D.C.
15. Repik, P., Flamand, A., Clark, H.F., Obijeski, J.F., Roy, P. and Bishop, D.H.L. (1974). *J. Virol.* 13, 250.
16. Schnitzlein, W.M. and Reichmann, M.E. (1976). *J. Mol. Biol.* 101, 307.
17. Schnitzlein, W.M. and Reichmann, M.E. (1977). *Virology* 77, 490.
18. Shapiro, A.L., Vinuela, E. and Maizel, J.V. (1967). *Biochem. Biophys. Res. Commun.* 28, 815.
19. Sonada, S. and Schlamowitz, M. (1970). *Immunochem.* 7, 885.
20. Stamminger, G.M. and Lazzarini, R.A. (1974). *Cell* 3, 85.

INVERTED COMPLEMENTARY TERMINAL SEQUENCES IN DEFECTIVE INTERFERING PARTICLE RNAs OF VESICULAR STOMATITIS VIRUS AND THEIR POSSIBLE ROLE IN AUTOINTERFERENCE

*J. PERRAULT, B.L. SEMLER, R.W. LEAVITT and J.J. HOLLAND

Department of Biology,
University of California, San Diego,
La Jolla, California 92093, U.S.A.

**Department of Microbiology and Immunology,*
Washington University School of Medicine,
St. Louis, Missouri 63110, U.S.A.

The widely occurring phenomenon of homologous virus autointerference first described succinctly for influenza virus by von Magnus (25) remains one of the most intriguing problems of molecular virology. The defective interfering virus particles (DI) which mediate this phenomenon have been most extensively characterized in the model vesicular stomatitis virus (VSV) system. VSV DI or T particles contain subgenomic RNA fragments ranging in size from 10-50% of the genome and specifically inhibit the replication of infectious virus at the replicase level (6).

Recently Holland *et al.* (3) have shown that VSV DI can attenuate viral infections *in vitro* and *in vivo*. Thus a thorough understanding of their mode of action is not only of potential therapeutic value but should also bring insight into the causes of slow, progressive, and persistent viral diseases.

Our more recent studies have focused on structural aspects of VSV DI particle RNAs (14, 16). Depending on the particular VSV T particle, these RNAs can be minus strands only (18, 23), unlinked plus and minus strands (11, 16, 17,19, 21) or covalently linked plus and minus strands also called snap-back (10, 14, 16).

The presence of inverted complementary terminal sequences has been reported for alphavirus genome and DI RNAs (5, 8) and for the paramyxovirus Sendai DI RNAs (9). Similar sequences have been suggested for the RNA from Uukuniemi virus (2), a Bunyavirus. In addition, similar sequences could explain the presence of circular ribonucleoproteins in cells infected by the paramyxovirus measles (24) and in both La Crosse (13) and

Lumbo (20) viruses, also members of the Bunyaviridae. Thus as suggested earlier by Kolakofsky (9) this structural feature may be common in animal virus RNAs. The inverted complementary terminal sequences appear to be ~150 base pairs long for Semliki Forest virus RNAs (8) and ~150-200 base pairs long for Sendai DI RNAs (9) representing 1-2% of the total RNA. This report describes complementary terminal sequences of VSV DI RNA.

CIRCULAR STRUCTURES IN VSV DI RNAs

When RNA from the short T particles (~1/10th genome size) generated from our stock of tsG31 VSV is spread under aqueous conditions and examined microscopically two types of double-stranded molecules can be observed (Fig.1A). If the preparation is melted and quick-cooled just before spreading

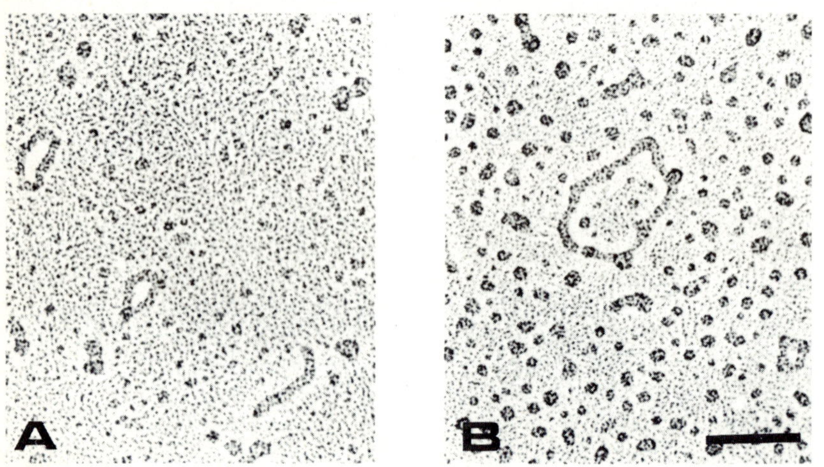

Fig.1. *Electron micrographs of tsG31 DI RNA. (A) Aqueous spreading showing the appearance of circular molecules after partial renaturation; (B) Formamide spreading showing two circular molecules, a monomer and an oligomer. The bar in the photographs represents 0.2 μm.*

essentially all the molecules are linear and homogenous in size (~0.23 μm). However, if the RNA is allowed to renature partially before spreading a variable number of circular (Fig. 1A) and multimeric structures (not shown) are generated. In experiments detailed elsewhere (16) we have shown that the linear molecules obtained without a renaturation protocol correspond to snap-back structures which are found frequently in various VSV T particle RNA preparations. These snap-back molecules all appear to conform to the same general hairpin structure with a very small single-stranded region linking the two complementary strands. The circular and multimeric structures, however, appear to be generated by concentration dependent annealing of identical size complementary strands and end-to-end annealing of inverted complementary terminal sequences. As many as 32% of the molecules appear circular when this same RNA is spread in formamide which unfolds single-stranded molecules but leaves snap-back molecules intact (Fig. 1B). Fig.2 summarizes our interpretation of the various structures seen in the electron microscope. Of particular significance is the finding that the snap-back DI RNA molecules contain inverted complementary sequences at *both* ends of *both* strands of the duplex since circles can be generated after RNase treatment (Fig.2D).

ISOLATION OF INVERTED COMPLEMENTARY TERMINAL SEQUENCES FROM DI RNA

Although our tsG31 VSV stock generates two T particles, the larger of which contains the snap-back molecules, it is possible to purify the smaller particle essentially free of the larger snap-back DI. When this shorter particle RNA (~350,000 daltons molecular weight) is melted and quick-cooled approximately 2-10% of the RNA reanneals in a concentration independent manner and can be isolated after RNase treatment. This RNA is essentially a single duplex species, approximately 60 bases in length (16). Similar small RNA fragments have been isolated from other VSV T particle sources including snap-back RNA from DI (unpublished observations). Many of the experiments reported below were carried out with this purified double-stranded fragment from tsG31 RNA which our evidence indicates corresponds to base-paired stems of single-stranded circles (16).

Fig.3A shows an electropherogram of stem RNA from a preparative acrylamide gel purification of a mixture of the two tsG31 T particles uniformly labelled *in vivo* with ^{32}P. The band near the top of the gel is nicked snap-back duplex RNA and its presence does not affect the yield or purity of stems from the single-stranded circles. Unless otherwise specified all hybridization experiments described below with unlabelled, ^{32}P-end-labelled, or uniformly ^{32}P-labelled stems were carried out with purified RNA eluted from similar discrete bands in preparative gels.

5' END-LABELLING OF STEMS AND STRAND SEPARATION

In order to obtain a ready source of labelled stems of high specific activity for hybridization studies we labelled the one available 5' hydroxyl end (generated by nuclease digestion) with phage T4 polynucleotide kinase and γ-^{32}P-labelled ATP (see illustration in Fig.4). The other strand corresponding

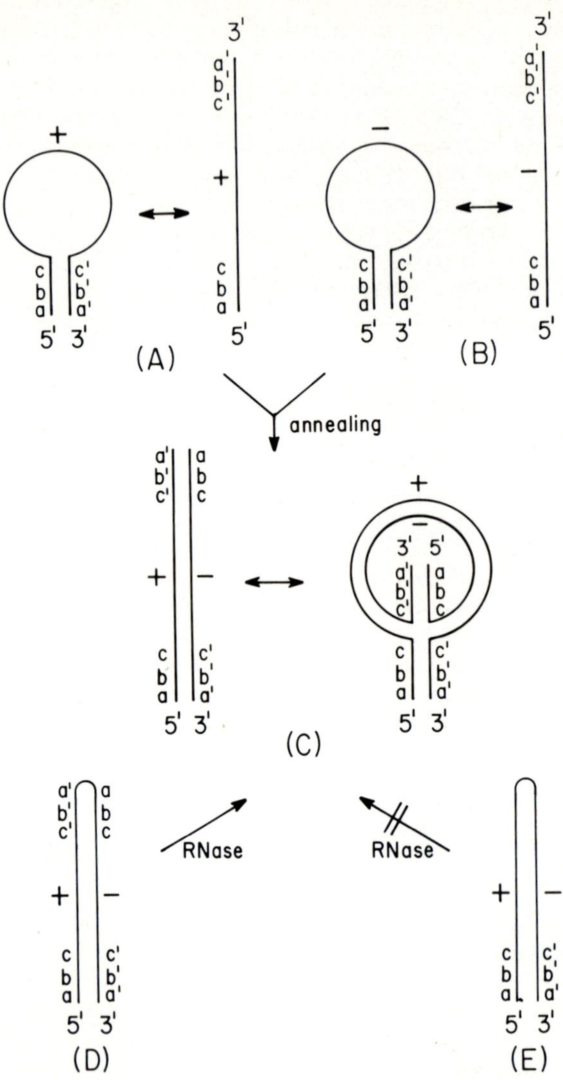

Fig.2. *Diagrammatic representation of structures proposed for the various DI RNA molecules observed microscopically. The letters abc and c'b'a' correspond to the proposed inverted complementary sequences at the ends of the RNA. (A) and (B), ssRNA of plus and minus strand polarity respectively in either linear or circular form; (C), linear and circular dsRNA resulting from intermolecular annealing of single-stranded plus and minus strands; (D) snap-back RNA structure with comple-*

to the original 5' end of the DI RNA molecule remains unlabelled presumably because it contains a phosphorylated terminus. The details of this procedure will be published elsewhere (Semler, Perrault, Abelson, and Holland, in preparation). Specific activities of the order of 100 x 10^6 c.p.m./μg were readily obtained. The end label in these duplex stems is resistant to RNases A and T_1 under high ionic strength conditions but is sensitive to these nucleases as well as to bacterial alkaline phosphatase following denaturation. Reannealing of the denatured single strands can be simply followed by increase in RNase resistance (see below).

Clearly the usefulness of both the end-labelled and uniformly-labelled materials as sensitive hybridization probes is greatly increased if the individual strands can be purified away from each other. This was carried out successfully by a modification of the Maxam and Gilbert procedure (12) for DNA strand separation which will be detailed elsewhere (Perrault, Semler and Holland, in preparation). Fig.3B shows an electropherogram of a preparative strand separation acrylamide gel of uniformly ^{32}P-labelled stems. Both the top band (slightly contaminated with undenatured duplex stems migrating just above it) and the bottom band were eluted from the gel and employed in the hybridization experiments described below. Strand separation of end-labelled stems gives rise to a labelled top band only.

FULL-SIZE GENOME RNA LACKS INVERTED COMPLEMENTARY TERMINAL SEQUENCES

The availability of both individual purified stem strands from tsG31 DI RNA allows us to test directly for the presence of these sequences in genome RNA by hybridization. The outcome of such an experiment (Table I) is clear. Only the top band (corresponding to the complementary sequence at the 3' end of the original DI RNA molecule) anneals to RNA from infectious virus. No sequences complementary to the 5' end of the DI RNA present in the stems are detected. This important fact, first observed with Sendai virus and its DI by Leppert *et al.* (see chapter 54), provides a basis for possible mechanisms of autointerference suggested below. As a test of the degree and extent of homology between complementary RNA molecules in our

---

mentary terminal sequences at both ends of the duplex which can generate circular dsRNA molecules after nicking with RNase; (E) hypothetical snap-back RNA structure with complementary sequences only at the open end of the duplex which cannot generate circles after nicking with RNase. VSV DI snap-back RNA molecules correspond to structure (D).

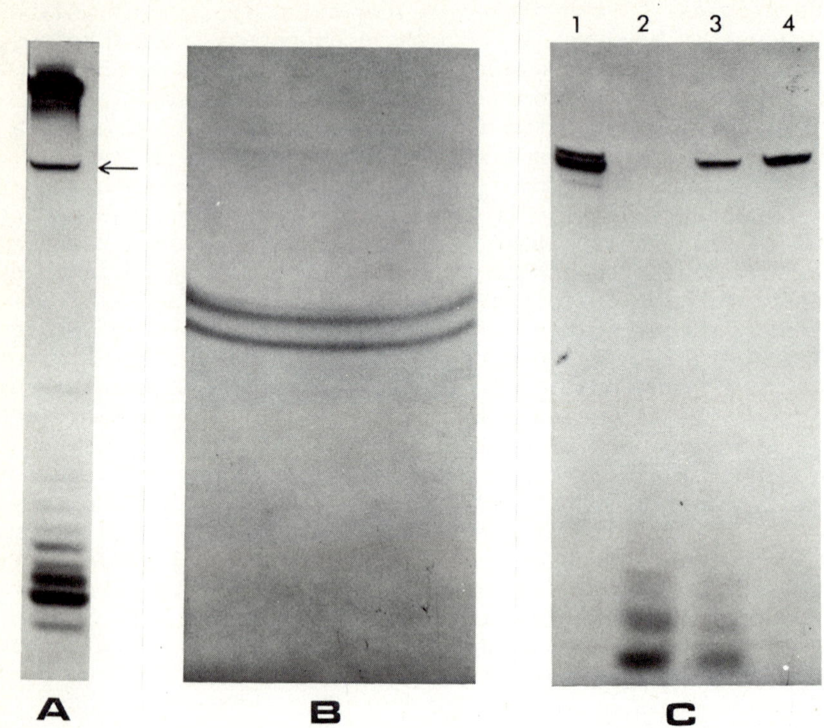

Fig.3. *Polyacrylamide gel electrophoretic analysis of ^{32}P-labelled tsG31 DI RNA stem structures. (A) 20% acrylamide preparative gel purification of uniformly ^{32}P-labelled duplex stems (arrow). (B) 8% acrylamide preparative gel separation of the individual strands from the duplex stem band shown in (A). (C) 20% acrylamide gel analysis of RNase resistant hybrids obtained from annealing reactions. Lane 1, ^{32}P-end labelled top band from tsG31 duplex stems eluted from a strand separation gel as in (B). The upper band in lane 1 represents undenatured duplex stems remaining in the sample; lane 2, RNase resistant fraction from ^{32}P-end labelled genome RNA self-annealed under the same conditions as the sample in lane 3; lane 3, RNase resistant fraction from ^{32}P-end labelled genome RNA annealed to an excess of unlabelled tsG31 DI RNA duplex stems; lane 4, RNase resistant fraction from the uniformly ^{32}P-labelled top band annealed to unlabelled genome RNA described in Table I. Experimental details of the gel techniques, autoradiography, and annealing procedures will be published elsewhere, (Perrault, Semler and Holland, in preparation).*

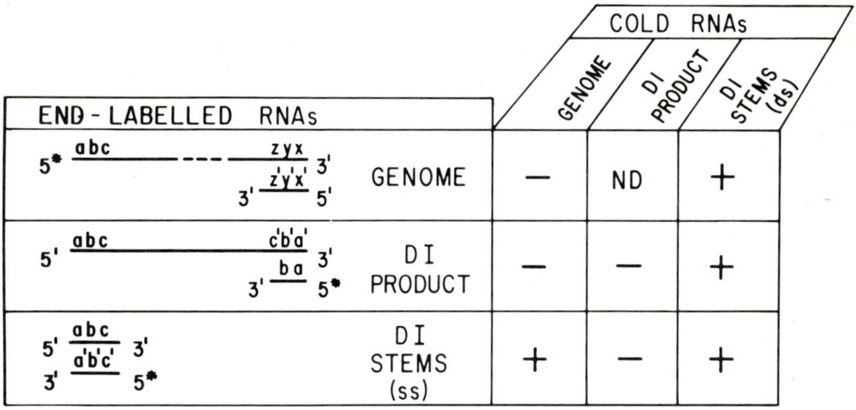

END-LABELLED RNAs		COLD RNAs: GENOME	COLD RNAs: DI PRODUCT	COLD RNAs: DI STEMS (ds)
5* abc --- zyx 3' 3' z'y'x' 5'	GENOME	−	ND	+
5' abc c'b'a' 3' 3' b a 5*	DI PRODUCT	−	−	+
5' abc 3' 3' a'b'c' 5*	DI STEMS (ss)	+	−	+

Fig.4. *Summary of annealing data using ^{32}P-end labelled RNAs. The diagrammatic structures shown are based on the results of these and other annealing data (see text). The * symbol denotes the position of the 5' end label on the various molecules. + and - signs indicate protection or no protection of the end label from RNase digestion after annealing. ND indicates not done. The details of these experiments will be published elsewhere, (Perrault, Semler and Holland, in preparation).*

hybridization experiments we routinely sized annealed duplexes on acrylamide gels after RNase treatment. This treatment is expected to digest not only non-complementary single-stranded molecules but also any extensive region of base-pairing mismatch in the duplexes which would give rise to smaller double-stranded fragments (some types of minor mismatching may not be detectable by this method). Fig.3C, lane 4, shows that essentially all of the RNase resistant material obtained by annealing the genome RNA to the uniformly ^{32}P-labelled top stem band migrates as a duplex identical in size to the original double-stranded stem. The background of RNase resistant material in the bottom band (Table I) is in part due to small poly A tracts and to small amounts of contaminating undenatured duplex RNA (data not shown). The above results therefore show that all or nearly all of the top band from the DI stem is complementary to a genome sequence.

CONSERVED GENOME 5' TERMINAL SEQUENCE IN DI RNAs AND DI POLYMERASE PRODUCT

In the experiment described above the top stem strand which anneals to genome RNA is also the one which becomes end-labelled with polynucleotide kinase. This provides strong evidence

TABLE I

Annealing of Separated Strands from Uniformly Labelled ^{32}P DI Stem RNA to Unlabelled VSV Genome RNA[α]

Source of RNA	Total c.p.m.	RNase Resistant c.p.m.	% Resistance
Top band, self-annealing	4,165	645	15
Top band plus genome RNA	22,880	20,460	90
Bottom band, self-annealing	3,280	1,184	36
Bottom band plus genome RNA	22,680	7,800	34

[α]Annealing reactions were carried out in 0.2 ml volumes of 0.6 M NaCl, 0.01 M Tris-HCl, pH 7.6, 0.001 M EDTA for 80 min at 70°. Conditions for *in vivo* labelling and particle RNA purifications were all described previously (16). The specific activity of the uniformly labelled ^{32}P top and bottom stem bands (eluted from the gel shown in fig.3B) was ~1 x 10^6 c.p.m./µg. Each band RNA sample alone or mixed with 12 µg of genome RNA was first denatured at 100° for 2 min in the buffer indicated above but without NaCl. Total c.p.m. and RNase resistant c.p.m. were determined from small portions of the total sample as described previously (16). The remaining portion from the top band plus genome RNA sample was RNase treated and the resistant RNA repurified before analyzing on the gel shown in fig.3C. The levels of RNase resistance obtained in the self-annealing reactions were verified to be saturating from independent experiments (not shown).

that the 5' end and not the 3' end of the DI RNA is conserved from the genome. This important conclusion was further tested and the location of this conserved sequence on the genome RNA was determined in a number of independent hybridization experiments which will be described in detail in a subsequent communication. A summary of these results as well as the relationship of the DI stems to the 5' end of the genome RNA and to the DI polymerase product (17) is illustrated in Fig.4. It should be mentioned here that this DI polymerase product RNA which is 45 nucleotides in length has been completely sequenced (Semler, Perrault, Abelson and Holland, in preparation) for two of our DI types. This sequence must be identical, or nearly so, to a stretch of 45 nucleotides at or very near the 5' end of the genome RNA, based on the annealing data summarized in Fig.4.

The structures depicted in Fig.4 are the only ones compatible with all the annealing data. In all combinations where hybridization did occur the size of the RNase resistant hybrids was determined on gels as in Fig.3C. In all cases the expected duplex size was obtained. In particular, the annealing of 5' end-labelled genome RNA to cold duplex DI stems gives rise to a labelled hybrid identical in size to the original DI stems (Fig.3C, lane 3). This critical finding confirms our assignments of polarities for these RNA molecules.

Lastly, similar hybridization experiments indicate that the 5' end sequence from tsG31 DI RNA is conserved not only in the Indiana serotype VSV (Mudd-Summers) genome but also in the VSV New Jersey (Ogden) genome. In addition, these sequences are conserved in DI derived from the VSV (Mudd-Summers) strain as well as in DI derived from persistently infected BHK-21 cells after three years of continuous carrier state *in vivo* (4). This conservation of stem sequences is not unexpected if they are involved as replicase recognition sites.

DISCUSSION

The above results show the relationship of the inverted complementary terminal sequences of VSV DI RNA to VSV genome RNA as well as to the DI polymerase product. First, using very sensitive hybridization probes from the separated strands of the inverted complementary terminal sequences of tsG31 DI RNA, we have shown that full-size VSV genome RNA lacks these inverted sequences. Only the 5' end (and not the 3' end) of VSV genome RNA is conserved in the DI RNA. In addition, the DI polymerase product was shown to be synthesized from the 3' end of the DI RNA. The fact that this product RNA did not anneal to genome RNA is further confirmation that the 3' end of genome RNA is not conserved in the DI RNA.

Evidence as to whether these complementary terminal sequences play a role in DI interference has not yet been obtained. However, it is an intriguing possibility that these sequences may contribute to the replicative advantage attributed to the DI RNA in autointerference (7, 15). If the stem sequences are involved as replicase binding and/or recognition sites, then the observation that they are at both ends of the DI RNA and only one end (the 5' end) of the genome RNA may allow the DI to inhibit or out-compete the replication of plus strand RNA (full-size RNA complementary to the genome). This assumes that for genome RNA the initial steps in replication, presumably recognition by a replicase of the sequence at the 3' end of minus strand virion RNA and subsequent synthesis of a plus strand complement, may utilize a different replicase (perhaps by adding or removing host cell or viral proteins) than the secondary steps in the replication scheme, which involve recognition of a different sequence at the 3' end of the newly-synthesized plus strand as template for the synthesis of progeny minus strands.

This scheme would be altered for DI RNAs containing inverted complementary terminal sequences because the sequence at the 3' end of the minus strand is identical to the sequence at the 3' end of the plus strand. For these RNAs, the synthesis of both plus and minus strands could be achieved by the same replicase. The replicative advantage would occur not only because genomes (unlike most DI) are tied up as templates for transcription, but also from a simple kinetic argument that the DI RNA need bind the replicase only once during the synthesis of a plus and minus strand whereas the genome RNA would require a separate binding event of one replicase for plus strand synthesis and another replicase for minus strand synthesis.

Furthermore, if the replicase complex which synthesizes genomic plus strand RNA using virion RNA as template is rate-limiting in infected cells and the other replicase complex (which synthesizes minus strand RNA) is not rate-limiting, then it is obvious that most DI RNAs would be selectively replicated since they have the same 3' sequences on both plus and minus strand RNA and can by-pass the rate limiting step. Schnitzlein and Reichmann (22) recently showed evidence from heterologous interference studies that replication of plus and of minus strands of RNA are distinguishably different processes. They also showed the importance of nucleocapsid protein type in replicase-template interactions so it must be kept in mind that this is a complex process involving much more than simple interactions between replicase and 3' termini of RNA. Full understanding of DI interference will require further studies, particularly since there is a rare class of VSV DI(HR-LT) with apparently different 3' termini than the majority class of DI studied above (1, 11, 21, 23).

The results above raise interesting questions regarding the mechanisms of generation of DI and many models could be advanced. Two related models have recently been proposed by D. Kolakofsky and his colleagues (see chapter 54) and by A.S. Huang (Eli Lilly Award Address, American Society for Microbiology Meeting,

New Orleans, La., 8-13 May 1977).

SUMMARY

The VSV DI RNAs we have examined contain inverted complementary terminal sequences approximately 60 nucleotides in length. Only the terminal sequence at the 5' end of these DI RNAs is found in the full size genome RNA. We discuss the implications for DI interference which arise from the observation that the 3' end of the genome RNA is not conserved in the majority of VSV DI RNAs.

ACKNOWLEDGEMENTS

We are grateful to Dr David Kohne for support and advice. We also thank Dr Daniel Kolakofsky for kindly communicating to us unpublished results and both him and Dr Laurent Roux for helpful discussions. We are grateful to Estelle Bussey for her excellent technical assistance. This investigation was supported by Public Health Service grants no. CA-10802 from the National Cancer Institute and no. NS-6-2336 from the National Institute of Central Nervous System Diseases and Stroke. R.W.L. is a Postdoctoral Fellow of the National Institute of Allergy and Infectious Diseases no. 1 F32 Al 05626-01. B.L.S. is supported by a U.S. Public Health Service Training Grant no. CA 05274.

REFERENCES

1. Colonno, R.J., Lazzarini, R.A., Keene, J.D. and Banerjee, A.K. (1977). *Proc. Natl. Acad. Sci. USA* 74, 1884.
2. Hewlett, M.J., Pettersson, R.F. and Baltimore, D. (1977). *J. Virol.* 21, 1085.
3. Holland, J.J., Villarreal, L.P., Breindl, M., Semler, B.L. and Kohne, D. (1976). *In* "Animal Virology", ICN-UCLA Symposia in Molecular and Cellular Biology (D. Baltimore, A.S. Huang and C.F. Fox, eds.), p.773.
4. Holland, J.J., Villarreal, L.P., Welsh, R.M., Oldstone M.B.A., Kohne, D., Lazzarini, R. and Scolnick, E. (1976). *J. Gen. Virol.* 33, 193.
5. Hsu, M.T., Kung, H-J. and Davidson, N. (1973). *Cold Spr. Har. Symp. Quant. Biol.* 38, 943.
6. Huang, A.S. (1973). *Ann. Rev. Microbiol.* 27, 101.
7. Huang, A.S. and Manders, E.K. (1972). *J. Virol.* 9, 909.
8. Kennedy, S.I.T. (1976). *J. Mol. Biol.* 108, 491.
9. Kolakofsky, D. (1976). *Cell* 8, 547.
10. Lazzarini, R.A. Weber, G.H., Johnson, L.D. and Stamminger, G.M. (1975). *J. Mol. Biol.* 97, 289.
11. Leamnson, R.N. and Reichmann, M.E. (1974). *J. Mol. Biol.* 85, 551.

12. Maxam, A.M. and Gilbert, W. (1977). *Proc. Nat. Acad. Sci. USA* 74, 560.
13. Obijeski, J.F., Bishop, D.H.L., Palmer, E.L. and Murphy, F.A. (1976). *J. Virol.* 20, 664.
14. Perrault, J. (1976). *Virology* 70, 360.
15. Perrault, J. and Holland, J.J. (1972). *Virology* 50, 159.
16. Perrault, J. and Leavitt, R.W. (1977). *J. Gen. Virol.* 38, 21.
17. Reichmann, M.E., Villarreal, L.P., Kohne, D., Lesnaw, J. and Holland, J.J. (1974). *Virology* 58, 240.
18. Roy, P. and Bishop, D.H.L. (1972). *J. Virol.* 9, 946.
19. Roy, P., Repik, P. Hefti, E. and Bishop, D.H.L. (1973). *J. Virol.* 11, 915.
20. Samso, A., Bouloy, M. and Hannoun, C. (1975). *C.r. Acad. Sci.*, Paris, Series D 280, 213.
21. Schnitzlein, W.M. and Reichmann, M.E. (1976). *J. Mol. Biol.* 101, 307.
22. Schnitzlein, W.M. and Reichmann, M.E. (1977). *Virology* 80, 275.
23. Stamminger, G. and Lazzarini, R.A. (1974). *Cell* 3, 85.
24. Thorne, H.V. and Dermott, E. (1976). *Nature* Lond. 264, 473.
25. von Magnus, P. (1954). *Adv. Virus. Res.* 2, 59.

A GENETIC MAP OF SENDAI VIRUS DI-RNAs

DANIEL KOLAKOFSKY, MARK LEPPERT and LILA KORT

Department of Microbiology,
University of Utah Medical Center,
Salt Lake City, Utah 84132, U.S.A.

When animal viruses are passaged at high multiplicities of infection, virus particles are generated which both interfere with the replication of standard virus and contain less genetic information than the standard virus (4). These defective-interfering (DI) particles are a common feature of most, if not all, animal virus systems, and Sendai virus, a member of the parainfluenza group, is no exception. Sendai DI-particles were first reported by Kingsbury, Portner, and Darlington (6) who demonstrated that these particles contained shorter genomic RNAs than standard virus particles, and appeared to interfere with standard virus infection at the level of replication.

We have recently isolated four Sendai DI-RNAs from four different stocks of Sendai virus (9). These DI-RNAs were found to be considerably shorter than non-defective (ND) genome RNA, representing from 8 to 30% the length of ND-genomes (15,000 nucleotides (N)). Hybridization studies demonstrated that these DI-RNAs are unique segments of the viral genome. A curious feature of these DI-RNAs is their ability to form circular structures with short panhandles (approximately 150 base pairs when viewed in the electron microscope). Since these panhandles are presumed to be the ends of the DI-RNAs, we concluded that the ends of the DI-RNA contain complementary sequences, i.e., each end of the DI-RNA is an inverted repeat of the other end.

As mentioned above, all four DI-RNAs we isolated contained complementary ends even though they ranged in size from 1200 to 4500 nucleotides long. The simplest model consistent with these results is that the ND-genome also contained complementary ends and that the DI-RNAs represent internal deletion mutations. In this model, the replicase and nucleocapsid protein recognition sites of the ND-genome, which are presumably at the ends of the genomic RNA, are conserved in the DI-RNAs and the DI-RNAs interfere with ND-genome replication by competing with the ND-genome for viral replicase.

In order to test this model, we have isolated the double-stranded RNA (ds-RNA) panhandles or stems from three of the four DI-RNAs previously described (9) and compared the sequences of these stems to the 5' end of the ND-genome by hybridization. The results of these studies indicate that the ends of all the DI-RNAs are very similar, if not identical, and that they all contain sequences homologous to the 5'-end of the minus-strand ND-genome. In addition, almost all the sequences present in the entire DI-RNAs are derived from sequences contiguous to the 5' end of the ND-genome. However, there appear to be no sequences present at its 5'-end, i.e., the ends of the ND-genome are not complementary. The complementary ends of the Sendai DI-RNAs therefore do not appear to have been generated by internal deletion of the ND-genome as previously proposed. A new model for the generation of the DI-RNAs is proposed.

SEQUENCE HOMOLOGY BETWEEN THE ds-STEMS AND THE 5'-END OF THE ND-GENOME

Since the Sendai DI-RNAs previously described all contain complementary ends (9), it seemed likely that the ends of the DI-RNAs derive from the ends of the 50s ND genome. To investigate this possibility, 50s RNA specifically labeled at either end would be most useful. The isolation of a soluble guanylyl-transferase from vaccinia virus which is capable of "capping" exogeneous large molecular weight RNAs has been described (13). Egg-grown Sendai 50s RNA (mostly minus strands) was "capped" using this guanylyl-transferase and α-^{32}P-GTP. The ^{32}P-capped-50s RNA was then characterized by enzymatic digestion and electrophoresis of the products on DEAE-paper at pH 3.5. As shown in Fig.1, when undigested RNA was electrophoresed, no radioactivity could be detected except at the origin. When the RNA was digested with ribonuclease T2, only a single spot of radioactivity migrating behind the ppppA marker is seen. Since the 5' end of Sendai 50s RNA is pppAp (Leppert and Kolakofsky, manuscript in preparation), the expected product of the T2 digestion of capped-50s RNA would be GpppAp. When the RNA was digested with Penicillium nuclease P1, almost all the radioactivity migrated as a single spot just slightly ahead of GTP (the expected product of GpppA), with a minor spot (<5% of the radioactivity) migrating at the position of pG. If alkaline phosphatase was included in the T2 digestion, the major spot now migrated in an identical manner as the major P1 digestion product (not shown) and there was no loss of radioactivity from the major spot. These experiments demonstrate that greater than 90% of the radioactivity introduced into Sendai 50s RNA is in an unmethylated 5' cap structure.

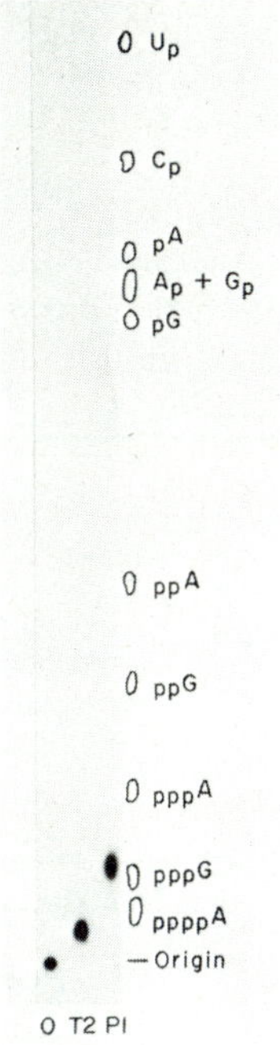

Up
Cp
pA
Ap + Gp
pG
ppA
ppG
pppA
pppG
ppppA
— Origin
O T2 PI

Fig.1. *Characterization of* ^{32}*P-capped 50s RNA*

Samples (4000 cpm cerenkov) of the ^{32}*P-RNA prepared as described (12) were digested with either 10 units of ribonuclease T2 plus 5 μg ribonuclease A in 15 μl of 10 mM sodium acetate pH 5.0, 1 mM EDTA, or 25 μg of Penicillium nuclease P1 in 10 μl of1mM sodium acetate pH 6.0, for 2 hr at 37°. The digestion mixes were spotted directly onto Whatman DE 81 paper along with various nucleotide markers. As a control, 2000 cpm (cerenkov) of undigested RNA was also spotted. Electrophoresis was carried out in pyridine-acetate buffer, pH 3.5., for 1.45 hr at 3 KV. The electrophoretogram was first exposed to X-ray film, and then washed in ethanol containing 1%* NH_4OH *and examined under uv light to locate the marker nucleotides whose position is indicated on the right-hand side of the figure.*

Double-stranded RNA stems which were 110, 135, and 150 base pairs (bp) long were isolated from the H-14s, J-15s, and R-205 circular DI-RNAs respectively as previously described (12). To determine whether these ds-stems contained sequences homologous to the 5' end of the 50s minus-strand genome, increasing amounts of each heat-denatured ^{3}H-stem was added to a constant amount of ^{32}P-capped 50s RNA and annealed for 45 min at 73°. As shown in Fig.2, if no stems were added, approximately 10% of the ^{32}P-radioactivity became resistant to ribonuclease digestion, indicating that the ^{32}P-capped RNA was composed almost entirely of minus-strands. When each of the ^{3}H-stems was annealed with the ^{32}P-capped 50s RNA, more than 90% of the ^{32}P-radioactivity became ribonuclease resistant. All three ds-stems therefore contain sequences homologous to the 5' end of the 50s minus-strand genome.

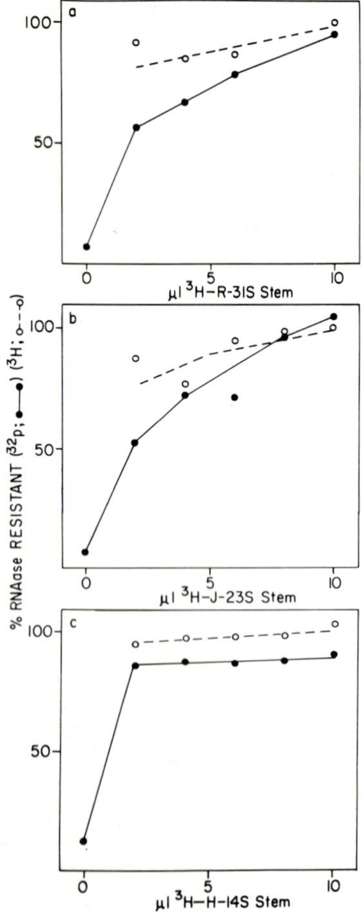

SEQUENCE HOMOLOGY BETWEEN WHOLE DI-RNAs AND THE 5' END OF THE ND-GENOME

Since some, if not all, the sequences present in the DI-stems are contiguous to the 5' end of the ND-genome, it seemed likely that most of the sequences present in the entire DI-RNAs would similarly derive from the 5' end of the ND-genome. To examine this possibility, ^{3}H-H-14s and R-20s DI-RNAs were isolated from preparative sucrose gradients as previously described (9), and one half of each preparation was coprecipitated with ^{32}P-capped 50s RNA while the other half of each preparation was ethanol precipitated by itself. The various RNAs were recovered by centrifugation, dissolved in 40 µl of 2 x buffer A containing 0.1% SDS, and annealed for 20 min at 73°. Those samples containing the ^{32}P-capped 50s RNA were digested with ribonuclease followed by pronase digestion and all samples were reprecipitated with ethanol, and then electrophoresed on a polyacrylamide slab gel along with ^{14}C-Reo ds-RNA as markers. That portion of the gel which contained the ^{32}P-capped 50s RNA samples was dried without PPO impregnation so that only ^{32}P- and ^{14}C-radioactivity would be seen by autoradiography, whereas that portion of the gel containing only the annealed ^{3}H-DI-RNAs was processed for fluorography. The results, presented in Fig.3, show that when the H-14s RNA is used to protect the ^{32}P-capped 50s RNA against ribonuclease digestion (slot b), a single major band of ^{32}P-radioactivity migrating at the level of the fastest Reo RNAs (s3 and s4 doublet) is seen. The control, ^{3}H-H-14s RNA which was annealed by itself without subsequent ribonuclease digestion (slot e),

---

Fig.2. *Annealing of heat-denatured ^{3}H-stems to ^{32}P-capped 50s RNA*

All ^{3}H-stems were first denatured by heating for 3 min at 99° in water followed by rapid cooling. Each annealing reaction contained 1500 cpm of ^{32}P-capped 50s RNA and increasing amount of either ^{3}H-R-31s stem (1000 cpm/µl, panel a), ^{3}H-J-23s stem (1000 cpm/µl, panel b), or ^{3}H-H-14s stem (1750 cpm/µl, panel c) in a total volume of 20 µl of 2 x buffer A. The contents of each annealing mix were sealed in glass capillaries and annealed for 45 min at 73°. The contents of each annealing reaction were then washed into 0.9 ml of 2 x buffer A containing 30 µg of ribonuclease A and digested for 30 min at 25°. Carrier rRNA (30 µg) was then added to each reaction and the remaining RNA was precipitated with 6% TCA and collected on millipore filters. Control samples were treated as above except that they were not digested with ribonuclease.

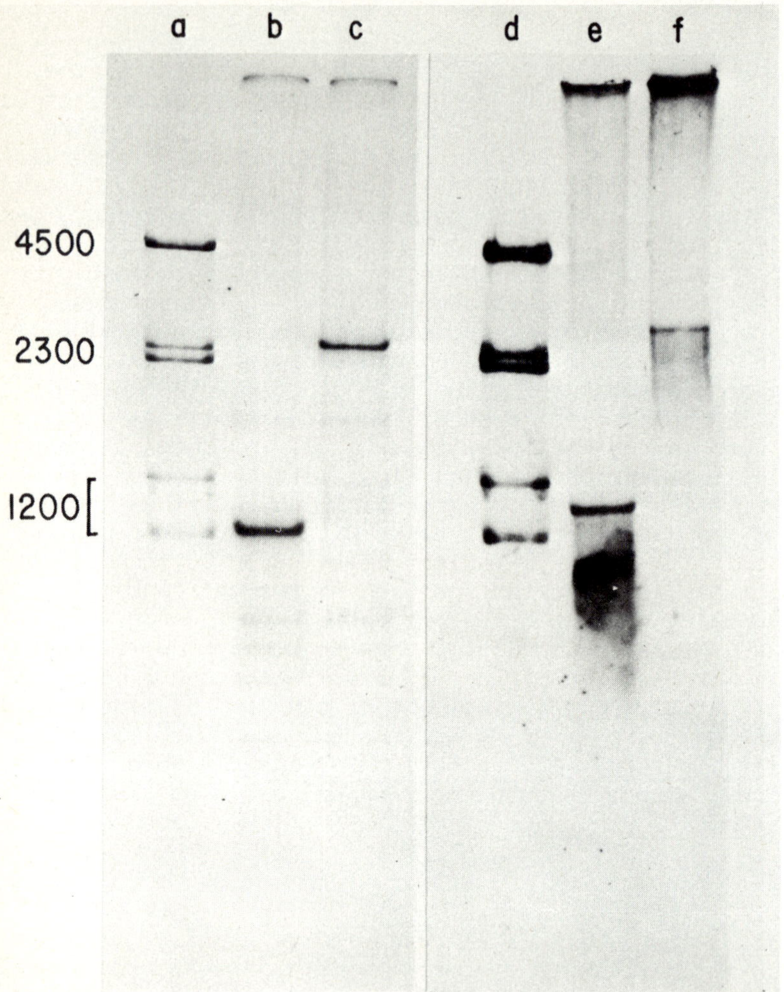

Fig.3. *Sequence homology between the DI-RNAs and the 5' end of the ND-genome by polyacrylamide gel electrophoresis*

^{3}H-H-14s and R-20s (20,000 cpm each) were both self-annealed by themselves (slots e and f) and in the presence of 5000 cpm of ^{32}P-capped 50s RNA (slots b and c, see text). After annealing, those samples containing the ^{32}P-RNA were digested with ribonuclease A (20 μg/μl) for 10 min at 25° in 0.3 M NaCl and then all the samples were recovered by ethanol precipitation and electrophoresed on a 2.2% polyacrylamide 0.5% agarose slab gel along with ^{14}C-Reo ds-RNAs (slots a and d). Slots a, b and c were autoradiographed, whereas slots d, e and f were

electrophoresed as a major band slightly above the ^{32}P-band in slot b. When R-20s RNA is used to protect the ^{32}P-capped 50s RNA against ribonuclease digestion, a single major band of ^{32}P-radioactivity is seen migrating just above the doublet of medium-size Reo RNAs (slot c). Similarly, annealed ^{3}H-R-20s RNA, run as a control (slot f) migrated as a single band just slightly slower than the ^{32}P-band in slot c. These results demonstrate that almost all the sequences in the H-14s and R-20s DI-RNAs are contiguous to the 5' end of the ND-genome. The small difference in mobility between the ^{32}P-bands and the ^{3}H-bands in each case (slot b versus slot 4 and slot c versus slot f) suggests that a small number of sequences (approximately 150 nucleotides)at the 5' end of each DI plus-strand (or the 3' end of each DI minus-strand) does not derive from the sequences adjacent to that portion of the ND-genome which is protected by the majority of each DI-RNA. That is to say, the majority, but not all, the sequences in the DI-RNAs are contiguous to the 5' end of the ND-genome.

THE ENDS OF THE ND-GENOME ARE NOT COMPLEMENTARY

We have so far shown that three DI-RNAs which range in size from 1200 to 2600 nucleotides all have identical complementary ends and that these complementary ends contain sequences homologous to approximately 150 nucleotides at the 5' end of the minus-strand genome. It therefore seemed likely that the ends of the ND-genome would also be complementary and that the DI-RNAs would be generated by internal deletions from the ND-genome. To test this model, we have performed three experiments to ask whether the ends of the ND-genome are complementary.

1. We have specifically labeled one end of the ND minus-strand genome by capping the 5' end. If the ends of the ND-genome are complementary, or, in fact, if there is a sequence anywhere in the ND-genome that is complementary to the 5' end, then self-annealing the capped 50s-RNA should cause the labeled 5' end to become resistant to ribonuclease digestion. The results of this experiment have already been presented in Fig.2. Even though the capped 50s-RNA has been self-annealed to a Crt of 1.5×10^{-1} (Crt ½ = 5×10^{-3}, (7)) only 10% of the cap structure became resistant to ribonuclease. This level of ribonuclease resistance is well within the amount expected due to the presence of minor amounts of 50s plus-strands in egg grown Sendai 50s RNA (15, 16).

---

fluorographed. The numbers at the left-hand side of the figure represent the average length (base pairs) of the small, medium, and large size classes of reovirus RNAs (5).

2. If there are any complementary sequences in the ND-genome large enough to form stable hybrids, self-annealing of the minus-strand genome followed by ribonuclease digestion should reveal these hybrids. Minus-strand RNA labeled with ^{3}H-uridine was prepared from intracellular nucleocapsid 50s RNA (for details see ref.12), self-annealed to 17 x Crt ½, and ds-RNA was isolated from the self-annealed RNA in an identical manner to that used for the isolation of the ds-stems from the DI-RNAs (12). Two percent of the ^{3}H-RNA (6000 cpm) was recovered as presumptive 50s-stems and electrophoresed on a polyacrylamide slab gel along with equal amounts of radioactivity of H-14s and J-15s stems. The gels were examined by fluorography and the results (Fig.4) show that the presumptive 50s-stem is composed only of very high molecular weight RNA which barely enters the gel. No bands in region of the H-14s and J-15s stems are visible even on prolonged exposure of the gel. This high molecular weight ds-RNA is most probably due to the presence of a minute amount of 50s plus-strands in the 50s minus-strand preparation (see ref.12).

3. The 5' end of the ND-genome has been determined to be pppAp (Leppert and Kolakofsky, manuscript in preparation). When this RNA is labeled with $^{32}PO_4$ and digested with ribonuclease T2, pppAp can easily be separated from the remainder of the digest by DEAE-paper electrophoresis and has been shown to represent 0.060% of the radioactivity (Table I). If there are any sequences present in 50s minus-strand RNA that are complementary to the 5' end of the RNA, self-annealing of this RNA followed by isolation of any ribonuclease-resistant RNA as described above should lead to a large enhancement of the fraction of radioactivity present in pppAp relative to the rest of the digest. Table I reports the percent of radioactivity in pppAp after ribonuclease T2 digestion of ^{32}P-H-50s and H-14s RNAs, H-14s stems, and the putative H-50s stems. If all the 5' ends of the 50s minus-strands had become ribonuclease-resistant due to a complementary sequence of the size of the H-14s stems present somewhere in 50s RNA, then pppAp should represent 2.3% of the ^{32}P-50s "stems". If, on the other hand, the ribonuclease resistance was due solely to the small amount of plus strands in the 50s minus-strand preparation, then pppAp should represent only 0.060% of the radioactivity. The fraction of radioactivity present as pppAp in H-50s "stem" was determined to be 0.065% (Table I).

It should be noted that in the above three experiments the radioactive 50s RNAs were annealed for a minimum of 45 min at 75°. Under these conditions, very few, if any, of the ND-genome RNAs remain intact as judged by sedimentation velocity. It is therefore unlikely that the absence of self-annealing to the 5'

Fig.4. *Polyacrylamide gel electrophoresis of putative 50s minus-strand stems.*

Sendai 50s minus-strands, labeled with ^{3}H-uridine and prepared as described in (17), was self-annealed to 17 x Crt ½ in 2 x buffer A at 73° and the remaining ribonuclease-resistant RNA was isolated as previously described (12) for the isolation of the DI-stems. Two percent of the radioactivity (6000 cpm) was found to be ribonuclease-resistant and was electrophoresed on a 6% polyacrylamide slab gel along with equal amounts of radioactivity of ^{3}H-H-14s and J-15s stems. The gel was then impregnated with PPO and fluorographed (1).

TABLE I

Percent pppAp in various Sendai RNA's

	Observed		Expected*
	%	(cpm)	%
H50s	0.060	(251)	0.046
H-14s	0.51	(9,228)	0.58
H-14s stem	1.3	(2,300)	2.3
H-50s "stem"	0.065	(20)	-

Double-stranded RNA was prepared from ^{32}P-50s minus strands (17) after exhaustive self-annealing as described in the legend to Fig.4. This ds-RNA, which represented 2% of the starting RNA was digested with 15 units of ribonuclease T2 plus 5 µg of ribonuclease A in a total volume of 20 ml of 10 mM sodium acetate pH 5.0 and 1 mM EDTA. As a control, a sample of ^{32}P-H14s stems was similarly digested. The digests were spotted directly on to Whatman DE 81 paper and electrophoresed for 1.75 hr at 2.5 KV. Radioactive spots were located by exposing the electrophoretogram to X-ray film, cut out, and counted by liquid scintillation.

---

end (experiments 1 and 3) or the generation of "stems" (experiment 2) was due to conformational restraints. Furthermore, fragmentation of the 50s RNA to approximately 15s by mild base hydrolysis before self-annealing of the RNA did not affect the results of experiment 2. Therefore, even though the above three experiments contain only negative evidence, they nevertheless strongly suggest that the ends of the ND-genome are not complementary. In addition, there do not appear to be any complementary sequences large enough to form hybrids which are stable to ribonuclease A digestion in 0.3 M NaCl at 25° anywhere in the ND-genome.

---

*The expected values for the fraction of radioactivity in pppAp relative to the total RNA digests have been determined from the relative specific activities of the snake venom phosphodiesterase and penicillium nuclease P1 digestion products of the eluted pppAp (Leppert and Kolakofsky, manuscript in preparation).

DISCUSSION

In this paper we have demonstrated that:

1. Although the complementary ends isolated from the three DI-RNAs are slightly different in length, they all contain sequences homologous to each other and to the 150 5'-terminal nucleotides of the ND-genome.
2. All the sequences in the DI-RNAs, except for a short stretch (approximately 150 nucleotides long) at the 3' end of each DI minus-strand, derive from sequences that are contiguous to the 5' end of the ND-genome.
3. Contrary to previous expectation, there does not appear to be any sequences present in the ND-genome (minus-strand) that are complementary to its 5' end, i.e., the ends of the ND-genome cannot be complementary.

A map of the DI-RNAs relative to the ND-genome is presented in Fig.5. In this map, the open squares represent the 5' terminal 150 nucleotides and the closed squares represent their complement. Since all three DI-RNAs contain both the 5' end of the ND-genome (50s minus-strand) and its complement as single strands (either plus or minus) whereas the ND-genome does not contain any sequences complementary to its 5' end, the complement to the 5' end of the ND-genome (closed squares) cannot have been derived from the ND-genome template but must have been derived from the nascent strand itself during replication. A model for the generation of DI-RNAs is presented in Fig.5b. In this model, DI-RNAs which map at the 5' end of the ND-genome are generated during replication of the ND-antigenome. Replication on the plus-strand template begins normally and the nascent minus-strand is covered with nucleocapisd protein so that the replicative intermediate (RI) contains a nascent minus-strand nucleocapsid tail. At some point during chain elongation, the nascent minus-strand nucleocapsid folds back on itself so that the 5' end of the nascent chain is closely associated with the viral replicase at the growing point of the RI. The replicase then leaves the ND-antigenome template without relinquishing the growing chain and attaches to the nascent minus-strand nucleocapsid at a point approximately 150 nucleotides from its 5' end. The replicase then continues its synthesis using the truncated minus-strand nucleocapsid as template for the short distance to the 5' end and terminates synthesis.

The product of this unusual replicative event is a short

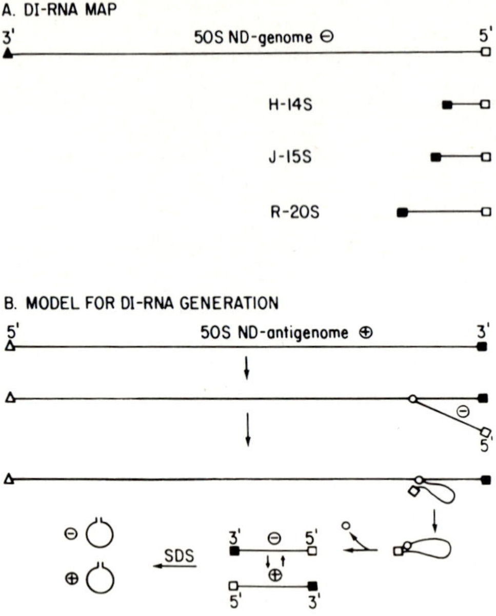

Fig.5. *A genetic map of the Sendai DI-RNAs and a model for their generation.*

(a) In this map, the ND-genome is 15,000 nucleotides long and the DI-RNAs are roughly drawn to scale. The open squares represent the 150 5'-terminal nucleotides of the ND-genome and the closed squares their complement. The closed triangles represent the 3'-terminal nucleotides of the ND-genome and, similarly the open triangles their complement.

(b) In this model, the open circle represents the viral replicase. Replicase recognition sites are denoted by closed triangles and squares.

minus-strand nucleocapsid which contains sequences identical to the 5' end of the 50s ND-genome (minus-strand), but in addition, contains sequences complementary to the 150 5' terminal nucleotides (closed squares) at its 3' end. Since these latter sequences are identical to the 3' end of the ND-antigenome, the DI-RNA minus strand thus contains the ND-antigenome replicase recognition site at its 3' end. Moreover, since the ends of the DI-RNA are complementary, both plus and minus DI-RNAs will contain this replicase recognition site at their 3' ends.

The mechanism which generates DI-RNAs therefore need not necessarily be one which retains both the original ends of the ND-genome, i.e., internal deletions. The generation of inverted terminal repeats of one end of the ND-genome is apparently sufficient to ensure that the DI-RNA can be replicated in the presence of helper ND-genomes. Those functions essential for replication e.g., sites which the replicase and the nucleocapsid proteins recognize and attach to, must therefore be contained within the inverted terminal repeats.

The unexpected finding that a parainfluenza virus does not contain genomes with complementary ends has important implications for the general mechanism of parainfluenza virus replication and, specifically, for the competitive advantage which DI-genomes possess. Since replication takes place on both genome and antigenome templates whereas transcription takes place only on genome templates, DI-genomes which derive only from the 5' end of the ND-genome will not contain a site (presumably at the 3' end of the ND-genome) required for transcription and will consequently be unable to transcribe. Such 5' end DI-minus-strands will therefore be used only as templates for replication. Since many rounds of replication take place during a single infection, DI-RNAs which do not transcribe will have a distinct competitive advantage due to template availability.

We have so far not been able to determine whether or not Sendai DI particles transcribe *in vitro*, since it is extremely difficult to purify these particles away from ND-particles (6) whose *in vitro* transcriptase activity is both unstable and only marginal in our hands. In the case of the several dozen VSV DI particles isolated to date, all except one contain sequences which derive from the 5' end of the minus-strand genome (11, 20) and these DI-particles do not transcribe *in vitro* even though they have been shown to contain a transcriptase which is perfectly competent if transferred to a ND-genome template (3). The one exception is the HR DI-particle which derives from the 3' end of the ND-genome (11, 20), which does transcribe, both in *in vitro* (A. Banerjee and R. Lazzarini, personal communication), and *in vivo* (2). If the model described for the generation of Sendai DI-RNAs is also applicable to VSV (another non-segmented negative strand virus whose mechanism of replication has so far proved very similar to that of Sendai virus), then VSV DI-RNAs should also contain complementary ends. Recent results (J. Perrault and R. Leavitt, personal communication), indicate that this is indeed so. Furthermore, when two or more VSV DI-particles containing different size RNAs are generated from a single clonal isolate of ND-VSV virions, all the sequences contained in the smallest DI-RNA are also present in the larger DI-RNAs (18, 19).

It is somewhat curious that all three Sendai DI-RNAs we have characterized contain rather short complementary ends of almost identical size (110-150 bp). The reasons for this are unclear in the general scheme presented in Fig.5b which predicts that DI-RNAs containing complementary ends of any length are possible. The only other Sendai or VSV DI-RNA whose secondary structure has been characterized is the "snap-back" VSV DI-RNA (10, 14) which contains complementary ends which make up the entire DI-genome. We have also recently observed what appears to be Sendai snap-back DI-RNAs. In the case of the VSV snap-back DI-RNAs, Perrault and Leavitt (personal communication) have suggested that these RNAs have the general structure abc------------c'b'a' abc------------c'b'a'. Such a structure can be generated in a similar manner as that presented in Fig. 5b if during the replication of a DI-RNA already containing the short complementary ends (abc:c'b'a'), the replicase switches from the template to the nascent chain at a point very close to the 5' end of the almost completed nascent chain ("rounds the corner") and continues synthesis back along itself. In both cases of DI-RNA generation described above, the replicase would switch from its template to its nascent chain near the ends of the nascent chain. Sites to which the replicase can switch may be available only at the ends of the nascent chains.

In conclusion, two qualifications of the discussion should be added. The first is that although the model we have presented for the generation of the DI-RNAs is the one we consider to be most likely, alternate models in which the nascent minus strand is first cut and then spliced to the 3' end of the plus strand template cannot be excluded. A second qualification concerns our conclusion that the ends of the ND-genome are not complementary. The methods we have used in an attempt to detect such complementarity depends on the stability of hybrids to ribonuclease digestion under our conditions of temperature and salt concentration. Small regions of complementarity (10-20 nucleotides long) at the ends of the ND-genome would probably not have been detected by this method. The presence or absence of such small regions of complementarity must await direct sequence determination.

REFERENCES

1. Bonner, W.M. and Laskey, R.A. (1974). *Eur. J. Biochem.* 46. 83.
2. Chow, J.M., Schnitzlein, W.M. and Reichmann, M.E. (1977). *Virology* 77, 579.
3. Emerson, S.U., and Wagner, R.R. (1972). *J. Virol.* 10, 279.
4. Huang, A.S. (1973). *Ann. Rev. Microbiol.* 27, 101.
5. Joklik, W.K. (1974). *In* "Comprehensive Virology" Vol.2, p.231. Plenum Press, N.Y.

6. Kingsbury, D.W., Portner, A. and Darlington, R.W. (1970). *Virology* 42, 857.
7. Kolakofsky, D., Boy de la Tour, E. and Bruschi, A. (1974). *J. Virol.* 14, 33.
8. Kolakofsky, D. and Bruschi, A. (1975). *Virology* 66, 185.
9. Kolakofsky, D. (1976). *Cell* 8, 547.
10. Lazzarini, R.A., Weber, G.H., Johnson, L.D. and Stamminger, G.M. (1975). *J. Mol. Biol.* 97, 289.
11. Leamson, R.N. and Reichmann, M.E. (1974). *J. Mol. Biol.* 85, 551.
12. Leppert, M., Kort, L. and Kolakofsky, D. (1977). *Cell* 12, 539.
13. Martin, S. and Moss, B. (1975). *J. Biol. Chem.* 250, 9330.
14. Perrault, J. (1976). *Virology* 70, 360.
15. Portner, A. and Kingsbury, D.W. (1970). *Nature* 228, 1196.
16. Robinson, W.S. (1970). *Nature* 225, 944.
17. Roux, L. and Kolakofsky, D. (1975). *J. Virol.* 16, 1426.
18. Schnitzlein, W.M. and Reichmann, M.E. (1976). *J. Mol. Biol.* 101, 307.
19. Schnitzlein, W.M. and Reichmann, M.E. (1977). *Virology* 77, 490.
20. Stamminger, G.M. and Lazzarini, R.A. (1974). *Cell* 1, 85.

SYNTHESIS OF BOTH + AND - STRAND VIRAL RNA IN CELLS INFECTED WITH *ONLY* VSV DI PARTICLES

L.D. JOHNSON and R.A. LAZZARINI

Laboratory of Molecular Biology,
Molecular Virology Section,
NINCDS, National Institutes of Health,
Bethesda, Maryland 20014, USA.

A number of different types of VSV defective interfering (DI) particles have been described in the recent literature (9, 13, 17, 18). One feature that is common to several of these particles, notably those derived from the 5' end of the VSV genome, is that by themselves they are without effect on host cells (2, 6, 8). They neither kill the cell nor express an appreciable amount of their genetic information, despite containing a full complement of VSV proteins. Also, compared to the infectious virus, they direct very little RNA synthesis in cell-free systems (3, 16, 19, 20). This deficiency lies in the ribonucleoprotein (RNP) template and not in the transcriptase (7). It has been proposed that the RNA of these DI particles lacks a 3' terminal transcription initiation sequence (5, 16).

A second type of particle, DI-LT, contains an RNA derived from the 3' half of the VSV genome (17). This DI RNA anneals to the leader sequence (the proposed complement of the transcription initiation sequence) and is transcribed into 12-17S mRNAs *in vitro* as efficiently as VSV (5). An examination of the biochemical capabilities of DI-LT in cultured cells is presented here.

In order to attribute molecular events taking place in infected cells to the infecting DI particles, the contribution of the contaminating standard particles must be negligible. A purified stock of DI-LT was prepared by routine methods (5) and the level of contamination with full-size VSV was estimated by procedures outlined in Table I. VSV and DI-LT particles were distinguishable by their size using the electron microscope. Samples of the DI preparation were centrifuged directly onto parlodion coated EM grids, and prepared for EM visualization as previously described (11). The concentration of DI-LT and contaminating full-size VSV was estimated from electron micrographic enlargements. Maximum and minimum concentrations of VSV were obtained by scoring either all aggregates and deformed particles large enough to be a VSV, or just clearly recognizable VSV particles. A second, independent estimate was derived from the infectivity (pfu/ml) of the preparation after correcting for plaque reduction due to autointerference and plaquing efficiency (9-20%) under our conditions. For the work presented here a conservative value of 1/1000 was chosen for the estimated VSV contamination in the DI-LT stock, based on physical particles. An estimate based on biological activities (plaque forming particles and interfering particles) indicate that there were 8000 times more interfering particles than plaque forming particles (21).

Since DI-LT and VSV have been shown to be equally efficient in *in vitro* transcription (5), it might be expected that the rates of primary transcription would be the same in cells infected with either DI-LT or VSV (fig.1). This expectation is borne out: when RNA synthesis is limited to primary transcription by blocking protein synthesis with cycloheximide, the rates are virtually the same in cells infected with 200 particles of either DI-LT or VSV. However, in the absence of cycloheximide, VSV directs RNA synthesis at a rate much faster than that observed with DI-LT, indicating more replication and amplification in the former infection. DI-LT synthesizes more than primary transcripts since in several experiments, more RNA was synthesized when no cycloheximide was added (data not shown). Infection with VSV at the level contaminating the DI-LT stock (0.2 physical particles/cell) results in slightly less RNA synthesis than is observed in the uninfected control. The incorporation shown in fig.1 has been corrected for the incorporation by the uninfected control. The magnitude of this correction can be seen in fig.2b where total incorporations are reported.

The synthesis of specific RNAs by DI-LT in the absence and presence of cycloheximide is shown in fig.2a. Three radioactive species are observed. The peak of radioactivity at the top of the gradient is also seen in uninfected host cells and unlike the other species is resistant to ribonuclease (RNase) even after heating in 0.1 M NaCl. It is slightly diminished by cycloheximide, contains no poly A sequences and does not appreciably anneal to excess 42S viral RNA (vRNA). The nature of this material was not further explored but we suspect from

TABLE I

Purity of DI-LT Inoculum

Assay	Titer ($x10^{-10}$/ml)
DI-LT physical particles	
electron micrographs*	95
VSV physical particles	
electron micrographs*	0.07-0.28
est. from pfu+	0.04-0.9
Plaque forming particles**	0.008

*Electron microscopic determinations were performed as described previously (11).

**VSV and DI-LT preparations were mixed prior to adsorption to determine plating efficiency. This value was used to correct the pfu/ml obtained directly with DI-LT for plaque reduction due to autointerference.

+The pfu/ml value in the last line was converted to physical particles (pp)/ml using a plaquing efficiency range of 9-20%.

its behavior with RNase that it is not RNA.

The middle peak migrates slightly slower than 18S rRNA and hence is an appropriate size for mRNAs produced by DI-LT. Also consistent with this identity is the insensitivity of its synthesis to cycloheximide, its polarity (75% anneals to excess vRNA) and polyadenylation (24%). This material shows little self-annealing (10%) indicating the presence of only traces of (-) strand RNA.

The fastest sedimenting peak migrates to the same position as 28S rRNA and its synthesis is sensitive to cycloheximide. The size and dependence on protein synthesis of the 28S RNA species indicate that it may result from replicative RNA synthesis, i.e. the synthesis of full length + or - RNA strands, and not from transcriptive synthesis. This material does not contain appreciable amounts of the 28S mRNA which codes for the viral L protein since it does not anneal to RNA derived from DI-T, an RNA which has been shown to contain part of the L-cistron (14, 22). Furthermore, the radiolabel at 28S is not due to an association of small nascent mRNAs with the input

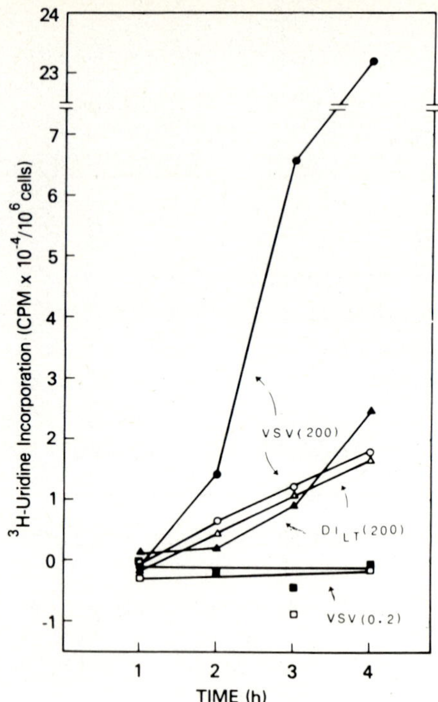

Fig.1. *RNA synthesis by DI-LT and VSV in the absence and presence of cycloheximide. BHK-21 suspension cells were prepared for infection as described previously (10). DI-LT was used at 200 physical particles/cell (Δ) and VSV at both 200 (O) and 0.2 (□) physical particles/cell. Parallel experiments were done with cycloheximide at 100 μg/ml (open symbols). At the indicated times, duplicate samples were TCA-precipitated and processed for liquid scintillation counting. Radiolabel incorporated by uninfected control cells has been subtracted at each time point.*

template RNA (transcriptive intermediates) since it does not contain detectable double stranded structures and sediments as a single, high molecular weight species in a denaturing sucrose gradient containing 80% DMSO. The fact that its synthesis is sensitive to cycloheximide further underscores this differentiation.

Replicative RNA synthesis can produce both (+) replicative RNA and (-) genomic RNA. The polarity of the newly synthesized 28S RNA was determined by annealing (fig.3). In the presence of excess unlabeled DI-LT RNA, 37% of the 28S RNA enters duplex structures and becomes resistant to RNase. Thus, 37% is (+) strand. When annealed with excess unlabeled mRNA, 61% forms

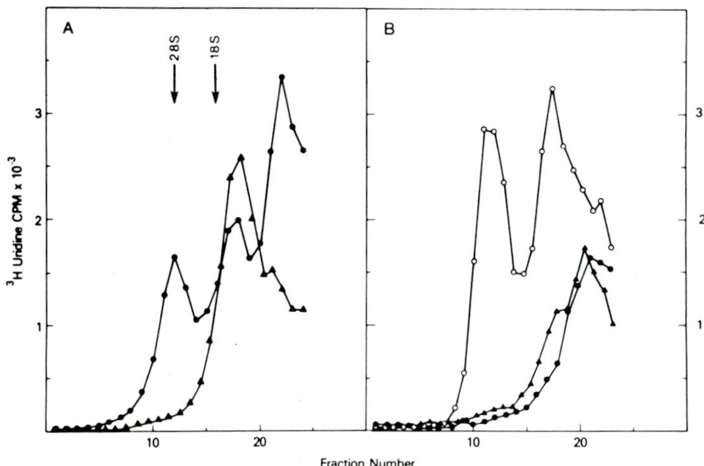

Fig.2. *a. Sucrose gradient separation of RNA made by DI-LT in the absence and presence of cycloheximide. BHK-21 suspension cells without (●) or with (Δ) cycloheximide were infected with DI-LT at 200 physical particles/cell. ^{3}H-uridine (20 μCi/10^6 cells) was added. At 4 hr, cells were harvested and RNAs separated on 10-30% (w/v) sucrose gradients (17). Radioactivity in 50 μl portions of each fraction was determined after TCA precipitation. Sedimentation of ribosomal RNA markers was determined by A_{260} measurement.*

b. Sedimentation of RNAs from uninfected control cells, and cells infected with VSV and DI-LT. Cells were infected with DI-LT at 200 (o), or with VSV at 0.2(▲) physical particles per cell in the absence of cycloheximide. The control (●) represents similarly treated uninfected cells. Other procedures were performed as in a.

hybrids indicating that this amount of the labeled RNA has a (-) polarity; i.e., is new genomic DI-LT RNA. These relative abundances of + and - strand were confirmed in self-annealing experiments where, in the absence of added RNA, the strand present in the lesser quantity, the (+) strand in the present case, governs the amount of self-annealing. Assuming (+) and (-) strands have equal radio-specific activity, self-annealing should drive 74% of the label into duplexes. The observed value of 77% is in excellent agreement. We conclude from these data that the 28S RNA is composed of genomic length RNA, 37% of which is (+) strand and 61% (-) strand.

In cells infected with VSV or VSV and DI particles, the respective full length (+) and (-) RNA species are present as characteristic RNP structures. These are easily identified by

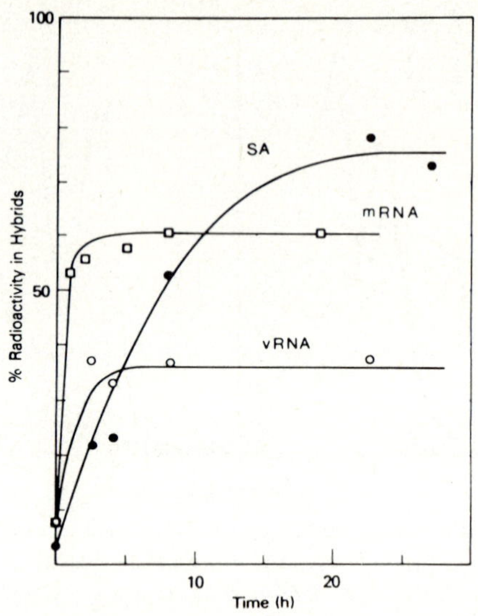

Fig.3. *Polarity of 28S RNA self-annealing at 60° was performed on 28S RNA after heating at 100° for 1 min, cooling and adjusting to 0.4 M NaCl-0.01 NaPi, pH 6.8 and a final volume of 0.25 ml (●). For determination of (+) strand RNA, 28S RNA was incubated with 7.38 μg of unlabeled DI-LT vRNA under the above conditions (O). The amount of (-) strand RNA was measured by incubating 28S RNA and an excess of viral mRNA which was prepared as described previously (10). At the indicated times, samples were removed for TCA precipitation before and after RNase treatment.*

their sedimentation rate in nondeproteinizing sucrose gradients and by their resistance to RNase. In the experiments shown in fig.4, the 28S RNA is present in structures whose sedimentation rate and RNase resistance indicates that they are RNP's. When equal amounts of cytoplasmic extracts are fractionated with and without deproteinization, slightly greater radiolabel is recovered in the RNP fractions than in 28S RNA (fig.4b). However, analysis of RNA subsequently prepared from the RNP peak indicates that some is nascent mRNA and that 90% of the 28S RNA synthesized in the DI-LT infected cell is present as an RNP.

The production of new mRNA species by pure DI-LT suggests that the coding potential of DI-LT, *i.e.* synthesis of N, NS, M and G proteins, may be realized *in vivo*. We have attempted to demonstrate directly viral protein synthesis by labeling

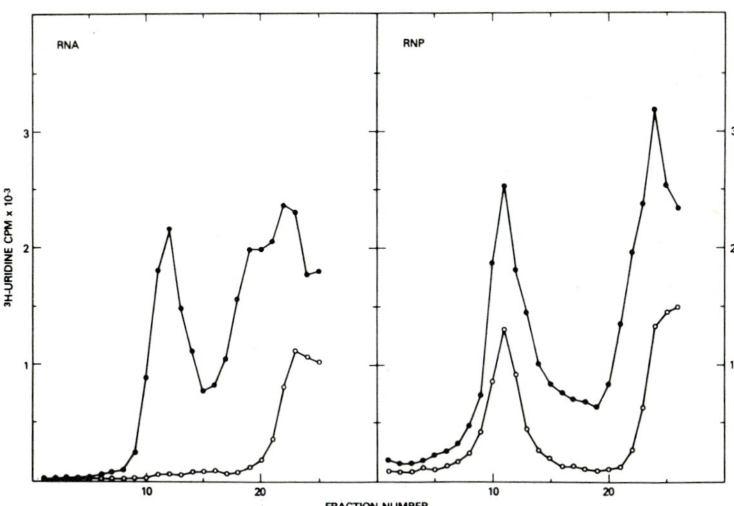

Fig.4. *Sedimentation of RNAs and RNPs isolated from DI-LT infected cells. Infection with DI-LT was performed as previously described and a cytoplasmic extract prepared (10).*
a. A sample was made 1% SDS and the RNA was separated on a sucrose gradient as described in fig.2.
b. A second portion of the same extract was analyzed on a 5-20% sucrose gradient in NEB containing 1M NH_4Cl for separation of RNP species (11). Both RNA and RNP profiles were determined by TCA precipitation of portions of the respective gradients (closed circles). RNase treatment (open circles) on the same volume was performed by incubation in 0.4 M NaCl-0.01 M Pi pH 6.8 for 30 min, 37°, with 50 μg/ml RNase A and 200 U/ml RNase T_1. The samples were then TCA precipitated and counted.

infected cultures with ^{35}S-methionine. An autoradiogram of an SDS polyacrylamide gel electrophoretogram of labeled extracts from DI-LT and VSV infected cells is shown in fig.5a. VSV at 200 physical particles/cell, stimulates the synthesis of all 5 viral proteins while VSV at the level contaminating the DI-LT does not stimulate detectable VSV protein synthesis. In cells infected with only DI-LT, M protein synthesis is readily observed. However, host protein synthesis is not shut off well and the synthesis of other viral proteins is obscured. We have attempted to circumvent this problem by purifying intracellular viral RNP structures away from the obscuring host proteins prior to electrophoresis. Emerson and Wagner (7) have shown that RNP isolated in high ionic strength, as in fig.4, contains only the N protein, while those isolated in low ionic strength contain N, NS and L proteins. Fig.5b shows that newly synthesized N protein is present in RNP isolated at high ionic strength from DI-LT infected cells. When the RNP structures are isolated in low-ionic strength gradients, both N and a small amount of a second protein which tracks with NS are distinguishable. The L protein, which is not coded in the DI-LT genome, is not synthesized. We have not, as yet, detected the synthesis of the G protein in cells infected with only DI-LT with ^{35}S-methionine. The demonstration of this synthesis (if present) may require more specific methods, such as labeling with radioactive carbohydrate precursors.

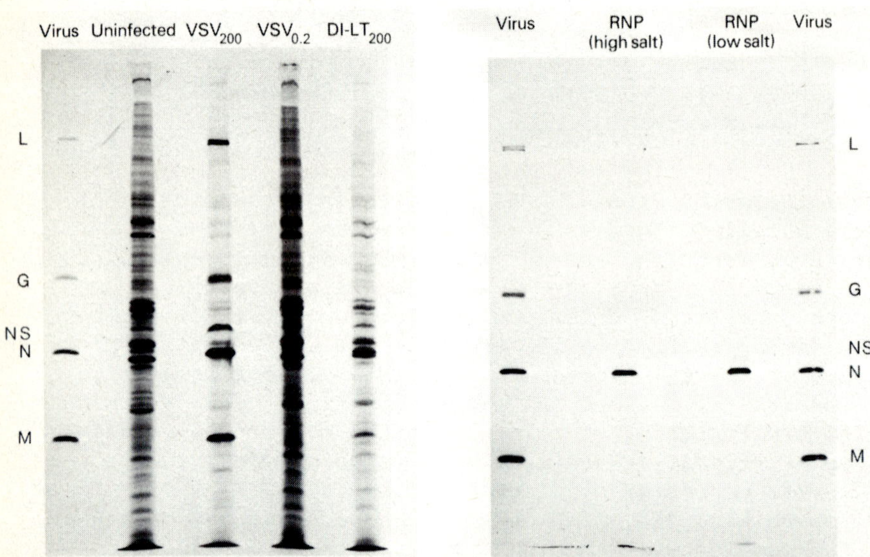

Fig.5 a and b. *SDS polyacrylamide gel electrophoresis of extracts from VSV and DI-LT infected cells. BHK-21 spinner cells were treated with actinomycin D (10 μg/ml) and infected with the indicated numbers of VSV or DI-LT particles as in fig.1 but with methionine free medium. At 3 hr post-infection ^{35}S-methionine was added to a final concentration of 2 μCi/10^6 cells. The cells were incubated for 1 hr, washed with PBS and dounce homogenized. Samples of the nuclei free extract (left panel) or RNP fractions prepared from them (right panel) were analyzed by slab - SDS polyacrylamide gel electrophoresis using a modification of the Laemmli procedure (12). ^{35}S-methionine labeled VSV was run for comparison. Radioautograms of the dried gels are shown.*

The biological activity of the long defective interfering particle of the HR strain, DI-LT, has been investigated in several laboratories but unfortunately only partial agreement of results has been obtained. Mori and Howatson (15) originally reported that this particle exhibited *in vitro* transcriptase activity equal to or better than that seen with the parent virus. Of particular importance was the fact that the product was comparable in size to VSV mRNA's. Reichmann *et al.* (19), while finding a small amount of activity with their preparations of DI-LT, showed that the transcription product was very much smaller. Recently, Colonno *et al.* (5) reported that the DI-LT not only synthesized RNA at a rate equal to that of the parental virus but also that the major products of this reaction were four full-size, capped and polyadenylated VSV mRNA's.

The explanation for these differences is not known for certain, but two points seem relevant. First, differences in the conditions used to culture and purify the DI particles may affect significantly the *in vitro* synthetic capacity of the DI. In this regard, it is interesting to note that host proteins have a profound affect on the *in vitro* transcription activity of VSV and DI-LT (1, 5). Under appropriate conditions the addition of cell extracts stimulates *in vitro* RNA transcritpion 5 or more fold. Second, it seems unlikely that the absence of *in vitro* transcriptase activity in the DI-LT used by Reichmann *et al.*,(19) is due to a defective template, since he and his colleagues have convincingly demonstrated that the genetic information of that particle is expressed *in vivo* during mixed infection with New Jersey serotype VSV (4).

Whatever the explanation for the reported differences, it is clear that the DI-LT particle contains information that can be expressed *in vivo*. The results reported here as well as our earlier results (10), show that this expression does not necessarily require coinfection with helper virus. These heretofore unrecognized abilities of DI-LT suggest that the effect of this particle on virus infection of animals might be quite complex. In addition to interfering with the replication of the infectious virus in mixedly infected cells, it can augment or complement the production of viral proteins. In cells singularly infected with only DI-LT, it can kill the host cell (21) and induce a limited synthesis of some viral proteins, phenomena which might be expected to trigger an enhanced immune response by the host.

ACKNOWLEDGEMENTS

We gratefully acknowledge the excellent technical assistance of Mitchel Binder in carrying out the SDS gel electrophoresis analyses.

REFERENCES

1. Ball, L.A. and White, C.N. (1976). *Proc. Nat. Acad. Sci. USA* 73, 442.
2. Baxt, B. and Bablanian, R. (1976). *Virology* 73, 370.
3. Bishop, D.H.L. and Roy, P. (1971). *J. Mol. Biol.* 58, 798.
4. Chow, J.M., Schnitzlein, W.M. and Reichmann, M.E. (1977). *Virology* 77, 579.
5. Colonno, R.J., Lazzarini, R.A., Keene, J.D. and Banerjee, A.K. (1977). *Proc. Nat. Acad. Sci. USA* 74, 1884.
6. Doyle, M. and Holland, J.J. (1973). *Proc. Nat. Acad. Sci. USA* 70, 2105.
7. Emerson, S.U. and Wagner, R.R. (1972). *J. Virol.* 10, 297.
8. Huang, A.S. and Manders, E.K. (1972). *J. Virol.* 9, 909.
9. Huang, A.S., Greenawalt, J.W. and Wagner, R.R. (1966). *Virology* 30, 161.
10. Johnson, L.D. and Lazzarini, R.A. (1977). *Proc. Nat. Acad. Sci. USA* 74, 4387.
11. Khan, S.R. and Lazzarini, R.A. (1977). *Virology* 77, 189.
12. Laemmli, U.K. (1970). *Nature* 222, 680.
13. Lazzarini, R.A., Weber, G.H., Johnson, L.D. and Stamminger, G.M. (1975). *J. Mol. Biol.* 97, 289.
14. Leamnson, R.N. and Reichmann, M.E. (1974). *J. Mol. Biol.* 85, 551.
15. Mori, H. and Howatson, A.F. (1973). *Intervirology* 1, 168.
16. Perrault, J. and Holland, J.J. (1972). *Virology* 50, 159.
17. Petric, M. and Prevec, L. (1970). *Virology* 41, 615.
18. Reichmann, M.E., Pringle, C.R., Follett, E.A.C. (1971). *J. Virol.* 8, 154.
19. Reichmann, M.E., Villareal, L.P., Kohne, D., Lesnau, J. and Holland, J.J. (1974). *Virology* 58, 240.
20. Roy, P. and Bishop, D.H.L. (1972). *J. Virol.* 9, 946.
21. Sekellick, M.J., Marcus, P.I., Johnson, L.D. and Lazzarini, R.A. (1977). *Virology* 82, 242.
22. Stamminger, G. and Lazzarini, R.A. (1974). *Cell* 3, 85.

RNA SYNTHESIS BY DEFECTIVE PARTICLES OF VESICULAR STOMATITIS VIRUS

SUZANNE U. EMERSON, PETER M. DIERKS, and J. THOMAS PARSONS

Department of Microbiology,
University of Virginia School of Medicine,
Charlottesville, Virginia 22901, U.S.A.

Vesicular stomatitis virus, a rhabdovirus, is a negative strand RNA virus, 1.85 nm in length, with a bullet shape morphology (22). Standard B virions contain a single piece of RNA which has a molecular weight of $\sim 3.6 \times 10^6$ daltons and codes for the five known viral proteins. During passage of the virus at high multiplicity of infection, shorter virus particles of similar morphology are generated and amplified (13). These truncated or T particles contain shorter pieces of viral RNA and are defective for replication in the absence of helper B particles. A large number of T particles of various lengths have been isolated. The particular clone of T particles obtained is characteristic for a specific virus stock and depends on the parent virus and the cells used to grow it (12, 18). The molecular basis of T particle formation is presently unknown.

Both B and T virions contain the same 5 viral proteins. Two of the proteins, G and M, are major structural components which with lipids comprise the virion envelope. The third major structural protein, N, is a nucleocapsid protein which binds tightly to the virion RNA to form an RNase resistant helical nucleocapsid. The remaining two proteins, L and NS, are minor components of the virion and comprise the viral transcriptase (9). *In vitro* and *in vivo* experiments have demonstrated that this enzyme uses the B particle N protein-RNA complex as a template to synthesize five monocistronic viral mRNAs which are capped with 7 mG at the 5' end and polyadenylated at the 3' end (2, 4, 21). In addition, *in vitro* studies have shown that a small "leader" RNA which contains free 5' phosphates and is not polyadenylated is also synthesized (6, 7). U.V. inactivation experiments indicate that the gene order along the B virion RNA is 3'-leader-N-NS-M-G-(L)-5' (1, 3).

The majority of T particles described contain RNA which is derived predominantly from the L gene region (19). These T particles do not synthesize mRNAs *in vitro* (8, 15, 17). The isolation of an active viral transcriptase from T particles which *in vitro* incorporate only very low levels of nucleotides compared to B particles, led us to originally argue that the low level of RNA synthesis represented B particle contamination and the T particle template was defective for transcription (8). Later reports have suggested that preparations of T particles synthesized RNA molecules smaller than 4S which were polyadenylated; however, since it was not shown that these molecules were coded for by the T particle RNA, the question of whether T particle templates can function *in vitro* was unresolved (15, 17). Because T particles do contain a virion polymerase and T particle nucleocapsids must function *in vivo* in order to serve as a template for their own replication, we reinvestigated the question of whether the small amount of RNA synthesized *in vitro* by T particle preparations was actually coded for by the T virion RNA.

The T particle we examined was derived from a stock of VSV Indiana obtained from R.R. Wagner and passaged initially on L cells and then on BHK-21 cells. Since the T particles generated during virus growth in BHK-21 cells have an RNA which migrates on gels with an apparent value of 30S compared to the B virion RNA which migrates as a 42S molecule, it is a long T or LT particle. 10% of the LT virion RNA self-anneals over a wide range of RNA concentrations suggesting some intrastrand base pairing can occur (Table I). The extent of annealing increases substantially when 28S viral mRNA, which codes for the L protein, is added, whereas addition of the 13-18S viral mRNAs, which code for the remaining four proteins, does not increase the RNase resistance significantly (Table I). Therefore, since the virion RNA is similar in size to the L protein mRNA and hybridizes to it, we conclude that the LT virion RNA contains most if not all of the L gene.

In order to compare the relative rates of transcription of B and LT virions we purified ^{14}C amino acid labeled B and LT virions from the same infection and tested each particle for its ability to synthesize RNA *in vitro*. Equal concentrations of B and LT particles were prepared (based on ^{14}C label) and four concentrations covering a 16-fold dilution range were tested in a standard transcription assay. The LT preparations incorporated 8.4 - 9.9% as much ^{3}H-UTP into RNA as did the B preparations (fig.1). Since the LT virions are approximately half the size of the B virions (S. Leech, personal communication), each reaction contained approximately twice as many LT virions as B virions at each dilution tested and therefore, per virion, the LT particles synthesize about 5% as much RNA as B particles do.

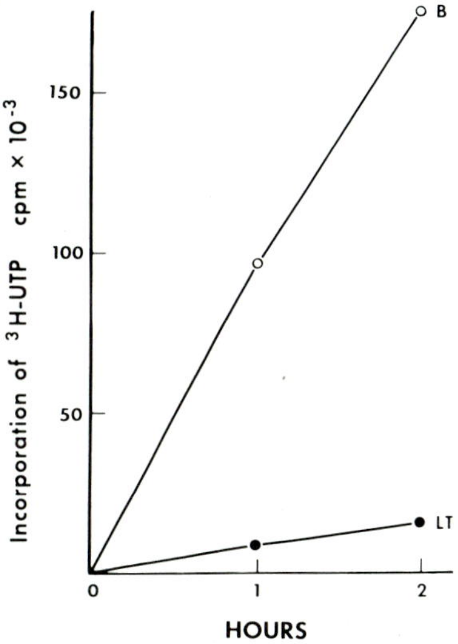

Fig.1. *In vitro RNA synthesis by purified B and LT virions. ^{14}C-amino acid labeled B and LT virions were purified by sucrose gradient centrifugation and diluted to contain equal amounts of ^{14}C-label. Four dilutions of each preparation were then assayed in a standard transcription system containing ^{3}H-UTP (19). The most concentrated virus sample is plotted. The background of 1000 cpm was substracted.*

Analysis of the *in vitro* RNA products by polyacrylamide gel electrophoresis demonstrates that the low level of UTP incorporated by the LT preparation is not due to contaminating B particles because the products of the LT and B reactions are quite different (fig.2). The majority of the B products migrate more slowly than a 4S marker generating a pattern characteristic of B virion mRNAs while the LT product migrates as a single band ahead of the 4S marker.

The LT product was checked for its ability to anneal to LT or B virion RNAs. Table II shows that the LT product labeled with radioactive UTP, CTP, or ATP is completely RNase sensitive prior to annealing and therefore does not contain extensive secondary structure or poly(A) sequences. If the LT product is annealed with intact LT or B virion RNAs very low levels of hybridization are observed. However, when the LT product was annealed with a 3-fold molar excess of alkali-fragmented LT

TABLE I

Characterization of LT Virion RNA by Annealing

Annealing volume	mRNA added[a]	TCA insoluble[b] cpm		RNase resistance %[c]
		-RNase	+RNase	
0.1 ml[d]	--	2741	299	10.91
0.5 ml[d]	--	2636	272	10.34
1.0 ml[d]	--	2521	265	10.51
2.0 ml[d]	--	1857	182	9.8
80 μl[e]	--	2488	238	9.57
80 μl[e]	10 μl (13-18S)	2726	312	11.46
80 μl[e]	25 μl (13-18S)	2584	269	10.41
80 μl[e]	10 μl (28S)	2372	986	41.57
80 μl[e]	25 μl (28S)	2471	2013	81.49

[a] ^{32}P-labeled viral mRNA was isolated from polysomes and purified by oligo(dT) cellulose chromatography and sucrose gradient sedimentation. The purity of the fractions was verified by electrophoresis on 2% polyacrylamide gels.

[b]Duplicates (0.5 ml) were incubated for 30 min at 37° with or without nine units each of RNase A and RNase T_1 and then precipitated with trichloroacetic acid, collected on cellulose acetate filters and counted.

[c]Calculated as (trichloroacetic acid-insoluble cpm after RNase/trichloroacetic acid-insoluble cpm before RNase) x 100.

[d] ^{3}H-uridine labeled LT virion RNA was dissolved in 0.3 M NaCl-0.3 M sodium citrate pH 7.0. Samples were boiled for 2 min, self-annealed at 68° for 15 min and then diluted to 2.2 ml with the same buffer.

[e] ^{3}H-uridine labeled LT virion RNA was mixed with ^{32}P-labeled mRNA in 0.4 M NaCl-10 mm Tris-HCl (pH 7.4). Samples were boiled for 30 sec then annealed at 68° for 2.5 hr and finally diluted to 2.1 ml with 0.15 M NaCl-10 mm Tris-HCl (pH 7.4).

TABLE II

Characterization of *in vitro* RNA by Annealing

in vitro product (label)	virion RNA added	TCA insoluble cpm[a]		RNase resistance of *in vitro* product[b] %
		-RNase	+RNase	
LT (^{3}H-ATP)[c]	--	4772	47	0.98
LT (^{3}H-UTP)[c]	--	1560	46	2.95
B (^{3}H-ATP)[c]	--	14870	1941	13.05
B (^{3}H-UTP)[c]	--	10585	49	0.46
LT (^{32}P-CTP)[d]	--	1821	48	2.66
LT (^{32}P-CTP)[d]	0.4 μg LT	1917	663	34.59
LT (^{32}P-CTP)[d]	1 μg LT	1756	1289	73.43
LT (^{32}P-CTP)[d]	0.6 μg B	1577	130	8.28
LT (^{32}P-CTP)[d]	1 μg B	1241	138	11.12

[a]Determined as in Table Ib

[b]Calculated as in Table Ic

[c]*In vitro* products from a 5 hr reaction were centrifuged to remove template, phenol extracted, and chromatographed on Sephadex columns to remove low molecular weight components. RNA was incubated at 72° for 2 hr in 100 μl of 0.4 M NaCl-10 mM Tris-HCl (pH 7.4) and then diluted to 2.2 ml with 0.15 M NaCl-10 mM Tris-HCl (pH 7.4).

[d]^{32}P-CTP labeled LT product purified as above was mixed with alkali fragmented ^{3}H-uridine labeled B or LT-virion RNA in 55 μl of 0.3 M NaCl-0.3 M sodium citrate, pH 7.0. Samples were boiled for 30 sec and incubated at 70° for 17.5 hr then samples were diluted to 2.1 ml with 0.3 M NaCl-10 mM Tris-HCl (pH 7.4).

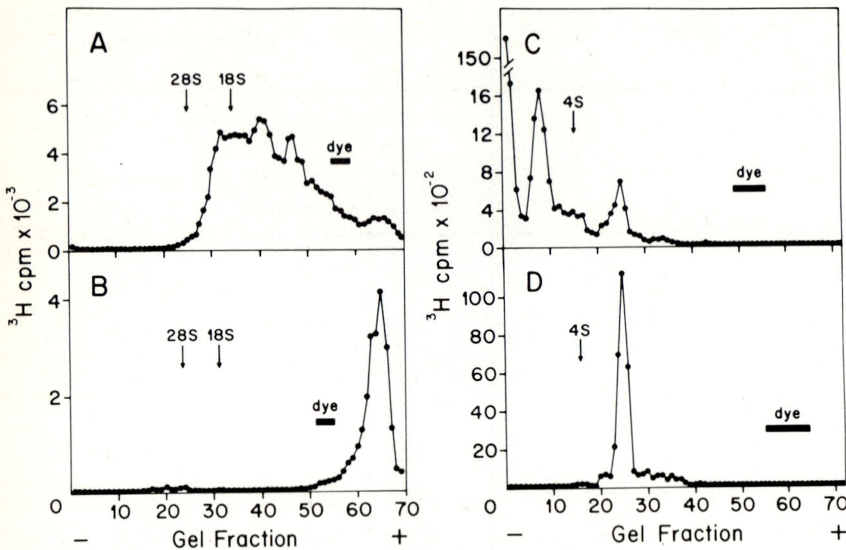

Fig.2. *Polyacrylamide gel profiles of RNAs synthesized in vitro by purified B and LT virions.* ^{3}*H-UTP labeled products from a 3 hr in vitro polymerase reaction were separated from template by centrifugation, phenol extracted, and separated from unincorporated nucleotides and small molecules by Sephadex chromatography before analysis on polyacrylamide gels. Arrows denote the positions of marker RNAs. (A) 2% gel, B-virion product; (B) 2% gel, LT-virion product; (C) 20% gel, B-virion product; (D) 20% gel, LT-virion product.*

virion RNA, 73% of the product hybridized. Therefore, the LT product is complementary to the LT genome and represents a template coded RNA. On the other hand, we have been unable to detect appreciable hybridization to either intact or alkali-fragmented B virion RNA.

Polyacrylamide gel electrophoresis of the LT product revealed a relatively homogeneous RNA species migrating more rapidly than a 4S marker. Partial sequence analysis has confirmed that the LT product was homogeneous and was composed of a single RNA species approximately 45 bases in length. The oligonucleotide fingerprints shown in fig.3 demonstrate that $\alpha^{32}P$-GTP labeled products yield only 6 oligonucleotides when digested with RNase T_1. If the product is treated with bacterial alkaline phosphatase prior to T_1 digestion, one oligonucleotide (No.T-6) disappears and a new oligonucleotide (No.T-6') appears. Thus, the presence of an alkaline phosphatase sensitive phosphate group in No.T-6 suggests that T-6 is the 5' terminus of the LT product. The nucleotide compositions and length of each T_1 oligonucleotide are shown in Table III. Although the product contains 57% A residues, the longest run of adenine residues is only 5-6 bases long.

DISCUSSION

Our data demonstrate that a T particle from the L gene region can function as a template for the virion polymerase. However, expression of the LT genome is limited to a small fraction of the template which codes for a unique RNA species. The LT particles incorporate only 5% as much UTP into RNA compared to B particles. However, since the LT product has approximately 5%-10% of the complexity of the B particle products, the polymerase appears to initiate synthesis on the B and LT particle templates with equal efficiency. The question of why the LT template transcription is limited to a specific nucleotide sequence has not been resolved. We have been unable to detect a larger product synthesized by LT particles at different temperatures or various ionic or detergent concentrations.

Our results are basically in agreement with those of Mori and Howatson and Reichman *et al.* who demonstrated that all T particles examined synthesized very small RNA molecules (15, 17). It is now crucial to compare the sequences of RNAs synthesized by a large number of different T particles since such comparisons will be extremely useful in defining how T particles are generated and function. Our prediction is that all T particles which can replicate will synthesize a small A-rich RNA with a sequence similar to LT product. There are a number of reasons for this prediction. First, parallels can be drawn between the LT product and the leader sequence studied by Colonno and Banerjee (6, 7). Both the LT product and leader are small RNA molecules, which are extremely rich in A and begin with 5'(p)ppApCpGp. Since the leader is complementary to the 3' end of the B virion genome, it seems most reasonable to assume this region represents a polymerase binding site. If so, then all T particles which transcribe or replicate would require a similar if not identical site on the 3' end. The first step in replication apparently involves a plus strand intermediate which should also require a polymerase site on its 3' end and this would be complementary to the 5' end of the T virion RNA. Therefore, the two ends of either B or T virion RNAs might be complementary. An observation consistent with this argument is that many T particle virion RNAs contain snapback RNA and some form circles with panhandles, indicative of complementary ends (14, 16). Terminal complementarity is

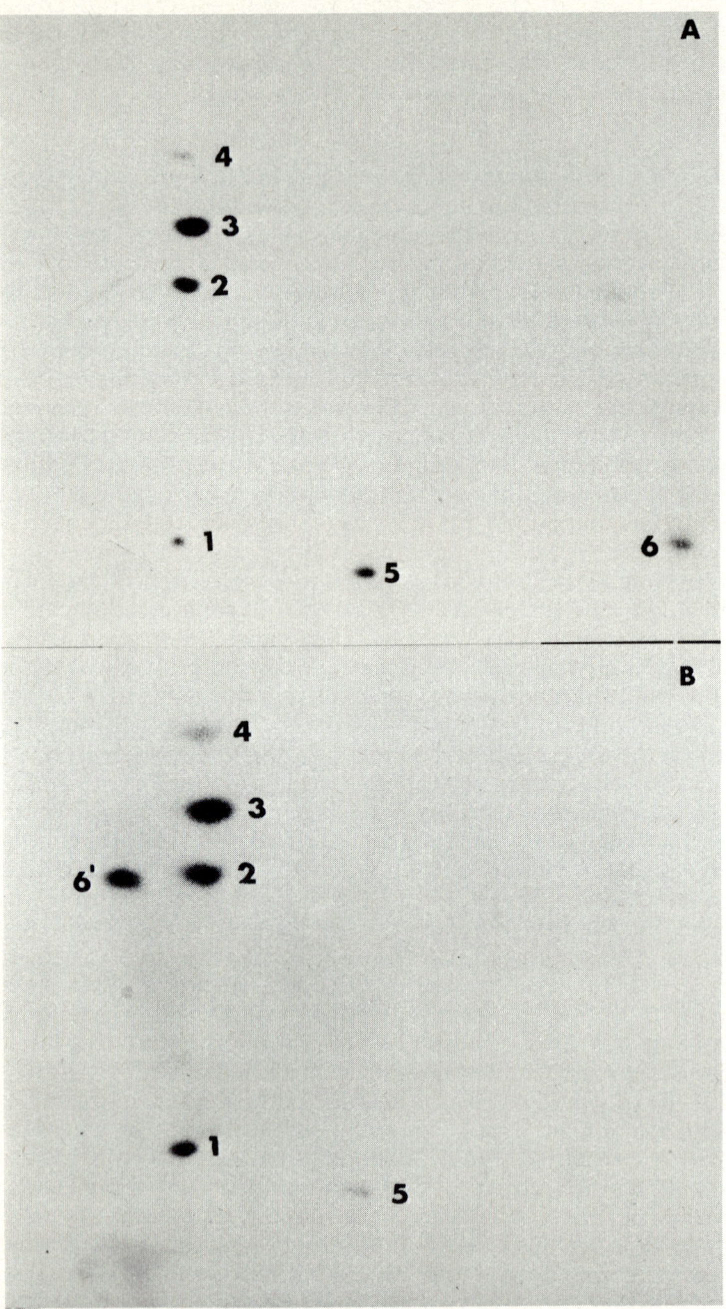

Fig. 3.

also supported by our observation that the extent of hybridization of the LT product to LT virion RNA increases dramatically if the LT virion RNA is first fragmented. Fragmentation would dissociate the two ends of the molecule and decrease the potential for self-annealing thus allowing the LT product to more effectively compete for its complementary sequence. Further sequence comparisons should demonstrate whether or not an LT product-like RNA is synthesized by all VSV T particles which replicate. If such a sequence is conserved and characteristic for all T particles, it must be required for either RNA synthesis or virion packaging.

Elucidation of how T particles acquire this sequence should now be possible. The LT particle we have studied contains L gene sequences yet the leader RNA template sequences and the L gene are thought to reside on opposite ends of the B virion RNA (1, 3). If the LT product and leader are indeed homologous, how did juxtaposition of the L gene and sequences coding for the leader-like sequences occur?

The LT particles characterized in this study contain L gene sequences as well as sequences that code for a leader-like RNA. We would like to consider the question of how T particles are generated and how they may synthesize a leader-like sequence that is not complementary to B virion RNA. We propose that the 3' and 5' ends of B particle RNA would have the following structure: (-) 3' *UGC[Xp]*$_{42}$... ...*[Yp]*$_{42}$ *GCA* 5'. A feature of this model is that while the terminal three nucleotides of the minus strand RNA are complementary (5, 11) the adjacent sequences contain little homology.

Replication of the minus strand by virion polymerase would yield a replicative intermediate with the following structure: (+) 5' *ACG[X'p]*$_{42}$.... ....*[Y'p]*$_{42}$*CGU* 3'. As suggested by both Kolakofsky and Huang most T particles would be generated from plus strand intermediates by a doubling back of the RNA transcriptase, utilizing the newly synthesized minus strand as a

---

Fig. 3. *Two-dimensional oligonucleotide fingerprints of* ^{32}*P-GTP labeled RNA synthesized in vitro by LT particles.* ^{32}*P-labeled product from a single preparation was divided into two equal portions. One portion (A) was digested with RNase* T_1 *and fingerprinted. The second portion (B) was first treated with bacterial alkaline phosphatase and then RNase* T_1*. The direction of electrophoresis in the first dimension on strips of cellulose acetate in 5 M urea, pH 3.5, is from left to right. Homochromatography in the second dimension on polyethyleneimine-cellulose thin layers at pH 7.5 is from bottom to top.*

TABLE III

Sequence Analysis of RNase T_1 Resistant Oligonucleotides

Oligonucleotide No.[a]	Complexity[b]	Relative Molar Yield[c]		Composition[d]
		Untreated	Phosphatase Treated	
T-1	13	.64	.51	$[(A_4C_2)\ (AC)\ (AC_2)]\ AG(A)$
T-2	3	1.00	1.00	$A_2G(A)$
T-3	2	.82	.76	AG(G)
T-4	1	.72	.71	G (G,U)[e]
T-5	20-21	.38	.20	$[(A_{5-6}U)\ (A_5C_2)(AU)(AC)]A_2G(A)$
T-6	3	.41	--	ppACG(A)[f]
T-6'	3	--	.79	ACG (A)

[a]See fig.3.

[b]Complexity was calculated from the composition and does not include the residue which nearest neighbor labels the guanosine residue.

[c]A_2G = 1.0 using $\alpha^{32}P$-GTP as source of label.

[d]Compositions were deduced from analysis of products obtained by digestion with RNase A and RNases T_1, T_2, and A (see ref.10). The last nucleotide in parenthesis was deduced to be the nearest neighbor to the 3' terminal guanosine residue. A detailed discussion of the sequence analysis will be presented elsewhere (manuscript in preparation).

[e]Nearest neighbor labelled with both $\alpha^{32}P$-GMP and $\alpha^{32}P$-UMP.

(continued next page)

(continued from previous page)
[f]Oligonucleotide 6, labelled with ^3H-ATP and α ^{32}P-CTP, released a product which contained both ^{32}P and ^3H labels and migrated on 3 MM paper (pH 3.5) as ppAp after digestion with RNases T_1, T_2 and A.

---

template. This would generate a T particle in which the ends are self-complementary: (-) 3' *UGC*$[Y'p]_{42}$... ...$[Yp]_{42}$ *GCA* 5'. *In vitro* transcription of such a T-particle would result in a T-product complementary to the 3' end of the T-particle but having little if any homology to B virion RNA.

This model for T-particle formation predicts that T virion RNAs would contain self complementary sequences. The extent of sequence complementarity may vary considerably among T-particle preparations as already shown by Perrault and Lazzarini (14, 16). Preparations of LT virion RNA have only limited complementarity since only about 10% of labeled LT virion RNA is resistant to RNase digestion after self-annealing. This model is further supported by the fact that the LT product transcribed from the LT particle, while containing a terminal ACG, differs from the leader sequence of B virion RNA (Colonno and Banerjee, personal communication). In addition we have been unable to demonstrate hybridization of LT product to B virion RNA. Therefore, in general, the small products synthesized *in vitro* by all T particles containing the L gene region should be identical and none should hybridize to B virion RNA. In addition the short sequence of approximately 45 nucleotides at the 3' terminus of T genomes should be complementary to the 5' terminus of B genomes. We are currently testing these predictions.

ACKNOWLEDGEMENTS

This research was supported by Public Health Service grants AI-11722 from the National Institute of Allergy and Infectious Diseases (S.U.E.) and CA-16553 from the National Cancer Institute (J.T.P.) and by grant VC-186 from the American Cancer Society (J.T.P.). S.U.E. is the recipient of Public Health Service Research Career Development Award AI-00013 from the National Institute of Allergy and Infectious Diseases.

REFERENCES

1. Abraham, G. and Banerjee, A.K. (1976). *Proc. Nat. Acad. Sci. USA* 73, 1504.
2. Abraham, G., Rhodes, D.P. and Banerjee, A.K. (1975). *Cell* 5, 51.
3. Ball, L.A. and White, C.N. (1976). *Proc. Nat. Acad. Sci.*

USA 73, 442.
4. Banerjee, A.K. and Rhodes, D.P. (1973). *Proc. Nat. Acad. Sci. USA* 70, 3566.
5. Banerjee, A., and Rhodes, D.P. (1975). *Biochem. Biophys. Res. Comm.* 68, 1387.
6. Colonno, R.J. and Banerjee, A.K. (1976). *Cell* 8, 197.
7. Colonno, R.J. and Banerjee, A.K. (1977). *Virology* 77, 260.
8. Emerson, S.U. and Wagner, R.R. (1972). *J. Virol.* 10, 297.
9. Emerson, S.U. and Yu, Y-H. (1975). *J. Virol.* 15, 1348.
10. Emerson, S.U., Dierks, P.M. and Parsons, J.T. (1977). *J. Virol.* 23, 708.
11. Hefti, E. and Bishop, D.H.L. (1975). *J. Virol.* 15, 90.
12. Holland, J.J., Villarreal, L.P. and Breindl, M. (1976). *J. Virol.* 17, 805.
13. Huang, A.S. (1973). *Ann. Rev. Micro.* 27, 101.
14. Lazzarini, R.A., Weber, G.H., Johnson, L.D. and Stamminger, G.M. (1975). *J. Mol. Biol.* 97, 289.
15. Mori, H. and Howatson, A.F. (1973). *Intervirology* 1, 168.
16. Perrault, J. (1976). *Virology* 70, 360.
17. Reichmann, M.E., Villarreal, L.P., Kohne, D., Lesnaw, J. and Holland, J.J. (1974). *Virology* 58, 240.
18. Reichmann, M.E., Pringle, C.R. and Follett, E.A. (1971). *J. Virol.* 8, 154.
19. Schnitzlein, W.M. and Reichmann, M.E. (1976). *J. Mol. Biol.* 101, 307.
20. Soria, M., Little, S.P. and Huang, A.S. (1974). *Virology* 61, 270.
21. Villarreal, L.P. and Holland, J.J. (1973). *Nature New Biol.* 246, 17.
22. Wagner, R.R. (1975). *In* "Comprehensive Virology" (H. Fraenkel-Conrat and R.R. Wagner, eds.), vol.4. p.1. Plenum Press, New York.

THE ROLE OF THE *ts*L AND P^- MUTATIONS OF VSV T1026 IN PERSISTENT INFECTION

C.P. STANNERS and T. LAM

The Ontario Cancer Institute and
Department of Medical Biophysics, University of Toronto,
Toronto, Canada

T1026 is a *ts* mutant of the HR strain of VSV-Indiana which, at semi-permissive temperatures, is much less cytocidal than its parent (1). The mutant produces a delayed neurological death in newborn hamsters (8). The *ts* mutation of T1026 was shown, by complementation, to be in Group I (8), i.e., in the L protein, which is the major determinant of the virion RNA-dependent RNA polymerase activity. Recently, we have shown that T1026 has a second non-*ts* mutation in P, the virus function which produces a decline in total protein synthesis of infected cells (7). P appears to act chiefly at the level of initiation of translation, and tends to be selective for the inhibition of host cell protein synthesis (7). The genotype of T1026 can thus be denoted *ts*L, P-.

Concerning persistence, we have drawn attention to the fact that many viruses emerging from persistently infected cultures are double mutants being both *ts* and small plaque (10), and the P^- mutation in T1026 correlates with failure to increase plaque size after several days (7). Thus T1026 could represent a model for a wide range of virus mutants capable of persistent infection. Logically the P^- mutation would be required for persistence to prevent cell death due to inhibition of cellular protein synthesis (3) by the P function, and the *ts*L mutation to prevent cell death due to inhibition of cellular protein synthesis by competition between viral and cellular mRNA for limited translational machinery. This paper represents a preliminary report on the role of the two types of mutation in T1026 in the establishment and maintenance of persistent infection.

Our earlier studies on persistence were done chiefly using

normal hamster embryo fibroblasts (6). Long term maintenance of persistence was difficult with these cells, as the requirement for regrowth after partial killing of the cultures exceeded their proliferative capacity. This difficulty was overcome by using STR-T, a polyoma virus-transformed line of hamster embryo fibroblasts (9) and L cells, both of which have unlimited proliferative capacity. A second difficulty concerned the nature of the particular *ts*L mutation in T1026. The expression of this mutation shows extreme dependence on temperature near its cut-off temperature (see Fig. 2), making temperature regulation over the relatively long periods of time involved in persistence experiments rather critical. To overcome this difficulty, a new *ts* mutation was introduced into R1, a *ts* revertant of T1026 (7) to give R1S1. The new *ts* mutation in R1S1 was shown by complementation to be Group I (Francoeur, unpublished), i.e., *ts*L, and its expression was found to be a more gentle function of temperature (Fig. 2). Persistent infection could be established and maintained much more easily with STR-T or L cells than with normal hamster embryo fibroblasts, and R1S1 was somewhat more consistent in producing persistence than T1026.

The results of experiments on persistence with STR-T and L cells are summarized in Fig. 1 and Table 1. Firstly, it can be seen that, using a wide range of input multiplicities, persistence could not be obtained with mutants carrying P^- mutations or *ts*L mutations alone. P^- mutations alone always give lysis, whereas *ts*L mutations alone gave either lysis or cure, depending on their cut-off temperatures (Fig. 1). *ts*L, P^- double mutants, on the other hand, could nearly always establish persistence, even at high m.o.i. Secondly, maintenance of per-

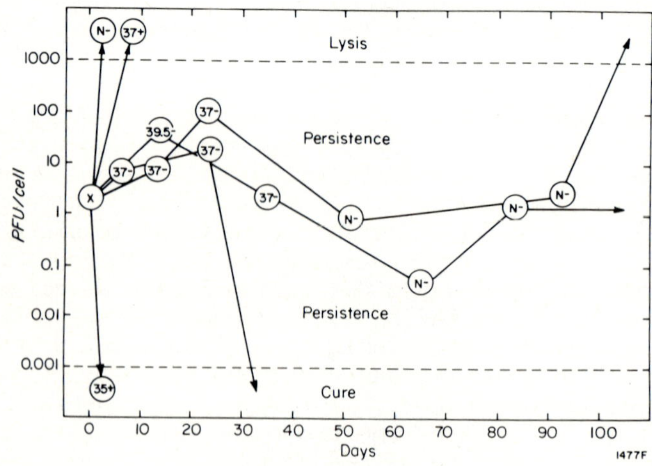

Fig.1

sistence for times exceeding about 3 weeks often involved an adjustment of the *ts*L mutation to lower temperature sensitivity or higher cut-off temperature (Fig. 2), though the P^- mutation was always retained.

This work, showing the necessity of both *ts*L and P^- mutations for establishment of persistent infection, is consistent with the model for persistence outlined above. Our virus lysates do not contain demonstrable interferon or DI particles, and it seems from the work of others that these genetic requirements can be relaxed if cultures are co-infected with large numbers of DI particles (2, 10) or treated with antiviral antibody (4). Concerning maintenance of persistent infection, it would appear that the mutations put the virus at a long term disadvantage, due presumably to delayed production of DI particles and/or interferon, so that the cultures can become cured. Curing is avoided by the emergence of P^- mutants with less stringent *ts*L mutations. In confirmation of these ideas, preliminary experiments indicate that the appearance of factors which interfere with virus development in culture fluids accompanies the observed shift in the *ts*L mutation. These adjustments in viral and cellular characteristics which accompany persistent infection are important in exploring possible mechanisms involved in persistence. It will be of great interest to determine whether they apply to other systems of viral persistence, and perhaps more importantly, to persistence *in vivo*.

Fig. 1. *The role of genotype in persistent infection. Cultures of STR-T cells in exponential phase at 2 x 10^6 cells per 4 oz medicine bottle (Brockway Glass Co.) in 10 ml α-MEM (5) with 10% fetal calf serum (FCS) at 38.5° were infected with various VSV mutants at 2 pfu/cell. The cultures were then incubated at 38.5° and subcultured every 3 days or when required by cell growth. At various times the total virus per cell was assayed by plaque assay (1) at 34° and by cell number determination with an electronic particle counter. Culture fluids were also assayed for virus type by plaque size at a series of temperatures to assess the ts mutation, and by the PSI assay (7) to assess the P mutation. The numbers in the circles representing each experimental point refer to the minimum temperature at which plaque size at 2-3 days was clearly less than the plaque size of non-ts virus. "N" represents non-detectable difference in plaque size relative to non-ts virus up to 39.5°. The + or - signs in the circles refer to the state of the P function. Cultures which entered the "cure" area could be subsequently cultured indefinitely with no evidence of infectious virus or virus products (6). Cultures which entered the "lysis" area showed no cell survival whatsoever.*

TABLE 1

Ability of VSV Mutants to Establish Persistence

Mutant	Genotype	m.o.i. pfu/cell	Cell type	Persistence[a]
HR	L^+, P^+	0.001	STR-T	-
		to		
		20	L	-
R1	L^+, P^-	0.001	STR-T	-
		2	"	-
T102[b]				
T128		0.001	STR-T	-
T129	*ts*L, P^+	2	"	-
T230				
T1026	*ts*L, P^-	0.001	STR-T	+, -
		0.1	"	-
		2	"	+, +
		0.001	L	+
		0.1	"	+
		2	"	+
R1S1	*ts*L, P^-	0.001	STR-T	+, +
		0.01	"	-
		0.1	"	-, +
		2	"	+, +, +
		5	"	+
		20	"	+
		0.001	L	+
		0.1	"	+
		2	"	+

[a] - denotes no survival of cells after 1 week or less.

+ denotes survival of infected cells for more than 2 weeks. Separate entries represent independent experiments.

[b] These are all *ts* group I mutants with high cut-off temperatures.

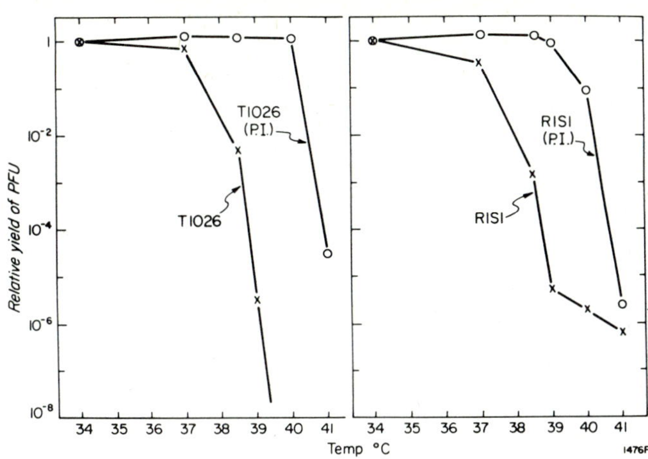

Fig. 2. *ts phenotype of input virus and output virus of two mutants used in persistence experiment of Fig. 1. The two infected cell lines which showed persistence for >80 days in Fig. 1 were originally infected with T1026 or R1S1. The virus emerging after 80 days, (last points in Fig. 1) denoted "T1026 P.I." and "R1S1 P.I.", together with the parental T1026 and R1S1 were used to infect cultures of 2 x 10^6 hamster embryo fibroblasts at 10^5 pfu per culture. After incubation for 18 hr at the indicated temperature (controlled by submersion of bottles in temperature-regulated water baths) in 10 ml α-MEM plus 2% FCS, the virus produced was assayed by plaque assay at 34°. The total yields of virus at each temperature are expressed as fractions of the yield at 34°.*

ACKNOWLEDGEMENTS

We thank W.A. Mehring for capable technical assistance. This work was supported by the Medical Research Council and the National Cancer Institute of Canada.

REFERENCES

1. Farmilo, A.J. and Stanners, C.P. (1972). *J. Virol.* 10, 605.
2. Holland, J.J. and Villarreal, L.P. (1974). *Proc. Nat. Acad. Sci. (USA)* 71, 2956.
3. Marvaldi, J.L., Lucas-Menard, J., Sekellick, M.J. and Marcus, P.I. (1977). *Virology* 79, 267.
4. Sinarachatanant, P. and Huang, A.S. (1977). *J. Virol.* 21, 161.
5. Stanners, C.P., Eliceiri, G.L. and Green, H. (1971). *Nature New Biol.* 230, 52.
6. Stanners, C.P., Farmilo, A.J. and Goldberg, V.J. (1975). *In* "Negative Strand Viruses" (Mahy, B.W.J. and Barry, R.D. eds), Academic Press, London, p.785.

7. Stanners, C.P., Francoeur, A.M. and Lam, T. (1977). *Cell* 11, 273.
8. Stanners, C.P. and Goldberg, V.J. (1975). *J. gen. Virol.* 29, 281.
9. Stanners, C.P., Till, J.E. and Siminovitch, L. (1963). *Virology* 21, 448.
10. Youngner, J.S., Dubovi, E.J., Quagliana, D.O., Kelly, M. and Preble, O.T. (1976). *J. Virol.* 19, 90.

INTERFERENCE AND PERSISTENCE IN MICE INFECTED WITH VESICULAR STOMATITIS VIRUS

JOAN CRICK and F. BROWN

Animal Virus Research Institute,
Pirbright, Woking, Surrey, GU ONF.

In 1970, Huang and Baltimore suggested that virus defective interfering (DI) particles may have a role in determining the outcome of disease in the intact animal (8), and the involvement of such particles in the establishment and maintenance of persistently infected cells has been shown for vesicular stomatitis virus (VSV) (6) rabies virus (10, 11) and for various other viruses (17). However, there are still only the experiments of von Magnus (21) with influenza virus and Mims (12) with Rift Valley fever virus to support the idea that DI particles are important in persistent infections *in vivo*.

Holland and his coworkers (3, 5, 7) achieved a significant prolongation of life and, in some cases, complete protection in mice when large amounts of VSV-Indiana DI particles were inoculated intracerebrally together with homologous virus. They concluded that their results were due to *in vivo* interference and distinguished the effect from the protection (presumably immunological) conferred by DI particles given 10 days before virus challenge.

The RNA of the DI particles must be functional for *in vitro* interference to be established (9) yet we have shown that two effects in mice similar to those observed by Holland and his colleagues may be obtained with inactivated DI particles. Since our results cannot be explained on the basis of interference we have suggested that the results of Holland and his coworkers may, like our own, be accounted for by host immunological responses.

Holland and Doyle (5) and Doyle and Holland (3) also pointed out that in their experiments mice which had received massive doses of DI particles and small amounts of virus, and which did not survive indefinitely, died from a slow, progressive wasting disease quite different from the rapid paralysis and death usually associated with the intracerebral inoculation of VSV. They regarded this 'delayed death' as also being due to interference, but the condition is not unlike that observed after the intracerebral administration of certain *ts* mutants of VSV (2, 16), and in our own experiments with inactivated DI particles. We have also noticed similar effects in mice infected with other rhabdoviruses and, very occasionally, with very small doses of VSV alone (1). Some mice recovered from the wasting disease and are still alive after 15-18 months in this condition although one or both hind legs have remained paralysed. How, therefore, is 'delayed death' caused in infected mice? Can it be related to a persistent infection and are DI particles involved? The following experiments were undertaken in an attempt to answer these questions.

EFFECT OF DI PARTICLES ON VIRUS TITRE IN 7 DAY OLD MICE

We first examined the effect of DI particles on the ability of VSV to infect 7 day old mice. Groups of mice were inoculated with serial dilutions of VSV-Indiana diluted in phosphate buffer or in a suspension of DI particles inactivated with acetylethyleneimine (AEI) (1). The mice were inoculated intracerebrally or intramuscularly in the right hind leg. Other mice were inoculated similarly with partially purified DI particles which had not been inactivated and which contained residual virus. The mice were kept for three weeks following inoculation.

The results in Table I show that there was a very slight depression of virus titre by inactivated DI particles if the intracerebral route of inoculation was used but a much greater depression when the injections were made intramuscularly. The mice also reacted more slowly after intramuscular inoculation of virus and there was even more delay if inactivated DI particles were also given.

When untreated DI particles were inoculated by the intracerebral route, the mice reacted more slowly than if the same dose of virus was given alone, but when dilutions of the same preparation were given intramuscularly very few mice reacted. Furthermore, there was a distinct prozone effect so that more mice reacted at higher dilutions (10^{-4} in the Table) than with more concentrated inocula. All the mice that reacted died within 11 days, but were paralysed for several days beforehand.

TABLE I

Effect of DI particles of VSV-Indiana on the titre of homologous virus in 7 day old mice

Inoculum	Titre (log ID_{50}/0.03 ml) i.c.	i.m.	
Virus alone	9.1 (2-5 days)[x]	8.5 (3-5 days)	
Virus + DI particles[xx]	8.3 (2-4 days)	5.7 (4-8 days)	
DI preparation containing residual virus	5.7 (2-8 days)	10^{-5} 2/5[xxx]	(4-11 days)
		10^{-4} 4/5	
		10^{-3} 1/5	
		10^{-2} 1/5	
		10^{-1} 0/10	
		undiluted 1/9	

[x]Time between first and last reaction

[xx]DI particles inactivated with AEI

[xxx]No. of mice reacting at different dilutions of inoculum

EFFECT OF AGE AND ROUTE OF INOCULATION ON THE OUTCOME OF INFECTION IN MICE

The results above suggested that the intramuscular route of inoculation might prove useful in interference experiments and in the development of subclinical or persistent infections. We have compared the titres of virus in groups of mice varying between 2 and 10 days old and between 3 and 8 weeks old. These ages were presumed to differentiate between reactions in mice which have different levels of immunological competence. The results in Table II show that there was very little difference in sensitivity between the two groups to virus given intracerebrally but a dramatic difference in response to virus given intramuscularly and, of those which did, most remained paralysed for more than 16 weeks, frequently showing the arched back and ruffled fur previously described (3, 5). There was no reaction at dilutions higher than 10^{-3}.

TABLE II

Effect of age and route of inoculation on the outcome of infection of mice with VSV-Indiana

Age of mice	Route of inoculation i.c.	i.m.	
2-10 days	7.5 (2-5 days) → death	6.7 (3-7 days) → death	
2 weeks-8 weeks	7.7 (2-12 days) →	10^{-3}	1/15[x]
	usually death,	10^{-2}	1/15
	occasionally paralysis	10^{-1}	1/15
	which lasts	undiluted	1/10
	indefinitely		

[x]These mice reacted 6-8 days after inoculation, most remained paralysed for more than 16 weeks.

When DI preparations containing residual virus were given to the older mice intracerebrally, death was delayed (cf.3) and mice which received the higher dilutions of virus died first. No mice showed any signs of disease when the DI preparation was inoculated intramuscularly.

CAN MICE BECOME PERSISTENTLY INFECTED AFTER INTRAMUSCULAR INOCULATION OF VIRUS?

In this series of experiments groups of fifteen 3 to 4 week old mice were given serial dilutions (undiluted up to 10^{-3}) of virus or virus inactivated with AEI. Twelve days after inoculation, 5 mice from each group were killed and their sera, after heat inactivation, used for the determination of serum neutralizing activity. The remaining mice in each group (apart from those which had received untreated virus and which were also paralysed) were challenged intracerebrally with 5-10 LD_{50} of homologous virus. The results in Table III show the superiority of the live virus over the inactivated virus in producing neutralizing activity and protecting against challenge. Very small doses of infectious virus stimulated the production of far greater amounts of serum neutralizing activity than undiluted, but inactivated virus. In more recent experiments, using dilutions of virus up to 10^{-8}, we have

TABLE III

Effect of intramuscular inoculation of 3-4 weeks old mice with VSV-Indiana on the production of serum neutralizing activity and protection against intracerebral challenge with homologous virus

Virus inoculum ($\log_{10}$ ID_{50} before inactivation)	Serum neutralizing activity on day 12[x]		Reactors[xx] following i.c. challenge on day 15	
	Untreated virus	Inactivated virus	Untreated virus	Inactivated virus
4.5	4.1	0.6	0/9	6/10
5.5	4.1	0.7	0/9	6/10
6.5	4.4	2.1	1/9	6/10
7.5	4.8	3.2	0/5	2/10

[x]Log_{10} depression of virus titre by 0.015 ml 1/10 serum. G.M. of 4 sera

[xx]Challenge dose 5-10 ID_{50}. Killed 10/10 control mice within 12 days.

detected significant neutralizing activity (1-2 logs in 0.015 1/10 serum) 7 days after the intramuscular inoculation of only 300 ID_{50} of virus while protection against homologous intracerebral challenge was conferred at 21 days by as little as 10 ID_{50} of virus. At this time the mice had serum neutralizing activities as high as 3 logs in 1/10 serum.

We also followed the development of protection and serum neutralizing activity in 3-4 weeks old mice given DI preparations containing relatively small amounts ($10^{5.3}ID_{50}$) of residual virus. At 21 days after inoculation the protective effect and the virus neutralizing activity stimulated by the untreated preparation (undiluted and diluted 1/10 was of the same order as that conferred by 10-30 ID_{50} of virus alone.

DISCUSSION

Our results suggest that inapparent infections may be established in 3-4 week old mice by intramuscular inoculation of VSV-Indiana. Protection against infection is less readily conferred and less neutralizing activity produced when substantial amounts of DI particles are added to the inoculum as would be expected if these inhibit virus replication. Large amounts of viral antigen must be produced to account for the massive immune response which we observe after the inoculation of small doses of virus. However, neither of these observations rules out the possibility that DI particles are produced as the virus replicates in the mouse. We know that the virus seed used throughout these experiments produces DI particles when passaged in BHK21 cells.

In preliminary experiments in 3-4 weeks old mice we have been able to show limited virus replication at the site of virus introduction (right hind leg) up to 4-5 days after infection, and subsequently in the brain, though we have failed to recover infective virus from the blood or the uninoculated hind leg, the only two tissues so far examined. These results agree with those of Falke and Rowe (4) who stated that in mice VSV becomes increasingly confined to the central nervous system with increasing age and in fact travels there not in the blood, but through the peripheral nerves, an interesting link with rabies virus which travels to the brain by the same route (13, 14).

Subclinical infections with VSV have been described earlier (19) and apparently persistent infections have also been obtained with *ts* mutants (2, 16, 20). Our results suggest that we may also have a model for persistent infection with VSV. Whether DI particles are involved remains an open question, which we are investigating.

ACKNOWLEDGEMENT

It is a pleasure to thank Mr R.J. Darnell for his help with the experiments.

REFERENCES

1. Crick, J. and Brown, F. (1977). *Inf. Immun.* 15, 354.
2. Dal Canto, M.C., Rabinowitz, S.G. and Johnson, T.C. (1976). *Br. J. Exp. Path.* 57, 321.
3. Doyle, M. and Holland, J.J. (1973). *Proc. Nat. Acad. Sci. USA* 70, 2105.
4. Falke, D. and Rowe, W.P. (1965). *Arch. Virusforschung* 17, 549.
5. Holland, J.J. and Doyle, M. (1973). *Infect. Immun.* 7, 526.
6. Holland, J.J. and Villareal, L.P. (1974). *Proc. Nat. Acad. Sci. USA* 71, 2956.
7. Holland, J.J., Villareal, L.P. and Etchison, J.R. (1974). *In* "Mechanisms of virus disease" (W.S. Robinson and C.F.Fox ed.) p.131. ICN-CLU Symposia on molecular and cellular biology, vol.1 W.A. Benjamin, In., Melno Park, Calif.
8. Huang, A.S. and Baltimore, D. (1970). *Nature* London 226, 325.
9. Huang, A.S. and Wagner, R.R. (1966). *Virology* 30, 173.
10. Kawai, A., and Matsumoto, S. (1977). *Virology* 76, 60.
11. Kawai, A., Matsumoto, S. and Tanabe, K. (1975). *Virology* 67, 520.
12. Mims, C.A. (1956). *Br. J. Exp. Path.* 37, 129.
13. Murphy, F.A., and Bauer, S.P. (1974). *Intervirol.* 3, 256.
14. Murphy, F.A., Bauer, S.P., Harrison, A.K., and Winn, W.C. Jr. (1973). *Lab. Invest.* 28, 361.
15. Olitsky, P.K., Sabin, A.B. and Cox, H.R. (1936). *J. Exp. Med.* 64, 723.
16. Rabinowitz, S.G., Dal Canto, M.C., and Johnson, T.C. (1976). *Inf. Immun.* 13, 1242.
17. Rima, B.K. and Martin, S.J. (1976). *Med. Microbiol. Immunol.* 162, 89.
18. Sabin, A.B. and Olitsky, P.K. (1937). *J. Exp. Med.* 66, 35.
19. Skinner, H.H. (1957). *J. Comp. Path.* 67, 69.
20. Stanners, C.P., Farmilo, A.J. and Goldberg, V.J. (1975). *In* "Negative Strand Viruses" (B.W.J. Mahy and R.D. Barry, eds.) p.785. Academic Press, London.
21. von Magnus, P. (1951). *Acta. Pathol. Microbiol. Scand.* 29, 156.

CHARACTERIZATION OF RABIES DEFECTIVE VIRUSES

SEIICHI MATSUMOTO and AKIHIKO KAWAI

*Institute for Virus Research,
Kyoto University,
Kyoto 606, Japan.*

Persistent infections in cultured cells have been established with many virulent viruses. It is known that biologically they are not produced by a common specific factor but rather different mechanisms such as interferon, temperature-sensitive mutations, proviral intermediates, or DI particles are involved in the establishment of persistence (1, 2, 6, 7). We reported that persistent rabies virus infection, with a cyclic rising and falling pattern, was readily produced in BHK-21 cells infected with the HEP-Flury strain, and synthesis of DI particles was implicated in the establishment and maintenance of this persistence (3). This communication deals with a concise summary of our observation of defective rabies viruses in studies carried out over the past few years.

DEFECTIVE VIRUS PRODUCED IN PERSISTENT INFECTION

When BHK-21 cells were infected with the virus at an m.o.i. of 0.01 pfu/cell, marked cpe occurred about 4 days post infection. In cases of the same virus infection at a high m.o.i. of 3 pfu/cell, cpe was slightly observed all over the cell sheet. The surviving cells of both cases which eventually grew to form a new monolayer were subcultured every 3 days. It was observed that the rabies virus persistent infection was established in all of those BHK-21 cell sublines. The amounts of released infective virus rose and fell cyclically and the appearance of inclusion (accumulation of viral nucleocapsid) positive cells also fluctuated from 100 to 0% with the same periodicity of infectivity.

TABLE I

Effect of the Cpe-suppressing Factor on the Replication of Homologous Virus

cpe-suppressing factor (units/culture)	Yield of LPV				
	pfu/ml at:				Ratio (%)
	15 hr	48 hr	72 hr	96 hr	at 72 hr
1.4×10^5	7.0×10^4	5.5×10^3	1.8×10^4	4.0×10^4	0.045
1.4×10^4	1.0×10^6	2.3×10^6	1.5×10^6	1.1×10^6	3.8
1.4×10^3	5.8×10^6	8.0×10^6	7.5×10^6	4.2×10^6	18.8
1.4×10^2	7.3×10^6	9.0×10^6	9.0×10^6	5.0×10^6	22.5
Control	9.8×10^6	3.0×10^7	4.0×10^7	4.5×10^7	100

BHK monolayers were co-infected with the HEP virus (LPV) at an m.o.i. of 5 pfu/cell and a series of 10-fold dilutions of the partially purified cpe-suppressing factor. After virus adsorption for 1 hr at 37°, cultures were washed twice, fresh growth medium was added and incubation was carried out at 37°. Culture fluids harvested at 15, 48, 72, 96 hr after infection were assayed for infectivity. The column on the extreme right shows ratios of the virus yield with that of the control at 72 hr after infection.

The infected cells were resistant against the homologous virus challenge, whereas they were susceptible to superinfection with a heterologous virus (VSV). This corresponded to the fact that interferon was not produced at all in the persistently infected cells.

Taking into consideration that the persistent rabies virus infection is readily established depending on the grade of cpe, cpe-suppressing activity was quantitatively assayed as follows. 5 x 10^5 BHK-21 cells were seeded into each well of multidish-trays. After 5 hr 0.2 ml of mixture of a series of 3 fold-diluted culture fluid and equal volume of rabies virus, sufficient to induce complete cytolysis, were added to each well. The final determination of cpe was done on the 7th day and the titer of the activity expressed as the reciprocal of the dilution causing lysis of 50% of cells. Table I shows that the culture fluid containing cpe-suppressing factor interferes with replication of homologous virus. This activity was associated with the virus particles. The culture fluid harvested at the time of a high level of persistent infection was used for virus purification. Two opalescent bands were observed after sucrose gradient centrifugation. The lower band, containing rabies virions having a size of 65 x 180 nm, coincided with the peak of infectivity. On the other hand, the interfering or cpe-suppressing activity was only recognizable in the upper band which contained shortened virions about 70 nm in length. Yields of such DI particles synthesized by persistently infected cells had the same cyclic pattern as the infectivity. It was noticeable that the cyclic production of DI particles regularly lagged behind that of infective viruses. This agrees with the suggestion by Palma and Huang (5) that cyclic production is the result of interdependence between defective viruses (helper dependent) and standard viruses (susceptible to inhibition by DI particles).

SMALL PLAQUE VIRUS PRODUCED IN PERSISTENT INFECTION

It was found that a small plaque virus (SPV) appeared in the virus population of this persistent infection and completely replaced the wild-type large plaque virus (LPV) after about 20 transfers of persistently infected BHK-21 cells (3, 4). A number of small plaque variants have been isolated from various persistent infections. They are generally considered to be attenuated viruses. In contrast, however, the SPV caused a marked cpe and did not produce viral persistence to the extent seen with the LPV. Thus, we carried out an investigation of the biological properties of SPV with regard to production of DI particles.

BHK-21 cells were infected with the SPV at a low m.o.i. and subcultured until inclusion positive cells reached about 100% and cpe became evident. Released viruses were then purified and two bands were obtained after sucrose density gradient centrifugation. Infectivity was found in the lower band, which contained only typical standard virions. The upper band contained truncated particles, 65 x 70-100 nm, which possessed neither infectivity nor suppressing activity against cpe produced by either the SPV or LPV. These defective particles also did not interfere with the growth of both infective viruses. When purified standard and defective non-interfering (DNI) viruses were inoculated at a high m.o.i., a large number of DNI particles was produced in 48 hr of the first passage and the quantitative ratio between standard and DNI particles as well as total volumes of each virus band were not influenced by addition of DNI particles to the inoculum. Therefore,

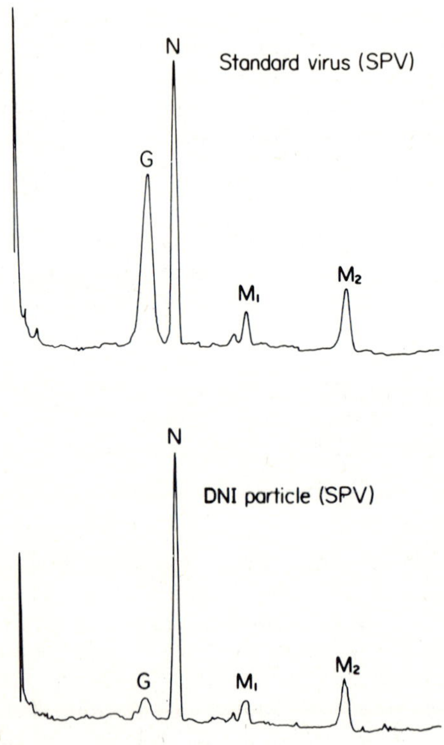

Fig. 1. *Comparison of electrophoretic patterns of polypeptides constituting standard and DNI particles of the small plaque virus.*

it can be concluded that in contrast to the wild-type virus (LPV), DNI particles were produced even from the first passage in BHK-21 cells, regardless of the multiplicity of infection. This is in contrast to the conventional view (von Magnus phenomenon) that several passages of undiluted virus in the host cell are required to generate defective viruses. In Fig. 1 showing polyacrylamide gel electrophoresis of polypeptides of standard and DNI particles of the SPV, respectively, the peak of glycoprotein of the latter is significantly lower than that of the former. This led to the inference that the DNI particles do not relate with the replication of progeny viruses. Whether or not the DNI particles could be adsorbed onto the host cell was examined. The ratios of cell associated viruses were sequentially compared between the standard and DNI viruses until 90 min after infection and a difference was observed in the adsorption rate of these labelled viruses. The value for DNI particles was very low. This result was coincident with the negative contrast figure of DNI particles in which the projecting spikes were observed to be somewhat obscure (Fig. 2).

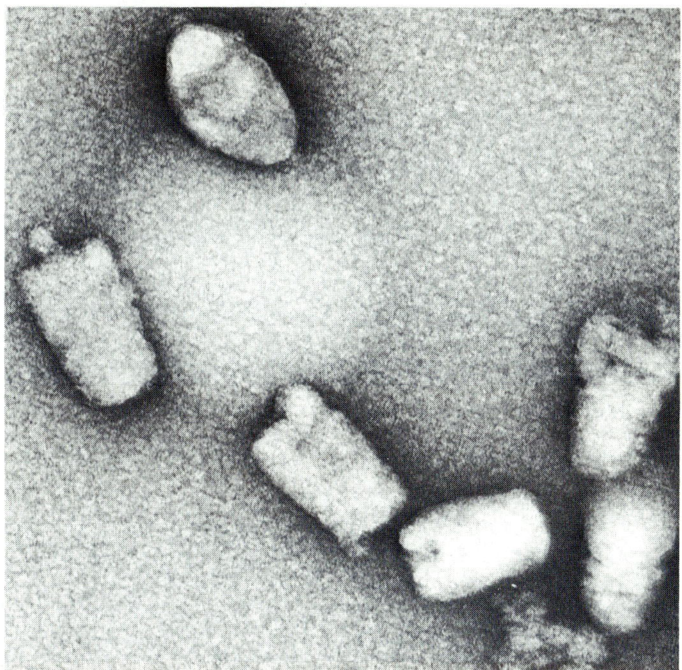

Fig. 2. *Negative contrast preparation from the upper band of SPV. Most of the particles are short and bullet-shaped. The figure of projecting spikes is obscure. x 170,000.*

Production of DNI particles may be involved in the mechanism by which the SPV caused a marked cpe and resulted in a decreased ability to induce persistent infection than did the LPV. We observed that when SPV completely replaced LPV after many transfers of infected cells, a certain grade of cpe occurred and the cyclic pattern of infection temporarily altered in irregular curves. It suggests that the spontaneous cpe is responsible for the disappearance of DI particles of LPV destined to displace DNI particles of SPV.

The results mentioned above seem to be contradictory to the evidence that the viral persistence continues not only after the replacement of virus population by the SPV but also is induced by undiluted passage of SPV, because DI particles played a major role in the establishment of rabies persistent infection. The following study was carried out to identify DI particles generated by the SPV. Culture fluids were harvested at high levels of cyclic infection and applied for virus purification. Besides two main bands, a relatively vague and broad band was visible at an upper position (Fig. 3). The lowest band coincided with the peak of infectivity

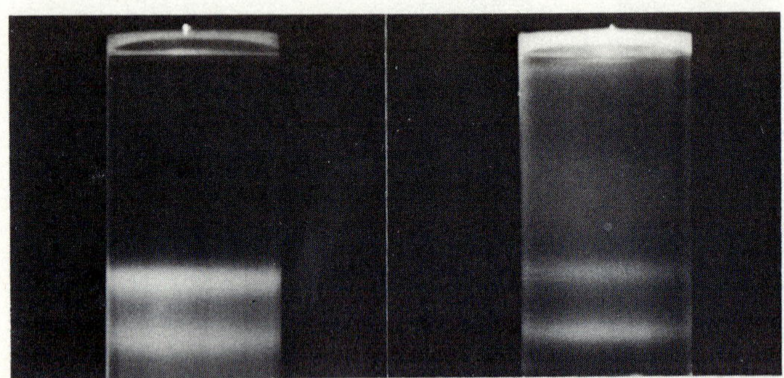

Fig. 3. *(a) Two bands in a linear sucrose density gradient of SPV collected at 6 days post infection at the m.o.i. of 0.001 pfu/cell. (b) Two sharp and one upper broad bands in a sucrose density gradient of SPV harvested from the culture fluid of persistent infection.*

and cpe-suppressing activity was not observed in the middle band. The top diffuse band exhibited cpe-suppressing activity against LPV as well as SPV. When cells were serially infected with undiluted SPV, cpe gradually became obscure after several passages and simultaneously a new band appeared in sucrose density gradient centrifugation, whereas the same phenomenon took place after fewer passages of LPV. The morphology of these particles was variable and the size was smaller than that of LPV-DI particles. Of special interest is the finding that DI particles generated by LPV did not interfere with the SPV. Namely, when the interfering activity was compared with LPV-DI and SPV-DI particles against homologous or heterologous standard viruses, it was found that only the LPV-DI particle did not interfere with the growth of heterologous standard SPV. Such data are expedient for understanding the stepwise transition from LPV to SPV in persistent infection. The SPV once evolved could be effectively selected for, since they are not susceptible to interference by the pre-existing DI particles of LPV.

REFERENCES

1. Holland, J.J. and Villarreal, L.P. (1974). *Proc. Nat. Acad. Sci.* 71, 2956.
2. Inglot, A.D., Albin, M. and Chudzio, T. (1973). *J. Gen. Virol.* 20, 105.
3. Kawai, A., Matsumoto, S. and Tanabe, K. (1975). *Virology* 67, 520.
4. Kawai, A. and Matsumoto, S. (1976). *Virology* 76, 60.
5. Palma, E.I. and Huang, A. (1974). *J. Inf. Dis.* 126, 402.
6. Simpson, R.W. and Iinuma, M. (1975). *Proc. Nat. Acad. Sci.* 72, 3230.
7. Youngner, J.S., Dubovi, E.J., Quagliana, D.O., Kelly, M. and Preble, O.T. (1976). *J. Virol.* 19, 90.

STUDY OF VIRULENT AND AVIRULENT RABIES VIRUSES AND THEIR DEFECTIVE RNA-CONTAINING PARTICLES

WILLIAM H. WUNNER and H. FRED CLARK

The Wistar Institute of Anatomy and Biology, 36th Street at Spruce Philadelphia, Pennsylvania 19104, USA.

Virulent rabies viruses kill adult mice following intracerebral (IC) inoculation according to a normal dose response curve. Avirulent or attenuated rabies virus strains fail to kill adult mice at high concentrations following IC inoculation (6). The pathogenic properties associated with a virulent strain such as CVS (challenge virus standard) appear to be lost in the attenuated Flury virus strain that has been adapted to replicate in eggs or in CVS that has been propagated in rabbit endothelial cells (7). The attenuated high egg passage (HEP) Flury strain of rabies virus was observed to interfere with a pathogenic rabies virus infection in WI-38 cells in these early experiments, but no attempts were made then to isolate and identify non-pathogenic or defective-interfering (DI) particles which could interfere with production of pathogenic virus particles. More recently Crick and Brown (4) examined the autointerference phenomenon caused by DI particles and showed that under suitable conditions BHK/21 cells infected with the low egg passage (LEP) Flury strain of virus produced a short RNA-containing particle which interferes with the multiplication of standard virus. Following this report, Holland and Villarreal (5) described how rabies standard (B) virions generated large amounts of DI (T) particles during a single passage in newborn mouse brains. Wiktor and co-workers in our laboratory further demonstrated that BHK cells persistently infected with ERA strain rabies virus produced large quantities of RNA-containing DI particles which were characteristically shorter than standard virus (8). We have been examining the biological phenomenon of killing by IC inoculation of fixed rabies virus strains (adapted to growth in laboratory animals, ref. 2) in adult mice with the aim of identifying some biochemical marker(s) associated with pathogenicity in virulent rabies virus infections.

TITRATION OF THREE STRAINS OF FIXED RABIES VIRUS IN MICE

Three rabies virus strains propagated in BHK/21 cells and injected intracerebrally into adult mice produce distinct dose response curves as shown in Fig. 1. The virulent CVS strain kills 100% of mice inoculated with undiluted virus (10^7 pfu/ml; dose = 0.03 ml) and serial ten-fold virus dilutions up to 10^{-5}. Mice inoculated with a standard laboratory stock of ERA virus (3) were protected at the undiluted dose

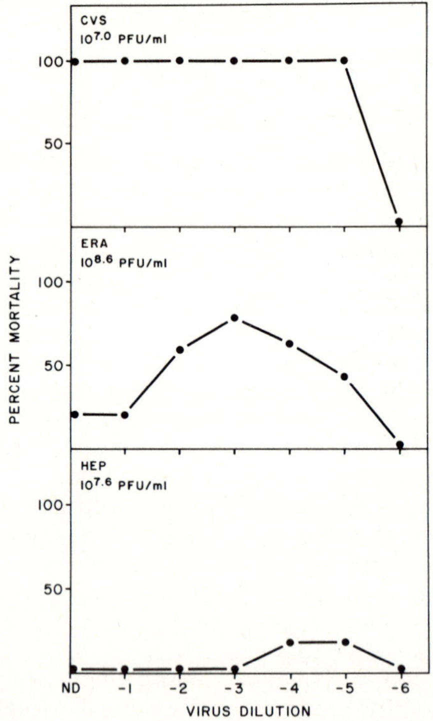

Fig. 1. *Dose response curves of cell culture propagated fixed rabies virus in adult mice. Virus stocks of CVS, ERA, and HEP Flury strains of rabies virus were prepared in BHK/21 cell monolayers infected at multiplicities of 0.1-1 pfu/cell. CVS and ERA viruses were titrated directly from culture fluid. HEP Flury was purified by twice banding B virions by velocity sedimentation in 5-30% sucrose density gradients. Serial 1:10 dilutions from non-diluted (ND) stocks were inoculated intracerebrally into five mice per dilution.*

of $10^{8.6}$ pfu/ml but frequently died following inoculation of lower concentrations of the virus, giving a dose response curve suggesting a typical autointerference phenomenon. A strong suggestion of autointerference was also observed with our laboratory stock of attenuated HEP Flury strain ($10^{7.6}$ pfu/ml, undiluted dose) which consistently caused a small number (10-40%) of rabies-specific deaths only near virus terminal dilutions. This dose response curve was consistent whether produced by inoculation with virus obtained directly from culture fluid or following a single step purification by sucrose density gradient centrifugation.

ANALYSIS OF VIRUS PARTICLES

Each of the virus stocks that were titrated directly in mice without prior purification was also propagated in BHK/21 cells in the presence of ^{3}H-uridine in order to characterize the typical population size distribution of virus particles and virion RNAs. Virus particles labelled with ^{3}H-uridine were purified as described in the legend of Fig. 2., and particle populations with different characteristic sediment-

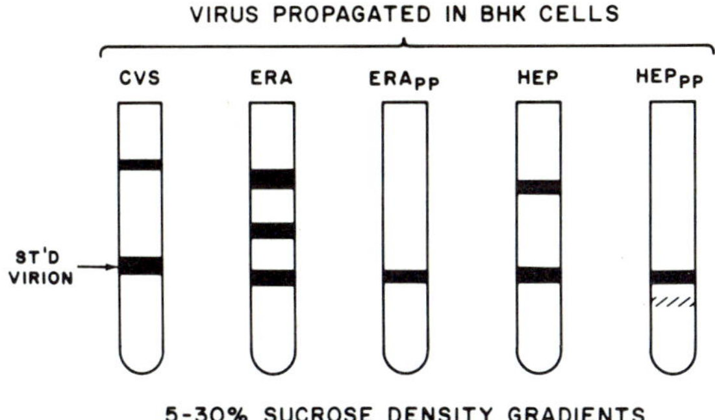

Fig. 2. *Velocity sedimentation gradients of rabies viruses propagated in BHK/21 cells. Monolayer cultures were infected with virus at a multiplicity of 0.1-1 pfu/cell and incubated for 4 to 6 days at 33° until maximum cytopathic effect was observed. Virus was pelleted from culture fluid by high speed centrifugation (35,000 g for 90 min) purified by velocity sedimentation in a 5-30% (w/v) sucrose density gradient centrifuged for 45 min at 54,000 g and 4°. The positions of visible bands in the gradients are indicated in the figure by solid lines; a faint band is indicated by the slashed lines. Rabies CVS, ERA (not recently plaque-purified), ERApp (plaque-purified), HEP (not recently plaque-purified) and HEPpp (plaque-purified) viruses are compared.*

ation profiles were titrated in mice by IC inoculation and analyzed for viral RNA size. RNA was isolated from both concentrated virus pellets analyzed prior to gradient purification by sedimentation velocity centrifugation and from gradient-purified virus preparations. The ^{3}H-RNA was sized by velocity sedimentation in sucrose density gradients and by SDS-polyacrylamide gel electrophoresis (SDS-PAGE) as described in the legend to Fig. 3.

The CVS, ERA and HEP virus stocks used in this study produced standard virions as shown in Fig. 2 which banded in sucrose density gradients in approximately the same position under identical centrifugation conditions. In addition to the major band of standard virions produced in BHK/21 cells, CVS stock virus produced a single slower sedimenting visible band that was positioned higher in the gradient than any visible band obtained from ERA or HEP Flury virus. The infectious titer (pfu) recovered from the CVS standard virion band was between 2 and 4.5 logs greater than that recovered from the slower sedimenting band after a single cycle purification step. The stock ERA virus (not recently plaque-purified) produced a middle (M) band in addition to the top (T) and standard virion or bottom (B) bands. The infectivity of T and M bands was ≤1% of the standard virion band. The T band was shown to be a double band by recycling T particles through a 5-30% sucrose density gradient for an extended centrifugation time. The HEP virus stock (not recently plaque-purified) produced a single slow sedimenting visible band which moved slightly faster than the T band of ERA. Both ERA and HEP viruses failed to produce any visible bands of slower sedimenting particles in the first few passages following renewed plaque-purification (3 to 5 times) of virus derived from standard stocks.

ANALYSIS OF PARTICLE-BOUND RNA

RNA was isolated from standard virions as well as top and middle band particles in these gradients and analyzed by sedimentation velocity in sucrose density gradients containing SDS. Standard virions of CVS, ERA and HEP contained a major ^{3}H-labelled RNA species of 42S with an apparent molecular

---

Fig. 3. *Polyacrylamide gel electrophoresis of rabies virus RNA (ERA strain) isolated from top (T), middle (M), and bottom (B) band particles separated by velocity sedimentation in sucrose density gradients. BHK cell monolayers were infected with the ERA strain of rabies at a multiplicity of 0.5 pfu/ cell and incubated at 33°. ^{3}H-uridine (10 μCi/ml) was added*

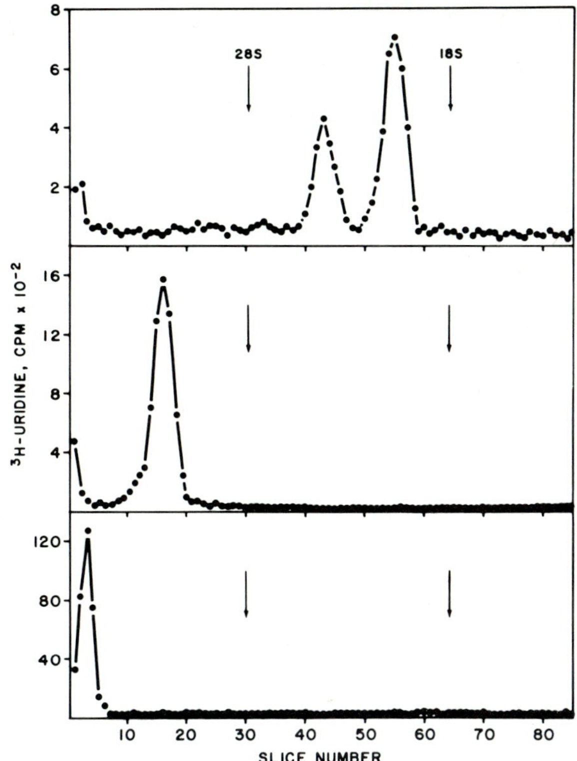

to the incubation medium 48 hr later. The culture fluid and extracellular label were removed at 72 hr and the infected cells were refed with fresh medium and incubated at 33° for a further 72 hr. Virus was then harvested, concentrated and purified as described in Fig. 2. The visible bands were taken from the centrifuge tube by side puncture, digested for 2 hr at 37° with proteinase K (100 µg/ml) in 10 mM Tris/HCl, pH 8, 10 mM NaCl, 10 mM EDTA and 0.5% SDS. The digest containing approximately 150,000 cpm of ^{3}H-RNA was layered onto a 15-30% (w/v) sucrose density gradient containing 1% SDS in 10 mM Tris/HCl, pH 7.2, 0.1 M NaCl and 10 mM EDTA and centrifuged for 16 hr at 21,000 g and 22° in an SW50.1 rotor. The peak fractions of ^{3}H-viral RNA from each gradient were precipitated in two volumes of ethanol prior to electrophoresis in SDS-polyacrylamide gel (1). 28S and 18S ribosomal RNA markers (arrows) were located with a Joyce-Loebl Scan 400. Gel slices 1 mm thick were counted for radioactivity in a liquid scintillation spectrometer. RNA from T, M and B band particles are shown from top to bottom.

weight of 4 x 10^6 as determined by SDS-PAGE. Each of the slower sedimenting virus particle bands contained at least a major ^{3}H-labelled RNA species which was smaller than standard virion RNA suggesting that these could be defective particles of the classical deletion mutant type. The RNA components isolated by sucrose gradient centrifugation from the three major bands (T, M and B) of ERA virus had sedimentation coefficients of approximately 18-20S, 30S and 42S for T, M and B, respectively. The individual RNA peak fractions from the three size classes were precipitated in ethanol and analyzed by SDS-PAGE. By the higher resolving power of SDS-PAGE it was possible to show that the ERA T band particles contained two RNA species. This is consistent with the observation that the T band of ERA is a double band. The major RNA peak from M band particles also migrated as a distinct single molecular species in polyacrylamide gel.

TITRATION OF DEFECTIVE PARTICLES AND PURIFIED VIRIONS IN MOUSE BRAIN

Both T and M band particles titrated in adult mice gave autointerfering-type death curves with approximately 60 to 80% killing at T band particle dilutions of 10^{-4} and 10^{-5} and approximately 40% killing at M band particle dilutions of 10^{-3} and 10^{-4}. Titration of standard virions (B band) showed complete protection at 10^{-2} and 10^{-3} dilutions before a biphasic curve was exhibited with 40% killing at 10^{-4}, no deaths at 10^{-5} and a steady rise to 80% deaths at 10^{-7} before gradually falling to end point beyond 10^{-8} dilution. Recently plaque-purified ERA (ERApp) virus produced only standard virions and ^{3}H-labelled standard RNA in BHK/21 cells. However, this virus continued to produce a rather bizarre autointerference-type death curve following IC inoculation in mice.

Standard virions and top band particles of CVS propagated in BHK/21 cells and separated by density gradient centrifugation (Fig. 2.) were titrated in adult mouse brains. The resulting percent mortality curves for each type of particle depicted a normal dose response similar to that shown for unpurified CVS in Fig. 1. While CVS top band particles contained a major 12-14S RNA species and a minor 20S RNA species but no detectable ^{3}H-labelled standard virion RNA, these particles did not appear to protect mice from virus killing at high concentrations.

Finally, HEP Flury virus that was recently plaque-purified (HEPpp) and grown in BHK/21 cells produced standard virions which banded in a single main band and a faint band sedimenting ahead of the main band. Analysis of the ^{3}H-labelled

RNA from main band particles and faint band particles in sucrose density gradients revealed that each contained both a major peak of 42S RNA and a minor heterogeneous peak of 28-30S RNA. The major RNA peak from the sucrose gradient was clearly standard virion RNA by SDS-PAGE, and the minor peak of 28-30S RNA was resolved into two smaller RNA species which migrated slower than 28S ribosomal RNA marker in SDS-polyacrylamide gel.

SUMMARY AND CONCLUSION

Three strains of fixed rabies virus produce particles other than standard virions that contain deletion-type RNA molecules. The CVS strain is the only one of these three stock viruses which presently kills adult mice when inoculated intracerebrally, showing a maximum response at the higher concentrations of virus inoculum. Both ERA and HEP Flury viruses exhibit an inhibition of killing with high concentrations of virus inoculum and a less efficient inhibition at lower concentration before end point dilution was reached. If the particles bearing substandard viral RNA molecules are truly DI particles capable of interfering with multiplication of virulent standard rabies virions, then it would not be surprising to observe an autointerfering-type mortality curve following IC inoculation of CVS in adult mice. Similarly, recently plaque-purified ERA and HEP viruses that do not produce detectable defective particles would be expected to express enhanced virulence. In neither situation could it be demonstrated that defective or DI particles altered the pathogenicity of rabies standard virions for adult mice. At present there is not a reliable interference assay for the DI particles *in vivo* and so their biological function remains obscure. It does not seem likely, however, that the virulence or avirulence of rabies viruses are mediated primarily by DI particles.

ACKNOWLEDGEMENTS

We thank Dr. Hilary Koprowski for his encouragement and Sally Shane and Nancy F. Parks for their technical assistance. This work was supported by USPHS research grants AI-09706 from the National Institute of Allergy and Infectious Diseases and RR-05540 from the Division of Research Resources.

REFERENCES

1. Bishop, D.H.L., Claybrook, J.R. and Spiegelman, S. (1967). *J. Mol. Biol.* 26, 373.
2. Clark, H.F. and Wiktor, T.J. (1972a). *In* "Strains of Human Viruses" (M. Majer and S.A. Plotkin, eds.) p. 177. Karger, Basel.

3. Clark, H.F. and Wiktor, T.J. (1972b). *J. Infect. Dis.* 125, 637.
4. Crick, J. and Brown, F. (1974). *J. Gen. Virol.* 22, 147.
5. Holland, J.J. and Villarreal, L.P. (1975). *Virology* 67, 438.
6. Koprowski, H. (1954). *Bull. Wld. Hlth. Org.* 10, 709.
7. Wiktor, T.J., Fernandes, M.V. and Koprowski, H. (1964). *J. Immun.* 93, 353.
8. Wiktor, T.J., Dietzschold, B., Leamnson, R.N. and Koprowski, H. (1977). *J. Virol.* 21, 626.

THE PATHOGENICITY OF NEWCASTLE DISEASE VIRUS AS DEPENDING ON THE STRUCTURE OF THE VIRAL GLYCOPROTEINS

HANS-DIETER KLENK, RUDOLF ROTT, YOSHIYUKI NAGAI* and WERNER BERK

Institut für Virologie der Justus-Liebig-Universität Giessen, Frankfurter Str. 107, 63 Giessen/Lahn, Germany

**Present address: Department of Virology, Cancer Research Institute, Nagoya University School of Medicine, Nagoya, Japan*

The envelope of paramyxoviruses contains 2 types of spikes: one type consists of glycoprotein HN which has hemagglutinating and neuraminidase activity (9, 10, 12, 14, 15), and the other one consists of glycoprotein F which induces cell fusion and hemolysis (2, 11, 13). It is generally agreed that these activities reflect the essential roles which both glycoproteins play in the initiation of infection: glycoprotein HN appears to be responsible for adsorption, i.e. the attachment of the virus to neuraminic acid-containing receptors on the host cell surface, whereas glycoprotein F appears to be involved in the penetration process which is assumed to take place by fusion of the viral envelope with the cell membrane.

Envelope maturation is a multistep process involving sequential incorporation of viral proteins into cellular membranes. Cell fractionation studies on NDV-infected cells (8) have shown that the polypeptide chains of the glycoproteins are synthesized on the rough endoplasmic reticulum. From there they migrate via the smooth endoplasmic reticulum and the Golgi apparatus to the plasma membrane. The M protein attaches to areas containing viral glycoproteins. Thus, patches of viral envelope are formed which are able to bind nucleocapsid strands floating in the cytoplasm. The mature virus particle is then released by budding. In the course of migration to the plasma membrane, the polypeptides of the glycoproteins undergo posttranslational modifications. These involve, as can be assumed by analogy with influenza virus (4), sequential glycosylation on the rough endoplasmic reticulum and on smooth internal membranes. Another type of modification is proteolytic cleavage which takes place at smooth internal membranes and at the plasma membrane. Thus, a larger precursor (F_0) to glycoprotein F has been observed with Sendai virus (2, 11) and with NDV (1). In both Sendai (14) and NDV virions (8), evidence has also been obtained for a smaller cleavage fragment, which appears to be bound to F by disulfide linkage.

This report is concerned with studies on NDV. The virus comprises a wide range of strains which differ markedly in pathogenicity for their natural host, the chicken (16). We will report here about studies which show (1) that precursors exist to both glycoproteins of NDV which are converted by proteolysis into the biologically active form, (2) that the NDV strains differ from each other with respect to cleavage of the glycoproteins, and (3) that these differences are an important factor underlying the variations in the pathogenicity of NDV.

STRAIN-DEPENDENT DIFFERENCES IN THE SUSCEPTIBILITY OF THE VIRAL GLYCOPROTEINS TO PROTEOLYTIC CLEAVAGE

A comparative analysis of the glycoproteins of 5 virulent (Italien, Herts, Field Pheasant, Texas, Warwick) and 5 avirulent strains (La Sota, B1, F, Queensland, Ulster) has been carried out. To investigate the formation of virus-specific glycoproteins, infected BHK21-F cells were pulse-labeled with radioactive glucosamine and subsequently subjected to polyacrylamide gel electrophoresis. Fig. 1 shows that 3 different glycoprotein patterns can be discriminated. The first type shows glycoprotein HN (MW 74,000) and glycoprotein F (MW 56,000). This type is demonstrated on the example of strains Italien and Herts and is typical for all virulent strains. The second type shows glycoprotein HN and the precursor glycoprotein F_0 (MW 68,000). This pattern has been found with the avirulent strains La Sota (Fig. 1), B_1, and F (not shown). The third type has been found with the avirulent strains Ulster (Fig. 1) and Queensland (not shown). It shows glycoproteins F_0 and HN_0. HN_0 has a molecular weight (82,000) slightly higher than that of HN and has not been observed before with paramyxoviruses. Essentially the same glycoprotein profiles have been observed when the viruses were grown in MDBK and chick embryo cells.

The precursor-product relationship of these glycoproteins has been clearly assessed by *in vitro* incubation of purified virus particles with proteases. For instance, when virions of strain Ulster derived from MBDK cells have been treated in this way glycoproteins HN_0 and F_0 were replaced by HN and F (7). These results together with an analysis of cyanogen bromide fragments (6) demonstrate that HN_0 and F_0 are precursors which are converted by tryptic cleavage into HN and F, respectively.

The findings described so far show that virulent strains

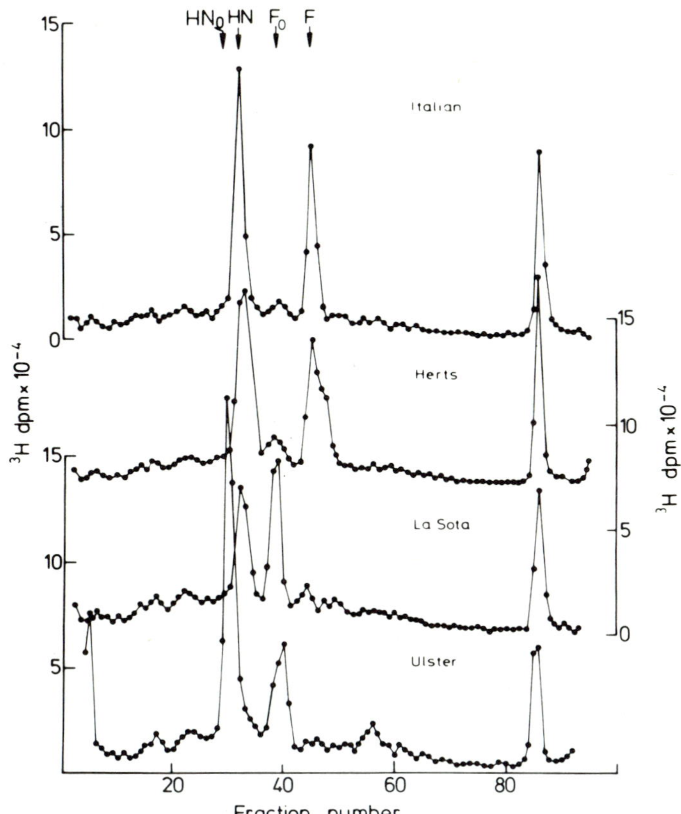

Fig. 1. *The glycoproteins synthesized in BHK21-F cells infected with strain Italien, Herts, La Sota, and Ulster. Cells were labeled at 5 hr p.i. for 3 hr with ^{3}H-glucosamine and subjected to polyacrylamide gel electrophoresis. For details see ref. 7.*

contain both glycoproteins in their cleaved form, whereas the avirulent ones contain one or both glycoproteins in the uncleaved form if they are grown in these cells.

The question arose now whether the differences in cleavage might be explained by structural variations of the glycoproteins resulting in different susceptibilities to proteolytic cleavage or whether they reflect differences in the level of proteolysis in infected cells. To discriminate between these possibilities double infection experiments have been carried out with virulent and avirulent strains. Analysis of the glycoprotein patterns synthesized under these conditions indicate that the relative resistance of the glycoproteins of the avirulent strains to cleavage is a genuine structural property of these molecules rather than the consequence of a low level of proteolysis in infected cells (7).

BIOLOGICAL ACTIVITIES OF VIRAL GLYCOPROTEINS AND THEIR DEPENDENCE ON PROTEOLYTIC CLEAVAGE

It was now of interest to see whether cleavage of the precursors of NDV glycoproteins was necessary for their biological activities as it has been found before with glycoprotein F of Sendai virus (2, 11). We have carried out, therefore, a comparative analysis of the effect of trypsin on virulent and avirulent strains grown in BHK21-F cells (Table 1). The virus was subjected to enzymatic treatment *in vitro* and then analyzed for hemagglutinating, neuraminidase, and hemolytic activity, and for infectivity. The titers of all activities obtained after trypsin treatment, which were similar with both virulent and avirulent strains, were compared to those of untreated samples, and the ratios were taken as a measure for activation. The ratios demonstrate that the virulent strains which contain both glycoproteins in the cleaved form are biologically fully active if grown in BHK21-F cells. In contrast, the avirulent strains La Sota, B1, and F which contain HN in the cleaved and F in the uncleaved form require trypsin for the activation of hemolysis. Strains Ulster and Queensland which have both glycoproteins in the uncleaved form require trypsin for the activation of hemolysis as well as hemagglutinating and neuraminidase activity.

The activation rates of the latter activities are rather low. This has probably to be explained by the fact that the experiment shown in Table 1 has been carried out on virus which usually contained a fair amount of cleaved and, thus, presumably active hemagglutinin-neuraminidase glycoprotein. It appeared, therefore, necessary to isolate HN_0 and to perform the activation on the purified glycoprotein. Under these conditions, hemagglutinating activity increased by a factor of more than 30 and neuraminidase activity by a factor of more than 15 (6). HN_0 is, therefore, biologically inactive.

These data demonstrate that cleavage is necessary for the biological activity of glycoprotein F as well as glycoprotein HN. Activity of both glycoproteins appears to be necessary for infectivity (Table 1): Virions containing F_0 (La Sota, B1, F) have a reduced infectivity. Infectivity is even lower if both glycoproteins are present in the uncleaved form (Ulster, Queensland). If these results are compared to the data on the mean death time of chick embryos after inoculation with the various strains (also shown in Table 1), a striking correlation emerges between infectivity in the non-activated state and virulence: strains Queensland and Ulster exhibit very low infectivity and very low virulence; strains Italien, Herts, Field Pheasant,

TABLE 1

Effect of Trypsin Treatment on the Biological Activities of NDV Grown in BHK21-F Cells

Strain	Virulence	Mean death hr[a]	Glycoprotein composition	Rate of activation by trypsin[b]			
				Hemagglutinin	Neuraminidase	Hemolysis	Infectivity
Italien	virulent	50	HN, F	1.0	1.01	0.62	1.13
Herts	virulent	49	HN, F	1.0	0.96	0.76	0.79
Field Pheasant	virulent	50	HN, F	1.0	1.10	0.87	1.35
Texas	virulent	50	HN, F	1.0	1.01	0.68	1.30
Warwick	virulent	50	HN, F	1.0	0.99	0.79	1.80
La Sota	avirulent	103	HN, F_o	1.0	1.02	5.00	5.55
B_1	avirulent	120	HN, F_o	1.0	0.97	5.34	7.10
F	avirulent	168	HN, F_o	1.0	1.00	5.58	-
Queensland	avirulent	∞	HN_o, F_o	2.0	2.03	5.34	18.10
Ulster	avirulent	∞	HN_o, F_o	2.0	2.22	9.20	144.00

[a]The data for strains Field Pheasant and Warwick are those of Moore and Burke (5), the data for all other strains are those of Waterson, Pennington, and Allan (16).

[b]Biological activities have been determined on cell-associated virus. Incubation with trypsin was carried out at a concentration of 2.5 μg/ml at 37° for 8 min. The activation rates are the ratios of the activities of trypsin-treated samples versus untreated controls. With strain Italien activities have been determined 9 hr p.i., with all other strains 13 hr p.i. For further details see ref. 7.

Texas, and Warwick are fully infectious and highly virulent; and strains La Sota, F and B1 are in between these extremes.

On the basis of these results it had to be expected that a virulent strain producing biologically active virus in a given cell culture should be able to undergo multiple replication cycles in this host. In contrast, if the same host would be infected with an avirulent strain producing inactive virus, infection should not proceed beyond the first cycle of replication. After substitution of trypsin, however, multiple cycle replication should also be possible in the case of the avirulent strains. This is indeed the case as shown elsewhere (7). The virulent strain Italien grew to the same titer in MDBK cells regardless of the multiplicity of infection and regardless as to whether trypsin was present or absent. On the other hand, the avirulent strain Ulster grew under multiple cycle conditions only if trypsin was present in the medium. By the same principle, avirulent strains produce plaques in MDBK and chick embryo cells only if trypsin is added to the overlay medium; whereas virulent strains can do so in the absence of the enzyme. These results demonstrate that avirulent strains in contrast to virulent ones require trypsin to undergo successive growth cycles in MDBK and chick embryo cells.

DIFFERENCES IN HOST RANGE OF VIRULENT AND AVIRULENT STRAINS AS DETERMINED BY THE SUSCEPTIBILITY OF THE ENVELOPE GLYCOPROTEINS TO CLEAVAGE

The inability of the avirulent strains to produce biologically active virus is a host-specific phenomenon. In certain hosts, such as chorio-allantoic membrane cells or the chick embryo, avirulent strains produce active virus containing cleaved glycoproteins as do the virulent strains (5, 7). These host-specific differences are probably due to the presence or absence of appropriate proteases. From these observations the following conclusions can be drawn, which are summarized in Fig. 2: only a few host systems are permissive for avirulent strains, i.e. they produce highly infectious virus which can undergo successive replication cycles in these cells; other systems are non-permissive, i.e. they produce defective virus; in contrast, all host systems are permissive for virulent strains. These observations demonstrate striking differences in host range between virulent and avirulent strains which are determined by the susceptibility of the envelope glycoproteins to proteolytic cleavage. It is a fair assumption that infection with a virulent strain which readily produces virus of full biological activity in a wide spectrum of different cells, spreads more rapidly in the organism than infection with an avirulent strain, which has a narrow host range.

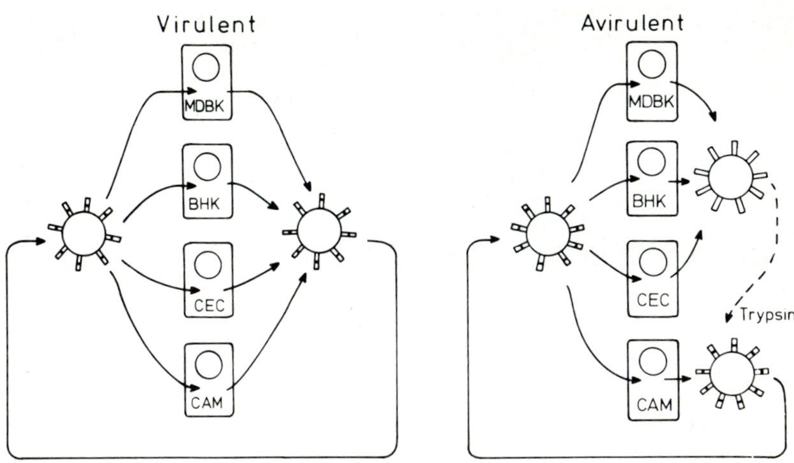

Fig. 2. *Host range of virulent and avirulent strains of NDV as determined by the susceptibility of the envelope glycoproteins to proteolytic cleavage. Avirulent strains produce fully infective progeny virus with cleaved glycoproteins in chorio-allantoic membrane (CAM) cells. In MDBK, BHK21-F, and chick embryo (CE) cells they produce inactive virus with uncleaved glycoproteins, which can be converted, however, into active virus by trypsin treatment. Virulent strains produce fully infectious virus with cleaved glycoproteins in all cells analyzed.*

CHANGE IN PROTEASE SENSITIVITY OF GLYCOPROTEIN F_0 BY MUTATION CAUSES CHANGE IN PATHOGENICITY

It is a reasonable assumption that the virulent and avirulent strains analyzed here arose from each other by spontaneous mutation which, if our concept were correct, should affect the susceptibility of the viral glycoproteins to proteases. It was, therefore, of high interest, if mutants could be obtained *in vitro* in which a change in protease susceptibility is paralleled by a change in pathogenicity.

Such experiments have been performed on the avirulent strain La Sota, and they have been designed to obtain mutants of higher pathogenicity. Wild type virus was exposed to nitrous acid, and we have searched for mutants which, unlike the avirulent wild type but like the virulent strains, were able to produce infectious virus and to undergo multiple replication cycles in MDBK and in CE cells. We have, thus, selected for a change in host range. One mutant has been obtained by this procedure which as a result of this change is able to produce plaques in chick embryo cells (Fig. 3). Furthermore, unlike the wild type and like virulent strains, the mutant induces cell fusion in BHK21-F cells and does not require trypsin treatment for full expression

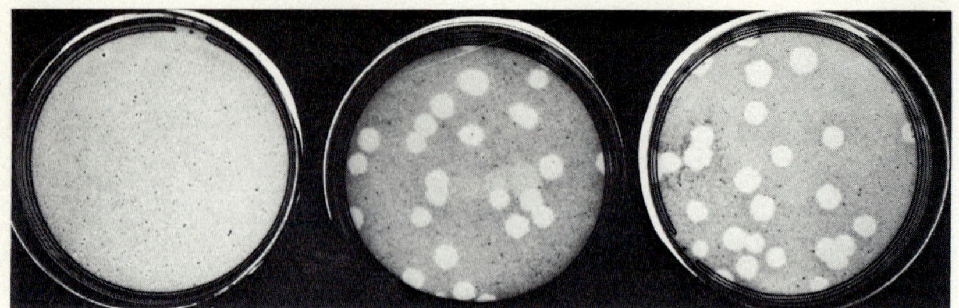

Fig. 3. *Plaque formation in chick embryo cells by La Sota-wild type (left), La Sota-mutant (middle), and Italien (right). Monolayers were inoculated with equal concentrations of egg-grown virus. The plaque assay was carried out by the procedure described previously (3). The overlay medium did not contain trypsin.*

of hemolytic activity and infectivity if grown in MDBK or chick embryo cells (data not shown).

^3H-Glucosamine-labeled virions of the mutant grown in MDBK cells have been isolated and the glycoproteins have been analyzed by polyacrylamide gel electrophoresis. Fig. 4 shows that, in contrast to wild type, the mutant contains glycoprotein F in the cleaved form as do virulent strains. The mutant has also been grown in CE cells, BHK cells, and in the embryonated egg, and was always found to contain the cleaved glycoprotein. Thus, cleavage occurs in all host cells analyzed as has been observed with virulent strains.

TABLE 2

Pathogenicity of La Sota (Mutant) as Compared to La Sota (Wild Type) and Italien

VIRUS STRAIN	Death time of chicks (days)[a]		Death time of chick embryos (hr)[b]	
	median	range	median	range
La Sota (wild type)	∞	-	142	89-184
La Sota (mutant)	9	5-12	99	65-141
Italien	5	4-6	71	65-112

[a]Out of 4 chicks infected with La Sota (wild type) all survived showing no symptoms, out of 9 infected with La Sota (mutant) 1 survived over 14 days showing severe disease, out of 7 infected with Italien none survived. Each animal was inoculated with 10^6 pfu of egg-grown virus.

[b]30 ten-day-old embryonated eggs were inoculated with La Sota (mutant), 10 eggs with La Sota (wild type), and 10 eggs with Italien. The inoculum was 10 pfu of egg-grown virus. All embryos died.

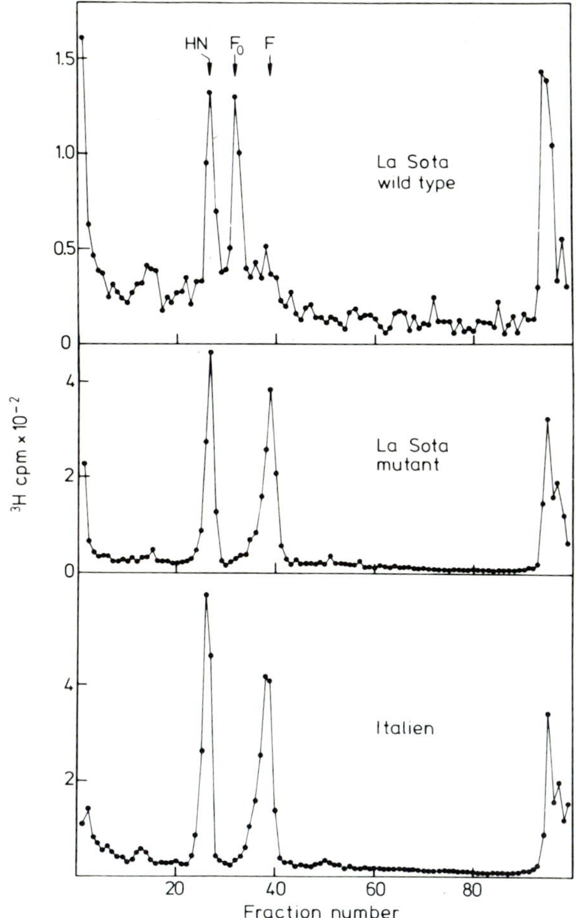

Fig. 4. *The glycoproteins of virions of strains La Sota (wild type), La Sota (mutant), and Italien grown in MDBK cells. Virus was labeled with ^{3}H-glucosamine and analyzed by polyacrylamide gel electrophoresis.*

The pathogenicity of the mutant has been compared to that of the wild type and of the virulent strain Italien by determining the mean death time for chick embryos (16) and for chicks (Table 2). Although the mutant appears to be not quite as virulent as strain Italien, there is a distinct increase in pathogenicity if compared to the wild type.

These findings taken together demonstrate that the susceptibility of glycoprotein F to proteases has been altered by mutation in such a way that cleavage, and thus activation of infectivity, can occur in a large spectrum of different host cells. The increase in host range is in turn paralleled by an increase in pathogenicity. The properties of the mutant provide, therefore, strong support to the concept that, with NDV, a relatively simple molecular mechanism, namely the susceptibility of the viral glycoproteins to proteolytic cleavage, is of high importance for such a complex phenomenon as pathogenicity.

ACKNOWLEDGEMENTS

This work was supported by the Deutsche Forschungsgemeinschaft (SFB 47, Virologie). Y.N. was a recipient of a research fellowship of the Alexander von Humboldt-Foundation.

REFERENCES

1. Hightower, L.E. and Bratt, M.A. (1974). *J. Virol.* 13, 788.
2. Homma, M. and Ohuchi, M. (1973). *J. Virol.* 12, 1457.
3. Klenk, H.-D., Rott, R., Orlich, M. and Blödorn, J. (1975). *Virology* 68, 426.
4. Klenk, H.-D., Schwarz, R.T., Schmidt, M.F.G. and Wöllert, W. (1977). *Topics in Infectious Disease* (in press)
5. Moore, N.P. and Burke, D.C. (1974). *J. Gen. Virol.* 25, 275.
6. Nagai, Y. and Klenk, H.-D. (1977). *Virology* 77, 125.
7. Nagai, Y., Klenk, H.-D. and Rott, R. (1976). *Virology* 72, 494.
8. Nagai, Y., Ogura, H. and Klenk, H.-D. (1976). *Virology* 69, 523.
9. Scheid, A., Caliguiri, L.A., Compans, R.W. and Choppin, P.W. (1972). *Virology* 50, 640.
10. Scheid, A. and Choppin, P.W. (1973). *J. Virol.* 11, 263.
11. Scheid, A. and Choppin, P.W. (1974). *Virology* 57, 475.
12. Seto, J.T., Becht, H. and Rott, R. (1973). *Med. Microbiol. Immunol.* 159, 1.
13. Seto, J.T., Becht, H. and Rott, R. (1974). *Virology* 61, 354.
14. Shimizu, K., Shimizu, Y.K., Kohama, T. and Ishida, N. (1974) *Virology* 62, 90.
15. Tozawa, H., Watanabe, M. and Ishida, N. (1973). *Virology* 55, 242.
16. Waterson, A.P., Pennington, T.H. and Allan, W.H. (1967). *Brit. Med. Bull.* 23, 138.

INHIBITION OF PROTEIN SYNTHESIS BY NEWCASTLE DISEASE VIRUS

C. WORTH CLINKSCALES* and MICHAEL A. BRATT

Department of Microbiology,
University of Massachusetts Medical School,
Worcester, Mass. 01605, USA.

* *Present address:*
Department of Microbiology and Immunology,
Oral Roberts University School of Medicine & Dentistry,
Tulsa, Oklahoma 74171, USA.

We are interested in the alterations of protein metabolism caused by paramyxovirus infection. Hightower and Bratt (5, 6) have shown that, following Newcastle disease virus (NDV) infection of secondary cultures of chicken embryo cells, total protein synthesis gradually decreases. By 6 hr post-infection it reaches approximately 50% of that in uninfected cells; it then remains constant before declining again as the cells die. The inhibition could not be explained by the gradual decline in uptake of precursor amino acids, nor was increased protein degradation observed. It is apparently due in large part to a decreased rate of protein synthesis, *per se*. Concomitant with the inhibition of protein synthesis, a gradual transition from host-specified to virus-specified polypeptide synthesis occurs.

Studies on Sendai virus infection of chicken embryo cells (8, 23) have provided no evidence for inhibition of protein synthesis (possible changes in amino acid uptake or protein degradation have not been reported). Instead, total protein accumulation increases and virus-specific protein accumulation is gradually superimposed on that of the host.

Our interest is in the molecular events associated with both the inhibition of total protein synthesis and the preferential synthesis of NDV-specific proteins. In these studies we have concentrated on infection of Chinese hamster ovary (CHO) cells. These cells have the advantage for cell fractionation studies of growing in suspension. Moreover, while virus production in them is similar (kinetics and amounts) to that in chicken embryo cells (unpublished data), both the inhibition of total cell protein accumulation and the conversion to virus-specific polypeptide synthesis are more rapid and extensive. In this case, also, the inhibition of protein accumulation cannot be attributed to decreased amino acid uptake. Preliminary pulse-chase studies also suggest that protein degradation is not increased during this period. Thus, in this system the inhibition of protein accumulation appears to result from a decreased rate of protein synthesis.

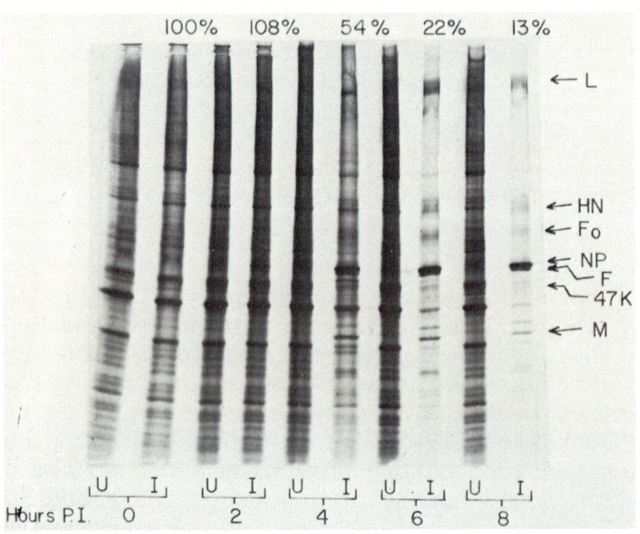

Fig.1. *Electrophoretic analysis of ^{35}S-methionine labeled polypeptides at various times after infection*

Media and cells were as described by Clinkscales et al. (4). At various times after infection at an moi of 5, and incubation at 38°, aliquots of 5 x 10^6 CHO cells were removed from suspension by centrifugation and resuspended in 1 ml balanced salts solution containing 5 μCi/ml of ^{35}S-methionine. After 30 min the samples were placed in balanced salt solution at 0°, centrifuged, washed, and solubilized in 0.2 ml of gel sample buffer. TCA-precipitable radioactivity was determined from duplicate samples and is expressed at the top of the figure as % of incorporation in uninfected cells. Aliquots were subjected to electrophoresis at a constant current of 30 mAmp through 10% SDS-polyacrylamide gels as described (4).

Fig.1 shows an autoradiogram of polypeptides from CHO cells labeled for 30 min with ^{35}S-methionine at various times after infection. The decreasing intensity of labeling in infected cells is an indication of the decline in rate of protein synthesis, as are the percentages indicating TCA-precipitable incorporation as a percent of incorporation in uninfected cultures (shown at the top for each time period); by 6 hr, there is approximately an 80% inhibition. Difference analyses (5) of densitometer tracings of these autoradiograms reveal that approximately 65% and 85% of the polypeptides synthesized at 6 and 8 hr, respectively, are virus-specified.

Fig.2 shows velocity sedimentation patterns of cytoplasmic extracts prepared from cells infected for various periods of time. A gradual shift of A_{260} absorbing material from the polyribosome region to a peak at 80S is observed. Whereas 90% of the A_{260} absorbing material is found in the polyribosome region before infection, by 6 hr post-infection 79% is found in the 80S region. The average size of the polyribosomes also diminishes.[1]

The distribution of NDV-specific RNA in these gradients at 6 hr post-infection is shown in fig.3; 60% of the rapidly sedimenting RNA is found in the 80S region (55S-85S). Elsewhere (2, 3), we have shown that the NDV-specific RNA which is released by EDTA from pooled fractions containing polyribosomes and the 80S region consists of the single stranded 35S mRNA and the 18S mRNAs (4, 12, 16). In the present study we could detect no differences in the distribution of NDV-specific RNAs from the polyribosomes and the 80S region by velocity sedimentation (data not shown), nor by electrophoresis in urea-containing polyacrylamide gels (procedure of Duncan McGeoch, personal communication). Moreover for both the RNA from the polyribosomes and the 80S region the latter procedure revealed 5 bands similar to those observed in formamide-containing polyacrylamide gels of the 18S mRNAs of NDV (22). Other possible physical differences such as the presence or absence of poly(A) (21, 22), or the presence or absence of capped and/or methylated 5' ends (15) on these RNAs have not yet been explored.

We next wanted to determine the extent to which protein synthesis is associated with both the polyribosomes and the 80S region. Our approach was to determine the distribution of nascent polypeptide chains by subjecting extracts of cells

---

[1]These effects are probably not the result of ribosomal run-off since cycloheximide was present throughout the preparation of the extracts. Nor does it appear to be due to enhanced polyribosomal degradation during preparation of the infected cell extracts, since a cytoplasmic extract of a mixture of infected and uninfected cells yields a composite of the two separate patterns (unpublished data.)

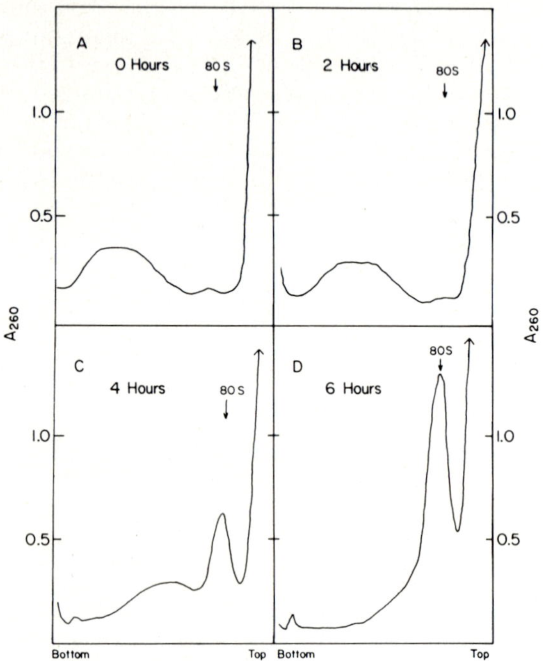

Fig.2. *Velocity sedimentation of cytoplasmic extracts at various times after infection*

Cycloheximide (100 μg/ml) was added to 2 x 10^7 cells at various times after infection. The cells were then diluted in 0° buffer (0.01M Tris, pH 8.4, 0.01 M NaCl, and 0.0015M $MgCl_2$) containing the same concentration of cycloheximide. After two centrifugation steps and resuspension in the same buffer, the cells were lysed by the addition of 1% (v/v) Triton X-100 and 0.5% (w/v) sodium deoxycholate. After removal of the nuclei by centrifugation the extracts were layered onto 10-40% sucrose gradients in the same buffer (lacking cycloheximide) and centrifuged for 1 hr in a Beckman SW41 rotor at 200,000 xg (40K) at 4°. Gradients were collected from the bottom, absorbance monitored at 260 nm, and fractions collected. The distribution at time zero was identical to that of uninfected cells throughout the experiment.

labeled with ^{35}S-methionine for 1 and 5 min (illustrated in figs.4A and 4B, respectively) to the velocity sedimentation procedures employed in figs.2 and 3. The distribution of

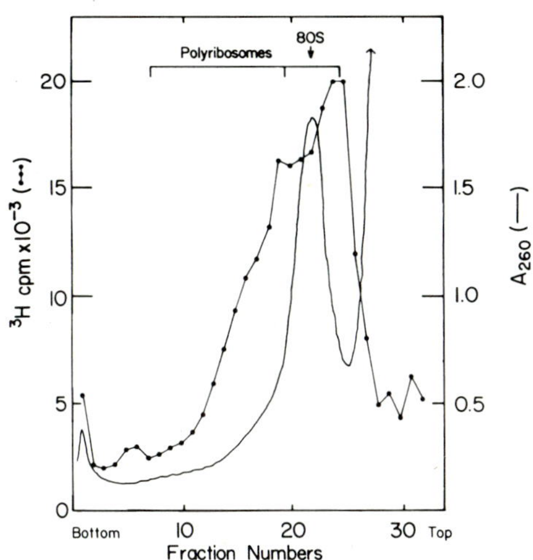

Fig.3. *Velocity sedimentation of cytoplasmic virus-specific RNA at 6 hours post infection*

2 x 10^7 cells infected at an moi of 5 were treated with 10 μg/ml Actinomycin D at 4.5 hr post infection, labeled with 25 μc/ml of ^3H-uridine at 5 hr post-infection and then at 6 hr extracted and fractionated exactly as described in fig.2. Samples of each fraction were TCA-precipitated prior to scintillation counting. Uninfected cells treated similarly showed no radioactivity above background.

nascent chain labeling for infected cells was similar to that for uninfected cells; surprisingly, even though 80% of the A_{260} material and 60% of the NDV-specific mRNA is found to sediment at 80S or less, only 22% of the nascent chains are found there. The remainder of the nascent chains are found in the polyribosomes, which by this criterion, also appear to be somewhat smaller than in the uninfected cell.[2]

---

[2]The similar levels of incorporation in the infected and uninfected cultures during these very short pulses would seem to be at odds with the greatly reduced levels of protein synthesis in infected cultures shown in fig.1. This discrepancy is also seen in similar studies on VSV infection (7). As yet we have no explanation for this finding. However, it seems reasonable to suppose that it may be related to the previously described difference in the time course of accumulation of viral and host

Finally, we have measured the ability of the RNAs in various fractions of these gradients to program cell-free protein synthesis in the rabbit reticulocyte system (13) and the wheat germ system (as modified (4)) from the procedures of Roberts and Paterson (14). RNA from the pooled regions of the gradient (as indicated in fig.3) was added to these cell-free extracts after deproteinization. Table I shows that uninfected cell RNA from both the polyribosomal and 80S regions promotes ^{35}S-methionine incorporation almost equally well in either the reticulocyte or wheat germ systems. In fact, the stimulatory effect was approximately proportional to the amount of RNA contained in each of several smaller pooled fractions throughout the gradient (data not shown). In contrast, the infected cell RNA in the 80S region was 2.78 and 38.1 times less effective in promoting incorporation in the reticulocyte and wheat germ systems, respectively. Other experiments in which the cell-free products were analyzed by SDS-polyacrylamide gel electrophoresis revealed that wherever incorporation was promoted by infected cell RNA, the products were mainly the NDV-specific polypeptides previously detected in the cell-free translation of unfractionated infected cell RNA (4). Whether the relative proportions of the individual polypeptides were identical in all cases has not yet been determined. The finding that infected cell RNA in the 80S region is less effective in promoting cell-free translation than is RNA from the polyribosomal region correlates well with the low protein synthetic activity associated with the 80S region RNA in the infected cell (fig.4).

DISCUSSION AND CONCLUSIONS

The inhibition of protein synthesis during viral infection might result from alterations to the cell's protein synthetic machinery -- e.g. effects on ribosomes, initiation factors etc. These alterations could be mediated directly by virus-specific products -- either specific proteins or the viral RNAs, themselves. Table II summarizes data which engender an

---

(contd. from previous page)

polypeptides during short pulses and/or the very unstable 50-60,000 dalton fraction found in both infected and uninfected cells including CHO cells (6) or possibly to differences in the rate of equilibration of amino acids in the soluble pools in infected and uninfected cells. In fact, the difference between incorporation in infected and uninfected cultures can be seen here to increase with an increase in pulse length from 1 to 5 min.

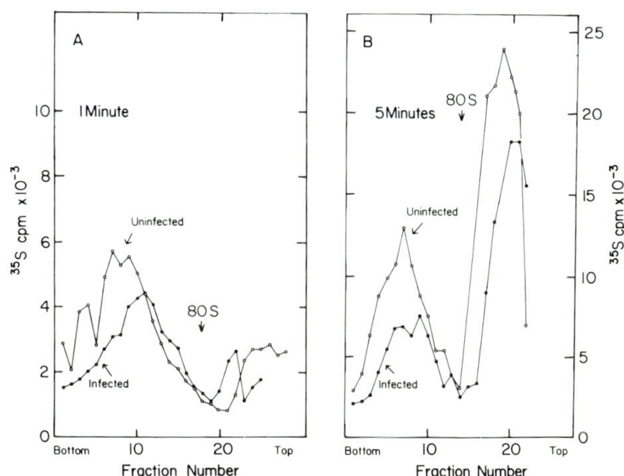

Fig.4. *Nascent chain labelled in infected and uninfected cells* *Uninfected cultures and cultures infected as in the previous figures, but in medium lacking L-methionine and containing dialysed fetal calf serum, were labeled for 1 or 5 min at 6 hr post-infection with 25 μCi/ml of ^{35}S-methionine. Cultures were then treated with cycloheximide and prepared and fractionated in the same way. Fractions were digested with 0.1 N NaOH, TCA-precipitated, and radioactivity determined.*

hypothesis invoking a sequestering of ribosomes as the cause of the inhibition of protein synthesis in NDV-infected CHO cells. At 6 hr post-infection total protein synthesis has decreased to 20% of the level in uninfected cells. This decrease is accompanied by a quantitative transfer of 80% of the A_{260} material from the polyribosomes to 80S structures. This material is presumably ribosomes because in the presence of EDTA it dissociates into 60S and 40S structures which contain 28S and 18S RNA, respectively (data not shown). At 6 hr post-infection 60% of the virus-specific mRNA also sediments in the 80S region. In contrast, the vast majority of protein synthetic activity (78% of the nascent chain labelling) is associated with the polyribosomes. Finally, the RNA in the 80S region is much less translatable -- particularly in the wheat germ system.[3]

Although we have not yet proven that the poorly functional mRNAs are physically associated with ribosomes it seems

TABLE I

Cell-free Protein Synthesis Directed by 80S and Polyribosomal RNAs

A. Reticulocyte Cell-Free System

	A	B	Ratio A/B
	RNA from Polyribosome region	RNA from 80S region	
Uninfected	16.1*	11.3*	1.42
Infected	16.4	5.9	2.78

B. Wheat Germ Cell-Free System

	A	B	Ratio A/B
Uninfected	19.1	18.1	1.06
Infected	100.3	2.7	38.1

*CPM x 10^{-3}

Each sample represents incorporation of ^{35}S-methionine into a TCA precipitable fraction. Reactions were done in duplicate and averaged. The background was subtracted and the radioactivity expressed as cpm per µg of RNA as determined by A_{260}.

(from previous page)

[3]Lodish and Rose (9) have shown that the removal of the 5' terminal 7-methylguanosine from vesicular stomatitis virus (VSV) mRNAs causes approximately 4-fold and 20-fold reductions in the ability of this mRNA to stimulate incorporation in cell-free extracts derived from reticulocytes and wheat germ, respectively. Our finding that RNAs from the polyribosomes and 80S region are also differentially translated in these same two systems could possibly be attributable to the same phenomenon.

TABLE II

Summary

	Uninfected		Infected (6 hours)	
Protein synthesis (% of uninfected)	100		20	
	Polyribosomes	80S	Polyribosomes	80S
A_{260} (% of rapidly sedimenting)	90	10	21	79
Virus-specific RNA (% of rapidly sedimenting)	--	--	40	60
Nascent chain labelling (% of rapidly sedimenting)	80	20	78	22
Efficiency of translation in reticulocyte system (% of polyribosomes)	100	70	100	36
Efficiency of translation in wheat germ system (% of polyribosomes)	100	95	100	2.7

reasonable to propose that the observed 80S structure may be an mRNA-monoribosome complex -- perhaps an aberrant initiation complex. The accumulation of ribosomes in such a complex could then account for the inhibition of total protein synthesis. Many questions remain to be answered in order for this hypothesis to be proven. Is there really a physical association of ribosomes and non-functional mRNAs or is their cosedimentation fortuitous? Is there enough of the non-functional mRNA -- at least one molecule per ribosome -- to account for this phenomenon? Is the non functional mRNA limited to the viral species we have detected or are non-functional host mRNAs also present? Is the poor translational ability of the mRNAs the cause of the accumulation of the accumulation of the 80S structures? Alternatively the inability of these mRNAs to be translated might be the result of some type of inactivation (by alteration of its 5' end, for instance) which could occur, for some other reason, when aberrant initiation complexes have formed. Finally, it will be necessary to eliminate the possibility that the poor translation of the RNA in the 80S region is due to the presence of an inhibitor.

An accumulation of ribosomes in the 80S region has also been seen in other systems including reovirus-infected cells (20) and VSV-infection (7, 17). As in the present study little protein synthetic activity was associated with the ribosomes. Whether viral mRNAs are also present was not determined.

Although the direct involvement of NDV mRNA in the inhibition of protein synthesis is as yet only hypothetical it is supported by several lines of evidence. For instance, chicken embryo cells infected with non-cytopathic mutants of the same virulent strain of NDV used in the present study or with naturally occurring avirulent strains of NDV (1, 10 chapter 72) accumulate less virus-specific RNA and less mRNA, in particular, than virulent virus. In addition, they also inhibit protein synthesis to a lesser extent. We have also found that all of the RNA temperature-sensitive mutants of NDV (18, 19) fail to inhibit protein synthesis at the non-permissive temperature. In contrast all of the RNA^+ mutants we have tested show an inhibition. However, while these findings are consistent with a mechanism of inhibition invoking a direct involvement of mRNA they do not rule out the involvement of a specific gene product; in each case the amount of this product could be reduced because of low levels of its mRNA.

Whether or not it is correct, the model we have suggested for the inhibition of protein synthesis should serve as a stimulus for further experimentation in this area. It will also be interesting to determine whether the mechanism of conversion of host to viral protein synthesis is related to this inhibition.

ACKNOWLEDGEMENTS

Supported by grants from the National Science Foundation (PCM76-84135) and the National Institute of Allergy and Infectious Disease (AI-12467).

C.W.C. was supported by a Public Health Service postdoctoral fellowship from the National Institute of Allergy and Infectious Disease (AI-00232).

We thank Helen Kotilainen, Pamela Chaitis, and Susan Stambler for excellent technical assistance and Kathryn Foley for help in preparation of this manuscript. We are also grateful to Charles H. Madansky and Trudy G. Morrison for numerous helpful discussions.

REFERENCES

1. Bratt, M.A. (1969). *Virology* 38, 485.
2. Clinkscales, C.W. and Bratt, M.A. (1975). *ASM Abstracts* p.251.
3. Clinkscales, C.W. and Bratt, M.A. (1977). *ASM Abstracts* p.283.
4. Clinkscales, C.W., Bratt, M.A. and Morrison, T.G. (1977). *J. Virol.* 22, 97.
5. Hightower, L.E. and Bratt, M.A. (1974). *J. Virol.* 13, 788.
6. Hightower, L.E. and Bratt, M.A. (1975). *J. Virol.* 15, 696.
7. Huang, A.S., Baltimore, D. and Stampfer, M. (1970). *Virology* 42, 946.
8. Lamb, R.A., Mahy, B.W.J. and Choppin, P.W. (1976). *Virology* 69, 116.
9. Lodish, H.F. and Rose, J.K. (1977). *J. Biol. Chem.* 252, 1181.
10. Lomniczi, C.H., Meager, A. and Burke, D.C. (1971). *J. Gen. Virol.* 13, 111.
11. Madansky, C.H. and Bratt, M.A. (1977). *Abst. Amer. Soc. Microbiol.* p.335.
12. Morrison, T.G., Weiss, S., Hightower, L., Spanier-Collins, B. and Bratt, M.A. (1975). *In* "*In vitro* transcription and translation of viral genomes" (A.L. Haenni and G. Beard, eds.), p.281. INSERM, Paris.
13. Pelham, H.R.B. and Jackson, R.J. (1976). *Eur. J. Biochem.* 67, 247.
14. Roberts, B.E. and Paterson, B.M. (1973). *Proc. Nat. Acad. Sci. U.S.A.* 70, 2330.
15. Rose, J.K. (1975). *J. Biol. Chem.* 250, 8098.
16. Spanier, B.B. and Bratt, M.A. (1977). *J. Gen. Virol.* 35, 439.
17. Stanners, C.P., Francoeur, A.M. and Lam, T. (1977). *Cell* 11, 273.
18. Tsipis, J.E. and Bratt, M.A. (1975). *In* "Negative Strand Viruses", vol.2 (B.W.J. Mahy and R.D. Barry, eds.), p.777. Academic Press, London.
19. Tsipis, J.E. and Bratt, M.A. (1976). *J. Virol.* 18, 848.
20. Warrington, R.C. and Wratten, N. (1977). *Virology* 81, 408.
21. Weiss, S.R. and Bratt, M.A. (1974). *J. Virol.* 13, 1220.
22. Weiss, S.R. and Bratt, M.A. (1976). *J. Virol.* 18, 316.
23. Zaides, V.M., Selimova, L.M., Zhirnov, O.P. and Bukrinskaya, A.G. (1975). *J. Gen. Virol.* 27, 319.

VARIATIONS IN PROTEINS SYNTHESIZED DURING THE ESTABLISHMENT OF PERSISTENT INFECTIONS WITH MEASLES VIRUS

B.K. RIMA, E.A. GOULD and S.J. MARTIN

Departments of Biochemistry and Microbiology,
Queen's University of Belfast
Belfast, N. Ireland.

Undiluted passage of measles virus allows the accumulation of defective interfering (DI) particles (6) which appear to be involved in the establishment of persistent infections of Vero cells by both measles virus (12) and canine distemper virus (15). In order to study the molecular events that occur during the transition from lytic to persistent infections we have investigated the effect of DI particles on polypeptide synthesis in measles virus-infected Vero cells. Also ten strains of measles virus, SSPE virus and canine distemper virus (CDV) have been compared to ascertain if there is a link between the polypeptide patterns and the biological characteristics of these viruses. These investigations have required us to reassess the nature and origin of the polypeptides associated with measles virus and CDV.

POLYPEPTIDES OF MEASLES VIRUS AND CANINE DISTEMPER VIRUS

Previous publications have reported 6 to 8 polypeptides in purified preparations of the morbilli virus group with molecular weights of 150-180,000; 69-80,000; 65-70,000; 58-61,000; 51-54,000; 42-46,000; 36-39,000; and 15-20,000 (1, 5, 10, 16). These polypeptides have been detected by Coomassie blue staining and by radioactive labelling of proteins.

During the present series of experiments we have detected a more complex pattern of proteins than previously reported and this was present in both virus released from cells and in virus released by sonication or by freezing and thawing of infected cultures. Furthermore, numerous attempts to purify the virus by repeated tartrate, sucrose or metrizamide gradient centrifugation failed to reduce the number of bands, some of which were glycoproteins (75,000-110,000 daltons) as they were stained with Schiff's base. As some of these polypeptides could be host contaminants or intermediates in the post-translational modifications that paramyxovirus glycoproteins have been reported to undergo (11), we have tried to determine which of the polypeptides present in purified virus are host or virus-induced proteins.

Monolayers of Vero cells were infected with the Edmonston B strain (Edm-1) of measles virus and labelled at various times post infection with ^{35}S-methionine for 30 min periods. The patterns of polypeptide synthesis were analysed by electrophoresis on polyacrylamide slab gels. Fig.1 shows that after development of c.p.e. at 17 hr p.i. a number of newly synthesized bands were detected in measles virus infected cells. Furthermore, the bands of host proteins decreased in intensity late after infection. Only five new bands were seen with mol. wts. of 69,000 (P1); 65,000 (P2); 58,000 (P3); 37,000 (P4) and 15,000 (P5). P3 had been identified as the nucleocapsid protein (1, 5, 10). Labelling late in infection with tritiated fucose, glucosamine, galactose or mannose showed that P1 was the only major glycopolypeptide of measles virus and of CDV, as had been shown previously for CDV (1). Mountcastle and Choppin (10) suggested that P2 of measles virus is associated with the nucleocapsid similarly to the P protein of Sendai Virus. We have not been able to ascribe a function to P4 and P5, although P4 may be the matrix protein of measles virus by analogy with the paramyxoviruses. In an earlier report (5) it was considered that the 42-46,000 mol. wt. polypeptide was the measles matrix protein, however,our present results show that this is a host component, perhaps actin. A protein of this size was also found in CDV (1), rinderpest virus (16) and measles virus (10).

In most preparations of purified virus a protein at 40,000 daltons is also present as recently described by Tyrrell and Norrby, (see chapter 14). However, no protein of this size was detected in uninfected or infected cells during pulse-labelling experiments (cf. fig.1). It is possible that the 40,000 dalton polypeptide is a cleavage product of a larger viral precursor protein comigrating with one of the other viral proteins.

Our results from pulse-labelling experiments are consistent with those of Follett *et al.*, (2) who found only four newly synthesized proteins in measles virus infected BS-C-1 cells, although they did not indicate the mol. wt. of the induced polypeptides nor detect the 15,000 dalton component.

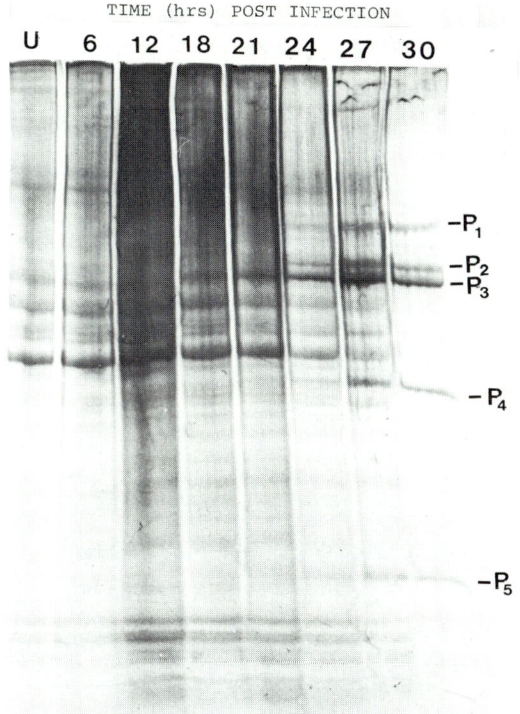

Fig.1. *Synthesis of polypeptides in measles infected Vero cells. Vero cell monolayers (45 mm petri dishes) were infected with measles virus (Edm-1) at a m.o.i. of 5 p.f.u./cell and pulse-labelled at various times post infection with 10 µCi (4 µCi/ml)* ^{35}S*-methionine for 30 min periods. The monolayers were then solubilized in 250 µl of 1% SDS, 1% mercaptoethanol, 1 M urea and boiled for 2 min before storage at -20°. Polyacrylamide slab gel electrophoresis was carried out on 8 to 15% gradient slabs at constant current. The buffers used were as described at Hall and Martin (5). Slabs were dried and autoradiographed on Kodak Royal X-Omat RP54 film.*

COMPARISON OF VARIOUS STRAINS OF MEASLES, SSPE AND CANINE DISTEMPER VIRUS

The polypeptide patterns of ten strains of CDV, measles and SSPE viruses grown in Vero cells labelled with ^{35}S-methionine have been compared. Virus was purified as described by Hall and Martin (5) after sonication of the infected monolayer. No differences were detected (fig.2) between an Edmonston strain with salt-dependent haemagglutinin (P9) and the Edm-1 strain

(14). The Onderstepoort strain of CDV and a large plaque variant of the Lec strain of SSPE (4) appear to differ from the other strains in the mobility of the P2 protein.

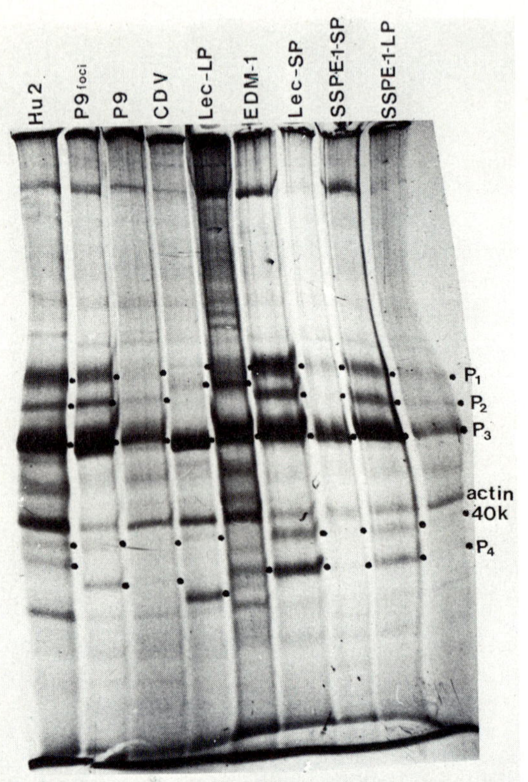

Fig.2. *Comparison of measles and canine distemper virus strains. Vero cell monolayers (75 cm) were infected at m.o.i. of 0.1-1.0 p.f.u./cell and labelled with 50 μCi ^{35}S-methionine (7.0 μCi/ml) after large fusing syncytia had developed. After 6 hr labelling the cell sheet was frozen and thawed, the supernatant and cell debris were sonicated and virus was purified on discontinuous sucrose and continuous tartrate gradients as described by Hall and Martin (5). Virus was pelleted at 200,000g for 1 hr and solubilized in 250 μl of 1% SDS, 1% β-mercaptoethanol and 1 M urea, by boiling for 2 min. Electrophoresis was carried out on 7.5% polyacrylamide gel slabs at constant current. Slabs were dried and autoradiographed on Kodak Royal X-Omat RP54 film.*

The Hu-2 strain of measles virus, described by Shirodaria *et.al.*, (14) as possibly derived from the Schwarz vaccine strain, and CDV, differ from the Edm-1 strain in the mobility of the P4 (M-protein). No differences were detected between the Edm-1, Hu-1 (data not shown), the small plaque variant of the Lec-SSPE strain, (4), a variant of P9 that gives rise to dense foci instead of large syncytia and the small and large plaque variants of SSPE-1. In contrast, Schluederberg *et al.*, (13) reported a difference in the mobility of the M-protein between the Edmonston strain and the SSPE-1 strain (7).

EFFECT OF DI PARTICLES ON MEASLES VIRUS POLYPEPTIDE SYNTHESIS

Vero cells infected with measles virus can produce 100-200 p.f.u./cell. However, stocks are easily contaminated with DI particles which can reduce the yield of infectious virus to less than 0.05 p.f.u./cell. These non-productive infections often lead to the survival of a substantial percentage of the cells and the establishment of a persistently infected culture (12). Even in non-productive infections, a considerable number of virus particles were released from the cells by sonication and these contained a higher proportion of glycoproteins with mol. wt. ranging from 75-110,000. There was also a decrease in the amount of viral protein (P4) and an increase in the actin-like component (fig.3A). Gels stained with Coomassie Blue (fig.3B) also showed that the nucleocapsid band (P3) was less abundant in these particles, perhaps caused by degradation.

These results indicate that during a non-productive infection there is an increase in the amount of host proteins associated with the released particles.

CONCLUDING REMARKS

Five virus-induced polypeptides were detected in Vero cells infected with measles virus. Only one of these was a glycopolypeptide (P1). The variation in the mobility of M protein of CDV and measles strains was a reproducible characteristic of these viruses. The exact nature of the P5 polypeptide remains to be established.

Our results indicate that measles virus contains a considerable number of specific host proteins, especially high mol. wt. glycoproteins (75-110,000), a glycoprotein with mol. wt. 51-54,000 and an actin-like component with mol. wt. 42-46,000. These proteins appear to be intimately associated with morbilli viruses. Recently, Mountcastle and Choppin (10) have described the maturation of measles virus as defective and our results would be consistent with their conclusion. Whether the host components are integrated into the virus envelope is not yet clear, but their association in the virus

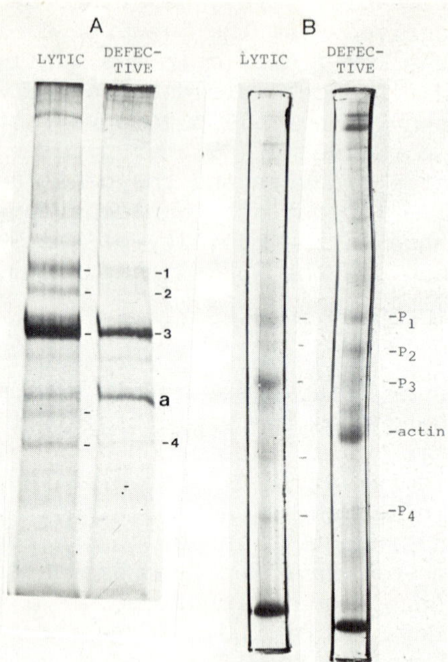

Fig.3. *Comparison of measles virus produced in non-productive (defective) and productive (lytic) infections. (a) Autoradiogram of a 7½% polyacrylamide slab gel of* ^{35}*S-labelled purified measles virus from a lytic and a defective infection labelled as in Fig.2. (b) Coomassie blue stained tubular 7½% polyacrylamide gels of purified measles virus from a lytic and a defective infection in Vero cells.*

---

particles appears to be non-random. A role for host envelope proteins in the formation of infectious measles virus particles may explain the inability of this virus to grow in enucleated cells and the particular sensitivity of measles virus growth to the condition of the host cell (2). When DI particles are produced, we observed a reduced amount of virus-specific envelope components in the released particles and we suggest that this is caused by the inhibition of secondary transcription by subgenomic material, that may be predicted from Kingsbury's model for the control of replication of paramyxoviruses (9). The transfer from a lytic to a persistent infection may occur when the amount of virus specific envelope proteins decreases below the level required to initiate the maturation process. This model is consistent with the general characteristics of persistently infected cells, where there is a decrease in viral surface antigens (3, 8).

ACKNOWLEDGEMENT

We thank the Medical Research Council (UK) for support.

REFERENCES

1. Bussell, R.H., Waters, D.J. and Seals, M.K. (1974). *Med. Microbiol. Immunol.* 160, 105.
2. Follett, E.A.C., Pringle, C.R., Pennington, T.H. and Shirodaria, P.V. (1976). *J. Gen. Virol.* 32, 163.
3. Fraser, K.B. and Martin, S.J. (1978). *In* "Measles Virus and its Biology", p.115. Academic Press, New York and London.
4. Gould, E.A., Cosby, S.L. and Shirodaria, P.V. (1976). *J. Gen. Virol.* 33, 139.
5. Hall, W.W. and Martin, S.J. (1973). *J. Gen. Virol.* 19, 175.
6. Hall, W.W., Martin, S.J. and Gould, E.A. (1974). *Med. Microbiol. Immunol.* 160, 155.
7. Horta-Barbosa, L., Fuccillo, D.A., Sever, J.L. and Zeman, W. (1969). *Nature* London 221, 974.
8. Joseph, B.S., Perrin, L.H. and Oldstone, M.B.S. (1976). *J. Gen. Virol.* 33, 107.
9. Kingsbury, D.W. (1974). *Med. Microbiol. Immunol.* 160, 73.
10. Mountcastle, W.E. and Choppin, P.W. (1977). *Virology* 78, 463.
11. Nagai, Y. and Klenk, H-D. (1977). *Virology* 77, 125.
12. Rima, B.K., Davidson, W.B. and Martin, S.J. (1977). *J. Gen. Virol.* 35, 89.
13. Schluederberg, A., Chavanich, S., Lipman, M.B. and Carter, C. (1974). *Biochem. Biophys. Res. Commun.* 58, 647.
14. Shirodaria, P.V., Dermott, E. and Gould, E.A. (1976). *J. Gen. Virol.* 33, 107.
15. ter Meulen, V. and Martin, S.J. (1976). *J. Gen. Virol.* 32, 431.
16. Underwood, B. and Brown, F. (1974). *Med. Microbiol. Immunol.* 160, 105.

HOST CELL-VIRUS RELATIONSHIPS IN A PERSISTENT INFECTION OF BHK-21 CELLS WITH MUMPS VIRUS

J.V. HALLUM and A.L. TRUANT

Department of Microbiology and Immunology, University of Oregon Health Sciences Center, Portland, Oregon 97201, USA.

Recently we reported the initiation and characterization of a persistent infection of BHK-21 cells by mumps virus (10). This persistence was established by inoculating cultures of BHK-21 cells with mumps virus at a multiplicity of infection of 0.01. The virus used for initiating the persistence (MuVo) was obtained from the Center for Disease Control, Atlanta, as a clinical isolate from a throat washing; it was not further adapted to BHK-21 cells. After inoculation, the cultures were incubated at 37°. Within 7 days, more than 99% of the cells had been destroyed. The few remaining cells repopulated the flasks, giving rise to the persistent infection (BHKpi). That the cultures were persistently infected was indicated by the presence of multinucleate giant cells, hemadsorption with chicken erythrocytes, and the continual release of mumps virus. The properties of the BHKpi cultures and the mumps virus released in the persistence (MuVpi) are summarized in Table I. Cultures of these cells have been carried in our laboratory for over four years with no significant variation in these properties.

In contrast to MuVo, MuVpi could establish a persistent infection directly in BHK-21 cells. When MuVpi was added at a moi of 0.01 to BHK-21 cells, little or no cell destruction was observed and the culture remained stable, showing properties similar to BHKpi cells by 10 days post infection.

The temperature sensitive nature of MuVpi was acquired early in the persistent infection; by passage 2 of the cells, the virus produced was clearly temperature-sensitive.

Because mumps virus has been implicated in such clinical

TABLE I

Summary of Properties of BHKpi and MuVpi

Properties of BHKpi	Properties of MuVpi
1. 85-95% carry mumps virus components	1. More heat-labile at 40° than MuVo
2. No detectable interferon produced	2. Slower replicative cycle than MuVo
3. Interference with NDV, VSV, and MuVo replication	3. Less virulent than MuVo
4. Produce 10^2 - 10^3 FFU/ml	4. Established a persistent infection with less cytopathology than MuVo
5. Approximately 5-10% cells form infectious centres	5. Temperature-sensitive
6. Cytoplasmic and nuclear immunofluorescence	
7. Doubling time shorter for BHKpi	
8. Cloning efficiency in soft agar higher for BHKpi	
9. Uninfected and infected clones isolated	

disorders as viral arthritis (6), diabetes (8), primary endocardial fibroelastosis (7), Hodgkins disease (9) and a number of other organ involvements (2, 5, 12) we began a study to investigate the mechanism by which mumps virus could persist in cell culture. The need for *in vitro* model systems for persistent infections *in vivo* has been recognized (1, 3, 4).

STUDIES ON THE MECHANISM OF MAINTENANCE OF THE BHKpi PERSISTENT INFECTION

The major factors described as contributing to the ability of a virus to persist in cell culture are interferon, the action of specific anti-viral antibodies, genetically resistant cells, defective interfering particles (DIPs), *ts* mutants and DNA intermediates or copies of the infecting viral genomes. To distinguish which, if any, of these possible mechanisms were functioning in BHKpi cultures, a series of cloning experiments were undertaken.

Two types of clones could be isolated from the parent BHKpi population (10). Uninfected BHK-21 cells were obtained which were as equally susceptible to MuVo as the original cells used to start the persistence. These clones did not contain viral antigens as tested by hemadsorption or immunofluorescence nor did they release infectious virus. The second type of clone obtained did not release infectious virus, but contained viral antigens as evidenced by immunofluorescence and their ability to adsorb chicken erythrocytes (10). These non-producer clones have been carried in our laboratory for over 40 passages with no change in the properties as summarized in Table II. The listed properties are consistent with a designation of these cultures as a latent-inducible infection designated as BHKli. The viral antigens have continued to be demonstrated in these cells, indicating their continual synthesis.

That no virus-producing cells could be isolated by cloning experiments suggests first that the replication of the *ts* mutant, MuVpi, is lethal for these cells. Second, it leads to the suggestion that at least two mechanisms are functioning in the maintenance of this persistence: one, a mechanism which is expressed as an equilibrium between the release of the lethal *ts* mutant and the rate of appearance of susceptible uninfected cells; the second mechanism is dependent upon the existence of the non-producer or latently infected clones.

If cultures of the latently infected cells are incubated at 33° instead of 37°, they begin to produce infectious virus within 10 min after exposure to the lower temperature (11). This appearance of infectious mumps virus cannot be blocked by pre-treatment of the cultures with actinomycin D, puromycin or cycloheximide (50 μg/ml). The released virus, which has been

TABLE II

Summary of properties of BHKli and MuVli

Properties of BHKli	Properties of MuVli
1. 50-80% express mumps virus antigens (Had or ImF).	1. Neutralized by MuVo antiserum.
2. At 37° do not release infectious mumps virus.	2. Fluorescent focus smaller than MuVo or MuVpi.
3. Release infectious mumps virus (10^3-10^4 FFU/ml) when downshifted to 33°.	3. Not temperature-sensitive.
4. Interference with MuVo, but not with NDV or VSV.	
5. Clone 3 formed approximately 25% infectious centers at 33° or 37°.	

designated MuVli, can be neutralized with mumps virus antiserum. MuVli cannot be obtained from BHKli cells by freezing or thawing. The properties of MuVli are listed in Table II.

MuVli was found not to be temperature-sensitive in growth when tested in HeLa cell cultures at 33° and 39°. The data confirming this conclusion are shown in Table III.

An additional property of MuVli is that it can form a latent infection in BHK-21 cells directly if the virus is added at 37°.

DISCUSSION

Previous reports from this laboratory have shown that persistently infected BHK-21 cells produce infectious, temperature-sensitive mumps virus (Table I). In this report we have shown that cell clones isolated from the parent BHKpi population were infected with a latent-inducible mumps virus (MuVli).

At least two mechanisms must be operating to maintain the persistent infection. The first of these is suggested to be an equilibrium between the appearance of susceptible, uninfected BHK-21 cells and the rate of appearance of the temperature-sensitive MuVpi. This conclusion is supported by (i) the finding of cloning experiments of uninfected, susceptible cells; (ii) the observation that no virus-producing clones could be isolated, suggesting that replication of MuVpi is lethal for its host cells even though the virus is less virulent than wild-type MuVo; (iii) treatment of BHKpi cultures with mumps antiserum does not

TABLE III

Efficiencies of plating of MuVli at 33° and 39°

	Released virus 33° FFU/ml	Released virus 39° FFU/ml	Ratio of plating efficiencies 39°/33°
MuVli p10*	3.0×10^4	2.5×10^4	0.83
MuVli p20	1.5×10^4	2.0×10^4	1.30
MuVli p25	3.7×10^4	5.1×10^3	0.18
MuVli p30	2.0×10^4	5.1×10^3	0.26

*Cell passage number from original clonal isolation

cure the cultures, but only eliminates virus production (unpublished observations).

The second mechanism is the latent mumps infection. These latently infected cells do not produce infectious virus, are resistant to challenge with MuVo and yet continue to produce mumps virus antigens which do not dilute out with subsequent cell passage.

Of the major factors, listed above as contributing to the maintenance of persistent infection *in vitro*, several can be eliminated as controlling BHKpi persistence. No interferon could be detected, even when the cell fluid was concentrated 100-fold. No antiserum was added to the BHKpi cells either to initiate or maintain the persistence. The cloning experiments eliminate genetically resistant cells as a contributing mechanism since the isolated, infected cells were as equally susceptible to mumps virus as the original BHK-21 cells.

If defective interfering particles, which would normally inhibit production of infectious virions, are responsible for maintaining the latent infection at 37°, it is difficult to understand the rapid appearance of MuVli following the downshift in incubation temperature (10 min). Also, if defective viruses are present in the population released in the persistent infection, they must be of the same density as either MuVo or MuVpi. Both MuVo and MuVpi show a single peak at 1.19 gm/cc in sucrose gradients and all of the infectious virus is contained in this single peak. However, these findings are only suggestive. Further studies on the virion RNA will be required to eliminate DIPs as a controlling factor in this persistence.

Temperature-sensitive mutants are clearly involved in the equilibrium mechanism mentioned above. These *ts* mutants, which can be isolated from the wild-type population perhaps are selected because of their slower rate of growth and their reduced yield per cell, properties which could permit an equilibrium with uninfected cells. However, *ts* mutants are clearly not involved in the latent infection.

Whether or not DNA copies of mumps virus genomic RNA are involved in maintaining the latent state of infection is unknown at this time.

These studies on a persistent infection of mumps virus have pointed out that a latent infection can be derived from this ubiquitous agent. The medical consequences of a latent infection of mumps virus are unknown at this time. However, the reported potential involvement of mumps in such conditions as viral arthritis, diabetes and Hodgkins disease indicate the need for continued studies of mumps virus in persistent infection.

ACKNOWLEDGEMENTS

This investigation was supported by a grant from the John A. Hartford Foundation, New York, N.Y.

REFERENCES

1. Ackermann, W.W. (1957). *Annals N.Y. Acad. Sci.* 67, 392.
2. Gray, J.A. (1973). *Brit. Med. J.* 1, 338.
3. Holland, J.J., Villarreal, L.P., Etchison, J.R. (1974). *In* "ICN-UCLA Symposium on Molecular and Cellular Biology", Vol.1, 131. W.A. Benjamin, Menlo Park, California.
4. Johnson, K.P. and Norrby, E. (1974). *Exp. and Molec. Path.* 21, 166.
5. Nicolis, G.L., Sabetzhadam, R., Hsu, C.S.C., Sohval, A.R. and Gabrilove, J.L. (1973). *JAMA* 223, 1033.
6. Smith, J.W. and Sanford, P. (1967). *Annals of Internal Medicine* 67, 651.
7. St Gene, J.W., Noren, S.R. and Adams, P. (1966). *New England J. Med.* 275, 339.
8. Sultz, H.A., Hart, B.A., Zielezny, M. and Schlessinger, E.R. (1975). *J. Pediatrics* 86, 654.
9. Truant, A.L. and Hallum, J.V. (1976). *Oncology* 33, 241.
10. Truant, A.L. and Hallum, J.V. (1977). *J. Med. Virol.* 1, 49.
11. Truant, A.L. and Hallum, J.V. (1977). Submitted to *Proc. Soc. Exptl. Biol. Med.*
12. Wood, C.B., Bradbrook, R.A. and Blumgart, L.H. (1974). *Brit. J. Clin. Prac.* 28, 67.

CYTOCIDAL AND PERSISTENT INFECTION OF BS-C-1 CELLS BY RESPIRATORY SYNCYTIAL VIRUS

C.R. PRINGLE, P. CASH, H.B. GIMENEZ and P.V. SHIRODARIA*

M.R.C. Virology Unit,
Institute of Virology,
Church Street, Glasgow, G11 5JR.

**Department of Microbiology,*
Queen's University,
Grosvenor Road, Belfast, BT12 6BN

Chronic non-lytic infection can be the normal outcome of infection of cultured cells by retroviruses and by certain cytoplasmic RNA-containing viruses, typical examples of the latter being the infection of rhesus monkey kidney cells by the paramyxovirus SV_5 (3) and infection of insect cells by arboviruses. Persistent infection of cultured cells has also been obtained experimentally with several viruses which are normally cytocidal. A recent analysis of persistent infection of BHK-21 cells by vesicular stomatitis virus (VSV) showed that the initiation of persistent infection was dependent on the presence of a specific defective interfering (DI) particle which was associated with temperature-sensitive (*ts*) mutants of complementation group III. Once initiated the persistent infection was maintained by a different DI particle which appeared during passage, and which could itself initiate persistent infection by wild type VSV (7). Similarly Youngner *et al*. (14) observed that DI particles were necessary for initiation, but not maintenance, of persistent infection in mouse L cells by VSV. Instead it was observed that *ts* mutants (all of which belonged to complementation group I), accumulated during serial passage. These *ts* mutants moderated the cytocidal activity of wild type virus and behaved functionally as conditional DI particles (15).

Other mechanisms of achieving persistence have been described which required the presence of inhibitors (*e.g.* antibody or interferon) in the culture medium, and conceivably more than one mechanism may operate in any particular situation. The idea that persistent infection might also be achieved by reverse transcription of conventional RNA viruses into the genome of the host cell has been explored by Zhdanov (17), and supported by varied experimental evidence from several virus-cell systems including tick-borne encephalitis virus in Hep-2 cells, measles virus in chick embryo cells, and Sindbis virus, SV_5 and VSV in mouse L cells. Independently Simpson and Iinuma (10) have challenged the concept that RNA tumour viruses and related viruses which possess a virion-associated reverse transcriptase are unique in their capacity to be transcribed into a DNA proviral form. These workers reported that they were able to recover infectious respiratory syncytial (RS) virus following infection of susceptible cells with nuclear DNA extracted from cells persistently infected by RS virus. Others have been unable to detect integration of viral genetic information, however, in cultures persistently infected by other RNA viruses using both biological and biochemical techniques (8). Also no evidence of proviral DNA has been detected in cultures which survived infection by VSV as a result of prolonged interferon treatment (6).

RS virus is believed to be a negative-strand virus, because of its morphological similarity to viruses of the paramyxovirus group, although definitive proof of this is lacking. The precise nature of the genome of RS virus is unknown, but it appears to be an unsegmented single-stranded RNA molecule with a possible minimum size of 2.5×10^6 daltons (13). The absence of genetic recombination (11) and the existence of at least six complementation groups (Gimenez and Pringle in preparation) support this inference. Although RS virus multiplies efficiently in enucleate BS-C-1 cells (5), it remains possible that RS virus is prone to integrate into nuclear DNA, because antigen has been detected in the nuclei of cells infected by certain *ts* mutants at restrictive temperature (4).

We have set out, therefore, to derive persistently infected cultures of BS-C-1 cells analogous to those described by Simpson and Iinuma (10) for HeLa and calf kidney cells to enable us to analyse the factors involved in establishment and maintenance of persistence, and perhaps confirm the existence of a proviral form of RS virus.

CYTOCIDAL INFECTION OF BS-C-1 CELLS BY RS VIRUS

Infection of monolayers of human diploid cells normally resulted in extensive syncytium formation and eventually their total degeneration. Infection of monolayers of an African Green Monkey Kidney cell line (BS-C-1) produced deeply stained aggregates of cells. This cytopathic effect (CPE) has been observed with eleven different isolates of human RS virus and two of bovine RS virus (1, 4). Despite the resemblance of these aggregates to transformation foci, no cells survived infection with wild type virus. Cells infected by *ts* mutants of RS virus, however, survived infection if incubated at the restrictive temperature (39°) and could be propagated for several transfers.

INITIATION OF PERSISTENT INFECTION

The *ts* mutants of the RSN-2 strain of RS virus have been classified into three groups according to the pattern of immunofluorescence observed at 39°, (4) and four complementation groups (provisionally designated B, D, E and F) are now recognised (Gimenez and Pringle, in preparation).

Several of these *ts* mutants representing the three immunofluorescence groups have been used in attempts to initiate persistent infection of BS-C-1 or HeLa cells. These experiments will be described in detail elsewhere (Pringle and Malloy, in preparation), and are summarised here only.

Monolayers of BS-C-1 cells were infected at approximately 1 p.f.u./cell and incubated for 4 days at 39°. Then the monolayers were scraped into suspension, reseeded after 1/2-1/4 dilution in fresh culture bottles and incubated a further 3-5 days at 39° until confluence was again achieved. These cycles of incubation and respreading were continued for as long as possible. The final outcome varied according to the mutant used to initiate infection; either wild type revertants appeared and destroyed the culture within a few transfers, or infectivity gradually diminished until the cultures were apparently cured of infection, or a state of stable chronic infection was achieved. Only the latter will be considered in detail since the cured cultures remained free of any trace of RS virus. Furthermore, the cured cultures were fully susceptible to superinfection and could be propagated indefinitely (> 100 transfers during 14 months continuous culture). RS virus infectivity could not be recovered either by co-cultivation with susceptible cells, or by exposure to halogenated pyrimidines, or by transfection of nuclear DNA to susceptible cells.

Stable chronic infection was attained through a sequence of cytopathic crises which gradually abated and thereafter the cultures could be maintained either at 39° or at 37°. Two of the *ts* mutants of the RSN-2 strain successfully initiated stable chronic infection. Both these mutants, *ts* 1, and *ts* 9, belong to immunofluorescence group II, which is characterised by considerable antigen production at 39°. In contrast reversion or abortive infection followed infection of BS-C-1 cells with five other *ts* mutants (*ts* 5, *ts* 15, *ts* 16, *ts* 17 and *ts* 18). These five mutants belong to immunofluorescence groups I or III (reduced flourescence at 39°). Mutants *ts* 15, *ts* 16, *ts* 17 and *ts* 18 are now provisionally classified in complementation group B and *ts* 1 in group D, suggesting that initiation of chronic infection may be mutant specific.

MAINTENANCE OF PERSISTENT INFECTION

Chronically infected cultures initiated by infection with mutant *ts* 1 have been maintained in continuous culture by more than 75 transfers with infectious virus continuously released into the medium (generally <1 p.f.u./cell). The number of virus-producing cells as measured by infectious centre assay varied from 1-100% in consecutive passages and by immunofluorescence from 3-58%.

An interferon-like activity was detected in the supernatant of the *ts* 1 - initiated line. Undiluted supernatant inhibited plaque formation by a number of viruses to varying extents, pseudorabies virus being the most sensitive (>95% inhibition) and RS virus the least (no inhibition). Defective interfering (DI) particles cannot be identified directly, because RS virus is pleomorphic in shape and difficult to purify. However, co-infection of normal BS-C-1 cells with wild type RS virus (or mutant *ts* 1) and medium from the persistently infected culture did not bring about any reduction in yield, indicating that DI particles were not present at high concentration in the culture medium. Therefore neither interferon nor DI particles appeared to be instrumental in maintaining the infection.

These results suggest that the persistent infection is a steady-state situation where all cells are infected and release virus at reduced rate, rather than an equilibrium between infected cells and cells made refractory by interferor or interference. However, discrimination between the steady-state and equilibrium hypotheses must await the outcome of attempts to clone the persistently infected cells in the presence of antiserum.

Simpson and Iinuma (10) reported that DNA extracted from persistently infected cultures of HeLa (virus-producing) and bovine kidney (non-producing) cells was infectious. However, our attempts to obtain infectious DNA from chronically infected (the *ts* 1-initiated line) or cured (the *ts* 5-, *ts* 15- and *ts* 18- initiated lines) BS-C-1 cells have not been successful (Pringle and Malloy, unpublished data). DNA extraction and transfection were attempted under a variety of conditions, including those which had been established as optimal for transfection of feline syncytium forming virus (FSFV), a feline "foamy" virus (2). The assay conditions for FSFV were sufficiently sensitive to detect 10-20 p.f.u. per μg DNA in FEA cells. However, no RS virus infectivity was observed following exposure of BS-C-1 cells to several 100 μg DNA under similar conditions. There was no direct evidence, therefore, to implicate a proviral DNA form of the RS virus genome in maintenance of the persistent infections established by the *ts* mutants of the RSN-2 strain of RS virus.

Present evidence favours the hypothesis that chronic infection is maintained by the replacement of the initiating mutant by a variant of RS virus endowed with reduced cytopathogenicity. The virus produced by the *ts* 1-initiated line now forms plaques in normal BS-C-1 cells at 39° as well as 37°. This variant did not differ markedly from *ts* 1 in growth cycle, thermal inactivation rate and polypeptide profile, although plaque size was reduced. Infection of normal BS-C-1 cells at 37° with this virus produced an initial CPE, which appeared to be self-limiting with rapid development of a state of chronic infection.

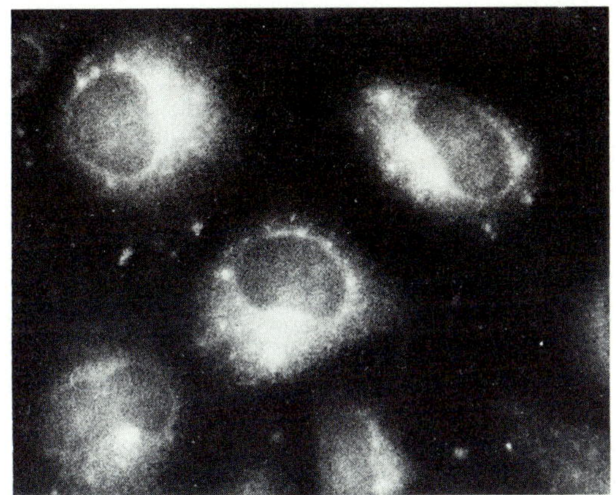

Fig.1. *Indirect immunofluorescence staining with bovine anti-RS virus serum and FITC-conjugated anti-bovine globulin. This culture had undergone 21 transfers at 39° followed by 26 transfers at 37°.*

SOME CHARACTERISTICS OF THE *ts* 1-INITIATED LINE OF CHRONICALLY INFECTED CELLS

Diffuse cytoplasmic fluorescence and cytoplasmic inclusions were observed in monolayers stained by fluorescent antibody (Fig.1). Faint nuclear antigen was detected occasionally, but the profusion of surface filaments characteristic of cytocidal RS virus infection (4), was absent. This pattern of fluorescence was similar to that observed in cells infected with the mutants classified in immunofluorescence group I by Faulkner *et al.* (4), but none of those tested (*ts* 5, *ts* 16, *ts* 17 and *ts* 18) were capable of initiating a chronic infection.

A major intracellular polypeptide, equivalent to VP 41 (the nucleoprotein), was the only RS virus polypeptide (1, 12, 13) which could be resolved clearly in polyacrylamide gel electrophoresis of extracts of chronically infected cells (Fig.2). There is evidently no build-up of viral protein in chronically infected cells, indicating a degree of regulation. Other experiments suggested a corresponding regulation of mRNA production, since no RS virus polypeptides were resolved by *in vitro* translation of total cytoplasmic RNA from chronically infected cells, although this was possible with RNA from lytically infected cells (Preston and Cash, unpublished data).

An increase in the agglutinability of cells by plant lectins accompanies transformation by oncogenic viruses and is observed in the initial stages of infection of cells by some non-oncogenic viruses (9, 16). Fig.3 shows that the *ts* 1-initiated line was agglutinated by a concentration of concanavalin A which had no effect on normal BS-C-1 cells. The virtual total involvement of the monolayer suggests there can be few, if any uninfected cells in the chronically infected culture.

CONCLUSION

Persistent infection of BS-C-1 cells has been obtained by propagation of *ts* mutant infected cells at 39°. The initiation of chronic infection appeared to be mutant-specific. Maintenance of chronic infection may be associated with appearance of a variant with reduced cytopathogenicity and does not appear to depend on temperature sensitivity, interferon production, DI particles, or formation of proviral DNA.

REFERENCES

1. Cash, P., Wunner, W.H. and Pringle, C.R. (1977). *Virology* 82, 369.
2. Chiswell, D.J. and Pringle, C.R. (1977). *J. Gen. Virol.* 36, 551.
3. Choppin, P.W. (1964). *Virology* 23, 224.
4. Faulkner, G.P., Shirodaria, P.V., Follett, E.A.C. and Pringle, C.R. (1976). *J. Virol.* 20, 487.
5. Follett, E.A.C., Pringle, C.R. and Pennington, T.H. (1975). *J. Gen. Virol.* 26, 183.
6. Friedman, R.M. and Costa, J.R. (1976). *Infection and Immunity* 13, 487.
7. Holland, J.J. and Villareal, L.P. (1974). *Proc. Nat. Acad. Sci. USA* 71, 2956.
8. Holland, J.J., Villareal, L.P., Welsh, R.M., Oldstone, M.B.A., Kohne, D., Lazzarini, R. and Scolnick, E. (1976).

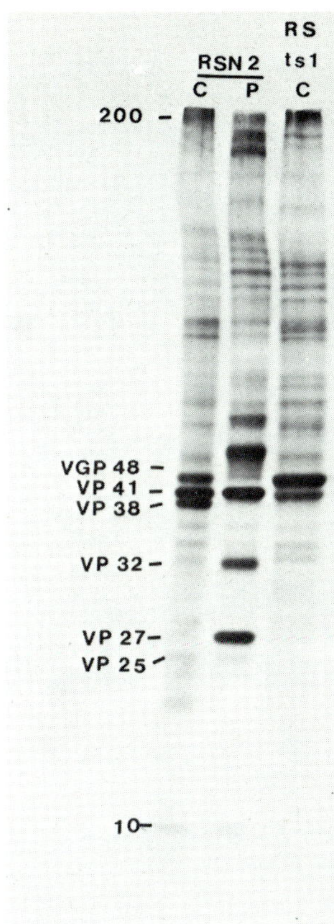

Fig.2. *Autoradiograph of ^{35}S-methionine-labelled polypeptides from BS-C-1 cells lytically and persistently infected with RS virus. Discontinuous SDS-polyacrylamide electrophoresis was carried out using gradient slab gels of 6-15% total acrylamide containing 0.1% (w/v) SDS and a constant proportion (5% w/w) of N, N'-methylene bisacrylamide as described by Cash et al. (1).*

C - *cell pellet extract*

P - *polyethylene glycol precipitate of clarified culture supernatant*

VP - *viral polypeptide*

VGP - *viral glycopolypeptide*

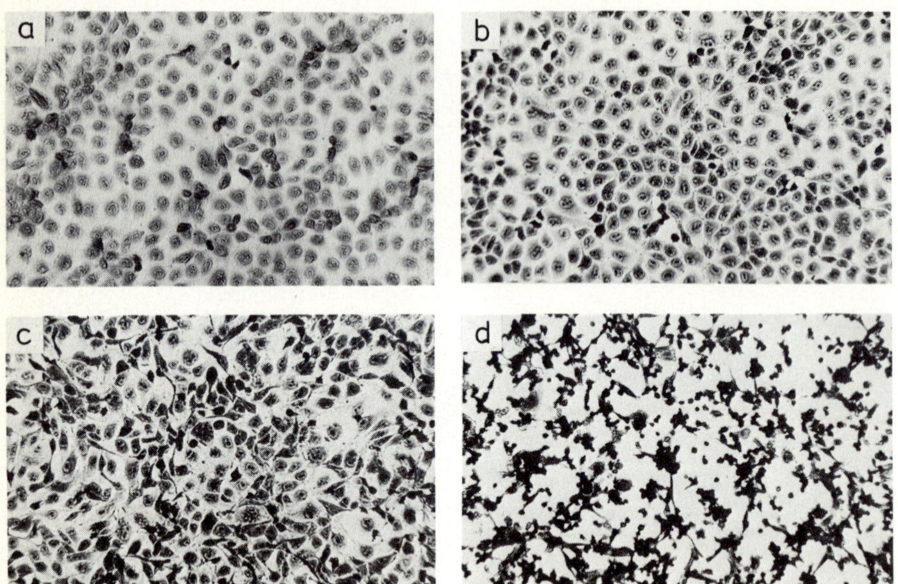

Fig.3. *The effect of exposure of normal and persistently infected BS-C-1 cells to 30 μg/ml concanavalin A for 2 hr at 31°.*

a *Normal BS-C-1 cells, untreated*
b *Normal BS-C-1 cells plus concanavalin A*
c *RS virus persistently-infected cells, untreated*
d *RS virus persistently-infected cells, plus concanavalin A*

J. Gen. Virol. 33, 193.

9. Poste, G. and Reeve, P. (1972). *Nature New Biol.* 237, 113.
10. Simpson, R.W. and Iinuma, M. (1975). *Proc. Nat. Acad. Sci. USA* 72, 3230.
11. Wright, P.F., Gharpure, M.A., Hodes, D.S. and Chanock, R.M. (1973). *Arch. ges. Virusforsch.* 41, 238.
12. Wunner, W.H., Faulkner, G.P. and Pringle, C.R. (1975). *In* "Negative Strand Viruses" (B.W.J. Mahy and R.D. Barry, eds.), Vol.1, p.193. Academic Press, London.
13. Wunner, W.H. and Pringle, C.R. (1976). *Virology* 73, 228.
14. Youngner, J.S., Dubovi, E.J., Quagliana, D.O., Kelly, M. and Preble, O.T. (1976). *J. Virol.* 19, 90.
15. Youngner, J.S. and Quagliana, D.O. (1976). *J. Virol.* 19, 102.
16. Zarling, J.M. and Tevethia, S.T. (1971). *Virology* 45, 313.
17. Zhdanov, V.M. (1975/76). *Intervirology* 6, 129.

STRUCTURE AND PATHOGENICITY OF INFLUENZA VIRUSES

R. ROTT, C. SCHOLTISSEK, H.-D. KLENK and M. ORLICH

Institut für Virologie, Justus-Liebig-Universität Giessen, Germany

The pathogenicity of a virus - that is its property to cause a disease in an organism - comprises a multitude of factors determined by the biological and biochemical characteristics of the virus on the one hand, and reactivity of the host on the other. It is obvious therefore that a molecular basis for viral pathogenicity is not easily defined.

In influenza viruses evidence has been presented for the polygenic nature of viral pathogenicity (1, 2). In the meantime, methods have become available which enable us to characterize more precisely the vague term polygenicity.

ACTIVATION OF THE HAEMAGGLUTININ

Proteolytic cleavage of the haemagglutinin (HA) glycoprotein is a prerequisite for the infectivity of influenza viruses, and is dependent on the host cell (4, 5). Whether proteolytic cleavage is possible in a given host cell is determined primarily by the individual structural characteristics of the HA rather than by the activation of cellular proteases by the infecting virus. This could be shown by double infection of chicken fibroblasts with fowl plague virus (FPV) and virus N. Under these conditions infectious FPV is formed whereas newly produced virus N, exhibiting uncleaved HA, is not infectious, the same result which was obtained when the cells were independently infected with the two viruses (3). The significance of the structural identity of HA for infectivity of the virus is underlined in a series of experiments with recombinant viruses where in all cells tested only those recombinant strains produce infectious progeny virus which possess the cleaved HA of FPV, regardless of the other structural components (4).

With strains, such as virus N, where non-infectious virus particles can be obtained with uncleaved HA and, thus, are accessible to proteolytic cleavage under *in vitro* conditions, some conclusions can be drawn regarding the enzymes active *in vivo*. Proteases with various substrate specificities are capable of producing cleavage products of HA which cannot be discriminated by their molecular weights from those obtained after trypsin treatment. However, only trypsin-like enzymes produce infectious virus particles *in vitro* (Table 1). It is remarkable that under *in vivo* conditions we have never observed virus particles which exhibit the cleaved HA but are non-infectious, as obtained after *in vitro* treatment with non-trypsin-like proteases such as thermolysine. This indicates that among the various proteases present in the cell only enzymes of trypsin specificity are able to activate virus infectivity. Our data show that certain cells contain such an activating enzyme whereas others do not.

SIGNIFICANCE OF VIRAL GLYCOPROTEINS FOR PATHOGENICITY

It seems certain that a large spectrum of host cells supporting replication of infectious virus helps the virus to be spread in the organism, and to infect cells of vital importance. There is no doubt, therefore, that pathogenicity of influenza viruses depends on the structure of the HA. It became clear, however, that viral glycoproteins alone do not determine pathogenicity (6, 8). We could show that back recombinants derived from antigenic hybrids used as parent strains were only infrequently pathogenic for chickens, although they carried both surface antigens of the highly pathogenic FPV (Table 2). In spite of the fact that the pathogenic and non-pathogenic back recombinants were indistinguishable with respect to the antigenic structure of both glycoproteins, only a few back recombinants obviously retained the whole set of the genes from the wild type FPV and were, therefore, again pathogenic (8).

INFLUENCE OF A SINGLE GENE EXCHANGE ON PATHOGENICITY

The question arose, therefore, which of the known genes other than the one coding for the HA, contributes decisively to the expression of pathogenicity. With the technique for analyzing the gene assignment of influenza virus recombinants which has been presented here already by C. Scholtissek *et al*. (Chapter 3) we studied first the influence of a single gene exchange on pathogenicity.

For the production of the necessary recombinants in which individual RNA segments were exchanged, primary chick fibroblasts were simultaneously infected at 40°, the non-permissive temperature, with corresponding *ts* mutants of FPV and another influenza

TABLE I

Effect of *in vitro* incubation with different proteases on cleavage of HA and activation of infectivity of virus N derived from chicken fibroblasts

Enzyme	Percentage of haemagglutinin present in cleaved form[a]	Virus activation[b]
None	0	-
Trypsin	100	+
Thermolysine	48	-
Chymotrypsin	45	-
Papain	34	-
Clostripain	15	-
Elastase	0	-

Virus released into the culture medium of chicken fibroblasts was subjected to *in vitro* incubation for 30 min at 37° with various proteases (10 µg/256 HA units). (a) The virus was purified, and the amount of cleaved glycoprotein was calculated from the polypeptide patterns obtained after polyacrylamide gel electrophoresis. (b) After treatment with proteases the virus was used for infecting chicken fibroblasts with a multiplicity of infection of 10 pfu/cell. 20 hr later the culture medium was tested for haemagglutinin (3).

---

virus prototype strain which does not form plaques on these host cells, and is non-pathogenic for chickens (10). Plaque forming viruses were selected. So far we have not yet succeeded in isolating recombinants in which only segment 7 is not derived from FPV. Adult chickens were infected intramuscularly with about 10^6 pfu of the characterized recombinant.

It could be shown that many of the FPV recombinants have a lower pathogenicity than the wild type. Infected chickens had either a delayed death time, or they recovered after signs of illness of varying degrees. Other recombinants did not produce any pathological signs at all (10). This means that an exchange of any RNA segment can modify pathogenicity which therefore cannot be confined to a particular gene. A degree of pathogenicity comparable to FPV wild type can be obtained by exchange of genes

TABLE II

Pathogenicity of back recombinants for chicken

Recombination pairs	Recombinants tested	Activities of the back recombinants grown in chicken eggs			
		HA units	pfu/ml x 10^7	μg of NANA/ml	Pathogenicity for chicken (mean death time in days)
FPV(H)Equine 2(N)	FPV(H)FPV(N)	512	1.4	72	-
x	FPV(H)FPV(N)	512	10	64	-
Equine 2(H)FPV(N)	FPV(H)FPV(N)	256	80	206	-
	FPV(H)FPV(N)	128	36	152	-
	FPV(H)FPV(N)	1024	160	208	+ (3.5)

Taken from Rott *et al*. (8)

TABLE III

Correlation between RNA segment composition of recombinants, in which a single RNA segment is exchanged, and their pathogenicity

RNA segment not derived from FPV	RNA segment derived from:						
	PR8	FM1	A2	A3	Equi 2	Swine	Virus N
1 (ts 3)	(1)-,-,- (5)-,-,-	(1)±,± (2)±,+ (3)±,± (4)-,±	(1)+,∓,+ (2)∓,+,+,+ (3)±,+,-	(1)-,-,- (2)+,+,+ (3)+,+,+,+ (4)-,-,∓,∓ (5)- (6)+,∓ (7)+,+,+	(1)+,+,+,± (2)±,-,+	(1)+,+ (2)+,±,±,+	(2)+,+,+ (3)±,±,∓
2 (ts 90)	(1)∓,-,+,-,+ (4)∓,+,-,+,+ (7)+,± (8)-,-	n.a.	(2)+,+,+ (3)+,±,+	(1)-,-,- (2)-	(6)±,±,±	(1)-,- (4)-,-,-	(10)+,+ (11)+,+

Numbers in parentheses are the numbers of individual isolates. Some of them have been described (9). Each symbol represents one chicken.

+ means dead after 2 to 3 days after infection; ± means delayed death after severe illness
∓ means significant illness and recovery; - means no signs of illness; n.a. = no recombinant available.

For further details see Scholtissek *et al.* (10).

TABLE IV

Pathogenicity of FPV recombinants in which more than one gene is replaced by the corresponding gene of another influenza virus prototype

Segments not derived from FPV	Segments derived from influenza A virus				
	Virus N	PR8	A2	A3	Equine 2
1,2			∓	+	
1,3		−			
1,6		∓	+	+	
1,8	−				
1,2,3			+		
1,3,6,8			+		
1,2,3,5,8	∓				
1,2,6,7,8	−				
2,3	+	∓			
2,5	±				
2,8	+				+
2,3,5	+				
2,3,6		±			
2,3,8	+	∓			
2,3,5,8	∓	−			
2,3,5,6,8	−				
2,3,4,5,8	−				
3,5	+				
3,6			+		
3,8	+				±
3,5,6	±				
3,6,8	−				
3,4,5,8	−				
3,5,6,8	−				
3,6,7,8	−				
5,8	−				−
5,6					−
6,8		∓			

For further details see Table 3

TABLE V

Gene constellation of mouse neurovirulent recombinants derived from the non-neurovirulent parent strains fowl plague virus (FPV) and influenza A England/1/61 (Engl.)

Virus recombinants[1]	Mouse neuro-virulence[2]	RNA segments derived from 1	2	3	4	5	6	7	8
V1 (Hav1N1)	+	Engl.	Engl.	FPV	FPV	FPV	FPV	FPV	FPV
V2 (Hav1N1)	-	FPV	Engl.	Engl.	FPV	Engl.	FPV	FPV	FPV
V3 (Hav1N1)	+	Engl.	Engl.	Engl.	FPV	Engl.	FPV	FPV	Engl.
L4a (Hav1N2)	+	Engl.	Engl.	Engl.	FPV	Engl.	Engl.	FPV	Engl.
L4b (Hav2N2)	-	FPV	Engl.	FPV	FPV	Engl.	Engl.	FPV	FPV
U1 (H2N1)	-	Engl.	Engl.	Engl.	Engl.	Engl.	FPV	Engl.	FPV
U2 (H2N1)	-	Engl.	Engl.	Engl.	Engl.	Engl.	FPV	Engl.	FPV

For assignment of the RNA segments see Scholtissek *et al.* (9).

[1]For more details see Vallbracht (11)

[2]- = negative

1, 2, 3, 5, 6, and 8. In any case the pathogenic properties of a recombinant are determined by the virus strain from which the corresponding gene was derived. For example, if the RNA segment 1 is taken from the influenza strain PR8 pathogenicity is not reconstituted, whereas segment 1 taken from swine influenza virus or virus N means that the recombinant is as pathogenic as the FPV wild type. If segment 2 on the other hand originates from swine influenza virus or from an A3-strain, no pathogenicity is recovered, whereas segment 2 from an A2-strain or virus N produces pathogenic virus (Table III).

DEFINED MULTIPLE GENE EXCHANGE

In a further set of experiments we started to select recombinants in which more than one gene of FPV was exchanged for genes from another influenza virus non-pathogenic for fowl. The few experimental data available so far indicate again that pathogenicity depends on the RNA segments of FPV which are exchanged as well as on the origin of these particular genes (Table IV). There is a trend that the probability of losing pathogenicity increases with the number of exchanged genes. However, we also made the interesting observation that a single gene exchange of RNA segment 1 or 2 from FPV by the Hong Kong strain led to complete loss of pathogenicity, while after exchange of both genes in one recombinant, pathogenicity was fully regained.

RECOMBINANTS EXHIBITING MOUSE NEUROVIRULENCE

An increase of pathogenicity of recombinants with FPV cannot be studied in fowl, because of the already extremely high pathogenicity of this virus. For such studies we have used neurovirulence for mice as a pathogenicity marker. The plaque-purified parent strain FPV (Hav1N1) and the influenza A strain England/1/61 (H2N2) which were used for recombination, both do not exhibit neurovirulence. However, as recently found by Vallbracht (11) 3 out of 7 recombinants examined killed suckling mice after intracerebral inoculation of 10^3-10^4 pfu within 6 days. In Table V the gene constellation of these recombinants are shown. As can be seen, the 3 neurovirulent strains have in common that RNA segments 1 and 2 are derived from the England strain whereas segment 4 comes from FPV. Since the other non-neurovirulent recombinants do not have this gene constellation it might be concluded that with this kind of recombination the virus particles are neurovirulent if at least the genes coding for the polymerase 1 and the transport protein are derived from the England strain and the HA gene from FPV. According to our previous experience one has to be cautious, however, to generalize from these results. It might well be that with other parent strains other gene constellations will also determine neurovirulence.

SUMMARY AND CONCLUSIONS

The available data suggests that cleavage of the HA is essential for the formation of a pathogenic virus. Besides this gene coding for HA any one of the other genes may contribute to pathogenicity, since their exchange may produce a virus with enhanced pathogenicity, or a less pathogenic, or totally non-pathogenic virus. Full reconstitution of pathogenicity of the wild type FPV largely depends on the parent strain used for the gene exchange and the particular gene which is re-assorted. There is a trend suggesting that infection with recombinants carrying defined multiple gene exchanges produces less severe signs of clinical illness the more segments that have been exchanged in the infecting virus. There is no experimental indication, therefore, that something like a virulence gene exists in influenza viruses. This is in contrast to NDV where pathogenicity appears to be determined primarily by the structure of the glycoproteins (7, see chapter 61). All our data point to the requirement for an optimal constellation of all genes in order to produce a highly pathogenic virus. Not every recombination will result in a functional genome (8, 9). The individual segments have to fit into a certain restricted pattern to yield a pathogenic virus evolving from a given recombination. In view of these results one might postulate analogous recombinations occurring in nature which result in field recombinants with low pathogenicity, most of which will pass unrecognized. However, one must also consider the possibility that on occasion recombinants more pathogenic than either parent may result.

ACKNOWLEDGEMENT

This work was supported by the Deutsche Forschungsgemeinschaft (Sonderforschungsbereich 47).

REFERENCES

1. Burnet, F.M. (1959). *In* "The Viruses" (Burnet, F.M. and Stanley, W.M. eds.), Vol.3, p.275. Academic Press, New York and London.
2. Kilbourne, E.D. (1963). *Progr. Medical Virology* 5, 81.
3. Klenk, H.-D., Rott, R. and Orlich, M. (1977). *J. Gen. Virol.* 36, 151.
4. Klenk, H.-D., Rott, R., Orlich, M. and Blödorn, J. (1975). *Virology* 68, 426.
5. Lazarowitz, S.G. and Choppin, P.W. (1975). *Virology* 68, 440.
6. Mayer, V., Schulman, J.L. and Kilbourne, E.D. (1973). *J.*

Virol. 11, 272.

7. Nagai, Y., Klenk, H.-D. and Rott, R. (1976). *Virology* 72, 494.
8. Rott, R., Orlich, M. and Scholtissek, C. (1976). *J. Virol.* 19, 54.
9. Scholtissek, C., Harms, E., Rohde, W., Orlich, M. and Rott, R. (1976). *Virology* 74, 332.
10. Scholtissek, C., Rott, R., Orlich, M., Harms, E. and Rohde, W. (1977). *Virology* 81, 74.
11. Vallbracht, A. (1977). Doctoral thesis, University of Tübingen.

BIOLOGICAL PROPERTIES OF RECOMBINANTS OF INFLUENZA A/HONG KONG AND A/PR8 VIRUSES: EFFECTS OF GENES FOR MATRIX PROTEIN AND NUCLEOPROTEIN ON VIRUS YIELD IN EMBRYONATED EGGS

JEROME L. SCHULMAN and PETER PALESE

Mount Sinai School of Medicine
of the City University of New York,
100th Street and Fifth Avenue,
New York City, New York, U.S.A.

Previously we described how the derivation of each gene of recombinant influenza A viruses could be determined by analysis of RNA migration patterns on urea-polyacrylamide gels (4). Subsequently this procedure, in concert with other techniques, permitted the establishment of a complete genetic map of influenza A viruses (4, 6, 7). We then speculated that these methods would facilitate the identification of genes involved in strain related differences in biological activity in different host cell systems (8). We proposed that comparison of viruses identical with respect to the derivation of all genes except one would provide a basis for determining the relationship of that gene product to differences in virulence.

Employing these techniques we observed that the distinctive capacity of influenza A/WSN/33 (HON1) virus to produce plaques in bovine kidney (MDBK) cells is related to neuraminidase (NA). Recombinant viruses which derived only the NA of WSN virus produced plaques whereas recombinant viruses deriving all of their genes from WSN virus except NA did not. Subsequent SDS polyacrylamide gel analysis of virus proteins obtained after infection of MDBK cells demonstrated predominantly cleaved hemagglutinin (HA) with WSN virus whereas under the same conditions predominantly uncleaved HA was observed in recombinants lacking the WSN NA. From these observations, we postulated that a previously unrecognized function of NA is to facilitate cleavage of HA by removing sialic acid residues which otherwise obstruct the cleavage site of the HA (9). The fact that among the influenza A viruses examined only WSN virus produces plaques in MDBK cells made the identification of the gene essential for plaque formation convenient to straightforward experimental approach.

NP AND M PROTEIN GENES IN HIGH YIELDING RECOMBINANTS

In the present communication, we discuss the results of experiments designed to elucidate the genes which contribute to strain related differences in growth capacity in embryonated chicken eggs. The use of genetic recombinants derived from PR8 virus to obtain high yielding recombinants was first suggested by Kilbourne and Murphy (1) and in recent years this method has been employed routinely with H3N2 viruses to obtain strains suitable for vaccine production (2). RNA analysis of one such recombinant, X-31, demonstrated that only the genes coding for HA and NA were derived from the low yielding A/Hong Kong/8/68 (H3N2) virus parent (Palese and Kilbourne, unpublished observations). Similar results were obtained following analysis of X-53 (Hsw1N1) derived by recombination of A/New Jersey/11/76 (Hsw1N1) and A/PR/8/34 (H0N1) viruses (5).

In an attempt to determine which particular PR8 genes are required for transfer of high yielding properties in embryonated chicken eggs we first examined the HA titers of 47 separate cloned recombinants of HK and PR8 viruses obtained after mixed infection in embryonated eggs or in cell culture. Each recombinant was assayed with respect to HA titers in allantoic fluids and was analyzed on RNA gels to determine the derivation of each gene following published procedures (4, 7, 8).

As summarized in Table I, analysis of these 47 recombinants provided preliminary evidence indicating that PR8 virus genes coding for NP and M proteins were required for the high yielding property. Thus, of the 34 recombinants which derived both of these genes from PR8 virus all but one replicated to HA titers of 1:2048 or greater in eggs. In contrast of the 13 recombinants which did not derive both NP and M protein genes from PR8 virus, 10 (91%) replicated to titers of 1:512 or less.

Although these data demonstrated an extremely close correlation between PR8 NP and M proteins and high yielding properties in embryonated eggs, interpretation of the data was complicated by the fact that 4 exceptions were observed; one recombinant deriving both NP and M protein genes from PR8 virus, did not replicate to high titer, and 3 recombinants deriving one or the other of these genes from HK virus did replicate to high titer. These discrepancies could not be explained on the basis of other genes or combination of genes, suggesting that variation associated with the selection of individual clones might influence the results.

CLONED RECOMBINANTS FROM SINGLE CYCLE YIELD

TABLE I

Genes for NP and M Protein in Recombinants of influenza A/PR/8/34 and A/HK/8/68 viruses

Derivation of genes			HA Titer	
NP	M	Number	≥2048	≤512
P	P	34	33	1
P	H	8	1	7
H	P	3	2	1
H	H	2	0	2
		47	36	11

P = derived from PR8 virus
H = derived from HK virus

In addition, we recognized that this preliminary analysis was compromised by the fact that most of the recombinants were obtained from embryonated eggs in different experiments following mixed infection and multicycle replication. We recognized that under these conditions in which PR8 virus replicates to much higher titer than HK virus it was to be expected that PR8 virus genes would predominate in the recombinants which were obtained. Therefore we focused on 14 recombinants obtained in one experiment from a single cycle yield following mixed infection of MDCK cells. Individual recombinant clones were selected in the presence of appropriate antiserum and purified by plaque to plaque passage as described previously (5, 8). As seen in Table II, analysis of the derivation of genes in the 14 recombinants revealed that with the possible exception of the NS gene, genes coding for the other 7 polypeptides were derived in approximately the same frequency from the 2 parents. Growth in embryonated eggs of each recombinant was determined by measuring HA titers of pooled allantoic fluids (4 eggs/each virus) 40 hr after infection with 100 EID_{50}.

Analysis of the gene composition and growth characteristics of the 14 recombinants revealed that all 6 of the high yielding recombinants derived both the genes coding for NP and M from

TABLE II

Relationship of yield in embryonated eggs and gene constellation of recombinants of PR8 and HK viruses[a]

Recombinant	Gene Derivation*								HA Titer
	P1	P2	P3	HA	NA	NP	M	NS	
1	P	P	P	P	H	P	P	P	16,384
2	H	H	H	P	H	P	P	P	8,192
3	P	P	P	P	H	P	P	P	8,192
4	P	P	P	H	P	P	P	P	8,192
5	P	P	P	H	P	P	P	P	8,192
6	H	H	P	H	P	P	P	H	4,096
7	H	H	H	H	H	H	H	P	256
8	H	H	H	P	P	P	H	P	512
9	H	P	H	P	H	P	P	P	256
10	H	H	H	P	H	H	H	H	256
11	H	H	P	P	H	H	P	P	512
12	P	P	P	H	P	P	H	P	128
13	H	H	P	H	P	H	H	P	32
14	H	P	H	H	P	P	H	P	256

[a]All cloned recombinants derived from a single mixture in MDCK cells

*H = gene from HK virus; P = gene from PR8 virus

PR8 virus (upper portion of Table II), whereas all 7 recombinants which derived either or both of these genes from HK virus replicated to low titer. However, one recombinant (no.9) which derived genes for NP and M from PR8 virus was a low yielding virus.

One interpretation of these results is that the property of high yield in embryonated eggs among recombinants of PR8 and HK viruses is associated with PR8 genes for NP and M protein, but that spontaneous mutations or particular combinations of of other genes may influence expression of the high yield characteristic.

"NATURAL SELECTION" OF RECOMBINANTS *IN OVO*

Recognizing the difficulties inherent in the analysis of individual clones, especially those obtained after multicycle replication in a host cell system in which one parent replicates much more efficiently than another, we concluded that a different approach was necessary to identify which PR8 genes are required to confer high yielding properties to recombinants of PR8 and HK viruses. We postulated that if embryonated eggs were inoculated at low dilution with a mixture of recombinant viruses containing equivalent proportions of HK and PR8 RNAs coding for each gene product, selection for recombinants with high yield properties would result in enrichment for specific PR8 virus gene(s) and elimination of corresponding HK virus gene(s). In other words, we proposed to inoculate embryonated eggs with a mixture containing as many recombinants as possible and then "let nature take its course". Since the virus yields obtained were not cloned we expected to avoid discrepancies attributable to clonal variation. The only restriction which we felt necessary to impose was to suppress high yielding wild type PR8 virus by mixing the inoculum with antiserum to PR8 (H0N1) or to PR8-HK (H0N2) virus.

An appropriate recombination mixture for such an experiment was already available, as evidenced by the analysis of the recombinants obtained from a single mixture, listed in Table II. Additional support for this premise was provided by infecting MDCK cells with that mixture and radioactively labelling the RNA in the presence of ^{32}P. The virus yield was purified, and the RNA extracted and analyzed on polyacrylamide-urea gels in accordance with previously described methods (4). As shown in Fig.1 both PR8 and HK genes coding for HA, NA, NP, M and NS can be seen and as far as it is possible to estimate by autoradiography HK and PR8 RNAs coding for these proteins are present in roughly equivalent proportions. Although the genes coding for P proteins are closely spaced, PR8 genes for P3 and P2 proteins and the HK gene for P2 can be discriminated, and it is likely that the other P genes are represented on this gel in the diffuse band below the PR8 P3 gene.

This recombinational mixture which contained approximately 10^7 EID_{50}/ml was serially diluted, mixed with antiserum to PR8 virus and inoculated into embryonated eggs. After determination of HA titers, individual allantoic fluids were used to infect MDCK cells for ^{32}P labelling of RNAs. Fig.2A demonstrates the RNA patterns of virus in 4 of these fluids along with the patterns of the parent viruses, PR8 virus

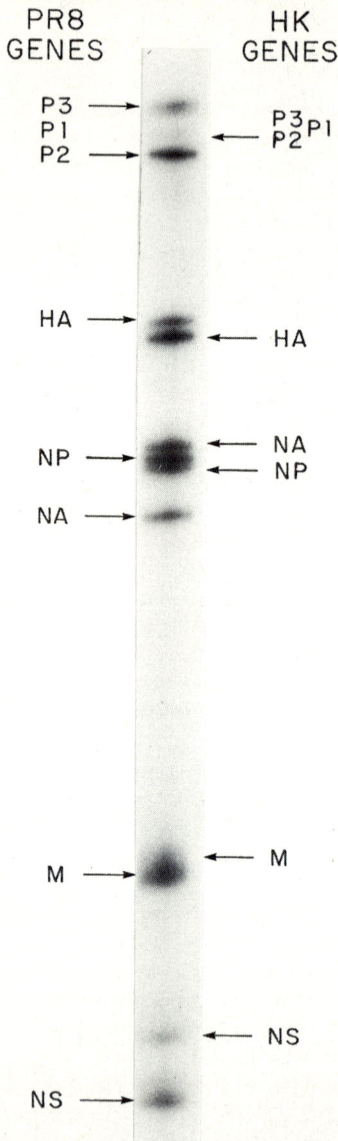

Fig.1. *Polyacrylamide gel analysis of the RNAs of a recombinant mixture obtained 8 hr after mixed infection of MDCK cells with influenza A/HK/8/68 and A/PR/8/34 viruses. Purification of virus and RNA analysis was done as described previously (4, 7). Arrows on the left indicate PR8 virus RNAs in the mixture; arrows on the right indicate HK virus RNAs.*

(lane 1) and HK virus (lane 5). It can be seen that viruses in all 4 recombination fluids derive genes coding for hemagglutinin (RNA 4) and neuraminidase (RNA 5) from HK virus. Yields a, b and c which were obtained after inoculation of eggs with a 1:10 dilution of the original recombinational mixture contain mixtures with respect to some genes. Multiple bands derived from both HK and PR8 viruses can be seen in the region of the RNAs coding for P proteins although RNAs 1 and 3 (P3 and P2 genes) from PR8 are the most intense. It should be noted that on this gel at least 5 of the 6 PR8 and HK genes coding for P proteins can be discriminated. In addition, bands corresponding to the NS genes of both PR8 and HK viruses can be seen at the bottom; a longer exposure of the film confirms the presence of RNA 8 from HK virus in virus in recombination fluid c (lane 4). However, only PR8 derived genes for NP (RNA 6) and M protein (RNA 7) can be seen. In summary, virus in these 3 fluids which were obtained in the presence of an antiserum screen to PR8, contained only HK genes for HA and NA, a mixture of HK and PR8 genes for P proteins and NS protein, but only PR8 virus derived genes for NP and M protein. As seen in Table III fluids a, b, c had high HA titers, and it is reasonable to speculate that the *in ovo* selection for high yielding virus was associated with the selection of virus particles containing PR8 genes for NP and M protein, and the exclusion of corresponding HK virus genes.

Fluid d (lane 6) also was obtained from an embryonated egg inoculated with a 1:10, dilution of the original mixture. The RNA pattern of virus in fluid d does not demonstrate the presence of a mixture. Genes coding for P1 (RNA 2), P2 (RNA 3), HA (RNA 4), NA (RNA 5) and M protein (RNA 7) are derived from HK virus and the remaining genes are derived from PR8 virus (Fig. 2A). As seen in Table III, fluid d has a much lower HA titer than the other fluids, an observation which is consistent with the hypothesis that the PR8 gene for M protein is an important factor contributing to the property of high yield in embryonated eggs. Although interference may provide a possible explanation, it is not clear why fluid d does not appear to be a mixture and why high yielding virus was not selected.

Fig.2B demonstrates the RNA patterns of virus in 2 other fluids obtained after infection of embryonated eggs with a 1:100 dilution of the original recombinational mixture. Neither of these yields appear to contain mixtures. Virus in fluid e (lane 2) derives all of its genes except HA (RNA 4) and NA (RNA 5) from PR8 virus. Fluid f (lane 3) contains virus which derives an additional gene, RNA 2 from HK virus. As seen in Table III, both of these recombination fluids have high HA titers.

Fig.3 shows the RNA patterns of 5 other recombinants obtained in embryonated eggs from the same original recombinational mixture. In this instance antiserum to PR8-HK (H0N2) virus was

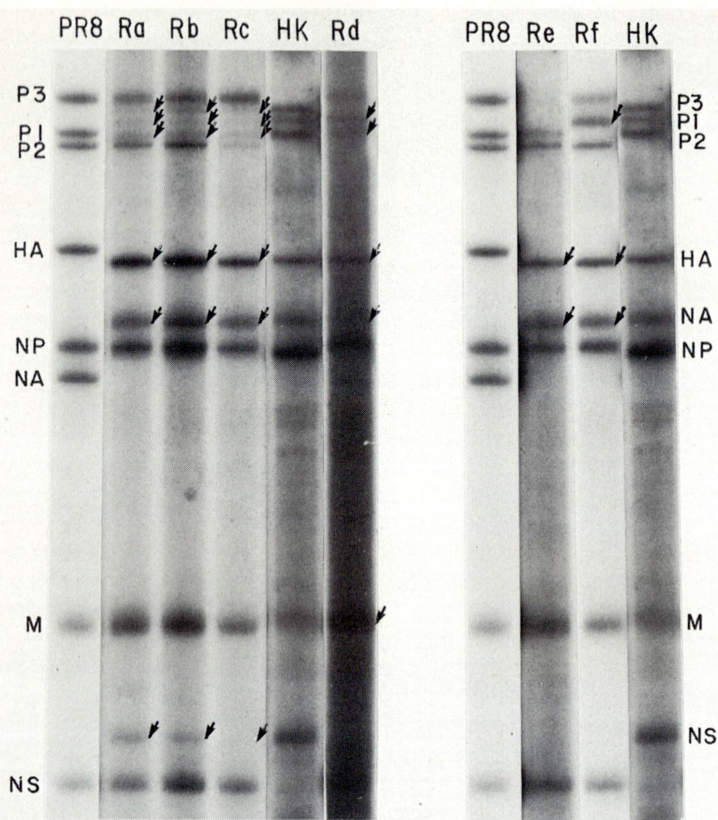

Fig.2A. *Analysis of the RNAs of PR8 and HK viruses and of virus yields from embryonated eggs infected with the same recombinational mixture shown in fig.1. Lane 1, RNA of PR8 virus; lane 2 RNAs of virus in fluid a; lane 3, RNAs of virus in fluid b; lane 4, RNAs of virus in fluid c; lane 5, RNA of HK virus; lane 6, RNAs of virus in fluid d. Fluids were derived after addition of antibody to PR8 (HON1) virus to the original recombinational mixture (see fig.1) prior to infection of embryonated eggs. Dilution of recombinational mixture for inoculation was 1:10. Conditions for extraction and analysis of RNA were the same as in fig.1.*

Fig.2B. *Analysis of the RNAs of PR8 and HK viruses and of virus yields from embryonated eggs infected with the same recombinational mixture shown in fig.1. Lane 1, RNA of PR8 virus; lane 2, RNAs of virus in fluid e; lane 3, RNAs of virus in fluid f; lane 4, RNA of HK virus. Fluids were derived after addition of* employed to select for high yielding HK-PR8 (H3N1) recombinants. It should be noted that the 5 recombinants shown in fig.3 all were derived from embryonated eggs inoculated with a 1:1000 dilution of the starting recombinational mixture. Fig.3 demonstrates that recombinant i (lane 4), recombinant j (lane 5) and recombinant k (lane 6) derive all of their genes from PR8 virus except RNA 4 (gene coding for HA). As seen in Table III all 3 are high yielding viruses. Recombinant h (lane 3) derives its NS gene (RNA 8) as well as RNA 4 from HK virus. Recombinant g (lane 2) derives 4 genes from HK virus, RNA 2 (gene for P1), RNA 4 (HA gene), RNA 7 (gene for M protein) and RNA 8 (gene for NS protein). As seen in Table III this is the only low yielding recombinant among the 5 HK-PR8 viruses analyzed. Thus as was seen with the HK-HK (H3N2) recombinants, the high yielding phenotype appears to require derivation of both NP and M protein genes from PR8 virus.

COMMENT

We believe that the data presented in this report are consistent with the hypothesis that in recombinants of PR8 and HK viruses, PR8 genes for NP and M proteins are required for high yielding properties. Although exceptions to this generalization have been noted and discussed, these are probably related to cloning procedures employed in selecting recombinants and clonal variation due to spontaneous mutation. In this respect the later experiments, in which "nature was allowed to take its course" in selecting high yielding recombinants, provided a more accurate representation of which genes were selected or excluded.

In advancing this hypothesis we do not mean to preclude the possibility that differences in other genes also may play a role. For example, analysis of the HA titers of recombinants included in Tables II and III suggests a relationship between low yield and derivation of the gene coding for P1 from HK virus, although this association is minimized when derivation of genes for NP and M protein is considered. Thus 3 out of 4 viruses which derive the P1 gene from HK virus and the genes for both NP and M from PR8 virus are high yielding. In addition the presence of HK genes coding for P1 in high titered mixtures (fluids a-c, table III) also indicates that the derivation of the P1 gene from HK virus does not necessarily

---

antibody to PR8 (H0N1) virus to the original recombinational mixture (see fig.1) prior to infection of embryonated eggs. Dilution of recombinational mixture for inoculation was 1:100. Conditions for extraction and analysis of RNA were the same as in fig.1.

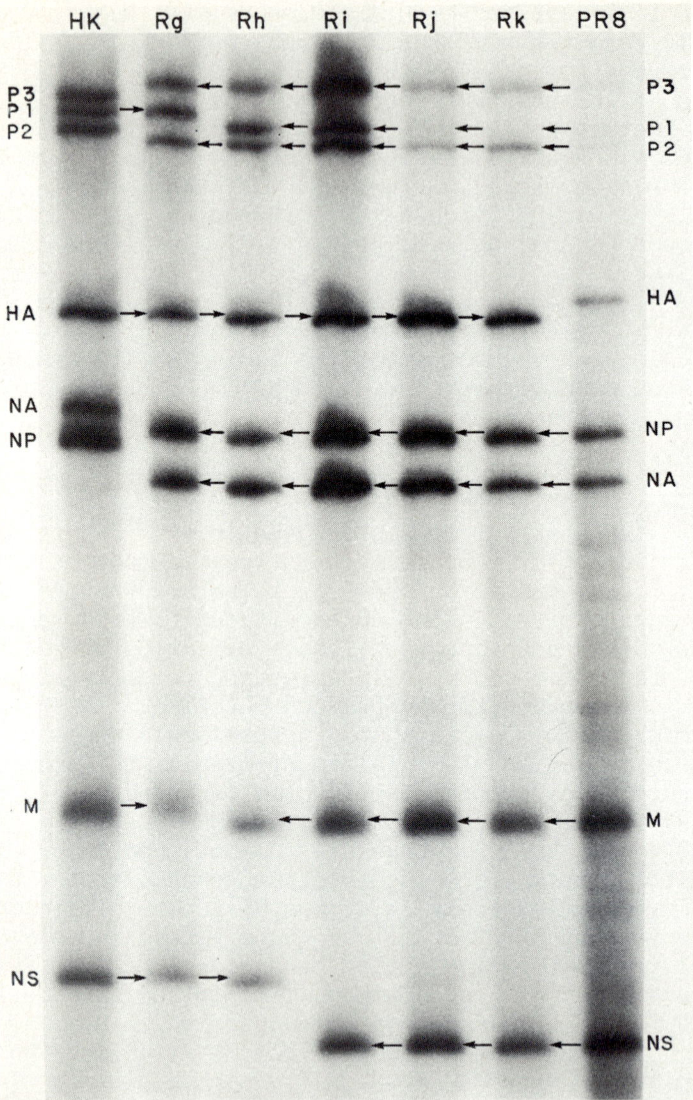

Fig.3. *Analysis of the RNAs of PR8 and HK viruses and of virus yields from embryonated eggs infected with the same recombinational mixture shown in fig.1. Lane 1, RNA of HK virus; lane 2, RNAs of virus in fluid g; lane 3, RNAs of virus in fluid h; lane 4, RNAs of virus in fluid i; lane 5, RNAs in fluid j; lane 6, RNAs of virus in fluid k; lane 7, RNA of PR8 virus. Fluids were derived after addition of antibody to PR8-HK (HON2) virus to the original recombinational mixture prior to infection*

TABLE III

Selection for PR8 genes for NP and M proteins in high yielding recombinants of PR8 and HK viruses*

Recombination fluid	Gene Derivation**								HA Titer
	P1	P2	P3	HA	NA	NP	M	NS	
a	Mix	Mix	Mix	H	H	P	P	Mix	4,096
b	Mix	Mix	Mix	H	H	P	P	Mix	8,192
c	Mix	Mix	Mix	H	H	P	P	Mix	4,096
d	H	H	P	H	H	P	H	P	512
e	P	P	P	H	H	P	P	P	8,192
f	H	P	P	H	H	P	P	P	8,192
g	H	P	P	H	P	P	H	H	256
h	P	P	P	H	P	P	P	H	4,096
i	P	P	P	H	P	P	P	P	4,096
j	P	P	P	H	P	P	P	P	8,192
k	P	P	P	H	P	P	P	P	8,192

*Selection of viruses was done in embryonated chicken eggs using a 1:10 (fluids a-d), 1:100 (fluids e, f) or 1:1000 (fluids g-h) dilution of the same recombinational mixture.

**H = gene derived from HK virus

P = gene derived from PR8 virus

Mix = RNAs from PR8 and HK viruses both present

produce low yielding virus.

Although a close relationship between PR8 virus genes for NP and M protein and high yield is demonstrated by these data, the mechanism for this relationship has not been established. It is possible that it is primarily the gene for M protein which determines whether or not a recombinant is high yielding and that selection for the PR8 gene for NP may be a secondary

of embryonated eggs. Dilution of recombinational mixture for inoculation was 1:1000. Conditions for extraction and analysis of RNA were the same as in fig.1.

effect reflecting more efficient interaction of NP and M proteins from the same virus. It has been shown previously that M protein is found in much greater abundance in virus particles than in infected cells which prompted speculation that synthesis of M protein may be a rate limiting step in influenza virus replication (3). The observations in the present report are in accord with such an hypothesis. Differences in the PR8 and HK genes for M protein could result in differences in the rate at which these RNAs are transcribed or translated. Alternatively, it is possible that differences in the gene product could influence the stability of the virus or the rate at which assembly takes place. Likewise, differences in genes for NP could hypothetically influence the rate of virus replication in a variety of ways.

ACKNOWLEDGEMENTS

We wish to thank Kaye Leitzinger and Marlene Lin for their expert technical assistance. This work was made possible by Public Health Service grants AI-11823, AI-09394, and AI-14053 from the National Institute of Allergy and Infectious Diseases; by grant PCM-76-11066 from the National Science Foundation and by a grant from the American Cancer Society, ACS (VC-234).

REFERENCES

1. Kilbourne, E.D. and Murphy, J.S. (1960). *J. Exp. Med.* 111, 387.
2. Kilbourne, E.D., Schulman, J.L., Schild, G.C., Schloer, G., Swanson, J. and Bucher, D. (1971). *J. Inf. Dis.* 124, 449.
3. Lazarowitz, S.G., Compans, R.W. and Choppin, P.W. (1971). *Virology* 46, 830.
4. Palese, P. and Schulman, J.L. (1976). *Proc. Natn. Acad. Sci. USA* 73, 2142.
5. Palese, P., Ritchey, M.B., Schulman, J.L. and Kilbourne, E.D. (1976). *Science* 194, 334.
6. Palese, P., Ritchey, M.B. and Schulman, J.L. (1977). *Virology* 76, 114.
7. Ritchey, M.B., Palese, P. and Schulman, J.L. (1976). *J. Virol.* 20, 307.
8. Schulman, J.L. and Palese, P. (1976). *J. Virol.* 20, 248.
9. Schulman, J.L. and Palese, P. (1977). *J. Virol.* 24, 170.

A SINGLE GENE CONTROLLING THE HOST RANGE OF FOWL PLAGUE VIRUS

J.W. ALMOND and R.D. BARRY

Division of Virology, Department of Pathology,
University of Cambridge,
Laboratories Block, Addenbrookes Hospital,
Hills Road, Cambridge CB2 2QQ.

The genetic elements which control and influence the diverse biological properties of influenza viruses are at present poorly understood. Of particular interest are the virus-coded factors which influence properties such as virulence, pathogenicity and host range (3, 22, 28), since they are probably of central importance in the evolution of new pandemic strains. Recent major advances in the characterisation of the genome of influenza viruses (1, 9, 16, 18, 20) facilitated by improved methods of polyacrylamide gel electrophoresis (4) and by the finding that the RNA segments of different strains show large differences in their migration patterns (1, 17) or base sequence homology (24), has opened up new possibilities for investigating which virus genes are associated with particular biological properties. The construction of interstrain recombinants for which the parental origin of every gene is known has already been exploited in assigning coding functions to each of the eight genome segments (1, 16, 18, 20), and in mapping *ts* mutations (1, 19, 21, 23). In a previous publication we described a method for mapping *ts* mutations which involved the construction of recombinants between two avian influenza virus strains (fowl plague virus), FPV-Rostock and FPV-Dobson (1). These experiments, together with many others, culminated in the production of a large number of Rostock-Dobson recombinants with different, but well-defined gene compositions. These recombinant strains readily lend themselves to investigation of the genetic basis of biological characteristics which differ between the two parental types. The simplest detectable difference between

FPV-Rostock and FPV-Dobson was in the range of host cells that they could infect *in vitro*. FPV-Dobson forms plaques on cells of mammalian origin, namely BHK-21 and L cell lines, whereas FPV-Rostock will not form plaques on either. Both strains however produce clear plaques on CEF. This paper describes an investigation into the genetic basis of this difference in host range.

Previous studies on the host cell range of influenza viruses, using plaque formation as a criterion for growth, implicated the viral haemagglutinin. For example, all recombinants from a cross between UV-killed FPV and the human strain A_2/Singapore/1/57 that were able to form plaques on CEF monolayers, possessed the haemagglutinin of FPV (29). FPV was the only parent capable of forming plaques in this cell type. Similar results were obtained from crossing a number of other human strains with FPV (15). It is known that cleavage of the viral haemagglutinin is necessary for the activation of progeny virions and hence for the multicycle growth required for plaque formation (11, 14). This observation, together with the finding that the degree of cleavage of the haemagglutinin varied between strains of virus in different cell types (13, 25), led Klenk and co-workers (11) to postulate that "the cleavage of the haemagglutinin may play an important role in the host range and spread of infection of these viruses". Consistent with this hypothesis is the observation that some strains of influenza viruses which do not normally form plaques on CEF monolayers will do so when a proteolytic enzyme such as trypsin is included in the overlay medium (2). Recently it has been shown that cleavage is dependent on the structure of the haemagglutinin *per se* and not on a capacity of the virus to induce any specific protease (12).

While the cleavage of the haemagglutinin is undoubtedly important in plaque formation, there may also be other viral factors which control host range. For example, Tobita (27) showed that the ability to form plaques in the human amnion-derived cell line (FL) segregated independently of the haemagglutinin and neuraminidase serotypes in recombinants between the plaque-forming WSN strain and the non plaque-forming A/HK/1/68. Further genetic investigation of plaque formation was not reported.

The Rostock-Dobson FPV recombinants provide suitable starting material for a thorough investigation of the genetic basis of host cell range because it is possible to determine the complete genotype of the recombinants. This in turn takes into consideration factors other than the surface antigens in the control of host range.

THE PARENTAL STRAINS

The Rostock strain of FPV used in this study was a plaque purified clone (S_3), that grew to high titre in eggs, caused rapid chick lethality and formed large clear plaques on CEF monolayers. No record of growth of the strain in mammalian cells could be traced.

The Dobson HRM (host range mutant) strain of FPV from which the clone 4H, used in this study was derived, has been studied previously (10). Dobson HRM is a host-range adapted variant of the Dobson wild-type strain of FPV. The adaptation process consisted of multiple passages and one plaque purification in BHK-21 cells (Zavada, personal communication).

The ability of Rostock S_3, Dobson HRM clone 4H and the Dobson wild type to form plaques on CEF, BHK-21 and L cell monolayers is summarised in Table I. We were surprised to find that wild type Dobson forms plaques as efficiently as Dobson HRM in BHK-21 cells and that the adaptation process extended the host range to L cells.

GENETIC BASIS OF HOST RANGE

The genetic determinants of plaque formation on BHK-21 and L cell monolayers, that is, the difference between Rostock (S_3) and Dobson HRM (4H) was investigated as follows. A number of Rostock-Dobson (S_3-4H) recombinants of known or partially-known genetic composition were tested for their ability to form plaques on L cell, BHK-21 cell and CEF monolayers. The results obtained, together with the genotypes of the recombinants, determined as described previously (1) are shown in Table II. Those recombinants which formed plaques in BHK-21 and L cells did so with an efficiency similar to that of Dobson 4H. The results indicate that plaque formation in both L and BHK-21 cells was dependent on the presence of the Dobson P_3 protein since this was the only common feature of all the viruses which exhibit this phenotype. Recombinants which were predominantly Dobson (e.g. Di44-7 and Dd45-e) but lacked Dobson P_3 were unable to form plaques, whereas those which were predominantly Rostock (e.g. DimN5c-1 and DimN5c-3) but possessed Dobson P_3, formed plaques in the same way as the Dobson 4H parent.

Because cleavage of the haemagglutinin is important in plaque formation (2, 11, 14) the possibility that the Dobson P_3 protein might somehow facilitate this process was investigated. Infected cells were pulse-labelled at 4 hr post-infection for 15 min with ^{35}S-methionine and subjected to electrophoresis as described elsewhere (8). A parallel set of samples was pulse-labelled and then chased for 45 min.

TABLE I

Plaque formation of FPV strains on three cell types

	Plaque titres*		
	CEF	BHK	L
FPV-Rostock, clone S_3	9.5×10^7	0	0
FPV-Dobson, wild type	2.7×10^9	9.0×10^8	0
FPV-Dobson, HRM clone 4H	1.4×10^9	8.0×10^8	1.8×10^8

*0 = no detectable plaques at any m.o.i.

The viruses investigated in this way are shown in Table III. The outcome of these experiments is shown in Figures 1 and 2. A reduction in the intensity of the haemagglutinin (HA) band, suggesting cleavage, was detected in all chased samples regardless of cell type or virus strain. It appeared that Rostock HA was cleaved in BHK-21 cells even though it could not form plaques (detection of the cleavage products, HA_1 and HA_2 was difficult in BHK-21 cells because host-cell protein synthesis was not shut off). This result, together with genetic evidence for the non-involvement of HA in host cell range control (e.g. R47i-2, that contains Dobson HA but does not form plaques, compared with DimN5c-1, containing Rostock HA but able to form plaques) led to the conclusion that the defect in plaque formation is due to an incorrect or incomplete functioning of the Rostock P_2* protein in the non-permissive cells.

DOBSON WILD TYPE

An important question is whether the functional equivalent of Dobson P_3 protein plays a role in the regulation of virus multiplication in various cell types for other influenza virus strains. Suggestive evidence that it does comes from consideration of the data presented in Table II, with respect to Dobson wild-type. Because all the recombinants tested which formed plaques in BHK-21 cells, also formed plaques in L cells, both properties must be carried by the gene which codes for the P_3 protein of Dobson HRM 4H. This implies that the mutation(s) responsible for the derivation of Dobson HRM from Dobson wild-type occurred in this same gene. The possibility cannot be ruled out however that, to grow in L cells, the Dobson HRM-P_3 protein requires cooperation with another gene product which can be provided by Rostock and Dobson HRM, but not by Dobson wild type. Construction of recombinants between Dobson wild type and Rostock would be required to provide conclusive evidence on this point.

Dobson HRM and Dobson wild type have been compared previously (10), and it was found that the HA of Dobson HRM migrated slightly faster than that of wild type on polyacrylamide gel electrophoresis. Also the RNA migration patterns showed a difference in the position of one of the bands. However, no data was presented which correlated these changes directly with the change in host range. Comparison of infected cell proteins of Dobson HRM with those of Dobson wild type (Fig.2) confirms the observation that the HA's migrate differently.

---

*Rostock P_2 is equivalent to Dobson P_3 see legend to Table II.

TABLE II

VIRUS	PROTEINS								PLAQUE TITRES		
	P_1	P_2 &	P_3	HA	NP	NA	M	NS	CEF	BHK	L
ROSTOCK	R	R^2	R^3	R	R	R	R	R	9.5×10^7	0	0
DOBSON	D	D^2	D^3	D	D	D	D	D	1.4×10^9	8.0×10^8	1.8×10^8
recombinants											
Di 47 c-1	R	R^2	R^3	R	R	R	R	D	7.5×10^7	0	0
Di US4-5	R	R^2	R^3	D	R	R	?	R	1.8×10^8	0	N.D.
Di US1c-2	R	R^2	R^3	R	D	R	?	R	1.3×10^8	0	N.D.
Di 39c-5	D	R^2	R^3	R	D	R	?	R	1.9×10^8	0	N.D.
Di mN5 c-1	D	D^3	R^3	R	R	R	?	R	2.6×10^8	1.4×10^8	3.7×10^7
Di mN5 c-3	R	D^3	R^3	R	D	R	?	R	1.9×10^8	1.5×10^8	1.9×10^7
Di 39 c-7	?	R^2	R^3	R	D	R	?	D	1.2×10^8	0	N.D.
Di 44 c-4	D	D^3	R^3	R	D	R	R	R	9.0×10^7	1.0×10^8	N.D.
Dd 45-h	R	D^2	R^2	D	D	?	?	R	4.3×10^7	0	N.D.
Di 44-2	R	D^3	R^3	R	D	R	?	D	2.9×10^8	1.6×10^8	6.7×10^7
Di 44 c-1	R	D^3	R^3	D	D	R	?	D	3.3×10^7	5.5×10^7	1.4×10^7
Di mN5 c-6	D	D^3	R^3	D	D	R	?	D	4.5×10^8	2.3×10^8	N.D.
R47 1-2	D	D^2	R^2	D	D	E	?	D	1.1×10^7	0	0
Dd 45-3	D	D^2	R^2	D	D	?	?	D	6.0×10^7	0	N.D.
Di 44-7	D	D^3	R^3	R	D	D	D	D	2.3×10^7	1.9×10^7	N.D.
Dd 45-e	D	D^2	R^2	D	D	R	D	D	2.4×10^7	0	N.D.

Genotypes, and plaque titres of FPV Rostock (S_3), Dobson (HRM-4H) and recombinants on CEF, BHK and L cell monolayers. 0 = no detectable plaques at any dilution. R = Rostock, D = Dobson, N.D. = not done, HA = haemagglutinin, NP = nucleoprotein, NA = neuraminidase, M = matrix protein, NS = non-structural protein.

P_1, P_2 and P_3, the polymerase associated proteins have been numbered purely on the order of their electrophoretic mobilities. Recombination studies have shown that Rostock P_2

Legend to Table II continued.

is equivalent to Dobson P_3 and similarly, Rostock P_3 is equivalent to Dobson P_2. Therefore recombinants involving these P proteins will contain either Rostock P_2 *and* Dobson P_2 or Rostock P_3 *and* Dobson P_3. Therefore R^3 = Rostock P_3, R^2 = Rostock P_2, D^3 = Dobson P_3, D^2 = Dobson P_2.

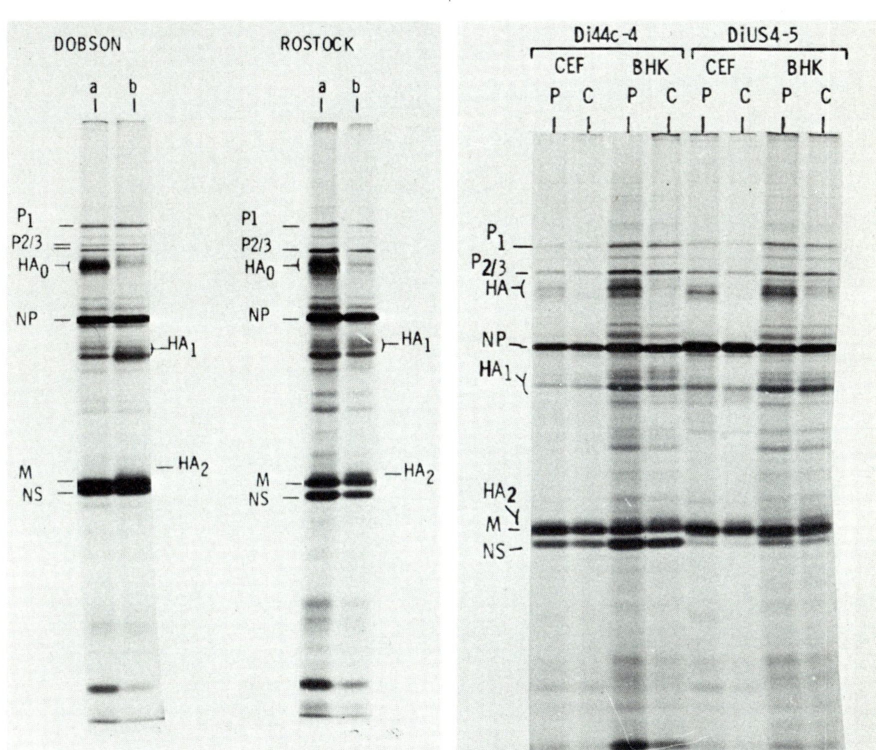

Fig.1. *Investigation of HA cleavage in CEF and BHK-21 cells. Infected cells were pulse-labelled with ^{35}S-methionine for 15 min at 4 hr post-infection, and either harvested, or chased for 45 min with cold methionine. Infected cells were prepared for electrophoresis as described in reference 8.*

a) Dobson HRM-4H and Rostock S_3 in BHK cells.
a = pulse, b = chase.

b) Recombinants in CEF and BHK cells.
P = pulse, C = chase.

In addition, a difference in the migration of the P_2 proteins was detected. A change in the migration of the putative host-range controlling protein P_3 of Dobson HRM,however, was not observed.

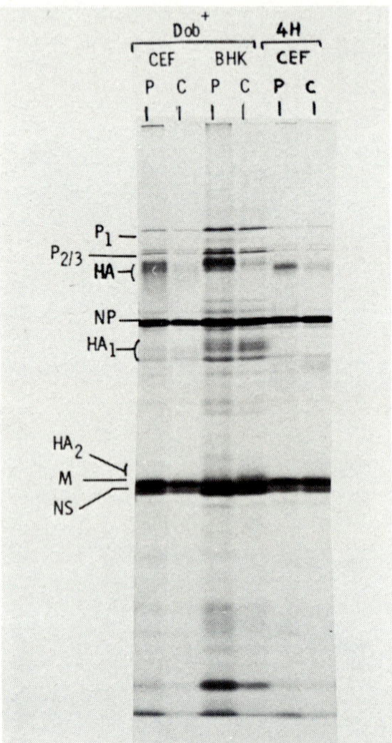

Fig.2. *Pulse-chase labelling of virus-induced proteins in cells infected with Dobson HRM (4H) and Dobson wild-type. Cells were pulse-labelled with ^{35}S-methionine for 15 min at 4 hr post-infection (P), or chased for 45 min with cold methionine (C). Samples were prepared and subjected to electrophoresis as described by Inglis et al. (8).*

CONCLUSION

The results presented here indicate that the Dobson P_3 protein or its equivalent can control the host cell range of FPV. This conclusion suggests that this protein depends on host cell factor(s) for its function.

The role of the Dobson P_3 protein in virus replication is unknown. Extrapolation from the results of other workers suggest that it may be involved in vRNA synthesis (19), or at some stage in the amplification of mRNA synthesis (7). Alternatively this protein may function in nuclear membrane transport, a role postulated for one of the P proteins by

TABLE III

Investigation of HA cleavage. Viruses examined

Virus	Cell type	Characteristics
Rostock S_3	BHK	Non-permissive
Dobson 4H	BHK	Permissive
Dobson wild type	BHK CEF	Permissive
Di44c-4	BHK CEF	Rostock HA with Dobson P_3 protein
DiUS4-5	BHK CEF	Dobson HA with Rostock P_2 protein

Scholtissek and Bowles (24). Studies on abortive infections support this latter explanation since a block at the stage of migration of RNP from nucleus to cytoplasm has been observed in non-permissive cells (5, 6, 26).

Further experiments are required to determine the precise mechanism by which the Dobson P_3 protein effects a control over host range; a comparison of the growth of Rostock and Dobson in CEF versus BHK or L cells may be valuable in this respect.

The significance of these results to other influenza virus strains is particularly important; it is possible that they merely indicate the host range limitation of FPV-Rostock and it may be found that mutations in other viral genes can also selectively affect multiplication in different cell types. On the other hand, if the P_3 protein is important in the control of host range in other influenza virus strains, its properties would be worthy of special attention in the construction of recombinant viruses for use as live vaccines in man. Work with other virus strains is required before a definite conclusion can be reached on the importance of this protein in the control of host range.

REFERENCES

1. Almond, J.W., McGeoch, D. and Barry, R.D. (1977). *Virology* 81, 62.
2. Appleyard, G. and Maber, H.B. (1974). *J. Gen. Virol.* 25, 351.
3. Burnet, F.M. and Lind, P.E. (1954). *Nature (London)* 173, 627.
4. Floyd, R.W., Stone, M.P. and Joklik, W.K. (1974). *Anal.*

Biochem. 59, 599.
5. Franklin, R.M. and Breitenfeld, P.M. (1959). *Virology* 8, 293.
6. Ghandi, S.S., Bell, H.B. and Burke, D.C. (1971). *J. Gen. Virology* 13, 423.
7. Hay, A.J., Skehel, J.J., Abraham, G., Smith, J.C. and Fellner, P. (1978). This volume, chapter 31.
8. Inglis, S.C., Carroll, A.R., Lamb, R.A. and Mahy, B.W.J. (1976). *Virology* 74, 489.
9. Inglis, S.C., McGeoch, D. and Mahy, B.W.J. (1977). *Virology* 78, 522.
10. Israel, A., Semmel, M. and Huppert, J. (1975). *Virology* 68, 503.
11. Klenk, H.D., Rott, R., Orlich, M. and Blödorn, J. (1975). *Virology* 68, 426.
12. Klenk, H.D., Rott, R. and Orlich, M. (1977). *J. Gen. Virol.* 36, 151.
13. Lazarowitz, S.G., Compans, R.W. and Choppin, P.W. (1973). *Virology* 52, 199.
14. Lazarowitz, S.G. and Choppin, P.W. (1975). *Virology* 68, 440.
15. McCahon, D. and Schild, G.C. (1971). *J. Gen. Virol.* 12, 207.
16. Palese, P. and Schulman, J.L. (1976). *Proc. Natl. Acad. Sci. USA* 73, 2142.
17. Palese, P. and Schulman, J.L. (1976). *J. Virol.* 17, 876.
18. Palese, P., Ritchey, M.B. and Schulman, J.L. (1977). *Virology* 76, 114.
19. Palese, P., Ritchey, M.B. and Schulman, J.L. (1977). *J. Virol.* 21, 1187.
20. Ritchey, M.B. and Palese, P., and Schulman, J.L. (1976). *J. Virol.* 20, 307.
21. Ritchey, M.B. and Palese, P. (1977). *J. Virol.* 21, 1196.
22. Rott, R., Orlich, M. and Scholtissek, C. (1976). *J. Virol.* 19, 54.
23. Scholtissek, C. and Bowles, A.L. (1975). *Virology* 67, 576.
24. Scholtissek, C., Harms, E., Rohde, W., Orlich, M. and Rott, R. (1976). *Virology* 74, 332.
25. Stanley, P., Gandhi, S.S. and White, D.O. (1973). *Virology* 53, 92.
26. ter Meulen, V. and Love, R. (1967). *J. Virol.* 1, 626.
27. Tobita, K. (1971). *Arch. Gesamte Virusforsch.* 34, 119.
28. Tuckova, E., Vonka, V. and Starek, M. (1968). *Acta Virol.* 12, 316.
29. Tumova, B. and Pereira, H.G. (1965). *Virology* 27, 253.

PHENOTYPIC PROPERTIES ASSOCIATED WITH INFLUENZA GENOME SEGMENTS

W.J. BEAN, Jr., and R.G. WEBSTER

Division of Virology,
St. Jude Children's Research Hospital,
332 N. Lauderdale, P.O. Box 318,
Memphis, Tennessee 38101, USA.

Strains of influenza A viruses vary greatly in their ability to produce disease in both man and animals. The pandemic of 1918-19 is believed to have been caused by a strain which was more virulent than those that have appeared subsequently, and while most avian influenzas cause relatively mild infections, a few, such as fowl plague virus, are highly virulent. Little is known about the genetic and biochemical factors that govern the properties of intrinsic virulence. Earlier work indicated that virulence is polygenic (6) and that it can be dissociated from hemagglutinin (HA) and neuraminidase (NA) activities (7).

The RNA of influenza viruses occurs as eight separate segments in the virion. Separation of the virion RNA segments has permitted genetic analysis of influenza virus and each segment can be ascribed to a particular protein (10). Association between virion RNA segments and biological properties of the virus can now be studied. In this report, we describe efforts to associate virulence with specific virus genome RNAs and to elucidate further the biological significance of cleavage of the HA protein into two subunits.

PROCEDURES

Viruses and recombinants

The virus strains used in this study were A/turkey/Ontario/7732/66 [Hav5 Nav6] (8) and A/WSN/33 [H0N1]. Turkey/Ontario is a highly virulent strain which causes a fatal viremia in turkeys and young chickens while WSN is avirulent in these species. Both strains form plaques with high efficiency on chick embryo fibroblast monolayers, and their plaque size and morphologies are readily distinguishable.

Recombinants were produced by mixedly infecting confluent chick embryo fibroblast monolayers (16 mm) with varying dilutions of the two parental strains and incubating the infected cells with antisera to select for viruses with the desired surface antigens. Virus yields were passaged twice in the presence of the same antisera and the putative recombinants were then grown in eggs and identified serologically. Those with the desired antigenic combinations were cloned twice, either at limit dilution in eggs or by plaque isolation in CEF. Backcrosses to the parental-type viruses were produced from pairs of recombinants. UV-inactivation of one parent was sometimes used to encourage production of recombinants with a preponderance of internal components from the non-UV-treated parent (11).

Assay for virulence of viruses

Virulence of the parental and recombinant influenza viruses was determined by injecting 10^4 PFU intramuscularly into 1-day-old chicks. For the purpose of this investigation, virulence is defined as the ability to produce morbidity and mortality in chicks without regard to other biological properties such as plaque size or ability to grow to high titers in eggs or tissue culture.

Analysis of Genome Composition

Virion RNA segments were either labeled with ^3H-uridine during replication in CEF monolayers or the RNA extracted from egg-grown virus was labeled with ^{125}I *in vitro*(Bean, in preparation). Iodination was done using a modification of the method described by Heiniger *et al.* (3). Separation of the virion RNA segments was accomplished using a formamide polyacrylamide gel system adapted from that described by Duesberg and Vogt, (2), with the formamide concentration reduced to allow the formation of some secondary structure in the RNA segments. A 65% formamide, 3% acrylamide gel was used to resolve segments 1 through 6 while a 50% formamide, 3% acrylamide gel was necessary to distinguish segments 7 and 8 (figs. 1 and 2). Electrophoresis was performed in a vertical slab apparatus with a 0.15 x 10 x 18 cm gel. All gels were processed for fluorography as described by Bonner and Laskey (1).

RESULTS

Analysis of recombinants with a defect in hemagglutinin cleavage

Proteolytic enzymes such as trypsin that cleave the HA into two subunits, HA_1 and HA_2 (4, 5) stimulate plaque formation in

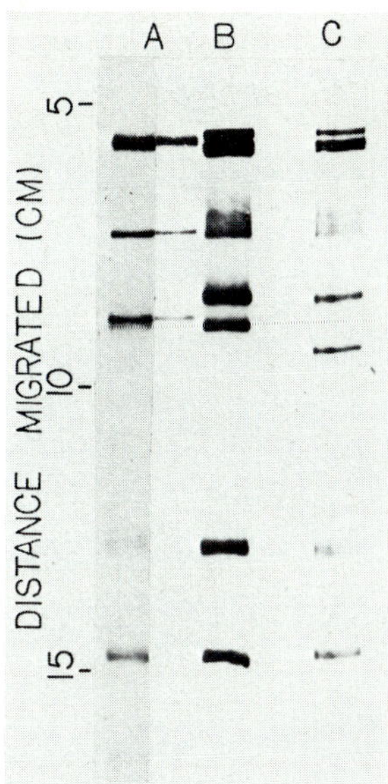

Fig.1. *Fluorogram of ^{3}H-uridine labeled viral RNA segments separated on a 65% formamide, 3% acrylamide gel*

Electrophoresis was for 14 hours at 170 volts with the temperature maintained at 15°.

Lane A - Turkey/Ontario, (left side is a 4 day exposure, right side is a 1 day exposure)

Lane B - Recombinant with all RNA segments derived from WSN, with the exception of segment 6 coding for the neuraminidase

Lane C - WSN

many strains of influenza virus. In the present study we will demonstrate that cleavage of the HA and plaque production are dependent on particular HA-NA combinations.

In our initial efforts to produce recombinants between turkey/Ontario and WSN we found that while it was relatively easy to obtain recombinants containing T/Ont$_{(H)}$-WSN$_{(N)}$, it was much more difficult to produce the reciprocal recombinant. The first recombinant isolated possessing WSN$_{(H)}$-T/Ont$_{(N)}$ did not form plaques on chick embryo fibroblast monolayers, despite the fact that both of the parental strains are plaque formers. The possibility that this recombinant was a non-plaque-forming mutant as well as a recombinant was essentially ruled out by the finding that backcrosses with reciprocal recombinants produced progeny of both parental types, all of which would plaque.

Analysis of the RNA composition of this recombinant indi-

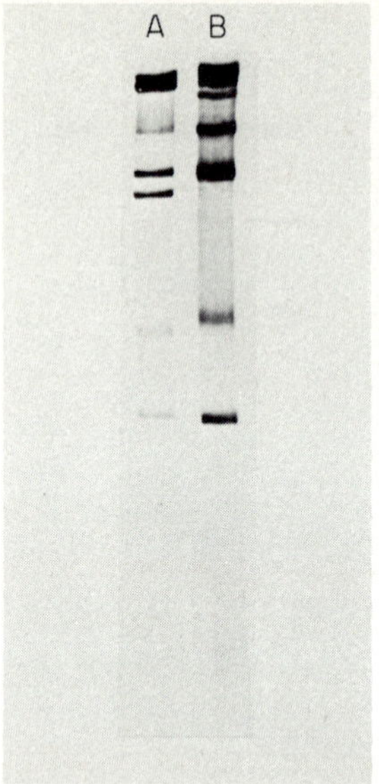

Fig.2. *Fluorogram of ^{3}H-uridine labeled viral RNA segments separated on a 50% formamide, 3% acrylamide gel*

Electrophoresis was for 14 hours at 150 volts with temperature maintained at 15°. Exposure 3 days.

Lane A - WSN

Lane B - Turkey/Ontario

cated that all of its genetic information was derived from the WSN parent with the exception of the RNA coding for NA. A second recombinant of this antigenic type was isolated and found to be phenotypically identical to the first isolate. However, analysis of its RNA composition indicated that all of its genetic information was derived from the turkey/Ontario parent with the exception of the RNA segment coding for the WSN HA. Further study revealed that both recombinants formed plaques if trypsin was incorporated into the agar overlay of the CEF monolayer (fig.3). Trypsin dependence for plaque formation was found to be a general phenomenon for seven recombinants with this combination of surface antigens. The remaining genetic constitution appeared to be irrelevant so long as the recombinant contained the HA from WSN and the NA from turkey/Ontario. In contrast, 49 other recombinants, including those of the reciprocal antigenic type and back-crosses having both surface proteins from one parent, all plaqued without trypsin.

In addition to the trypsin-dependent characteristics, all of the non-plaquing recombinants are avirulent in chickens.

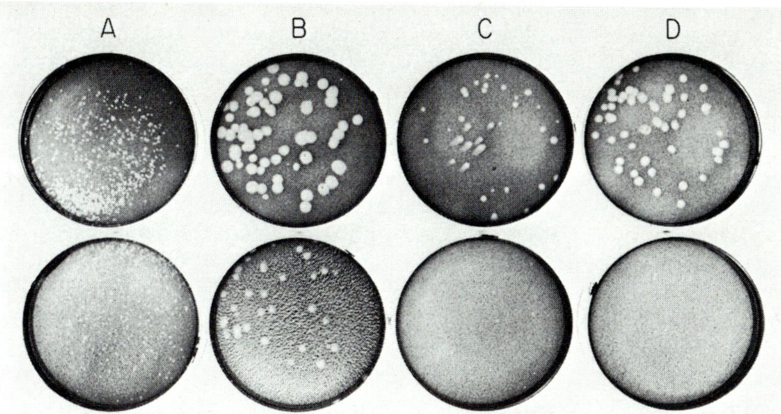

Fig.3. *Trypsin dependent plaque formation by WSN(HA)-Turkey/Ontario(NA) recombinants in CEF monolayers*

Upper Row - 5 μg trypsin/ml
Lower Row - No trypsin
A - Turkey/Ontario
B - WSN
C - Recombinant containing all RNA segments from turkey/Ontario except that coding for the WSN hemagglutinin
D - Recombinant containing all RNA segments from WSN except that coding for the turkey/Ontario neuraminidase

This avirulence is not due simply to the presence of the WSN HA since, as discussed below, virulent recombinants with both WSN surface antigens can be produced.

These results indicate that there is a basic incompatibility between the WSN HA and the turkey/Ontario NA in that when they are combined on a single particle, a biologically crippled virus is produced. This suggests that there is a functional relationship between the NA and the HA, in that a virion must have the proper NA to produce a cleaved HA.

Analysis of intrinsic virulence of influenza virus

To determine which gene segments of influenza virus are responsible for virulence, a large number of recombinants between the virulent and avirulent strains of influenza virus described above, were studied. The genetic composition of the parental and recombinant viruses was determined by polyacrylamide gel electrophoresis and correlated with their virulence for 1-day-old chicks. Fifty-six recombinants could be assigned to 3 groups; Group 1 viruses were identical to turkey/Ontario (all chicks being killed within 3 days). Group 2 recombinants were also virulent; these chicks stopped eating and became listless with ruffled feathers. Forty to 90% died 2 to 6 days after infection while the remaining birds recovered. Group 3 recombinants were identical to WSN and were completely avirulent in chickens. Table I lists the genetic composition of 11 recombinants grouped according to virulence. All three combinations of surface proteins, other than the $WSN_{(H)}$-$T/Ont_{(N)}$ group discussed above, are represented in both the virulent and avirulent groups. This indicates that with these two virus strains the surface proteins are not critical in determining virulence. However, two of the RNA segments coding for internal proteins are consistently linked with virulence. All recombinants that were virulent for chickens contained RNA segments 3 and 5 from the virulent parent, while avirulent recombinants derived these segments from WSN. RNA segment 5 codes for the nucleoprotein, while segment 3 has been reported to code for protein P2 of PR8 and Hong Kong (9). In all recombinants so far isolated and characterized, segments 3 and 5 have remained "linked". Although more recombinants will have to be characterized to determine if these segments can be segregated, the data suggest that the proteins coded by these RNAs may have a close functional relationship. The two strains used in this study may not be related closely enough for these proteins to be compatible.

One recombinant, Rec 47 (with both surface antigens from turkey/Ontario), has an RNA segment (No.1) that migrates more slowly in a 65% formamide gel than the corresponding segment of either parent. This may represent a mutation which has caused a change in the secondary structure of the RNA molecule. However, the possibility that a change may have arisen via a true recombinational event cannot be ruled out and is currently being investigated.

DISCUSSION AND CONCLUSIONS

These studies show that there is a functional relationship between the HA and NA of influenza viruses necessary for cleavage of the HA polypeptide and plaque formation. The WSN HA is incompatible with the NA of turkey/Ontario in that virions with this combination of surface antigens require trypsin for plaque formation. These viruses appear to be biological cripples, since they are also avirulent. Virions with any other combination of surface proteins of these two strains may be either virulent or avirulent depending on their internal components. Two RNA segments were consistently linked to each other and to virulence. In all virulent recombinants, segments 3 and 5 were derived from turkey/Ontario while in all

TABLE I

Comparison of Genome Composition with Virulence

Virus	Surface Antigen HA	Surface Antigen NA	Origin of RNA Segments 1	2	3	4	5	6	7	8	Virulence Group[a]
Parent	T	T	T	T	T	T	T	T	T	T	1
Rec 49	T	T	T	T	T	T	T	T	W	T	
Rec 6	T	W	T	T	T	T	T	W	T	T	
Rec 47	W	W	?[b]	T	T	W	T	W	W	T	
Rec 37	T	T	W	T	T	T	T	T	T	T	2
Rec 22	W	W	T	T	T	W	T	W	T	W	
Rec 16	W	W	T	W	T	W	T	W	T	T	
Rec 14	W	W	T	T	T	W	T	W	W	W	
Rec 42	W	W	W	T	W	W	W	W	W	W	3
Rec 43	W	W	W	W	W	W	W	W	W	T	
Rec 48	T	T	T	T	W	T	W	T	W	T	
Rec 56	T	W	W	W	W	T	W	W	W	W	
Parent	W	W	W	W	W	W	W	W	W	W	

[a]Group 1 viruses were identical to turkey/Ontario, all chicks dead within 3 days. Group 2 recombinants were also virulent; all infected chicks ceased eating and became listless with ruffled feathers and 40 to 90% died 2 to 6 days after infection while the remaining birds recovered. Group 3 recombinants were identical to WSN; and were completely avirulent.

[b]RNA segment does not co-migrate with corresponding segment of either parent.

avirulent recombinants these segments were from WSN.

The theoretical number of possible recombinants between any two influenza viruses with eight RNA segments is 256. However, the actual number may be much lower due to the incompatibility of certain combinations. Additional recombinants are being selected to confirm the results obtained with the limited number of recombinants listed above. In addition, other biological properties such as plaque morphology, host range, and virion morphology are also being studied in an attempt to find a genetic basis for these properties.

ACKNOWLEDGEMENTS

The authors wish to thank Diana Jones for excellent technical assistance. This work was supported by Research Grant AI 08831 from the National Institute of Allergy and Infectious Diseases, Childhood Cancer Center Grant CA 08480 from the National Cancer Institute and by ALSAC.

REFERENCES

1. Bonner, W.M. and Laskey, R.A. (1974). *Eur. J. Biochem.* 46 83.
2. Duesberg, P.H. and Vogt, P.K. (1973). *J. Virol.* 12, 594.
3. Heiniger, H.J., Chen, H.W., and Commerford, S.L. (1973). *Journal of Applied Radiation and Isotopes*, Vol. 24, 425.
4. Klenk, H-D., Rott, R., Orlich, M. and Blödorn, J. (1975). *Virology* 68, 426.
5. Lazarowitz, S.G. and Choppin, P.W. (1975). *Virology* 68, 440.
6. McCahon, D. and Schild, G.C. (1972). *J. Gen. Virol.* 15, 73.
7. Mayer, V., Schulman, J.L. and Kilbourne, E.D. (1973). *J. Virol.* 11, 272.
8. Narayan, O., Lang, G., and Rouse, B.T. (1969). *Arch. ges. Virusforsch.* 26, 149.
9. Palese, P., Ritchie, M.B., and Schulman, J.L. (1977). *Virology* 76, 114.
10. Palese, P. (1977). *Cell* 10, 1.
11. Schulman, J.L. and Palese, P. (1976). *J. Virol.* 20, 248.

INFLUENZA A VIRUS RECOMBINANTS OF DEFINED VIRULENCE FOR MAN: DIFFERENCES IN VIRUS RNA, POLYPEPTIDES AND TYPE-SPECIFIC NP ANTIGENS

J.S. OXFORD, D.J. McGEOCH[*], G.C. SCHILD and A.S. BEARE[**]

Division of Virology,
National Institute for Biological Standards and Control,
Holly Hill, London, NW3 6RB.

[*]*MRC Virology Unit,*
Institute of Virology,
Church Street, Glasgow.

INTRODUCTION

Recent studies have indicated that the genome of the avian influenza virus A/Fowl Plague/27 consists of 8 pieces of single stranded RNA (9). Each of these RNAs represents the gene for one virus-coded polypeptide and RNA - polypeptide assignments have been made for several strains of influenza A viruses (2, 7, 12).

Such detailed analysis of the virus genome offers a method for characterising the genome of attenuated influenza viruses to be used as vaccines in man. Palese and Schulman (12) have demonstrated that under suitable conditions the RNA fragments of different influenza A viruses migrate at different rates in polyacrylamide slab gels containing 6M urea. In addition, differences in the polypeptide migration rates can be measured (2, 13). Thus, in recombinant influenza viruses, which are commonly used as candidate attenuated vaccine strains, (reviewed in chapter 75) the parental origin of each RNA gene can now be established. In the present study we have analysed 6 potential recombinants of influenza A/PR/8/34 (H0N1) and A/England/69 (H3N2) viruses, obtained from a mixed infection in eggs by cloning in the presence of specific antibody. The parent, A/PR/8/34 virus had a high yield of virus haemagglutinin (HA) in embryonated hens eggs and was attenuated in virulence for man (3, 5) whilst the A/England/69 parent was

[**]*MRC Common Cold Research Unit, Harvard Hospital, Salisbury, Wiltshire.*

virulent for man, and provided a low yield of HA in eggs. The recombinants were either virulent or attenuated for man. Therefore by complete analysis of the genome of the recombinants we attempted to answer questions concerning relatedness of possible gene groups or constellations to attenuation. We recognized that the recombinant set was small but considered that the previous characterisation of their human virulence rendered this examination worthwhile.

METHODS

Viruses

The recombinant series used in the study was originally isolated and characterised biologically by MacCahon and Schild (8). Clinical studies were carried out using volunteers as described by Beare and Reed (4). The viruses were cloned by terminal dilution and titration in 10 day old embryonated hens eggs. For polypeptide or RNA analysis, virus was purified by differential and rate zonal centrifugation as described previously (14). Virus concentrates of 15 mg/ml protein were stored at -70°.

Immunological detection of NP antigen

Influenza nucleoprotein antigen (NP) was isolated from the recombinant X-42 virus (A/equine/Prague/56 Heq1-A/Port Chalmers/1/73 N2) by electrophoresis on cellulose acetate strips at pH 6.6 (11) using purified and concentrated virus disrupted with 1% w/v sodium sarcosyl detergent (SS or Ciba-Geigy NL97). The NP antigen of this recombinant is derived from the A/Port Chalmers/1/73 parent (unpublished data). Antiserum was produced in goats and rabbits by two serial i.m. injections of 40 µg protein in Freunds complete adjuvant. Immunodouble diffusion tests were carried out in 1% agarose as described previously (14).

Polypeptide analysis

MDCK cells were infected with allantoic fluid suspensions of virus to give a multiplicity of approximately 10 pfu/cell. After 6 hr or 18 hr incubation at 37° the cells were pulsed for 30 min with 20 µCi per 60 mm petri dish in 0.2 ml Geys medium of ^{35}S-methionine. The cells were then washed and lysed with a mixture of 2% w/v SDS and 0.6% v/v β-mercaptoethanol, heated at 100° for 2 min, and 40 µl volumes layered onto slab gels (15) of 17.5% polyacrylamide using discontinuous buffers (1) and electrophoresed for 18 hr at 40 mA. Gel slabs were dried and autoradiographed by standard procedures as described previously (10).

Gel electrophoresis of virus RNAs

RNA was extracted from two preparations of each virus as described previously (7) and fractionated by electrophoresis through 2.6% and 2.8% polyacrylamide slab gels containing 6M urea. These gels were 22 cm long x 15 cm x 1.5 mm and were prepared as described by Palese and Schulman (12). 2-5 µg of each preparation were loaded in 1 cm slots and electrophoresis carried out for 27 hr at 110V, at room temperature and with tank buffer recirculation. The gel slabs were then stained for 30 min with 0.2% toluidine blue in 40% methoxyethanol and destained with several changes of 30% methoxyethanol until the stained RNA bands became visible.

RESULTS

Analysis of RNA extracted from the recombinants indicated the presence of 8 bands in 6M urea-acrylamide gels. RNAs with coding potential for 7 of the 8 virus polypeptides could be identified by migration rate differences as originating from either the A/PR/8/34 or the A/England/69 parent viruses (Table I). Of particular interest was the gene composition of clones 7 and 6, both virulent viruses for man. Clone 7 was established to be identical to the A/England/69 parent in all 7 genes examined by the above technique. Clone 6, in contrast had received at least 6 of the 8 genes from the avirulent A/PR/8/34 parent and yet was virulent for man. When the gene composition of the remaining four attentuated recombinants was examined, (Table I) no particular constellation or group of genes was characteristic of all viruses which could have resulted in the attenuation property. All the attenuated recombinants possessed genes 5 or 3 derived from the A/PR/8/34 parent but the virulent clone 6 virus also possessed those 2 genes.

Confirmation of the parental origin of the NP, HA and NS1 genes established above by direct analysis of virus RNA could be obtained by investigation of migration rate differences of the polypeptides. Thus, clear differences (Fig.1) were detected in the rates of migration of the NP and NS1 polypeptides. NS2 polypeptide was detected in virus infected MDCK cells but no differences were detected between the migration rate of this polypeptide from the two parental viruses in the 17.5% polyacrylamide gels used in these experiments. In further experiments purified virus preparations were solubilised and electrophoresed in 17.5% slab polyacrylamide gels and the polypeptides stained with Coomassie blue. Differences in the migration rates of HA1 and HA2 polypeptides could be detected between H0 and H3 HA. The relatively small differences in the rate of migration of M protein did not allow the parental origin of gene 7 to be determined. The data for the parental origin of the NP and HA

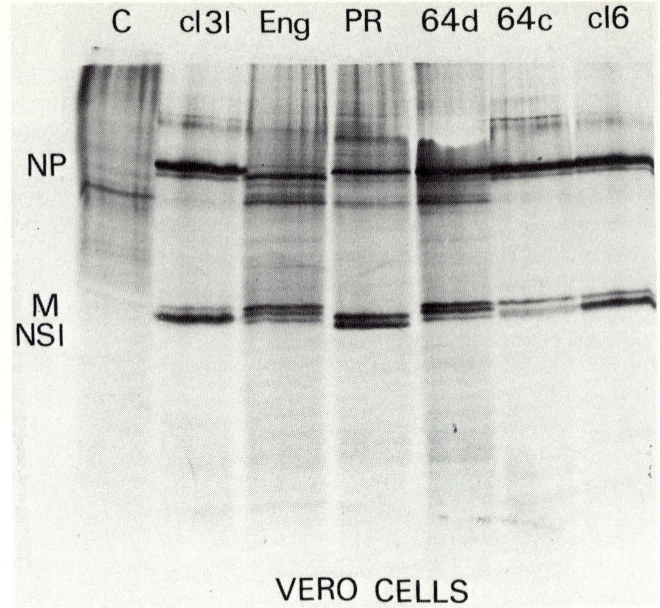

Fig.1. *Comparison of rates of migration of ^{35}S-methionine labelled virus polypeptides of recombinants*

Vero cells were pulse labelled for 30 min at 6 hr post infection with ^{35}S-methionine and the SDS lysed cell extracts electrophoresed in 17.5% polyacrylamide gel slabs with discontinuous buffer systems.

was identical to that obtained using direct analysis of RNA except that the NP of clone 31 was considered to be of A/PR/8 origin (Table I). At present we have been unable to distinguish clear migration rate differences in the three P polypeptides between the two parents.

Immunological analysis to determine the parental origin of the HA and NA genes was carried out using immunodouble diffusion tests. Furthermore, differences in antigenic determinants between the NP antigen of A/PR/8/34 and A/England/69 parents were detected. A common determinant was detected in all the A viruses tested but the marked spur formation indicated antigenic differences between the two parental viruses (Fig.2). Therefore the parental origin of the gene coding for the NP antigen could be deduced. The results were identical to those of Table 1; only clone 7 possessed the gene coding for the A/England/69 type of NP.

DISCUSSION

Direct analysis of RNA isolated from purified or semipurified

TABLE I

Comparison of migration rates of virus RNA's of recombinants in polyacrylamide gels containing 6M urea

gene	recombinant					
	clone 6	clone 7	64c	64d	clone 31	64b
1	P	E	E	P	P	P
2	P	E	E	P	P	P
3	P	E	P	P	P	P
4 (HA)	E	E	E	E	E	P
5 (NP)	P	E	P	P	?	P
6 (NA)	E	E	E	E	P	E
8 (NS1)	P	E	P	E	P	P

P gene originating from the A/PR/8/34 parent virus
E gene originating from the A/England/69 parent

Analysis was performed generally on 2 preparations of virus RNA.

Note that RNA 5 of clone 31 migrated with the corresponding A/PR/8/74 band in 2.6% gels but in a position *between* the two parental bands on 2.8% gels and so could not be clearly assigned.

---

viruses in 6M urea-acrylamide gels demonstrated the presence of 8 RNA pieces confirming earlier studies with A/FPV (9). Palese and Schulman (12) using ^{32}P-labelled virus RNA also detected 8 bands in polyacrylamide gels with A/PR/8/34 virus. Analysis of gene products of the recombinants by immunodouble diffusion and ^{35}S-methionine labelling and subsequent electrophoresis in analytical SDS polyacrylamide gels provided confirmative data on the genes coding for NP, NS1, HA and NA proteins, except for that of the origin of gene 5 in clone 31.

Virulence of influenza viruses is an important biological attribute with no known mechanism in molecular terms at present. Cleavage of virus HA may be an important factor whilst experiments with A/FP virus have suggested that gene P3 may control virus host range and hence contribute to virus virulence (1).

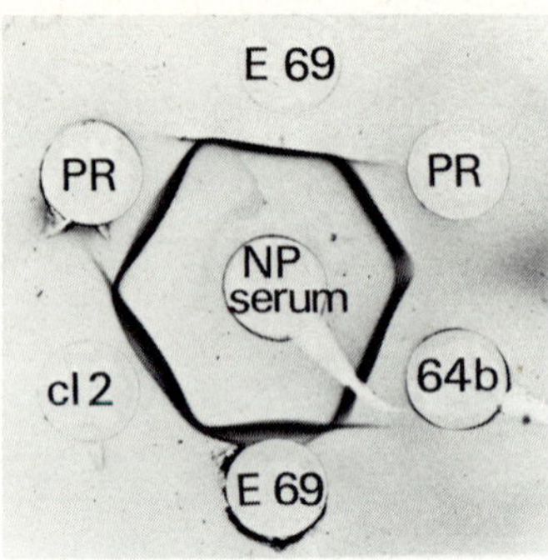

Fig.2. *Immunodouble diffusion analysis of NP antigens from influenza A recombinants*

Potent monospecific antiserum prepared against the NP of recombinant virus (X-42) was tested against 1% SS disrupted concentrates of the different viruses. Spur formation indicates antigenic differences in the NP antigen.

Recent studies with recombinants of avian influenza virus indicated that virulence was polygenic and linked with genes coding for NP and P polypeptides (chapter 69). With the cold-adapted attenuated vaccine viruses, attenuation appears to be correlated with transfer of genes 3, 5 and 7 (chapter 74). We analysed our data to determine whether there is any association of particular genes with growth ability in eggs and with human virulence. Excepting clone 7, which may be a mutant of A/England/69, all the viruses contain gene 3 and the NP gene from A/PR/8/34. Thus these genes correlate with the ability of the recombinants to grow to high titre in eggs like the A/PR/8/34 parent. This association could have other origins: for instance recombinants with A/PR/8/34 genes other than these two might be nonviable without P1 and NP from A/PR/8/34. It was of interest that clone 7 consisted of at least seven, and possibly eight, genes from the A/England/69 parent and therefore may be the result of mutation of this parent to allow good growth in eggs rather than a reassorted form. With clone 31, the NP polypeptide behaved like A/PR/8/34 parent in gel mobility or antigenicity. However, under some conditions, the corresponding RNA gene had mobility intermediate to the parental types. We presume this anomaly to be the result of mutation.

Recombinant clone 6 had 6 genes originating from the A/PR/8/34 parent and only the genes coding for HA and NA glycoproteins originated from the virulent A/England/69 parent. However, the recombinant was virulent for man. In an investigation of recombinants made with viruses appearing in 1972, 1973 and 1974,

Florent *et al.* (6) related virulence in man to the extent to which recombinant v RNA hybridized with A/PR/8/34 cRNA and considered that a figure of 55% or above signified attenuation. It would seem that the test cannot be universally applicable since clone 6 should theoretically give a hybridization figure of 75%.

Relatively few recombinants were examined in the present series, but all viruses had been characterised completely in volunteers. The difficulty of providing suitable viruses for analysis is aggravated by continuous antigenic drift in nature and the rapid acquisition of HA and NA antibodies by potential volunteers. By the use of a virulent H0 N1 virus to which few young people will be resistant, it is hoped to produce a series of recombinants suitable for long term studies (chapter 75). Whether virulence mechanisms will prove the same for all influenza viruses remains to be determined.

ACKNOWLEDGEMENTS

We would like to thank R. Newman for the preparation of purified influenza viruses, Terry Corcoran for technical assistance and B. Watts for photography.

REFERENCES

1. Almond, J.W. (1977). *Nature*, London 270 617.
2. Almond, J.W., McGeoch, D. and Barry, R.D. (1977). *Virology* 81, 62.
3. Beare, A.S. and Hall, T.S. (1971). *Lancet* ii, 1271.
4. Beare, A.S. and Reed, S.E. (1977). *In* "Chemoprophylaxis and Viral Infections of the Respiratory Tract". (J.S. Oxford, ed.). Chapter 7. CRC Press, Cleveland, Ohio, USA.
5. Beare, A.S., Schild, G.C. and Craig, J.W. (1975). *Lancet* ii, 729.
6. Florent, G., Lobmann, M., Beare, A.S. and Zygraich, N. (1977). *Archives of Virology* 54, 19.
7. Inglis, S.C., McGeoch, D. and Mahy, B.W.J. (1977). *Virology* 78, 522.
8. McCahon, D., and Schild, G.C. (1972). *J. Gen. Virol.* 15, 73.
9. McGeoch, D., Fellner, P. and Newton, C. (1976). *Proc. Nat. Acad. Sci. USA* 73, 3045.
10. Oxford, J.S. (1975). *J. Gen. Virol.* 28, 409.
11. Oxford, J.S. and Schild, G.C. (1976). *Virology* 74, 394.
12. Palese, P. and Schulman, J.L. (1976). *J. Virol.* 17, 876.
13. Palese, P., Ritchey, M.B. and Schulman, J.L. (1977). *Virology* 76, 114.
14. Schild, G.C., Oxford, J.S., and Virelizier, J.L. (1976).

In "The role of Immunological Factors in Infectious, Allergic and Autoimmune Processes". (R.F. Beers and E.C. Bassett, eds.). Raven Press, New York.

15. Skehel, J.J. and Schild, G.C. (1971). *Virology* 44, 396.
16. Studier, F.W. (1973). *J. Mol. Biol.* 79, 237.

ISOLATION AND PRELIMINARY CHARACTERIZATION OF *ts* MUTANTS OF RABIES VIRUS

FRANÇOISE BUSSEREAU and ANNE FLAMAND

Bât.400 - Universite Paris-Sud, 91405 Orsay - Cedex, France.

In an attempt to improve our understanding about the rabies virus growth cycle, a genetic study of the virus was undertaken. This paper describes conditions of isolation and preliminary characterization of induced thermosensitive (*ts*) mutants.

ISOLATION OF INDUCED *ts* MUTANTS

The CVS strain of rabies virus was employed as wild-type virus for the isolation of *ts* mutants. Preliminary studies have shown that this strain grows equally well over the temperature range of 33°-39.6° on BHK cells - 10^8 plaque forming units (pfu) per ml. The cycle lasts about 30 hours. The rabies virus plates on CER monolayers between 33° and 38.5° but not at higher temperatures. These findings led to the use of 33° as the permissive temperature (PT) and 38.5° or 39.6° as the non-permissive temperature (NPT) depending on conditions (either growth or titration).

Mutagenesis by 5-fluorouracil (5-FU): the wild type virus was mutagenized by growth in presence of the base analogue 5-FU (50 μg/ml). The BHK cells were infected at a multiplicity of infection (moi) equal to 5 pfu/cell. In presence of the drug, virus production was depressed by one log. A second growth cycle in the absence of the drug was permitted in order to allow the eventual mutation to be expressed. The mutagenized stock was frozen at -70°.

Mutagenesis by nitrous acid: the wild type virus was incubated at room temperature with 0.03M $NaNO_2$ in 0.25 M citrate-phosphate buffer (pH 4.5 or pH 4.2). At 10 min intervals aliquots were removed and frozen at -70°. Samples were assayed for infectivity. After 10 min treatment 75% of the virus was inactivated at pH 4.5 and 99% at pH 4.2.

The mutagenized stock (5-FU or nitrous acid) was then plated at 33° on CER cells. After 5 days of incubation, well isolated plaques were picked up and titrated at 33° and 38.6°. Clones which plated at 33° but not 38.6° were grown on BHK cells. (A plaque contains about 5×10^4pfu). If, for a clone, thermosensitivity was confirmed, a stock was made and kept at -70°.

Of the 2,346 clones tested from two different 5-FU mutageneses 70 were thermosensitive, which represents 3% of *ts* mutant. No *ts* mutants among 450 clones tested was found at pH 4.5 (20 and 40 min treatment) and only 2 *ts* mutants among 675 clones tested were found at pH 4.2 (10 min treatment).

Nitrous acid was less efficient than 5-FU in inducing *ts* mutations in the rabies genome.

COMPLEMENTATION TEST

During preliminary experiments, mixed as well as single infections were performed with 9 *ts* mutants and the complementation index was measured as the ratio:

$$C = \frac{\text{yield in mixed infection after 24 hr at 39.6°}}{\text{sum of the yields in single infections after 24hr at 39.6°}}$$

In each cross C was $\leqslant 2$, and therefore considered as negative (3). As those 9 mutants failed to complement each other, they seemed to be in the same complementation group. Further crosses between two of them and 38 new mutants gave only a negative result ($C \leqslant 2$). At this step of the investigation three alternative hypotheses could be formulated:

1. the mutants are not in the same group and conditions have to be worked out so as to reveal complementation.
2. all the mutants were in the same complementation group, which is not completely unlikely since in the case of VSV, 80% of the *ts* mutants fell into complementation group I (3, 7).
3. there is no complementation in the rabies subgroup unlike the VSV subgroup.

In order to test the first hypothesis conditions of complementation were changed, using a few examples of *ts* mutants to test all the other mutants. Harvests were taken at different times after infection (18, 20, 24, and 30 hr) and incubation at PT for various lengths of time (6, 8, 10 and 12 hr) followed by a shift at NPT were tried, without any success. Therefore, at present there is no evidence of complementation among the rabies *ts* mutants.

VIRAL INTRA-CELLULAR SYNTHESIS

RNA synthesis

Viral RNA synthesis in BHK cells infected with the wild type strain was shown to be maximum between 8 and 12 hr post-infection at 37° (2). Experiments were undertaken on CER infected cells with the wild type strain at 33° and 39.6° by labelling the cells with ^{3}H-uridine in presence of actinomycin D (15 µg/ml). The same results as for BHK cells were obtained in CER cells. Under those conditions (see Table I) uridine incorporation into acid-insoluble material was measured.

Uridine incorporation at NPT was compared with that obtained at PT for the wild type as well as for 35 *ts* mutants (Table I). Nineteen mutants synthesized significant amounts of RNA at NPT (more than 50% of that synthesized at PT). For 16 mutants the synthesis was less than 50% and for two of them no incorporation was detected at NPT.

Uridine incorporation by 24 of the 35 *ts* mutants previously analysed was compared with that of the wild type (Fig.1). At PT, incorporations showed a broad distribution although usually lower than the wild type probably due to the moi used. At NPT, incorporations were always lower than the wild type; however, 4 mutants synthesized more than 50% compared to the wild type and 8 mutants less than 10%.

Putative RNA^{+} mutants were retained on the basis that their RNA synthesis was the same as for the wild type at PT, and at NPT was significant, although less than the wild type. Putative RNA^{-} mutants were retained on the basis that RNA synthesis at NPT was less than 10% of the wild type and normal at PT.

Protein synthesis

Cells infected with the wild type and treated with fluorescent antibody directed against the nucleocapsid were brightly fluorescent after 20 hr at PT or NPT (Table I). All the *ts* mutants induced similar bright fluorescence at PT. At NPT 7 mutants, including those which were affected in RNA synthesis, failed to develop any distinct fluorescence in infected cells. For the other mutants fluorescence at NPT was normal.

Protein synthesis in cells infected with the wild type or *ts* mutants was studied under hypertonic conditions (5). Results are shown in Fig.2. The rabies polypeptides G, N, M_1 and M_2 were synthesized in cells infected with the wild type at PT and NPT and with *ts* 16 at PT. At NPT no protein synthesis could be detected in cells infected with this mutant. This result is coherent with the fact that the mutant induced no fluorescence in cells incubated at NPT and that RNA synthesis was inhibited at this temperature. Experiments are in progress to study protein synthesis in cells infected with mutants which induced normal fluorescence and RNA synthesis at non-permissive temperature.

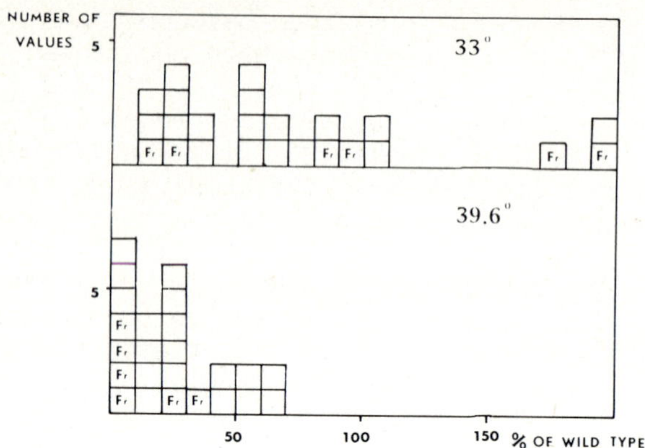

Fig.1. *Histogram of ^{3}H-uridine incorporation by 24 ts mutants. For each ts mutant the ratio:*

$$\frac{\textit{Incorporation in ts infected cells-incorporation in mock infected cells}}{\textit{Incorporation in wild type infected cells-incorporation in mock infected cells}}$$

was calculated for 33° and 39.6°.

Fr = mutant for which fluorescence is reduced or absent at the non-permissive temperature.

DISCUSSION

A total of 72 induced (5FU and nitrous acid) *ts* mutants have been isolated. During preliminary studies no positive complementation could be observed. The hypothesis that the mutants would be affected in the same function is unlikely since at least some of them could be distinguished on physiological criteria, although one has to interpret the data with care. For instance induction of fluorescence in infected cells at non-permissive temperature does not necessarily mean that secondary transcription has occurred. The fact that cells infected with the wild type start to fluoresce 4 hr after infection (T. Wiktor, personal communication) at a period where almost no amplification of RNA synthesis has occurred (1) illustrates this possibility. For early periods of incubation as well as for RNA^{-} mutants at NPT (at least those which would be able to perform primary transcription), the intensity of fluorescence will also depend on the multiplicity of infection. Therefore the presence or absence of detectable fluorescence at NPT cannot be equated with the RNA^{+} or RNA^{-} phenotype. Careful experiments at early times after infection, controlling the moi, are necessary to clarify this point. On the other hand the RNA and protein synthesis induced by rabies virus has a low efficiency (2, 6). The fact that the virus does not inhibit cellular synthesis forces the use of inhibitors which in turn may affect viral synthesis (4, 6). This probably explains why the results are not as clear as with VSV.

Our failure to detect complementation is likely to be due to the experimental conditions rather than to physiological characteristics of the viral growth cycle. Experiments are in progress, using mutants that can be distinguished in protein and RNA synthesis at non-permissive temperature, in order to answer this question.

ACKNOWLEDGEMENTS

We are grateful to Ph. Vigier for helpful discussions and J. Crick for reading the manuscript.

We thank D. Pese, J. Benejean and F. Aguero for excellent technical assistance and M. Dahuron for typing the manuscript.

Studies on protein synthesis in infected cells were done by one of us (A.F.) at the Wistar Institute. The advice of T. Wiktor, P. Madore, B. Dietzchold and J. England is greatly acknowledged. We also thank T. Wiktor for CER cells.

This research was supported by the "Centre National de la Recherche Scientifique" through the L.A. 40086, the NATO grant 994 and by the Commissariat a l'Energie Atomique (Saclay).

REFERENCES

1. Bishop, D.H.L. and Flamand, A. (1975). *In* "Control Processes in Virus Multiplication". Cambridge University Press.
2. Ermine, A. and Flamand, A. (1977). *Ann. Microbiol.* (Inst. Pasteur) 128A, 477.
3. Flamand, A. (1970). *J. Gen. Virol.* 8, 187.
4. Flamand, A., Pese, D. and Bussereau, F. (1977). *Virology* 78, 323.
5. Madore, H.P. and England, J.M. (1975). *J. Virol.* 16, 1351.
6. Madore, H.P. and England, J.M. (1977). *J. Virol.* 22, 102.
7. Pringle, C.R. (1970). *J. Virol.* 5, 559.

For Figure 2 and Table I - see over.

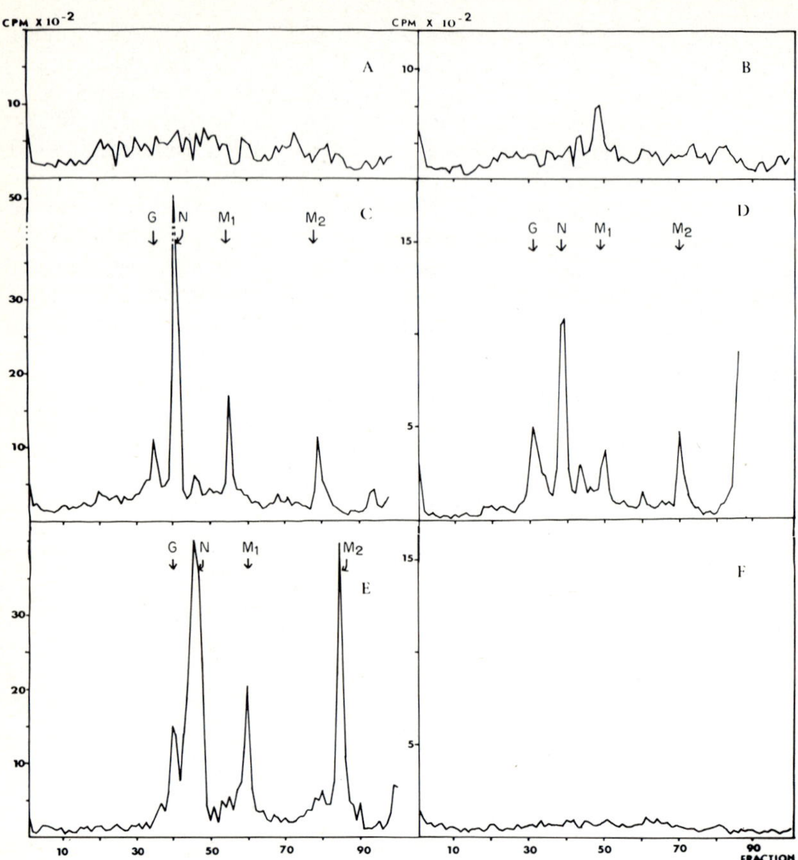

Fig.2. *Comparison of protein synthesis in BHK cells, infected with the wild type or ts 16 mutant, or mock infected cells, under hypertonic conditions.*

mock infected cells: panel A = 33°, panel B = 39.6°
wild type infected cells: panel C = 33°, panel D = 39.6°
ts 16 mutant infected cells: panel E = 33°, panel F = 39.6°

Confluent monolayers of BHK cells (3 x 10^6 cells per plate) were infected (moi of 5) or mock infected. After 24 hr of incubation at 33° or 39.6° they were washed twice with 5 ml of Earle's medium containing 10% foetal calf serum and incubated in 600 mOsM medium at 33° or 39.6°. After 30 min, medium was removed and replaced by 1 ml of 600 mOsM medium containing 20 µC of ^{3}H-leucine. After 2 hr of incubation at 33° or 39.6° the cells were washed three times and resuspended in 2 ml of ice-cold NTE buffer (0.13 M NaCl, 0.05 M Tris, pH 7.8, 0.001 M EDTA). After sonication and alcohol precipitation,

TABLE I

Behaviour of 35 *ts* mutants of rabies virus at 33° and 39.6°

Number of mutants	Fluorescence in infected cells after 24 hr		RNA synthesis in infected cells:
	33°	39.6°	^{3}H-uridine incorporation 39.6°/33°
7	bright	greatly reduced or absent	≤ 0.5
9	bright	bright	≤ 0.5
19	bright	bright	> 0.5
wild type	bright	bright	≥ 2

For Legend to Table I see over page.

cellular extracts were solubilized into 0.2 ml of buffer containing 0.0625 M Tris HCl pH 6.8, 2% SDS, 10% glycerol and 5% mercaptoethanol. 100 μl of the suspension were layered onto 11 cm 10% Laemmli gels. ^{14}C-leucine-labelled virus was subjected to co-electrophoresis with the cell lysates. The arrows refer to the ^{14}C-rabies proteins.

Table I: Behaviour of 35 *ts* mutants of rabies virus at 33° and 39.6°

FLUORESCENCE:

Slides of BHK cells were infected with wild type or *ts* mutants (moi between 1 and 5). After 20 hr of incubation cells were washed, fixed with acetone and treated with fluorescent antibody directed against the N protein (Institut Pasteur production - Paris).

URIDINE INCORPORATION

Monolayer cultures of CER cells (3×10^6 cells per 50 mm petri dish) were infected with rabies virus wild type or *ts* mutants (moi between 1 and 5 depending on each viral stock) and then overlayed with 3 ml of culture medium (2, 4). At 7.5 hr post-infection (time 0 refers to the end of adsorption period) culture medium was replaced by 1 ml of prewarmed medium containing 15 µg of actinomycin D (Sigma). 30 min later radioactive medium was added (60 µC of ^{3}H-uridine; specific activity 30 Ci/mM - Saclay). At 12 hr monolayers were washed rapidly with saline buffer and the cells disrupted in 1.5 ml of an extraction buffer (0.4 M NaCl, 0.01 M Tris pH 4.2, 0.002 M EDTA) containing SDS 0.1%. The extracts were precipitated with TCA 5% and collected on membrane filters (Selectron 0.45 µ) to determine the total TCA insoluble counts (2).

Control cells were mock infected and treated with actinomycin D and ^{3}H-uridine exactly as the infected cells. Under those conditions incorporation with the wild type was 45,325 cpm at 33° and 124,711 cpm at 39.6°, whereas in mock infected cells incorporation was 9,072 cpm at 33° and 17,015 cpm at 39.6°.

Relative incorporation was equated by the ratio:

$$\frac{\text{Incorporation in infected cells at } 39.6° - \text{incorporation in mock infected cells at } 39.6°}{\text{Incorporation in infected cells at } 33° - \text{incorporation in mock infected cells at } 33°}$$

NON-CYTOPATHIC MUTANTS OF NEWCASTLE DISEASE VIRUS: COMPARISON WITH NATURALLY OCCURRING AVIRULENT STRAINS

CHARLES H. MADANSKY and MICHAEL A. BRATT*

Department of Microbiology and Molecular Genetics,
Harvard Medical School,
Boston, Mass. 02115, USA.

**Department of Microbiology,*
University of Massachusetts Medical School,
Worcester, Mass. 01605, USA.

Newcastle Disease Virus (NDV), a paramyxovirus causing disease in birds, is represented in nature by many strains which differ in the organ systems they affect and the severity of the symptoms they produce (8). These strains vary in virulence from those which produce asymptomatic infections to those which cause death within a few days. It is possible to classify strains according to their virulence in the chicken by measuring the time it takes a standard dose of virus to kill ten day old embryos at 38° (22).

Investigators have traditionally compared the phenotypes of various strains of NDV in tissue culture, occasionally noting consistent differences between virulent and avirulent strains. It has long been observed, for example, that virulent strains invariably cause plaques in chicken embryo cell cultures, while avirulent strains consistently fail to do so (18). Several groups noted that avirulent strains failed to cause fusion from within (FFWI), while this property varied among virulent strains (4, 15). Recently, evidence was presented suggesting that, in certain cell types, the F_0 polypeptide of avirulent strains of NDV is not cleaved into the F polypeptide, thus preventing FFWI and causing the production of non-infectious virus particles (14). In these instances, therefore, avirulent strains are unable to spread and go through the multiple cycles of re-infection necessary for plaque formation. A further consistent difference is the finding that during infection of chicken embryo cells, avirulent strains synthesize less NDV-specific RNA; particularly noteworthy is the reduction in 18S mRNA relative to 50S RNA (2, 11).

We are interested in the molecular basis of cell killing by NDV, and the relationship between cytopathogenicity and virulence. On the basis of strain comparison alone, however, it is impossible to establish a causal relationship between any one phenotype, and cell killing in tissue culture or virulence in the intact animal. We therefore undertook the selection of mutants of a virulent strain of NDV which possess a non-cytopathic phenotype. The character we chose was plaque formation. We have isolated six independent, non-conditional, non-cytopathic mutants of the Australia-Victoria strain of NDV (AV-WT) which produce infectious virus but do not form plaques (12). These mutants exhibit several striking similarities to naturally occurring avirulent strains.

The mutants were selected as follows: a cloned stock of AV-WT was mutagenized with N-methyl, N-nitrosoguanidine (20). Cultures were infected with a few infectious particles, overlaid with agar and incubated for two to three days to allow plaques to form. Plaques were marked, the plates and agar were marked with ink to allow subsequent realignment, and the agar was then removed, intact. The monolayers were then flooded with chicken erythrocytes at 0°C, and hemadsorption-positive spots which had not previously registered as plaques were identified. The agar previously over these spots was then picked, diluted and used to reinitiate infection; virus was cloned several times in the same manner.

Table I summarizes several of the properties of the 6 mutants thus selected. Each mutant makes characteristic and reproducible hemadsorbing spots. 5 of the 6 mutants grow to titers similar to wild-type virus (measured as spot-forming units versus plaque-forming units), but mutant nc7 grows to titers 10-fold lower. Several mutants are normal for FFWI at 36° or 40°. 5 of the 6 are altered in this property at 41.8°. All mutants failed to cause cytopathic effects visible through light microscopy during single cycle growth. In contrast to wild-type virus, which dramatically inhibits total protein accumulation relative to uninfected cells, the mutants either show no effect (nc9, nc12) or are intermediate (nc7, nc17, nc4, nc16) in their ability to inhibit total protein accumulation. Mutants nc9, nc12 and nc17 give mean embryo death times of 60 hrs, compared to 40 hrs for AV-WT. Mutants nc7, nc4 and nc16 give mean embryo death times of 80 hrs, approaching values obtained with naturally occurring avirulent strains.

RNA accumulation in mutant-infected cells was measured as actinomycin D-resistant incorporation of ^{3}H-uridine into TCA-precipitable material between 4 and 10 hrs post-infection. Table II shows that the mutants accumulate 15-30% the levels of RNA accumulated in wild-type infection. This is similar to the levels accumulated in cells infected with the naturally occurring avirulent strains, NJ-La Sota and B1-Hitchner. When this RNA was phenol extracted and subjected to velocity sedimentation, the patterns shown in Fig.1 were obtained.

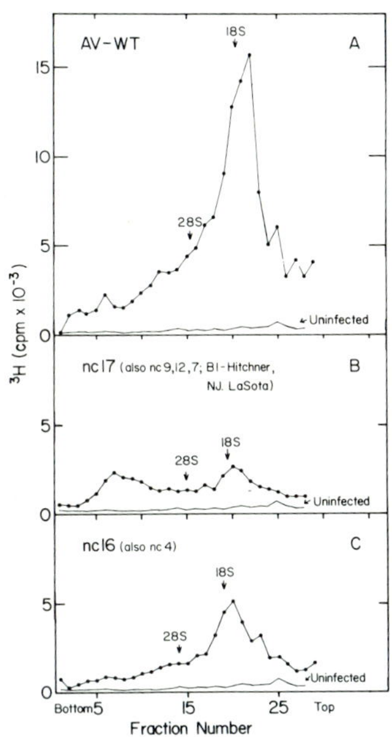

Fig.1. *Velocity sedimentation analysis of virus-specific RNAs*

Cells infected at a moi of 5 and incubated at 40° were labeled with ^3H-uridine between 4 and 10 hr post infection. Cells were solubilized with SDS-containing buffer (5). The RNA was extracted with chloroform-isoamyl alcohol-phenol (7), precipitated twice in ethanol, layered onto 15-30% SDS-sucrose gradients, and centrifuged at 180,000 x g for 4 hr.

In panel A (AV-WT), the 18S peak represents exclusively single-stranded, poly A-containing mRNA which codes for all of the known virus-specific polypeptides (7), except the L polypeptide, presumably involved in RNA transcription and most likely encoded by a 35S mRNA (13, 19). The ratio of 18S to 50S RNA is dramatically altered in cells infected by 4 out of the 6 mutants; they and the avirulent strains have essentially the same pattern shown in Panel B. Mutants nc4 and nc16 (Panel C) accumulate somewhat more 18S and less 50S RNA, though the ratio still differs from wild-type values.

Fig.2 shows an autoradiogram of a 7% SDS polyacrylamide

TABLE I

Properties of Non-cytopathic Mutants

	Spot Size[a]	Growth[b] (% WT)	Fusion from[c] within 36°	40°	42°	CPE[d] (light micro.)	Inhibition of[e] total protein accumulation (% uninfected)	Mean Embryo[f] Death Time (hrs)
AV-WT (Parental strain)	○	100	+	+	+	+++	20	40
nc 9	⬤	100	-	-	-	-	100	60
nc 12	⬤	100	-	-	-	-	100	60
nc 7	•	10	-	-	-	-	60	80
nc 17	●	50	+	+	+	-	44	60
nc 4	●	80	+	+	±	-	44	80
nc 16	●	80	+	+	±	-	40	80

a. Relative size of hemadsorption spots was determined by infecting chick embryo cell cultures at limiting dilution, overlaying with media containing agar, and incubating at 40° for 48 hrs at pH 7.2. The agar overlay was then removed and cells washed with phosphate-buffered saline and flooded with a 2% solution of chicken RBCs at 4° for 20 min.

b. Figures represent the average titer of virus released at 6, 9 and 12 hr post-infection at 40° from cells infected at an moi of 5.

c. FFWI was measured at 12 hr p.i. on non-confluent monolayers at pH 8.4 as previously described (3).

TABLE I continued.

d. Cytopathic effects measured by light microscopy include the following criteria: parallel array, spindle shape, presence of vacuoles, and number of cells attached to plate after 12 hr in liquid media. Cells were infected at an moi of 5. Cells were fixed with 95% methanol and visualized after giemsa staining.

e. Cells infected at an moi of 5 were labeled for 30 min at 12 hr post-infection (40°) with medium containing a ^{3}H-amino acid mixture as previously described (9). Total protein accumulation is corrected for differences in protein content and amino acid uptake and expressed as a percentage of uninfected cultures.

f. Mean embryo death time was measured as the time required to kill half of the 10 day old embryos infected with 10^3 infectious units and incubated at 38°.

TABLE II

NDV-specific RNA Synthesis

	cpm $^{3}H \times 10^{-3}$*	Percentage of AV-WT
Uninfected	9.3	3
AV-WT	295.9	100
nc 9	59.2	20
nc 7	62.1	21
nc 12	53.3	18
nc 17	44.4	15
nc 4	91.7	31
nc 16	79.9	27
NJ-La Sota	47.3	16
B1-Hitchner	29.6	10

*Radioactivity incorporated into TCA-precipitable material in cells infected at a moi of 5 and incubated at 40°, and labeled with ^{3}H-uridine in the presence of 5 μg/ml actinomycin D between 4 and 10 hr post-infection. (Average of 3 experiments.)

gel of ^{35}S-methionine-labeled polypeptides synthesized in mutant-infected chick embryo cells. Densitometer tracings of these gels reveal that all the viral polypeptides appear to accumulate at 30-50% of wild-type levels (unpublished data). An exception is the L polypeptide which seems to accumulate at levels less than 5% of those observed in wild-type infection. Mutants nc4 and nc16 (which show greater 18S RNA synthesis in Fig.1) accumulate slightly more L than the other mutants. We have been unable to detect increased turnover of this polypeptide in mutant-infected cells by pulse-chase analysis (unpublished data) and must therefore seek other explanations for the low levels of intracellular L.

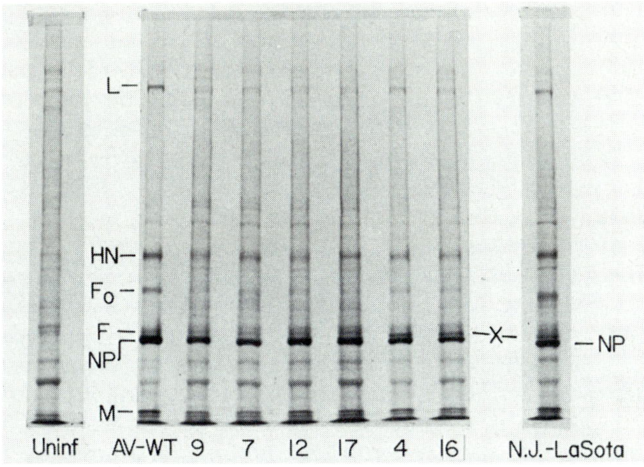

Fig.2. *Electrophoretic analysis of intracellular polypeptides*

Cells infected at an moi of 5 and incubated at 36° were labeled with ^{35}S-methionine in Hanks BSS for 30 min at 12 hr post-infection. Cells were solubilized and samples electrophoresed on a 7% SDS-polyacrylamide gel (7) with a 5% stacking gel (120V constant voltage). The gels were dried and autoradiography performed using Royal X-Omat x-ray film.

Several other features of the autoradiogram in Fig.2 are noteworthy. Mutants nc4 and nc16 accumulate relatively large amounts of a polypeptide (X) which migrates between NP and F. Tryptic peptide analysis reveals differences between this polypeptide and other viral polypeptides including F_0, as does analysis by limited proteolysis with chymotrypsin (data not shown: procedures ref.6 and 10). However, polypeptide X does appear to be related to the band migrating slightly behind NP in the pattern generated by the avirulent NJ-LaSota strain. (NP of both the NJ-LaSota and B1-Hitchner strains migrates faster than NP of AV-WT). Thus both of the mutants and two avirulent strains appear to accumulate a polypeptide not present in AV-WT-infected cells. This polypeptide is incorporated into both mutant and avirulent virions (data not shown); mutant virions contain the F polypeptide as well. Recently, a preliminary comparison by limited chymotrypsin digestion of F_o synthesized in a cell-free reaction and polypeptide X has suggested that these two polypeptides might be related (Hightower, personal communication). It is therefore unclear whether X is an entirely new viral product or an aberrant intermediate in the processing of F_o. In the latter case, the apparent differences in the tryptic peptide patterns of F_o synthesized *in vivo* and X (several peptides present in X are not present in F_o and vice-versa) might be explained by differences in post-translational modification. Further and more detailed tryptic peptide analysis will be necessary to firmly establish the identity of the X polypeptide.

Fig.3 illustrates the results of a four minute pulse with ^{35}S-methionine-containing medium followed by a 30 min chase with unlabeled methionine for cells infected with either AV-WT, NJ-LaSota, or B1-Hitchner. Polypeptide X (which migrates slightly slower in B1-Hitchner-infected cells) is present in significant amounts in the pulse only in cells infected with the avirulent strains, and does not appear to be unstable. However, a polypeptide migrating in the region of F_o in the pulse disappears in the chase in the avirulent strains; a new band (F_A), which migrates more rapidly than F_o then appears. Limited chymotrypsin digestion suggests that both F_o and F_A from avirulent strains are related to F_o from AV-WT-infected cells, with some sequences missing from F_A. We therefore believe that F_A is an aberrant cleavage product of F_o. In contrast with the report by Nagai and Klenk (14) we find F_A rather than F_o in avirulent virions (data not shown).

Fig.4 shows an autoradiogram of a polyacrylamide gel of ^{35}S-methionine-labeled polypeptides in virions released from cells infected with the non-cytopathic mutants. Only mutant nc7 (which grows to characteristically lower titers) shows an altered pattern by its inclusion of a polypeptide migrating in the region of F_o, in addition to F. (Protein X is not resolved in this 9% gel). This mutant can be activated by exogenous trypsin both to cause FFWI and to form plaques in chicken embryo cell cultures in a manner similar to that reported for avirulent strains (14).

In order to analyze these mutants genetically, we have taken several approaches. The first is complementation for the enhancement of RNA synthesis with RNA^-temperature-sensitive mutants derived from the same parental strain (20, 21).

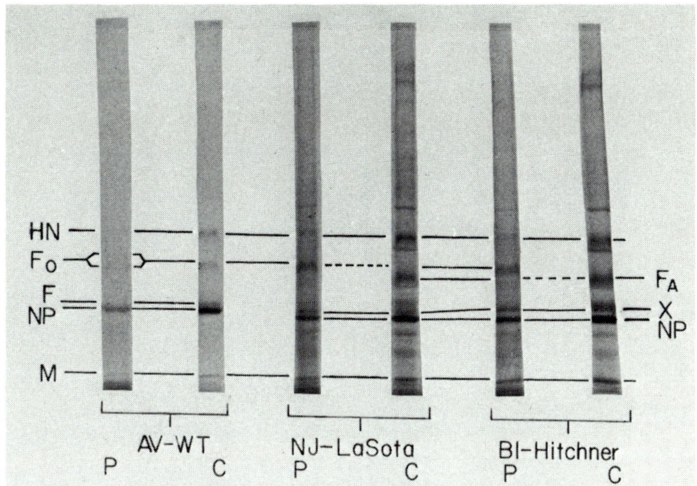

Fig.3. *Pulse-chase analysis of intracellular polypeptides*

Cells infected at an moi of 5 and incubated at 39.5° were pulsed for 4 min with ^{35}*S-methionine in Hanks BSS or pulsed and then chased for 30 min with medium containing a 1000-fold excess of unlabeled methionine. Cells were washed and solubilized, and samples electrophoresed as in Fig.2.*

Results indicate that the non-cytopathic mutants will complement for RNA synthesis with mutants in complementation group E but not with mutants in Group A (data not shown).

Another approach has been to select independent, plaque-forming revertants from individual clones of each mutant. By analyzing these revertants with respect to RNA synthesis, FFWI and protein species we hope to determine whether the lesions in the non-cytopathic mutants are single-step mutations with pleiotropic effects, or multi-step mutations, and determine thereby which phenotypes relate directly to cell killing.

DISCUSSION

The fact that mutants selected to be less cytopathic than their virulent parent exhibit phenotypes similar to naturally occurring avirulent strains, in addition to increased mean embryo death times, suggests 1) that these common phenotypes render avirulent strains less cytopathic and, 2) that reduced cytopathogenicity may account, in part, for their avirulence *in vivo*. Furthermore, since only one of the non-cytopathic mutants accumulates F_0 in the virion, we conclude that other phenotypes (*e.g.* RNA accumulation, defects in F_0 processing)

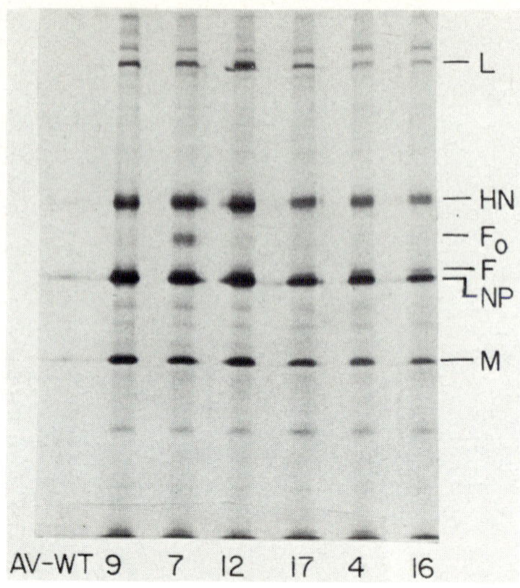

Fig.4. *Polypeptides in virions*

Cells infected at a moi of 5 and incubated at 39.5° were labeled from 4-16 hr post-infection with ^{35}S-methionine-containing medium. Virus-containing medium was centrifuged at 12,000 x g for 10 min and the supernatant sedimented through 20% sucrose onto a 65% sucrose-D_2O pad. The peak of virus was pooled and pelleted at 60,000 x g for 1 hr. The virus pellet was solubilized and electrophoresed through a 9% resolving gel as described in Fig.2. Comparable amounts of protein were electrophoresed for all mutants, but only 10% as much was used for AV-WT. Both AV-WT and mutant virions are labeled to an equivalent extent and released from infected cells in similar quantities (unpublished results).

can also contribute to avirulence. It is necessary to distinguish between those contributions due to the production of non-infectious virus during a single cycle of growth in non-permissive tissues, as proposed by Nagai and Klenk (14) and those due to infectious virus which does not kill or in some way alter infected cells.

Several lines of evidence point to accumulation of mRNA as a critical point in the determination of cytopathic effects. Blair and Robinson (1) suggested that differences in accumulation of 18S mRNA might account for differences in cytopathic effects during Sendai virus and NDV infection of chicken embryo cells. Preble and Youngner (16) found that a variant of NDV obtained from persistently infected cells was defective in RNA synthesis. During infection of CHO cells by the same virulent strain of NDV used in the present studies, 18S mRNA accumulates in 80S structures which are inactive in protein synthesis (7). This mRNA translates much less efficiently in cell-free systems than the mRNA from polyribosomes. Thus mRNA itself might lead to one cytopathic effect - inhibition of total protein synthesis.

Of course some product(s) of the mRNA may also be involved in cytopathology. In this regard the significance of the selection of 5 of 6 mutants with some alteration in the ability to cause FFWI is as yet unclear. Reeve *et. al.*(17) suggested that the balance between synthesis and utilization of virus-specific products in infected cells may be a determinant of cytopathology. This notion is reinforced by the non-cytopathic mutants which produce infectious virus in amounts similar to AV-WT even though they accumulate both less RNA and less protein.

Virulence *in vivo* is clearly a complex phenomenon, including not only the variables of each virus-cell interaction - *e.g.* tissue tropisms, ability to produce infectious virus, and ability to kill or otherwise alter the host cell - but also a variety of protective responses of the infected animal. Our intention in selecting mutants with a non-cytopathic phenotype was to establish the relationships between specific viral phenotypes in tissue culture and aspects of virulence in the whole animal, as well as to discover the ways in which virulent strains of NDV kill cells. Hopefully, this approach will be successful and we can begin to apply our wealth of knowledge about NDV in tissue culture to the problems of pathogenesis *in vivo*.

ACKNOWLEDGEMENTS

Supported by grants from the National Science Foundation (PCM 76-84135) and The National Institute of Allergy and Infectious Disease (AI-12467).

C.H.M. is a predoctoral trainee of the National Institute of General Medical Sciences (GM-07196).

We thank Helen Kotilainen and Susan Stambler for excellent technical assistance and Kathryn Foley for help in preparation of this manuscript. We are also grateful to C. Worth Clinkscales and Trudy G. Morrison for numerous helpful discussions.

REFERENCES

1. Blair, C.D. and Robinson, W.S. (1968). *Virology* 35, 537.
2. Bratt, M.A. (1969). *Virology* 38, 485.
3. Bratt, M.A. and Gallaher, W.R. (1969). *Proc. Natl. Acad. Sci.*, 64, 536.
4. Bratt, M.A. and Gallaher, W.R. (1972). *In* "Membrane Research"

(C.F. Fox, ed.), p.383, Academic Press, New York.
5. Bratt, M.A. and Robinson, W.S. (1967). *J. Mol. Biol.* 23, 1.
6. Cleveland, D.W., Fischer, S.G., Kirschner, H.W. and Laemmli, U.K. (1977). *J. Biol. Chem.* 252, 1102.
7. Clinkscales, C.W., Bratt, M.A. and Morrison, T.G. (1977). *J. Virol.* 22, 97.
8. Hanson, R.P. (1963). *In* "Newcastle Disease Virus: An Evolving Pathogen", The University of Wisconsin Press, Madison.
9. Hightower, L.E. and Bratt, M.A. (1974). *J. Virol.* 13, 788.
10. Hightower, L.E., Morrison, T.G. and Bratt, M.A. (1975). *J. Virol.* 16, 1599.
11. Lomniczi, B., Meager, A. and Burke, D.C. (1971). *J. Gen. Virol.* 13, 111.
12. Madansky, C.H. and Bratt, M.A. (1977). *Abst. Amer. Soc. Microbiol.* p.335.
13. Morrison, T.G., Weiss, S.R., Hightower, L.E., Collins, B. Spanier and Bratt, M.A. (1975). *In* "*In Vitro* Transcription and Translation of Viral Genomes". (A.L. Haenni and G. Beaud, eds.), p.281-289. INSERM 47 Institute National de la Sante et de la Recherche Medicale, Paris.
14. Nagai, Y. and Klenk, H.D. (1977). *Virology* 77, 125.
15. Poste, G. and Waterson, A.P. (1975). *In* "Negative Strand Viruses" (B.W.J. Mahy and R.D. Barry, eds.), p.905-922. Academic Press, London.
16. Preble, O.T. and Youngner, J.S. (1973). *J. Virol.* 12, 472.
17. Reeve, P., Rosenblum, M. and Alexander, D. (1970). *J. Hyg.* 68, 61.
18. Schloer, G.M. and Hanson, R.P. (1968). *J. Virol.* 2, 40.
19. Spanier, B.B. and Bratt, M.A. (1977). *J. Gen. Virol.* 35, 439.
20. Tsipis, J.E. and Bratt, M.A. (1975). *In* "Negative Strand Viruses", (B.W.J. Mahy and R.D. Barry, eds.) p.777-784. Academic Press, London.
21. Tsipis, J.E. and Bratt, M.A. (1976). *J. Virol.* 18, 848.
22. Waterson, A.P., Pennington, T.H. and Allan, W.H. (1967). *British Med. Bull.* 23, 138.

BIOLOGIC CHARACTERISTICS OF INFLUENZA VIRUS RECOMBINANTS DERIVED AT SUBOPTIMAL TEMPERATURES

[1]H.F. MAASSAB, [2]S.B. SPRING, [3]A.P. KENDAL and [1]A.S. MONTO

[1]*University of Michigan, School of Public Health, Ann Arbor, Michigan 48104, USA.*

[2]*Laboratory of Infectious Diseases, National Institutes of Health, Bethesda, Maryland 20014, USA.*

Adaptation of influenza viruses of types A and B to grow at suboptimal temperature (25°) has provided cold mutants which have the ability to give high yield and to form plaques at 25°. In addition, these cold mutants were shown to be temperature-sensitive (ts) and genetically stable, to share lesion(s) with ts mutants derived by mutagenesis, and to be attenuated for man (3, 4, 8, 9).

Recombination-genetic reassortment at 25° can be used to transfer cold gene(s) from a cold variant to recent antigenic variants of types A and B influenza viruses. Clones isolated from this interaction were also found to be cold-adapted, to share lesion(s) with ts mutants, and to grow to a high yield at 25°. All had the surface antigen (HA) of the wild type, while some had the neuraminidase antigen characteristics of the cold variant. All showed loss of virulence when tested in animals and man (1, 4, 9). The percentage of the clones possessing the surface antigen(s) of the wild type ranged from 20 to 40 percent. A set of clones will be analyzed to demonstrate 1) that the lesion(s) responsible for cold adaptation and/or ts properties can reside in viral genes other than those coding for hemagglutinin or neuraminidase, 2) the clones derived are also the product of recombination rather than the result of spontaneous emergence of a cold mutant, 3) cold adaptation can be segrated from the temperature-sensitivity,

[3]*World Health Organization Influenza Center, Center for Disease Control, Atlanta, Georgia 30333, USA.*

and 4) cold gene(s) can bc transferred at the permissive temperature of 33°.

DERIVATION OF COLD RECOMBINANTS

Development and characterization of cold-adapted recombinants have been performed throughout these studies in two host systems, primary chick kidney cultures and embryonated eggs. The derivation of cold-adapted recombinants was accomplished using the genetically stable cold variant "Master Strain" A/AA/6/60-H3N2 as donors of cold gene(s) and a variety of recent wild type parents as illustrated in the flow diagram.

FLOW DIAGRAM
Derivation of Cold Recombinant

Treatment of "Wild Parent"	Passage of the Recombinant	Treatment of "Cold Variant"
"Wild" epidemic strain	"Wild" strain x cold variant "Master strain"	A/AA/6/60 - H2N

Passage 1 at 25° in primary chick kidney tissue culture (PCKTC) (48-72 hr)

Passage 2 and 3 at 25° in PCKTC with anti A/AA/6/60 immune serum (5 days each)

Passage 4 in embryonate eggs at 25° without antiserum (72 hr)

Passage 5-plaque titration at 25° in PCKTC (5 days)

No yield *3 x 10^{-3} PFU/ml* *No yield*

Passage 6-plaques (20)picked for separate passage in embryonate eggs at 33° (72 hr)

Selection of vaccine candidate after biologic and immunologic characterization of yields.

For the cold recombinant - (AA-CR18) - The "wild" parent strain is A/Scot./840/74-H3N2 (E4)

For the cold recombinant - (AA-CR19) - The "wild" parent strain is A/Vict./3/75-H3N2

For the cold recombinant - (AA-CR22) - The "wild" parent strain is A/Vict./3/75-H3N2 (Lot E-81)

For the cold recombinant - (AA-CR24) - The "wild" parent strain is A/NJ/8/75-HswN1 (E2)

E = Designates passage in embryonated eggs.

The wild type parent of AA-CR18-H3N2 is the A/Scotland/840/74-E4; the one of AA-CR19-H3N2 is A/Victoria/3/75-E5; and the one of AA-CR22-H3N2 is A/Vict/3/75 (Lot No.-81). The procedure of derivation of cold recombinant clones of the above viruses has been described elsewhere (4, 6). In general, mixed infection of PCKC was initiated at 25° using 5 PFU/cell with the attenuated cold variant A/AA/6/60-H3N2 and the wild type virus. After 72 hr, second and third passages were made at 5 day intervals with antisera directed against the cold variant: A fourth passage without antiserum was carried out in embryonated eggs held at 25°. Parallel passage of wild and cold parents failed to yield virus, demonstrating that the antiserum employed had been effective in suppressing growth of the cold parent and that the final yield of the mixed infection series was the product of genetic recombination rather than the result of spontaneous emergence of a cold mutant (9). The fourth passage of the mixed infection was passed without antiserum, and the fifth passage was titrated in PCKC at 25° and 20-30 clones isolated for biologic and immunologic characterizations. Recent studies with other strains have shown unequivocally that this technique offers independent assortment of the surface antigens from the biologic properties as a result of such a cross and suggest that recombination rather than selection was occurring in the dually infected cells (5).

GENETIC ANALYSIS OF COLD RECOMBINANT

In a collaborative study with Dr S.B. Spring in Dr Chanock's laboratory genetic analysis of 5 cold-adapted influenza viruses and 12 cold recombinants was performed by the plate-recombination technique to determine to which complementation group(s) they belonged. The nine 5-FU mutants previously assigned to 7 complementation groups were used as prototype strains (8, 9). The data are summarized in Table I. The results showed that the cold-adapted variants and recombinants which were temperature-sensitive (ts) were shown to share a lesion with only one of the 7 complementation groups of the 5-FU mutants. Several of the cold variants like the A/AA/2/67, A/Aichi/2/68 and A/AA/1/70 also had additional lesions. The cold variants A/AA/Marton/43, and A/FM/1/47 strains could not be evaluated since they were found to be less than 100-fold restricted at 39°. Likewise, a cold recombinant prepared from a cross between A/AA/6/60 cold variant and A/Dunedin/4/73 strain had a shut-off temperature greater than 39°. These 3 viruses were, therefore, not suitable for genetic analysis. Mutants derived by the procedure of cold adaptation are then amenable to genetic analysis and for some, they share lesion(s) with the 5-FU mutants.

TABLE I

ts Lesions in Cold Variants and Cold Recombinants

Virus	Method of Derivation	Shared Lesions (Complementation Group)[1]	Shut-off Temperature[2]	Antigenic Subtype
Cold Variants				
A/AA/Marton/43-HON1	Stepwise		>39	$H0_{43}N1_{43}$
A/FM/1/47-H1N1	Stepwise		>39	$H1_{47}N1_{47}$
A/AA/2/65-H2N2	Plaque selection	1	37	$H2_{65}N2_{65}$
A/AA/2/67-H2N2	Plaque selection	1,3,5	37	$H2_{67}N2_{67}$
A/Aichi/2/68-H3N2	Plaque selection	1,6	38	$H3_{68}N2_{68}$
A/AA/1/70-H3N2	Plaque selection	1,3,6		$HE_{68}N2_{68}$
P17-A/AA/60/60-H2N2	Stepwise	1	37	$H2_{60}N2_{60}$
Cold Recombinants				
CR6	P17 X A/Queen/6/72	1	37	$H3_{72}N2_{72}$
CR12	P17 X A/AA/9/73	1	38	$H3_{73}N2_{73}$
CR13	P17 X Dunedin/4/73		>39	$H3_{73}N2_{73}$
CR18 - Clone 1	P17 X A/Scot/840/74	1	38	$H3_{74}N2_{74}$

TABLE I (Contd.)

Clone 2	1		$H3_{74}N2_{74}$
Clone 3	1		$H3_{74}N2_{60}$
Clone 4	1		$H3_{74}N2_{60}$
Clone 5	1	38	$H3_{74}N2_{60}$
Clone 6	1	37	$H3_{74}N2_{74}$
Clone 7	1		$H3_{74}N2_{74}$
Clone 8	1		$H3_{74}N2_{60}$
Clone 9	1	37	$H3_{74}N2_{60}$
Clone 10	1	38	$H3_{74}N2_{60}$

[1]Determined by complementation recombination assay on RMK monolayers using 5-FU Hong Kong-*ts* mutants and recombinants (8, 9).

[2]Taken as temperature at which there is a 100-fold loss in pfu.

TABLE II

Dissociation of cold-adapted property

A. Properties of the Cold Recombinant AA-CR13-H3N2 *in vitro* and *in vivo*

Virus	Infectivity titer in pfu/ml		Clinical Signs*	Antibody Response	Temperature Sensitivity
	25°	39°			
Wild parent - A/Dun/4/73	0	8×10^7	High Fever-Coryza	1024	WT-NCA
Attenuated Parent-A/AA/6/60	8×10^7	0	None	512	ts-CA
Recombinant-AA-CR13	5×10^6	10^6	None	512	$\overset{+}{ts}$-CA

*Comparable dose for each of the viruses listed (10^6 pfu/ml) were given intranasally to each of the 2 ferrets.

+Antibody titer by HI 21 days after infection. Ferrets were challenged 21 days after infection with 1000 tissue culture dose of the homologous "wild parent". Clinical manifestations were absent and complete protection was evident.

WT = Wild type; NCA = Not Cold-adapted; CA-Cold-adapted; $\overset{+}{ts}$ = phenotypically wild type; ts-temperature sensitive.

TABLE II (contd.)

B. Characterization of ts^+ Revertants Isolated from Hamsters Infected with CR12 Virus[a]

Plaque No.	Day	Hamster	Organ	Titer: pfu/ml Chick Kidney		Rhesus Monkey Kidney		Temperature Sensitivities[b]	
				25°	33°	33°	39°		
1	2	3	Nasal turb	$3x10^4$	$3x10^6$	$4x10^6$	$4x10^6$	ts^+	CAP
2	2	3	Nasal turb	$6x10^5$	$1.4x10^6$	$4x10^6$	$8x10^5$	ts^+	CA
3	2	3	Nasal turb	$1x10^4$	$4x10^5$	$4x10^6$	$4x10^5$	ts^+	CAP
1	3	5	Nasal turb	$1.3x10^4$	$3x10^5$	$4x10^6$	$2x10^6$	ts^+	CAP
1	2	9	Lung	$2x10^3$	$3x10^4$	$6x10^4$	$3x10^4$	ts^+	CA
Parental Viruses									
CR12						$2x10^6$	.0001	ts	CA
A/AA/9/73 - (CR12 wt parent)				0	$3x10^5$	$6x10^6$	$6x10^6$	wt	CAP

[a]RMK isolates were plated on RMK monolayers at 33° and 39°. Plaques appearing at 39° were picked, diluted 1:10 and inoculated into embryonated eggs. These allantoic pools were then studied; the data is presented in this table.

[b]ts = temperature sensitive, ts^+ = phenotypically wild type; CA = cold adapted; CAP = Cold-adapted properties; Plaques at 25° appeared late, were small and did not increase appreciably in size.

DISSOCIATION OF THE COLD-ADAPTED PROPERTY

Recent studies of ts properties of different animal viruses have shown the tendency of dissociation between the ts properties and virulence. Stanners and Goldberg (10) for instance, presented evidence that certain groups of mutants of vesicular stomatitis virus (VSV) were less virulent than would be expected from its extreme leakiness at body temperature. Preble and Youngner (7) have also demonstrated that reversion of the ts variant to wild type was not accompanied by reversion of persistence to virulence. The data shown in Table II provide evidence of such dissociation. In Table II part A the cold recombinant AA-CR13 did not exhibit the 100-fold reduction in infectivity when compared to the wild type, so was not temperature-sensitive. However, it did have the cold-adapted property with high yield and plaque formation at 25°. In addition, when this line was inoculated into ferrets, it was avirulent and immunogenic when compared to its wild type parent. Another example of dissociation between the cold-adapted and ts phenotypes in certain cold recombinant lines of influenza virus is shown in Table II B. It is evident that upon infection of hamsters with the cold recombinant CR12, ts+ revertants were isolated which accounted for 0.1 to 1 percent of the total population. However, these plaques when titrated for infectivity in primary chick kidney cells (PCKC) appeared to retain their cold-adapted property to a certain degree. One plaque was definitely cold-adapted, while the other three were intermediate and designated as possessing the cold-adapted property (CAP), since the plaques at 25° appeared late, were small and did not increase appreciably in size. Further insight into the roles of the ts and cold-adaptation properties might be obtained by comparing the level of replication of the ts+ revertants appearing in the hamster and ferrets lungs with that of its ts and cold recombinant parent (AA-CR12). Thus, this observation appears to contradict our earlier conclusion which emphasized uniformly that cold-adaptation and temperature-sensitivity of cold mutants cannot be dissociated (4).

These results also tend to suggest that the cold adaptation and ts properties play a role in the regulation of growth of these viruses since AA-CR12 grew better in the nasal turbinates of both ferrets and hamsters than did the wild type parent (Spring and Maassab, unpublished results).

TRANSFER OF COLD GENE(S) AT PERMISSIVE TEMPERATURE (33°)

A study was undertaken to demonstrate that the acquisition of "cold" genes in influenza virus is a stable genetic property which can be transferred by recombination at 33°. A cross between the "Master Strain" cold variant A/AA/6/60 and the wild type A/Victoria/3/75 (Lot E-81) was performed in PCKC at MOI of 2 pfu/cell at 33°. After suppression of the cold mutant with antiserum, a pool was made from the third passage level and plaqued at 33°. Table III A shows the titer of the infectivity titer of this line at different temperatures. From this titration, 10 clones were picked and a pool of each was made in the embryonated eggs at 33° for characterization. It is also of interest to note that the line was cold-adapted and temperature-sensitive (ts). Characterization of 6 of the 10 clones is shown in Table III B. It is evident that the clones exhibit different biological properties, but they were all temperature-sensitive with a shut-off temperature of 38° or lower. Two clones were cold-adapted, three had the cold-adapted properties, while the sixth was ts but not cold-adapted. Thus, we were able to isolate in one cross at 33° clones of different biological properties. Other clones from the same cross are being evaluated under the same conditions. Detailed characterization of these clones *in vitro* and *in vivo* is in progress and will be reported elsewhere. These findings demonstrate that stable cold variants can transfer cold gene(s) and segregate different markers by recombination at the permissive temperature of 33° instead of the usual incubation at 25° in PCKC.

ATTENUATION FOR MAN

Certain of these cold recombinants were administered to human volunteers intranasally as attenuated live influenza virus vaccines, and were found to be acceptable and immunogenic (1, 2, 5).

SUMMARY AND CONCLUSION

Derivation and characterization of new cold recombinants with surface antigens selected on the basis of contemporary epidemiological relevance have been accomplished. Analysis of a large number of clones has provided evidence that cold adaptation is a genetic property, and that the cold mutants isolated have the following characteristics: 1) The clones were derived by recombination-genetic reassortment at the suboptimal temperature of 25° in primary chick kidney cells (PCKC), 2) All the clones achieved a high yield and formed plaques at 25° and 33°, and 3) The cold recombinants were attenuated when tested in an animal model system such as the ferret. When the temperature sensitivity of these cold-adapted lines was studied, cut-off temperatures of 37°, 38° and 39° were obtained. Hence, it was postulated that the cold mutants were temperature-sensitive (ts). However, recent studies *in vitro* and *in vivo* have demonstrated that cold

TABLE III

Development and Characterization of the Recombinant AA-CR23 Derived at 33°

A. Derivation of clones

Virus	Antigenicity	Infectivity titer* (pfu/ml) at				Cut-off+ Temperature
		25°	33°	38°	39°	
Wild parent-A/Vict./3/75	H3N2	0	8×10^7	5×10^7	10^7	> 39°
Attenuated parent-A/AA/6/60	H2N2	8×10^7	6×10^8	10^2	0	38° or lower
3rd Passage-AA-CR23	H3N2	8×10^3	5×10^4**	0	0	38° or lower

* - Infectivity in primary chick kidney cells (PCKC)

** - From this passage level, 10 clones were isolated for characterization

+ - Defined as a 100-fold or greater reduction in plaquing efficiency on primary chick kidney cells (PCKC).

TABLE III (contd.)

B. Characterization of the Recombinant Clones

Recombinant Clones	Antigenicity	Infectivity titer* pfu/ml in PCKC			Properties of the Clones
		25°	33°	39°	
A	H3N2	4×10^5	3×10^6	0	CAP+
B	H2N2**	2×10^6	10^7	0	CA
C	H3N2	3×10^5	2×10^6	0	CAP
D	H3N2	3×10^5	8×10^6	0	CAP
E	H3N2	4×10^5	6×10^6	0	CA
BIII	H3N2	0	5×10^5	0	ts

* - In primary chick kidney cells (PCKC)

** - This clone is A/AA/6/60 - H2N2

\+ - CAP = Cold-adapted property - (Smaller in size and delayed in appearance).
CA = Cold-Adapted.

adaptation is a separate property of influenza virus which can be dissociated from the ts marker. Certain cold recombinants grew equally well at 39° and 25°, and also retain loss of virulence when administered to ferrets and mice. Cold genes can be transferred to new strains at 33°.

ACKNOWLEDGEMENTS

We thank Eve Bingham for expert technical assistance. Recombinants were prepared with financial support from the U.S. Army Medical Research and Development Command, contract No. DADA 17-73-C-3060 and National Institutes of Health, contract No. I-Al72521. We would like also to acknowledge the financial support of Philips-Duphar of Weesp, Holland.

REFERENCES

1. Davenport, F.M., Hennessy, A.V., Maassab, H.F., Minuse, E., Clark, L., Abrams, G.D., and Mitchell, J.R. (1977). *J. Infect. Dis.* 136, 17.
2. Hraber, A., Vodopija, I., André, F.E., Mitchell, J.R., Maassab, H.F., Hennessy, A.V., and Davenport, F.M. (1977). *In* "International Symposium on Influenza Immunization". Karger, Basel. In press.
3. Maassab, H.F. (1969). *J. of Immunol.* 102, 728.
4. Maassab, H.F. (1975). *In* "Negative Strand Viruses". (Mahy, B.W.J. and Barry, R.D., eds.) p.755. Academic Press, New York.
5. Maassab, H.F., Cox, N.J., Murphy, B.R. and Kendal, A.P. (1977). *In* "International Symposium on Influenza Immunization" Karger, Basel. In press.
6. Maassab, H.F., Kendal, A.P. and Davenport, F.M. (1972). *Proc. Soc. Expl. Biol. Med.* 139, 768.
7. Preble, O.T. and Youngner, J.S. (1973). *J. Virol.* 12, 481.
8. Spring, S.B., Maassab, H.F., Kendal, .P., Murphy, B.R. and Chanock, R.M. (1977). *Virology* 77, 337.
9. Spring, S.B., Maassab, H.F., Kendal, A.P., Murphy, B.R., and Chanock, R.M. (1977). *Arch. Virol.* 55, 233
10. Stanners, C.P. and Goldberg, V.J. (1975). *J. Gen. Virol.* 29, 281.

BIOCHEMICAL CHARACTERISTICS OF RECOMBINANT VIRUSES DERIVED AT SUB-OPTIMAL TEMPERATURES: EVIDENCE THAT *ts* LESIONS ARE PRESENT IN RNA SEGMENTS 1 AND 3, AND THAT RNA 1 CODES FOR THE VIRION TRANSCRIPTASE ENZYME

A.P. KENDAL, N.J. COX, S.B. SPRING[1], and H.F. MAASSAB[2]

World Health Organization Influenza Center, Center for Disease Control, Atlanta, Georgia 30333, USA.

[1]*Laboratory of Infectious Diseases, National Institutes of Health, Bethesda, Maryland 20014, USA.*

The basis for attenuation of cold-adapted (c.a.) viruses may be elucidated by determining the nature of their genetic lesions, and by preparing recombinant viruses having defined mixtures of mutant and wild-type genes so that their biological properties can be evaluated. We previously reported that the prototype c.a., temperature-sensitive (*ts*) influenza virus A/Ann Arbor/6/60 contained lesions affecting the pattern of its haemagglutinin (HA) synthesis at elevated temperatures (4). We have also shown, however, that recombinant viruses may be produced which derive c.a. and *ts* properties from the A/Ann Arbor/6/60 mutants, but their HA and neuraminidase (NA) from wild-type (w.t.) H3N2 strains (9, chapter 73). This suggests that multiple-step lesions were introduced into the genes of A/Ann Arbor/6/60 during the initial process of cold-adaptation, and that the lesions most important for the c.a. and *ts* properties are in genes other than the HA gene. To identify such genes we have determined the genetic composition of recombinant c.a. and *ts* viruses and have examined their biochemical properties. In this report we describe the particular results which indicate that A/Ann Arbor/6/60 mutant contains *ts* lesions in its RNA 1 and 3, and that RNA 1 is the gene coding for the virion transcriptase.

[2]*University of Michigan School of Public Health, Ann Arbor, Michigan 48104, USA.*

MUTANT VIRUSES

Derivation of c.a. viruses and recombinants have been described (8, 9 chapter 73). WSN *ts* mutants were obtained from Dr. Palese and Dr. Simpson. Seeds were prepared in eggs, and plaque titrations were done in MDCK cells as previously described (2). Water-jacketed incubators were used to maintain temperatures of 33° and 39°, and by modification of the temperature regulation system one incubator was maintained at 39° ± 0.1°. Our WSN *ts* mutant stocks had titers of $\geq 0.5 \times 10^6$ pfu/ml at 34° and $<0.5 \times 10^3$ pfu/ml at 39°. WSN *ts* 6, 53 and 15 belong to recombination groups I, II and III respectively (20, 21). Krug *et al.*,(6) showed that these mutants have *ts* lesions affecting synthesis of cRNA, vRNA and cRNA respectively, and Palese and co-workers (13, 14) have shown that each lesion is in a large RNA segment of the mutants.

PROCEDURES FOR BIOCHEMICAL IDENTIFICATION OF GENES IN A/Ann Arbor/6/60 AND H3N2 VIRUSES AND ASSIGNMENT OF THEIR FUNCTIONS

Eight single-stranded RNAs of influenza virus A/Ann Arbor/6/60 can be resolved by electrophoresis in warmed 3% acrylamide gels in the absence of any chemical denaturing agents (Fig.1a). In contrast to results of electrophoresis in the presence of urea (10), in the absence of denaturing agents differences may be observed in the migration rates of 7 RNAs, 1, 2, 3, 4, 5, and 6 (Fig.1b) and 8 (5) of A/Ann Arbor/6/60 and H3N2 viruses. We have previously shown that RNAs 7 and 8 code respectively for M and NS proteins, with M protein being identified by electrophoresis of viral protein (5). To assign functions to RNAs 4, 5 and 6 we have examined recombinants that contain either neuraminidase (NA) or both nucleoprotein (NP) and NA of A/Ann Arbor/6/60 but HA of a wild-type H3N2 parent. Under conditions of electrophoresis similar to those described in the legend of Fig.1, RNAs 4, 5 and 6 code respectively for HA, NA and NP in the H2N2 and H3N2 viruses. As shown below, RNA 3 of A/Ann Arbor/6/60 (as defined for our electrophoretic system) corresponds functionally with the RNA 2 of WSN-*ts* 15 virus (i.e. codes for a polypeptide involved in *in vivo* cRNA synthesis), and RNA 1 is the gene coding for the virion transcriptase. By elimination, RNA 2 therefore codes for the polypeptide which together with NP is involved in *in vivo* vRNA synthesis (6).

IDENTIFICATION OF RNA3 of A/Ann Arbor/6/60 AS THE GENE CONTAINING A CONDITIONALLY-LETHAL *ts* LESION

Using the above described electrophoresis procedures we showed that all c.a. and *ts* recombinant viruses so far examined derived their RNA 3, as well as NP and M proteins,

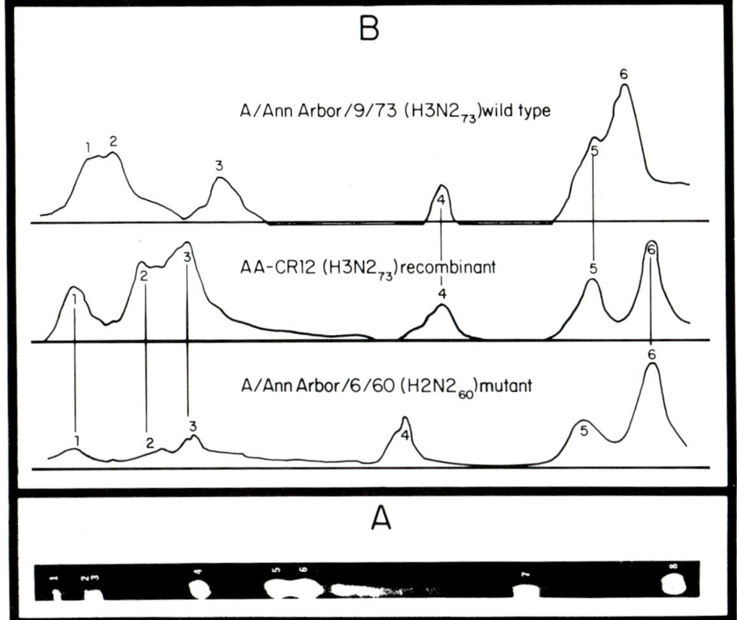

Fig.1. *Electrophoresis of single-stranded virion RNAs in warm acrylamide gels in the absence of denaturing agents. Viruses were grown in chick kidney cell monolayers in the presence of ^{3}H-uridine and were purified by standard density gradient centrifugation techniques (5). Techniques for electrophoresis were modified from Cox and Kendal (1). Gels containing 3% acrylamide, 0.15% bisacrylamide, and 0.5% agarose were prepared in electrophoresis buffer containing 30 mM sodium dihydrogen phosphate, 30 mM tris base and 0.9 mM EDTA, pH 7.8, and 0.2% SDS. The gels were cast as 30 cm long x 14 cm wide x 3 mm thick slabs in a vertical apparatus (Bio-Rad Laboratories, Richmond, Ca.). Purified virus in 10% running buffer without SDS was disrupted by heating to 56° for 2 min with 2 volumes of electrophoresis buffer containing 20% sucrose, loaded into sample slots and electrophoresed for 20 hr at 100 mA. Bands of viral RNA were detected by autoradiography with the procedure of Laskey and Mills (7). A. Autoradiogram of RNA from influenza virus A/Ann Arbor/6/60. B. Densitometer tracing of autoradiograms from a different gel containing two parent viruses, A/Ann Arbor/6/60 mutant and wild-type A/Ann Arbor/9/73, and their cold-adapted recombinant AA-CR12. Despite some band-splitting of RNA 2, it is clear that the recombinant derives RNAs 1, 2, and 3, in addition to RNA 6, from the mutant parent, and only RNAs 4 and 5 from its w.t. parent. Since antigenic analysis showed the recombinant derived HA and NA*

/continued

exclusively from the mutant parent A/Ann Arbor/6/60. In contrast all other genes of the mutant parent, including RNAs 1 and 2 have segregated independently of the c.a. and *ts* properties (Table Ia). To determine whether RNA 3 contained a *ts* lesion we investigated whether ts^+ recombinants could be produced using as parent viruses mutant A/Ann Arbor/6/60 and mutants of A/WSN/33 virus that had been shown to each contain a single lesion in a different one of their 3 largest RNA segments. When MDCK cells were mixedly infected with mutant A/Ann Arbor/6/60 and either WSN-*ts6* group I mutant) or WSN-*ts*53 (group II mutant), ts^+ recombinants were produced, whereas ts^+ recombinants were not recovered from a mixed infection of mutant A/Ann Arbor/6/60 and WSN-*ts* 15 (group III mutant), (Table Ib). These results indicate that A/Ann Arbor/6/60 mutant contains a conditionally lethal *ts* lesion in its RNA 3 which must code for the same gene product as WSN-*ts* 15 RNA 2, as resolved in a different electrophoresis system (13).

TEMPERATURE-SENSITIVITY OF VIRION TRANSCRIPTASE IN A/Ann Arbor/6/60 MUTANT AND RECOMBINANT VIRUSES

Because the lesion in WSN-*ts*15 virus affects *in vivo* cRNA synthesis (6) we investigated whether the transcriptase of mutant A/Ann Arbor/6/60 was *ts*. We found that although transcriptase activity of w.t. A/Ann Arbor/6/60 virions was reduced by about half at 40° as compared to that observed at 31° during the first 10 min of enzyme reaction, transcriptase activity of the mutant A/Ann Arbor/6/60 was reduced to a greater extent (about fourfold at 40° as compared to 31°) (Fig.2a). We verified the *ts* nature of the mutant transcriptase by examining the recombinant virus AA-CR12, which (Fig.1b) derives its RNAs 1, 2 and 3, (as well as NP) from A/Ann Arbor/6/60 mutant. The characteristics of AA-CR12 transcriptase were clearly similar to those of its mutant parent A/Ann Arbor/6/60 rather than to those of its w.t. parent A/Ann Arbor/9/73 (Fig.2b).

When we investigated the c.a. recombinants AA-CR19 and AA-CR22, however, they were found to have transcriptases that were less *ts* than that of their mutant parent A/Ann Arbor/6/60 (Fig.2c). Both AA-CR19 and AA-CR22 contain RNA 3 of their mutant parent (Table Ia) and exhibit a conditionally-lethal *ts* phenotype in their growth patterns (not shown). These recombinants both differ genetically from mutant A/Ann Arbor/

---

from the w.t. parent, it may be concluded that under these conditions of electrophoresis, RNA 6 of the recombinant corresponds to the NP gene of the mutant parent. RNAs 7 and 8 were not scanned, but RNA 8 of the recombinant was observed to correspond in migration rate to that of the mutant parent, and to differ from that of the w.t. parent.

6/60 and from recombinant AA-CR12 in that their RNA 1 is derived from a wild-type H3N2 virus (Table Ia). Therefore we conclude that the *ts* nature of the virion transcriptase observed for A/Ann Arbor/6/60 mutant and AA-CR12 recombinant is probably caused by lesions in their RNA 1, and that the *ts* lesion in RNA 3 affects some other function necessary for *in vivo* RNA synthesis.

SUMMARY AND CONCLUSIONS

By comparing the electrophoretic migration rates of virion RNAs in warmed 3% acrylamide gels we showed that genes transferred from the c.a. and *ts* mutant A/Ann Arbor/6/60 to c.a. and *ts* H3N2 recombinants always include RNA 3, whereas the other large RNAs, segments 1 and 2, have segregated independently of the c.a. and *ts* markers. Because the A/Ann Arbor/6/60 mutant fails to recombine with a WSN group III single-mutant which has a *ts* lesion in one of its three large RNA segments but under the same growth conditions, does recombine with WSN group I and II mutants which have *ts* lesions in the other two large RNA segments, we have concluded that the A/Ann Arbor/6/60 mutant contains a *ts* lesion in its RNA 3 segment which affects *in vivo* cRNA synthesis and is responsible for the conditionally-lethal *ts* phenotype of the virus.

Additionally, by *in vitro* assays of the transcriptases in recombinant viruses which derived different combinations of RNAs 1, 2, and 3 from the A/Ann Arbor/6/60 mutant and w.t. parents, we showed that the mutant probably also has a lesion in its RNA 1 that increases the temperature-sensitivity of its virion transcriptase compared to w.t. virus. The existence of a *ts* lesion in the RNA 1 of A/Ann Arbor/6/60 mutant was not, however, detected in recombination experiments with WSN mutant *ts*6 (Table Ib). We suggest that this is because cells were mixedly-infected at the permissive temperature, and because recombinants produced at 34° which derive RNA 1, but not RNA 3, from A/Ann Arbor/6/60 mutant may have a sub-lethal *ts* phenotype when subsequently grown at 39°. Presence of a *ts* lesion in RNA 1 of the A/Ann Arbor/6/60 mutant is consistent with previous results of Spring *et al.* (18, 19), who showed that the virus failed to complement the replication at 39° of the NIH mutant R 1 which contains a lesion in its RNA 1 (12). To obtain further information about the nature of *ts* lesions in the RNA segments 1 and 3 of mutant A/Ann Arbor/6/60 we must now study recombination at 34° between this virus and the NIH mutants, similarly to the recombination studies with WSN mutants. This appears necessary since the previously described complementation analyses using the NIH mutants were done at 39°, and results may accordingly have been affected by the

TABLE I

Genetic properties of cold-adapted, temperature-sensitive mutant A/Ann Arbor/6/60 and its recombinants.

A. Gene composition of influenza virus recombinants deriving cold-adaptation and temperature-sensitive properties from mutant A/Ann Arbor/6/60

Recombinants					Derivation of genes for internal protein					
Expt.	Wild-type parent	Clone	HA	NA	RNA 1	RNA 2	RNA 3	NP	M	NS
AA-CR6	A/Queensland/6/72	0	w[1]	w	[2]		μ[3]	μ	μ	μ
AA-CR12	A/Ann Arbor/9/73	0	w	w	μ	μ	μ	μ	μ	μ
AA-CR13[4]	A/Dunedin/4/73	5,9	w	w		w	μ	μ	μ	μ
AA-CR18	A/Scotland/840/74	3,4,5	w	μ		μ	μ	μ	μ	w
AA-CR18	A/Scotland/840/74	7	w	w		μ	μ	μ	μ	w
AA-CR19	A/Victoria/3/75	0	w	w	w	w	μ	μ	μ	μ
AA-CR22	A/Victoria/3/75	1	w	w	w	μ	μ	μ	μ	μ
AA-CR22	A/Victoria/3/75	17	w	μ	w	μ	μ	μ	μ	μ

[1] w indicates gene derived from wild-type parent
[2] blank indicates not done
[3] μ indicates gene derived from mutant parent
[4] AA-CR13 clones 5 and 9 have a c.a. and *ts* phenotype, although a different clone of AA-CR13 is not *ts* (19).

B. Production of ts^+ virus from mixed infections of mutant A/Ann Arbor/6/60 and WSN *ts* mutants of recombination groups I, II and III

Cells infected at 34° with		Infectivity titer at 39° (pfu/ml) of 24 hr harvest from infected cells
A/Ann Arbor/6/60	WSN mutant	
+[a]	-	$<10^2$
-[b]	*ts* 6 (I)[c]	$<10^2$
+	*ts* 6 (I)	2.5×10^4[d]
-	*ts* 53 (II)	3×10^2
+	*ts* 53 (II)	5.5×10^3[d]
-	*ts* 15 (III)	$\leq 10^2$
+	*ts* 15 (III)	$<10^2$

[a] + = 1 to 5 pfu/cell

[b] - = virus omitted

[c] Virus added at 1 to 5 pfu/cell. Number in parentheses = complementation/recombination groups of virus (from 20, 21).

[d] Plaques appearing at 39° were picked and shown on subsequent plaque titration to have a ts^+ phenotype as predicted for true recombinants.

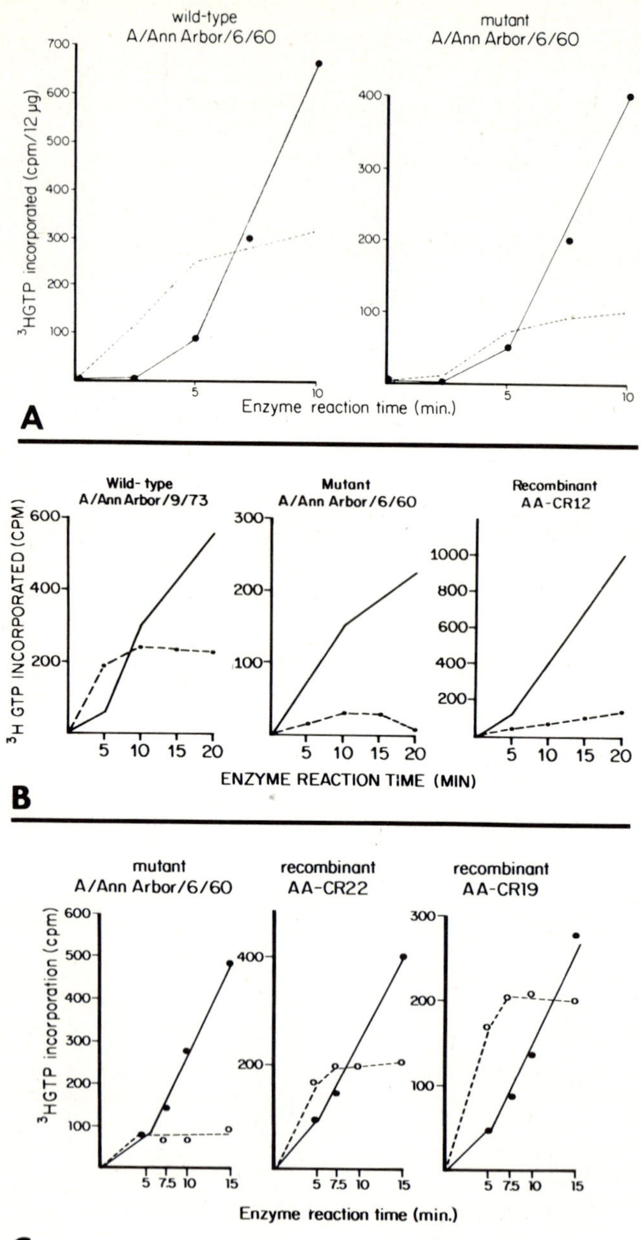

Fig.2. /For Legend see opposite page.

degree of leakiness of different *ts* lesions, as well as by the nature of interactions between the mutant gene products.

Defects attributable to inhibition of RNA synthesis of the mutant A/Ann Arbor/6/60 at elevated temperatures were not observed in our previous study probably because we infected cells at high multiplicities (4). It is known that under these conditions significant levels of RNA synthesis can occur with mutants that otherwise exhibit an RNA$^-$ phenotype when cells are infected at lower multiplicity (3, 6).

Temperature-sensitive transcriptases have been observed rarely for influenza *ts* mutants, a more common finding being temperature-sensitivity of RNA synthesis *in vivo* (3, 6, 15-17). Ghendon *et al.* (3) did, however, observe a *ts* polymerase in one *ts* mutant of fowl plague virus that was also RNA$^-$ for *in vivo* synthesis at elevated temperatures, whereas mutants of three other complementation/recombination groups with RNA synthesis defects did not have *ts* virion transcriptases. Assignment of RNAs to the different genetic groupings has not been reported for this set of mutants. Similarly to the results of Ghendon *et al.* (3), virus from only one of the WSN *ts* mutant groups believed to have defects in RNA synthesis (6, 21) has a *ts* virion transcriptase (11) and in this case it is known that the relevant gene is RNA 1 (12).

The finding of a virion transcriptase associated with

Fig.2. *Virion transcriptase activity of cold-adapted viruses at 31° and 40°. Virus samples were treated with 0.5% Triton N101 at 40° for 5 min, and then added to prewarmed reaction mixtures containing in 1 ml the following components: Tris-HCl pH 8.2, 50 mM; KCl, 150 mM; dithiothreitol, 5 mM; $MgCl_2$, 4 mM; ATP, 2 μm; CTP and UTP, 0.4 μM; ^{3}HGTP, 0.1 μM, specific activity 80 mCi/mMol; GpG, 0.25 μM. Aliquots were withdrawn at the times shown and dispensed into 5 ml of cold 5% TCA. After standing for 30 min, RNA was recovered on Whatman GFC filters, and washed 3 times with 5% TCA and 10 times with cold ethanol. Filters were oven dried and radioactivity was counted in a toluene-based cocktail. All results are the means of duplicate samples, after correction for the background observed in samples withdrawn immediately after addition of virus. A. Wild-type and mutant A/Ann Arbor/6/60. B. Mutant A/Ann Arbor/6/60, wild-type A/Ann Arbor/9/73, and their cold-adepted recombinant AA-CR12 that derives its RNAs 1, 2, and 3 from the mutant. C. Mutant A/Ann Arbor/6/60 and its recombinants AA-CR19 and AA-CR22 that derive RNA 3 from the mutant but RNA 1 from a wild-type H3N2 virus.*

lesions in only one of the genes implicated in RNA synthesis *in vivo*, and the consistency between our results and those of Mowshowitz and Ueda (11) that the responsible gene is RNA 1, suggests that the gene product of RNA 1 is the true virion transcriptase. Functions for the other gene affecting *in vivo* cRNA synthesis may include interaction with the host cell nucleus to permit amplification of cRNA synthesis early in infection.

Additional subjects of investigation with the A/Ann Arbor/6/60 mutant are the nature of lesions permitting growth at sub-optimal temperatures. The role of the mutant virus NP and M protein genes in determining the viral phenotype will be examined because all recombinants produced at 25° that we have so far studied derived genes for these proteins, in addition to their RNA 3, exclusively from the A/Ann Arbor/6/60 parent virus.

ACKNOWLEDGEMENTS

We thank Eve Bingham, Judy Galphin and Donna Sasso for expert technical assistance. Recombinants were prepared with financial support from the US Army Medical Research and Development Command, contract no.DADA 17-73-C-3060 and National Institutes of Health, contract no. I-A172521.

Use of trade names is for identification purposes only and does not constitute endorsement by the Public Health Service or the US Department of Health, Education, and Welfare.

REFERENCES

1. Cox, N.J. and Kendal, A.P. (1976). *Virology* 74, 239.
2. Cox, N.J., O'Neill, M.C. and Kendal, A.P. (1977). *J. Gen. Virol.* 37, 161.
3. Ghendon, Y.Z., Markushin, S.G., Blagovezhenskaya, O.V. and Genkina, D.B. (1975). *Virology* 66, 454.
4. Kendal, A.P., Kiley, M.P. and Maassab, H.F. (1973). *J. Virol.* 12, 1503.
5. Kendal, A.P., Cox, N.J., Murphy, B.R. Spring, S.B. and Maassab, H.F. (1977). *J. Gen. Virol.* 37, 145.
6. Krug, R.M., Ueda, M. and Palese, P. (1975). *J. Virol.* 16, 790.
7. Laskey, R.A. and Mills, A.D. (1975). *Europ. J. Biochem.* 56, 335.
8. Maassab, H.F. (1967). *Nature, Lond.* 213, 612.
9. Maassab, H.F. (1975). *In* "Negative Strand Viruses", (Mahy, B.W.J. and Barry, R.D., eds.), p.755. Academic Press, New York.
10. Maassab, H.F., Cox, N.J., Murphy, B.R. and Kendal, A.P.

(1977). *In* "International Symposium on Influenza Immunization". Karger, Basel.
11. Mowshowitz, S.L. and Ueda, M. (1976). *Arch. Virol.* 52, 135.
12. Palese, P. and Ritchey, M.B. (1977). *Virology* 78, 183.
13. Palese, P., Ritchey, M.B. and Schulman, J.L. (1977). *J. Virol.* 21, 1187.
14. Ritchey, M.B. and Palese, P. (1977). *J. Virol.* 21, 1196.
15. Scholtissek, C. and Bowles, A.L. (1975). *Virology* 67, 576.
16. Scholtissek, C., Kruczinna, R., Rott, R. and Klenk, H.-D. (1974). *Virology* 58, 317.
17. Scholtissek, C., Harms, E., Rohde, W., Orlich, M. and Rott, R. (1976). *Virology* 74, 332.
18. Spring, S.B., Maassab, H.F., Kendal, A.P., Murphy, B.R. and Chanock, R.M. (1977). *Virology* 77, 337.
19. Spring, S.B., Maassab, H.F., Kendal, A.P., Murphy, B.R. and Chanock, R.M. (1977). *Arch. Virol.* 55, 233.
20. Sugiura, A., Tobita, K. and Kilbourne, E.D. (1972). *J. Virol.* 10, 639.
21. Sugiura, A., Ueda, M., Tobita, K. and Enomoto, C. (1975). *Virology* 65, 363.

RECOMBINANT INFLUENZA VIRUSES FROM PARENTS OF KNOWN VIRULENCE FOR MAN

A.S. BEARE

MRC Common Cold Unit,
Harvard Hospital,
Salisbury, Wilts. SP2 8BW, U.K.

The scientific basis of virulence is not known and most of the experiments designed for its study have concentrated on its genetic character and on its variability under different conditions. Conditional lethal mutants have especially been used as experimental probes, and in the case of influenza, it had been hoped that temperature-sensitive (*ts*) viruses, or recombinants derived from them, would emerge as live vaccine by-products for use in man (9, 10, 13, 16). The difficulties of applying laboratory research in this field to the mass immunization of man have, however, proved very great. None of the parent viruses chosen have been ideal, and the Bethesda group in the U.S.A. after much ingenious work with *ts*-mutants of a plaque-forming H2 N2 virus (which could not itself be studied in man because of a lack of volunteers free of anti-H2 and anti-N2 antibodies), concluded that the cut-off temperature alone was important in attenuation and not the sites of the defects in the RNAs. Beare *et al.* (5) using the older virus A/PR/8/34 (H0 N1), a host-range mutant that had lost its human infectivity, thought that the inclusion of genes from this virus in recombinants derived from itself and virulent viruses commonly signified a tendency towards reduced virulence. The effect seemed quantitative and related to the number of genes which had been incorporated. This view received support from Florent *et al.* (8) who found that increased hybridization of ^{3}H-cRNA of A/PR/8/34 to recombinant vRNA corresponded to increased 'attenuation' of the recombinants. This paper will examine the results obtained with A/PR/8/34 and other similar mutants, the problems that have made experimentation on human virulence so slow and so difficult, and a method by which it may now be expedited. A companion paper (chapter 70) describes interim results on the analysis by polyacrylamide gel electrophoresis of RNAs of parent viruses and recombinants of known virulence for man.

VIRUSES

Most viruses were either newly isolated or obtained from the World Influenza Centre, Mill Hill, London. Their origin and subsequent histories have been described elsewhere (3-5, 11). Propagation was by allantoic inoculation of 11-day embryonated hens' eggs by standard methods. Viruses used for repeated human inoculation were usually reisolated from nasal washings and passed thereafter in specific pathogen-free (SPF) eggs.

Pairs of viruses from which recombinants were selected, commonly consisted of wild-type H3 N2 type A strains and host-range mutants known to be attenuated or non-infectious for man (see Tables I-IV). Type B recombination pairs were chosen on similar principles (Table V). Evidence that recombination had occurred was mostly inferred from biological properties of parents and progeny, usually growth properties, high haemagglutination (HA) titre, virulence for animals, nature of envelope antigens, and human virulence (3, 12). Influenza A virus recombination was performed by mixed infection of embryonated eggs followed by subsequent cloning by terminal dilution, while influenza B recombinants were obtained in bovine kidney cell monolayers and cloned by repeated plaque selection (4).

The virus A/USA/43 (H0 N1) was kindly provided by Dr. J.J. Skehel as a freeze-dried seed. It was said to have had 9 egg passes only. After inoculation into antibody-free volunteers it was treated in 3 different ways, (1) passed as infected nasal washings to other volunteers, recovered in SPF eggs, and cloned by triple passage at limit dilution in SPF eggs (2) passed from nasal washings into SPF eggs and cloned as in (1), and (3) cloned in eggs without preliminary passage in man. The intervening human passages between the egg passes were designed to provide alternative seeds suitable for virulence experiments, as man-to-man passage was known to produce rapid enhancement of human virulence (2). Anti-HA and anti-neuraminidase (NA) antibodies in volunteer sera were measured by standard methods (2, 7). Selection of volunteers, clinical surveillance and grading of reactions were as described by Beare and Reed (4).

EXPERIMENTAL TRIALS

The 2 attenuated viruses, A/PR/8/34 (H0 N1) and A/Okuda/57, and 8 wild-type H3Na strains are shown in Table I. Most of the H3 N2 viruses seemed to have a standard type of virulence

TABLE I

Influenza A viruses (or comparable serotypes) of defined human virulence used in the study or from which recombinants were made

Name	Surface antigens	Origin and passage	Whether tested in human trials	Known or presumed virulence
A/PR/8/34	H0 N1	Unrecorded passages	Yes	0
A/Okuda/57	H2 N2	280 egg passes	No	+
A/Hong Kong/1/68	H3 N2	few egg passes	Yes	+++
A/England/939/69	H3 N2	isolate from clinical influenza	No	+++
A/England/878/69	H3 N2	few egg passes	Yes	+++
A/England/42/72	H3 N2	" " "	Yes	+
A/Port Chalmers/1/73	H3 N2	" " "	Yes	+++
A/Finland/4/74	H3 N2	" " "	Yes	+++
A/Scotland/840/74	H3 N2	" " "	Yes	+++
A/Victoria/3/75	H3 N2	" " "	No	+++

Testing in human trials consisted of inoculation into antibody-free volunteers and observation for clinical effects, virus excretions and antibody rises. Virulence: +++ = capable of causing influenzal symptoms; ++ = local symptoms and constitutional symptoms; + = local symptoms only; ± = overattenuation; 0 = non-infectious. Only mild effects were observed with wild A/England/42/72, but it was difficult to find antibody-free volunteers.

(1) signified here by the symbol +++, or the ability to induce influenza-like illness in antibody-free volunteers. The one exception to this generalization was A/England/42/72 which was given with little clinical effect to many volunteers, but at a time when it was difficult to find seronegatives. A/PR/8/34 (H0 N1) was non-infectious for young adults free of anti-H0 and anti-N1 antibodies who were readily available among the volunteers attending the Unit (5) but none could be found who were free of anti-H2 and anti-N2 antibodies, and A/Okuda/57, therefore remained uncharacterized. It was presumed however, that this virus which had been used as a live vaccine by Okuno and Nakamura (18) was attenuated but still infectious. The effects of attenuating the wild virus were consistently similar. The influenza syndrome of fever, headache, joint pains, etc. shaded into coryza or minimal local symptoms in the respiratory tract. Adequate 'attenuation' is denoted in the tables by + and low infectivity by ±.

The effect of crossing the non-infectious virus A/PR/8/34 and the epidemic strain A/England/939/69 are shown in Table II. The 6 viruses shown were presumed to be recombinants on the basis of biological characteristics, but the particular genes that had been inherited from the individual parents could not of course be distinguished other than those coding for the HA and N proteins which were identified by immunological tests. Three of the viruses, 64c, (H3 N2), 64d (H3 N2) and 31 (H3 N1), were 'attenuated' but fully infectious; one, no.6 (H3 N2), was possibly slightly less virulent than the wild parent; one, no. 7 (H3 N2), showed no evidence of attentuation; and an H3 N2 virus, X-31, with a similar parentage (A/PR/8/34 and A/Aichi/2/68) was partially attenuated. The H0 N2 hybrid, 64b, was of low infectivity. None of the presumptive recombinants, however, resembled A/PR/8/34 in its total lack of infectivity for man.

Recombinants made with A/PR/8/34 and later wild-type H3 N2 serotypes are shown in Table III. Some of the results obtained with A/England/42/72 gave probelms in interpretation. The 'wild' strain did not produce influenza in people apparently free of anti-H3 antibody, and MRC 5 (H3 N2) and MRC 8 (H3 N1) were therefore more virulent than the wild parent. MRC2 and MRC7 were, however, attenuated, infectious and antigenic. MRC3 (H0 N2) like its counterpart 64b (H0 N2) in the A/England/939/69 series, was of low infectivity. Five more H3 N2 recombinants selected from A/PR/8/34 and 2 later variants of the Hong Kong subtype were all infectious for man and of varying degrees of virulence.

Fewer recombinants were made with A/Okuda/57 and wild viruses. McCahon *et al.* (11) tested 4 recombinants made with the strain

TABLE II

The human virulence of influenza A virus recombinants prepared from A/PR/8/34 and early A/Hong Kong/68 - like wild strains

Virus	Surface antigens	Virulence in human trials
A/PR/8/34 - England/939/69:		
clone 6	H3 N2	++
clone 7	H3 N2	+++
clone 64c	H3 N2	+
clone 64d	H3 N2	+
clone 31	H3 N1	+
clone 64b	H0 N2	±
A/PR/8/34 - Aichi/2/68:		
X-31	H3 N2	++

A/Finland/4/74 (Table IV), of which 2 were attenuated, 1 possibly overattenuated, and 1 virulent. Because of the rapidly changing immunological state of the population at the time the study was carried out, finding volunteers of comparable susceptibility to infection for each of the trials proved extremely difficult.

A later study showed that the same methods could be applied to the selection of attenuated influenza B viruses (Table V). The recombination technique now used was, however, different. All influenza B viruses plaqued efficiently (but with a similar plaque morphology) in monolayers of bovine kidney, and there was evidence of a high frequency of recombination and a modification of virulence corresponding to that which had been observed with the influenza A viruses (4).

From the data given, there can be little indication of the relative importance for human virulence of individual influenza virus genes. However, if one accepts a premise that virulence

TABLE III

The virulence for man of influenza A recombinants prepared from A/PR/8/34, and later wild-type H3 N2 strains

Virus	Surface antigens	Virulence in human trials
A/PR/8/34 - England/42/72:		
MRC 2	H3 N2	+
MRC 5	H3 N2	++++
MRC 7	H3 N2	+
MRC 8	H3 N1	+++
MRC 3	H0 N2	±
A/PR/8/34 - P.C./1/73:		
MRC 9	H3 N2	+++
MRC 11*	H3 N2	+++
A/PR/8/34 - Scotland/840/74:		
MRC 12a	H3 N2	++
MRC 12b	H3 N2	++
MRC 12c	H3 N2	+

*MRC 11 was a recombinant of A/PR/8/34 and MRC 9.

and infectivity are simply different degrees of the same thing, it would seem that virulence can be quantitatively modified by an admixture of non-virulence genes to virulence genes. There is no apparent qualitative difference between a virulent and an attenuated virus, as between a pneumo- or neurotropic virus in mice. Also none of the recombinants containing the H3 haemagglutinin were as attenuated as those containing the H0 haemagglutinin, which might suggest a significant role for the HA gene in the determination of virulence. Florent *et al.*(8) found a

TABLE IV

The virulence for man of influenza A recombinants prepared from A/Okuda/57 and later wild-type H3 N2 strain

Virus	Surface antigens	Virulence in human trials
A/Okuda/57 - Finland/4/74:		
WRL 100	H3 N2	+++
WRL 94	H3 N2	±
WRL 95	H3 N2	+
WRL 105	H3 N2	+

correlation between the homology of A/PR/8/34 RNA with that of recombinants derived from it, and the human virulence of those recombinants (Table VI). A problem arose in the observation that A/PR/8/34 and wild H3 N2 strains produced basic hybridization figures of 35% apparently due to partial homology of individual RNA segments. Florent (personal communication) was unable to apply his test to the A/PR/8/34-England/939/69 recombinants because of technical problems but found 40% hybridization between the RNAs of the parental viruses. Oxford *et al.* (chapter 70) have now attempted to analyse the RNA segments and polypeptides of these viruses and of the recombinants derived from them and to consider their findings in relation to the virulence of the viruses.

DISCUSSION

Many attempts have been made in recent years to apply genetic techniques to the study of influenza virus virulence for man. The virus contains 8 segments of RNA, each with a separate gene function. Theoretically, the number of complementation groups should correspond, and Murphy *et al.* (14) in a review of different *ts*-mutants, concluded that the degree of thermal sensitivity and not the biochemical defects at non-permissive temperature determined the attenuation. They suggested that the virus replicated at the temperature prevailing in the upper respiratory tract, but failed to do so at that which was found in the lung and hence infection produced strictly local effects. The actual gene containing the defect which rendered the virus

TABLE V

The virulence for man of influenza B viruses and their recombinants

Virus	Human virulence
B/Lee/40	+
B/Setagaya/3/56	not tested; presumed +
B/Hannover/1/70	+++
B/Hong Kong/8/73	++
B/Lee 40 - Hannover/1/70:	
clone 232	++
clone 233	+
clone 234	not diagnosable
clone 235	+++
B/Setagaya/3/56 - Hong Kong /8/73:	
clone 101	+
clone 102	+++
clone 103	+++
clone 104	++
clone 105	+

ts was relatively unimportant, provided it was not one of those coding for HA or N, since that would make it impossible to use the virus for making recombinants with wild-type HA and N polypeptides. The results obtained with our own systems must obviously be examined from a different angle. A/PR/8/34 is non-infectious for man and replicates at very high temperatures, a property which it transmits to some of its recombinants. When the genes of virulent viruses are diluted with those of

TABLE VI

Relationship between human virulence and degrees of hybridization between A/PR/8/34 cRNA and recombinant vRNA. All recombinants were prepared from A/PR/8/34 (non-infectious for man) and wild-type H3 N2 strains (fully virulent for man). The technique used in the hybridization analysis has been described(8)

Virus	Virulence for man	% hybridization between PR8 cRNA and recombinant vRNA
A/PR/8/34	non-infectious (control)	100
A/England/42/72 and later serotypes	mostly +++	35
A/PR/8/34 - England/42/72:		
MRC 2 (H3 N2)	+	55
MRC 5 (H3 N2)	+++	31
MRC 7 (H3 N2)	+	45
MRC 8 (H3 N1)	+++	50
A/PR/8/34 - Port Chalmers/1/73:		
MRC 11 (H3 N2)	+++	35
A/PR/8/34 - Scotland/840/74:		
RIT 4022 (H3 N2)	+++	47
RIT 4023 (H3 N2)	+	62
RIT 4025 (H3 N2)	+	67
AM 37 (H3 N1)	+	77
A/PR/8/34 - Victoria/3/75:		
RIT 4050 (H3 N2)	+	56

non-virulent viruses, the hybrid is relatively attenuated. The identity of the virulence genes has not been established, but there is a suggestion that that governing the HA may have a more important role to play than any of the others. No H3 N2 recombinants made from A/PR/8/34 and wild H3 N2 viruses resembled A/PR/8/34 in its lack of infectivity for man, and H0 N2 recombinants appeared more attenuated than H3 N1 recombinants (Tables II-III). Further, a virus A/equine/1/56 (Heq1) - England/42/72 (N2) proved non-infectious for people without anti-Heq1 antibodies, and this appeared to reflect the lack of infectivity for man of A/equine/1/56 (not shown in the tables). Chapter 70 describes attempts to identify by polyacrylamide gel electrophoresis the genes contained in the recombinants listed in Table II.

Although influenza viruses have a wide host range, there is no animal substitute for man himself in the study of human virulence, in spite of much experimental work that has been done in ferrets and mice. As much data as possible must therefore be accumulated from epidemiological and experimental sources whenever practicable. However, the immunological state of man is not static, and only after major antigenic changes in nature are people available in a fully susceptible state for strictly controlled scientific experiments with the new viruses. Such experiments cannot afterwards be repeated and hence information is collected piecemeal and at irregular intervals. Our own investigation has already stretched over the entire 9 years of the A/Hong Kong/68 subtype era, even though volunteers were always available in adequate numbers. Now that methods of analysing RNAs by polyacrylamide gel electrophoresis, by RNA hybridization and by combinations of the two, are rapidly improving it has become urgently necessary to expedite the human studies. We should ourselves like to suggest the use of obsolete virulent serotypes in isolated volunteers free of anti-haemagglutinin and anti-neuraminidase antibody. A virus, A/USA/43 (H0 N1), has been used in this way to a limited extent in Salisbury and recombinants developed from this strain could be studied concurrently in volunteers and in the laboratory. The investigation would thus be insulated from the effects on man of natural antigenic drift. Whether findings from individual virus strains would be widely applicable to the group as a whole, now and in the future, is still a matter for conjecture.

ACKNOWLEDGEMENTS

I am grateful to my collaborators, Dr. G.C. Schild and Dr. D. McCahon, many of whose experiments I have described, and to the technical staff at Salisbury, Mrs. K. Callow, Mrs. B. Head and Mr. J. Sherwood.

REFERENCES

1. Beare, A.S. (1975). *Progr. Med. Virol.* 20, 49.
2. Beare, A.S., Bynoe, M.L. and Tyrrell, D.A.J. (1968). *Brit. Med. J.* 4, 482.
3. Beare, A.S. and Hall, T.S. (1971). *Lancet ii* 1271.
4. Beare, A.S. and Reed, S.E. (1977). *In* "Chemoprophylaxis and Viral Infections of the Respiratory Tract". (J.S. Oxford, ed.), Chapter 7. Chemical Rubber Company, Cleveland, Ohio, USA.
5. Beare, A.S., Schild, G.C. and Craig, J.W. (1975). *Lancet ii* 729.
6. Beare, A.S., Sherwood, J.E., Callow, K.A. and Craig, J.W. (1977). *Infect. Immun.* 15, 347.
7. Callow, K.A. and Beare, A.S. (1976). *Infect.Immun.* 13, 1.
8. Florent, G., Lobmann, M., Beare, A.S. and Zygraich, N. (1977). *Archives of Virology* 54, 19.
9. Ghendon, Y.Z., Marchenko, A.T., Markushin, S.G., Ghenkina, D.B., Mikhejeva, A.V. and Rozina, E.E. (1973). *Archives of Virology* 42, 154.
10. Mackenzie, J.S. (1969). *Brit. Med. J.* 3, 757.
11. McCahon, D., Beare, A.S. and Stealey, V.M. (1976). *Post-grad. Med. J.* 52, 389.
12. McCahon, D. and Schild, G.C. (1972). *J. Gen. Virol.* 15, 73.
13. Murphy, B.R., Chalhub, E.G., Nusinoff, S.R. and Chanock, R.M. (1972). *J. Infect. Dis.* 126, 170.
14. Murphy, B.R., Richman, D.D., Spring, S.B. and Chanock, R.M. (1976). *Postgrad. Med. J.* 52, 381.
15. Okuno, Y. and Nakamura, K. (1966). *Biken J.* 9, 89.
16. Spring, S.B., Nusinoff, S.R., Mills, J., Richman, D.D., Tierney, E.L., Murphy, B.R. and Chanock, R.M. (1975). *Virology* 66, 522.

FURTHER STUDIES ON THE EFFECT OF COMPLEMENT ON THE MEMBRANE OF NEWCASTLE DISEASE VIRUS (NDV)

M.I. SAWA and K. APOSTOLOV

Department of Virology,
Royal Postgraduate Medical School,
London, England.

The incubation of paramyxoviruses with erythrocytes at 37° leads to three phenomena which appear at first to be unrelated. However, there is evidence that they have a common mechanism.

The first was reported by Burnet and Anderson (5). They showed that incubation of erythrocytes with Newcastle disease virus (NDV) became agglutinable by heterophile antibody. The second discovery, by Morgan *et al.* (11), was the capacity of mumps virus to lyse erythrocytes. Finally, Okada (12) discovered the capacity of Sendai virus to fuse cells in culture. Subsequently it was shown that NDV and Sendai can also fuse erythrocytes. In each case, attachment of the virus to the erythrocytes is necessary. Howe and Morgan (9) demonstrated that during virus-erythrocyte interaction, fusion of the viral envelope with the erythrocyte membrane occurs.

In 1972 a theory was put forward based on electron micrograph negative staining that the fusing agent is the viral membrane (nanogranular layer) of the viral envelope, and the haemolysis is a sequential process culminating in the escape of haemoglobin from that part of the virus which is integrated into the erythrocyte membrane (1, 2). The following steps were identified in the haemolysis process: attachment of the virus by the tips of the spikes; fusion of the viral membrane (nanogranular layer) with the erythrocyte membrane; the spikes (projections) float off; and finally, the creation of a break in the integrated viral membrane and escape of haemoglobin. This concept of haemolysis was utilized to explain the phenomenon associated with the interaction of paramyxoviruses with erythrocytes and to design experiments to support it. In a recent paper we reported that heterophile antibody and antibody to the host in which the virus grows, in the presence of complement, enhances the haemolytic property of the virus several-fold (3). In this paper we report additional observations concerning this phenomenon.

KINETICS OF HAEMOLYSIS BY VIRUSES PRE-TREATED WITH COMPLEMENT

The Queensland strain of NDV was obtained from Dr. D.J. Alexander, Central Veterinary Laboratory, Weybridge. The virus was grown, harvested, concentrated and purified by the same method as previously described for Sendai virus (Apostolov and Almėida, (1).

As shown in Fig.1 there is no significant difference in the kinetics of haemolysis with pre-adsorbed and non-adsorbed virus. This finding suggests that the fusion of the viral membrane to the erythrocyte membrane follows the attachment without delay. It can also be seen that most of the haemolysis is over after 10 min, and most of the delay is probably due to the time taken for haemoglobin to diffuse out through complement holes. The lytic event and the release of haemoglobin from erythrocytes by immune haemolysis requires osmotic swelling of the cells. It has been shown by Green *et al* (7) that the initial lesion is too small to allow protein to pass, and complement-lysed cells lose their intracellular potassium but do not rupture or release haemoglobin until swelling of the cell and expansion of the complement holes occurs.

KINETICS OF HAEMOLYSIS BY SINGLE AND MULTIPLE CYCLE VIRUSES

Viruses harvested 18 hr after inoculation of chick chorioallantoic membrane are called single cycle, and have been shown to have very little haemolytic activity compared to viruses harvested after 48 hr, which are called multiple cycle viruses (8).

The results of an experiment in which we examined their relative haemolytic properties are shown in Fig.2. Primary haemolysis in iso-osmotic KCl is higher with multiple cycle virus. Haemolysis after complement treatment is about the same. Haemolysis by the two viruses is different when the erythrocytes from primary haemolysis are washed and subjected to treatment with unheated human plasma (4), then reincubated for 30 min at 37°. In this case, the additional haemolysis obtained is higher with the single cycle virus.

These results suggest that virus harvested under single cycle conditions has a more stable membrane than virus harvested after 48 hr. It also suggests that the total of fused membranes in the same for both viruses, as seen from the curve for complement-treated virus.

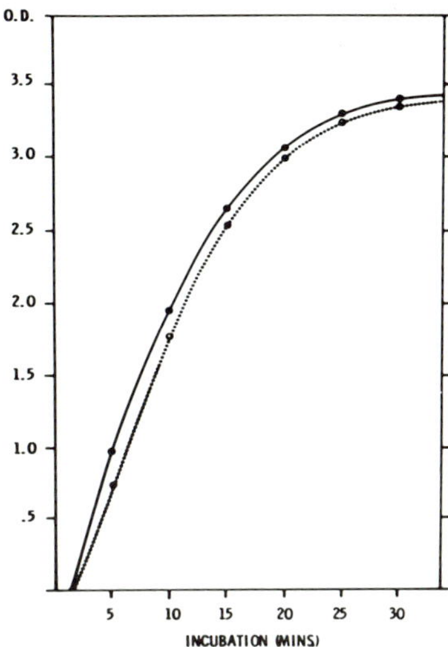

Fig.1. *Virus (NDV) treated with unheated plasma for 3 min at 37° as a source of heterophile antibody and complement (3) was adsorbed to the erythrocytes for 30 min at 4°. The erythrocytes were then washed in iso-osmotic KCl and incubated at 37° for 30 min (●——●). In another set of tubes the virus pre-treated with unheated human plasma for 3 min was added at zero time to erythrocytes and incubated at 37° for 30 min (o---o). Samples were taken at the appointed time and haemoglobin was measured as previously described (3).*

STABILIZING EFFECT OF HETEROPHILE ANTIBODY ON THE VIRUS MEMBRANE

In order to examine the effect of heterophile antibody on the haemolytic property of the virus, we have compared primary haemolysis by complement-treated virus with and without EDTA, and the results are shown in Fig.3. It can be seen that heterophile antibody reduces the amount of primary haemolysis. The effect is not due to inhibition of attachment, since the haemagglutination titre of the virus is the same. However, incubation of the virus for 3 min in unheated plasma leads to enhancement of the haemolytic property of the virus due to activation of complement and production of complement holes (10). That the effect is due to complement is supported by the observation that the enhancement is inhibited in the presence of EDTA (6). Calcium and magnesium ions are needed for activation of complement, and they are complexed with EDTA, so the complement is not activated. The enhancement of haemolysis in NDV and Sendai virus (unpublished) due to com-

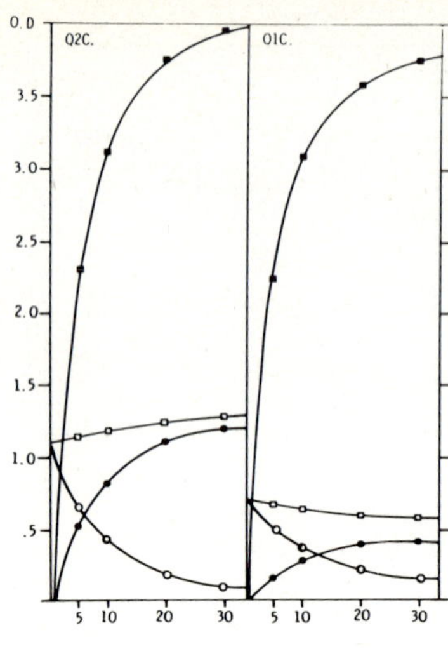

Fig.2. *The Queensland strain of NDV was harvested 18 hr after inoculation chorio-allantoic membranes (Q1C) and after 48 hr (Q2C). After purification both populations were adjusted to the same HA titre (HAU 256). Primary haemolysis was determined after incubation at 37° in iso-osomotic KCl (●———●), the same amount of virus (0.2 ml) was incubated with an equal amount of unheated human plasma for 3 min at 37°. For both populations (■———■) the kinetics of primary and complement-treated virus were performed as in Graph 1. In addition, the erythrocytes from primary haemolysis, at the appointed time intervals, were washed and reincubated as packed erythrocytes (mixed later) with 0.2 ml plasma for 30 min at 37°, and the released haemoglobin was plotted (o———o). Primary haemolysis plus haemolysis additional to primary after complement treatment was also plotted (□———□).*

plement is illustrated in Fig.4. There are clearly visible multiple complement holes of approximately 10 nm in diameter in the negatively stained particles obtained after treatment with unheated plasma.

CONCLUSION

The results presented in this paper support the concept that the three phenomena associated with the interaction of paramyxoviruses with erythrocytes are aspects of the main event in haemolysis. The phenomenon of agglutination of paramyxovirus-treated erythrocytes depends on the availability of the virus membrane on the surface of the erythrocytes after the spikes have floated off. The phenomenon of haemolysis

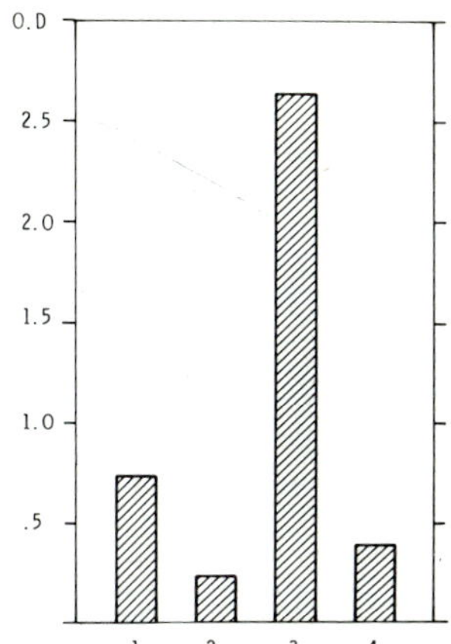

Fig.3. *Virus was distributed in equal amounts (0.2 ml) in tubes at 4°, unheated plasma was added in the tubes represented in columns 2, 3 and 4. After adsorption at 4° for 30 min 0.2 ml EDTA (100 mM) was added to the preparation in the 4th column, and the preparation from columns 3 and 4 incubated for 30 min at 37°. The erythrocytes were added to all tubes and adsorbed for 30 min at 4°, washed once and re-suspended to standard volume. After incubation at 37° for 30 min, haemolysis was determined.*

depends on the stability of the viral membrane before and after fusion with the erythrocyte membrane, and, finally, the fusion of the viral membrane and its integration into the erythrocyte membrane. In our opinion, the phenomenon of fusion of erythrocytes is initiated by bridging the erythrocytes with the virus membrane.

ACKNOWLEDGEMENTS

We thank Professor A.P. Waterson for discussion on the present work and for checking the manuscript.

REFERENCES

1. Apostolov, K. and Almeida, J.D. (1972). *J. Gen. Virol.* 15, 227.
2. Apostolov, K. and Poste, G. (1972). *Microbios,* 10A, 139.
3. Apostolov, K. and Sawa, M.I. (1976). *J. Gen. Virol.* 33, 459.
4. Apostolov, K. and Waterson, A.P. (1974). *Microbios,* 11A, 85.
5. Burnet, F.M. and Anderson, S.G. (1946). *Brit. J. Exp. Path.* 27, 236.
6. Frank, M.M., Rapp, H.J. and Borsos, T. (1964). *J. Immunology* 93, 409.

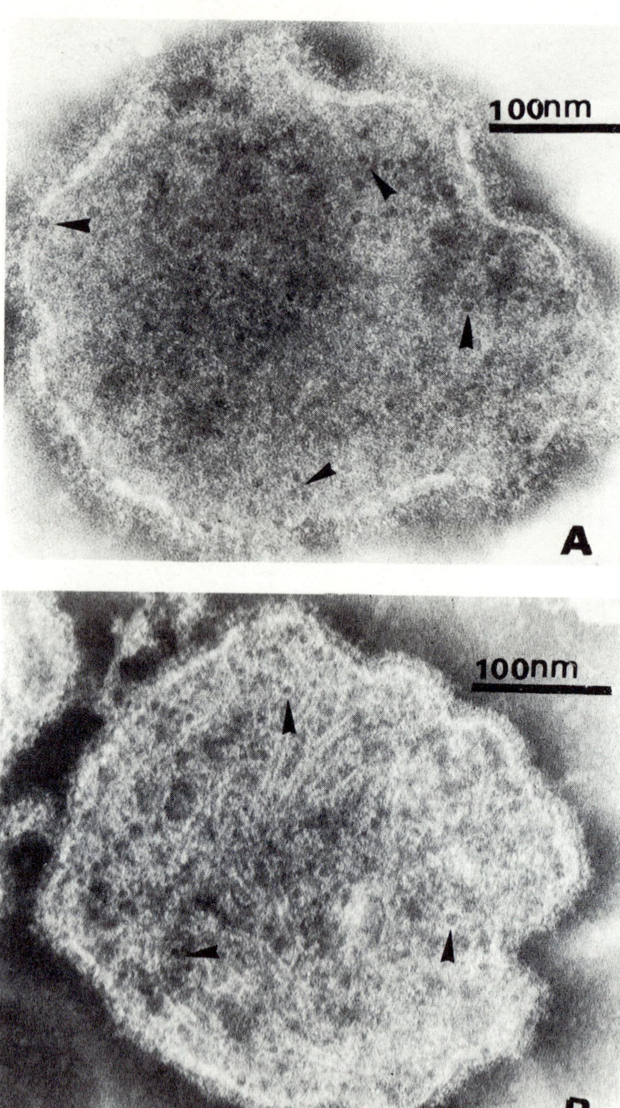

Fig.4. *Virus treated with unheated human plasma for 10 min at 37° was washed by centrifugation at 30,000 g and re-suspended in distilled water and negatively stained by 3% PTA (1). The complement holes on a particle of NDV (arrows) are shown in* a *and on Sendai virus in* b. *(x 215,000).*

7. Green, H., Barrow, P. and Goldberg, B. (1959). *J. Exp. Med.* 110, 699.
8. Homma, M., Shimizu, K., Shimizu, Y.K. and Ishida, N. (1976). *Virology* 71, 41.
9. Howe, C. and Morgan, C. (1969). *J. Virol.* 3, 70.
10. Humphrey, J.H. and Dourmashkin, R.R. (1965). *In* "Ciba Foundation Symposium" (G.E.W. Wolstenholme and Julie Knight, eds.), p.175. J.A. Churchill Ltd.
11. Morgan, H.R., Enders, J.F. and Wagley, P.F. (1948). *J. Exp. Med.* 88, 503.
12. Okada, Y. (1958). *Biken's Journal* 1, 103.

ISOLATION AND PROPERTIES OF THE MEMBRANE (M) PROTEIN OF MEASLES VIRUS

W.W. HALL, K. NAGASHIMA, W.R. KIESSLING and V. TER MEULEN

Institute of Virology and Immunobiology,
University of Würzburg,
Versbacher Landstrasse 7, 8700 Würzburg, W. Germany.

The membrane (M) protein of measles virus is believed to be the smallest major protein of the virus particle having a mol. weight of approximately 37,000 daltons. By analogy with the paramyxoviruses it is likely that the protein is located beneath the lipid membrane and may be linked to the internal nucleocapsid structures (2, 3, 6). Since the M protein of the paramyxoviruses apparently only plays a structural role and is devoid of any biological activity only a few studies have appeared describing its biochemical properties. In the present report we describe the isolation of the M protein of measles virus and an analysis of its chemical, physical and antigenic properties.

PURIFICATION OF M PROTEIN

M protein was isolated from purified measles virus which had been grown in Vero cells using an adaption of the method originally devised for the paramyxoviruses SV5 and NDV (4, 5). Purified virus was disrupted with Triton-X-100 in 0.01M PO_4 buffer containing 1M KCl. Nucleocapsids and partially disrupted viruses were removed by centrifugation and the supernatant fluids were dialysed for 2-3 days against de-ionised distilled water. The dialysate containing the precipitated M protein (4) was centrifuged through a 10 ml cushion of 18% (W/V) sucrose at 18,000 r.p.m. for 45 min in a SW 27 rotor. The pellet containing partially purified M protein was resuspended in Triton-X-100-1M KCl; vigorously stirred for 1 hr at room temperature; re-dialysed and centrifuged through a sucrose cushion as described above. After 3 to 5 re-cycles of this procedure pure preparations of M protein could be obtained.

PHYSICAL AND CHEMICAL PROPERTIES

M protein isolated as described above was examined for its purity by electrophoresis on 15% polyacrylamide-SDS gels. Fig.1 shows that only a single protein was present. The mol. wt. was estimated to be approximately 38,000 daltons. However, when high concentrations of Triton (> 10%) were used during purification the protein band tended not to be sharp and also appeared to have a variable and sometimes increased mol. wt. (up to 42,000). It is thought that these observations are due to some types of interaction between the detergent and the hydrophobic protein. A similar spreading of the Sendai virus M protein after Triton extraction can also be observed in the publication of Hewitt and Nermut (2).

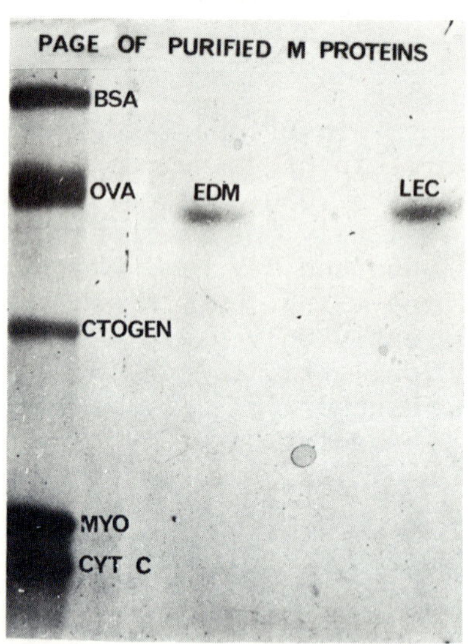

Fig.1. *SDS polyacrylamide gel electrophoresis of purified Edmonston (EDM) and SSPE (Lec) virus M proteins. The single protein band of the two M proteins has a mol. wt of approx. 38,000 daltons.*

The amino acid composition of the purified M protein is shown in Table I. A comparison of the data obtained with that previously reported for the M proteins of influenza (1) and SV5 (3) viruses shows a wide variation between all 3. However, polar amino acids constituted 45% of the total in the measles M protein giving a further indication of its hydrophobic nature.

Electron microscopic studies using negative staining

TABLE I

Amino Acid Composition of Membrane Protein

Amino Acid	Residues per 100 Residues		
	Measles	SV5*	Influenza**
Aspartic Acid	7.4	7.8	7.5
Threonine	1.8	7.1	6.9
Serine	8.6	7.6	5.9
Glutamic Acid	11.0	11.0	13.0
Proline	4.2	6.5	2.9
Glycine	14.2	9.9	6.8
Alanine	10.7	5.0	10.6
Valine	3.9	7.3	6.6
Methionine	0.84	2.1	3.2
Isoleucine	3.6	5.7	4.3
Leucine	7.0	8.1	11.1
Tyrosine	2.7	2.6	2.2
Phenylalanine	7.2	3.8	3.4
Histidine	2.4	2.0	2.0
Lysine	8.3	6.8	5.6
Arginine	5.6	4.8	7.1

* From McSharry *et al.* (3).

**From Gregoriades (1).

Legend to Table I

The amino acid composition was kindly carried out by Dr Schiltz, Department of Physiological Chemistry, University of Würzburg, using a Beckman "Multichrom" two column system.

revealed a variation in its structure which seemed to be dependent on salt concentration. Under conditions of high salt (1M KCl; Fig.2a) subunit structures, some of which had aggregated into "fibre-like" particles were observed. In contrast under low salt conditions (distilled water; Fig.2b) the protein had assembled into organised tubular or cylindrical structures. The basic tubular structure had a diameter of approximately 14 nm and varied widely in length. A second type of structure was also observed having a diameter of 22 nm, and it is felt that this is composed of the basic 14 nm particle with an additional "coat" or protein. Similar types of tubular structures have been reported for the M protein of Sendai virus (2) and in agreement with these authors we feel certain that this form adapted by the M protein is not what would be found in the virus particle. It would seem likely that it has arisen from a tight "wrapping up" of the hydrophobic subunits.

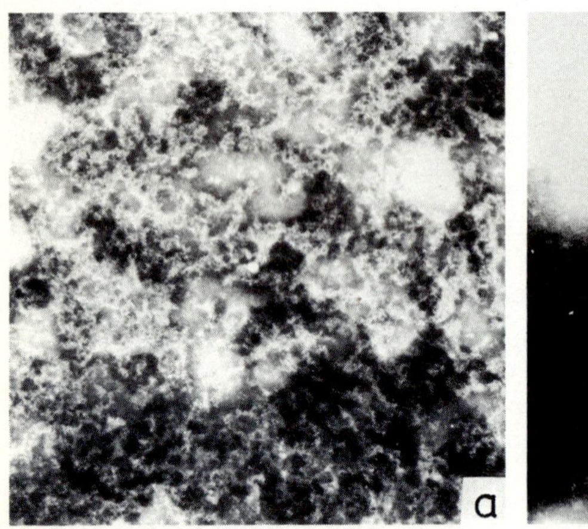

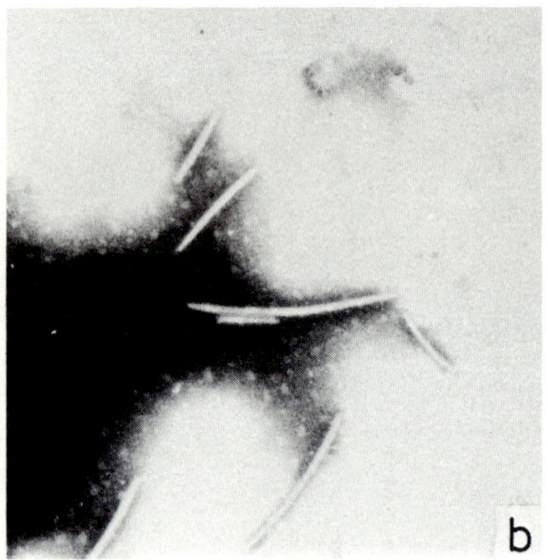

Fig.2. *M-protein of measles virus Edmonston negatively stained with 2% phosphotungstic acid (pH 7.0) and air dried x 65,000.*

a) M protein from a 1 M KCl solution
b) M protein from a low salt suspension.

PREPARATION OF ANTISERUM TO M PROTEIN

Purified M protein was suspended in sterile water containing 0.2% sodium deoxycholate (DOC), mixed with Freund's adjuvant and inoculated into rabbits. A second inoculation in the absence of adjuvant was given after a 4 week interval and the animals were finally bled after a further 2 weeks. Immunodiffusion tests revealed a single precipitin line between antigen and antiserum whereas no precipitation was observed when disrupted Vero cells were used as antigen. It was also possible to titre the antisera using an indirect radioimmunoassay (IRIA) test (7). In addition antisera were assayed for their ability to inhibit the biological properties of whole virus particles. It could be demonstrated that the antiserum had no neutralising, HI or HLI activities.

Attempts were also made to use the antisera to determine the intracellular localisation of the M protein. Indirect immunofluorescence revealed only a hardly visible cytoplasmic staining indicating that either only a small amount of protein was present or alternatively demonstrating the difficulty of detecting the presence of such a small protein.

ANTIGENIC COMPARISON OF THE M PROTEINS OF MEASLES AND SUBACUTE SCLEROSING PANENCEPHALITIS (SSPE) VIRUSES

Using the techniques described above M proteins and their corresponding rabbit antisera were produced from Edmonston strain measles virus and from the Lec strain of SSPE virus. Immunodiffusion studies surprisingly revealed only precipitation lines in the homologous system and no cross reaction was observed. Because the test system used can at times be relatively insensitive we employed an immunoelectrophoresis test. Fig.3 demonstrates that once again precipitation was only observed in the homologous system indicating that the M proteins of the two viruses are antigenically different.

The above observation of distinct antigenic differences between SSPE and measles virus raises an interesting point in the consideration of the etiology of SSPE. We have shown (chapter 15) that the viruses isolated from patients (SSPE viruses) may be derived from the normal measles virus as a result of a mutation or modification in the gene region coding for the M protein. The present data corroborates this hypothesis and demonstrates for the first time the exact differences between the two viruses. It cannot at present be ascertained whether the mutation occurs in the host or that the patient has been infected with both the wild type and mutant virus. However, it seems feasible to speculate that in the host the changes involved during such a mutation could result in the production of a membrane protein which could be incapable in maintaining its normal function in virus assembly. It is possible that such a breakdown of membrane protein function could result in block in the assembly of virus particles causing a non-productive or persistent infection.

ACKNOWLEDGEMENT

This work has been supported by the Deutsche Forschungsgemeinschaft "Schwerpunkt Multiple Sklerose und verwandte Erkrankungen" Az.: Me 270/16.

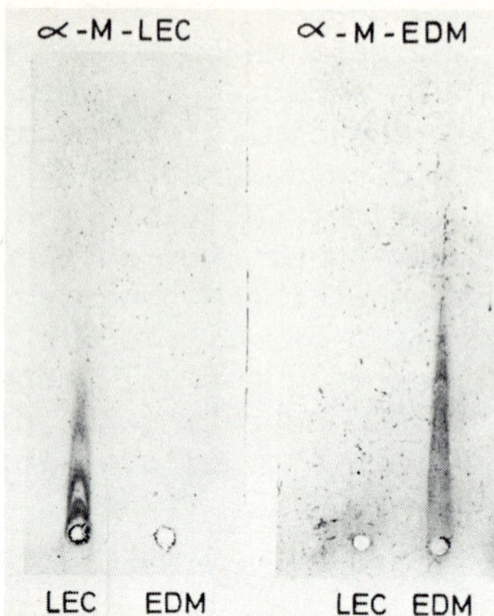

Fig.3. *Antigenic electrophoresis in antibody containing agarose gel.*

α-*M-Lec*	=	*anti-M-protein serum from rabbits immunized with purified M protein from SSPE virus Lec.*
α-*M-EDM*	=	*anti-M-protein serum from rabbits immunized with purified M protein from measles virus Edmonston*
Lec	=	*purified M protein sonicated and suspended in 0.4% DOC*
EDM	=	*purified M protein sonicated and suspended in 0.4% DOC.*

REFERENCES

1. Gregoriades, A. (1973). *Virology* 54, 369.
2. Hewitt, J.A. and Nermut, M.V. (1977). *J. Gen. Virol.* 34, 127.
3. McSharry, J.J., Compans, R.W., Lacland, H. and Choppin, P.W. (1975). *Virology* 67, 365.
4. Scheid, A., Caliguiri,L.A., Compans, R.W. and Choppin, P.W. (1972). *Virology* 50, 640.
5. Scheid, A. and Choppin, P.W. (1973). *J. Virol.* 11, 263.
6. Shimizu, K. and Ishida, N. (1975). *Virology* 67, 427.
7. Yung, LLL., Loh, W. and ter Meulen, V. (1977). *Med. Microbiol. Immunol.* 163, 111.

DISTRIBUTION AND RELATIONSHIPS OF ANTIGENS OF MEASLES VIRUS MEMBRANE COMPLEX IN THE INFECTED CELL AND THE NATURE OF THE HAEMOLYSIN ANTIGEN

K.B. FRASER, M. GHARPURE, P.V. SHIRODARIA, M.A. ARMSTRONG, A. MOORE and E. DERMOTT

Department of Microbiology and Immunobiology, The Queen's University of Belfast, Belfast, N. Ireland.

Antigens of influenza virus (7), of paramyxoviruses (10) and of oncornaviruses (1) have been identified as individual polypeptides with some precision. The number of virus-specific proteins in measles virus-infected cells is still uncertain, and even the structural polypeptides of the virion have not yet been exactly identified as familiar antigens (9). We have, therefore, adopted the alternative method of recognising and tracing virus antigens by means of antibodies prepared against semi-purified virus materials (11, 13) intending to relate them eventually to single polypeptides separated by electrophoresis. The site of origin and distribution in the infected cell of 4 structural antigens of measles virus, haemagglutinin (HA), haemolysin (HL), nucleocapsid (NC) and an antigen of the membrane protein complex (MP) are described here.

Vero cells were grown in 199 medium with 5% calf serum and HEp2 cells in Eagle's medium with 10% calf serum. The 243 strain of Edmonston measles virus (4) was grown from diluted inocula in Vero cells maintained in 199 medium with 1% calf serum. Stocks of titre not less than 2×10^7 p.f.u./ml., tested on Vero cells, were kept at -70°. When semi-confluent monolayers of Hep 2 cells were infected with 243 virus at a moi of 9, new infectious virus appeared in the cells at 8 hr and released virus by 14 hr.

FUNCTIONAL SPECIFICITY OF ANTISERA

The crude, unabsorbed antisera showed remarkable functional specificity (Table I). The anti-HL activity of anti-HA was a secondary effect, for after complete absorption of anti-HA and

TABLE I

Specificity of antisera to measles virus

Serum[‡] ANTI		Virus functional test and titre				Infected cell
		ANTI-HA	ANTI-HL	CFT HA	CFT NC	FA*
-HA		16,000	5,120	1,024	<10	8,000
-HL		<10	10,240	10	<10	4,000
-NC		<10	<20	ND	5,120	3,000
-MP	1.	NIL	NIL	NIL	NIL	300
	2.	160	300	ND	ND	1,000

[‡] All antisera except one rabbit anti-MP serum were produced in guinea-pigs. 3 deep s.c. injections were given, with intervals of 3 weeks then 10 days, of antigens in complete Freund's adjuvants at doses stated below.

HA - Haemagglutinin from tween-80-ether-treated virus, released from infected Vero cells. Purified on CsCl discontinuous gradient. Band at density 1.23 to 1.25 used as antigen (300 μg/ml). No haemolytic activity.

HL - Spikeless, trypsin treated, virions released from infected Vero cells (12). Protein content 330 μg/ml.

NC - Nucleocapsid from 1% NP40 - disrupted infected Vero cells. Twice purified in CsCl gradient. Band at density 1.31 used as antigen 250 mg/ml.

TABLE I continued.

MP - (1) Ether-precipitated protein from acid chloroform-methanol solution (6) of measles virus-infected cell pellet.
(2) Similar but not precipitated in KCl; semi-purified protein, MW 36,000 on acrylamide gradient gel (8). Used as antigen at 455 μg.

Anti-HA Anti-HL - 4 units whole virus in 0.25 ml equal vol. 2% green monkey erythrocytes and equal volume serum dilution (14).

CFT - 3 units complement, 4 units antigen overnight fixation 4°.

FA - See Table II.

*Absorbed sera, no staining of uninfected cells.

ND = Not done.

anti-HL properties with acetone-treated virus, which does not absorb anti-HL (Armstrong, unpublished) there was no independent anti-HL globulin present. The serum did not react with NC. The anti-HL serum did not react with HA or NC nor did anti-NC react with HA or HL. Of the 2 anti-MP sera, one (guinea-pig serum) was non-reactive in functional tests. The other had moderate amounts of anti-HA and anti-HL reactivity, but only up to 10% of the fluorescent antibody titre as measured before absorption of sera for fluorescent staining. The anti-HA and HL activity of anti-MP could be absorbed by para-formaldehyde-fixed virus with subsequent loss of fluorescent antibody titre.

DISTRIBUTION OF ANTIGEN

The fluorescent-staining titres of the 4 antisera allowed 15 to 20 fluorescent staining doses of each to be used to trace antigen with no non-specific staining and very small chances of cross-reaction. The following descriptions are summarized in Table II.

HA - the HA antigen appeared first at 8 to 10 hr after infection as faintly fluorescing dots or threads in a paranuclear position, exactly as described by Norrby (11). It appeared externally on unfixed cells at 12 to 14 hr after infection. The antigen was resistant to acetone but was removed by 0.025% and 0.004% trypsin.

HL - the anti-HL serum revealed an antigen which had at first a much more diffuse and homogeneous fluorescence than HA, appeared at 9 to 11 hr after infection and formed specifically stained dots or globules amongst the homogeneous material 4 or 5 hr later. It apparently reached the exterior of the cell 13 to 15 hr after infection, but these short differences in time may represent time to reach detectable concentrations rather than time of arrival or time of formation at the membrane. The surface HL was distinctly sensitive to acetone fixation but was preserved after trypsin treatment of carrier cells. It will be shown presently to be 2 different antigens.

NC - the formation of stainable NC antigen in fluorescent dots or circular inclusions in the cytoplasm again followed Norrby's description, except that, at our multiplicity of infection, they were visible from the second hour of the growth cycle onwards. At 10 hr antigen began to spread diffusely from the inclusion bodies to the membrane, which it reached at 14 hr after infection. It was acetone-resistant and never situated on the surface of the cell.

It is probable that there are more virus antigens than one in inclusion bodies. Immuno-electron microscopy showed that anti-NC reacted with the actual nucleocapsid structures, (Fig. 1, g) but not with nucleocapsids of Sendai virus. Anti-HA did

not react with measles virus nucleocapsid.
MP - the guinea-pig anti-MP had no complement-fixing activity against crude measles antigens but the rabbit had some, to a titre of 40; both gave unexpected fluorescent-staining reactions. They each stained in the cytoplasm the same aggregates of nucleocapsid as were stained by anti-NC. This was shown by double immunofluorescent staining; at 8 hr and later no inclusion body failed to fluoresce red (anti-NC) and green (anti-MP) under appropriate filter systems. Spread through the cytoplasm followed the course of NC staining until the membrane was reached at 15 hr when the MP staining was external, acetone-resistant and trypsin-sensitive.

Because the MP staining was so unlike that of the precisely known distribution of Sendai MP (16) further tests were made. These showed: 1. That anti-MP blocked its own staining at membrane specifically but was not strong enough to block staining of IBs. 2. Anti-NC fails to block membrane staining by anti-MP but completely blocked staining of IBs by anti-MP; anti-HA did not block surface staining but anti-HL blocked it partially. 3. Absorption of anti-MP with purified nucleocapsid removes the reactivity of anti-MP with cytoplasmic IBs but did not affect the membrane staining. 4. Anti-MP caused agglutination and globulin adherence when added to nucleocapsid, but this reactivity was lost when MP was absorbed with nucleocapsid (Fig.1, h,i). The titre of membrane-staining remained unreduced. Clearly anti-MP had some anti-NC which was removable.

Re-examination of antigen formation in the growth cycle with anti-MP serum that had been absorbed with acetone-treated, infected cells, now showed intracytoplasmic and surface staining similar in distribution to that given by guinea-pig anti-HL serum and no staining of inclusion bodies. This did not establish its specificity.

THE SPECIFICITY OF ANTI-HL

The preceding findings raised doubts about the specificity of anti-HL; that is, anti-spikeless virion. Carrier 34 cells provided evidence for the existence of 2 antigenic determinants. The number of cells in the line carrying virus antigens at the surface varied with the pass used from 10% to 35%. Acetone treatment reduced the number of cells showing specific staining by anti-HL at the surface to about one half (*e.g.*, unfixed 17.7%, fixed 7.3%). On 34 cells trypsin treatment also reduced the number of stained cells (*e.g.*, 17.7% untreated, 9.3% treated). Both treatments brought the number of surface-stainable cells to less than 1% (unpublished data). Accordingly, anti-HL and 5 human sera containing measles antibody were each absorbed with acetone-treated virus until all anti-HA and

TABLE II

Specific reactions with antisera to measles virus

Serum ANTI	Fluorescence staining				GEL-diffusion
	Pattern	Time (PI) antigen		Properties of antigen at membrane	Against antigens HA and NC
		1st seen (hr)	At membrane (hr)		
-HA	Paranuclear mesh	8-10	13-14	External Acetone-R Trypsin-S	Single line of "identity" sera HL, NC, MP
-HL	Cytoplasmic, diffuse and blobs	9-11	14-15	External Acetone-S Trypsin-R	Single line of "identity" sera HA, NC, MP
-NC	Minute dots in cytoplasm grow to large IC and IN aggregates	2-4	14	Internal Acetone-R	Two lines One of "identity" sera HA, HL, MP
-MP	Coincidental with HL	10-11	15	Acetone-PS Trypsin-R	Single line of "identity" sera HA, HL, NC

TABLE II continued

Sera - as in Table I

Fluorescence staining	(1) Indirect with FITC-conjugated anti-globulins. Unfixed for external antigen. Acetone-fixed 8 min room temperature for internal antigen.
	(2) Direct simultaneous staining of separate antigens with FITC, or Rhodamine-B sulphonyl-chloride-conjugated sera.
Trypsin-sensitivity	Tested on persistently-infected HEp2 cells (5) bearing virus antigen on surface of 37% cells and internal antigen in all, including >90% of cell nuclei, 0.004% crystalline trypsin 37° 40 min followed by ice-cold 4% para-formaldehyde 4 min (2, 3).

PI = post infection. R = resistant. S = sensitive. PS = partially sensitive

IC = Intra-cytoplasmic. IN = Intranuclear

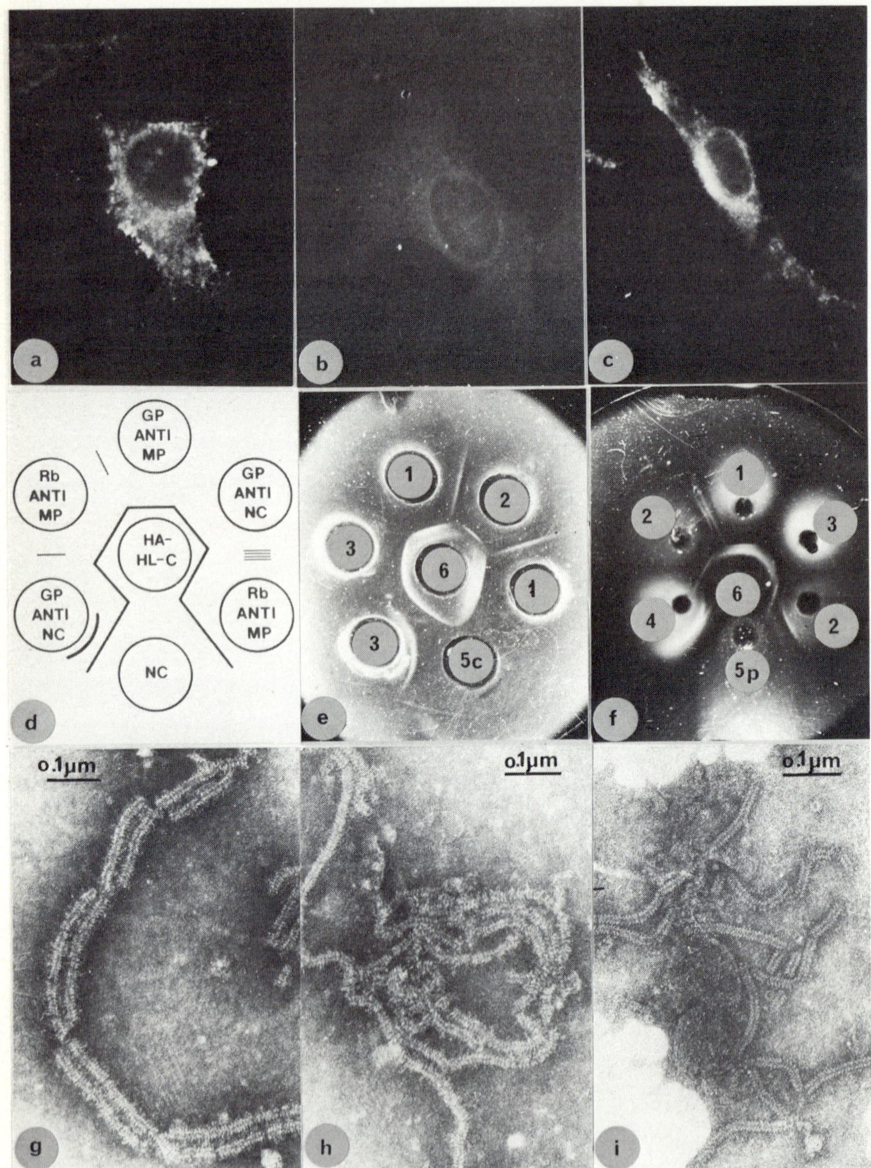

Fig.1. *a-c, Indirect immunofluorescence staining of HEP_2 cells infected with measles virus; a, unfixed cells, b, acetone fixed cells, stained with human anti-HL serum and FITC-conjugated anti-IgG, c, cells fixed by drying and stained similarly, d-f, Immunodiffusion reactions (13,15) of HA-HL complex and*

anti-NC reactivity was removed and only functional anti-HL remained; they were then tested for fluorescent staining on virus infected cells, unfixed or fixed in acetone. Only the unfixed cells showed surface fluorescence. Two of the human sera, applied to the fluorescent antibody testing of the growth cycle, showed that HL appeared in the cytoplasm at 10 hr in cells fixed without acetone, and on the surface of the unfixed cells at the same time. HL antigen was, therefore, completely acetone-sensitive in contrast to HA, NC and MP (Fig.1, a,b,c). Its distribution was peri-nuclear, almost ring-like, when first seen. This growth curve had been accelerated by high virus input.

DOUBLE IMMUNO-DIFFUSION

Reactions with Tween-20-treated virus (HA) showed, firstly, one line of identity, using anti-HA, anti-HL, and anti-MP but not anti-NC. None of the antisera reacted with detergent-treated normal, or Sendai virus-infected, vero cells. Secondly, the same line of identity between HA and MP and crude, sucrose gradient-separated NC, was shown by anti-MP but the precipitin line was absent from NC twice-purified on caesium chloride gradients (Fig.1, d,e,f). Thirdly, a single polypeptide, M.W. 36,000 eluted after electrophoresis from polyacrylamide gel reacted faintly with anti-MP and anti-HA, but strongly with anti-HL diluted to 1:10.

We can conclude:

1. That sera anti-HA and anti-NC have a predominant specificity corresponding to a characteristic staining pattern.
2. Anti-HL has 2 specificities. The true HL antigen is recognised also by human anti-HL, is acetone sensitive and recognisable on the surface of trypsin-treated infected cells.
3. MP, as used, was poorly antigenic and anti-MP, was contaminated with other antibody so that the presence of MP at the cell surface could not be substantiated, but anti-

---

unpurified nucleocapsid; 1, guinea-pig anti-HA, 2, rabbit anti-MP, 3, guinea-pig anti-HL, 4, guinea-pig anti-NC, 5_C*, crude preparation of NC treated with SDS,* 5_P*, purified preparation of NC treated with SDS, 6, Tween-20 treated HA. g-i, Electron micrographs of NC reacted with antisera; equal volumes of NC and antibodies were incubated 1 hr at 37° and the product applied to carbon-coated grids with 2% phosphotunstate contrast stain. g, NC with guinea-pig anti-NC, h, NC with unabsorbed rabbit anti-MP, i, NC with absorbed rabbit anti-MP.*

HL, made against spikeless virions, has a precipitin reacting with MP and with a 36,000 M.W. polypeptide isolated from crude MP.

4. Non-functioning anti-virus antibody can be detected by the fluorescent antibody method.

REFERENCES

1. Bolognesi, D.P. (1974). *Adv. Virus Res.* 19, 315.
2. Ehrnst, A. and Sundqvist, K-G. (1975). *Cell* 5, 351.
3. Fraser, K.B., Shirodaria, P.V. and Harie, M. (1974). *Med. Microbiol. Immunol.* 160, 221.
4. Gould, E. (1974). *Med. Microbiol. Immunol.* 160, 211.
5. Gould, E. and Linton (1975). *J. Gen. Virol.* 28, 21.
6. Gregoriades, A. (1973). *Virology* 54, 369.
7. Kilbourne, E.D., Choppin, P.W., Schultze, I.T., Scholtissek, C. and Bucher, D.L. (1972). *J. Infect. Dis.* 125, 447.
8. Marsden, H.S., Crombie, I.K. and Subak-Sharpe, J.H. (1976). *J. Gen. Virol.* 31, 347.
9. Mountcastle, W.E. and Choppin, P.W. (1977). *Virology* 78, 463.
10. Mountcastle, W.E., Compans, R.W. and Choppin, P.W. (1971). *J. Virol.* 7, 47.
11. Norrby, E. (1972). *Microbios.* 5, 31.
12. Norrby, E. and Gollmar, Y. (1975). *Infect. Immun.* 11, 231.
13. Norrby, E. and Hammarskjold, B. (1972). *Microbios.* 5, 17.
14. Saburi, Y. and Matsumoto, M. (1966). *Arch. ges. Virusforsch.* 17, 29.
15. Wadsworth, C. (1957). *Int. Arch. Allergy* 10, 355.
16. Yoshida, T., Yoshiyuki, N., Yoshii, S., Maeno, K. and Matsumoto, T. (1976). *Virology* 71, 143.

MEMBRANE FLUIDITY AND CELL FUSION INDUCED BY SENDAI VIRUS

C.A. HART*, D. FISHER, T. HALLINAN and J.A. LUCY

Department of Biochemistry and Chemistry,
Royal Free Hospital School of Medicine, University of London,
8 Hunter Street, London WC1N 1BP.

**Present address:*
Department of Medical Microbiology,
The New Medical School, Liverpool University,
Pembroke Place, P.O. Box 147, Liverpool L69 3BX

The concept of membrane fluidity has greatly enlarged our understanding of the structure and function of membranes. Much knowledge of membrane fluidity has arisen from studies on simple artificial membranes (6), though recently more information is being gathered from studies on biological membranes (23). When an artificial membrane, composed solely of one type of phospho-lipid, is gradually warmed, a sharp endothermic change occurs at a particular temperature known as the transition temperature (Tc). At the Tc, the lipids undergo a phase change from a gel to a liquid crystalline phase. Below the Tc the lipids form a regular rigid array with the hydrocarbon tails closely packed and relatively immobile. Above the Tc the rigid array has melted and the hydrocarbon tails are less closely packed and hence the membrane is more fluid. The fluidity of a membrane at a particular temperature depends upon several factors. These include the nature of the lipid headgroup, the length and degree of saturation of the hydrocarbon tails, and the presence of cholesterol, of water, and of hydrated ions such as Ca^{2+} (6). In addition a variety of proteins can alter membrane fluidity (21) and it has been suggested that extrinsic structures underlying biological membranes such as microtubules or microfilaments may also alter membrane fluidity (3). Biological membranes are much more complex and only rarely show sharp phase changes of the type described above, because they are composed of many different lipids, including cholesterol, and contain many proteins. At physiological temperatures it has been calculated that approximately 70% of the lipids of the sarcoplasmic reticulum are in liquid crystalline or fluid state (19).

The fluidity of a membrane is known to affect many of its properties. The mobility of a variety of antigens on the cell surface increases with increasing membrane fluidity (9, 15). Membrane fluidity seems to have an important role in the fusion of both artificial (5, 20) and biological membranes (1, 4, 8, 13). Indeed some chemicals which are known to increase membrane fluidity are themselves chemical fusogens (11, 13).

We have investigated the role of membrane fluidity in the Sendai virus-induced cell fusion of chicken erythrocytes.

Low levels of Sendai virus, grown *in ovo*, were used to induce fusion in a suspension of chicken erythrocytes as described previously (10).

THE EFFECT OF TEMPERATURE

Sendai virus-induced cell fusion was measured at a variety of temperatures. Under the conditions used, cell fusion

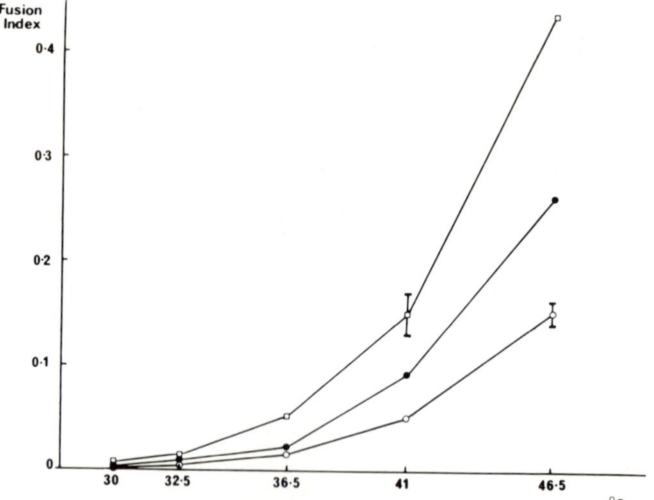

Fig.1. *The effect of temperature on the fusion of chicken erythrocytes induced by Sendai virus*

Chicken erythrocytes were prepared as described previously (10) and were incubated at 37° for 30 min. On cooling, Sendai virus (○ *400 HAU;* ● *600 HAU;* □ *800 HAU) and* Ca^{2+} *(4 mM) were added. The suspension was then kept on ice for 20 min with frequent mixing. The suspension was placed in a water bath at the required temperature and treated as described previously. Each point represents the mean of duplicate samples assayed within one experiment with error bars to show the difference from the mean (where applicable).*

increased with increasing temperature above 25°, and fusion was not detected below 25° (Fig. 1). An Arrhenius plot was constructed by plotting Log_{10} Fusion Index (F.I.) against the reciprocal of the absolute temperature (Fig. 2). The slopes of the straight lines obtained by this method are proportional to the activation energy.

The activation energy which is obtained from an Arrhenius plot is equal to 2.303 R x Slope ($Kcal.mol^{-1}$), where R = gas constant (R = 1.987 $cal.mol^{-1}$). The mean activation energy for Sendai virus-induced fusion of chicken erythrocytes obtained from sixteen separate experiments was 50.7 ± 8.1 Kcal. mol^{-1}.

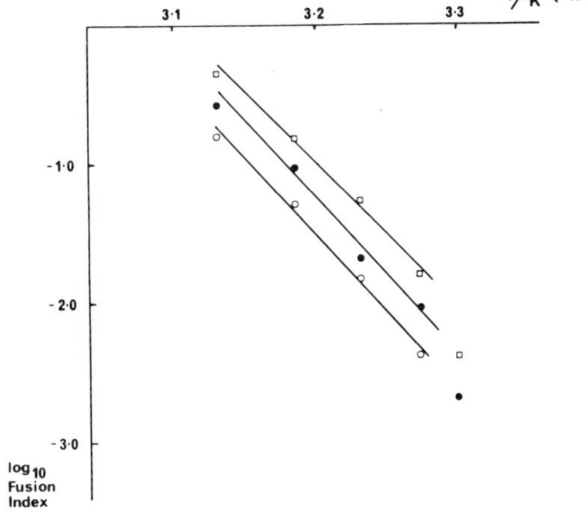

Fig.2. *The effect of temperature on the fusion of chicken erythrocytes induced by Sendai virus*

Legend is as for Fig. 1. except that the results are plotted as Log_{10} Fusion Index against the reciprocal of the absolute temperature (1/°K). (An Arrhenius plot).

This activation energy is more than twice that reported for several other types of membrane fusion. Myoblast fusion has an activation energy of 18-22 $Kcal.mol^{-1}$ (4). The activation energy for pinocytosis was 17.6 $Kcal.mol^{-1}$ (24) and the activation energy for the fusion of chicken erythrocytes induced by either retinol or glycerol mono-oleate was 20 $Kcal.mol^{-1}$ (8).

It is possible to obtain an activation energy for three other actions of Sendai virus upon cell membranes. The haemolysis of chicken erythrocytes by Sendai virus has an activation energy of 22.6 $Kcal.mol^{-1}$ (18). An activation energy of 19.4 $Kcal.mol^{-1}$ for Sendai virus-induced leak of choline from cultured cells can be calculated from the work of

Pasternak and Micklem (22). Also, Maeda et al. (16) found that if spin-labelled Sendai virus was incubated with human erythrocytes then the spin-label was diluted among the phospho-lipids of the erythrocyte membranes. This dilution was temperature dependent and its activation energy can be calculated as approximately 24 $Kcal.mol^{-1}$. If, as is likely, each of these phenomena is a reflection of the initial fusion of the viral envelope with the cell membranes, then cell fusion induced by Sendai virus appears to have a step with an activation energy that is more than twice that necessary for the initial fusion. The high activation energy for virus-induced cell fusion might then be considered as evidence against the hypothesis that cell fusion occurs via the formation of cell-virus-cell bridges. Another important aspect of an Arrhenius plot is the presence or absence of discontinuities in the slopes. The points at which the lines of different gradient meet are thought to be temperatures at which transitions from one phase to another are occurring. In the present study, no discontinuities were observed between 48° and 30°. Below 30° the rate of fusion was too slow to be reliably measured and hence the presence or absence of discontinuities below this temperature could not be investigated. Li et al. (14) have shown that the optimal temperature for the infectivity of Newcastle disease was 15° and these workers have related this temperature to phase changes in the membrane of the infected cells.

FLUIDISING AGENTS

Many chemicals are known to increase the fluidity of both artificial and biological membranes (12). Among these are benzyl alcohol (7, 17) and glycerol mono-oleate (11). Both glycerol mono-oleate (0.28 mM) and benzyl alcohol (10 mM) greatly enhanced cell fusion in the presence of Sendai virus. The most obvious explanation for these findings would be that both benzyl alcohol and glycerol mono-oleate were increasing membrane fluidity and thus enhancing virus induced cell fusion. However under other conditions both benzyl alcohol (Galloway and Fisher, unpublished) and glycerol mono-oleate (2) are themselves chemical fusogens. Nevertheless, control experiments showed that under the conditions used neither benzyl alcohol nor glycerol mono-oleate caused cell fusion in the absence of Sendai virus, nor in cells agglutinated by non-fusogenic influenza virus.

A method for distinguishing between the alternatives of chemical-enhancement of virus-induced fusion and viral-enhancement of chemically-induced fusion is offered by a study of the effect of Ca^{2+} on the two systems. Chemically-induced cell fusion has a requirement for Ca^{2+} (2) whereas cell fusion induced by low levels of Sendai virus is maximal in the absence of Ca^{2+} (10). If the action of glycerol mono-oleate were simply to increase membrane fluidity, glycerol mono-oleate should enhance Sendai virus-induced cell fusion whether Ca^{2+} was present or not. As can be seen from Fig. 3. glycerol mono-oleate did not increase the extent of Sendai virus-induced fusion of chicken erythrocytes unless Ca^{2+} was present. It seems therefore that glycerol mono-oleate acts mainly as a chemical fusogen, and that its action is enhanced by the presence of the virus.

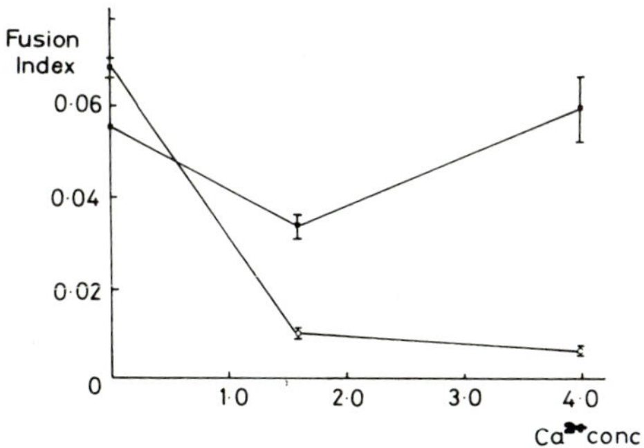

Fig.3. *The effect of glycerol mono-oleate and* CA^{2+} *on the fusion of chicken erythrocytes induced by Sendai virus*

Chicken erythrocytes were prepared as described previously, and were incubated at 37° for 30 min. On cooling, Sendai virus (128 HAU), the required amount of Ca^{2+} *alone (□) or in the presence of GMO (0.28 mM) (■) were added. The suspension was kept on ice for 20 min with frequent mixing. The suspension was then treated as described previously. Each point represents the mean of duplicate samples examined within our experiment with error bars (where applicable) to show the difference from the mean.*

C.A.H. was a Wellcome Research Training Scholar.

REFERENCES

1. Ahkong, Q.F., Cramp, F.C., Fisher, D., Howell, J.I., Tampion, W., Verrinder, W. and Lucy, J.A. (1973). *Nature (New Biol.)* 242, 215.
2. Ahkong, Q.F., Fisher, D., Tampion, W. and Lucy, J.A. (1973). *Biochem. J.* 136, 147.
3. Berlin, R.D. (1975). *Ann. N.Y. Acad. Sci.* 253, 445.
4. Bosch, J. van der, Schudt, C. and Pette, D. (1973). *Exp. Cell Res.* 82, 433.

5. Breisblatt, W. and Ohki, S. (1975). *J. Memb. Biol.* 23, 385.
6. Chapman, D. (1974). *In* "Fundamentals of Lipid Chemistry" (R.M. Burton and F.C. Guerra, eds.) Ch. 23, p. 565. Bi-Science, Missouri.
7. Colley, C.M. and Metcalfe, J.C. (1972). *Febs Letters* 24, 241.
8. Fisher, D. (1975). *In* "Biomembranes, Lipids and Receptors" (R.M. Burton and L. Packer, eds.) Ch. 5, p. 75. Bi-Science, Missouri.
9. Frye, C.D. and Edidin, M. (1970). *J. Cell Sci.* 7, 319.
10. Hart, C.A., Fisher, D., Hallinan, T. and Lucy, J.A. (1976). *Biochem. J.* 158, 141.
11. Kennedy, A. and Rice-Evans, C. (1976). *Febs Letters* 69, 45.
12. Kosower, E.M., Kosower, N.S., Faltin, Z., Diver, A., Saltoun, G. and Frensdorff, A. (1974). *Biochim. Biophys. Acta* 363, 261.
13. Kosower, N.S., Kosower, E.M. and Wegman, P. (1975). *Biochim. Biophys. Acta* 401, 530.
14. Li, J.K-K., Williams, R.E. and Fox, C.F. (1975). *Biochem. Biophys. Res. Comms.* 62, 470.
15. Lustig, S., Pluznik, D.H., Kosower, N.S. and Kosower, E.M. (1975). *Biochim. Biophys. Acta* 401, 458.
16. Maeda, T., Asano, A., Ohki, K., Okada, Y. and Ohnishi, S. (1975). *Biochemistry* 14, 3736.
17. Metcalfe, J.C., Seeman, P. and Burgen, A.S.V. (1968). *Mol. Pharmacol.* 4, 87.
18. Neurath, A.R. (1965). *Acta Virol.* 9, 34.
19. Oldfield, E. and Chapman, D. (1972). *Febs Letters* 23, 285.
20. Papahadjopoulos, D., Poste, G., Schaffer, B.E. and Vail, W.J. (1974). *Biochim. Biophys. Acta* 352, 10.
21. Papahadjopoulos, D., Moscarello, M., Eylar, E.H. and Isac, T. (1975). *Biochim. Biophys. Acta* 401, 317.
22. Pasternak, C.A. and Micklem, K.J. (1973). *J. Memb. Biol.* 14, 293.
23. Singer, S.J. (1975). *In* "Advances in Experimental Medicine and Biology" (R.W. Meints and E. Davis, eds.) 62, 181. Plenum. N.Y. and London.
24. Steinman, R.M., Silver, J.M. and Cohn, Z.A. (1974). *J. Cell Biol.* 63, 949.

ENVELOPED VIRUSES: STRUCTURE AND INTERACTION WITH CELL MEMBRANES

F.W. LANDSBERGER, D.S. LYLES and P.W. CHOPPIN

The Rockefeller University, New York, N.Y. 10021, U.S.A.

The lipid composition of the envelope of budding viruses such as myxo-, paramyxo-, and rhabdoviruses largely reflects that of the plasma membranes while the protein composition is determined by the viral genome. A lipid bilayer appears to be a ubiquitous feature of enveloped virions. The outer surface of the enveloped virions is covered by glycoprotein projections or "spikes" which possess, depending on the virion, a variety of functions including haemagglutination, neuraminidase, cell fusion, and haemolysis. Associated with the inner surface of the lipid bilayer is a non-glycosylated membrane (M) protein.

We have been using several approaches with a special emphasis on spin label electron spin resonance (ESR) techniques to investigate the molecular architecture of enveloped viruses and have also begun to investigate the molecular events in the membrane-membrane interactions associated with virus adsorption, hemolysis, fusion, and penetration.

SPIN LABELLING AS A METHOD

We have applied spin labelling techniques with the fundamental strategy of incorporating nitroxide derivatives of such molecules as stearic acid

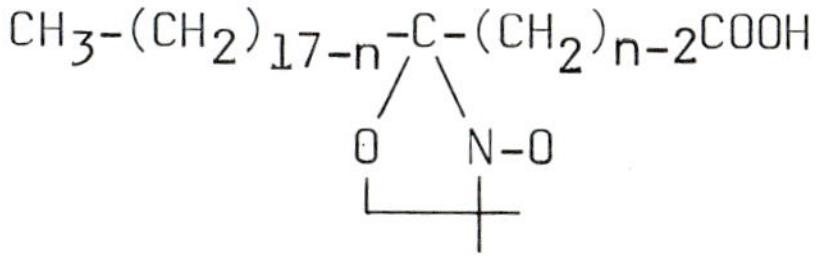

```
CH3-(CH2)17-n-C-(CH2)n-2COOH
              / \
             O   N-O                C ,
             |___|                   n
                 |
```

phosphatidylcholine

```
                                           O          j=10, k= 3: PC5
                                           ||
          ┌───┼─                     O   CH2-O-C-R    j=10, k= 5: PC7
          O   N-O                    ||   |
           \ /                       ||   |           j= 5, k=10: PC12
CH3(CH2)j-C-(CH2)k-C-O-CH           O
                                         |    ||      j= 1, k=14: PC16
                                         |    ||             +
                                       CH2-O-P-O-(CH2)2-N(CH3)3
                                              |
                                              O-
```

or phosphatidylethanolamine

```
                                           O
                                           ||
          ┌───┼─                     O   CH2-O-C-R              PE12
          O   N-O                    ||   |
           \ /                       ||   |
CH3(CH2)5-C-(CH2)10-C-O-CH          O
                                         |    ||           +
                                       CH2-O-P-O-(CH2)2-NH3
                                              |
                                              O-
```

into the lipid phases of either virions or plasma membranes. The stearic acid spin labels are incorporated such that the carboxyl moiety is at the approximate depth in the lipid bilayer of the phosphate group of the phospholipids (7). Presumably, the phosphatidylcholine and phosphatidyl-ethanolamine spin label derivatives behave similarly to the native lipids. As implied by the structures, the nitroxide moiety can be located at a variable distance from the polar functions.

The ESR spectra of the lipid spin labels reflect properties of the local environment into which they are inserted. One property of the lipid phase which can be observed by spin labels is the fluidity of the local environment. The distance between the outermost peaks (cf. $2A'_{zz}$ in Fig.1) of the ESR spectrum decreases as the degree of motion of the spin label (which must reflect the motion of the neighboring molecules) increases (11, 13).

STRUCTURE OF THE VIRAL LIPID BILAYER

For each of the viral envelopes examined, the region in the vicinity of the glycerol backbone of the phospholipids probed by the C_5 spin label is considerably more rigid than that of the deep hydrocarbon phase probed by the C_{16} label. Such a rigidity gradient is characteristic of a lipid bilayer and has been used as a diagnostic assay. Table I gives an

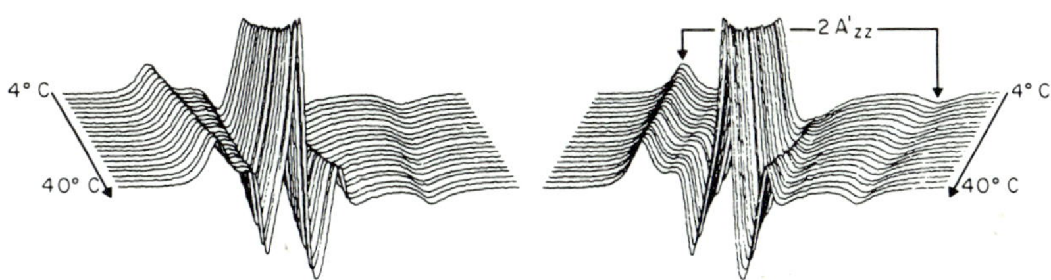

Fig.1. *Change in the electron spin resonance (ESR) spectrum of VSV labelled with the C_5 spin label derivative of stearic acid. The splitting between the outermost peaks in a spectrum is defined as $2A'_{zz}$. As temperature increases, the motion of the lipid fatty acyl chain increases. In addition to a change in $2A'_{zz}$ as the motion of the spin label is altered, the entire lineshape also changes. The vertical scale is the same for all the spectra.*

inclusive list of the systems examined. An important question is to what extent the structure of the viral bilayer is determined by its lipid and protein components.

An increase in the cholesterol-phospholipid ratio of influenza, SV5, and vesicular stomatitis virions causes an increase in the rigidity of the viral lipid bilayer (20, 21) in accord with the results of model membrane studies (7).

The viral lipid bilayer is considerably more rigid than that of the host plasma membrane, (2, 21, 33). Since the lipid compositions of the two systems are essentially the same (16, 18) this indicates that viral protein has a significant role in determining the organization of the viral bilayer. Lipid membranes prepared from extracted viral lipids are more fluid than the native viral envelope as assayed by both spin label (1, 33) and ^{13}C NMR methods (35). However, the phospholipid classes are asymmetrically distributed about the viral bilayer (6, 24) and it is unlikely that this symmetry is precisely maintained in the preparation of bilayers from the extracted lipids. These experiments do provide support for the view that the viral proteins are a determining factor in the observed rigidity of the viral bilayer.[1]

---

[1]Using a fluorescent probe, Barenholz *et al.* (2) did not observe a difference between the intact VSV virion and the bilayer formed from the extracted lipids. The reason for this difference in the experimental results is not clear but it may reflect incorporation of the fluorescent probe into regions of the lipid bilayer not assayed by the other techniques.

TABLE I

Experimental Evidence for a Lipid Bilayer Structure in Envelopes of Animal Viruses

Spin Label ESR	Ref.
1. Myxoviruses	
Influenza A_0 (WSN)	19
2. Paramyxoviruses	
SV5	20
Sendai	Landsberger, Scheid & Compans unpublished
Newcastle disease*	18
3. Rhabdoviruses	
Vesicular stomatitis	21
4. Togaviruses	
Sindbis	33
Venezuelan equine encephalitis	12
5. Oncornaviruses	
Rauscher murine leukemia	22
Murine mammary tumor	Landsberger, Yagi & Compans unpublished
6. Arenaviruses	
Pichinde	Landsberger, Gard & Compans unpublished
X-ray Diffraction	
1. Togaviruses	
Sindbis	9

*Spin label: 2,2-dimethyl, 4-butyl, 4-pentyl, N-oxyloxazolidine.

By comparing the rigidity of the viral lipid bilayer of different virions grown in the same cell type, the differential effect of viral proteins on the lipid bilayer can be assessed. The rigidity of the bilayers of influenza and SV5 virions could not be distinguished (20).[2] In contrast, the range of bilayer rigidity resulting from growth in MDBK versus BHK cells is less for VSV than that for either influenza or SV5 virions (21). These results also imply that viral proteins may in fact have a significant role in defining the structure of the viral bilayer.

It is of interest to define whether either the glycoproteins or the M-protein is involved in determining the rigidity of the viral bilayer. One approach has been to modify the glycoproteins proteolytically (Table II). Removal of the glycoprotein spikes from the virion surface causes either no change in the viral bilayer or a small increase in the fluidity of the lipid fatty acyl chains (19, 21, 33). The HA of influenza B virus grown in HKCC cells can be cleaved with a resulting increase in infectivity (23). Cell-fusing and haemolytic activity of Sendai virus grown in MDBK cells can be activated by cleavage of a glycoprotein (31). Spin label studies have not detected a change in the viral bilayer upon proteolytic activation of either virus preparation. These results indicate that either complete or limited proteolysis has either no or a small effect on the interaction between the viral lipids and the hydrophobic feet of the glycoproteins which putatively penetrate the lipid bilayer. An alteration in the amount of glycoprotein per influenza virion does not detectably alter the bilayer fluidity (24), suggesting that the bilayer fluidity is not primarily determined by the hydrophobic pieces of the viral glycoprotein.

We have recently made a simple calculation which suggests that the amount of glycoprotein present with the bilayers of influenza virions is not enough to contribute significantly to the bilayer rigidity. The proteolytically inaccessible hydrophobic pieces of the glycoproteins are presumably about 5,000-6,000 daltons in size (28, 32, 34, 37), and could possibly form an α-helix spanning the lipid bilayer. From reconstitution of the cytochrome oxidase system, it has been found that there is a lipid layer one molecule thick which sheaths the protein and is highly immobilized (14). If a similar monolayer of lipid exists around a glycoprotein α-helical segment spanning the

---

[2]We have recently repeated the experiment of 1973 comparing SV5 and influenza virions. The previous results were confirmed over a temperature range of 13°-37° under conditions of considerably higher signal-to-noise than were previously possible.

TABLE II

Proteolytic Modification of Envelope Glycoproteins

Virus	Enzyme	Result of Enzyme Treatment	Effect on Lipid Bilayer Structure	Ref.
Influenza A_o (WSN)	Chymotrypsin Bromelain	"Spikeless" particles	None	19,21
Vesicular stomatitis	Trypsin	"Spikeless" particles	More fluid	21
Sindbis	Pronase	"Spikeless" particles	More fluid	27,33
Influenza B (1760) (HKCC grown)	Trypsin	HA cleavage	None	Landsberger & Choppin, unpublished
Sendai (MDBK grown)	Trypsin	Activate cell-fusing and hemolytic activities	None	Landsberger, Scheid, and Choppin, unpublished

bilayer, the lipid in such a layer would be only a very small fraction of the total virion lipid.

Although a direct evaluation of the depth to which the glycoproteins penetrate the bilayer and of the degree of lipid-glycoprotein interaction has not yet been fully possible, the presently available data indicate that the glycoproteins do not determine the lipid rigidity, suggesting that the M-protein may be involved in the observed greater structural rigidity of the viral lipid bilayer in comparison with that of the host plasma membrane. This postulate is consistent with the observed close promixity of the M-protein to the lipid bilayer (4). Furthermore, the M-protein is essential for assembly (4). Thus it is not unreasonable to speculate that the strong interaction between the M-protein and the lipid bilayer may be a significant factor in virion maturation. The M-protein upon arriving at the cytoplasmic surface of the plasma membrane may possibly increase the local lipid rigidity over that of the bulk membrane and thereby perhaps be a factor in regulating the proteins present in the assembly area.

Newly isolated influenza virus strains have a filamentous shape, but on continuous passage a virus with a spherical shape is selected (5). Furthermore, filamentous versus spherical morphology has been shown to be genetically determined (15). These results suggest that a virus-specified protein is involved in determining the shape of the virus and the M-protein would seem to be the logical candidate. Studies to date have revealed no differences which appear to be significant in the amounts or the electrophoretic mobility of the virion proteins (Choppin, unpublished experiments), but detailed studies of the proteins from virions of different morphology are continuing. We have found evidence, however, which indicates differences in lipid-protein interaction in the two kinds of virions. Spin label studies have shown that the lipid bilayer of the filamentous version is more fluid than that of the spherical virion (Fig.2). Further studies of the individual protein should provide new insight into lipid-protein interactions and their possible role in determining the morphology of membrane-enclosed viruses.

ENVELOPED VIRUSES - CELL SURFACE INTERACTIONS

We have approached the problem of enveloped virus-cell surface interaction by extensively using the Sendai virus system to investigate the separate processes of adsorption, haemolysis, fusion, and penetration. Sendai virus grown in MDBK cells has an active HA but a precursor glycoprotein (F_0) which lacks haemolytic and cell-fusing activities. Upon *in vitro* trypsinization of MDBK-grown Sendai virus, the F_0 glyco-

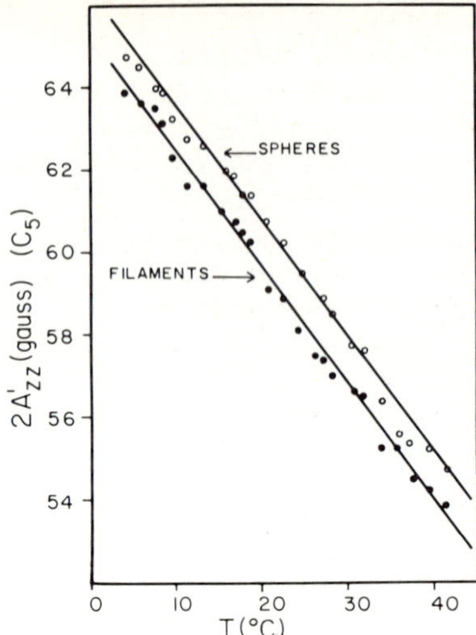

Fig.2. *Comparison of the lipid bilayers of the filamentous and spherical forms of influenza virus using the C_5 spin label. The lipid bilayer of the filamentous form is more fluid (smaller $2A'_{zz}$) than that of the spherical form at each temperature examined. A similar linear dependence of $2A'_{zz}$ on temperature has also been observed with SV5 and VSV (cf. Fig.1). No apparent phase "transitions" or "discontinuities" are observed in any of these systems with the C_5 label.*

protein is cleaved into the active form F resulting in a virion which also exhibits haemolytic and cell-fusing activities (31). Sendai virus grown in embryonated eggs and harvested after 48 hr is infectious and possesses haemolytic and cell-fusing activities. However, when egg-grown Sendai virus is harvested early (24 hr after infection), it contains the F glycoprotein in the active form and can induce membrane fusion but has very little haemolytic activity, which can be induced by several techniques including freezing and thawing (10).

ENVELOPED VIRUS ADSORPTION TO CELL SURFACES

The attachment of influenza and Sendai viruses grown in MDBK cells as well as the lectins con A and wheat germ agglutinin, results in an increase in bilayer fluidity of the chicken erythrocyte plasma membrane which is sensitive to microtubule-disrupting drugs such as colchicine, vinblastine, and tetracaine as detected by a variety of spin labels including derivatives of stearic acid, phosphatidylcholine, and phosphatidylethanolamine (25). As can be seen from Table III, avian and amphibian erythrocytes, which are nucleated and

TABLE III

Species Differences in Effect on Agglutination on ESR Spectrum of C_5 Label in Erythrocyte Membranes

	Haemagglutination (HAU)[1]		Δ (gauss)[2]	
Species	Influenza virus	Con A	Influenza virus	Con A
Chicken	2000	500	1.1	1.0
Pigeon	2000	4000	1.0	1.7
Duck	2000	0	1.0	-0.4
Goose	500	30	1.1	-0.1
Frog	--[3]	--[3]	1.7	1.7
Human	2000	0	-0.1	-0.1
Monkey	500	4000	0.2	0.0
Guinea pig	1000	16,000	0.0	0.3
Rabbit	0	8000	0.1	0.2
Sheep	250	16,000	-0.2	-0.3
Bovine	250	0	0.3	0.3

[1]HA assays were performed using a 0.5% (vol/vol) suspension of erythrocytes of each species to titer a standard suspension of virus (2000 HAU/ml) or con A (5 mg/ml). HAU = haemagglutinating units, the reciprocal of the lowest dilution that will agglutinate an equal volume of an 0.5% suspension of a particular species of erythrocyte.

[2] $\Delta = 2A'_{zz}$ (no additions) - $2A'_{zz}$ (agglutinin added). In general, differences as small as 0.3 gauss are not considered significant.

[3]Not determined. Frog erythrocytes were agglutinated by both influenza virus and Con A.

contain microtubules, exhibit a structural change in the lipid bilayer upon agglutination whereas the mammalian erythrocytes, which lack microtubules, do not. The generality of this dichotomy and the observed drug sensitivity supports the hypothesis that the structural change in the lipid bilayer resulting from virus adsorption involves microtubules on the cytoplasmic surface of the membrane.

The observed structural changes in the plasma membrane suggest that a redistribution of the receptor molecules within the plane of the membrane may occur as a result of cross-linking of the receptors following virus attachment. The HA glycoprotein isolated from influenza virus by bromelain digestion (3) binds to erythrocytes without cross-linking the receptors and has no effect on the bilayer of the cell. In contrast, the influenza virus HA isolated by treatment with Triton X-100 followed by detergent removal results in protein aggregates which are multivalent (30), can haemagglutinate, and do induce an increase in the fluidity of the chicken erythrocyte bilayer. Likewise, the succinylated derivative of con A, which is not capable of cross-linking receptors (8) attaches to chicken erythrocytes but does not agglutinate them and does not result in a change in the lipid bilayer. However, when anticon A antibodies are added to chicken erythrocytes pretreated with succinyl-con A, the cells agglutinate and an increase in bilayer fluidity is induced. Therefore, the structural change in the plasma membrane upon virus attachment involves the cross-linking of virus receptors. The apparent involvement of microtubules and the cross-linking of receptors in the bilayer structural changes suggests that they may be factors in the early events of virus penetration.

STRUCTURAL EVENTS IN THE ERYTHROCYTE BILAYER UPON VIRUS-INDUCED HAEMOLYSIS

Sendai virus-induced haemolysis of chicken or human erythrocytes results in a structural change in the erythrocyte bilayer, reducing the difference in fluidity that exists in the intact erythrocyte between the environment of PC12 and PE12 spin labels such that both labels exist in regions of similar fluidity in the haemolyzed erythrocyte[3] (26). Similar structural changes have been observed upon osmotic haemolysis (36). In the absence of Mg^{++} and Ca^{++}, the structural changes in the erythrocyte bilayer reflect the degree of haemolysis.

---

[3]Due to the agglutination effect on chicken erythrocytes, for simplicity most experiments have been done with human erythrocytes. Similar effects have been observed with both systems.

The haemolysis-induced structural changes appear to involve erythrocyte protein located on the cytoplasmic surface (36).

Sendai virions grown in MDBK cells or grown in eggs and harvested early have very low haemolytic activity and do not cause a structural change in the lipid bilayer. The ability to induce the structural change in the bilayer as well as to haemolyze is activated by freezing and thawing of the early harvest virions and by *in vitro* trypsinization of the MDBK-grown virions. Therefore, not only must the Sendai virions have the F glycoprotein in the active cleaved form, but they must also be haemolytically active to induce the fluidity change in the erythrocyte bilayer (26).

KINETICS OF VIRUS-INDUCED HAEMOLYSIS AND CELL FUSION

The kinetics of fusion of Sendai virions with erythrocyte membranes has been measured using virus labeled at low levels with phospholipid derivative spin labels. As fusion occurs, the viral and erythrocyte lipid bilayers mix, and the ESR spectrum changes with time from that characteristic of the virus to that of the cell membranes. The kinetics of the structural change in the erythrocyte bilayer resulting from Sendai virus-induced haemolysis can be measured by recording the ESR spectrum of spin labeled erythrocytes as a function of time after adsorption of the virions.

In Table IV the results of such kinetics experiments are given. The results are independent of the spin label used. The rates of fusion of early and late harvest Sendai virions with erythrocytes are the same indicating that the absence of haemolytic activity does not affect the rate of virus fusion. Fusion of the viral envelope with the erythrocyte bilayer occurs at a rate similar to that of the change induced by haemolysis. These results suggest that fusion may be the rate-limiting step under conditions of active haemolysis.

ACKNOWLEDGEMENTS

This work was supported by Grants PCM76-09962 and PCM76-0993 from the National Science Foundation and AI-05600 from the National Institutes of Health. F.R.L. is an Andrew W. Mellon Fellow. D.S.L. is a National Institutes of Health Postdoctoral Fellow.

REFERENCES

1. Alstiel, L.D., Landsberger, F.R. and Compans, R.W. (1975). *Abs. Ann. Meeting ASM*, 238.
2. Barenholz, Y., Moore, N.F. and Wagner, R.R. (1976). *Biochemistry* 16, 3563.
3. Brand, C.M. and Skehel, J.J. (1972). *Nature New Biol.*

TABLE IV

Time for the Completion of Half of the Reaction of Virus Fusion and Haemolysis

1. Virus Fusion with Human Erythrocytes
(Spin-labeled virus added to unlabeled cells)

Virus	Spin Label	Erythrocytes Intact	Erythrocytes Ghosts
Sendai (early harvest)	PE12	7 min	--
	PC7	--	7 min
Sendai (late harvest)	PE12	7 min	--
	PC5, PC7, PC12	--	7 min
Influenza	PC7	ND[1]	ND[1]

2. Virus-induced Haemolysis of Human Erythrocytes
(Unlabeled virus added to spin-labeled cells)

Virus	Spin Label	Erythrocytes Intact
Sendai (late harvest)	PE12, PC12	7 min
Sendai (early harvest)	PE12	25 min[2]
Influenza	PE12	ND[1]

Virus (0.5 mg) was mixed with erythrocytes or ghosts (50 µl of packed cells), adsorbed at 0° for 10 min, and then pelleted. The ESR spectrum of the pellet was recorded as a function of time at 37°. The reaction half-times were determined from the slope of the data plotted according to the first order rate equation: $\ln [2A'_{zz}(\text{final}) - 2A'_{zz}(t)]$ vs. t; the half-time is $-\ln 2/\text{slope}$.

[1] ND = No change in the ESR spectrum detected after 1 hr at 37°.

[2] Some haemolysis by early harvest virus occurs in these experiments due to the large quantities of virus used, which are an order of magnitude higher than the amount of late harvest virus required to give significant haemolysis. The amount of virus used in these experiments is necessary to obtain reasonable ESR spectra from labeled virus.

238, 145.
4. Choppin, P.W. and Compans, R.W. (1975). *In* "Comprehensive Virology" (H. Fraenkel-Conrat and R.R. Wagner, eds.), vol. 4, p.95. Plenum Press, New York.
5. Choppin, P.W., Murphy, J.S. and Tamm, I. (1960). *J. Expl. Med.* 112, 945.
6. Fong, B.S., Hunt, R.C. and Brown, J.C. (1976). *J. Virology* 20, 658.
7. Godici, P.E. and Landsberger, F.R. (1975). *Biochemistry* 14, 3927.
8. Gunther, G.R., Wang, J.L., Yahara, I., Cunningham, B.A. and Edelman, G.M. (1973). *Proc. Nat. Acad. Sci. USA* 70, 1012.
9. Harrison, S.C., David, A., Jumblatt, J. and Darnell, J.E. (1971). *J. Mol. Biol.* 60, 523.
10. Homma, M., Shimizu, K., Shimizu, Y.K. and Ishida, N. (1976). *Virology* 71, 41.
11. Hubbell, W.L. and McConnell, H.M. (1971). *J. Amer. Chem. Soc.* 93, 314.
12. Hughes, F. and Pedersen, C.E. (1975). *Biochim. Biophys. Acta* 394, 102.
13. Jost, P., Libertini, L.J., Hebert, V.C. and Griffith, O.H. (1971). *J. Mol. Biol.* 59, 77.
14. Jost, P.C., Griffith, O.H., Capaldi, R.A. and Vanderkooi, G. (1973). *Proc. Nat. Acad. Sci. USA* 70, 480.
15. Kilbourne, E.D. and Murphy, J.S. (1960). *J. Expl. Med.* 111, 387.
16. Klenk, H.-D. and Choppin, P.W. (1969). *Virology* 38, 255.
17. Klenk, H.-D. and Choppin, P.W. (1970a). *Virology* 40, 939.
18. Klenk, H.-D. and Choppin, P.W. (1970b). *Proc. Nat. Acad. Sci. USA* 66, 57.
19. Landsberger, F.R., Lenard, J., Paxton, J. and Compans, R.W. (1971). *Proc. Nat. Acad. Sci. USA* 68, 2579.
20. Landsberger, F.R., Compans, R.W., Choppin, P.W. and Lenard, J. (1973). *Biochemistry* 12, 4498.
21. Landsberger, F.R. and Compans, R.W. (1976). *Biochemistry* 15, 2356.
22. Landsberger, F.R., Compans, R.W., Paxton, J. and Lenard, J. (1972). *J. Supramol. Struct.* 1, 50.
23. Lazarowitz, S.G. and Choppin, P.W. (1975). *Virology* 68, 440.
24. Lenard, J., Tsai, D.K., Compans, R.W. and Landsberger, F.R. (1976). *Virology* 71, 389.
25. Lyles, D.S. and Landsberger, F.R. (1976). *Proc. Nat. Acad. Sci. USA* 73, 3497.
26. Lyles, D.S. and Landsberger, F.R. (1977). *Proc. Nat. Acad. Sci. USA* 74, 1918.

27. Moore, N.F., Barenholz, Y. and Wagner, R.R. (1976). *J. Virol.* 19, 126.
28. Mudd, J.A. (1974). *Virology* 62, 573.
29. Rothman, J.E., Tsai, D.K., Dawidowicz, E.A. and Lenard, J. (1976). *Biochemistry* 15, 2361.
30. Scheid, A. and Choppin, P.W. (1973). *J. Virol.* 11, 263.
31. Scheid, A. and Choppin, P.W. (1974). *Virology* 57, 475.
32. Schloemer, R.H. and Wagner, R.R. (1975). *J. Virol.* 16, 237.
33. Sefton, B.M. and Gaffney, B.J. (1974). *J. Mol. Biol.* 90, 343.
34. Skehel, J.J. and Waterfield, M.D. (1975). *Proc. Nat. Acad. Sci. USA* 72, 93.
35. Stoffel, W., Bister, K., Schreiber, C. and Tunggal, B. (1976). Hoppe-Seyler's *Z. Physiol. Chem.* 357, 905.
36. Tanaka, K.-I. and Ohnishi, S.-I. (1976). *Biochim. Biophys. Acta* 426, 218.
37. Utermann, G. and Simons, K. (1974). *J. Mol. Biol.* 85, 569.
38. Wisnieski, B.J., Parkes, J.G., Huang, Y.O. and Fox, C.F. (1974). *Proc. Nat. Acad. Sci. USA* 71, 4381.

ORTHO- AND PARAINFLUENZA VIRUS-INDUCED HOST CELL MEMBRANE MODULATION IN CELL-MEDIATED CYTOTOXICITY

Y. HOSAKA, T. SAKURAI, K.K. KO AND K. FUKAI

Research Institute for Microbial Diseases,
Osaka University, Suita City,
Osaka, Japan.

Virus-induced cell-mediated immunity is important both for the defence mechanism against viral diseases and for the physiological functions of the immune surveillant, since virus-infected cells provide "altered-self" antigens (5). Many reports have been published concerning the recognition of virus-infected targets by immune T lymphocytes. However, few reports are available dealing with host cell membrane modulations in virus-infected cells. Ada *et al* found that ultraviolet (UV)-irradiated ectromelia virus-inoculated cells are susceptible to cell-mediated lysis (1) and suggested that glycoprotein synthesis is a prerequisite for ectromelia virus or lymphocytic choriomeningitis (LCM) virus-infected cells to become susceptible to cell-mediated lysis (15). Anti-viral antiserum did not inhibit cell-mediated lysis of Sendai virus (8), ectromelia virus (9) or Coxsackievirus B-3 infected cells (20).

This investigation deals with the requirements for the induction of susceptibility to cytotoxic lymphocytes (CL) in Sendai virus- or influenza virus-infected cells and the possible role of CL in inhibiting growth of influenza virus in murine cell cultures.

The results suggest that when the cell mediated immune response is compared in mice infected with normal or UV-irradiated Sendai or influenza viruses, CL are manifested only with infectious virus. Additional studies were carried out to determine whether inactive Sendai virus produced from L929 cells (13) could induce formation of CL before and after inactivation by trypsin (11).

Unless otherwise mentioned, egg-grown Sendai virus, Z strain, and influenza virus, PR8 strain (HONI) were used for mouse immunization, and for *in vitro* infection of established murine cell lines, L929(H-2^k) and MC57G(H-2^b) cells were used as target cells.

EFFECT OF UV IRRADIATION OF SENDAI OR PR8 VIRUSES ON THEIR CAPACITY TO INDUCE TARGET CELL SUSCEPTIBILITY

Fig.1a shows the effect of UV-irradiation of Sendai virus on infectivity and their capacity to render virus-infected L929 cells susceptible to cell-mediated lysis (CML). Fig.1b shows similar curves for UV irradiation of PR8 influenza virus.

In both cases, infectivity was rapidly inactivated but the capacity to induce the CML antigens was reduced more slowly. Thus, UV-irradiated instead of active virus can be used for studies on virus-induced cell-mediated immunity *in vivo* and *in vitro*. This has the advantage of being a simpler system, eliminating the effects of virus replication.

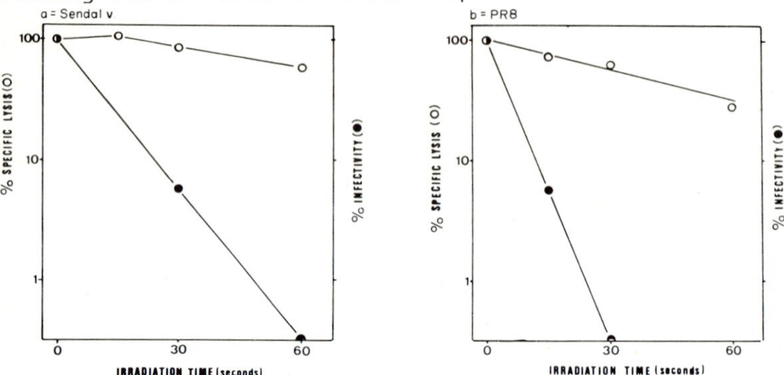

Fig.1. *Inactivation curves by UV-irradiation of viral infectivity and the capacity to induce susceptibility of infected L929 cells to CML. a: Sendai virus, b: PR8. Viruses were irradiated with UV light 20 cm distance from the light source (irradiated volumes were different for the two viruses). 10 μl and 100 μl aliquots were removed at intervals and assayed for their infectivity and the capacity to induce the susceptibility, respectively.*

Moi; 30 EID_{50}/cell, if virus was not irradiated. 6 day immune spleen cell to target ratio = 50. Incubation with the immune spleen cells; 12 hr. Cell-mediated lysis (or the susceptibility to CML) was expressed as follows:-

$$\text{Specific lysis } (\%) = \frac{{}^{51}Cr \text{ release of targets with immune spleen cells} - {}^{51}Cr \text{ release with normal spleen cells}}{\text{Total } {}^{51}Cr \text{ release with Triton X-100} - {}^{51}Cr \text{ release with normal spleen cells}} \times 100$$

CELL-MEDIATED IMMUNE RESPONSE IN MOUSE WITH LIVE AND UV-IRRADIATED VIRUSES

C3H/he(H-2^k) and C57BL/6(B6)(H-2^b) mice were inoculated intraperitoneally (ip) with 10-fold serial dilutions of live and UV-irradiated Sendai or PR8 viruses (B6 mice only with PR8). Their spleen cell cytotoxicities to virus-infected syngeneic targets were measured 6 days later.

Fig.2a shows that the minimum doses of live and UV-irradiated Sendai virus needed to evoke positive cytotoxicity in C3H mice were 10^{-5} and 10^{-1} HA, respectively. In terms of infectivity, both doses were almost the same, $10^{1.0-2.0}EID_{50}$. Fig.2b shows that the minimum dose of active and UV-irradiated PR8 needed to evoke cytotoxicity in C3H mice were 10^{-4} and 10 HA, respectively. Again, in terms of infectivity, both doses were almost the same, $10^{2.0-2.5}EID_{50}$. This finding with live Sendai virus is analogous to the report (8) that 10^1EID_{50} of Sendai virus stimulated cytotoxic spleen cell formation in C3H mice.

The figures also show that treatment of the cytotoxic spleen cells with anti-theta serum and complement abolished their cytotoxicity but treatment with anti-mouse IgG rabbit serum and complement did not. This finding suggests that T cells are responsible for the cytotoxicity. In B6 mice (Fig.2c), generation of cytotoxic spleen cells with PR8 virus was dependent on inoculation of a higher virus dose.

Possible reasons for the apparent inability of UV-killed virus to evoke spleen cell cytotoxicity were considered. First, most of the ip inoculated UV-killed virus was phagocytosed, rendering it incapable of sensitization of spleen cells. Hence, the effects of ip inoculation into C3H mice of *in vitro* UV virus-inoculated L929 targets was investigated. However, this immunization evoked spleen cell cytotoxicity against normal L929 cells. Thus, we were unable to measure a virus-specific cell-mediated immune response. Second, the antigenic structure of virus-infected target cells is in some ways different from that of UV virus-inoculated cells, and this might be responsible for the difference in the ability of the two viruses to evoke cytotoxicity. Therefore, target differences between active and UV-irradiated viruses, were investigated.

COMPARISON OF COMPETITION FOR CML BY LIVE AND UV VIRUS-INOCULATED COLD TARGETS

L929 cells were infected either with live Sendai or PR8 virus

---

In this figure, it was expressed as relative values (%) of specific lysis of targets with irradiated virus to that with unirradiated virus.

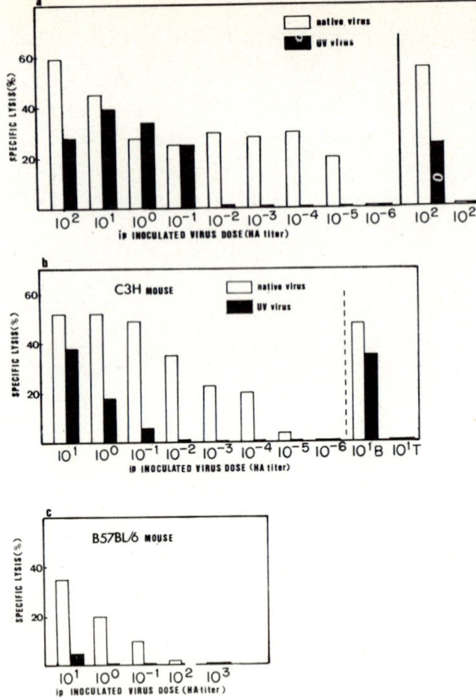

Fig.2. *Comparison of cell-mediated immune responses (spleen cell cytotoxicity) with native and UV-virus.*
a: Sendai virus in C3H mice
b: PR8 in C3H mice
c: PR8 in C57BL/6 mice
0.5 ml of 10-fold serial dilutions of native or UV-virus were ip inoculated into mice and 6 days later mouse splenocyte cytotoxicity to virus-infected syngeneic targets were examined. CML assay was same as in Fig.1.

at a multiplicity of 10 EID_{50}/cell or with UV-irradiated Sendai or UV-PR8 virus (1 min irradiated) which was equivalent to a multiplicity of 4 x 10 EID_{50}/cell. These conditions of infection gave maximum lysis (60-70%) of virus-inoculated targets. ^{51}Cr-labelled targets infected with live virus were added to unlabelled cells at target ratios of 1, 3, and 9 at 4 hr post-infection, and subsequently immune spleen cells were added to the mixtures. CML was assayed after 15 hr incubation. The inhibition patterns of CML by the non-radioactive target cells were almost the same for active and UV-irradiated virus with both Sendai and PR8 viruses.

No differences in competition by cold targets between active and UV-Sendai or between active and UV-PR8 virus were found which suggests that no differences in recognition pattern by cytotoxic T cells exists between active and UV virus-inoculated targets.

EFFECTS OF ACTINOMYCIN D, 2-DEOXY-D-GLUCOSE, CYCLOHEXIMIDE AND ANTI-VIRAL SERUM ON INDUCTION OF THE SUSCEPTIBILITY OF TARGETS TO CML

Actinomycin D (AMD) suppresses influenza virus growth by inhibiting complementary (messenger) RNA synthesis (16) but it does not affect Sendai virus growth (3). Therefore, it was of interest to investigate whether the drug displays a similar inhibitory pattern to induction of the susceptibility of Sendai and PR8-infected targets to CML. Figs.3a and 3b show that induction of the susceptibility in PR8-infected L929 cells is inhibited by the presence of the drug (1-4 μg/ml) but no inhibition is detected in Sendai-infected cells. The figures also show that when UV-virus is used, induction of the susceptibility of both PR8 and Sendai-inoculated L929 cells is inhibited by the drug. The reason for this difference in sensitivity to the drug between active and UV-Sendai virus infected targets remains to be determined.

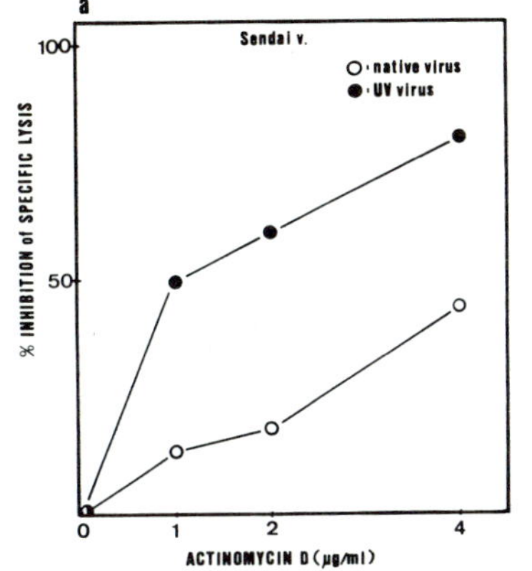

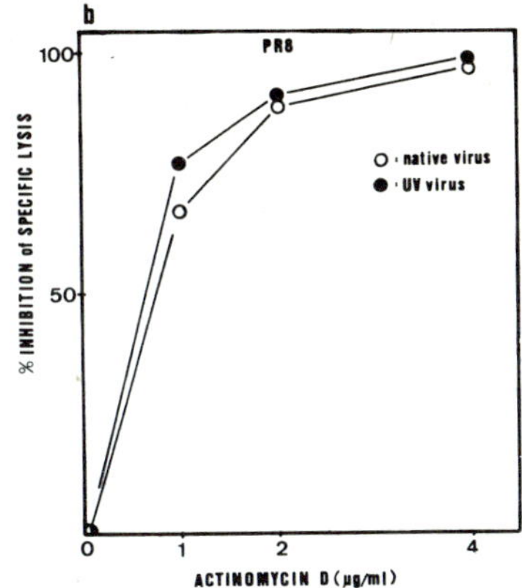

Fig.3. *Effect of AMD on cell-mediated lysis of native and UV-virus-inoculated L929 cells. a: Sendai virus. b: PR8. AMD was present throughout the incubation time of infected target cultures after 1 hr adsorption. Anti-virus immune C3H spleen cells were added to the target cultures 4 hr p.i. and incubated at 34° for 6-7 hr. UV virus: 1 min irradiated, moi, 10 EID_{50}/cell for native virus. 4 x 10 EID_{50}/cell for UV-virus, if it were not irradiated.*

$$\frac{\text{control lysis (no AMD)-lysis in the presence of AMD}}{\text{control lysis (no AMD)}} \times 100$$

Inhibition of PR8-induced susceptibility by AMD was small in comparison to the complete inhibition of PR8 growth in L929 cells by the drug (1 μg/ml) observed when measured by hemagglutinin production.

2-deoxy-D-glucose (2dG) inhibits influenza (14) and parainfluenza virus (10) growth by blocking glycoprotein synthesis. This compound also inhibits formation of virus-induced susceptibility to CML of ectromelia or LCM virus-infected targets, which suggests that glycoprotein synthesis is necessary for CML antigen susceptibility (15). Since the drug concentrations used do not inhibit attachment to ^{51}Cr-labelled immune spleen cells, we measured its effect on CML of Sendai and PR8 infected L929 cells. 5 mM 2dG inhibits about 50% of active and UV-irradiated PR8 inoculated targets while 20 mM 2dG results in 90% inhibition. UV-irradiated Sendai-inoculated targets are almost identical to PR8 in their sensitivity to the compound but active Sendai-infected targets show some resistance. These concentrations of 2dG inhibit PR8 and Sendai virus replication in L929 cells with Sendai exhibiting a greater sensitivity.

In the presence of cycloheximide (5-10 μg/ml), CML of active and UV-irradiated Sendai or PR8 virus is completely inhibited. This reagent is thought to suppress the cytotoxic action of immune spleen cells (4) in addition to inhibiting virus-induced protein synthesis.

Anti-PR8 guinea pig serum (HI titer 2560/ml), anti-Sendai guinea pig (HI titer 215/ml) and rabbit hyperimmune serum (HI titer 1600/ml) did not inhibit the CML of PR8 and Sendai virus-infected L929 cells or the CML of UV-irradiated Sendai virus-infected L929 cells, respectively.

REQUIREMENTS FOR INDUCTION OF THE SUSCEPTIBILITY OF SENDAI OR INFLUENZA VIRUS-INFECTED TARGETS TO CML

Exposure of Sendai or influenza viruses to 1 min irradiation does not abolish their ability to induce susceptibility to CML in virus-inoculated targets. Perhaps with these RNA viruses, only a part of the viral genome is necessary for susceptibility of infected target cells to CML. Since almost full expression of the susceptibility of Sendai targets to CML occurs in the presence of AMD there is probably no requirement for the synthesis of the bulk of host proteins. On the other hand, the inhibition of susceptibility of PR8 targets to CML in the presence of the drug suggests that expression requires transcription of the influenza virus genome.

The inhibition by 2dG and cycloheximide supports the hypothesis that expression of susceptibility requires glycoprotein synthesis in the target cells, as was previously shown in the case of ectromelia virus-infected cells (15). Since anti-viral serum did not interfere with cell-mediated cytolysis, this glycoprotein is not an envelope protein.

These results, therefore, are not consistent with the hypothesis (18) that cytotoxic T cells recognise a combination of preformed virus antigen and H-2-histocompatibility antigen, which implies that there is no requirement for protein synthesis in the system with UV-irradiated Sendai virus inoculated targets.

We were not able to find any difference in the susceptibility of target cells to CML which would explain the difference in capacity to evoke spleen cell cytotoxicity between active and UV-irradiated viruses.

It is possible that the antigenic structures determining susceptibility of both native and UV-irradiated virus targets to CML are similar, but the structures responsible for sensitisation of spleen cells are somehow different.

INHIBITION OF PR8 VIRUS GROWTH BY H-2 SPECIFIC IMMUNE SPLEEN CELLS

Zinkernagel and Althage (21) reported that syngenic immune spleen cells inhibit vaccinia virus growth, if they are added during the early stages of virus replication. We investigated whether the growth of an enveloped RNA virus, such as PR8 in L929 and MC57G cells, was inhibited by syngeneic and allogeneic immune spleen cells. Since non-infectious virus and HA of PR8 are produced in L929 and MC57G cells we measured virus growth by HA production. Fig.4 shows that only syngeneic immune spleen cells inhibit PR8 growth in these cells, and further that treatment of immune spleen cells with anti-theta serum and complement, but not treatment with anti-mouse IgG and complement, abolishes their inhibitory action. This finding suggests that the inhibition is T-cell dependent.

INHIBITION OF INFLUENZA VIRUS GROWTH IN L929 CELLS BY CROSS-REACTIVE IMMUNE SPLEEN CELLS

CL generated with serologically distinct strains of influenza A type attack all influenza A virus-infected targets (6, 7, 22), but homologous interactions show a stronger response. In our studies we obtained evidence that these cross reactive, syngeneic immune spleen cells inhibit the growth of serologically distinct influenza A virus in L929 cells, with inhibition being dependent on cell-mediated lysis.

The above results indicate that in humans, CL generated by influenza A virus infection may function in the defence mechanism against serologically distinct influenza virus infections.

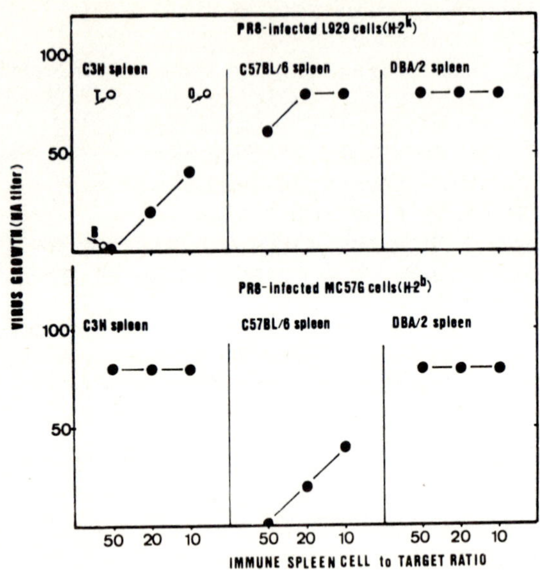

Fig.4. *Inhibition of PR8 growth in L929 ($H-2^k$) and MC57G ($H-2^b$) cells by syngeneic immune spleen cells.*

Anti-PR8 6 day immune spleen cells from C3H ($H-2^k$), C57BL/6 ($H-2^b$) or DBA/2 ($H-2^d$) mice were added to the PR8-infected target cultures, 5 hr p.i. at indicated ratio of spleen to target cells and incubated at 34° for 12 hr. Then the cells were washed once and cell-associated hemagglutinin determined. CML assay was same as in Fig.1.

INTERACTION OF VIRUS-INFECTED L929 CELLS AND IMMUNE C3H SPLEEN CELLS OBSERVED BY ELECTRON MICROSCOPY

Electron microscopy was used to observe the stages of virus replication blocked by immune spleen cells. Influenza A, WSN strain (HONI), was used instead of PR8, since WSN, unlike PR8, infection induces the formation of dense bodies, surrounded by fine granules, in the cytoplasm (cf Fig.5a). After incubation with immune spleen cells for 4.5 hr (Fig.5b), the dense bodies can be seen. After incubation for 16 hr, many disrupted cells and cell debris are seen. The dense bodies looked rather smaller and no budding virus was observed (Fig.5c).

It appears, therefore, that immune spleen cells, added to the target cells at 6 hr postinfection, inhibit budding or an assembly system for budding and also inhibit the formation of dense bodies. The biological significance of these dense bodies is, as yet, unknown.

Fig.5d shows the interaction of Sendai infected L929 cells and immune C3H spleen cells after a 6 hr incubation from 6 hr postinfection. A break on a target cell membrane in the region of its contact with a lymphocyte is seen. Fine dense materials are seen attaching to the area of the lymphocyte membrane just facing the break. One can speculate that the break in the target cell is just at the point where the lymphocyte attacks. The cytoplasm looks loosely organized which may be the initial response to cell-mediated lysis.

CELL-MEDIATED IMMUNE RESPONSES WITH UNTREATED AND TRYPSINIZED L929 CELL-GROWN SENDAI VIRUS (L-SENDAI)

We found that when L-Sendai which had a 10^{3-4} ratio of infectivity (HAD foci units) to hemagglutinin, was inoculated onto L929 cells (10^6 cells), the cells were lysed by 6 day immune C3H spleen cells. When L-Sendai virus (in MEM with 5% fetal bovine serum) was treated with 0.1% trypsin at 35° for 10 min, its infectivity was increased 40 times and trypsinized L-Sendai-infected L929 cells were lysed by the same immune cells.

We measured virus dose-dependency of the cell-mediated immune response in C3H mice with untreated and trypsinized L-Sendai. Intraperitoneal inoculation of the crude L-Sendai evoked spleen cell cytotoxicity to normal L cells and we could not assess the virus-specific immune response. Intranasal inoculation (IN) of crude L-Sendai, on the other hand, evoked virus-specific spleen cell cytotoxicity. Intraperitoneal inoculation of normal L929 cells or medium of L929 cell cultures into C3H mice evoked similar cytotoxicity to normal L929 cells. Since L cells originate from C3H mouse heart, the current L cell surfaces appear to have been altered to enable them to evoke cell-mediated immune response in syngeneic mice. If this immune response was caused by an endogenous C type virus, then C3H mice must not be tolerant to the virus.

We purified L-Sendai by adsorption-elution onto chicken red blood cells, followed by a differential centrifugation. The optimal concentration (0.01%) of trypsin needed to activate the purified virus (5000 HA/ml) was smaller than that (0.1%) needed to activate the crude virus. We found that C3H mice generated virus-specific spleen cell cytotoxicity selectively, when they were inoculated with the purified virus by either route.

Fig.6 shows that inhalation of the purified L-Sendai virus evoked splenocyte cytotoxicity more efficiently than ip inoculation and *in vitro* trypsinization increased the ability of the virus to evoke splenocyte cytotoxicity at least 1000-fold, although its infectivity, measured by HAD foci production, only

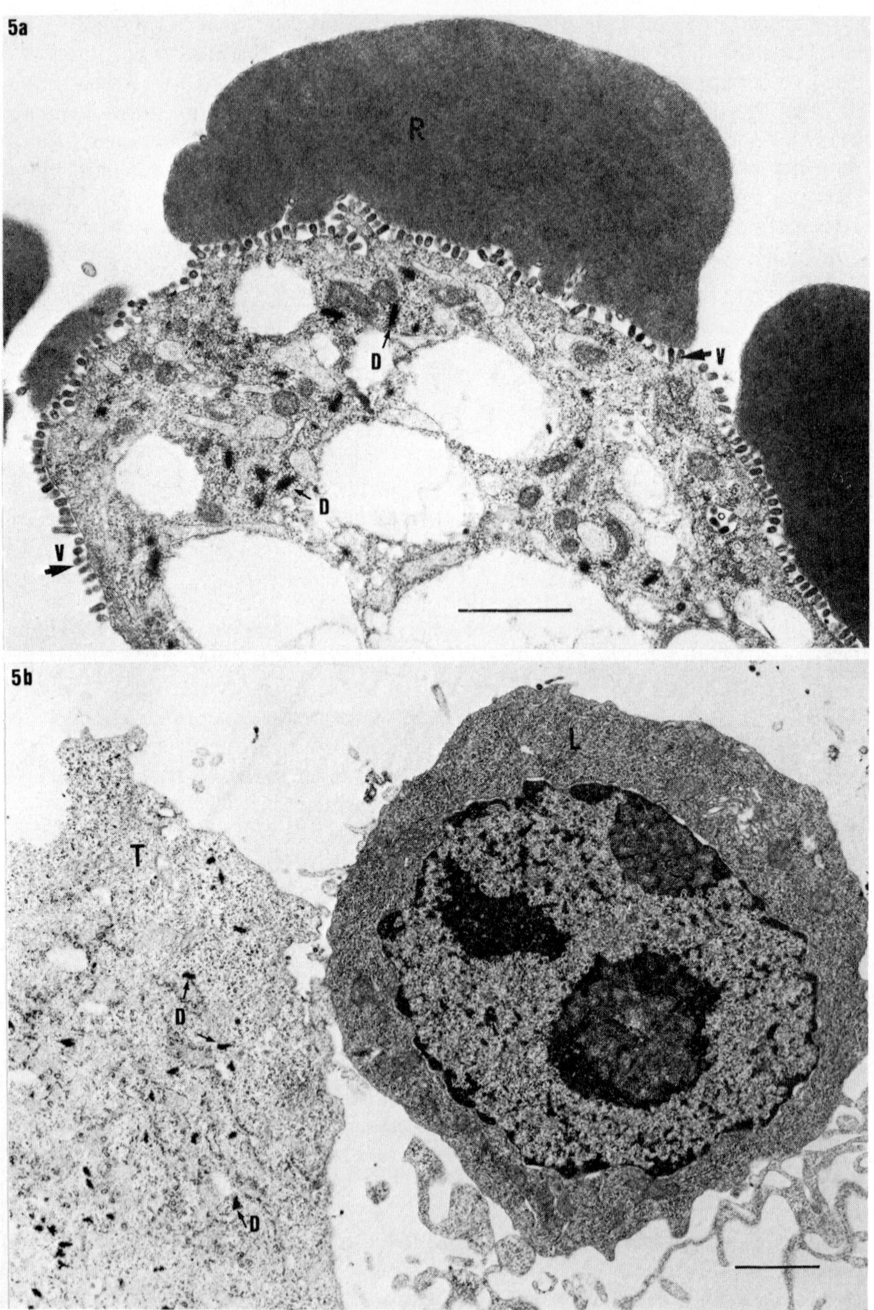

Fig.5a and 5b

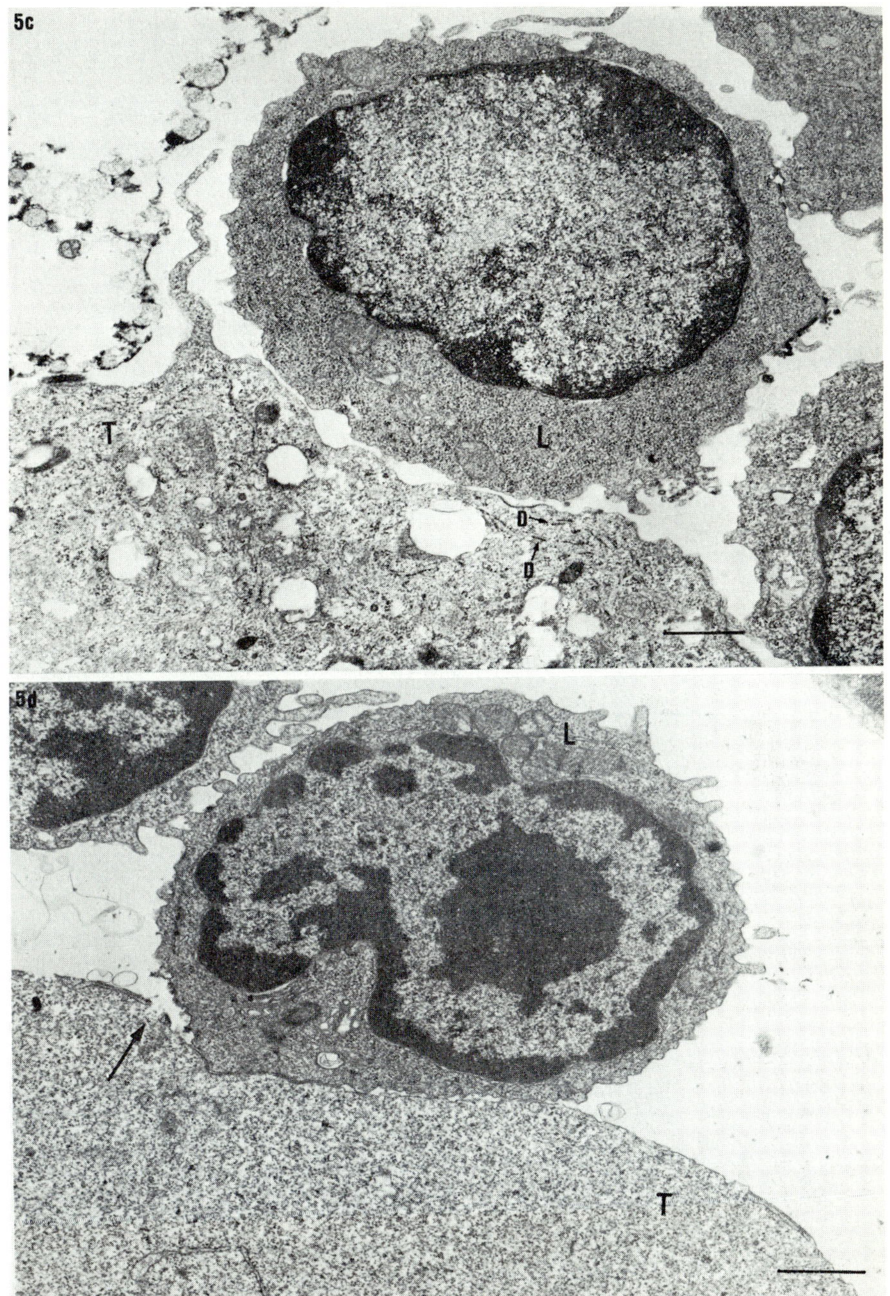

Fig.5c and 5d

Fig.5a. *L929 cells infected with WSN virus, 20 hr after infection (moi 5 EID_{50}/cell). No spleen cells but guinea pig red blood cells (RBC) were added. Many virus particles budding on cell membranes are seen, and guinea pig RBC are hemadsorbed. In the cytoplasm, virus-induced dense bodies are seen. R: RBC, V: virus particles, D: virus-induced dense bodies. Scale: 1 μm.*

Fig.5b and 5c. *L929 cells infected with WSN virus, incubated with immune C3H spleen cells for 4.5 hr and 16 hr from 6 hr postinfection, respectively. In Fig.5b, many fine virus-induced dense bodies are seen, and no morphological sign of presence of disrupted cells are observed at this stage. In Fig.5c, dissociated cell materials possibly derived from disrupted cells are seen in the left side. No budding virus is seen at 22 hr postinfection. Possibly degenerated dense bodies are seen in the cytoplasm. Dense materials are attaching to some parts of the lymphocyte cell membrane. T: infected target, L: lymphocyte.*

Fig.5d. *L929 cells infected with Sendai virus, incubated with immune C3H spleen cells for 6 hr from 4 hr postinfection. An arrow shows a break of cell membrane of the infected target, probably attacked by a cytotoxic lymphocyte. Fine dense materials are seen attaching to the part of the lymphocyte membrane just facing the break. Organization of the cytoplasm is looking loose. All the cells were fixed with glutaraldehyde-osmium tetroxide and sections stained with lead and uranyl acetate.*

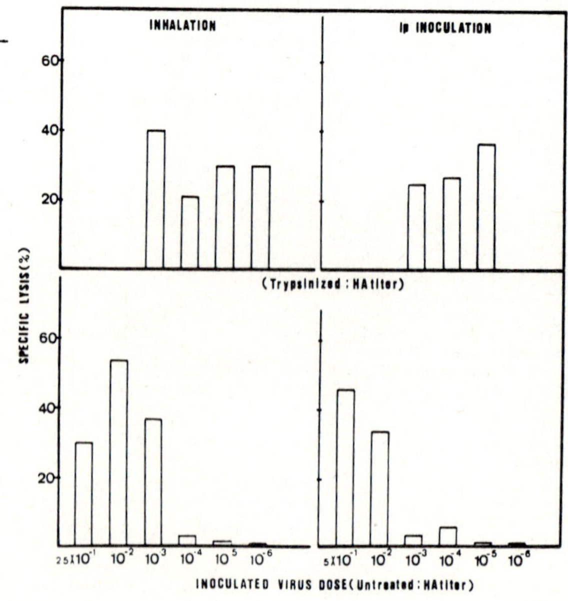

Fig.6

increased 20 fold.

DISCUSSION ON CELL-MEDIATED IMMUNE RESPONSE WITH L-SENDAI VIRUS

It is interesting to note that 10^2 infectious doses of L-Sendai was the minimum dose needed to evoke splenocyte cytotoxicity by ip inoculation. This suggests that non-infectious hemagglutinin of L-Sendai could not evoke splenocyte cytotoxicity indicating that, *in vivo*, activation of the virus by ip or respiratory proteolytic enzymes does not occur. The reason why inhalation of L-Sendai was a more efficient route for evoking a cell-mediated response is probably that the virus can replicate better in mouse lungs than in peritoneal cavity.

A fraction of the trypsinized L-Sendai, which could not produce *in vitro* HAD foci but evoked *in vivo* splenocyte cytotoxicity might be further exposed to activation by *in vivo* proteolytic enzymes, which act on the trypsinized L-Sendai virus but not the untreated virus.

CONCLUSION

1. Expression of susceptibility to CML of Sendai or influenza virus-infected targets seems to require new glycoprotein synthesis.
2. The new glycoprotein synthesis seems to be directed by a part of the viral genome RNA.
3. Homologous and heterologous immune spleen cells of influenza A viruses inhibited growth of an influenza A virus.
4. Electron microscopic study of the interaction of virus-infected targets and immune spleen cells shows that the immune spleen cells inhibited the budding or assembly system for budding of influenza virus.

Fig.6. *Cell-mediated immune responses (splenocyte cytotoxicity) in C3H mice with untreated and trypsinized L-Sendai virus. upper: untreated virus. lower: trypsinized virus.*

L929 cells were infected with Sendai virus at moi 10 EID_{50}. 2 days later, the culture medium was harvested. L-Sendai virus in the medium was purified by adsorption-elution onto chicken RBC, followed by a differential centrifugation. The purified virus (2.6 x 10^7 HAD foci units/ml, 5000 HA/ml) was treated with 0.01% trypsin at 35° for 10 min. The infectivity was increased by 20 times.

25 μl and 500 μl aliquots of 10-fold serial dilutions of untreated or treated L-Sendai were inhaled and ip inoculated into C3H mice, respectively. 6 days later splenocyte cytotoxicity of the immunized mice were assayed to Sendai virus-infected L929 targets. The cytotoxicity assay method was the same as in Fig.1.

5. Trypsinization of L929 cell-grown Sendai virus activated the capacity of the virus to evoke mouse splenocyte cytotoxicity.

ACKNOWLEDGEMENTS

During the course of this work T. Sakurai was JICA Fellow from Instituto Adolfo Lutz in Sao Paulo. K.K. Ko was a Taniguchi Scholarship Fellow from Catholic Medical College, Seoul, Korea.

REFERENCES

1. Ada, G.L., Jackson, D.C., Blanden, R.V., Tha Hla, R. and Bowern, N.A. (1976). *Scand. J. Immunol.* 5, 23.
2. Bean, W.J. and Simpson, R.W. (1973). *Virology* 56, 646.
3. Blair, C.D. and Robinson, W.S. (1968). *Virology* 35, 537.
4. Brunner, K.T., Mauel, J., Cerottini, J.-C. and Chapuis, B. (1968). *Immunol.* 14, 181.
5. Doherty, P.C., Blanden, R.V. and Zinkernagel, R.M. (1976). *Transplant. Rev.* 29, 89.
6. Doherty, P.C., Effros, R.B. and Bennink, J. (1977). *Proc. Nat. Acad. Sci. USA,* 74, 1209.
7. Effros, R.B., Doherty, P.C., Gerhard, W. and Bennink, J. (1977). *J. Exp. Med.* 145, 557.
8. Ertl, H. and Koszinowski, U. (1976). *Z. Immun. Forsch.* 152, 128.
9. Gardner, I., Bowern, N.A. and Blanden, R.V. (1974). *Europ. J. Immunol.* 5, 122.
10. Hodes, D.S., Schnitzer, T.J., Kalica, A.R., Camargo, E. and Chanock, R.M. (1975). *Virology* 63, 201.
11. Homma, M. (1971). *J. Virol.* 8, 619.
12. Homma, M. and Ouchi, M. (1973). *J. Virol.* 12, 1457.
13. Ishida, N. and Homma, M. (1961). *Virology* 14, 486.
14. Klenk, H.-D., Scholtissek, C. and Rott, R. (1972). *Virology* 49, 723.
15. Jackson, D.C., Ada, G.L., Hapel, A.J. and Dunlop, M.B. (1976). *Scand. J. Immunol.* 5, 1021.
16. Pons, M.W. (1975). *In* "Influenza Viruses and Influenza" (E.D. Kilbourne, ed.)., p.145. Academic Press, New York.
17. Scheid, A. and Choppin, P.W. (1974). *Virology* 57, 475.
18. Schrader, J.W. and Edelman, G.M. (1977). *J. Exp. Med.* 145, 523.
19. Smith, R.T. and Landy, M. (1971). Eds. "Immune Surveillance". Academic Press, New York.
20. Wong, C.Y., Woodruff, J.J. and Woodruff, J.F. (1977). *J. Immunol.* 118, 1159.
21. Zinkernagel, R.M. and Althage, A. (1977). *J. Exp. Med.* 145, 644.
22. Zweerink, H.J., Courtneidge, S.A., Skehel, J.J., Crumpton, M.J..and Askonas, B.A. (1977). *Nature* 267, 354.

INTERACTION OF LIPID VESICLES CONTAINING PARAMYXOVIRUS ENVELOPE GLYCOPROTEINS WITH CULTURED CELLS

GEORGE POSTE, *PETER REEVE and *HELMUT BACHMEYER

Department of Experimental Pathology, Roswell Park Memorial Institute, Buffalo, New York 14263, U.S.A.

**Sandoz Forschungsinstitut, A-1235, Vienna, Austria.*

Studies in several laboratories have shown that lipid vesicles (liposomes) can fuse with the plasma membrane of cells cultured *in vitro* (6). Since a variety of hydrophobic membrane proteins and glycoproteins can be incorporated into vesicle membranes, fusion of vesicles with the cellular plasma membrane offers a potential method for introducing specific proteins into the plasma membrane of cultured cells. To test the feasibility of this method, we have examined the ability of lipid vesicles to serve as carrier vehicles to introduce Newcastle disease virus (NDV) envelope glycoproteins into the plasma membrane of mammalian and avian cells. Viral glycoproteins were chosen for these experiments since they can be reliably distinguished from cell components and, in the case of NDV and other paramyxoviruses, their ability to induce hemagglutination and cell fusion (7) can be used to determine if vesicle-derived proteins retain their functional properties when assimilated into the plasma membrane. A final useful feature is that the uptake, subcellular distribution and eventual fate of paramyxovirus envelope glycoproteins introduced into cells by vesicles can be directly compared with events in normal virus infection in which envelope glycoproteins are incorporated into the plasma membrane by fusion of the virus envelope with the plasma membrane (2).

Many hydrophobic membrane proteins and glycoproteins have now been successfully incorporated into vesicles of widely differing lipid composition, size and structure (unilammelar versus multilamellar) (6). The recently developed class of large unilamellar vesicles (LUV) (3) were used in the present experiments since the efficiency of incorporation of paramyxovirus envelope glycoproteins into these vesicles is significantly higher than for large multilamellar vesicles or small unilamellar vesicles of similar composition. This may reflect the fact that during formation of LUV by fusion of small unilamellar vesicles (SUV) hydrophobic regions of the lipid bilayer become exposed in the SUV (3) and this may facilitate association of hydrophobic membrane proteins with the lipid bilayer.

The association of vesicles containing ^{125}I-labeled envelope glycoproteins from NDV and influenza virus with baby hamster kidney (BHK) is shown in Table I. The results indicate that incorporation of viral glycoproteins into lipid vesicles significantly increases their ability to associate with cells compared with free glycoproteins or free glycoproteins incubated with cells in the presence of pure phospholipid vesicles. Similar results have also been obtained with chick embryo cells (not shown).

These results do not, however, establish whether the vesicle-derived viral glycoproteins have been inserted into the plasma membrane or are merely bound to the surface. Cells were therefore incubated with trypsin to see if the cell-associated glycoproteins could be eluted. As shown in Table I, enzymatic modification of the cell surface caused a marked reduction (approx. 90%) in cell-associated radioactivity in cells that had been incubated at 37° with vesicles containing glycoproteins from NDV strain Ulster and influenza virus. This suggests that the majority of glycoproteins were associated with vesicles adsorbed to protease-susceptible cell surface components. The same enzyme treatment was less effective, however, in removing vesicle-derived glycoproteins from cells incubated at 37° with vesicles containing glycoproteins from NDV strain Herts (Table I). The vesicle-derived glycoproteins which resist elution by trypsin in this system could either be associated with vesicles adsorbed to enzyme-resistant cell components or have become incorporated into the plasma membrane as a result of vesicle-plasma membrane fusion. To distinguish between these possibilities, cells were incubated at 4° with vesicles containing NDV strain Herts glycoproteins before being exposed to trypsin since vesicle-plasma membrane fusion does not occur at this temperature (4). Under these conditions, trypsinization induced elution of approximately 90% of the vesicle-derived glycoproteins (Table I). This indicates that the enzyme-resistant fraction of vesicle-derived glycoproteins detected at 37° cannot be attributed to adsorption of vesicles to enzyme-resistant cellular components and enzyme-resistance is due instead to incorporation of the glycoproteins into the plasma membrane via vesicle fusion. Conversely, the elution of vesicles containing glycoproteins from NDV strain Ulster and influenza virus from cells by trypsin at both 37° and 4° (Table I) suggests that these vesicles are unable to fuse with the plasma membrane.

Vesicle-derived glycoproteins from NDV strain Herts which persist on the surface of the trypsinized cells exhibit hemadsorption activity, can be capped by anti-NDV antibodies and the capping process is inhibited by cytoskeletal-disruptive drugs (cytochalasin B; dibucaine HCl) which block ligand-induced capping of normal plasma membrane components. These findings, which will be described in detail in a subsequent publication, provide further evidence to suggest that the viral glycoproteins have been incorporated into the lipid bilayer of the plasma membrane via vesicle fusion.

The ability of NDV to induce cell fusion is determined by the presence of a specific glycoprotein (F) in the virus envelope and NDV strains which cannot fuse cells contain a different glycoprotein (F_0) (7). The latter is a structural precursor of the F glycoprotein and brief trypsinization of F_0-containing virions converts F_0 to F with accompanying acquisition of the ability to fuse cells. This difference in the cell fusion activity of the F_0 and F glycoproteins is also exhibited when the proteins are incorporated into vesicles and interacted with cells. As shown in Table II, vesicles containing the F glycoprotein (Herts and trypsinized Ulster) induce significant cell fusion, but vesicles containing the F_0 precursor (NDV strain Ulster) are ineffective. In addition to offering a new experimental approach for defining the functional properties of the F and F_0 glycoproteins, these results also indicate that integral membrane proteins introduced into cells using lipid vesicles as carriers can retain their functional activity.

The usefulness of vesicle-mediated transfer of proteins into cellular membranes as an experimental tool not only demands that the introduced proteins be functional, but also that they persist in the plasma membrane for sufficient time to enable experiments to be done. The persistence of vesicle-derived NDV glycoproteins in the plasma membrane of "acceptor" BHK and chick embryo cells is shown in Table III. The results indicate that "clearance" of viral glycoproteins is relatively rapid, but is no faster than that for viral glycoproteins introduced into the plasma membrane during normal virus infection (Table III). The fate of viral glycoproteins removed from the plasma membrane is not known.

The present experiments provide the first demonstration of the use of lipid vesicles as carrier vehicles to introduce novel hydrophobic membrane proteins into the plasma membrane

TABLE I

Association of ^{125}I-Labeled Virus Envelope Glycoproteins with BHK Cell Monolayers

Source of Envelope Glycoproteins	Sample[a,b]	Cell-associated Radioactivity (cpm x 10^{-3} per 10^6 cells)[c] untreated cells 37°	untreated cells 4°	trypsin-treated cells[d] 37°	trypsin-treated cells[d] 4°
NDV (Herts)	Vesicle-associated glycoproteins	8.59	5.26	4.64	0.63
	free glycoproteins	0.39	0.24	0.04	0
	free glycoproteins + vesicles	0.61	0.37	0.05	0
NDV (Ulster)	vesicle-associated glycoproteins	6.19	4.16	0.80	0.67
	free glycoproteins	0.42	0.27	0.06	0
	free glycoproteins + vesicles	0.58	0.33	0.05	0
Influenza (A/Scot/74)	vesicle-associated glycoproteins	10.64	6.28	1.28	0.75
	free glycoproteins	0.83	0.42	0.12	0
	free glycoproteins + vesicles	1.15	0.64	0.18	0

Footnotes to Table I (see opposite page)

a = Envelope glycoproteins were incorporated into large unilamellar vesicles (LUV) composed of phosphatidylserine (PS) produced by Ca^{2+}-induced fusion of small unilamellar vesicles (SUV) as described elsewhere (3). Viral glycoprotein (50-100μg) was added to SUV (total 2.5-5 mg lipid) before the addition of Ca^{2+} (4mM) to induce fusion of SUV to form LUV. The efficiency of association of envelope glycoproteins with LUV was determined by the binding of ^{125}I-labeled viral glycoproteins to the vesicles, by direct assay of protein content. and by assay of haemagglutination (HA) activity (10,000 HAU for vesicles containing glycoproteins corresponds to a viral protein concentration of approximately 320 μg; this compares with 200/μg viral protein for a similar HA activity in intact virions).

b = Vesicles containing ^{125}I-labeled viral glycoproteins (10,000 HAU) were incubated with monolayer cultures of BHK cells in 60 mm diameter plastic Petri dishes in serum-free, low calcium (0.5 mM) Dulbecco's modified Eagles medium for 2 hr at 37° or 4° as previously described (4). The total amount of ^{125}I-radioactivity added to cell cultures (cpm x 10^{-5}) was: NDV (Herts) 2.86; NDV (Ulster) 2.06; and influenza virus (A/Scotland/74/H3N2) 2.81. Replicate cell cultures were incubated under similar conditions with identical concentrations of glycoproteins without vesicles (free glycoproteins) or glycoproteins together with pure lipid vesicles (free glycoproteins + vesicles).

c = Results represent mean values from three separate experiments.

d = Cell monolayers incubated with 250 μg/ml crystalline (x2) trypsin (Worthington) for 15 min at 37° after incubation with the indicated samples for 2 hr at 37° or 4°.

---

of cultured cells. In addition to demonstrating the feasibility of using lipid vesicles as carriers to induce defined alterations in plasma membrane composition, the methods described here can also be used to obtain information on the functional properties of specific virus envelope glycoproteins.

ACKNOWLEDGEMENT

This work was supported in part by Grants CA13393 and CA-18260 from the National Institutes of Health (G.P.).

TABLE II

Fusion of BHK Cells Following Interaction with Lipid Vesicles Containing Newcastle Disease Virus Envelope Glycoproteins

Treatment[a]	% Cell Fusion[b]		
	Herts (HN and F)	Ulster (HN and F_o)	Ulster (Trypsinized)[c] (HN and F)
untreated control	2.7	2.4	2.1
intact UV-inactivated virus	69.2	2.9	35.3
vesicle-associated glycoproteins	32.6	7.2	19.4
free glycoproteins	4.7	6.9	3.2
free glycoproteins + vesicles	8.1	7.6	8.8

Footnotes.
a = Cells (4×10^6) were incubated with intact virus (10,000 HAU or a similar HAU titer of: 1. isolated glycoproteins incorporated into phosphatidylserine LUV as described in footnote (a) in Table I; 2. free glycoproteins; or 3. free glycoproteins together with pure phospholipid vesicles.

b = Cells were exposed to the treatments described in footnote (a) for 2 hr at 37°, after which cells were washed to remove virus or the various forms of isolated glycoproteins and the % cell fusion determined after a further incubation at 37° for 16 hr as previously described (5). The results represent mean values derived from three separate experiments.

(continued next page)

(Footnotes to Table II, continued)

c = Virions were incubated with 20 μg/ml crystalline (x3) trypsin for 15 min at 37°. Isolated glycoproteins were obtained by detergent-extraction of trypsinized virions rather than trypsination after extraction.

HN = hemagglutinin/neuraminidase, mol. wt. 70,000; F_o = F glycoprotein precursor, mol. wt. 65,000; and F = "fusion" protein, mol. wt. 54,000.

REFERENCES

1. Bachmeyer, H. (1975). *Intervirology* 5, 260.
2. Bächi, T., Deas, J.E. and Howe, C. (1977). *In* "Virus Infection and the Cell Surface". (G. Poste and G.L. Nicolson, eds.), p.83. Elsevier-North Holland, Amsterdam.
3. Papahadjopoulos, D., Vail, W.J., Jacobson, K. and Poste, G. (1975). *Biochim. Biophys. Acta.* 394, 483.
4. Poste, G. and Papahadjopoulos, D. (1976). *Proc. Nat. Acad. Sci. USA* 73, 1603.
5. Poste, G., Waterson, A.P., Terry, G., Alexander, D.J. and Reeve, P. (1972). *J. Gen. Virol.* 16, 95.
6. Poste, G., Papahadjopoulos, D. and Vail, W.J. (1976). *In* "Methods in Cell Biology" (D.M. Prescott, ed.), p.33 Academic Press, New York.
7. Rott, R. and Klenk, H-D. (1977). *In* "Virus Infection and the Cell Surface". (G. Poste and G.L. Nicolson, eds.), p.47. Elsevier-North Holland, Amsterdam.

(For Table III see over page)

TABLE III

Persistence of Newcastle Disease Virus Envelope Glycoprotein Antigens in the Plasma Membrane of BHK and Chick Embryo (CE) Cells as Determined by Binding of ^{125}I-labeled Anti-NDV Antibodies

Treatment[a]	Cell Type	Ratio of Antigen Content at Indicated Time to Initial Antigen Content[c]					
		1 hr	2 hr	4 hr	6 hr	8 hr	12 hr
Vesicle-associated glycoproteins (NDV-Herts)[a]	BHK	0.87	0.61	0.31	0.15	0	0
NDV-Herts (natural infection)[b]	BHK	0.73	0.46	0.17	0	0	0
Vesicle-associated glycoproteins (NDV-Herts)[a]	CE	0.65	0.37	0.19	0	0	0
NDV-Herts (natural infection)[b]	CE	0.69	0.48	0.12	0	0	0

Footnotes to Table III

a = Cells were incubated with LUV containing solubilized envelope glycoproteins (5,000 HAU) for 2 hr at 37° as described in the footnotes to Table I. The cells were then briefly trypsinized as described in Table I to elute adsorbed vesicles, washed with prewashed phosphate buffered saline and then incubated with ^{125}I-labeled anti-NDV antibodies for 30 min at 37° to determine antibody binding.

b = Cells were incubated with 5,000 HAU of intact virus for 1 hr at 37° and then treated identically to vesicle-treated cells in footnote (a).

c = Initial antigen content determined immediately after incubation with vesicles for 2 hr or intact virus for 1 hr.

THE INTERACTION OF VESICULAR STOMATITIS VIRUS WITH PHOSPHATIDYLCHOLINE LIPOSOMES

N.F. MOORE*, E.J. PATZER and R.R. WAGNER

Department of Microbiology, University of Virginia, School of Medicine, Charlottesville, Virginia 22901, U.S.A.

*Present address:
Unit of Invertebrate Virology, South Parks Road, Oxford OX1 3RB, U.K.

In previous studies we have attempted to determine the location of cholesterol in the VS viral membrane by exposure of intact virions to a combination of cholesterol oxidase and phospholipases (10). Cholesterol in the viral membrane is not accessible to the action of cholesterol oxidase alone (which oxidizes the β-OH of cholesterol to give cholest-4-en-3-one), but with the addition of phospholipase C or detergents the enzyme can act on the cholesterol. By comparison, phospholipase A_2 digestion of the viral membrane does not result in exposure of the cholesterol to cholesterol oxidase, which suggests that the phospholipid headgroups sterically hinder enzyme action.

Stoffel and Bister (13) have suggested that the high content of cholesterol in the viral membrane may provide some restriction on the mobility of the choline group of 3-en-phosphatidylcholine and sphingomyelin; however, our evidence suggests that the cholesterol in VS viral membrane has little effect on the mobility of the phosphate headgroup of the phospholipids (11). While these approaches have led to some suggestions as to how the cholesterol is interacting with the phospholipids of the viral membrane, it has told us little about the distribution of cholesterol on either face of the viral membrane bilayer and nothing at all about the function of the cholesterol in the membrane.

A number of studies have demonstrated that it is possible to exchange cholesterol out of various membranes into cholesterol-PC and PC liposomes; this exchange is a two-phase reaction, which suggests that the cholesterol in the outer face of a membrane is being exchanged first (4, 5, 7, 8, 12). The purpose of the present study was two-fold: (i) to investigate the relative amounts of cholesterol on the two faces of the VS_{Ind} viral membrane; (ii) to explore the possibility that removing cholesterol from the viral membrane by interaction with PC liposomes may affect viral infectivity.

CELLS AND VIRUS

The Indiana serotype of vesicular stomatitis (VS_{Ind}) virus was grown on confluent monolayers of BHK-21 cells. The virus was purified by sequential differential, velocity and equilibrium centrifugation (2). Plaque assays were performed by plating on monolayer cultures of L-929 cells (14).

ASSAYS

Total phospholipid phosphorus was determined in the lipid extract by the method of Bartlett (3). Protein was quantitated by the method of Lowry *et al*. (9). Cholesterol was determined using modifications of a published procedure (1).

PREPARATION OF LIPOSOMES AND CHOLESTEROL EXCHANGE REACTIONS

Single-lamellar liposomes were prepared from chromatographically pure egg PC in the presence and absence of 50 mol percent cholesterol by high intensity ultrasonic irradiation of the lipids for 5-10 min under nitrogen at 4-10° using a Heat Systems W-350 Sonifier (6). Radioactive lipids, including cholesterol and trace amounts of high specific activity cholesterol oleate, were incorporated into the liposomes. VS_{Ind} virus was interacted with a 50-fold excess of PC liposomes (based on a molar ratio of viral cholesterol to vesicle lipid). Following reaction for various times, liposomes and virus were separated by equilibrium centrifugation at 100,000 x g through a 20% sucrose pad; the liposomes remained in the supernatant while the virus pelleted. Alternatively, virus was banded to equilibrium on 0-40% tartrate:0-20% glycerol gradients where the liposomes remained at the top of the gradient.

RESULTS AND DISCUSSION

Cholesterol was found to be depleted from the viral membrane when VS_{Ind} virus was interacted with a fifty-fold excess of PC liposomes at 37° and then separated from the liposomes by pelleting at 100,000 x g for 1 hr through a pad of 20% sucrose. Cholesterol depletion was found to occur with both intact virions and virions which had been treated with trypsin or thermolysin to remove the surface spike glycoprotein (G). The kinetics of cholesterol removal with intact and spikeless particles was very similar, which demonstrates that the G protein is not sterically hindering the interaction of vesicle and viral membranes. In both cases the removal of cholesterol was a two-phase reaction; 60-70% of the cholesterol was removed in an initial fast phase (Fig.1), the remaining 30% being more difficult to remove. This is unlike the situation with influenza virus where prior removal of the glycoprotein spikes is necessary for interaction of virions and liposomes (8) to effect cholesterol exchange between virions and cholesterol-containing liposomes.

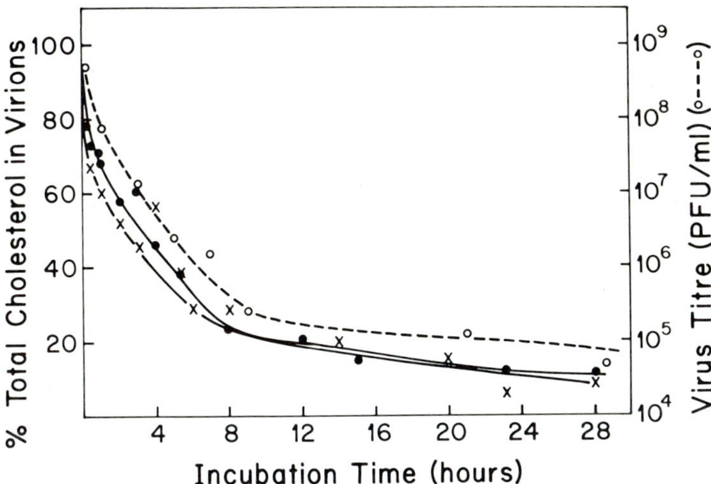

Fig.1. *The depletion of cholesterol from intact (●——●) and spikeless (X-X) VS virus membrane and the loss of viral infectivity (0-0) with intact virions caused by interaction of virions with phosphatidylcholine liposomes. Following interaction of VS virus with approximately 50-fold excess liposomes (original viral cholesterol to liposomal lipid molar ratio) the two components were separated by pelleting virions through a 20% sucrose pad at 100,000 x g. Viral infectivity was determined by plaque assay and the residual cholesterol content of the virions was determined enzymatically using cholesterol oxidase and quantitated as a percentage of the original cholesterol content.*

Paralleling the decrease in cholesterol content of the VS virus membrane was a decrease in infectivity of virus reacted with PC vesicles (Fig.1). This also occurred in a two-phase manner with little further decrease in viral infectivity after the first phase of cholesterol removal. This seems to suggest that the removal of cholesterol is directly related to loss of viral infectivity.

When VS_{Ind} virus was incubated with a similar fifty-fold excess of PC liposomes at 4°, the virus retained a much greater level of infectivity (Table I) which was apparently paralleled by a much smaller loss of viral membrane cholesterol. This appeared to support the theory that the cholesterol level is proportional to the infectivity of the virus.

An obvious experiment to test conclusively the relationship of cholesterol depletion to viral infectivity is to replenish cholesterol in cholesterol-depleted virions with cholesterol-containing PC-liposomes. Virions were first incubated with PC liposomes for 10 hr resulting in approximately 3 logs decrease in viral infectivity and the loss of 75% of the cholesterol. These depleted virions were then incubated with PC liposomes containing 50 mol percent cholesterol for various periods of time. Using radioactively labelled cholesterol and the sensitive cholesterol oxidase assay (controlled by trace amounts of high specific activity [^{14}C]cholesterol oleate to correct for adherent liposomes in the calculations), it was possible to demonstrate net exchange of cholesterol back to the viral membrane. However, no restoration in viral infectivity occurred (Table I). The exchange of cholesterol back to the viral membrane tended to be small and irreproducible.

To further investigate the mechanism of cholesterol exchange between viral membranes and liposomes, intact virions were incubated with cholesterol containing PC liposomes as previously described. A substantial reduction in infectivity also occurred, although the loss was somewhat less than incubation with liposomes made from phosphatidylcholine alone (Table I). When trypsinized virions were included with these liposomes, a two-phase exchange of cholesterol occurred with the first 60% of the viral cholesterol being exchanged rapidly. Spikeless virions provide a more satisfactory means to study the exchange of radioactive cholesterol from liposomes to virions because the amount of "sticking" of liposomes is approximately 50% of that with intact virions.

In conclusion, we have demonstrated that both cholesterol exchange and depletion occur by two-phase reactions with the cholesterol in the outer face of the viral membrane being removed rapidly. It seems risky to postulate that 60-70% of the cholesterol is located on the outer face of the bilayer. A component of the cholesterol located in the inner membrane layer could be in a position which allows it to be exchanged (flip-flop) more rapidly to the outer layer.

The loss of viral infectivity does not appear to be related directly to cholesterol depletion from the viral membrane

TABLE I

Effect of PC and PC-Cholesterol Liposomes on Infectivity of Intact and Cholesterol-Depleted VS Virions

Virus + Liposomes	Incubation Conditions		Infectivity (pfu/ml)	
	Temp (°C)	Time (hr)	Control	+ Liposomes
Intact VSV	4°	6	3.8×10^9	8.9×10^8
+ PC [a]	4°	14	3.8×10^9	4.5×10^8
Intact VSV	37°	4	8.0×10^8	3.0×10^6
+ PC [a]	37°	9	8.0×10^8	2.0×10^5
Intact VSV	37°	8	$\sim 10^{10}$	3.0×10^9
+ PC-Chol [b]	37°	17	7.0×10^9	6.7×10^7
Chol-depleted VSV	37°	2	5.6×10^6	5.9×10^6
+ PC-Chol [c]	37°	15	7.0×10^5	9.1×10^5

[a] VS_{Ind} virions were reacted with a 50-fold excess of PC vesicles

[b] Intact VS_{Ind} virions were reacted with excess PC liposomes containing 50 mol percent cholesterol

[c] VS_{Ind} virions previously depleted by PC liposomes of 80% cholesterol were reacted with excess PC liposomes containing 50 mol percent cholesterol.

since the incubation of virions with cholesterol containing PC liposomes also results in a substantial, albeit reduced, loss in infectivity. However, PC-cholesterol vesicles probably also produce net cholesterol depletion from the virus membrane due to the higher molar ratio of cholesterol in the viral membranes than in the liposomes. The study is further complicated by a low background exchange of viral phosphatidylcholine for the egg-PC of the liposomes with an unknown effect on viral infectivity. Thin layer chromatography of separated liposomes after interaction with VS_{Ind} virus demonstrated the presence of only PC and exchanged cholesterol. However, we cannot exclude the possibility of small amounts of other viral lipids being exchanged into the liposomes.

Electron microscopy of virus particles after long periods

of interaction with PC liposomes demonstrated the attachment of many liposomes to the viral membranes (Fig.2). This was further demonstrated by the apparently irreversible association of [^{14}C] cholesterol oleate counts (derived from the liposomes) with the viral membranes even after many conventional virus purification steps. While infectivity and cholesterol losses were reduced at lower temperature, sticking of liposomes did occur which argues against sticking being the only cause of infectivity loss.

Viral particles apparently remained fairly intact even after extensive cholesterol depletion as demonstrated by equilibrium banding, polyacrylamide gels of the separated virions and electron microscopy. Therefore the infectivity loss cannot be explained simply by dissolution of virion membranes.

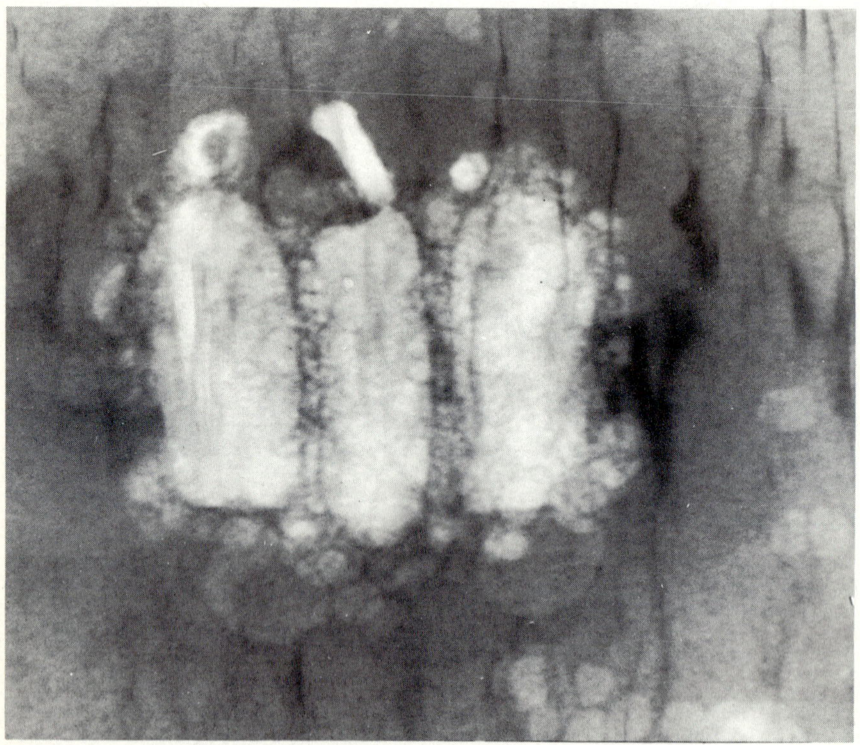

Fig.2. *Electron micrograph illustrating "sticking" of phosphatidylcholine liposomes to VS virus after interaction at 37° for 16 hr. VS virions were separated from the vast majority of liposomes by pelleting through a 20% sucrose pad; some free liposomes are still apparent.*

REFERENCES

1. Allain, C.C., Poon, L.S., Chan, C.S.G., Richmond, W. and Fu, P.C. (1974). *Clin. Chem.* 20, 470.
2. Barenholz, Y., Moore, N.F. and Wagner, R.R. (1976). *Biochemistry* 15, 3563.
3. Bartlett, G.R. (1959). *J. Biol. Chem.* 234, 466.
4. Bell, F.P. and Schwartz, C.J. (1971). *Biochim. Biophys. Acta* 231, 553.
5. Grunze, M. and Dueticke, B. (1974). *Biochim. Biophys. Acta* 356, 125
6. Huang, C. and Thompson, T.E. (1974). *Methods in Enzymol. Biomembranes* 32, 485.
7. Inbar, M. and Shinitzky, M. (1974). *Proc. Nat. Acad. Sci. USA* 71, 2128.
8. Lenard, J. and Rothman, J.E. (1975). *Proc. Nat. Acad. Sci. USA* 73, 391.
9. Lowry, O.H., Rosebrough, N.J., Farr, A.L. and Randall, R.J. (1951). *J. Biol. Chem.* 193, 265.
10. Moore, N.F., Patzer, E.J., Barenholz, Y. and Wagner, R.R. (1977). *Biochemistry* 16, 4708.
11. Moore, N.F., Patzer, E.J., Wagner, R.R., Yeagle, P.L., Hutton, W.C. and Martin, R.B. (1977). *Biochim. Biophys. Acta* 464, 234.
12. Pozansky, M. and Large, Y. (1976). *Nature* 259, 420.
13. Stoffel, W. and Bister, K. (1975). *Biochemistry* 14, 2841.
14. Wagner, R.R., Levy, A.H., Snyder, R.M., Ratcliffe, G.A. Jr., and Hyatt, D.F. (1963). *J. Immunol.* 91, 112.

MODIFICATIONS INDUCED BY VSV AND *TS* MUTANTS IN CHICK EMBRYO CELL PERMEATION

N. GENTY

Université Paris XI,
Institut de Microbiologie,
Centre d'Orsay, 91405 Orsay-Cedex, France.

Many lytic viruses inhibit macromolecular synthesis of their host cells (5, 6). The experimental procedure commonly used to follow this "shut off" phenomenon is the incorporation of labelled precursors into acid-precipitable material.

In a previous study (3) we observed that the "shut off" of RNA synthesis in chick embryo cells (CEC) after infection with vesicular stomatitis virus (VSV) is partly due to a reduced capacity of the infected cells to label the acid-soluble pool with exogenous uridine. Experiments with both temperature-sensitive mutants of group III (altered in the M protein of the viral envelope) and with M-depleted wild type VSV (by treatment with phospholipase C) suggest a specific role of the M protein in a sequence of events modifying uridine uptake.

In this report we describe experiments which will answer a few of the questions concerning the diminution of uridine accumulation in the acid-soluble pool of infected cells.

DOES THE URIDINE NUCLEOTIDE POOL LEAK OUT OF THE CEC AFTER INFECTION BY VSV?

The decreased labelling of the nucleotide pool by exogeneous uridine in VSV-infected CEC could be explained by a lowered effectiveness of the transport system for this nucleoside, but could also be a consequence of a leakage of the pool from the infected cells. It is known that nucleotides are normally trapped inside the cells (8). Since viral infection may alter membrane properties (10) it was decided to determine whether VSV-infected CEC are able to retain phosphorylated uridine compounds.

The uridine nucleotide pool of CEC was labelled at low temperature (11). The prelabelled CEC were infected with VSV or mock-infected. On rewarming them to 37° part of the labelled nucleotide pool was incorporated into acid-insoluble material. From Fig.1 it can be seen that: (i) there is no leakage of the acid soluble pool during the adsorption period; (ii) when rewarming to 37° after the adsorption period there is a decreased loss of label from the acid soluble pool in infected cells compared to non-infected control cells; (iii) during the first 3 hr after infection the total radioactivity associated with the cells does not decrease any more rapidly in the infected cells than in the control cells. In that experiment (Fig.1) 3 hr p.i., the total amount of radioactivity per monolayer of infected cells was 1.22×10^6 dpm and 1.26×10^6 dpm in the control monolayers. Since uridine derivatives are not released from the CEC after infection with VSV we can say that in our previous experiments it is the uptake of uridine by CEC that seems to be inhibited after infection by VSV.

From Fig.1 we can also conclude that the transfer of labelled nucleotides from the acid-soluble pool into RNA is rapidly inhibited to infected cells. This probably reflects an inhibition of the absolute rate of RNA synthesis in infected cells.

ARE CARRIER-MEDIATED TRANSPORT AND SIMPLE DIFFUSION BOTH ALTERED IN CEC INFECTED BY VSV?

Uridine enters the cell by a saturable transport reaction as well as by simple diffusion (9). The transport reaction is by far the more important process under physiological conditions where there is a low exogenous nucleoside concentration. Simple diffusion becomes the predominant mode of entry into the cells at high extracellular concentration, when the carrier mediated transport is saturated.

The fundamental cause of uridine transport changes in CEC may be assessed by comparing initial rates of nucleoside uptake as a function of exogenous concentration. Typical dependence of uptake velocity on substrate concentration is shown in Fig.2. It demonstrates that VSV infection markedly reduces the capacity of the cells to take up uridine. This effect seems to be due to an inactivation of the transport system. Only the saturable transport component of the uptake curve (i.e. less than 3×10^{-4} M exogenous uridine) seems to be affected by viral infection. The parallel nature of the uptake curves for infected and non-infected cells above 3×10^{-4} M of exogenous uridine suggests that the uptake of uridine by simple diffusion is not affected after VSV infection. We have estimated the rate of uridine uptake due to simple diffusion as a function of exogenous uridine concentration by drawing a line parallel to

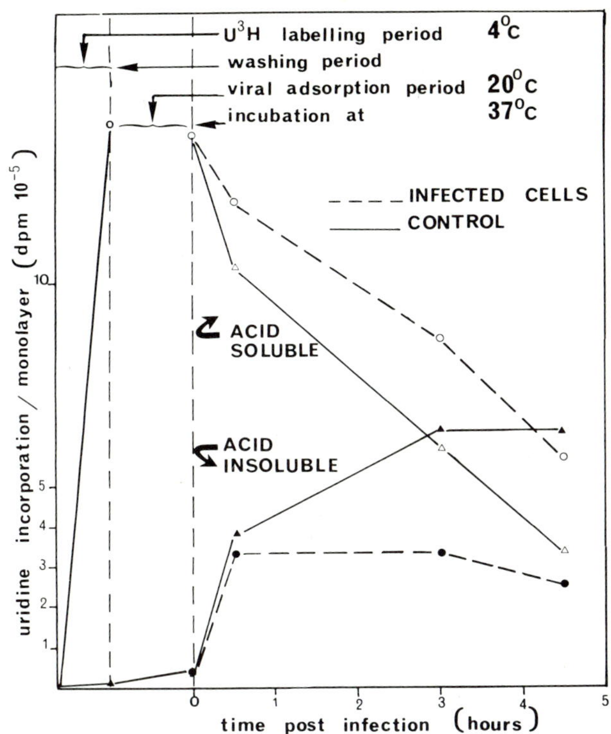

Fig.1. *Labelling of CEC with ^{3}H-uridine at 4°, chased at 37° after infection with VSV.*

Monolayers of CEC were supplemented with ^{3}H-uridine (1 μCi/ml 23 Ci/mmole) for 30 min at 4°. The cultures were then washed three times with Tris buffered saline pH 7.4, and infected with wild type VSV (m.o.i. = 5 pfu/cell) or mock-infected at room temperature. After 40 min of adsorption, monolayers were covered with 2 ml of fresh Eagle's medium and incubated at 37°. At intervals thereafter the acid-soluble pool and acid-insoluble material were tested for radioactivity as described elsewhere (3). All values represent averages of duplicate monolayers.

the straight line portion of the uptake curve which passes through the origin. Subtraction of the estimated simple diffusion rate from the overall rate of uptake, provides an estimate of the uptake due to the transport reaction. The values give curves typical of a saturable process highly inhibited in infected cells. Lineweaver-Burk plots of the initial rate of transport corrected for simple diffusion show similar apparent K_m values for uridine carrier mediated transport but a decreased apparent V_{max} due to infection. This is evidence of a non competitive type of inhibition.

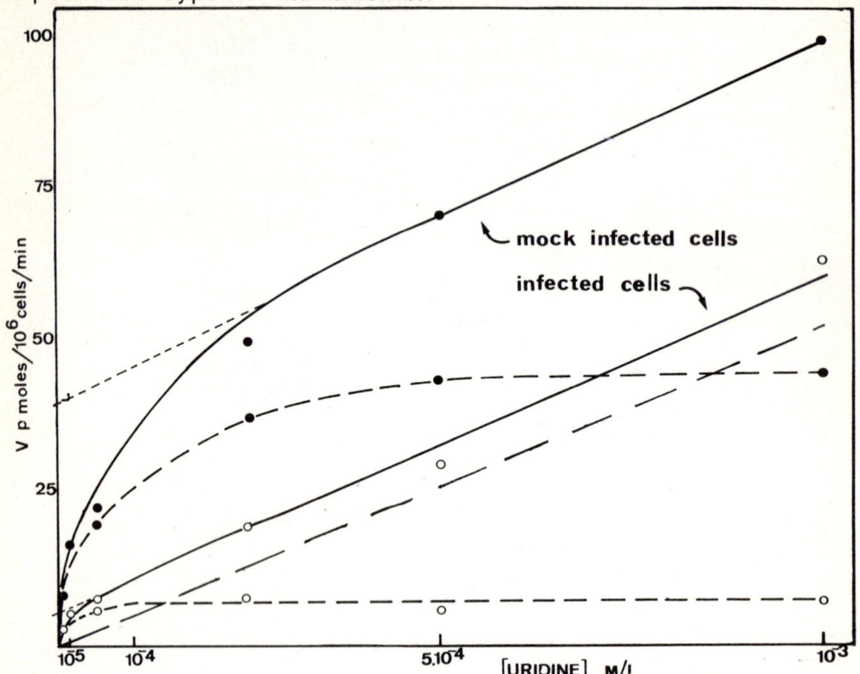

Fig.2. *Effect of VSV infection on uridine uptake by CEC*

CEC monolayers were infected with VSV wild type at a m.o.i. of 5, or mock infected. Cultures were then incubated in a constant temperature water bath (39.5°). To determine uridine uptake, at 2.5 hr post-infection, the 2 ml culture medium were supplemented with 0.5 ml of ^{3}H-uridine (5 µCi/ml, 23 Ci/mmole) diluted in cold uridine solutions to get the appropriate uridine concentrations. During this labelling monolayers were gently rocked. In order to determine the initial rate of uridine uptake, the duration of the labelling was 2 min. For washing, monolayers were placed on ice, rocked gently, and rinsed six times with ice cold buffer. The acid soluble pool was then extracted with cold 5% TCA. Since experiments were performed at a constant total radioactivity of uridine, the measured uptake as dpm, multiplied by the substrate concentration gives the amount of uridine and uridine nucleotides trapped into the cells. Uridine uptake by infected cells (—O——O—), by mock infected cells (—●——●—). Contribution of simple diffusion (— —— —). Uptake due to transport reaction by infected cells (---O---O---), by mock infected cells (--●---●--).

DOES VSV INFECTION CAUSE A GENERALIZED PERMEABILITY CHANGE?

We wondered whether the permeation of other metabolites could be affected after infection of CEC by VSV. We compared the uptake of different substrates during a 2 min pulse in monolayers of CEC infected at low m.o.i. by VSV, or mock infected, and cultured at 39.5° for 2.5 hr after viral infection.

The uptake of cytidine and uridine were equally affected (60% decrease), thymidine and guanosine uptake was weakly inhibited (30% decrease), while the transport of adenosine was not at all affected. The fact that the uptake of uridine and cytidine show the same kinetics of inhibition when CEC are infected with VSV is in agreement with the suggestions of Scholtissek (12) who showed that these two ribonucleosides are taken up at the same plasma membrane site of CEC while the site of transport of adenosine is different.

The effect of infection on orotic acid incorporation was investigated. During the first few hours post infection there is no decrease in the labelling for the nucleotide pool by exogenous ^{3}H-orotic acid but there is a decrease, although much slower than the one with uridine, of incorporation into TCA-precipitable material. The residual incorporation of these precursors into acid precipitable material of infected cells (in experimental conditions cited) is 58% for orotic acid and 32% for uridine of the incorporation in the control.

It is generally admitted that orotic acid penetrates cells by simple diffusion. Its penetration into CEC is one hundred fold less effective than uridine penetration. The simple diffusion of the precursor (like the simple diffusion of uridine) is not altered in VSV-infected CEC. The reduction in incorporation of orotic acid into acid-insoluble material of VSV infected CEC probably represents the actual diminution of rate of RNA synthesis.

IS THE DECLINE OF URIDINE TRANSPORT IN VSV INFECTED CEC SECONDARY TO THE INHIBITION OF CELL RNA SYNTHESIS AFTER INFECTION?

We have shown (Fig.1) that infection markedly inhibits the rate of incorporation of labelled uridine nucleotides into acid-precipitable material. The data are interpreted to indicate that VSV rapidly inhibits cell RNA synthesis.

Results from Fig.2 show that there is a failure of the infected cells to transport uridine into the acid soluble pool.

To determine whether these two virus-induced phenomena are interdependent we have analyzed the effect of different *ts* mutants at non permissive temperature or M-depleted VSV on uridine uptake and on actual RNA synthesis of CEC (treated or not with cycloheximide). Table I demonstrates the following

TABLE I

Shut Off of Uridine Incorporation in VSV-Infected CEC

Virus	Cycloheximide	1st Mechanism inhibition of uridine uptake	2nd Mechanism inhibition of RNA synthesis
W.T. ts I (1,5,7,53) ts II (52) ts IV (100) ts V (45)	None	+	+
	5 μg/ml	+	0
ts III (23,76,89)	None	0	+
M depleted VSV	5 μg/ml	0	0
ts 1026 ts 82	None	0	0

Inhibition of uridine incorporation in acid-precipitable material during a 30 min pulse in CEC treated or not with cycloheximide, infected for 2.5 hr with different types of viruses, at non-permissive temperature.

1st mechanism: The virus induces inhibition of exogenous uridine uptake; this causes a reduction of labelled nucleotides in the acid soluble pool which explains the diminution of labelling of acid insoluble material.

2nd mechanism: Diminution of the rate of RNA synthesis; (uridine incorporation in acid insoluble material has been corrected for differences in uridine uptake).

+ is assigned when the ratio $\frac{\text{infected cells}}{\text{mock infected cells}}$ is ~50%

0 is assigned when it is ⩾ 90%

types of response to viral infection: (i) the systems in which both permeability and RNA synthesis are affected; (ii) the systems in which only the uptake of uridine or the synthesis of RNA is altered; (iii) the systems in which absolutely no inhibition is apparent.

We conclude that the modifications of uridine uptake and RNA synthesis are unrelated. The M protein, which plays a role in the sequence of events that modify uridine transport (3), does not seem to be implicated in the inhibition of RNA synthesis.

CONCLUSIONS

Cell surface alterations have often been invoked as a result of viral infection. Since one of the important functions of the cell membrane is the uptake of precursors, it is not surprising that viral infection induces modifications of permeability (1, 2, 4, 7). We have determined that the diminution of the labelling of the acid-soluble pool with exogenous uridine in VSV-infected CEC is not due to a general breakdown of the cellular permeability barriers since: (i) infection does not release the acid soluble pool; (ii) under our experimental conditions the inhibition of penetration is specific for uridine and cytidine; (iii) the carrier mediated transport, but not simple diffusion, is restricted.

Additional experiments must determine whether the saturable system which is inhibited in VSV infected CEC is the facilitated diffusion of uridine, its phosphorylation by uridine kinase, or a transphosphorylating reaction (12).

REFERENCES

1. Carrasco, L. and Smith, A.E. (1976). *Nature* 264, 807.
2. Fuchs, P. and Giberman, E. (1973). *F.E.B.S. Letters* 31, 127.
3. Genty, N. (1975). *J. Virol.* 15, 8.
4. Hand, R. (1976). *J. Virol.* 19, 801.
5. Kohn, A. and Fuchs, P. (1973). *In* "Advances in Virus Research" 18, 159.
6. Martin, E.M. and Kerr, I.M. (1968). *In* "The Molecular Biology of Viruses", (L.V. Crawford and M.G.P. Stoker, eds.), p.15. Cambridge University Press.
7. Negreanu, Y., Reinhertz, Z. and Kohn, A. (1974). *J. Gen. Virol.* 22, 265.
8. Plagemann, P.G.W. and Roth, M.F. (1969). *Biochem.* 8, 4782.
9. Plagemann, P.G.W. (1970). *J. Cell Physiol.* 77, 213.
10. Rott, R., Becht, H., Hammer, G., Klenk, H.-D. and Scholtissek, C. (1975). *In* "Negative Strand Viruses" (B.W.J. Mahy and R.D. Barry, eds.), p.843. Academic Press.

11. Scholtissek, C. (1967). *Biochim. Biophys. Acta* **145**, 228.
12. Scholtissek, C. (1968). *Biochim. Biophys. Acta* **158**, 435.

85

THERMOLABILITY AND SOME PHYSICAL PROPERTIES OF TEMPERATURE-SENSITIVE MUTANTS OF VESICULAR STOMATITIS VIRUS (VSV) WITH ALTERED MEMBRANE PROTEINS

P. KELLER[ab], D.H. CLUXTON[c], E.E. UZGIRIS[d], and J. LENARD[a]

[a]*Department of Physiology, CMDNJ-Rutgers Medical School, Piscataway, New Jersey, U.S.A.*

Present address:

[b]*Merck, Sharp & Dohme Research Laboratories, West Point, Pennsylvania, U.S.A.*

Enveloped viruses have for several years attracted interest as simple membrane systems derived from the plasma membrane of their host cell by budding. VSV is unique in possessing only a single glycoprotein in its envelope. This fact, in addition to the existence of two well-characterized classes of temperature-sensitive mutants containing lesions in each of the two viral envelope proteins makes VSV a suitable choice for studying the role that each of these proteins plays in the initiation of viral infection, and in viral budding. This paper describes surface changes which occur in purified thermolabile group V (G protein) mutant virions upon exposure to non-permissive temperatures.

THERMOLABILITY

The thermolability of a number of group III and group V VSV mutants has previously been reported (2, 3, 5). These results have been confirmed and extended in our laboratory using purified virions grown at the permissive temperature of 31° (Figs. 1a and b). Both group III and group V mutants demonstrate a wide range of rates of inactivation at 39°, whereas none of the mutants were inactivated when held at 31° in phosphate buffered saline (PBS, data not shown).

[c]*Russell Sage College, Troy, New York, U.S.A.*

[d]*General Electric Research and Development Center, Schenectady, New York, U.S.A.*

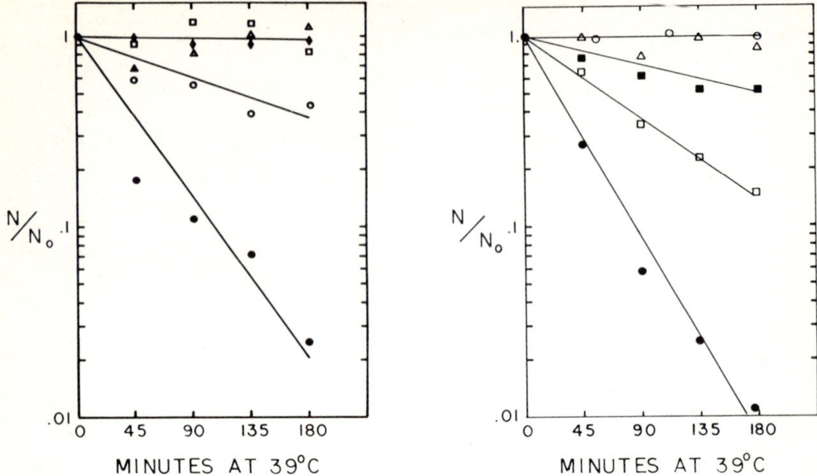

Fig.1a and b. *Thermolability of temperature sensitive mutants incubated at 39° in PBS. Virions grown from plaque purified stock were grown at the permissive temperature and purified by polyethylene glycol pptn. followed by a sucrose gradient. The bands were collected, dialyzed against PBS, incubated at 39° or 31° and plated on BHK cells at 31°.*
A. Group III virions ● *ts 31,* O *ts 33,* ♦ *ts 23,* Δ *ts 89,* □ *WT*
B. Group V virions ● *ts 45,* □ *ts 110,* ■ *ts 57,* Δ *ts 44,* O *WT*

Presumably inactivation of the thermolabile mutants results, either directly or indirectly, from a conformational change (or changes) in the protein affected by the mutation. Since the conformation of a protein is stabilized by its environment, it should be possible to alter the thermolability of any given mutant by altering the environment during exposure to the non-permissive temperature. We have demonstrated that this is indeed the case for one group V mutant, ts 45. The thermolability of ts 45 is greatly enhanced by incubation in PBS and 2% ethanol, or 0.1X PBS, while the presence of additional 0.60 M NaCl in PBS (total [NaCl] = 0.75M) protects the virus from inactivation (Fig.2). These conditions do not affect ts 45 at the permissive temperature, or wild type (WT) at permissive or non-permissive temperatures.

Similar experiments were performed both with a number of group III (M protein; ref.4) and group V (G protein) mutants (Table I). Only the group V mutants are altered in their thermolability by varying the NaCl concentration, or by adding 2% ethanol to PBS. None of the mutants were inactivated when incubated at the permissive temperature under these conditions.

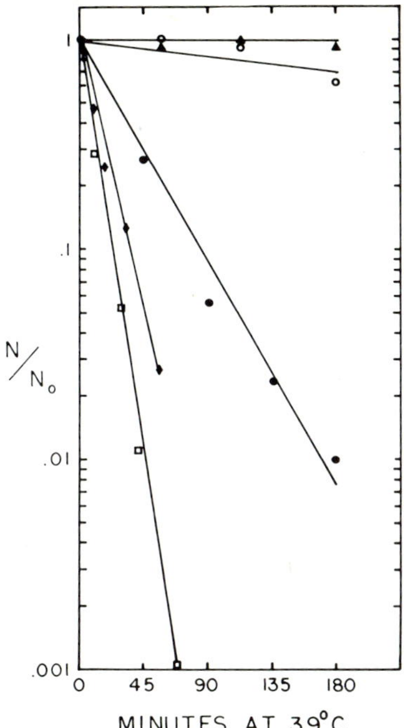

Fig.2. *Thermolability of ts 45. The experimental procedure was the same as described in Fig.1 except the buffer was varied as noted below during exposure to 39°.*
● *PBS,* □ *PBS & 2% Ethanol,* ♦ *0.1 PBS, O PBS & 0.06 M NaCl,* ▲ *WT PBS*

VIRAL INTEGRITY

It has been reported that the spikes and envelopes of ts 45 dissociate from the virion upon incubation at the non-permissive temperature of 39° (2). However, this did not occur under our experimental conditions, since no radioactivity was found at the top of a gradient after sedimenting ^{3}H-leucine labelled ts 45 which had been incubated at 39° (Fig.3). This suggested that all the viral proteins sedimented with the viral particles, a finding which was confirmed by SDS-polyacrylamide gel electrophoresis (data not shown). Interestingly, ts 45 sedimented more rapidly after incubation at 39°, implying either an increase in particle density or aggregation of the virions (Fig. 3).

MEASUREMENTS OF PARTICLE SIZE

To investigate the basis for the alteration in sedimentation behaviour another experimental approach was utilized. ts 45

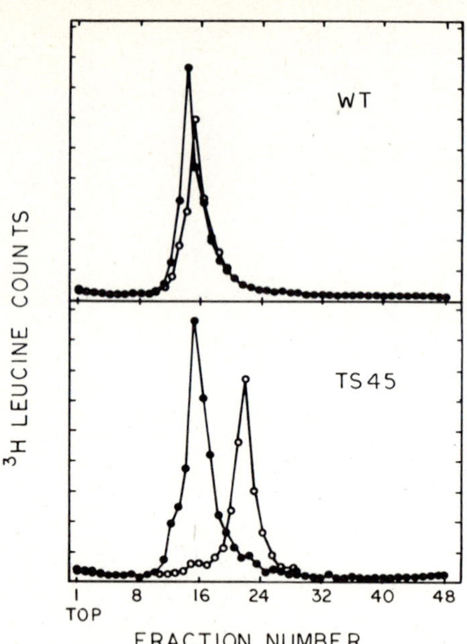

Fig.3. *Sucrose gradients of ts 45 and WT virions. ^{3}H Leucine labeled virions were purified as described in Fig.1, incubated at 31° or 39° at a concentration of about 1 mg/ml for 180 min, layered on a 15-60% sucrose gradient and run for 60 min at 22 k in a SW 41 rotor. ● 31° O 39°*

mutants were incubated at both the permissive and non-permissive temperatures, and their average particle size measured as a function of time by quasi-elastic light scattering (6).

ts 45 in PBS aggregated rapidly upon incubation at 39°, whereas aggregation did not occur upon incubation at 39° in PBS containing a total NaCl concentration of 0.75 M (Fig.4). Aggregation of ts 45 at the permissive temperature was not observed under any incubation condition, nor has it been observed on incubating any of the group III mutants, or the non-thermolabile group V mutant ts 44, at 39° (Fig.4).

When ts 45 was incubated at 39° and then shifted down to 31°, aggregation continued although at a different rate (Fig.5). This change in rate is not understood, but it is clear that a permanent alteration of the viral surface has occured which permits aggregation to continue.

The rate of aggregation of ts 45 was found to vary with particle concentration (Fig.6), as might be expected for a phenomenon which depends upon interaction between particles. On the other hand, the rate of loss of infectivity was inde-

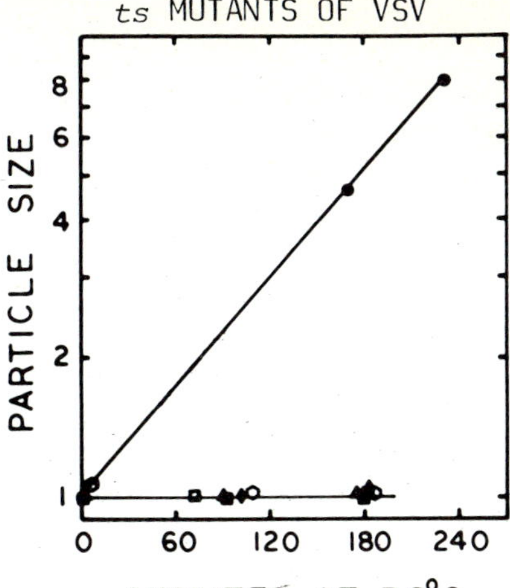

Fig.4. *Time course of aggregation of mutant virions. Virions were purified as in Fig.1, placed in appropriate buffer at a concentration of about 1 mg/ml, and their size measured using laser self-beat spectroscopy. Particle size refers to the mean particle size relative to particle size at t = 0 as defined by diffusion rate.*
● *ts 45 PBS,* ○ *ts 45 PBS & 0.60 M NaCl,* ♦ *wt PBS,* △ *ts 44 PBS,* □ *ts 89 PBS,* ■ *ts 33 PBS*

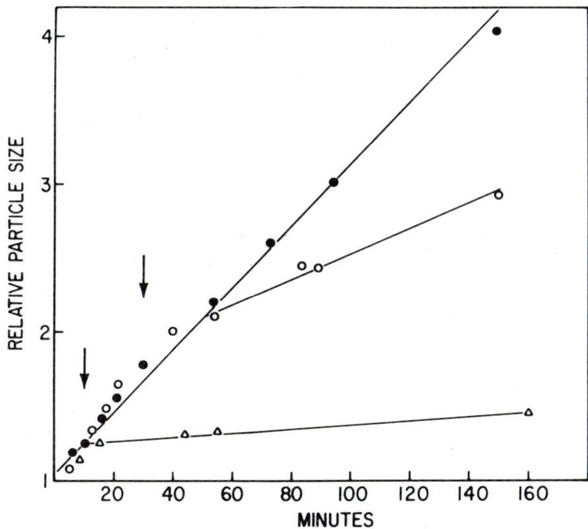

Fig.5. *Shift-down experiments with ts 45. Aggregation was measured as in Fig.4. ts 45 was incubated at 0.6 mg/ml in PBS at 39°. After 68 min (O), or 15 min (Δ) of exposure to 39°, the temperature was shifted down to 31° and measurements continued.*

pendent of particle concentration (data not shown). This shows that aggregation is not the primary cause of thermolability of ts 45, but is a secondary event associated with conformational change in the viral glycoprotein.

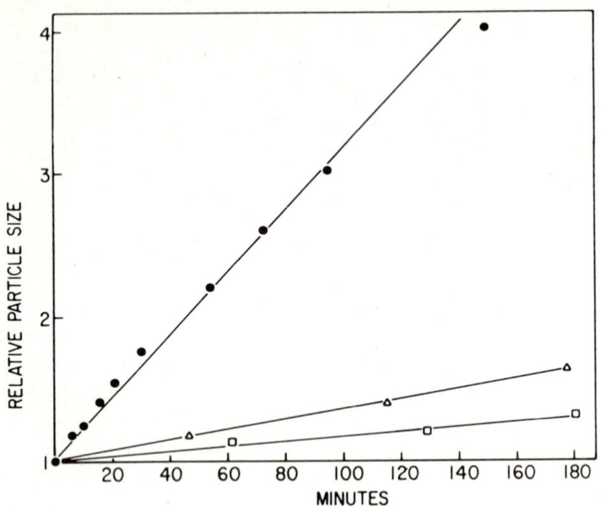

Fig.6. *Concentration dependence of aggregation of ts 45 at 39° in PBS.* ● *0.6 mg/ml;* △ *0.3 mg/ml;* □ *0.12 mg/ml*

DISUCSSION

The data presented in this paper indicate that thermolabile group V mutants are characterized by a change in conformation of the glycoprotein upon incubation at the non-permissive temperature. The effects of this change are progressive with time, and are influenced by environmental factors such as salt concentration and presence of small amounts of ethanol. Since ts 45 aggregates upon exposure to the non-permissive temperature it is clear that this mutant has altered surface properties, presumably as a result of changes in glycoprotein conformation.

These changes in the glycoprotein conformation upon incubation at the non-permissive temperature must underlie both the loss of infectivity and the aggregation of ts 45. However, aggregation cannot be the primary cause of the loss in infectivity because the rate of aggregation is dependent upon particle concentration (Fig.6), while the rate of inactivation is not. The conformational change induced at the non-permissive temperature is at least partly irreversible, since aggregation continues after shiftdown to the permissive temperature (Fig.5) and infectivity is not regained. Both inactivation and aggre-

TABLE I

THERMOLABILITY OF GROUP III AND GROUP V MUTANTS AT 39°

Mutant	Group	Thermo-lability in PBS[A]	Effect of 0.60 M NaCl*	Effect of 2% Ethanol*	Effect of 0.1 x PBS*
ts 31	III	Very	No effect	No effect	No effect
ts 33	III	Somewhat	No effect	No effect	-----
ts 23	III	None	No effect	No effect	No effect
ts 89	III	None	No effect	No effect	No effect
ts 45	V	Very	Protects	Increased	Increased
ts 110	V	Very	Protects	Increased	Increased
ts 57	V	Somewhat	Protects	Increased	Increased
ts 44	V	None	No effect	Slight increase	Increased
Wt		None	No effect	No effect	No effect

[A]see Figs.1 and 2

*Thermolability at 39° relative to thermolability at 39° in PBS

gation are prevented if incubation at the non-permissive temperature occurs in the presence of high salt concentration (Figs.2, 4).

Under the experimental conditons used, ts 45 does not lose its spikes at the non-permissive temperature, in contrast to a previous report. Also, preliminary evidence, based on changes in light scattering properties in environments of different osmotic strength (1) suggest that the osmotic response of the thermolabile group V virions is not lost upon incubation at the non-permissive temperature (data not shown). These observations are also consistent with an altered surface structure arising from a conformational change in the viral glycoprotein.

The availability of mutants with altered surface properties should provide a powerful tool in studying the initial events in virus infection.

ACKNOWLEDGEMENTS

Paul Keller was a Damon Runyon-Walter Winchell Cancer Fund Fellow during the course of this work. This study was supported in part by NSF grant PCN 76-10020 and NIH grant AI 13003-01. The authors are indebted to Roger Vanderoef for excellent technical assistance.

REFERENCES

1. Bittman, R., Majuk, Z., Honig, D.S., Compans, R.W. and Lenard, J. (1976). *Biochim. Biophys. Acta* 433, 63.
2. Deutsch, V. and Berkaloff, A. (1971). *Ann. Inst. Pasteur* 121, 101.
3. Lafay, F. (1974). *J. Virol.* 14, 1220.
4. Pringle, C.R. (1975). *Current Topics in Microbiology and Immunology* 69, 85.
5. Pringle, C.R. and Duncan, I.B. (1971). *J. Virol.* 8, 56.
6. Pusey, P.N., Koppel, D.E., Schaefer, D.W., Camerini-Otero, R.D. and Koenig, S.H. (1974). *Biochemistry* 13, 952.

SUBJECT INDEX